BRIDGE ENGINEERING HANDBOOK

BRIDGE ENGINEERING HANDBOOK

VOLUME II

EDITED BY
WAI-FAH CHEN and LIAN DUAN

CRC Press
Taylor & Francis Group
Boca Raton London New York

CRC Press is an imprint of the
Taylor & Francis Group, an **informa** business

First published 1999 by CRC Press
Taylor & Francis Group
6000 Broken Sound Parkway NW, Suite 300
Boca Raton, FL 33487-2742

Reissued 2019 by CRC Press

A Library of Congress record exists under LC control number:

ISBN 13: 978-0-367-26344-7 (set)
ISBN 13: 978-0-367-25329-5 (hbk)
ISBN 13: 978-0-367-25331-8 (pbk)
ISBN 13: 978-0-429-28724-4 (ebk)

Visit the Taylor & Francis Web site at http://www.taylorandfrancis.com and the
CRC Press Web site at http://www.crcpress.com

Foreword

Among all engineering subjects, bridge engineering is probably the most difficult on which to compose a handbook because it encompasses various fields of arts and sciences. It not only requires knowledge and experience in bridge design and construction, but often involves social, economic, and political activities. Hence, I wish to congratulate the editors and authors for having conceived this thick volume and devoted the time and energy to complete it in such short order. Not only is it the first handbook of bridge engineering as far as I know, but it contains a wealth of information not previously available to bridge engineers. It embraces almost all facets of bridge engineering except the rudimentary analyses and actual field construction of bridge structures, members, and foundations. Of course, bridge engineering is such an immense subject that engineers will always have to go beyond a handbook for additional information and guidance.

I may be somewhat biased in commenting on the background of the two editors, who both came from China, a country rich in the pioneering and design of ancient bridges and just beginning to catch up with the modern world in the science and technology of bridge engineering. It is particularly to the editors' credit to have convinced and gathered so many internationally recognized bridge engineers to contribute chapters. At the same time, younger engineers have introduced new design and construction techniques into the treatise.

This Handbook is divided into seven sections, namely:

- Fundamentals
- Superstructure Design
- Substructure Design
- Seismic Design
- Construction and Maintenance
- Special Topics
- Worldwide Practice

There are 67 chapters, beginning with bridge concepts and aesthestics, two areas only recently emphasized by bridge engineers. Some unusual features, such as rehabilitation, retrofit, and maintenance of bridges, are presented in great detail. The section devoted to seismic design includes soil-foundation-structure interaction. Another section describes and compares bridge engineering practices around the world. I am sure that these special areas will be brought up to date as the future of bridge engineering develops.

May I advise each bridge engineer to have a desk copy of this volume with which to survey and examine both the breadth and depth of bridge engineering.

T. Y. Lin
Professor Emeritus, University of California at Berkeley
Chairman, Lin Tung-Yen China, Inc.

Preface

The *Bridge Engineering Handbook* is a unique, comprehensive, and state-of-the-art reference work and resource book covering the major areas of bridge engineering with the theme "bridge to the 21st century." It has been written with practicing bridge and structural engineers in mind. The ideal readers will be M.S.-level structural and bridge engineers with a need for a single reference source to keep abreast of new developments and the state-of-the-practice, as well as to review standard practices.

The areas of bridge engineering include planning, analysis and design, construction, maintenance, and rehabilitation. To provide engineers a well-organized, user-friendly, and easy-to-follow resource, the Handbook is divided into seven sections. *Section I, Fundamentals*, presents conceptual design, aesthetics, planning, design philosophies, bridge loads, structural analysis, and modeling. *Section II, Superstructure Design*, reviews how to design various bridges made of concrete, steel, steel-concrete composites, and timbers; horizontally curved, truss, arch, cable-stayed, suspension, floating, movable, and railroad bridges; and expansion joints, deck systems, and approach slabs. *Section III, Substructure Design*, addresses the various substructure components: bearings, piers and columns, towers, abutments and retaining structures, geotechnical considerations, footings, and foundations. *Section IV, Seismic Design*, provides earthquake geotechnical and damage considerations, seismic analysis and design, seismic isolation and energy dissipation, soil–structure–foundation interactions, and seismic retrofit technology and practice. *Section V, Construction and Maintenance*, includes construction of steel and concrete bridges, substructures of major overwater bridges, construction inspections, maintenance inspection and rating, strengthening, and rehabilitation. *Section VI, Special Topics*, addresses in-depth treatments of some important topics and their recent developments in bridge engineering. *Section VII, Worldwide Practice*, provides the global picture of bridge engineering history and practice from China, Europe, Japan, and Russia to the U.S.

The Handbook stresses professional applications and practical solutions. Emphasis has been placed on ready-to-use materials, and special attention is given to rehabilitation, retrofit, and maintenance. The Handbook contains many formulas and tables that give immediate answers to questions arising from practical works. It describes the basic concepts and assumptions, omitting the derivations of formulas and theories, and covers both traditional and new, innovative practices. An overview of the structure, organization, and contents of the book can be seen by examining the table of contents presented at the beginning, while an in-depth view of a particular subject can be seen by examining the individual table of contents preceding each chapter. References at the end of each chapter can be consulted for more-detailed studies.

The chapters have been written by many internationally known authors from different countries covering bridge engineering practices, research, and development in North America, Europe, and the Pacific Rim. This Handbook may provide a glimpse of a rapidly growing trend in global economy in recent years toward international outsourcing of practice and competition in all dimensions of engineering. In general, the Handbook is aimed toward the needs of practicing engineers, but materials may be reorganized to accommodate undergraduate and graduate level bridge courses. The book may also be used as a survey of the practice of bridge engineering around the world.

The authors acknowledge with thanks the comments, suggestions, and recommendations during the development of the Handbook by Fritz Leonhardt, Professor Emeritus, Stuttgart University, Germany; Shouji Toma, Professor, Horrai-Gakuen University, Japan; Gerard F. Fox, Consulting Engineer; Jackson L. Durkee, Consulting Engineer; Michael J. Abrahams, Senior Vice President, Parsons, Brinckerhoff, Quade & Douglas, Inc.; Ben C. Gerwick, Jr., Professor Emeritus, University of California at Berkeley; Gregory F. Fenves, Professor, University of California at Berkeley; John M. Kulicki, President and Chief Engineer, Modjeski and Masters; James Chai, Senior Materials and Research Engineer, California Department of Transportation; Jinrong Wang, Senior Bridge Engineer, URS Greiner; and David W. Liu, Principal, Imbsen & Associates, Inc.

We wish to thank all the authors for their contributions and also to acknowledge at CRC Press Nora Konopka, Acquiring Editor, and Carol Whitehead and Sylvia Wood, Project Editors.

Wai-Fah Chen
Lian Duan

Editors

Wai-Fah Chen is a George E. Goodwin Distinguished Professor of Civil Engineering and Head of the Department of Structural Engineering, School of Civil Engineering at Purdue University. He received his B.S. in civil engineering from the National Cheng-Kung University, Taiwan, in 1959, M.S. in structural engineering from Lehigh University, Bethlehem, Pennsylvania in 1963, and Ph.D. in solid mechanics from Brown University, Providence, Rhode Island in 1966.

Dr. Chen's research interests cover several areas, including constitutive modeling of engineering materials, soil and concrete plasticity, structural connections, and structural stability. He is the recipient of numerous engineering awards, including the AISC T.R. Higgins Lectureship Award, the ASCE Raymond C. Reese Research Prize, and the ASCE Shortridge Hardesty Award. He was elected to the National Academy of Engineering in 1995, and was awarded an Honorary Membership in the American Society of Civil Engineers in 1997. He was most recently elected to the Academia Sinica in Taiwan.

Dr. Chen is a member of the Executive Committee of the Structural Stability Research Council, the Specification Committee of the American Institute of Steel Construction, and the editorial board of six technical journals. He has worked as a consultant for Exxon's Production and Research Division on offshore structures, for Skidmore, Owings and Merril on tall steel buildings, and for World Bank on the Chinese University Development Projects.

A widely respected author, Dr. Chen's works include *Limit Analysis and Soil Plasticity* (Elsevier, 1975), the two-volume *Theory of Beam-Columns* (McGraw-Hill, 1976–77), *Plasticity in Reinforced Concrete* (McGraw-Hill, 1982), *Plasticity for Structural Engineers* (Springer-Verlag, 1988), and *Stability Design of Steel Frames* (CRC Press, 1991). He is the editor of two book series, one in structural engineering and the other in civil engineering. He has authored or coauthored more than 500 papers in journals and conference proceedings. He is the author or coauthor of 18 books, has edited 12 books, and has contributed chapters to 28 other books. His more recent books are *Plastic Design and Second-Order Analysis of Steel Frames* (Springer-Verlag, 1994), the two-volume *Constitutive Equations for Engineering Materials* (Elsevier, 1994), *Stability Design of Semi-Rigid Frames* (Wiley-Interscience, 1995), and *LRFD Steel Design Using Advanced Analysis* (CRC Press, 1997). He is editor-in-chief of *The Civil Engineering Handbook* (CRC Press, 1995, winner of the Choice Outstanding Academic Book Award for 1996, *Choice Magazine*), and the *Handbook of Structural Engineering* (CRC Press, 1997).

Lian Duan is a Senior Bridge Engineer with the California Department of Transportation, U.S., and Professor of Structural Engineering at Taiyuan University of Technology, China.

He received his B.S. in civil engineering in 1975, M.S. in structural engineering in 1981 from Taiyuan University of Technology, and Ph.D. in structural engineering from Purdue University, West Lafayette, Indiana in 1990. Dr. Duan worked at the Northeastern China Power Design Institute from 1975 to 1978.

Dr. Duan's research interests cover areas including inelastic behavior of reinforced concrete and steel structures, structural stability and seismic bridge analysis and design. He has authored or coauthored more than 60 papers, chapters, and reports, and his research has focused on the development of unified interaction equations for steel beam-columns, flexural stiffness of reinforced concrete members, effective length factors of compression members, and design of bridge structures.

Dr. Duan is also an esteemed practicing engineer. He has designed numerous building and bridge structures. Most recently, he has been involved in the seismic retrofit design of the San Francisco-Oakland Bay Bridge West spans and made significant contributions to the project. He is coeditor of the *Structural Engineering Handbook* CRCnetBase 2000 (CRC Press, 2000).

Contributors

Michael I. Abrahams
Parsons, Brinckerhoff, Quade & Douglas, Inc.
New York, New York

Mohamed Akkari
California Department of Transportation
Sacramento, California

Fadel Alameddine
California Department of Transportation
Sacramento, California

Masoud Alemi
California Department of Transportation
Sacramento, California

S. Altman
California Department of Transportation
Sacramento, California

Rambabu Bavirisetty
California Department of Transportation
Sacramento, California

David P. Billington
Department of Civil Engineering and Operations Research
Princeton University
Princeton, New Jersey

Michael Blank
U.S.Army Corps of Engineers
Philadelphia, Pennsylvania

Simon A. Blank
California Department of Transportation
Walnut Creek, California

Michel Bruneau
Department of Civil Engineering
State University of New York
Buffalo, New York

Chun S. Cai
Florida Department of Transportation
Tallahassee, Florida

James Chai
California Department of Transportation
Sacramento, California

Hong Chen
J. Muller International, Inc.
Sacramento, California

Kang Chen
MG Engineering, Inc.
San Francisco, California

Wai-Fah Chen
School of Civil Engineering
Purdue University
West Lafayette, Indiana

Nan Deng
Bechtel Corporation
San Francisco, California

Robert J. Dexter
Department of Civil Engineering
University of Minnesota
Minneapolis, Minnesota

Ralph J. Dornsife
Washington State Department of Transportation
Olympia, Washington

Lian Duan
California Department of Transportation
Sacramento, California

Mingzhu Duan
Quincy Engineering, Inc.
Sacramento, California

Jackson Durkee
Consulting Structural Engineer
Bethlehem, Pennsylvania

Marc O. Eberhard
Department of Civil and Environmental Engineering
University of Washington
Seattle, Washington

Johnny Feng
J. Muller International, Inc.
Sacramento, California

Gerard F. Fox
HNTB (Ret.)
Garden City, New York

John W. Fisher
Department of Civil Engineering
Lehigh University
Bethlehem, Pennsylvania

Kenneth J. Fridley
Washington State University
Pullman, Washington

John H. Fujimoto
California Department of Transportation.
Sacramento, California

Mahmoud Fustok
California Department of Transportation
Sacramento, California

Ben C. Gerwick, Jr.
Ben C. Gerwick, Inc.
Consulting Engineers
San Francisco, California

Mahmoud Fustok
California Department of Transportation
Sacramento, California

Ben C. Gerwick, Jr.
Ben C. Gerwick, Inc.
Consulting Engineers
San Francisco, California

Chao Gong
ICF Kaiser Engineers
Oakland, California

Frederick Gottemoeller
Rosales Gottemoeller & Associates, Inc.
Columbia, Maryland

Fuat S. Guzaltan
Parsons, Brickerhoff, Quade & Douglas, Inc.
Princeton, New Jersey

Danjian Han
Department of Civil Engineering
South China University of Technology
Guangzhou, China

Ikuo Harazaki
Honshu–Shikoku Bridge Authority
Tokyo, Japan

Lars Hauge
COWI
Consulting Engineers and Planners
Lyngby, Denmark

Oscar Henriquez
Department of Civil Engineering
California State University
Long Beach, California

Susan E. Hida
California Department of Transportation
Sacramento, California

Dietrich L. Hommel
COWI
Consulting Engineers and Planners
Lyngby, Denmark

Ahmad M. Itani
University of Nevada
Reno, Nevada

Kevin I. Keady
California Department of Transportation
Sacramento, California

Michael D. Keever
California Department of Transportation
Sacramento, California

Sangjin Kim
Kyungpook National University
Taeg, South Korea

F. Wayne Klaiber
Department of Civil Engineering
Iowa State University
Ames, Iowa

Michael Knott
Moffatt & Nichol Engineers
Richmond, Virginia

Steven Kramer
University of Washington
Seattle, Washington

Alexander Krimotat
SC Solutions, Inc.
Santa Clara, California

John M. Kulicki
Modjeski and Masters, Inc.
Harrisburg, Pennsylvania

John Kung
California Department of Transportation
Sacramento, California

Farzin Lackpour
Parsons, Brickerhoff, Quade & Douglas, Inc.
Princeton, New Jersey

Don Lee
California Department of Transportation
Sacramento, California

Fritz Leonhardt
Stuttgart University
Stuttgart, Germany

Fang Li
California Department of Transportation
Sacramento, California

Guohao Li
Department of Bridge Engineering
Tongji University
Shanghai, People's Republic of China

Xila Liu
Department of Civil Engineering
Tsinghua University
Beijing, China

Luis R. Luberas
U.S.Army Corps of Engineers
Philadelphia, Pennsylvania

M. Myint Lwin
Washington State Department of Transportation
Olympia, Washington

Jyouru Lyang
California Department of Transportation
Sacramento, California

Youzhi Ma
Geomatrix Consultants, Inc.
Oakland, California

Alfred R. Mangus
California Department of Transportation
Sacramento, California

W. N. Marianos, Jr.
Modjeski and Masters, Inc.
Edwardsville, Illinois

Brian Maroney
California Department of Transportation
Sacramento, California

Serge Montens
Jean Muller International
St.-Quentin-en-Yvelines
France

Jean M. Muller
Jean M. Muller International
St.-Quentin-en-Yvelines
France

Masatsugu Nagai
Department of Civil and
Environmental Engineering
Nagaoka University of Technology
Nagaoka, Japan

Andrzej S. Nowak
Department of Civil and
Environmental Engineering
University of Michigan
Ann Arbor, Michigan

Atsushi Okukawa
Honshu–Shikoku Bridge Authority
Kobe, Japan

Dan Olsen
COWI
Consulting Engineers and Planners
Lyngby, Denmark

Klaus H. Ostenfeld
COWI
Consulting Engineers and Planners
Lyngby, Denmark

Joseph Penzien
International Civil Engineering
Consultants, Inc.
Berkeley, California

Philip C. Perdikaris
Department of Civil Engineering
Case Western Reserve University
Cleveland, Ohio

Joseph M. Plecnik
Department of Civil Engineering
California State University
Long Beach, California

Oleg A. Popov
Joint Stock Company
Giprotransmost (Tramos)
Moscow, Russia

Zolan Prucz
Modjeski and Masters, Inc.
New Orleans, Louisiana

Mark L. Reno
California Department of
Transportation
Sacramento, California

James Roberts
California Department of
Transportation
Sacramento, California

Norman F. Root
California Department of
Transportation
Sacramento, California

Yusuf Saleh
California Department of
Transportation
Sacramento, California

Thomas E. Sardo
California Department of
Transportation
Sacramento, California

Gerard Sauvageot
J. Muller International
San Diego, California

Charles Scawthorn
EQE International
Oakland, California

Charles Seim
T. Y. Lin International
San Francisco, California

Vadim A. Seliverstov
Joint Stock Company
Giprotransmost (Tramos)
Moscow, Russia

Li-Hong Sheng
California Department of
Transportation
Sacramento, California

Donald F. Sorgenfrei
Modjeski and Masters, Inc.
New Orleans, Louisiana

Jim Springer
California Department of
Transportation
Sacramento, California

Shawn Sun
California Department of
Transportation
Sacramento, California

Shuichi Suzuki
Honshu-Shikoku Bridge Authority
Tokyo, Japan

Andrew Tan
Everest International Consultants,
Inc.
Long Beach, California

Man-Chung Tang
T. Y. Lin International
San Francisco, California

Shouji Toma
Department of Civil Engineering
Hokkai-Gakuen University
Sapporo, Japan

M. S. Troitsky
Department of Civil Engineering
Concordia University
Montreal, Quebec
Canada

Keh-Chyuan Tsai
Department of Civil Engineering
National Taiwan University
Taipei, Taiwan
Republic of China

Keh-Chyuan Tsai
Department of Civil Engineering
National Taiwan University
Taipei, Taiwan
Republic of China

Wen-Shou Tseng
International Civil Engineering Consultants, Inc.
Berkeley, California

Chia-Ming Uang
Department of Civil Engineering
University of California
La Jolla, California

Shigeki Unjoh
Public Works Research Institute
Tsukuba Science City, Japan

Murugesu Vinayagamoorthy
California Department of Transportation
Sacramento, California

Jinrong Wang
URS Greiner
Roseville, California

Linan Wang
California Department of Transportation
Sacramento, California

Terry J. Wipf
Department of Civil Engineering
Iowa State University
Ames, Iowa

Zaiguang Wu
California Department of Transportation
Sacramento, California

Rucheng Xiao
Department of Bridge Engineering
Tongji University
Shanghai, China

Yan Xiao
Department of Civil Engineering
University of Southern California
Los Angeles, California

Tetsuya Yabuki
Department of Civil Engineering and Architecture
University of Ryukyu
Okinawa, Japan

Quansheng Yan
College of Traffic and Communication
South China University of Technology
Guangzhou, China

Leiming Zhang
Department of Civil Engineering
Tsinghua University
Beijing, China

Rihui Zhang
California Department of Transportation
Sacramento, California

Ke Zhou
California Department of Transportation
Sacramento, California

Contents

SECTION III Substructure Design

Section III

Substructure Design

26

Bearings

Johnny Feng
J. Muller International, Inc.

Hong Chen
J. Muller International, Inc.

26.1 Introduction

Bearings are structural devices positioned between the bridge superstructure and the substructure. Their principal functions are as follows:

1. To transmit loads from the superstructure to the substructure, and
2. To accommodate relative movements between the superstructure and the substructure.

The forces applied to a bridge bearing mainly include superstructure self-weight, traffic loads, wind loads, and earthquake loads.

Movements in bearings include translations and rotations. Creep, shrinkage, and temperature effects are the most common causes of the translational movements, which can occur in both transverse and longitudinal directions. Traffic loading, construction tolerances, and uneven settlement of the foundation are the common causes of the rotations.

Usually a bearing is connected to the superstructure through the use of a steel sole plate and rests on the substructure through a steel masonry plate. The sole plate distributes the concentrated bearing reactions to the superstructure. The masonry plate distributes the reactions to the substructure. The connections between the sole plate and the superstructure, for steel girders, are by bolting or welding. For concrete girders, the sole plate is embedded into the concrete with anchor studs. The masonry plate is typically connected to the substructure with anchor bolts.

26.2 Types of Bearings

Bearings may be classified as fixed bearings and expansion bearings. Fixed bearings allow rotations but restrict translational movements. Expansion bearings allow both rotational and translational movements. There are numerous types of bearings available. The following are the principal types of bearings currently in use.

0-8493-7434-0/00/$0.00+$.50

26.2.1 Sliding Bearings

A sliding bearing utilizes one plane metal plate sliding against another to accommodate translations. The sliding bearing surface produces a frictional force that is applied to the superstructure, the substructure, and the bearing itself. To reduce this friction force, PTFE (polytetrafluoroethylene) is often used as a sliding lubricating material. PTFE is sometimes referred to as Teflon, named after a widely used brand of PTFE, or TFE as appeared in AASHTO [1] and other design standards. In its common application, one steel plate coated with PTFE slides against another plate, which is usually of stainless steel.

Sliding bearings can be used alone or more often used as a component in other types of bearings. Pure sliding bearings can only be used when the rotations caused by the deflection at the supports are negligible. They are therefore limited to a span length of 15 m or less by ASHTTO [1].

A guiding system may be added to a sliding bearing to control the direction of the movement. It may also be fixed by passing anchor bolts through the plates.

26.2.2 Rocker and Pin Bearings

A rocker bearing is a type of expansion bearing that comes in a great variety. It typically consists of a pin at top that facilitates rotations, and a curved surface at the bottom that accommodates the translational movements (Figure 26.1a). The pin at the top is composed of upper and lower semi-circularly recessed surfaces with a solid circular pin placed between. Usually, there are caps at both ends of the pin to keep the pin from sliding off the seats and to resist uplift loads if required. The upper plate is connected to the sole plate by either bolting or welding. The lower curved plate sits on the masonry plate. To prevent the rocker from walking, keys are used to keep the rocker in place. A key can be a pintal which is a small trapezoidal steel bar tightly fitted into the masonry plate on one end and loosely inserted into the recessed rocker bottom plate on the other end. Or it can be an anchor bolt passing through a slotted hole in the bottom rocker plate.

A pin bearing is a type of fixed bearings that accommodates rotations through the use of a steel pin. The typical configuration of the bearing is virtually the same as the rocker described above except that the bottom curved rocker plate is now flat and directly anchored to the concrete pier (Figure 26.1b).

Rocker and pin bearings are primarily used in steel bridges. They are only suitable for the applications where the direction of the displacement is well defined since they can only accommodate translations and/or rotations in one direction. They can be designed to support relatively large loads but a high vertical clearance is usually required when the load or displacement is large. The practical limits of the load and displacement are about 1800 kN and ±100 mm, respectively, and rotations of several degrees are achievable [3].

Normally, the moment and lateral forces induced from the movement of these bearings are very small and negligible. However, metal bearings are susceptible to corrosion and deterioration. A corroded joint may induce much larger forces. Regular inspection and maintenance are, therefore, required.

26.2.3 Roller Bearings

Roller bearings are composed of one or more rollers between two parallel steel plates. Single roller bearings can facilitate both rotations and translations in the longitudinal direction, while a group of rollers would only accommodate longitudinal translations. In the latter case, the rotations are provided by combining rollers with a pin bearing (Figure 26.1c).

Roller bearings have been used in both steel and concrete bridges. Single roller bearings are relatively cheap to manufacture, but they only have a very limited vertical load capacity. Multiple roller bearings, on the other hand, may be able to support very large loads, but they are much more expensive.

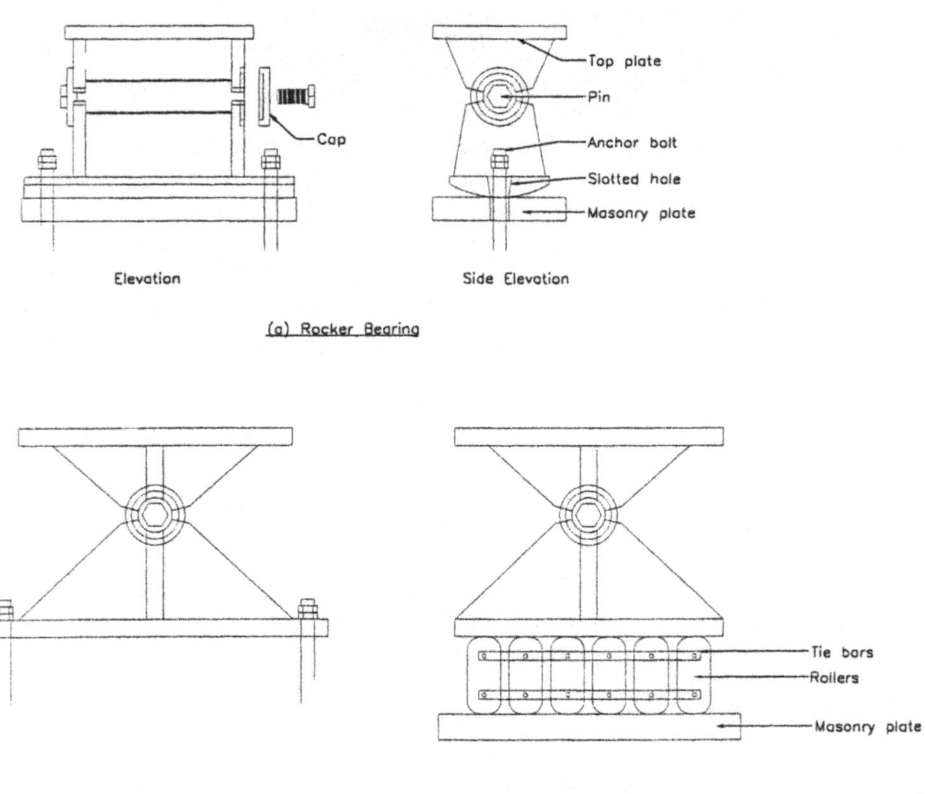

FIGURE 26.1 Typical rocker (a), pin (b), and roller bearings (c).

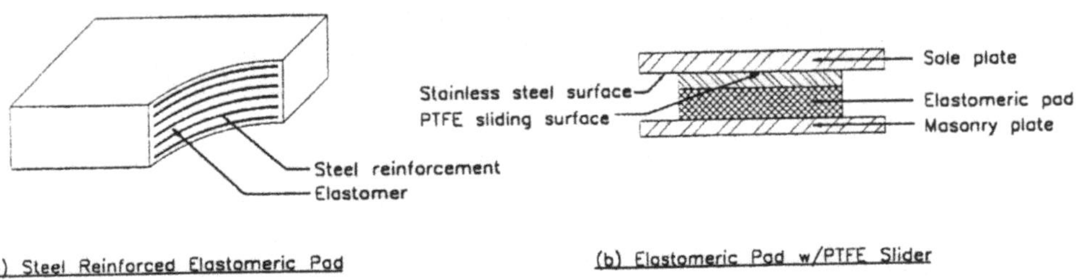

FIGURE 26.2 Elastomeric bearings. (a) Steel-reinforced elastomeric pad; (b) elastomeric pad with PTFE slider.

Like rocker and pin bearings, roller bearings are also susceptible to corrosion and deterioration. Regular inspection and maintenance are essential.

26.2.4 Elastomeric Bearings

An elastomeric bearing is made of elastomer (either natural or synthetic rubber). It accommodates both translational and rotational movements through the deformation of the elastomer.

Elastomer is flexible in shear but very stiff against volumetric change. Under compressive load, the elastomer expands laterally. To sustain large load without excessive deflection, reinforcement is used to restrain lateral bulging of the elastomer. This leads to the development of several types of elastomeric bearing pads — plain, fiberglass-reinforced, cotton duck–reinforced, and steel-reinforced elastomeric pads. Figure 26.2a shows a steel-reinforced elastomeric pad.

Plain elastomeric pads are the weakest and most flexible because they are only restrained from bulging by friction forces alone. They are typically used in short- to medium-span bridges, where bearing stress is low. Fiberglass-reinforced elastomeric pads consist of alternate layers of elastomer and fiberglass reinforcement. Fiberglass inhibits the lateral deformation of the pads under compressive loads so that larger load capacity can be achieved. Cotton-reinforced pads are elastomeric pads reinforced with closely spaced layers of cotton duck. They display high compressive stiffness and strength but have very limited rotational capacities. The thin layers also lead to high shear stiffness, which results in large forces in the bridge. So sometimes they are combined with a PTFE slider on top of the pad to accommodate translations (Figure 26.2b). Steel-reinforced elastomeric pads are constructed by vulcanizing elastomer to thin steel plates. They have the highest load capacity among the different types of elastomeric pads, which is only limited by the manufacturer's ability to vulcanize a large volume of elastomer uniformly.

All above-mentioned pads except steel-reinforced pads can be produced in a large sheet and cut to size for any particular application. Steel-reinforced pads, however, have to be custom-made for each application due to the edge cover requirement for the protection of the steel from corrosion. The steel-reinforced pads are the most expensive while the cost of the plain elastomeric pads is the lowest.

Elastomeric bearings are generally considered the preferred type of bearings because they are low cost and almost maintenance free. In addition, elastomeric bearings are extremely forgiving of loads and movements exceeding the design values.

26.2.4 Curved Bearings

A curved bearing consists of two matching curved plates with one sliding against the other to accommodate rotations. The curved surface can be either cylindrical which allows the rotation about only one axis or spherical which allows the bearing to rotate about any axis.

Lateral movements are restrained in a pure curved bearing and a limited lateral resistance may be developed through a combination of the curved geometry and the gravity loads. To accommodate lateral movements, a PTFE slider must be attached to the bearings. Keeper plates are often used to keep the superstructure moving in one direction. Large load and rotational capacities can be designed for curved bearings. The vertical capacity is only limited by its size, which depends largely on machining capabilities. Similarly, rotational capacities are only limited by the clearances between the components.

Figure 26.3a shows a typical expansion curved bearing. The lower convex steel plate that has a stainless steel mating surface is recessed in the masonry plate. The upper concave plate with a matching PTFE sliding surface sits on top of the lower convex plate for rotations. Between the sole plate and the upper concave plate there is a flat PTFE sliding surface that will accommodate lateral movements.

26.2.5 Pot Bearings

A pot bearing comprises a plain elastomeric disk that is confined in a shallow steel ring, or pot (Figure 26.3b). Vertical loads are transmitted through a steel piston that fits closely to the steel ring (pot wall). Flat sealing rings are used to contain the elastomer inside the pot. The elastomer behaves like a viscous fluid within the pot as the bearing rotates. Because the elastomeric pad is confined, much larger load can be carried this way than through conventional elastomeric pads.

Translational movements are restrained in a pure pot bearing, and the lateral loads are transmitted through the steel piston moving against the pot wall. To accommodate translational movement, a PTFE sliding surface must be used. Keeper plates are often used to keep the superstructure moving in one direction.

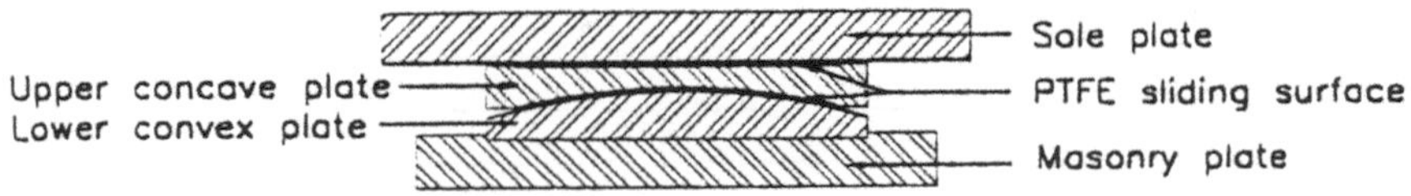

(a) Spherical Bearing

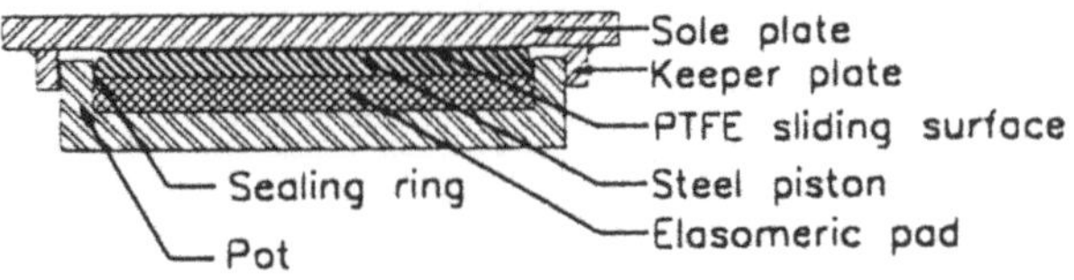

(b) PotBearing

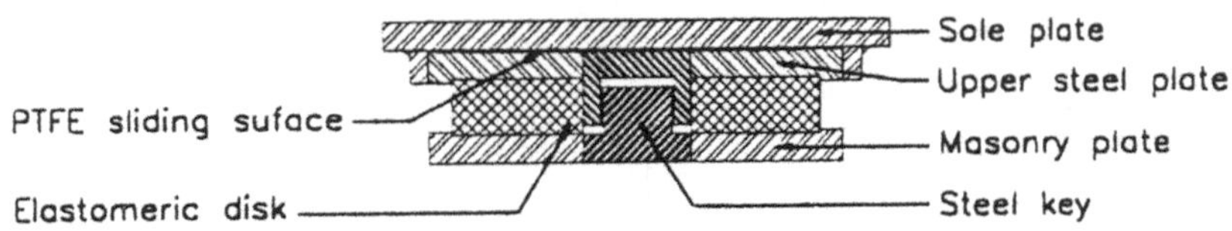

(c) Disk Bearing

FIGURE 26.3 Typical spherical (a), pot (b), and disk (c) bearings

26.2.6 Disk Bearings

A disk bearing, as illustrated in Figure 26.3c, utilizes a hard elastomeric (polyether urethane) disk to support the vertical loads and a metal key in the center of the bearing to resist horizontal loads. The rotational movements are accommodated through the deformation of the elastomer. To accommodate translational movements, however, a PTFE slider is required. In this kind of bearings, the polyether urethane disk must be hard enough to resist large vertical load without excessive deformation and yet flexible enough to accommodate rotations easily.

26.3 Selection of Bearings

Generally the objective of bearing selection is to choose a bearing system that suits the needs with a minimum overall cost. The following procedures may be used for the selection of the bearings.

26.3.1 Determination of Functional Requirements

First, the vertical and horizontal loads, the rotational and translational movements from all sources including dead and live loads, wind loads, earthquake loads, creep and shrinkage, prestress, thermal and construction tolerances need to be calculated. Table 26.1 may be used to tabulate these requirements.

TABLE 26.1 Typical Bridge Bearing Schedule

Bridge Name of Reference					
Bearing Identification mark					
Number of bearings required					
Seating Material		Upper Surface			
		Lower Surface			
Allowable average contact pressure (PSI)		Upper Surface	Serviceability		
			Strength		
		Lower Surface	Serviceability		
			Strength		
Design Load effects (KIP)	Service limit state		Vertical	max.	
				perm	
				min.	
			Transverse		
			Longitudinal		
	Strength limit state		Vertical		
			Transverse		
			Longitudinal		
Translation	Service limit state	Irreversible	Transverse		
			Longitudinal		
		Reversible	Transverse		
			Longitudinal		
	Strength limit state	Irreversible	Transverse		
			Longitudinal		
		Reversible	Transverse		
			Longitudinal		
Rotation (RAD)	Service limit state	Irreversible	Transverse		
			Longitudinal		
		Reversible	Transverse		
			Longitudinal		
	Strength limit state	Irreversible	Transverse		
			Longitudinal		
		Reversible	Transverse		
			Longitudinal		
Maximum bearing dimensions (IN)	Upper surface		Transverse		
			Longitudinal		
	Lower surface		Transverse		
			Longitudinal		
	Overall height				
Tolerable movement of bearing under transient loads (IN)			Vertical		
			Transverse		
			Longitudinal		
Allowable resistance to translation under service limit state (KIP)			Transverse		
			Longitudinal		
Allowable resistance to rotation under service limit state (K/FT)			Transverse		
			Longitudinal		
Type of attachment to structure and substructure			Transverse		
			Longitudinal		

Source: AASHTO, *LRFD Bridge Design Scecifications*, American Association of State Highway and Transportation Officials, Washington, D.C.

TABLE 26.2 Summery of Bearing Capacities [3,5]

	Load		Translation		Rotation	Costs	
Bearing Type	Min. (KN)	Max. (KN)	Min. (mm)	Max. (mm)	Max. (rad)	Initial	Maintenance
Elastomeric pads							
Plain	0	450	0	15	0.01	Low	Low
Cotton duck reinforced	0	1,400	0	5	0.003	Low	Low
Fiberglass reinforced	0	600	0	25	0.015	Low	Low
Steel reinforced	225	3,500	0	100	0.04	Low	Low
Flat PTFE slider	0	>10,000	25	>100	0	Low	Moderate
Disk bearing	1,200	10,000	0	0	0.02	Moderate	Moderate
Pot bearing	1,200	10,000	0	0	0.02	Moderate	High
Pin bearing	1,200	4,500	0	0	>0.04	Moderate	High
Rocker bearing	0	1,800	0	100	>0.04	Moderate	High
Single roller	0	450	25	>100	>0.04	Moderate	High
Curved PTFE bearing	1,200	7,000	0	0	>0.04	High	Moderate
Multiple rollers	500	10,000	100	>100	>0.04	High	High

26.3.2 Evaluation of Bearings

The second step is to determine the suitable bearing types based on the above bridge functional requirements, and other factors including available clearance, environment, maintenance, cost, availability, and client's preferences. Table 26.2 summarizes the load, movement capacities, and relative costs for each bearing type and may be used for the selection of the bearings.

It should be noted that the capacity values in Table 26.2 are approximate. They are the practical limits of the most economical application for each bearing type. The costs are also relative, since the true price can only be determined by the market. At the end of this step, several qualified bearing systems with close cost ratings may be selected [5].

26.3 Preliminary Bearing Design

For the various qualified bearing alternatives, preliminary designs are performed to determine the approximate geometry and material properties in accordance with design specifications. It is likely that one or more of the previously acceptable alternatives will be eliminated in this step because of an undesirable attribute such as excessive height, oversize footprint, resistance at low temperature, sensitivity to installation tolerances, etc. [3].

At the end of this step, one or more bearing types may still be feasible and they will be included in the bid package as the final choices of the bearing types.

26.4 Design of Elastomeric Bearings

26.4.1 Design Procedure

The design procedure is according to AASHTO-LRFD [1] and is as follows:

1. Determine girder temperature movement (Art. 5.4.2.2).
2. Determine girder shortenings due to post-tensioning, concrete shrinkage, etc.
3. Select a bearing thickness based on the bearing total movement requirements (Art. 14.7.5.3.4).
4. Compute the bearing size based on bearing compressive stress (Art. 14.7.5.3.2).
5. Compute instantaneous compressive deflection (Art. 14.7.5.3.3).
6. Combine bearing maximum rotation.
7. Check bearing compression and rotation (Art. 14.7.5.3.5).
8. Check bearing stability (Art. 14.7.5.3.6).
9. Check bearing steel reinforcement (Art. 14.7.5.3.7).

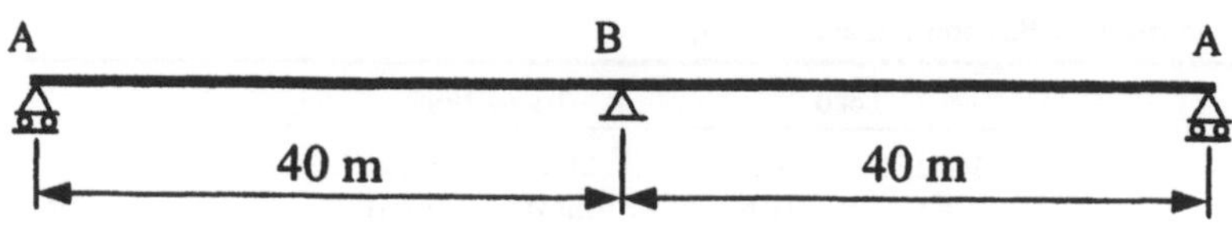

FIGURE 26.4 Bridge layout

26.4.2 Design Example (Figure 26.4)

Given

L	= expandable span length	= 40 m
R_{DL}	= DL reaction/girder	= 690 kN
R_{LL}	= LL reaction (without impact)/girder	= 220 kN
θ_s	= bearing design rotation at service limit state	= 0.025 rad
ΔT	= maximum temperature change	= 21°C
Δ_{PT}	= girder shortening due to post tensioning	= 21 mm
Δ_{SH}	= girder shortening due to concrete shrinkage	= 2 mm
G	= shear modulus of elastomer	= 0.9 ~ 1.38 MPa
γ	= load factor for uniform temperature, etc.	= 1.2
ΔF_{TH}	= constant amplitude fatigue threshold for Category A	= 165 MPa

Using 60 durometer reinforced bearing:

F_y	= yield strength of steel reinforcement	= 350 MPa

Sliding bearing used:

1. **Temperature Movement**
   From Art. 5.4.2.2, for normal density concrete, the thermal coefficient α is

$$\alpha = 10.8 \times 10^{-6}/°C$$

$$\Delta_{TEMP} = (\alpha)(\Delta T)(L) = (10.8 \times 10^{-6}/°C)(21°C)(40{,}000 \text{ mm}) = 9 \text{ mm}$$

2. **Girder Shortenings**

$$\Delta_{PT} = 21 \text{ mm and } \Delta_{SH} = 2 \text{ mm}$$

3. **Bearing Thickness**
   h_{rt} = total elastomer thickness
   h_{ri} = thickness of ith elastomeric layer
   n = number of interior layers of elastomeric layer
   Δ_S = bearing maximum longitudinal movement = $\gamma \cdot (\Delta_{TEMP} + \Delta_{PT} + \Delta_{SH})$
   Δ_S = 1.2 × (9 mm + 21 mm + 2 mm) = 38.4 mm
   h_{rt} = bearing thickness $\geq 2\Delta_S$ (AASHTO Eq. 14.7.5.3.4-1)
   h_{rt} = 2 × (38.4 mm) = 76.8

   Try h_{rt} = 120 mm, h_{ri} = 20 mm and n = 5

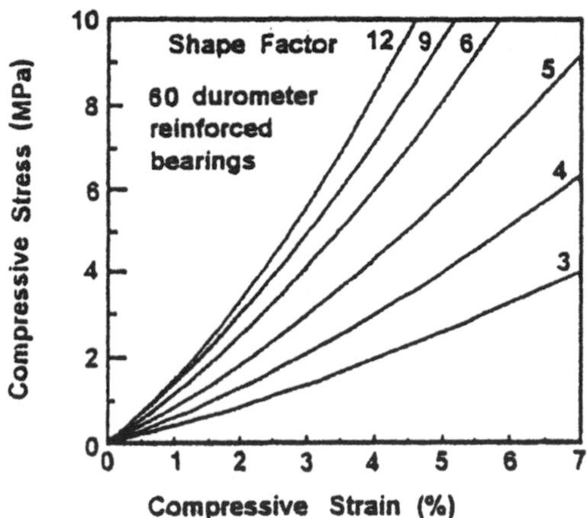

FIGURE 26.5 Stress–strain curves. (From AASHTO, Figure C14.7.5.3.3.1.)

4. **Bearing Size**

L = length of bearing
W = width of bearing
S_i = shape factor of thickness layer of the bearing = $\dfrac{LW}{2h_{ri}(L+W)}$

For a bearing subject to shear deformation, the compressive stresses should satisfy:

σ_S = average compressive stress due to the total load $\leq 1.66GS \leq 11$ (AASHTO Eq. 14.7.5.3.2-1)
σ_L = average compressive stress due to the live load $\leq 0.66\ GS$ (AASHTO Eq. 14.7.5.3.2-1)

$$\sigma_s = \frac{R}{LW} = \frac{1.66GLW}{2h_{ri}(L+W)}$$

Assuming σ_S is critical, solve for L and W by error and trial.

$$L = 300 \text{ mm and } W = 460 \text{ mm}$$

$$S = \frac{LW}{2h_{ri}(L+W)} = \frac{(300\text{ mm})(460\text{ mm})}{2(20\text{ mm})(300\text{ mm}+460\text{ mm})} = 4.54$$

$$\sigma_L = \frac{R_L}{LW} = \frac{(200{,}000\text{ N})}{(300\text{ mm})(460\text{ mm})} = 1.6\text{ MPa} \qquad \text{OK}$$

$$\leq 0.66\ GS = 0.66(1.0\text{ MPa})(4.54) = 3.0\text{ MPa}$$

5. **Instantaneous Compressive Deflection**

For $\sigma_S = 6.59$ MPa and $S = 4.54$, one can determine the value of ε_i from Figure 26.5:

$$\varepsilon_i = 0.062$$

$$\delta = \sum \varepsilon_i h_{ri} \qquad \text{(AASHTO Eq. 14.7.5.3.3-1)}$$

$$= 6(0.062)(20\text{ mm}) = 7.44\text{ mm}$$

6. Bearing Maximum Rotation

The bearing rotational capacity can be calculated as

$$\sigma_{\text{capacity}} = \frac{2\delta}{L} = \frac{2(7.44\ \text{mm})}{300\ \text{mm}} = 0.05\ \text{rad} < \theta_{\text{design}} = 0.025\ \text{rad} \qquad \text{OK}$$

7. Combined Bearing Compression and Rotation

a. *Uplift requirement* (AASHTO Eq. 14.7.5.3.5-1):

$$\sigma_{\text{s,uplift}} = 1.0GS\left(\frac{\theta_{\text{design}}}{n}\right)\left(\frac{L}{h_{\text{ri}}}\right)^2$$

$$= 1.0(1.2)(4.54)\left(\frac{0.025}{5}\right)\left(\frac{300}{20}\right)^2 = 6.13\ \text{MPa} < \sigma_s = 6.59\ \text{MPa} \qquad \text{OK}$$

b. *Shear deformation requirement* (AASHTO Eq. 14.7.5.3.5-2):

$$\sigma_{\text{s,shear}} = 1.875GS\left(1 - 0.20\left(\frac{\theta_{\text{design}}}{n}\right)\left(\frac{L}{h_{\text{ri}}}\right)^2\right)$$

$$= 1.875(1.0)(4.54)\left(1 - 0.20\left(\frac{0.025}{5}\right)\left(\frac{300}{20}\right)^2\right) = 6.60\ \text{MPa} > \sigma_s = 6.59\ \text{MPa} \qquad \text{OK}$$

8. Bearing Stability

Bearings shall be designed to prevent instability at the service limit state load combinations. The average compressive stress on the bearing is limited to half the predicted buckling stress. For this example, the bridge deck, if free to translate horizontally, the average compressive stress due to dead and live load, σ_s, must satisfy:

$$\sigma_s \leq \frac{G}{2A - B} \qquad \text{(AASHTO Eq. 14.7.5.3.6-1)}$$

where

$$A = \frac{1.92\frac{h_{rt}}{L}}{S\sqrt{1 + \frac{2.0\,L}{W}}} \qquad \text{(AASHTO Eq. 14.7.5.3.6-3)}$$

$$= \frac{1.92\frac{(120\ \text{mm})}{(300\ \text{mm})}}{(4.54)\sqrt{1 + \frac{2.0(300\ \text{mm})}{(460\ \text{mm})}}} = 0.11$$

$$B = \frac{2.67}{S(S+2.0)\sqrt{1+\frac{L}{4.0W}}}$$

(AASHTO Eq. 14.7.5.3.6-4)

$$= \frac{2.67}{(4.54)(4.54+2.0)\sqrt{1+\frac{(300\text{ mm})}{4.0(460\text{ mm})}}} = 0.08$$

$$\frac{G}{2A-B} = \frac{(1.0\text{ MPa})}{2(0.11)-(0.08)} = 6.87 > \sigma_s$$ OK

9. **Bearing Steel Reinforcement**
The bearing steel reinforcement must be designed to sustain the tensile stresses induced by compression of the bearing. The thickness of steel reinforcement, h_s, should satisfy:
a. *At the service limit state:*

$$h_s \geq \frac{3h_{max}\sigma_s}{F_y}$$ (AASHTO Eq. 14.7.5.3.7-1)

$$= \frac{3(20\text{ mm})(6.59\text{ MPa})}{(350\text{ MPa})} = 1.13\text{ mm}$$ (governs)

b. *At the fatigue limit state:*

$$h_s \geq \frac{2h_{max}\sigma_L}{\ddot{A}F_y}$$ (AASHTO Eq. 14.7.5.3.7-2)

$$= \frac{2(20\text{mm})(1.6\text{MPa})}{(165\text{MPa})} = 0.39\text{mm}$$

where h_{max} = thickness of thickest elastomeric layer in elastomeric bearing = h_{ri}.

Elastomeric Bearings Details
Five interior lays with 20 mm thickness each layer
Two exterior lays with 10 mm thickness each layer
Six steel reinforcements with 1.2 mm each
Total thickness of bearing is 127.2 mm
Bearing size: 300 mm (longitudinal) × 460 mm (transverse)

References

1. AASHTO, *LRFD Bridge Design Specifications*, American Association of State Highway and Transportation Officials, Washington, D.C., 1994.
2. AASHTO, *Standard Specifications for the Design of Highway Bridges*, 16th ed. American Association of State Highway and Transportation Officials, Washington, D.C., 1996.
3. Stanton, J. F., Roeder, C. W., and Campbell, T. I., High Load Multi-Rotational Bridge Bearings, NCHRP Report 10-20A, Transportation Research Board, National Research Council, Washington, D.C., 1993.
4. Caltrans, Memo to Designers, California Department of Transportation, Sacramento, 1994.
5. AISI, Steel bridge bearing selection and design guide, *Highway Structures Design Handbook*, Vol. II, American Iron and Steel Institute, Washington, D.C., 1996, chap. 4.

27

Piers and Columns

Jinrong Wang
URS Greiner

27.1 Introduction

Piers provide vertical supports for spans at intermediate points and perform two main functions: transferring superstructure vertical loads to the foundations and resisting horizontal forces acting on the bridge. Although piers are traditionally designed to resist vertical loads, it is becoming more and more common to design piers to resist high lateral loads caused by seismic events. Even in some low seismic areas, designers are paying more attention to the ductility aspect of the design. Piers are predominantly constructed using reinforced concrete. Steel, to a lesser degree, is also used for piers. Steel tubes filled with concrete (composite) columns have gained more attention recently.

This chapter deals only with piers or columns for conventional bridges, such as grade separations, overcrossings, overheads, underpasses, and simple river crossings. Reinforced concrete columns will be discussed in detail while steel and composite columns will be briefly discussed. Substructures for arch, suspension, segmental, cable-stayed, and movable bridges are excluded from this chapter. Chapter 28 discusses the substructures for some of these special types of bridges.

27.2 Structural Types

27.2.1 General

Pier is usually used as a general term for any type of substructure located between horizontal spans and foundations. However, from time to time, it is also used particularly for a solid wall in order to distinguish it from columns or bents. From a structural point of view, a column is a member that resists the lateral force mainly by flexure action whereas a pier is a member that resists the lateral force mainly by a shear mechanism. A pier that consists of multiple columns is often called a *bent.*

There are several ways of defining pier types. One is by its structural connectivity to the superstructure: monolithic or cantilevered. Another is by its sectional shape: solid or hollow; round, octagonal, hexagonal, or rectangular. It can also be distinguished by its framing configuration: single or multiple column bent; hammerhead or pier wall.

0-8493-7434-0/00/$0.00+$.50

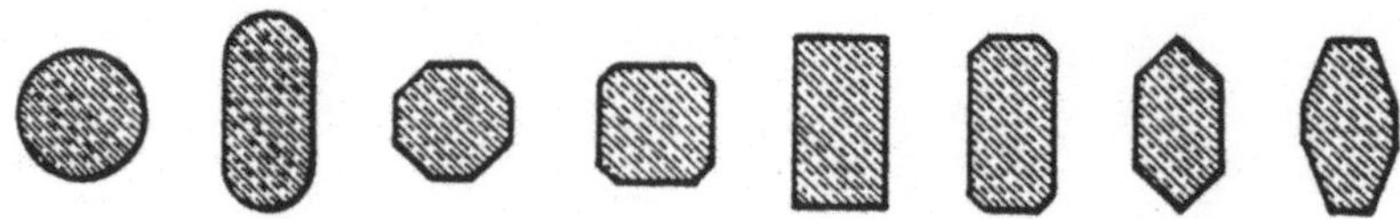

FIGURE 27.1 Typical cross-section shapes of piers for overcrossings or viaducts on land.

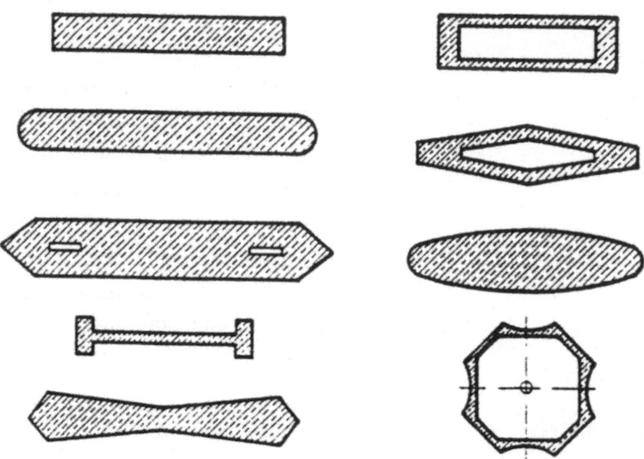

FIGURE 27.2 Typical cross-section shapes of piers for river and waterway crossings.

27.2.2 Selection Criteria

Selection of the type of piers for a bridge should be based on functional, structural, and geometric requirements. Aesthetics is also a very important factor of selection since modern highway bridges are part of a city's landscape. Figure 27.1 shows a collection of typical cross section shapes for overcrossings and viaducts on land and Figure 27.2 shows some typical cross section shapes for piers of river and waterway crossings. Often, pier types are mandated by government agencies or owners. Many state departments of transportation in the United States have their own standard column shapes.

Solid wall piers, as shown in Figures 27.3a and 27.4, are often used at water crossings since they can be constructed to proportions that are both slender and streamlined. These features lend themselves well for providing minimal resistance to flood flows.

Hammerhead piers, as shown in Figure 27.3b, are often found in urban areas where space limitation is a concern. They are used to support steel girder or precast prestressed concrete superstructures. They are aesthetically appealing. They generally occupy less space, thereby providing more room for the traffic underneath. Standards for the use of hammerhead piers are often maintained by individual transportation departments.

A column bent pier consists of a cap beam and supporting columns forming a frame. Column bent piers, as shown in Figure 27.3c and Figure 27.5, can either be used to support a steel girder superstructure or be used as an integral pier where the cast-in-place construction technique is used. The columns can be either circular or rectangular in cross section. They are by far the most popular forms of piers in the modern highway system.

A pile extension pier consists of a drilled shaft as the foundation and the circular column extended from the shaft to form the substructure. An obvious advantage of this type of pier is that it occupies a minimal amount of space. Widening an existing bridge in some instances may require pile extensions because limited space precludes the use of other types of foundations.

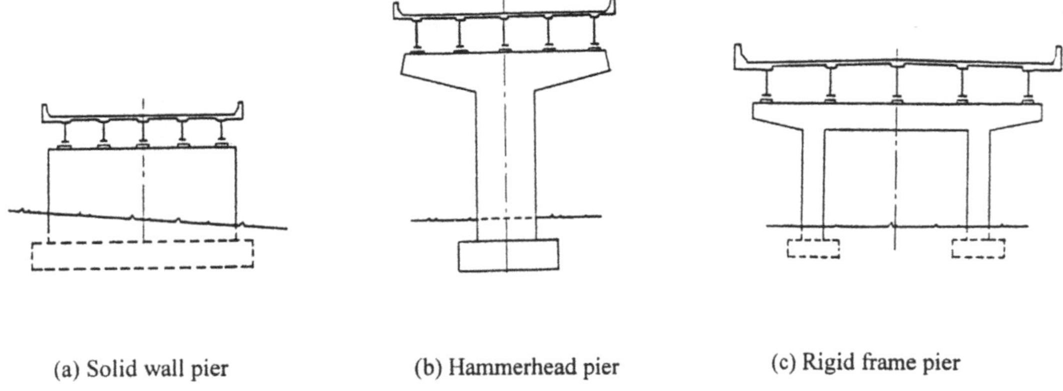

(a) Solid wall pier (b) Hammerhead pier (c) Rigid frame pier

FIGURE 27.3 Typical pier types for steel bridges.

Selections of proper pier type depend upon many factors. First of all, it depends upon the type of superstructure. For example, steel girder superstructures are normally supported by cantilevered piers, whereas the cast-in-place concrete superstructures are normally supported by monolithic bents. Second, it depends upon whether the bridges are over a waterway or not. Pier walls are preferred on river crossings, where debris is a concern and hydraulics dictates it. Multiple pile extension bents are commonly used on slab bridges. Last, the height of piers also dictates the type selection of piers. The taller piers often require hollow cross sections in order to reduce the weight of the substructure. This then reduces the load demands on the costly foundations. Table 27.1 summarizes the general type selection guidelines for different types of bridges.

27.3 Design Loads

Piers are commonly subjected to forces and loads transmitted from the superstructure, and forces acting directly on the substructure. Some of the loads and forces to be resisted by the substructure include:

- Dead loads
- Live loads and impact from the superstructure
- Wind loads on the structure and the live loads
- Centrifugal force from the superstructure
- Longitudinal force from live loads
- Drag forces due to the friction at bearings
- Earth pressure
- Stream flow pressure
- Ice pressure
- Earthquake forces
- Thermal and shrinkage forces
- Ship impact forces
- Force due to prestressing of the superstructure
- Forces due to settlement of foundations

The effect of temperature changes and shrinkage of the superstructure needs to be considered when the superstructure is rigidly connected with the supports. Where expansion bearings are used, forces caused by temperature changes are limited to the frictional resistance of bearings.

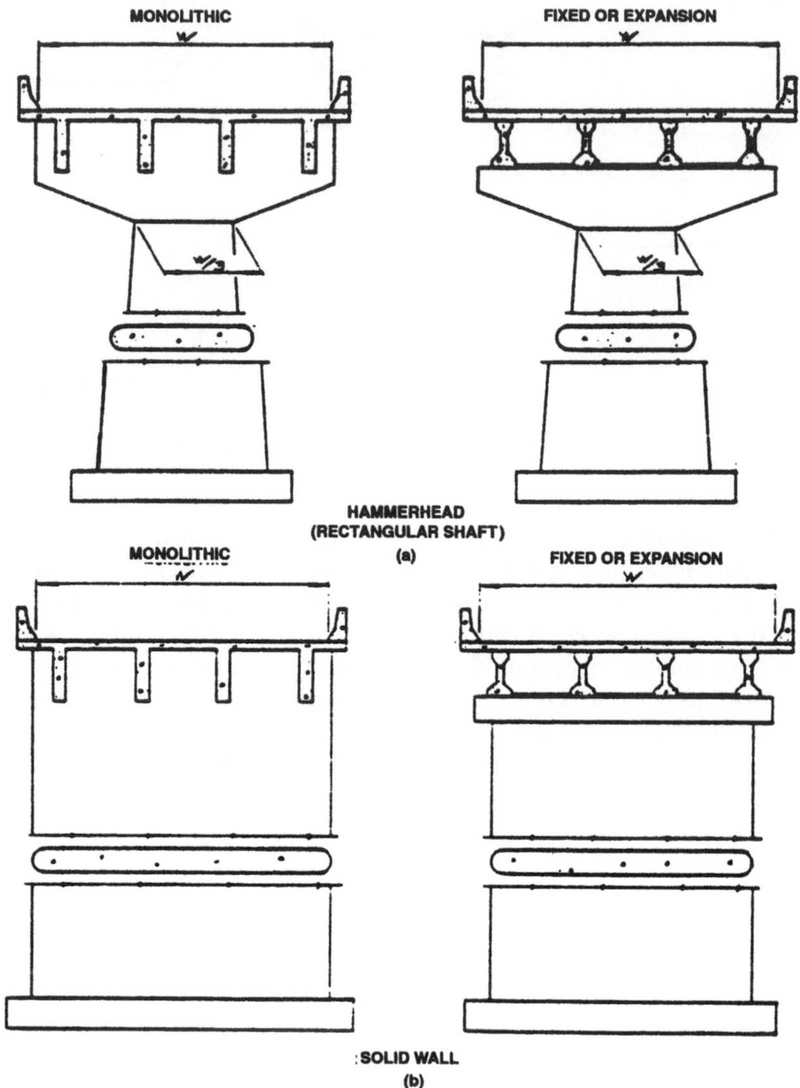

FIGURE 27.4 Typical pier types and configurations for river and waterway crossings.

Readers should refer to Chapters 5 and 6 for more details about various loads and load combinations and Part IV about earthquake loads. In the following, however, two load cases, live loads and thermal forces, will be discussed in detail because they are two of the most common loads on the piers, but are often applied incorrectly.

27.3.1 Live Loads

Bridge live loads are the loads specified or approved by the contracting agencies and owners. They are usually specified in the design codes such as AASHTO LRFD Bridge Design Specifications [1]. There are other special loading conditions peculiar to the type or location of the bridge structure which should be specified in the contracting documents.

Live-load reactions obtained from the design of individual members of the superstructure should not be used directly for substructure design. These reactions are based upon maximum conditions

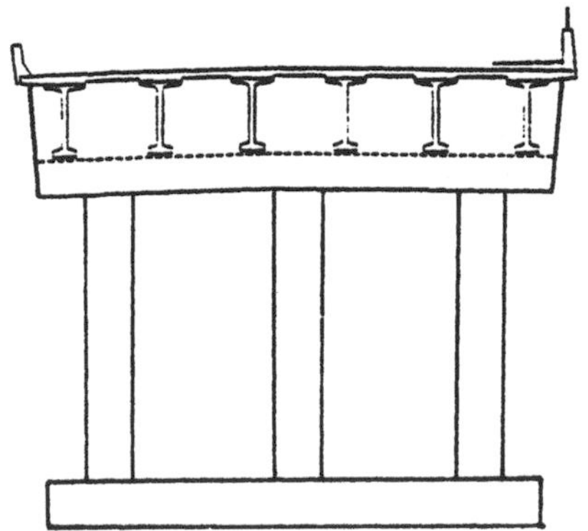

(a) Bent for precast girders

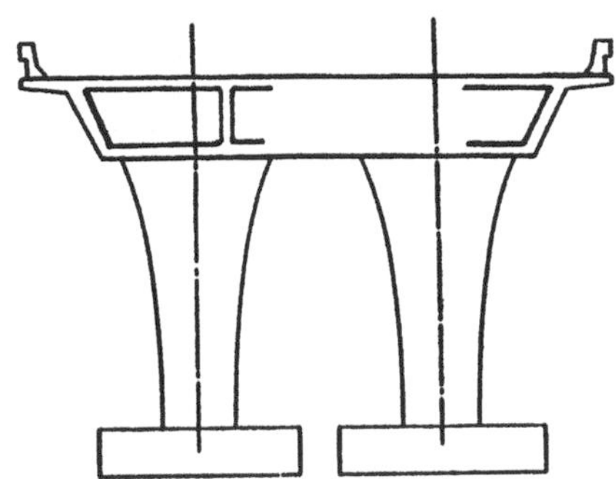

(b) Bent for cast-in-place girders

FIGURE 27.5 Typical pier types for concrete bridges.

TABLE 27.1 General Guidelines for Selecting Pier Types

		Applicable Pier Types
		Steel Superstructure
Over water	Tall piers	Pier walls or hammerheads (T-piers) (Figures 27.3a and b); hollow cross sections for most cases; cantilevered; could use combined hammerheads with pier wall base and step tapered shaft
	Short piers	Pier walls or hammerheads (T-piers) (Figures 27.3a and b); solid cross sections; cantilevered
On land	Tall piers	Hammerheads (T-piers) and possibly rigid frames (multiple column bents)(Figures 27.3b and c); hollow cross sections for single shaft and solid cross sections for rigid frames; cantilevered
	Short piers	Hammerheads and rigid frames (Figures 27.3b and c); solid cross sections; cantilevered
		Precast Prestressed Concrete Superstructure
Over water	Tall piers	Pier walls or hammerheads (Figure 27.4); hollow cross sections for most cases; cantilevered; could use combined hammerheads with pier wall base and step-tapered shaft
	Short piers	Pier walls or hammerheads; solid cross sections; cantilevered
On land	Tall piers	Hammerheads and possibly rigid frames (multiple column bents); hollow cross sections for single shafts and solid cross sections for rigid frames; cantilevered
	Short piers	Hammerheads and rigid frames (multiple column bents) (Figure 27.5a); solid cross sections; cantilevered
		Cast-in-Place Concrete Superstructure
Over water	Tall piers	Single shaft pier (Figure 27.4); superstructure will likely cast by traveled forms with balanced cantilevered construction method; hollow cross sections; monolithic; fixed at bottom
	Short piers	Pier walls (Figure 27.4); solid cross sections; monolithic; fixed at bottom
On land	Tall piers	Single or multiple column bents; solid cross sections for most cases, monolithic; fixed at bottom
	Short piers	Single or multiple column bents (Figure 27.5b); solid cross sections; monolithic; pinned at bottom

for one beam and make no allowance for distribution of live loads across the roadway. Use of these maximum loadings would result in a pier design with an unrealistically severe loading condition and uneconomical sections.

For substructure design, a maximum design traffic lane reaction using either the standard truck load or standard lane load should be used. Design traffic lanes are determined according to AASHTO

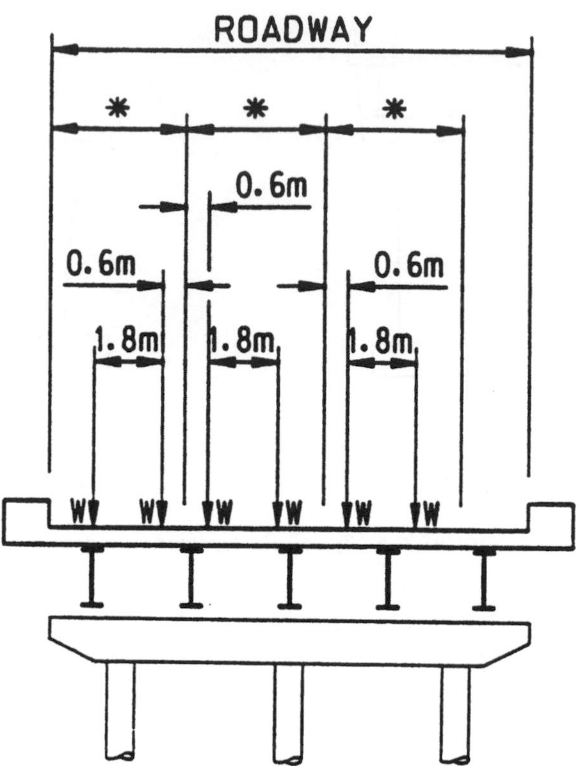

* DESIGN TRAFFIC LANE = 3.6m
NO. OF LANES = ROADWAY ÷ 3.6
REDUCED TO NEAREST WHOLE NUMBER

WHEEL LOADING $W = \frac{R_2}{2}$

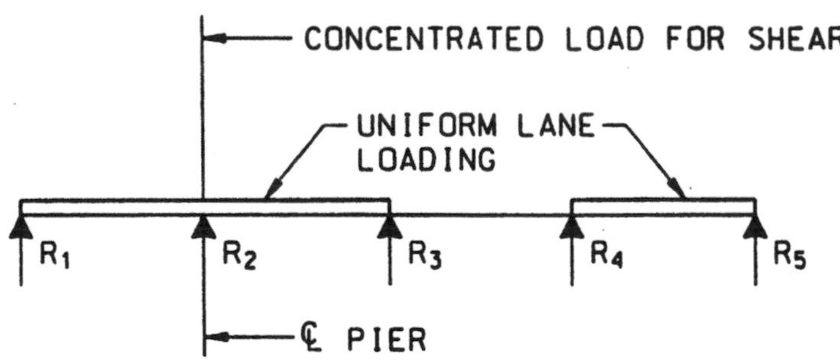

FIGURE 27.6 Wheel load arrangement to produce maximum positive moment.

LRFD [1] Section 3.6. For the calculation of the actual beam reactions on the piers, the maximum lane reaction can be applied within the design traffic lanes as wheel loads, and then distributed to the beams assuming the slab between beams to be simply supported. (Figure 27.6). Wheel loads can be positioned anywhere within the design traffic lane with a minimum distance between lane boundary and wheel load of 0.61 m (2 ft).

The design traffic lanes and the live load within the lanes should be arranged to produce beam reactions that result in maximum loads on the piers. AASHTO LRFD Section 3.6.1.1.2 provides load reduction factors due to multiple loaded lanes.

TABLE 27.2 Dynamic Load Allowance, IM

Component	IM
Deck joints — all limit states	75%
All other components	
• Fatigue and fracture limit state	15%
• All other limit states	33%

Live-load reactions will be increased due to impact effect. AASHTO LRFD [1] refers to this as the *dynamic load allowance, IM.* and is listed here as in Table 27.2.

27.3.2 Thermal Forces

Forces on piers due to thermal movements, shrinkage, and prestressing can become large on short, stiff bents of prestressed concrete bridges with integral bents. Piers should be checked against these forces. Design codes or specifications normally specify the design temperature range. Some codes even specify temperature distribution along the depth of the superstructure member.

The first step in determining the thermal forces on the substructures for a bridge with integral bents is to determine the point of no movement. After this point is determined, the relative displacement of any point along the superstructure to this point is simply equal to the distance to this point times the temperature range and times the coefficient of expansion. With known displacement at the top and known boundary conditions at the top and bottom, the forces on the pier due to the temperature change can be calculated by using the displacement times the stiffness of the pier.

The determination of the point of no movement is best demonstrated by the following example, which is adopted from Memo to Designers issued by California Department of Transportations [2]:

Example 27.1

A 225.55-m (740-foot)-long and 23.77-m (78-foot) wide concrete box-girder superstructure is supported by five two-column bents. The size of the column is 1.52 m (5 ft) in diameter and the heights vary between 10.67 m (35 ft) and 12.80 m (42 ft). Other assumptions are listed in the calculations. The calculation is done through a table. Please refer Figure 27.7 for the calculation for determining the point of no movement.

27.4 Design Criteria

27.4.1 Overview

Like the design of any structural component, the design of a pier or column is performed to fulfill strength and serviceability requirements. A pier should be designed to withstand the overturning, sliding forces applied from superstructure as well as the forces applied to substructures. It also needs to be designed so that during an extreme event it will prevent the collapse of the structure but may sustain some damage.

A pier as a structure component is subjected to combined forces of axial, bending, and shear. For a pier, the bending strength is dependent upon the axial force. In the plastic hinge zone of a pier, the shear strength is also influenced by bending. To complicate the behavior even more, the bending moment will be magnified by the axial force due to the P-Δ effect.

In current design practice, the bridge designers are becoming increasingly aware of the adverse effects of earthquake. Therefore, ductility consideration has become a very important factor for bridge design. Failure due to scouring is also a common cause of failure of bridges. In order to prevent this type of failure, the bridge designers need to work closely with the hydraulic engineers to determine adequate depths for the piers and provide proper protection measures.

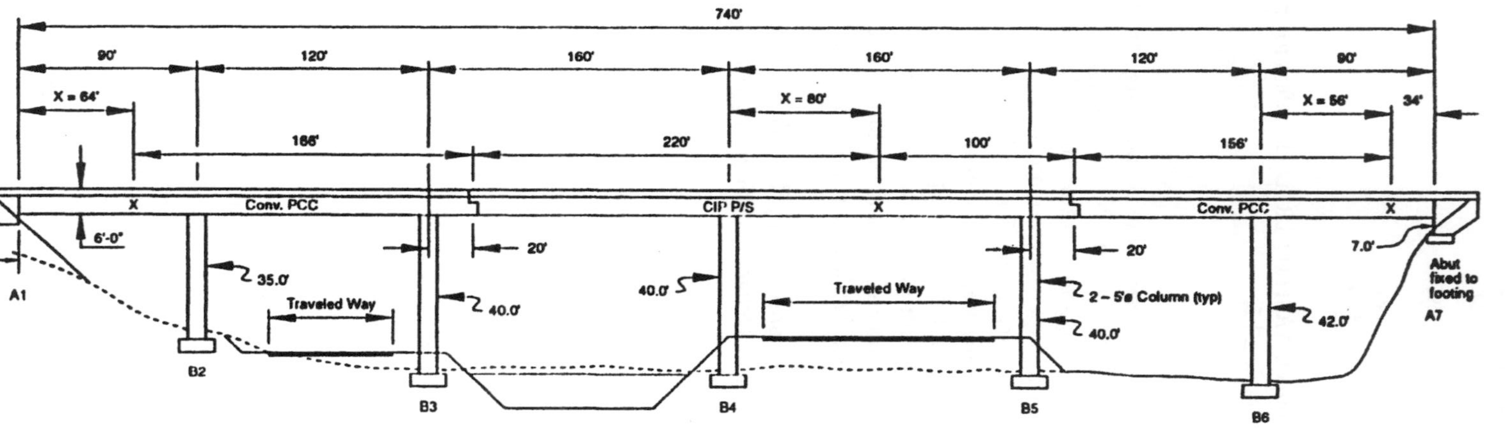

	A1	B2	B3		B4	B5		B6	A7	
I (Ft)^4	1.38	61.36	61.36		61.36	61.36		61.36	102	
L (Ft)	5.50	35.0	40.0	Sum	40.0	40.0	Sum	42.0	7.0	Sum
P (kips) @ 1" side sway	1200 +	618 +	415 =	2,233	415 +	415 =	830	359 Will slide +	600 =	959
D (distance from 1st member of frame)	0	90	210		0	160		0	90	
P x D / 100	0 +	556 +	872 =	1,428	0 +	664 =	664	0 +	540 =	540
$X = \frac{\Sigma(P \times D)/100}{\Sigma P}(100) = \frac{1,428}{2,233}(100) = \underline{64'}$					$\frac{664}{830}(100) = \underline{80'}$			$\frac{540}{959}(100) = \underline{56'}$		

Notes:
- Width of Structure = 78'
- Diameter of Column = 5'-0"
- K/Pile @ 1" deflection = 100 kips
- Point of No Movement = X
- Refer to Properties/Piles Table.

Assumptions:
1. Super str. inf. rigid
2. Columns fixed top and bottom
3. Abutment footing will slide @ a force equal to D.W.
4. E (piles) = 4 × 10⁶ psi
   E (columns) = 3 × 10⁶ psi

Fixed/Fixed Condition

$P\text{ (Col.)} = 12EI\frac{\Delta}{L^3}$

$@\ 1\text{" defl.} = \frac{432\,I}{\left(\frac{L}{10}\right)^3}$

Pinned/Fixed Condition

$P\text{ (Col.)} = 3EI\frac{\Delta}{L^3}$

$@\ 1\text{" defl.} = \frac{108\,I}{\left(\frac{L}{10}\right)^3}$

D.W. Abut 7 = 600 k (assume linear up to 1" deflection)

$I\text{ (abut)} = \frac{78}{12}(2.5)^3 = 102$

FIGURE 27.7 Calculation of points of no movement.

27.4.2 Slenderness and Second-Order Effect

The design of compression members must be based on forces and moments determined from an analysis of the structure. Small deflection theory is usually adequate for the analysis of beam-type members. For compression members, however, the second-order effect must be considered. According to AASHTO LRFD [1], the second-order effect is defined as follows:

> The presence of compressive axial forces amplify both out-of-straightness of a component and the deformation due to non-tangential loads acting thereon, therefore increasing the eccentricity of the axial force with respect to the centerline of the component. The synergistic effect of this interaction is the apparent softening of the component, i.e., a loss of stiffness.

To assess this effect accurately, a properly formulated large deflection nonlinear analysis can be performed. Discussions on this subject can be found in References [3,4] and Chapter 36. However, it is impractical to expect practicing engineers to perform this type of sophisticated analysis on a regular basis. The moment magnification procedure given in AASHTO LRFD [1] is an approximate process which was selected as a compromise between accuracy and ease of use. Therefore, the AASHTO LRFD moment magnification procedure is outlined in the following.

When the cross section dimensions of a compression member are small in comparison to its length, the member is said to be slender. Whether or not a member can be considered slender is dependent on the magnitude of the slenderness ratio of the member. The slenderness ratio of a compression member is defined as, KL_u/r, where K is the effective length factor for compression members; L_u is the unsupported length of compression member; r is the radius of gyration = $\sqrt{I/A}$; I is the moment of inertia; and A is the cross-sectional area.

When a compression member is braced against side sway, the effective length factor, $K = 1.0$ can be used. However, a lower value of K can be used if further analysis demonstrates that a lower value is applicable. L_u is defined as the clear distance between slabs, girders, or other members which is capable of providing lateral support for the compression member. If haunches are present, then, the unsupported length is taken from the lower extremity of the haunch in the plane considered (AASHTO LRFD 5.7.4.3). For a detailed discussion of the K-factor, please refer to Chapter 52.

For a compression member braced against side sway, the effects of slenderness can be ignored as long as the following condition is met (AASHTO LRFD 5.7.4.3):

$$\frac{KL_u}{r} < 34 - \left(\frac{12M_{1b}}{M_{2b}}\right) \tag{27.1}$$

where

M_{1b} = smaller end moment on compression member — positive if member is bent in single curvature, negative if member is bent in double curvature

M_{2b} = larger end moment on compression member — always positive

For an unbraced compression member, the effects of slenderness can be ignored as long as the following condition is met (AASHTO LRFD 5.7.4.3):

$$\frac{KL_u}{r} < 22 \tag{27.2}$$

If the slenderness ratio exceeds the above-specified limits, the effects can be approximated through the use of the moment magnification method. If the slenderness ratio KL_u/r exceeds 100, however, a more-detailed second-order nonlinear analysis [Chapter 36] will be required. Any detailed analysis should consider the influence of axial loads and variable moment of inertia on member stiffness and forces, and the effects of the duration of the loads.

The factored moments may be increased to reflect effects of deformations as follows:

$$M_c = \delta_b M_{2b} + \delta_s M_{2s} \tag{27.3}$$

where

M_{2b} = moment on compression member due to factored gravity loads that result in no appreciable side sway calculated by conventional first-order elastic frame analysis, always positive

M_{2s} = moment on compression member due to lateral or gravity loads that result in side sway, Δ, greater than $L_u/1500$, calculated by conventional first-order elastic frame analysis, always positive

The moment magnification factors are defined as follows:

$$\delta_b = \frac{C_m}{1 - \dfrac{P_u}{\phi P_c}} \geq 1.0 \tag{27.4}$$

$$\delta_s = \frac{1}{1 - \dfrac{\Sigma P_u}{\phi \Sigma P_c}} \geq 1.0 \tag{27.5}$$

where

P_u = factored axial load

P_c = Euler buckling load, which is determined as follows:

$$P_c = \frac{\pi^2 EI}{(KL_u)^2} \tag{27.6}$$

C_m, a factor which relates the actual moment diagram to an equivalent uniform moment diagram, is typically taken as 1.0. However, in the case where the member is braced against side sway and without transverse loads between supports, it may be taken by the following expression:

$$C_m = 0.60 + 0.40\left(\frac{M_{1b}}{M_{2b}}\right) \tag{27.7}$$

The value resulting from Eq. (27.7), however, is not to be less than 0.40.

To compute the flexural rigidity EI for concrete columns, AASHTO offers two possible solutions, with the first being:

$$EI = \frac{\dfrac{E_c I_g}{5} + E_s I_s}{1 + \beta_d} \tag{27.8}$$

and the second, more-conservative solution being:

$$EI = \frac{\dfrac{E_c I_g}{2.5}}{1 + \beta_d} \tag{27.9}$$

where E_c is the elastic modulus of concrete, I_g is the gross moment inertia, E_s is the elastic modules of reinforcement, I_s is the moment inertia of reinforcement about centroidal axis, and β is the ratio of maximum dead-load moment to maximum total-load moment and is always positive. It is an approximation of the effects of creep, so that when larger moments are induced by loads sustained over a long period of time, the creep deformation and associated curvature will also be increased.

27.4.3 Concrete Piers and Columns

27.4.3.1 Combined Axial and Flexural Strength

A critical aspect of the design of bridge piers is the design of compression members. We will use AASHTO LRFD Bridge Design Specifications [1] as the reference source. The following discussion provides an overview of some of the major criteria governing the design of compression members.

Under the Strength Limit State Design, the factored resistance is determined with the product of nominal resistance, P_n, and the resistance factor, ϕ. Two different values of ϕ are used for the nominal resistance P_n. Thus, the factored axial load resistance ϕP_n is obtained using $\phi = 0.75$ for columns with spiral and tie confinement reinforcement. The specifications also allows for the value ϕ to be linearly increased from the value stipulated for compression members to the value specified for flexure which is equal to 0.9 as the design axial load ϕP_n decreases from $0.10 f_c' A_g$ to zero.

Interaction Diagrams

Flexural resistance of a concrete member is dependent upon the axial force acting on the member. Interaction diagrams are usually used as aids for the design of the compression members. Interaction diagrams for columns are usually created assuming a series of strain distributions, and computing the corresponding values of P and M. Once enough points have been computed, the results are plotted to produce an interaction diagram.

Figure 27.8 shows a series of strain distributions and the resulting points on the interaction diagram. In an actual design, however, a few points on the diagrams can be easily obtained and can define the diagram rather closely.

- Pure Compression:

The factored axial resistance for pure compression, ϕP_n, may be computed by:

For members with spiral reinforcement:

$$P_r = \phi P_n = \phi 0.85 P_o = \phi 0.85 \left[0.85 f_c' \left(A_g - A_{st} \right) + A_{st} f_y \right] \tag{27.10}$$

For members with tie reinforcement:

$$P_r = \phi P_n = \phi 0.80 P_o = \phi 0.80 \left[0.85 f_c' \left(A_g - A_{st} \right) + A_{st} f_y \right] \tag{27.11}$$

For design, pure compression strength is a hypothetical condition since almost always there will be moments present due to various reasons. For this reason, AASHTO LRFD 5.7.4.4 limits the nominal axial load resistance of compression members to 85 and 80% of the axial resistance at zero eccentricity, P_o, for spiral and tied columns, respectively.

- Pure Flexure:

The section in this case is only subjected to bending moment and without any axial force. The factored flexural resistance, M_r, may be computed by

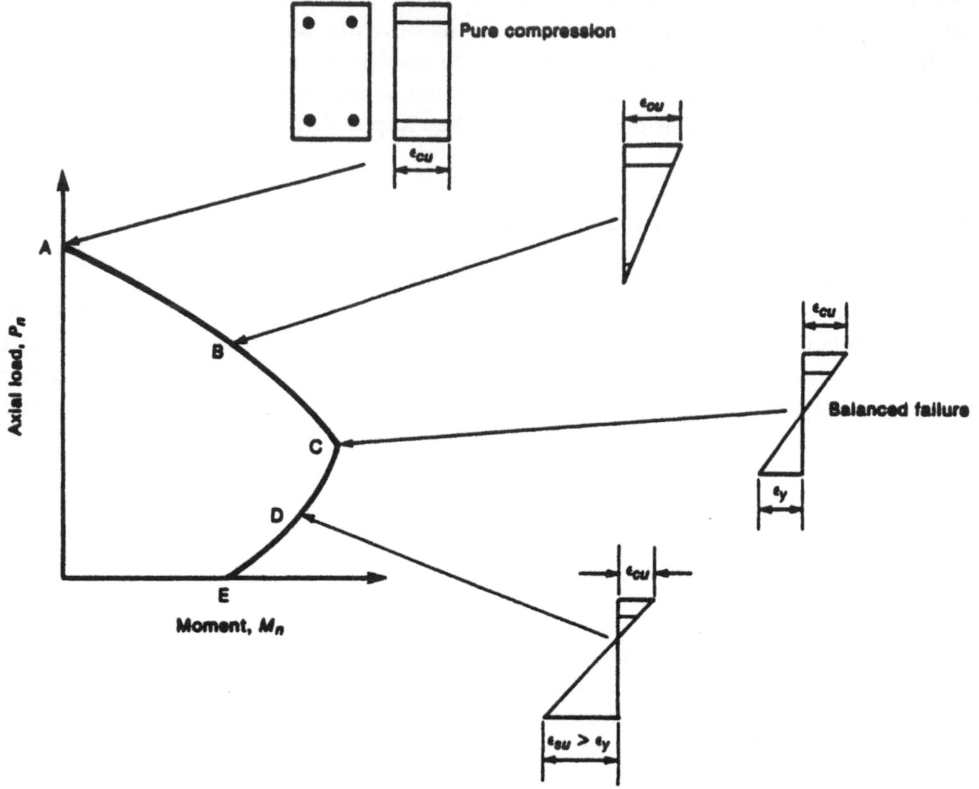

FIGURE 27.8 Strain distributions corresponding to points on interaction diagram.

$$M_r = \phi M_n = \phi\left[A_s f_y d\left(1 - 0.6\rho\frac{f_y}{f_c'}\right)\right]$$
$$= \phi\left[A_s f_y\left(d - \frac{a}{2}\right)\right] \tag{27.12}$$

where

$$a = \frac{A_s f_y}{0.85 f_c' b}$$

- **Balanced Strain Conditions:**

Balanced strain conditions correspond to the strain distribution where the extreme concrete strain reaches 0.003 and the strain in reinforcement reaches yield at the same time. At this condition, the section has the highest moment capacity. For a rectangular section with reinforcement in one face, or located in two faces at approximately the same distance from the axis of bending, the balanced factored axial resistance, P_r, and balanced factored flexural resistance, M_r, may be computed by

$$P_r = \phi P_b = \phi\left[0.85 f_c' b a_b + A_s' f_s' - A_s f_y\right] \tag{27.13}$$

and

$$M_r = \phi M_b = \phi\left[0.85 f_c' b a_b \left(d - d'' - a_b/2\right) + A_s' f_s' \left(d - d' - d''\right) + A_s f_y d''\right] \tag{27.14}$$

where

$$a_b = \left(\frac{600}{600 + f_y}\right)\beta_1 d$$

and

$$f_s' = 600\left[1 - \left(\frac{d'}{d}\right)\left(600 + \frac{f_y}{600}\right)\right] \le f_y$$

where f_y is in MPa.

Biaxial Bending

AASHTO LRFD 5.7.4.5 stipulates that the design strength of noncircular members subjected to biaxial bending may be computed, in lieu of a general section analysis based on stress and strain compatibility, by one of the following approximate expressions:

$$\frac{1}{P_{rxy}} = \frac{1}{P_{rx}} + \frac{1}{P_{ry}} - \frac{1}{P_o} \tag{27.15}$$

when the factored axial load, $P_u \ge 0.10 \phi f_c' A_g$

$$\frac{M_{ux}}{M_{rx}} + \frac{M_{uy}}{M_{ry}} \le 1 \tag{27.16}$$

when the factored axial load, $P_u < 0.10 \phi f_c' A_g$

where

P_{rxy} = factored axial resistance in biaxial flexure

P_{rx}, P_{ry} = factored axial resistance corresponding to M_{rx}, M_{ry}

M_{ux}, M_{uy} = factored applied moment about the *x*-axis, *y*-axis

M_{rx}, M_{ry} = uniaxial factored flexural resistance of a section about the *x*-axis and *y*-axis corresponding to the eccentricity produced by the applied factored axial load and moment, and

P_o = $0.85 f_c' (A_g - A_s) + A_s f_y$

27.4.3.2 Shear Strength

Under the normal load conditions, the shear seldom governs the design of the column for conventional bridges since the lateral loads are usually small compared with the vertical loads. However, in a seismic design, the shear is very important. In recent years, the research effort on shear strength evaluation for columns has been increased remarkably. AASHTO LRFD provides a general shear equation that applies for both beams and columns. The concrete shear capacity component and the angle of inclination of diagonal compressive stresses are functions of the shear stress on the concrete and the strain in the reinforcement on the flexural tension side of the member. It is rather involved and hard to use.

Alternatively, the equations recommended by ATC-32 [5] can be used with acceptable accuracy. The recommendations are listed as follows.

Except for the end regions of ductile columns, the nominal shear strength provided by concrete, V_c, for members subjected to flexure and axial compression should be computed by

$$V_c = 0.165\left(1+(3.45)\left(10^{-6}\right)\frac{N_u}{A_g}\right)\sqrt{f_c'}A_e \quad \text{(MPa)} \tag{27.17}$$

If the axial force is in tension, the V_c should be computed by

$$V_c = 0.165\left(1+(1.38)\left(10^{-5}\right)\frac{N_u}{A_g}\right)\sqrt{f_c'}A_e \quad \text{(MPa)} \tag{27.18}$$

(note that N_u is negative for tension),

where

A_g = gross section area of the column (mm²)
A_e = effective section area, can be taken as $0.8A_g$ (mm²)
N_u = axial force applied to the column (N)
f_c' = compressive strength of concrete (MPa)

For end regions where the flexural ductility is normally high, the shear capacity should be reduced. ATC-32 [5] offers the following equations to address this interaction.

With the end region of columns extending a distance from the critical section or sections not less than $1.5D$ for circular columns or $1.5h$ for rectangular columns, the nominal shear strength provided by concrete subjected to flexure and axial compression should be computed by

$$V_c = 0.165\left(0.5+(6.9)\left(10^{-6}\right)\frac{N_u}{A_g}\right)\sqrt{f_c'}A_e \quad \text{(MPa)} \tag{27.19}$$

When axial load is tension, V_c can be calculated as

$$V_c = 0.165\left(1+(1.38)\left(10^{-5}\right)\frac{N_u}{A_g}\right)\sqrt{f_c'}A_e \quad \text{(MPa)} \qquad (27.18)$$

Again, N_u should be negative in this case.

The nominal shear contribution from reinforcement is given by

$$V_s = \frac{A_v f_{yh} d}{s} \quad \text{(MPa)} \tag{27.20}$$

for tied rectangular sections, and by

$$V_s = \frac{\pi}{2}\frac{A_h f_{yh} D'}{s} \tag{27.21}$$

for spirally reinforced circular sections. In these equations, A_v is the total area of shear reinforcement parallel to the applied shear force, A_h is the area of a single hoop, f_{yh} is the yield stress of horizontal reinforcement, D' is the diameter of a circular hoop, and s is the spacing of horizontal reinforcement.

27.4.3.3 Ductility of Columns

The AASHTO LRFD [1] introduces the term *ductility* and requires that a structural system of bridge be designed to ensure the development of significant and visible inelastic deformations prior to failure.

The term *ductility* defines the ability of a structure and selected structural components to deform beyond elastic limits without excessive strength or stiffness degradation. In mathematical terms, the ductility μ is defined by the ratio of the total imposed displacement Δ at any instant to that at the onset of yield Δ_y. This is a measure of the ability for a structure, or a component of a structure, to absorb energy. The goal of seismic design is to limit the estimated maximum ductility demand to the ductility capacity of the structure during a seismic event.

For concrete columns, the confinement of concrete must be provided to ensure a ductile column. AASHTO LRFD [1] specifies the following minimum ratio of spiral reinforcement to total volume of concrete core, measured out-to-out of spirals:

$$\rho_s = 0.45\left(\frac{A_g}{A_c} - 1\right)\frac{f'_c}{f_{yh}} \tag{27.22}$$

The transverse reinforcement for confinement at the plastic hinges shall be determined as follows:

$$\rho_s = 0.16\frac{f'_c}{f_y}\left(0.5 + \frac{1.25P_u}{A_g f'_c}\right) \tag{27.23}$$

for which

$$\left(0.5 + \frac{1.25P_u}{A_g f'_c}\right) \geq 1.0$$

The total cross-sectional area (A_{sh}) of rectangular hoop (stirrup) reinforcement for a rectangular column shall be either

$$A_{sh} = 0.30ah_c\frac{f'_c}{f_{yh}}\left(\frac{A_g}{A_c} - 1\right) \tag{27.24}$$

or,

$$A_{sh} = 0.12ah_c\frac{f'_c}{f_y}\left(0.5 + \frac{1.25P_u}{A_g f'_c}\right) \tag{27.25}$$

whichever is greater,

where

a = vertical spacing of hoops (stirrups) with a maximum of 100 mm (mm)
A_c = area of column core measured to the outside of the transverse spiral reinforcement (mm^2)
A_g = gross area of column (mm^2)
A_{sh} = total cross-sectional area of hoop (stirrup) reinforcement (mm^2)
f'_c = specified compressive strength of concrete (Pa)
f_{yh} = yield strength of hoop or spiral reinforcement (Pa)
h_c = core dimension of tied column in the direction under consideration (mm)
ρ_s = ratio of volume of spiral reinforcement to total volume of concrete core (out-to-out of spiral)
P_u = factored axial load (MN)

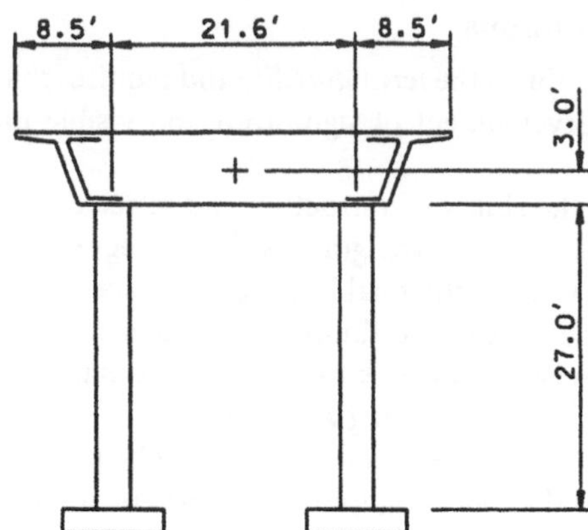

FIGURE 27.9 Example 27.2 — typical section.

TABLE 27.3 Column Group Loads — Service

	Dead Load	Live Load + Impact Case 1 Trans $M_{y\text{-max}}$	Case 2 Long $M_{x\text{-max}}$	Case 3 Axial N-max	Wind	Wind on LL	Long Force	Centrifugal Force-M_y	Temp.
M_y (k-ft)	220	75	15	32	532	153	208	127	180
M_x (k-ft)	148	67	599	131	192	86	295	2	0
P (k)	1108	173	131	280	44	17	12	23	0

TABLE 27.4 Unreduced Seismic Loads (ARS)

	Case 1 Max. Transverse	Case 2 Max. Longitudinal
M_y — Trans (k-ft)	4855	3286
M_x — Long (k-ft)	3126	3334
P — Axial (k)	–282	–220

Example 27.2 Design of a Two-Column Bent

Design the columns of a two-span overcrossing. The typical section of the structure is shown in Figure 27.9. The concrete box girder is supported by a two-column bent and is subjected to HS20 loading. The columns are pinned at the bottom of the columns. Therefore, only the loads at the top of columns are given here. Table 27.3 lists all the forces due to live load plus impact. Table 27.4 lists the forces due to seismic loads. Note that a load reduction factor of 5.0 will be assumed for the columns.

Material Data

f'_c = 4.0 ksi (27.6 MPa) $\quad$ E_c = 3605 ksi (24855 MPa)

E_s = 29000 ksi (199946 MPa) $\quad$ f_y = 60 ksi (414 MPa)

Try a column size of 4 ft (1.22 m) in diameter. Provide 26-#9 (26-#30) longitudinal reinforcement. The reinforcement ratio is 1.44%.

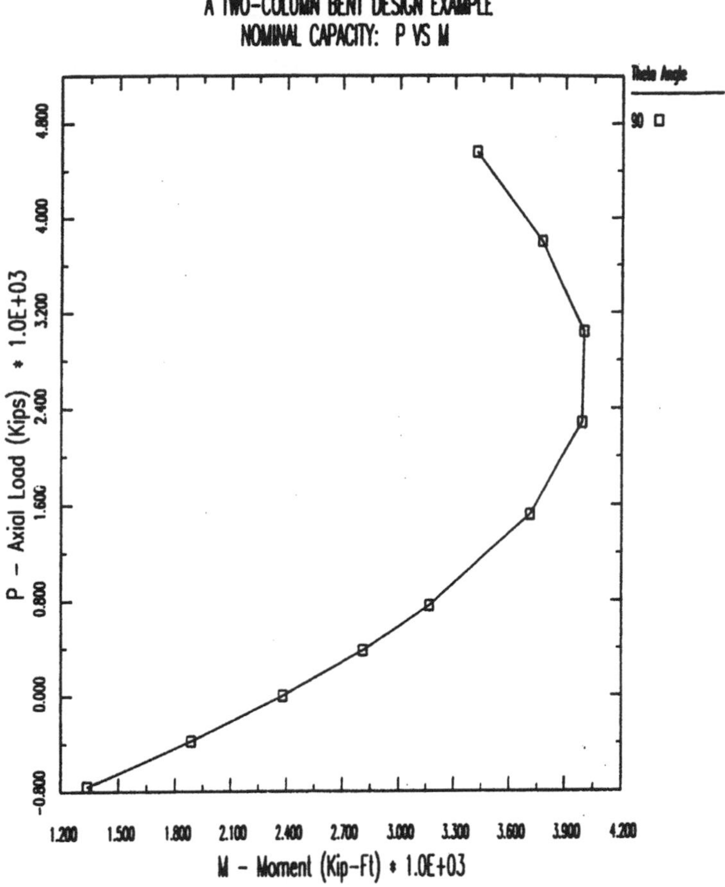

FIGURE 27.10 Example 27.2 — interaction diagram.

Section Properties

A_g = 12.51 ft² (1.16 m²) A_{st} = 26.0 in² (16774 mm²)

$I_{xc} = I_{yc}$ = 12.46 ft⁴ (0.1075 m⁴) $I_{xs} = I_{ys}$ = 0.2712 ft⁴ (0.0023 m⁴)

The analysis follows the procedure discussed in Section 27.4.3.1. The moment and axial force interaction diagram is generated and is shown in Figure 27.10.

Following the procedure outlined in Section 27.4.2, the moment magnification factors for each load group can be calculated and the results are shown in Table 27.5.

In which:

$$K_y = K_x = 2.10$$

$$K_y L/R = K_x\ L/R = 2.1 \times 27.0/(1.0) = 57$$

where R = radio of gyration = $r/2$ for a circular section.

$$22 < KL/R < 100 \quad \therefore \ \underline{\text{Second-order effect should be considered}}.$$

TABLE 27.5 Moment Magnification and Buckling Calculations

Load	P (k)	Moment Magnification			Cracked Transformed Section		Critical Buckling		Axial Load
Group	Case	Trans. M_{agy}	Long M_{agx}	Comb. M_{ag}	E^*I_y (k-ft²)	E^*I_x (k-ft²)	Trans. P_{cy} (k)	Long P_{cx} (k)	P(k)
I	1	1.571	1.640	1.587	1,738,699	1,619,399	5338	4972	1455
I	2	1.661	1.367	1.384	1,488,966	2,205,948	4571	6772	1364
I	3	2.765	2.059	2.364	1,392,713	1,728,396	4276	5306	2047
II		1.337	1.385	1.344	1,962,171	1,776,045	6024	5452	1137
III	1	1.406	1.403	1.405	2,046,281	2,056,470	6282	6313	1360
III	2	1.396	**1.344**	1.361	1,999,624	**2,212,829**	6139	**6793**	1305
III	3	1.738	1.671	1.708	1,901,005	2,011,763	5836	6176	1859
IV	1	1.437	1.611	1.455	1,864,312	1,494,630	5723	4588	1306
IV	2	1.448	1.349	1.377	1,755,985	2,098,586	5391	6443	1251
IV	3	1.920	1.978	1.936	1,635,757	1,585,579	5022	4868	1805
V		1.303	1.365	1.310	2,042,411	1,776,045	6270	5452	1094
VI	1	1.370	1.382	1.373	2,101,830	2,056,470	6453	6313	1308
VI	2	1.358	1.327	1.340	2,068,404	2,212,829	6350	6793	1256
VI	3	1.645	1.629	1.640	1,980,146	2,011,763	6079	6176	1788
VII	1	1.243	1.245	1.244	2,048,312	2,036,805	6288	6253	826
VII	2	1.296	1.275	1.286	1,940,100	2,053,651	5956	6305	888

Note: Column assumed to be unbraced against side sway.

The calculations for Loading Group III and Case 2 will be demonstrated in the following:

Bending in the longitudinal direction: M_x

$$\text{Factored load} = 1.3[\beta_D D + (L + I) + CF + 0.3W + WL + LF]$$

$\beta_D = 0.75$ when checking columns for maximum moment or maximum eccentricities and associated axial load. β_d in Eq. (27.8) = max dead-load moment, M_{DL}/max total moment, M_t.

$$M_{DL} = 148 \times 0.75 = 111 \text{ k-ft (151 kN·m)}$$

$$M_t = 0.75 \times 148 + 599 + 0.3 \times 192 + 86 + 295 + 2 = 1151 \text{ k-ft (1561 kN·m)}$$

$$\beta_d = 111/1151 = 0.0964$$

$$EI_x = \frac{\frac{E_c I_g}{5} + E_s I_s}{1+\beta_d} = \frac{\frac{3605 \times 144 \times 12.46}{5} + 29{,}000 \times 144 \times 0.2712}{1 + 0.0964} = 2{,}212{,}829 \text{ k-ft}^2$$

$$P_{cx} = \frac{\pi^2 EI_x}{(KL_u)^2} = \frac{\pi^2 \times 2{,}212{,}829}{(2.1 \times 27)^2} = 6793 \text{ kips } (30{,}229 \text{ kN})$$

$$C_m = 1.0 \text{ for frame braced against side sway}$$

$$\delta_s = \frac{1}{1 - \frac{\Sigma P_u}{\phi \Sigma P_c}} = \frac{1}{1 - \frac{1305}{0.75 \times 6793}} = 1.344$$

The magnified factored moment = 1.344 × 1.3 × 1151 = 2011 k-ft (2728 kN·m)

TABLE 27.6 Comparison of Factored Loads to Factored Capacity of the Column

		Applied Factored Forces (k-ft)				Capacity (k-ft)			
Group	Case	Trans. M_y	Long M_x	Comb. M	Axial P (k)	ϕM_n	ϕ	Ratio M_u/M	Status
I	1	852	475	975	1455	2924	0.75	3.00	OK
I	2	566	1972	2051	1364	2889	0.75	1.41	OK
I	3	1065	981	1448	2047	3029	0.75	2.09	OK
II		1211	546	1328	1137	2780	0.75	2.09	OK
III	1	1622	1125	1974	1360	2886	0.75	1.46	OK
III	2	1402	2011	2449	1305	2861	0.75	1.17	OK
III	3	1798	1558	2379	1859	3018	0.75	1.27	OK
IV	1	1022	373	1088	1306	2865	0.75	2.63	OK
IV	2	813	1245	1487	1251	2837	0.75	1.91	OK
IV	3	1136	717	1343	1805	3012	0.75	2.24	OK
V		1429	517	1519	1094	2754	0.75	1.81	OK
VI	1	1829	1065	2116	1308	2864	0.75	1.35	OK
VI	2	1617	1905	2499	1256	2842	0.75	1.14	OK
VI	3	2007	1461	2482	1788	3008	0.75	1.21	OK
VII	1	1481	963	1766	826	2372	0.67	1.34	OK
VII	2	1136	1039	1540	888	2364	0.65	1.54	OK

Notes:

1. Applied factored moments are magnified for slenderness in accordance with AASHTO LRFD.
2. The seismic forces are reduced by the load reduction factor $R = 5.0$.

$L = 27.00$ ft, $f'_c = 4.00$ ksi, $F_y = 60.0$ ksi, $A_{st} = 26.00$ in.^{2}

The analysis results with the comparison of applied moments to capacities are summarized in Table 27.6.

Column lateral reinforcement is calculated for two cases: (1) for applied shear and (2) for confinement. Typically, the confinement requirement governs. Apply Eq. 27.22 or Eq. 27.23 to calculate the confinement reinforcement. For seismic analysis, the unreduced seismic shear forces should be compared with the shear forces due to plastic hinging of columns. The smaller should be used. The plastic hinging analysis procedure is discussed elsewhere in this handbook and will not be repeated here.

The lateral reinforcement for both columns are shown as follows.

For left column:

V_u = 148 kips (659 kN) (shear due to plastic hinging governs)

ϕV_n = 167 kips (743 kN) ∴ No lateral reinforcement is required for shear.

Reinforcement for confinement = $\rho_s = 0.0057$ ∴ Provide #4 at 3 in. (#15 at 76 mm)

For right column:

V_u = 180 kips (801 kN) (shear due to plastic hinging governs)

ϕV_n = 167 kips (734 kN)

ϕV_s = 13 kips (58 kN) (does not govern)

Reinforcement for confinement =$\rho_s = 0.00623$ ∴ Provide #4 at 2.9 in. (#15 at 74 mm)

Summary of design:

4 ft (1.22 m) diameter of column with 26-#9 (26-#30) for main reinforcement and #4 at 2.9 in. (#15 at 74 mm) for spiral confinement.

FIGURE 27.11 Typical cross sections of composite columns.

27.4.4 Steel and Composite Columns

Steel columns are not as commonly used as concrete columns. Nevertheless, they are viable solutions for some special occasions, e.g., in space-restricted areas. Steel pipes or tubes filled with concrete known as composite columns (Figure 27.11) offer the most efficient use of the two basic materials. Steel at the perimeter of the cross section provides stiffness and triaxial confinement, and the concrete core resists compression and prohibits local elastic buckling of the steel encasement. The toughness and ductility of composite columns makes them the preferred column type for earthquake-resistant structures in Japan. In China, the composite columns were first used in Beijing subway stations as early as 1963. Over the years, the composite columns have been used extensively in building structures as well as in bridges [6–9].

In this section, the design provisions of AASHTO LRFD [1] for steel and composite columns are summarized.

Compressive Resistance

For prismatic members with at least one plane of symmetry and subjected to either axial compression or combined axial compression and flexure about an axis of symmetry, the factored resistance of components in compression, P_r, is calculated as

$$P_r = \phi_c P_n$$

where

P_n = nominal compressive resistance

ϕ_c = resistance factor for compression = 0.90

The nominal compressive resistance of a steel or composite column should be determined as

$$P_n = \begin{cases} 0.66^{\lambda} F_e A_s & \text{if} \quad \lambda \le 2.25 \\ \dfrac{0.88 F_e A_s}{\lambda} & \text{if} \quad \lambda > 2.25 \end{cases} \tag{27.26}$$

in which

For steel columns:

$$\lambda = \left(\frac{KL}{r_s}\pi\right)^2 \frac{F_y}{E_e} \tag{27.27}$$

For composite column:

$$\lambda = \left(\frac{KL}{r_s} \pi \right)^2 \frac{F_e}{E_e} \tag{27.28}$$

$$F_e = F_y + C_1 F_{yr} \left(\frac{A_r}{A_s} \right) + C_2 f_c \left(\frac{A_c}{A_s} \right) \tag{27.29}$$

$$Ee = E \left[1 + \left(\frac{C_3}{n} \right) \left(\frac{A_c}{A_s} \right) \right] \tag{27.30}$$

where

A_s = cross-sectional area of the steel section (mm^2)
A_c = cross-sectional area of the concrete (mm^2)
A_r = total cross-sectional area of the longitudinal reinforcement (mm^2)
F_y = specified minimum yield strength of steel section (MPa)
F_{yr} = specified minimum yield strength of the longitudinal reinforcement (MPa)
f_c' = specified minimum 28-day compressive strength of the concrete (MPa)
E = modules of elasticity of the steel (MPa)
L = unbraced length of the column (mm)
K = effective length factor
n = modular ratio of the concrete
r_s = radius of gyration of the steel section in the plane of bending, but not less than 0.3 times the width of the composite member in the plane of bending for composite columns, and, for filled tubes,

$$C_1 = 1.0; \quad C_2 = 0.85; \quad C_3 = 0.40$$

In order to use the above equation, the following limiting width/thickness ratios for axial compression of steel members of any shape must be satisfied:

$$\frac{b}{t} \leq k \sqrt{\frac{E}{F_y}} \tag{27.31}$$

where

k = plate buckling coefficient as specified in Table 27.7
b = width of plate as specified in Table 27.7
t = plate thickness (mm)

Wall thickness of steel or composite tubes should satisfy:

For circular tubes:

$$\frac{D}{t} \leq 2.8 \sqrt{\frac{E}{F_y}}$$

TABLE 27.7 Limiting Width-to-Thickness Ratios

	k	b
		Plates Supported along One Edge
Flanges and projecting leg or plates	0.56	Half-flange width of I-section Full-flange width of channels Distance between free edge and first line of bolts or welds in plates Full-width of an outstanding leg for pairs of angles on continuous contact
Stems of rolled tees	0.75	Full-depth of tee
Other projecting elements	0.45	Full-width of outstanding leg for single-angle strut or double-angle strut with separator Full projecting width for others
		Plates Supported along Two Edges
Box flanges and cover plates	1.40	Clear distance between webs minus inside corner radius on each side for box flanges Distance between lines of welds or bolts for flange cover plates
Webs and other plates elements	1.49	Clear distance between flanges minus fillet radii for webs of rolled beams Clear distance between edge supports for all others
Perforated cover plates	1.86	Clear distance between edge supports

For rectangular tubes:

$$\frac{b}{t} \leq 1.7\sqrt{\frac{E}{F_y}}$$

where
D = diameter of tube (mm)
b = width of face (mm)
t = thickness of tube (mm)

Flexural Resistance

The factored flexural resistance, M_r, should be determined as

$$M_r = \phi_f M_n \tag{27.32}$$

where
M_n = nominal flexural resistance
ϕ_f = resistance factor for flexure, $\phi_f = 1.0$

The nominal flexural resistance of concrete-filled pipes that satisfy the limitation

$$\frac{D}{t} \leq 2.8\sqrt{\frac{E}{F_y}}$$

may be determined:

$$\text{If } \frac{D}{t} < 2.0\sqrt{\frac{E}{F_y}}, \text{ then } M_n = M_{ps} \tag{27.33}$$

$$\text{If } 2.0\sqrt{\frac{E}{F_y}} < \frac{D}{t} \leq 8.8\sqrt{\frac{E}{F_y}}, \text{ then } M_n = M_{yc} \tag{27.34}$$

where
M_{ps} = plastic moment of the steel section
M_{yc} = yield moment of the composite section

Combined Axial Compression and Flexure

The axial compressive load, P_u, and concurrent moments, M_{ux} and M_{uy}, calculated for the factored loadings for both steel and composite columns should satisfy the following relationship:

$$\text{If } \frac{P_u}{P_r} < 0.2, \text{ then } \frac{P_u}{2.0P_r} + \left(\frac{M_{ux}}{M_{rx}} + \frac{M_{uy}}{M_{ry}}\right) \leq 1.0 \tag{27.35}$$

$$\text{If } \frac{P_u}{P_r} \geq 0.2, \text{ then } \frac{P_u}{P_r} + \frac{8.0}{9.0}\left(\frac{M_{ux}}{M_{rx}} + \frac{M_{uy}}{M_{ry}}\right) \leq 1.0 \tag{27.36}$$

where
P_r = factored compressive resistance
M_{rx}, M_{ry} = factored flexural resistances about x and y axis, respectively
M_{ux}, M_{uy} = factored flexural moments about the x and y axis, respectively

References

1. AASHTO, *LRFD Bridge Design Specifications,* 1st ed., American Association of State Highway and Transportation Officials, Washington, D.C., 1994.
2. Caltrans, Bridge Memo to Designers (7-10), California Department of Transportation, Sacramento, 1994.
3. White, D. W. and Hajjar, J. F., Application of second-order elastic analysis in LRFD: research to practice, *Eng. J.*, 28(4), 133, 1994.
4. Galambos, T. V., Ed., *Guide to Stability Design for Metal Structures,* 4th ed., the Structural Stability Research Council, John Wiley & Sons, New York, 1988.
5. ATC, Improved Seismic Design Criteria for California Bridges: Provisional Recommendations, Applied Technology Council, Report ATC-32, Redwood City, CA, 1996.
6. Cai, S.-H., Chinese standard for concrete-filled tube columns, in *Composite Construction in Steel and Concrete II,* Proc. of an Engineering Foundation Conference, Samuel Easterling, W. and Kim Roddis, W. M., Eds, Potosi, MO, 1992, 143.
7. Cai, S.-H., Ultimate strength of concrete-filled tube columns, in *Composite Construction in Steel and Concrete,* Proc. of an Engineering Foundation Conference, Dale Buckner, C. and Viest, I. M., Eds, Henniker, NH, 1987, 703.
8. Zhong, S.-T., New concept and development of research on concrete-filled steel tube (CFST) members, in *Proc. 2nd Int. Symp. on Civil Infrastructure Systems,* 1996.
9. CECS 28:90, *Specifications for the Design and Construction of Concrete-Filled Steel Tubular Structures,* China Planning Press, Beijing [in Chinese], 1990.

10. AISC, *Load and Research Factor Design Specification for Structural Steel Buildings and Commentary*, 2nd ed., American Institute of Steel Construction, Chicago, IL, 1993.
11. Galambos, T. V. and Chapuis, J., LRFD Criteria for Composite Columns and Beam Columns, Revised Draft, Washington University, Department of Civil Engineering, St. Louis, MO, December 1990.

28

Towers

Charles Seim
T. Y. Lin International

28.1 Introduction

Towers are the most visible structural elements of long-span bridges. They project above the superstructure and are seen from all directions by viewers and by users. Towers give bridges their character and a unifying theme. They project a mnemonic image that people remember as a lasting impression of the bridge itself. As examples of the powerful imagery of towers, contrast the elegant art deco towers of the Golden Gate Bridge (Figure 28.1) with the utilitarian but timeless architecture of the towers of the San Francisco–Oakland Bay Bridge (Figure 28.2). Or contrast the massive, rugged stone towers of the Brooklyn Bridge (Figure 28.3) with the awkward confusing steel towers of the Williamsburg Bridge in New York City (Figure 28.4).

Towers can be defined as vertical steel or concrete structures projecting above the deck, supporting cables and carrying the forces to which the bridge is subjected to the ground. By this definition, towers are used only for suspension bridges or for cable-stayed bridges, or hybrid suspension–cable-stayed structures. The word *pylon* is sometimes used for the towers of cable–stayed bridges. Both *pylon* and *tower* have about the same meaning — a tall and narrow structure supporting itself and the roadway. In this chapter, the word *tower* will be used for both suspension and for cabled-stayed bridges, to avoid any confusion in terms.

Both suspension and cable-stayed bridges are supported by abutments or piers at the point where these structures transition to the approach roadway or the approach structure. Abutments are discussed in Chapter 30. Piers and columns that support the superstructure for other forms of bridge structures such as girders, trusses, or arches, usually do not project above the deck. Piers and columns are discussed in Chapter 27.

The famous bridges noted above were opened in 1937, 1936, 1883, and 1903, respectively, and, if well maintained, could continue to serve for another 100 years. Bridge engineers will not design structures like these today because of changing technologies. These bridges are excellent examples of enduring structures and can serve to remind bridge engineers that well-designed and maintained structures do

0-8493-7434-0/00/$0.00+$.50

FIGURE 28.1 Golden Gate Bridge, San Francisco. (Courtesy of Charles Seim.)

last for 150 years or longer. Robust designs, durable materials, provisions for access for inspection and maintenance, and a well-executed maintenance program will help ensure a long life. The appearance of the bridge, good or bad, is locked in for the life of the facility and towers are the most important visual feature leading to the viewer's impression of an aesthetic structure.

28.2 Functions

The main structural function of the towers of cable-stayed and suspension bridges is carrying the weight of the bridge, traffic loads, and the forces of nature to the foundations. The towers must perform these function in a reliable, serviceable, aesthetic, and economical manner for the life of the bridge, as towers, unlike other bridge components, cannot be replaced. Without reliability, towers may become unsafe and the life of the entire bridge could be shortened. Without serviceability being designed into the structure, which means that it is designed for access and ease of maintenance, the bridge will not provide continuing long service to the user. The public demands that long-span bridges be attractive, aesthetic statements with long lives, so as not to be wasteful of public funds.

28.3 Aesthetics

While the main function of the towers is structural, an important secondary function is visual. The towers reveal the character or motif of the bridge. The bridges used as examples in the introduction are good illustrations of the image of the structure as revealed by the towers. Indeed, perhaps they are famous because of their towers. Most people visualize the character of the Brooklyn Bridge by the gothic, arched, masonry towers alone. The San Francisco–Oakland Bay Bridge and the Golden Gate Bridge give completely different impressions to the viewer as conveyed by the towers. Seim [7] measured the ratios of the visible components of the towers of the latter two bridges and found important, but subtle, diminution of these ratios with height above the tower base. It is the subtle changes in these ratios within the height of the towers that produce the much-admired proportions of these world-renowned bridges. The proportions of the towers for any new long-span bridge

FIGURE 28.2 San Francisco–Oakland Bay Bridge. (Courtesy of Charles Seim.)

should be carefully shaped and designed to give the entire bridge a strong — even robust — graceful, and soaring visual image. The aesthetics of bridges are discussed in greater detail in Chapters 2 and 3 of this volume.

The aesthetics of the array of cables many times are of secondary importance to the aesthetics of the towers. However, the array or form of the cables must be considered in the overall aesthetic and structural evaluation of the bridge. Main cables of suspension bridges always drape in a parabolic curve that most people instinctively enjoy. The large diameter of the cables makes them stand out as an important contribution to the overall visual impression as the supporting element of the roadway.

The cables of cable-stayed bridges are usually of small diameter and do not stand out visually as strongly as do the cables of suspension bridges. However, the array of the stays, such as harp, fan, radiating star, or others, should be considered in context with the tower form. The separated, parallel cables of the harp form, for example, will not be as obtrusive to the towers as will other arrangements. However, the harp cable form may not be appropriate for very long spans or for certain tower shapes. The cables and the towers should be considered together as a visual system.

Billington [2] presents an overview of the importance of the role of aesthetics in the history of the development of modern bridge design. Leonhardt [5] presents many examples of completed bridges showing various tower shapes and cable arrangements for both suspension and cable-stayed bridges.

FIGURE 28.3 Brooklyn Bridge, New York. (Courtesy of Charles Seim.)

28.4 Conceptual Design

Perhaps the most important step in the design of a new bridge is the design concept for the structure that ultimately will be developed into a final design and then constructed. The cost, appearance, and reliability and serviceability of the facility will all be determined, for good or for ill, by the conceptual design of the structure. The cost can be increased, sometimes significantly, by a concept that is very difficult to erect. Once constructed, the structure will always be there for users to admire — or to criticize. The user ultimately pays for the cost of the facility and also usually pays for the cost of maintaining the structure. Gimsing [4] treats the concept design issues of both cable-stayed and suspension bridges very extensively and presents examples to help guide designers.

A proper bridge design that considers the four functions of reliability, serviceability, appearance, and cost together with an erectable scheme that requires low maintenance, is the ideal that the design concept should meet.

A recent trend is to employ an architect as part of the design team. Architects may view a structure in a manner different from engineers, and their roles in the project are not the same. The role of the engineer is to be involved in all four functions and, most importantly, to take responsibility for the structural adequacy of the bridge. The role of the architect generally only involves the function of aesthetics. Their roles overlap in achieving aesthetics, which may also affect the economy of the structure. Since both engineers and architects have as a common objective an elegant and economical bridge, there should be cooperation and respect between them.

Occasional differences do occur when the architect's aesthetic desires conflict with the engineer's structural calculations. Towers, as the most visible component of the bridge, seem to be a target for this type of conflict. Each professional must understand that these differences in viewpoints will occur and must be resolved for a successful and fruitful union between the two disciplines.

While economy is usually important, on occasions, cost is not an objective because the owner or the public desires a "symbolic" structure. The architect's fancy then controls and the engineer can only provide the functions of safety and serviceability.

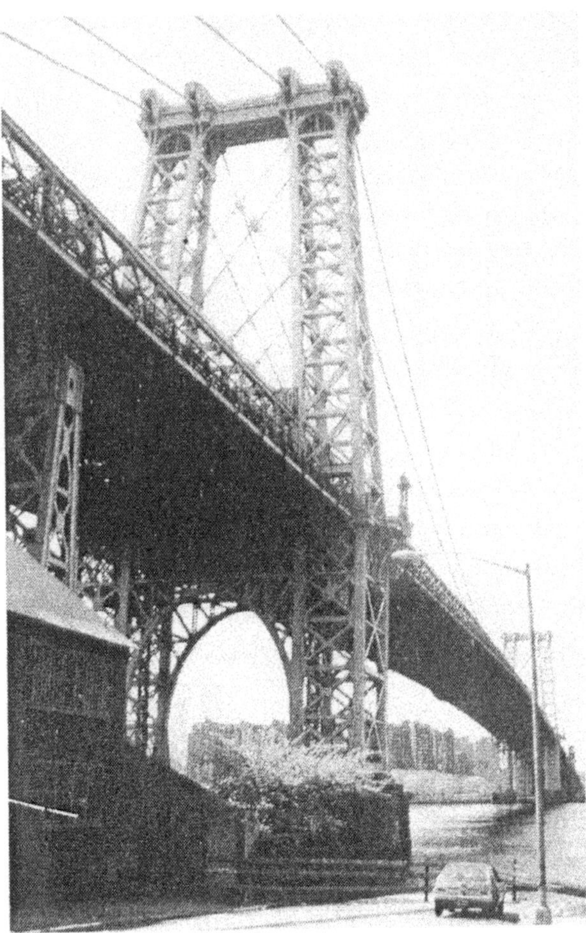

FIGURE 28.4 Williamsburg Bridge, New York. (Courtesy of Charles Seim.)

28.4.1 Materials

Until the 1970s steel was the predominant material used for the towers of both cable-stayed and suspension bridges. The towers were often rectangular in elevation with a cross-sectional shape of rectangular, cruciform, tee, or a similar shape easily fabricated in steel. Examples of suspension bridge steel tower design are the plain, rectangular steel towers for the two Delaware Memorial Bridges; the first constructed in 1951 and the parallel one in 1968 (Figure 28.5). An example of a cable-stayed bridge that is an exception to the rectangular tower form is the modified A-frame, weathering-steel towers of the Luling Bridge near New Orleans, 1983 (Figure 28.6).

The cross sections of steel towers are usually designed as a series of adjoining cells formed by shop-welding steel plates together in units from 6 to 12 m long. The steel towers for a suspension bridge, and for cable-stayed bridges with stays passing over the top of the tower in saddles, must be designed for the concentrated load from the saddles. The steel cellular towers for a cable-stayed bridge with cables framing in the towers must be designed for the local forces from the numerous anchorages of the cables.

Since the 1970s, reinforced concrete has been used in many forms with rectangular and other compact cross sections. Concrete towers are usually designed as hollow shafts to save weight and to reduce the amount of concrete and reinforcing bars required. As with steel towers, concrete towers must

FIGURE 28.5 Delaware Memorial Bridges. (Courtesy of D. Sailors.)

FIGURE 28.6 Luling Bridge, New Orleans, Louisiana. (Courtesy of Charles Seim.)

be designed for the concentrated load from the saddles at the top, if used, or for the local forces from the numerous anchorages of the cables framing into the tower shafts

Towers designed in steel will be lighter than towers designed in concrete, thus giving a potential for savings in foundation costs. Steel towers will generally be more flexible and more ductile and can be erected in less time than concrete towers. Steel towers will require periodic maintenance painting, although weathering steel can be used for nonmarine environments.

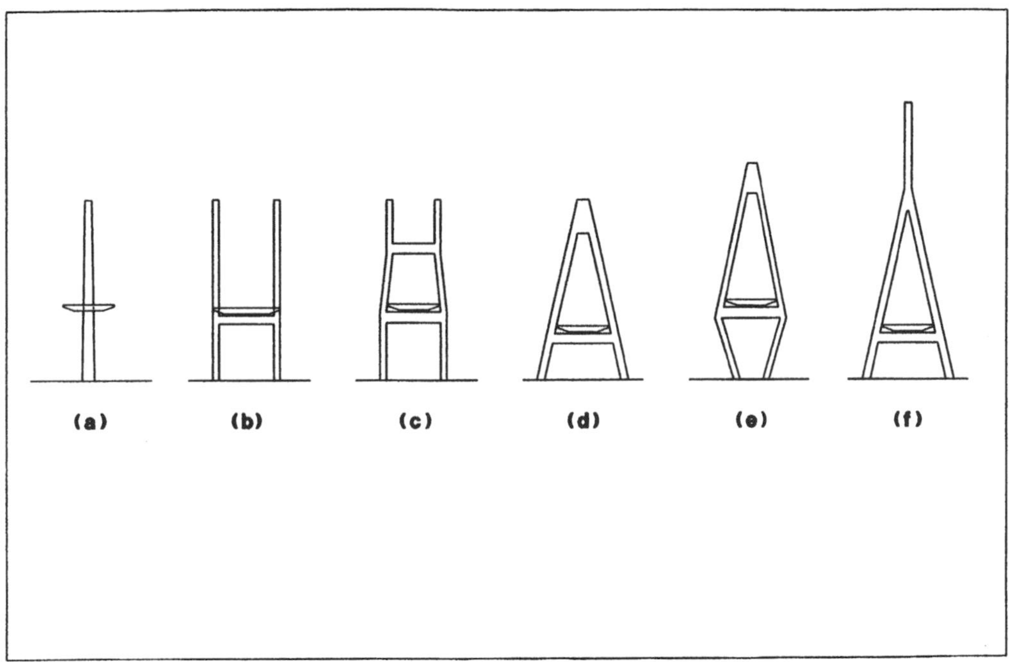

FIGURE 28.7 Generic forms for towers of cable-stayed bridges. (a) Single tower, I; (b) double vertical shafts, H; (c) double cranked shafts; (d) inclined shafts, A; (e) inclined shafts, diamond; (f) inverted Y.

The cost of steel or concrete towers can vary with a number of factors so that market conditions, contractor's experience, equipment availability, and the design details and site-specific influences will most likely determine whether steel or concrete is the most economical material. For pedestrian bridges, timber towers may be economical and aesthetically pleasing.

During the conceptual design phase of the bridge, approximate construction costs of both materials need to be developed and compared. If life-cycle cost is important, then maintenance operations and the frequencies of those operations need to be evaluated and compared, usually by a present-worth evaluation.

28.4.2 Forms and Shapes

Towers of cable-stayed bridges can have a wide variety of shapes and forms. Stay cables can also be arranged in a variety of forms. See Chapter 19. For conceptual design, the height of cable-stayed towers above the deck can be assumed to be about 20% of the main span length. To this value must be added the structural depth of the girder and the clearance to the foundation for determining the approximate total tower height. The final height of the towers will be determined during the final design phase.

The simplest tower form is a single shaft, usually vertical (Figure 28.7a). Occasionally, the single tower is inclined longitudinally. Stay cables can be arranged in a single plane to align with the tower or be splayed outward to connect with longitudinal edge beams. This form is usually employed for bridges with two-way traffic, to avoid splitting a one-way traffic flow. For roadways on curves, the single tower may be offset to the outside of the convex curve of the roadway and inclined transversely to support the curving deck more effectively.

Two vertical shafts straddling the roadway with or without cross struts above the roadway form a simple tower and are used with two planes of cables (Figure 28.7b) The stay cables would incline inward to connect to the girder, introducing a tension component across the deck support system; however, the girders are usually extended outward between the towers to align the cables vertically

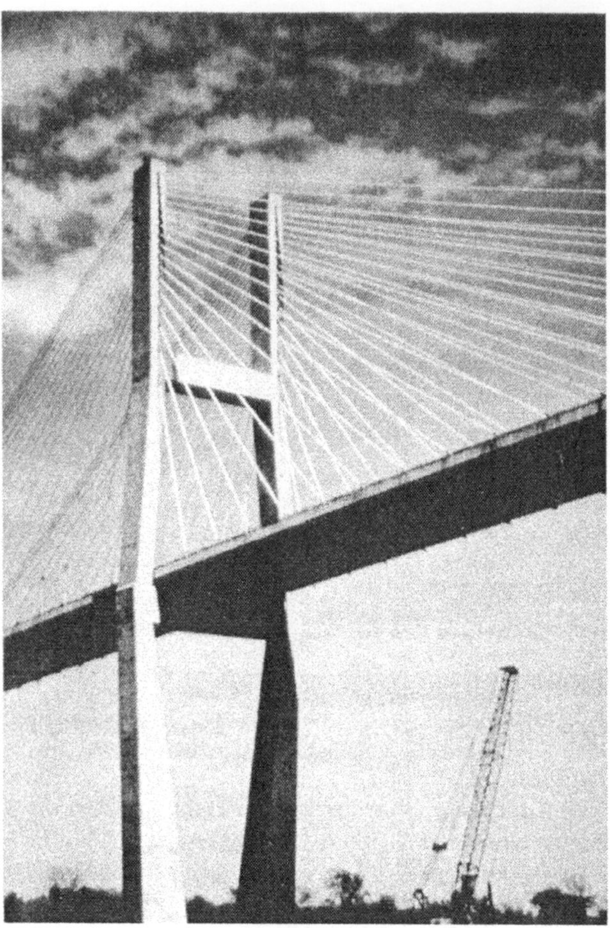

FIGURE 28.8 Talmadge Bridge, Georgia. (Courtesy of T. Y. Lin International.)

with the tower shafts. The tower shafts can also be "cranked" or offset above the roadway (Figure 28.7c). This allows the cables to be aligned in a vertical plane and to be attached to the girder, which can pass continuously through the towers as used for the Talmadge Bridge, Georgia (Figure 28.8). A horizontal strut is used between the tower shafts, offset to stabilize the towers.

The two shafts of cable-stayed bridges can be inclined inward toward each other to form a modified A-frame, similar to the Luling Bridge towers (Figure 28.6), or inclined to bring the shafts tops together to form a full A-frame (Figure 28.7d). The two planes of stay cables are inclined outward, producing a more desirable compression component across the deck support system.

The form of the towers of cable-stayed bridge below the roadway is also import for both aesthetics and costs. The shafts of the towers for a modified A-frame can be carried down to the foundations at the same slope as above the roadway, particularly for sites with low clearance. However, at high clearance locations, if the shafts of the towers for a full A-frame or for an inverted Y-frame are carried down to the foundations at the same slope as above the roadway, the foundations may become very wide and costly. The aesthetic proportions also may be affected adversely. Projecting the A-frame shafts downward vertically can give an awkward appearance. Sometimes the lower shafts are inclined inward under the roadway producing a modified diamond (Figure 28.7e), similar to the towers of the Glebe Island Bridge, Sidney, Australia (Figure 28.9). For very high roadways, the inward inclination can form a full diamond or a double diamond as in the Baytown Bridge, Texas (Figure 28.10). For very long spans requiring tall towers, the A-frame can be extended with a single vertical shaft forming an inverted Y shape (Figure 28.7f) as in the Yang Pu Bridge, China

FIGURE 28.9 Glebe Island Bridge, Sidney, Australia. (Courtesy of T. Y. Lin International.)

(Figure 28.11). This form is very effective for very long spans where additional tower height is required and the inclined legs add stiffness and frame action for wind resistance.

The number of shafts or columns within the towers of cable-stayed bridges can vary from one to four. Three-shaft towers generally are not used for cable-stayed bridges except for very wide decks. Four-shaft towers can be used best to support two separate structures instead of a single wide deck. The towers could share a common foundation or each have its own foundation depending on the cost.

Suspension bridges can have from one to four cables depending on structural or architectural needs. Only a few single-cable suspension bridges have been designed with an A or inverted Y form of towers. Usually towers of suspension bridges follow a more traditional design using two vertical shafts and two planes of cables, as illustrated by the steel towers for the Delaware Memorial Bridges (see Figure 28.5). However, concrete towers have recently proved to be economical for some bridges. The very long span (1410 m) Humber Bridge, England, 1983, used uniformly spaced, multi-strut concrete towers (Figure 28.12). The crossing of the Great Belt seaway in Denmark (Figure 28.13), opening in 1999, has concrete towers 254 m high with two struts, one near the midheight and one at the top.

For conceptual designs, the height of suspension bridge towers above the deck depend on the sag-to-span ratio which can vary from about 1:8 to 1:12. A good preliminary value is about 1:10. To this value must be added the structural depth of the deck and the clearance to the foundations to obtain the approximate total tower height. The shafts are usually connected together with several

FIGURE 28.10 Baytown Bridge, Texas. (Courtesy of T. Y. Lin International.)

FIGURE 28.11 Yang Pu Bridge, China. (Courtesy of T. Y. Lin International.)

struts or cross-bracing along the height of the tower, or the shafts are connected at the top with a large single strut. Some form of strut is usually required for suspension bridges as the large cables carry lateral wind and seismic loads to the tops of the tower shafts, which then need to be braced against each other with cross struts to form a tower-frame action.

28.4.3 Erection

During the concept design phase, many different tower forms may be considered, and preliminary designs and cost estimates completed. Each alternative considered should have at least one method

FIGURE 28.12 Humber Bridge, England. (Courtesy of Charles Seim.)

of erection developed during the concept design phase to ensure that the scheme under consideration is feasible to construct. The cost of unusual tower designs can be difficult to estimate and can add significant cost to the project.

28.5 Final Design

The AASHTO Standard Specifications for Highway Bridges [1] apply to bridges 150 m or less in span. For important bridges and for long-span cable-supported bridge projects, special design criteria may have to be developed by the designer. The special design criteria may have to be also developed in cooperation with the owners of the facility to include their operations and maintenance requirements and their bridge-performance expectations after large natural events such as earthquakes. See Chapter 18 for suspension bridge design and Chapter 19 for cable-stayed bridge design. Troitsky [8], Podolny and Salzi [6], and Walther [9] present detailed design theory for cable-stayed bridges.

Design methodology for the towers should follow the same practice as the design methodology for the entire bridge. The towers should be part of a global analysis in which the entire structure is treated as a whole. From the global analyses, the towers can be modeled as a substructure unit with forces and deformations imposed as boundary conditions.

Detailed structural analyses form the basis for the final design of the tower and its components and connections. Both cabled-stayed and suspension bridges are highly indeterminate and require careful analysis in at least a geometric nonlinear program.

28.5.1 Design Loads

The towers are subject to many different loading cases. The towers, as well as the entire structure, must be analyzed, designed, and checked for the controlling loading cases. Chapter 6 presents a detailed discussion of bridge loading.

The weight of the superstructure, including the self-weight of the towers, is obtained in the design process utilizing the unit weights of the materials used in the superstructure and distributed to the tower in accordance with a structural analysis of the completed structure or by the erection equipment during the construction phases.

FIGURE 28.13 Great Belt Bridge, Denmark. (Courtesy of Ben C. Gerwick, Inc.)

Loads from traffic using the bridge such as trains, transit, trucks, or pedestrians are usually prescribed in design codes and specifications or by the owners of the facility. These are loads moving across the bridge and the forces imparted to the towers must be obtained from a structural analysis that considers the moving loading. These are all gravity effects that act downward on the structure, but will induce both vertical and horizontal forces on the towers.

A current trend for spanning wide widths of waterways is to design multispan bridges linked together to form a long, continuous structure. With ordinary tower designs, the multispan cable-stayed girders will deflect excessively under live loads as the towers will not be sufficiently stiffened by the cable stays anchored within the flexible adjacent spans. For multispan suspension bridges with ordinary tower designs, the same excessive live-load deflection can also occur. Towers for multispan cable-supported bridges must be designed to be sufficiently rigid to control live-load deflections.

Towers are also subject to temperature-induced displacements, both from the superstructure and cable framing into the towers, and from the temperature-induced movement of the tower itself. Towers can expand and contract differentially along the tower height from the sun shining on them from morning until sunset. These temperature effects can cause deflection and torsional twisting along the height of the tower.

Wind blowing on the towers as a bluff shape induces forces and displacements in the tower. Forces will be induced into the cables by the pressure of wind on the superstructure, as well as by the wind forces on the cables themselves. These additional forces will be carried to the towers.

For long-span bridges and for locations with known high wind speeds, wind should be treated as a dynamic loading. This usually requires a wind tunnel test on a sectional model of the superstructure in a wind tunnel and, for important bridges, an aeroelastic model in a large wind tunnel. See Chapter 57. Under certain wind flows, the wind can also excite the tower itself, particularly if the tower is designed with light steel components. In the rare instances in which wind-induced excitation of the tower does occur, appropriate changes in the cross section of the tower can be made or a faring can be added to change the dynamic characteristics of the tower.

Forces and deformations of long-span structures from earthquakes are discussed in Chapter 40. The seismic excitation should be treated as dynamic inertia loadings inducing response within the structure by exciting the vibrational modes of the towers. Induced seismic forces and displacement can control the design of towers in locations with high seismic activity. For locations with lower seismic activity, the tower design should be checked at least for code-prescribed seismic loadings. The dynamic analysis of bridges is discussed in Chapter 34.

A full analysis of the structure will reveal all of the forces, displacements, and other design requirements for all loading cases for the final tower design.

28.5.2 Design Considerations

Suspension bridge cables pass over cable saddles that are usually anchored to the top of the tower. A cable produces a large vertical force and smaller, but important, transverse and longitudinal forces from temperature, wind, earthquake, or from the unbalanced cable forces between main and side spans. These forces are transmitted through the cable saddle anchorage at each cable location to the top of the tower. The towers and the permanent saddle anchorages must be designed to resist these cable forces.

The erection of a suspension bridge must be analyzed and the sequence shown on the construction plans. To induce the correct loading into the cables of the side span, the erection sequence usually requires that the saddles be displaced toward the side spans. This is usually accomplished for short spans by displacing the tops of the towers by pulling with heavy cables. For long spans, the saddles can be displaced temporarily on rollers. As the stiffening deck elements are being erected into position and the cable begins to take loads, the towers or saddles are gradually brought into final vertical alignment. After the erection of the stiffening deck elements are completed, the saddles are permanently fastened into position to take the unbalanced cable loads from the center and the side spans.

At the deck level, other forces may be imposed on the tower from the box girder or stiffening truss carrying the roadway. These forces depend on the structural framing of the connection of the deck and tower. Traditional suspension bridge designs usually terminate the stiffening truss or box girder at the towers, which produces transverse, and longitudinal, forces on the tower at this point. Contemporary suspension bridge designs usually provide for passing a box girder continuously through the tower opening which may produce transverse forces but not longitudinal forces. For this arrangement, the longitudinal forces must be carried by the girder to the abutments.

The most critical area of the tower design is the tower-to-foundation connection. Both shear forces and moments are maximum at this point. Anchor bolts are generally used at the base of steel towers. The bolts must be proportioned to transfer the loads from the tower to the bolts. The bolts must be deeply embedded in the concrete footing block to transfer their loads to the footing reinforcement. Providing good drainage for the rainwater running down the tower shafts will increase the life of the steel paint system at the tower base and provide some protection to the anchor bolts.

Concrete towers must be joined to the foundations with full shear and moment connections. Lapped reinforcing bars splices are usually avoided as the lapping tends to congest the connections, the strength of the bars cannot be developed, and lapped splices cannot be used for high seismic areas. Using compact mechanical or welded splices will result in less congestion with easier placement of concrete around the reinforcement and a more robust tower-to-footing connection.

Careful coordination between the foundation designers and tower designers is required to achieve a stable, efficient, and reliable connection.

The cable arrangements for cable-stayed bridges are many and varied. Some arrangements terminate the cables in the tower, whereas other arrangements pass the cable through the tower on cable saddles. Cables terminating in the tower can pass completely through the tower cross section and then anchor on the far side of the tower. This method of anchoring produces compression in the tower cross section at these anchorage points. Cables can also be terminated at anchors within the walls of the tower, producing tension in the tower cross section at the anchorage points. These tension forces require special designs to provide reliable, long-life support for the cables.

Just as for suspension bridges, the erection of cable-stayed bridges must be analyzed and the sequence shown on the construction plans. The girders, as they are erected outward from the towers, are very vulnerable. The critical erection sequence is just before closing the two arms of the girders at the center of the span. High winds can displace the arms and torque the towers, and heavy construction equipment can load the arms without benefit of girder continuity to distribute the loads.

28.6 Construction

Towers constructed of structural steel are usually fabricated in a shop by welding together steel plates and rolled shapes to form cells. Cells must be large enough to allow welders and welding equipment, and if the steel is to be painted, painters and cleaning and painting equipment inside each cell.

The steel tower components are transported to the bridge site and then erected by cranes and bolted together with high-strength bolts. The contractor should use a method of tensioning the high-strength bolts to give constant results and achieve the required tension. Occasionally, field welding is used, but this presents difficulties in holding the component rigidly in position while the weld is completed. Field welding can be difficult to control in poor weather conditions to achieve ductile welds, particularly for vertical and overhead welds. Full-penetration welds require backup bars that must be removed carefully if the weld is subject to fatigue loading.

Towers constructed of reinforced concrete are usually cast in forms that are removed and reused, or jumped to the next level. Concrete placing heights are usually restricted to about 6 to 12 m to limit form pressure from the freshly placed concrete. Reinforcing bar cages are usually preassembled on the ground or on a work barge, and lifted into position by crane. This requires the main load-carrying reinforcing bars to be spliced with each lift. Lapped splices are the easiest to make, but are not allowed in seismic areas.

Slip forming is an alternative method that uses forms that are pulled slowly upward, reinforcing bars positioned and the concrete placed in one continuous operation around the clock until the tower is completed. Slip forming can be economical, particularly for constant-cross-section towers. Some changes in cross section geometry can be accommodated. For shorter spans, precast concrete segments can be stacked together and steel tendons tensioned to form the towers.

Tower designers should consider the method of erection that contractors may use in constructing the towers. Often the design can reduce construction costs by incorporating more easily fabricated and assembled steel components or assembled reinforcing bar cages and tower shapes that are easily formed. Of course, the tower design cannot be compromised just to lower erection costs.

Some engineers and many architects design towers that are not vertical but are angled longitudinally toward or away from the main span. This can be done if such a design can be justified structurally and aesthetically, and the extra cost can be covered within the project budget. The difficulties of the design of longitudinally inclined towers must be carefully considered as well as the more expensive and slower erection, which will create additional costs.

Many towers of cable-stayed bridges have legs sloped toward each to form an A, an inverted Y, a diamond, or similar shapes. These are not as difficult to construct as the longitudinally inclined tower design. The sloping concrete forms can be supported by vertical temporary supports and cross struts that tie the concrete forms together. This arrangement braces the partly cast concrete tower legs against each other for support. Some of the concrete form supports for the double-diamond towers of the Baytown Bridge are visible in Figure 28.9.

As the sloped legs are erected, the inclination may induce bending moments and lateral deflection in the plane of the slope of the legs. Both of these secondary effects must be adjusted by jacking the legs apart by a calculated amount of force or displacement to release the locked-in bending stresses. If the amount of secondary stress is small, then cambering the leg to compensate for the deflection and adding material to lower the induced stress can be used.

The jacking procedure adds cost but is an essential step in the tower erection. Neglecting this important construction detail can "lock-in" stresses and deflections that will lower the factor of safety of the tower and, in an extreme case, could cause a failure.

Tower construction usually requires special equipment to erect steel components or concrete forms to the extreme height of the tower. Suspension bridges and some cable-stayed bridges require cable saddles to be erected on the tower tops. Floating cranes rarely have the capacity to reach to the heights of towers designed for long spans. Tower cranes, connected to the tower as it is erected, can be employed for most tower designs and are a good choice for handling steel forms for the erection of concrete towers. A tower crane used to jump the forms and raise materials can be seen in Figure 28.9. Occasionally, vertical traveling cranes are used to erect steel towers by pulling themselves up the face of the tower following the erection of each new tower component.

The erection sequence for a suspension bridge may require that the towers be pulled by cables from the vertical toward the sides spans or that the cable saddles be placed on rollers and displaced toward the side spans on temporary supports. The tower restraints are gradually released or the rollers pushed toward their final position as the erection of the deck element nears completion. This operation is usually required to induce the design forces into the cables in the side spans. The cable saddles then are permanently anchored to the towers.

Because the tower erection must be done in stages, each stage must be checked for stability and for stresses and deflections. The specifications should require the tower erection to be checked by an engineer, employed by the contractor, for stability and safety at each erection stage. The construction specifications should also require the tower erection stages to be submitted to the design engineer for an evaluation. This evaluation should be thorough enough to determine if the proposed tower erection staging will meet the intent of the original design, or if it needs to be modified to bring the completed tower into compliance.

28.7 Summary

Towers provide the visible means of support of the roadway on which goods and people travel. Being the most visible elements in a bridge, they give the bridge, for good or for ill, its character, its motif, and its identifying aesthetic impression. Towers usually form structural portals through which people pass as they travel from one point to another. Of themselves, towers form an aesthetic structural statement.

Towers are the most critical structural element in the bridge as their function is to carry the forces imposed on the bridge to the ground. Unlike most other bridge components, they cannot be replaced during the life of the bridge. Towers must fulfill their function in a reliable, serviceable, economical, and aesthetic manner for the entire life of the bridge. Towers must also be practicable to erect without extraordinary expense.

Practicable tower shapes for cable-stayed bridges are many and varied. Towers can have one or several legs or shafts arrayed from vertical to inclined and forming A- or inverted Y-shaped frames. Suspension bridge towers are usually vertical, with two shafts connected with one or several struts.

The conceptual design is the most important phase in the design of a long-span bridge. This phase sets, among other items, the span length, type of deck system, and the materials and shape of the towers. It also determines the aesthetic, economics, and constructibility of the bridge. A conceptual erection scheme should be developed during this phase to ensure that the bridge can be economically constructed.

The final design phase sets the specific shape, dimensions, and materials for the bridge. A practical erection method should be developed during this phase and shown on the construction drawings. If an unusual tower design is used, the tower erection should also be shown. The specifications should allow the contractor to employ an alternative method of erection, provided that the method is designed by an engineer and submitted to the design engineer for review. It is essential that the design engineer follow the project into the construction stages. The designer must understand each erection step that is submitted by the contractor in accordance with the specifications, to ensure the construction complies with the design documents. Only by this means are owners assured that the serviceability and reliability that they are paying for are actually achieved in construction.

The successful design of a cable-stayed or a suspension bridge involves many factors and decisions that must be made during the planning, design, and construction phases of the project. Towers play an important role in that successful execution. The final judgment of a successful project is made by the people who use the facility and pay for its construction, maintenance, and long-life service to society.

References

1. AASHTO, *Standard Specifications for Highway Bridges*, American Association of State Highway and Transportation Officials, Washington, D.C., 1994.
2. Billington, D. P., *The Tower and the Bridge, The New Art of Structural Engineering*, Basic Books, New York, 1983.
3. Cerver, F. A., *New Architecture in Bridges*, Muntaner, Spain, 1992.
4. Gimsing, N. J., *Cable-Supported Bridges — Concept and Design*, John Wiley & Sons, New York, 1997.
5. Leonhardt, F., *Bridges, Aesthetics and Design*, MIT Press, Cambridge, MA, 1984.
6. Podolny, W. and Scalzi, J. B., *Construction and Design of Cable Stayed Bridges*, 2nd ed., John Wiley & Sons, New York, 1986.
7. Seim, C., San Francisco Bay's jeweled necklace, *ASCE Civil Eng.*, 66(1), 14A, 1996.
8. Troitsky, M. S., *Cable Stayed Bridges*, Van Nostrand Reinhold, 1988.
9. Walter, R., *Cable Stayed Bridges*, Thomas Telford, U.K., 1988.

29

Abutments and Retaining Structures

Linan Wang
California Transportation Department

Chao Gong
ICF Kaiser Engineers, Inc.

29.1 Introduction

As a component of a bridge, the abutment provides the vertical support to the bridge superstructure at the bridge ends, connects the bridge with the approach roadway, and retains the roadway base materials from the bridge spans. Although there are numerous types of abutments and the abutments for the important bridges may be extremely complicated, the analysis principles and design methods are very similar. In this chapter the topics related to the design of conventional highway bridge abutments are discussed and a design example is illustrated.

Unlike the bridge abutment, the earth-retaining structures are mainly designed for sustaining lateral earth pressures. Those structures have been widely used in highway construction. In this chapter several types of retaining structures are presented and a design example is also given.

29.2 Abutments

29.2.1 Abutment Types

Open-End and Closed-End Abutments

From the view of the relation between the bridge abutment and roadway or water flow that the bridge overcrosses, bridge abutments can be divided into two categories: open-end abutment, and closed-end abutment, as shown in Figure 29.1.

For the open-end abutment, there are slopes between the bridge abutment face and the edge of the roadway or river canal that the bridge overcrosses. Those slopes provide a wide open area for the traffic flows or water flows under the bridge. It imposes much less impact on the environment

0-8493-7434-0/00/$0.00+$.50

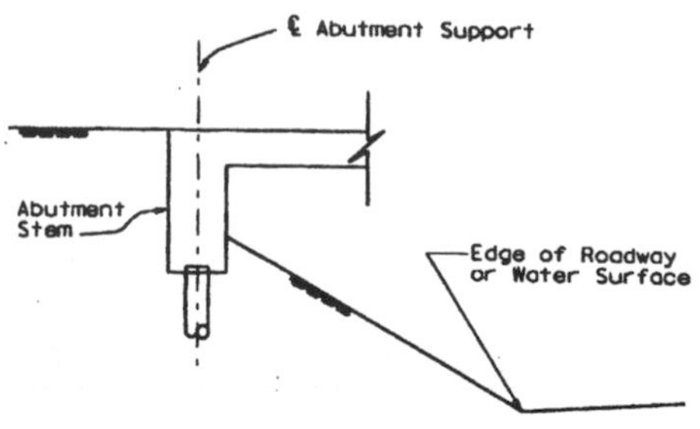

(a) Open End, Monolithic Type Abutment

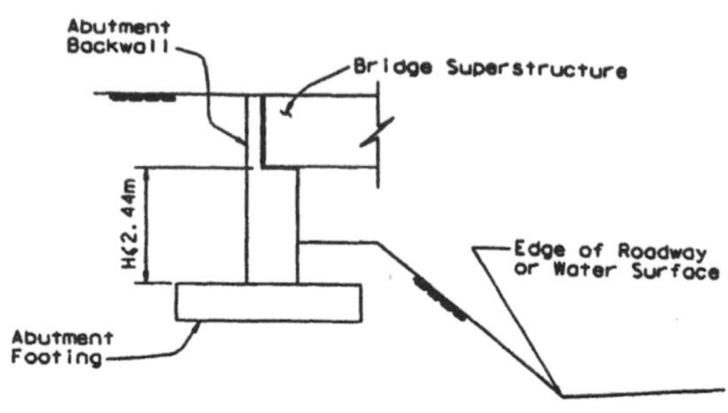

(b) Open End, Short Stem Seat Type Abutment

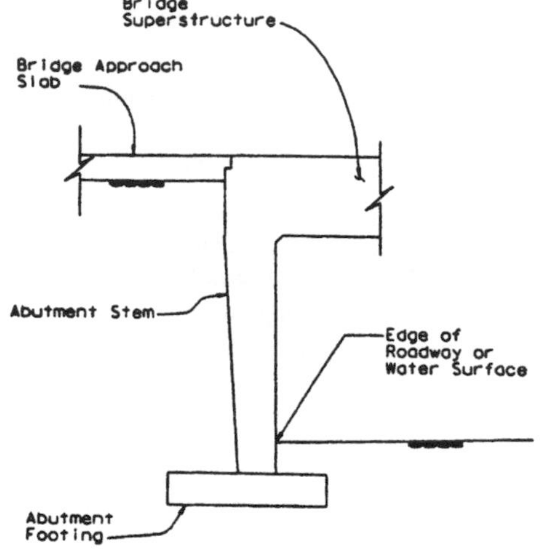

(c) Close End, Monolithic Type Abutment

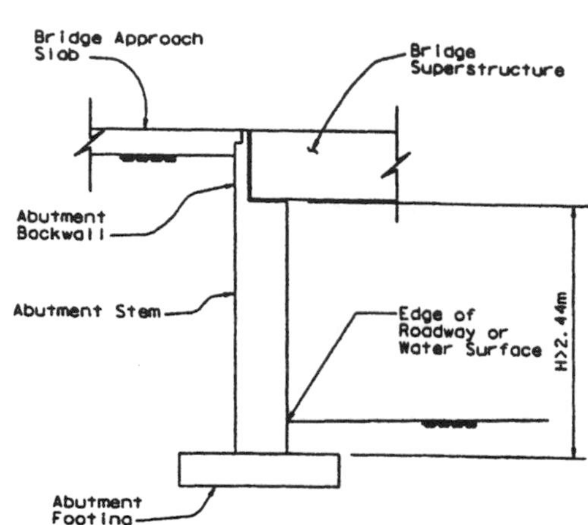

(d) Close End, High Stem Seat Type Abutment

FIGURE 29.1 Typical abutment types.

and the traffic flows under the bridge than a closed-end abutment. Also, future widening of the roadway or water flow canal under the bridge by adjusting the slope ratios is easier. However, the existence of slopes usually requires longer bridge spans and some extra earthwork. This may result in an increase in the bridge construction cost.

The closed-end abutment is usually constructed close to the edge of the roadways or water canals. Because of the vertical clearance requirements and the restrictions of construction right of way, there are no slopes allowed to be constructed between the bridge abutment face and the edge of roadways or water canals, and high abutment walls must be constructed. Since there is no room or only a little room between the abutment and the edge of traffic or water flow, it is very difficult to do the future widening to the roadways and water flow under the bridge. Also, the high abutment walls and larger backfill volume often result in higher abutment construction costs and more settlement of road approaches than for the open-end abutment.

Generally, the open-end abutments are more economical, adaptable, and attractive than the closed-end abutments. However, bridges with closed-end abutments have been widely constructed in urban areas and for rail transportation systems because of the right-of-way restriction and the large scale of the live load for trains, which usually results in shorter bridge spans.

Monolithic and Seat-Type Abutments

Based on the connections between the abutment stem and the bridge superstructure, the abutments also can be grouped in two categories: the monolithic or end diaphragm abutment and the seat-type abutment, as shown in Figure 29.1.

The monolithic abutment is monolithically constructed with the bridge superstructure. There is no relative displacement allowed between the bridge superstructure and abutment. All the superstructure forces at the bridge ends are transferred to the abutment stem and then to the abutment backfill soil and footings. The advantages of this type of abutment are its initial lower construction cost and its immediate engagement of backfill soil that absorbs the energy when the bridge is subjected to transitional movement. However, the passive soil pressure induced by the backfill soil could result in a difficult-to-design abutment stem, and higher maintenance cost might be expected. In the practice this type of abutment is mainly constructed for short bridges.

The seat-type abutment is constructed separately from the bridge superstructure. The bridge superstructure seats on the abutment stem through bearing pads, rock bearings, or other devices. This type of abutment allows the bridge designer to control the superstructure forces that are to be transferred to the abutment stem and backfill soil. By adjusting the devices between the bridge superstructure and abutment the bridge displacement can be controlled. This type of abutment may have a short stem or high stem, as shown in Figure 29.1. For a short-stem abutment, the abutment stiffness usually is much larger than the connection devices between the superstructure and the abutment. Therefore, those devices can be treated as boundary conditions in the bridge analysis. Comparatively, the high stem abutment may be subject to significant displacement under relatively less force. The stiffness of the high stem abutment and the response of the surrounding soil may have to be considered in the bridge analysis. The availability of the displacement of connection devices, the allowance of the superstructure shrinkage, and concrete shortening make this type of abutment widely selected for the long bridge constructions, especially for prestressed concrete bridges and steel bridges. However, bridge design practice shows that the relative weak connection devices between the superstructure and the abutment usually require the adjacent columns to be specially designed. Although the seat-type abutment has relatively higher initial construction cost than the monolithic abutment, its maintenance cost is relatively lower.

Abutment Type Selection

The selection of an abutment type needs to consider all available information and bridge design requirements. Those may include bridge geometry, roadway and riverbank requirements, geotechnical and right-of-way restrictions, aesthetic requirements, economic considerations, etc. Knowledge of the advantages and disadvantages for the different types of abutments will greatly benefit the bridge designer in choosing the right type of abutment for the bridge structure from the beginning stage of the bridge design.

29.2.2 General Design Considerations

Abutment design loads usually include vertical and horizontal loads from the bridge superstructure, vertical and lateral soil pressures, abutment gravity load, and the live-load surcharge on the abutment backfill materials. An abutment should be designed so as to withstand damage from the Earth pressure, the gravity loads of the bridge superstructure and abutment, live load on the superstructure or the approach fill, wind loads, and the transitional loads transferred through the connections between the superstructure and the abutment. Any possible combinations of those forces, which produce the most severe condition of loading, should be investigated in abutment design. Meanwhile, for the integral abutment or monolithic type of abutment the effects of bridge superstructure deformations, including bridge thermal movements, to the bridge approach structures must be

TABLE 29.1 Abutment Design Loads (Service Load Design)

Abutment Design Loads	Case I	II	III	IV	V
Dead load of superstructure	X	X	—	X	X
Dead load of wall and footing	X	X	X	X	X
Dead load of earth on heel of wall including surcharge	X	X	X	X	—
Dead load of earth on toe of wall	X	X	X	X	—
Earth pressure on rear of wall including surcharge	X	X	X	X	—
Live load on superstructure	X	—	—	X	—
Temperature and shrinkage	—	—	—	X	—
Allowable pile capacity of allowable soil pressure in % or basic	100	100	150	125	150

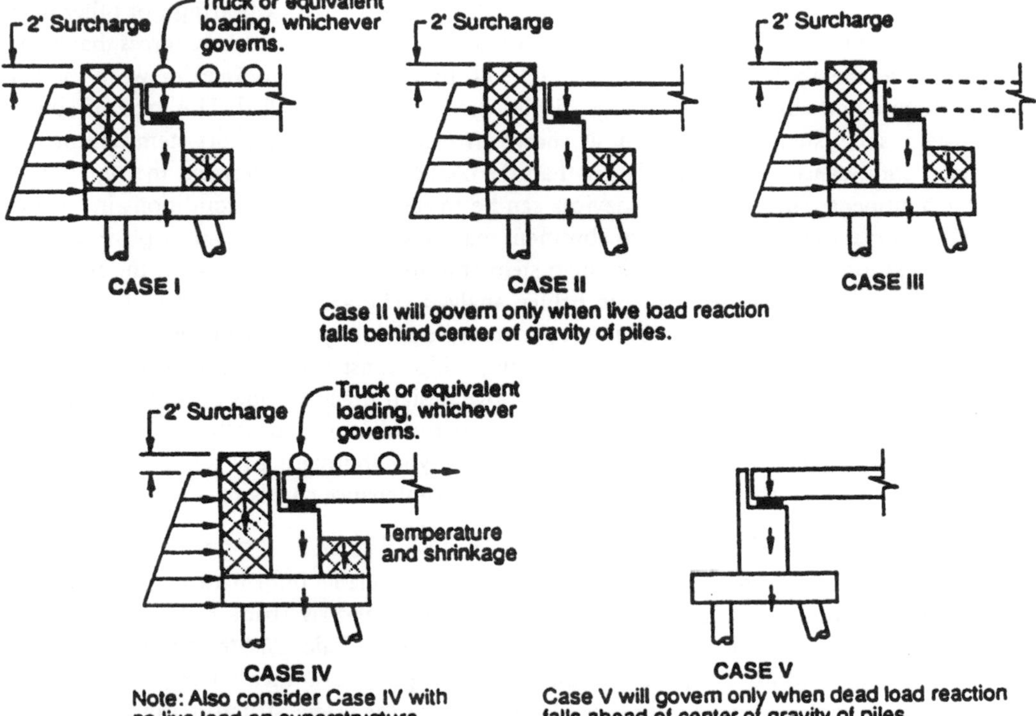

FIGURE 29.2 Configuration of abutment design load and load combinations.

considered in abutment design. Nonseismic design loads at service level and their combinations are shown in Table 29.1 and Figure 29.2. It is easy to obtain the factored abutment design loads and load combinations by multiplying the load factors to the loads at service levels. Under seismic loading, the abutment may be designed at no support loss to the bridge superstructure while the abutment may suffer some damages during a major earthquake.

The current AASHTO Bridge Design Specifications recommend that either the service load design or the load factor design method be used to perform an abutment design. However, due to the uncertainties in evaluating the soil response to static, cycling, dynamic, and seismic loading, the service load design method is usually used for abutment stability checks and the load factor method is used for the design of abutment components.

The load and load combinations listed in Table 29.1 may cause abutment sliding, overturning, and bearing failures. Those stability characteristics of abutment must be checked to satisfy certain restrictions. For the abutment with spread footings under service load, the factor of safety to resist sliding should be greater than 1.5; the factor of safety to resist overturning should be greater than 2.0; the factor of safety against soil bearing failure should be greater than 3.0. For the abutment with pile support, the piles have to be designed to resist the forces that cause abutment sliding, overturning, and bearing failure. The pile design may utilize either the service load design method or the load factor design method.

The abutment deep shear failure also needs to be studied in abutment design. Usually, the potential of this kind of failure is pointed out in the geotechnical report to the bridge designers. Deep pilings or relocating the abutment may be used to avoid this kind of failure.

29.2.3 Seismic Design Considerations

Investigations of past earthquake damage to the bridges reveal that there are commonly two types of abutment earthquake damage — stability damage and component damage.

Abutment stability damage during an earthquake is mainly caused by foundation failure due to excessive ground deformation or the loss of bearing capacities of the foundation soil. Those foundation failures result in the abutment suffering tilting, sliding, settling, and overturning. The foundation soil failure usually occurs because of poor soil conditions, such as soft soil, and the existence of a high water table. In order to avoid these kinds of soil failures during an earthquake, borrowing backfill soil, pile foundations, a high degree of soil compaction, pervious materials, and drainage systems may be considered in the design.

Abutment component damage is generally caused by excessive soil pressure, which is mobilized by the large relative displacement between the abutment and its backfilled soil. Those excessive pressures may cause severe damage to abutment components such as abutment back walls and abutment wingwalls. However, the abutment component damages do not usually cause the bridge superstructure to lose support at the abutment and they are repairable. This may allow the bridge designer to utilize the deformation of abutment backfill soil under seismic forces to dissipate the seismic energy to avoid the bridge losing support at columns under a major earthquake strike.

The behavior of abutment backfill soil deformed under seismic load is very efficient at dissipating the seismic energy, especially for the bridges with total length of less than 300 ft (91.5 m) with no hinge, no skew, or that are only slightly skewed (i.e., <15°). The tests and analysis revealed that if the abutments are capable of mobilizing the backfill soil and are well tied into the backfill soil, a damping ratio in the range of 10 to 15% is justified. This will elongate the bridge period and may reduce the ductility demand on the bridge columns. For short bridges, a damping reduction factor, D, may be applied to the forces and displacement obtained from bridge elastic analysis which generally have damped ARS curves at 5% levels. This factor D is given in Eq. (29.1).

$$D = \frac{1.5}{40\,C + 1} + 0.5 \tag{29.1}$$

where C = damping ratio.

Based on Eq. (29.1), for 10% damping, a factor $D = 0.8$ may be applied to the elastic force and displacement. For 15% damping, a factor $D = 0.7$ may be applied. Generally, the reduction factor D should be applied to the forces corresponding to the bridge shake mode that shows the abutment being excited.

The responses of abutment backfill soil to the seismic load are very difficult to predict. The study and tests revealed that the soil forces, which are applied to bridge abutment under seismic load, mainly depend on the abutment movement direction and magnitude. In the design practice, the Mononobe–Okabe method usually is used to quantify those loads for the abutment with no restraints on the top. Recently, the "near full scale" abutment tests performed at the University of California at Davis show a nonlinear relationship between the abutment displacement and the backfill soil reactions under certain seismic loading when the abutment moves toward its backfill soil. This relation was plotted as shown in Figure 29.3. It is difficult to simulate this nonlinear relationship between the abutment displacement and the backfill soil reactions while performing bridge dynamic analysis. However, the tests concluded an upper limit for the backfill soil reaction on the abutment. In design practice, a peak soil pressure acting on the abutment may be predicted corresponding to certain abutment displacements. Based on the tests and investigations of past earthquake damages, the California Transportation Department suggests guidelines for bridge analysis considering abutment damping behavior as follows.

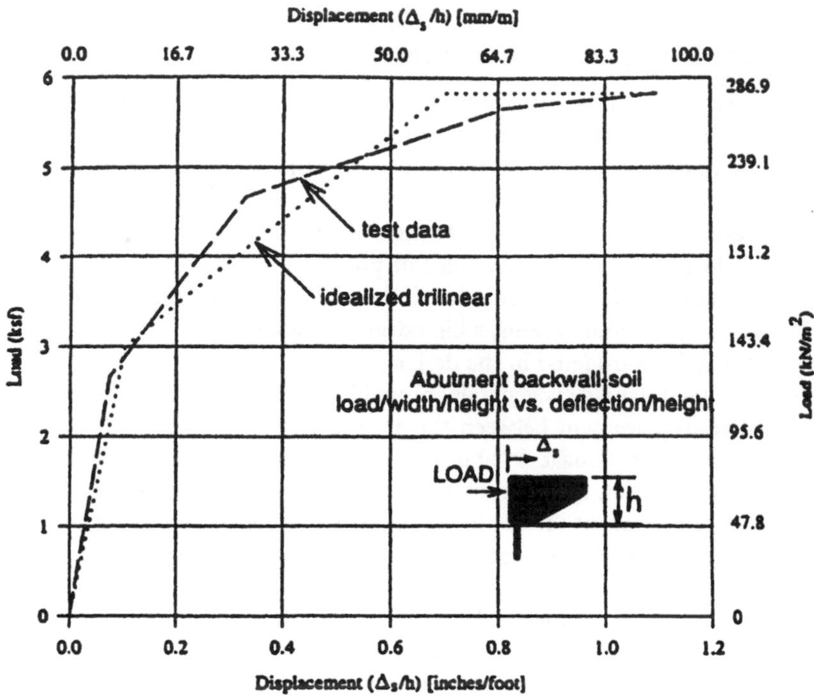

FIGURE 29.3 Proposed characteristics and experimental envelope for abutment backfill load–deformation.

By using the peak abutment force and the effective area of the mobilized soil wedge, the peak soil pressure is compared to a maximum capacity of 7.7 ksf (0.3687 MPa). If the peak soil pressure exceeds the soil capacity, the analysis should be repeated with reduced abutment stiffness. It is important to note that the 7.7 ksf (0.3687 MPa) soil pressure is based on a reliable minimum wall height of 8 ft (2.438 m). If the wall height is less than 8 ft (2.438 m), or if the wall is expected to shear off at a depth below the roadway less than 8 ft (2.438 m), the allowable passive soil pressure must be reduced by multiplying 7.7 ksf (0.3687 MPa) times the ratio of ($L/8$) [2], where L is the effective height of the abutment wall in feet. Furthermore, the shear capacity of the abutment wall diaphragm (the structural member mobilizing the soil wedge) should be compared with the demand shear forces to ensure the soil mobilizations. Abutment spring displacement is then evaluated against an acceptable level of displacement of 0.2 ft (61 mm). For a monolithic-type abutment this displacement is equal to the bridge superstructure displacement. For seat-type abutments this displacement usually does not equal the bridge superstructure displacement, which may include the gap between the bridge superstructure and abutment backwall. However, a net displacement of about 0.2 ft (61 mm) at the abutment should not be exceeded. Field investigations after the 1971 San Fernando earthquake revealed that the abutment, which moved up to 0.2 ft (61 mm) in the longitudinal direction into the backfill soil, appeared to survive with little need for repair. The abutments in which the backwall breaks off before other abutment damage may also be satisfactory if a reasonable load path can be provided to adjacent bents and no collapse potential is indicated.

For seismic loads in the transverse direction, the same general principles still apply. The 0.2-ft (61-mm) displacement limit also applies in the transverse direction, if the abutment stiffness is expected to be maintained. Usually, wingwalls are tied to the abutment to stiffen the bridge transversely. The lateral resistance of the wingwall depends on the soil mass that may be mobilized by the wingwall. For a wingwall with the soil sloped away from the exterior face, little lateral resistance can be predicted. In order to increase the transverse resistance of the abutment, interior supplemental shear walls may be attached to the abutment or the wingwall thickness may be increased, as shown in Figure 29.4. In some situations larger deflection may be satisfactory if a reasonable load path can be provided to adjacent bents and no collapse potential is indicated. [2]

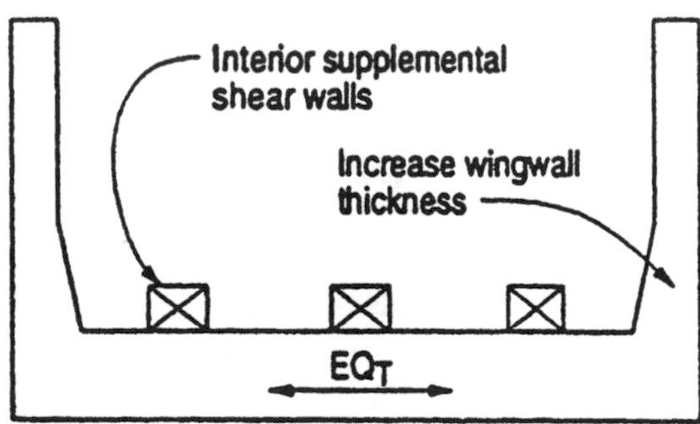

FIGURE 29.4 Abutment transverse enhancement.

Based on the above guidelines, abutment analysis can be carried out more realistically by a trial-and-error method on abutment soil springs. The criterion for abutment seismic resistance design may be set as follows.

Monolithic Abutment or Diaphragm Abutment (Figure 29.5)

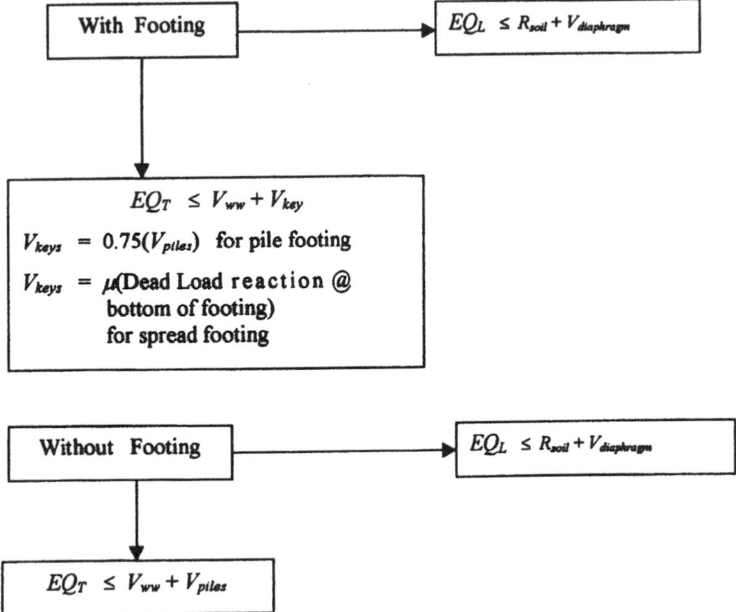

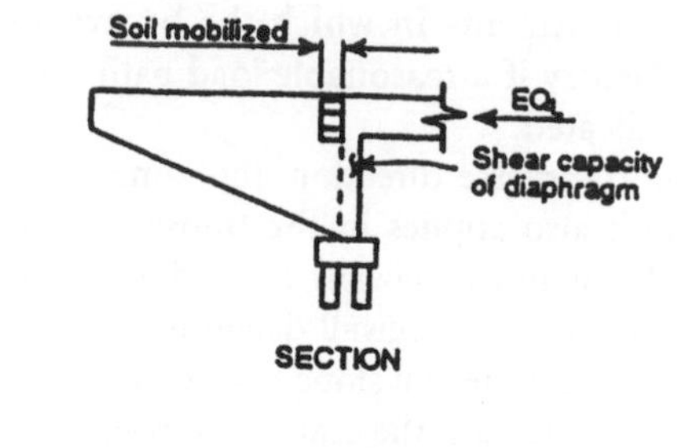

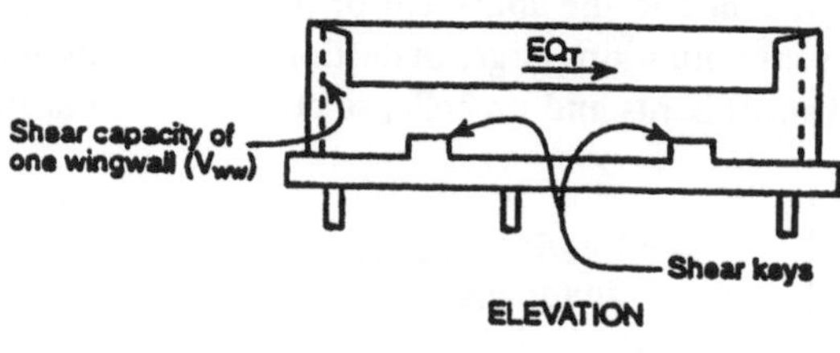

(a)with footing

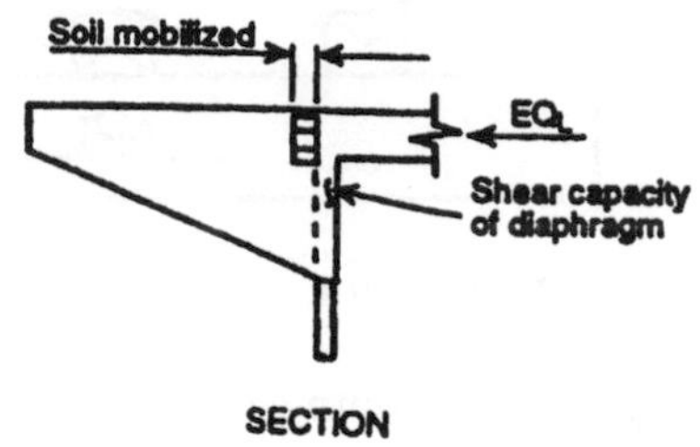

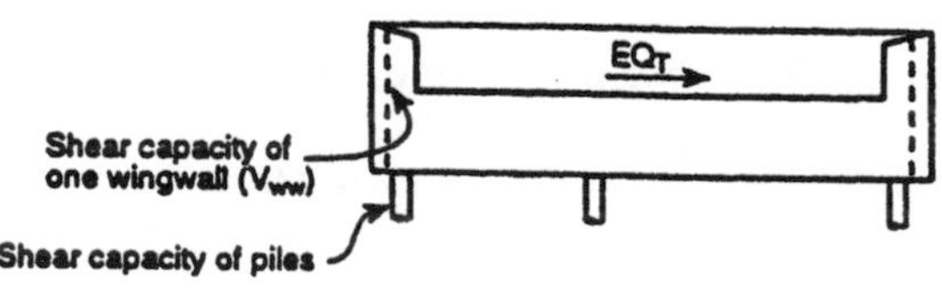

(b)without footing

FIGURE 29.5 Seismic resistance elements for monolithic abutment.

Seat-Type Abutment (Figure 29.6)

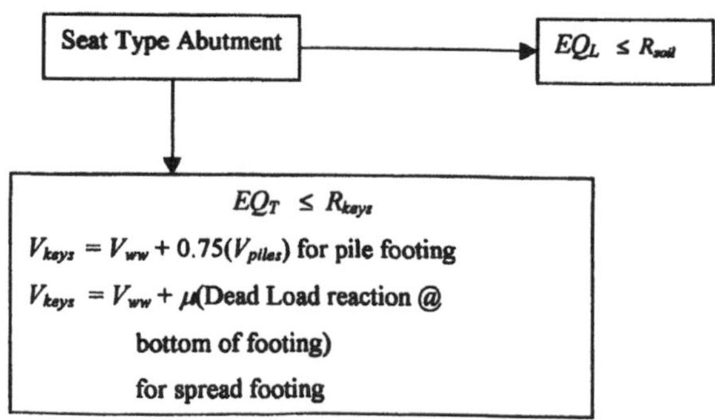

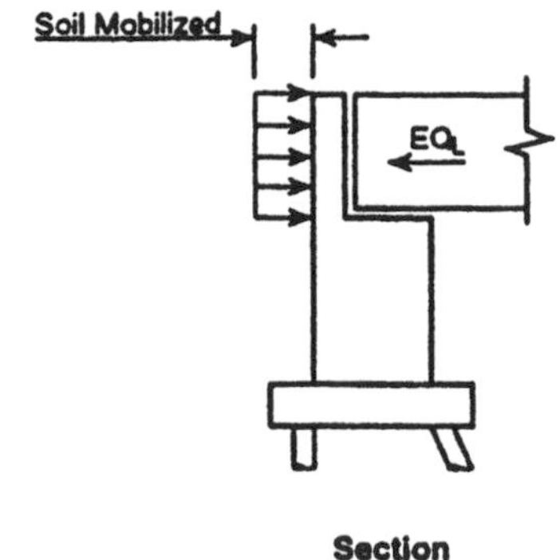

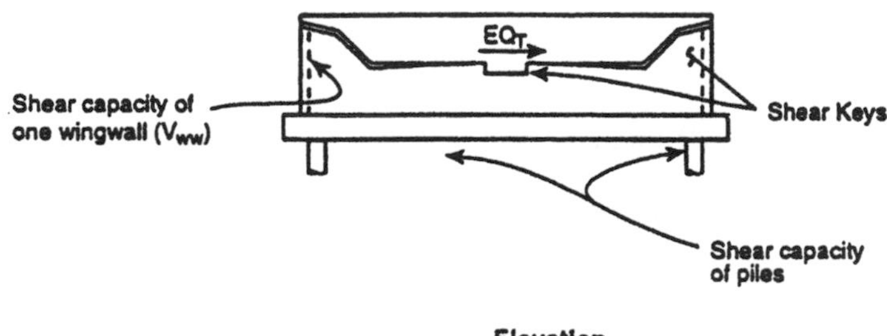

FIGURE 29.6 Seismic resistance elements for seat-type abutment.

where

EQ_L = longitudinal earthquake force from an elastic analysis
EQ_T = transverse earthquake force from an elastic analysis
R_{soil} = resistance of soil mobilized behind abutment
$R_{diaphragm}$ = φ times the nominal shear strength of the diaphragm
R_{ww} = φ times the nominal shear strength of the wingwall
R_{piles} = φ times the nominal shear strength of the piles
R_{keys} = φ times the nominal shear strength of the keys in the direction of consideration
φ = strength factor for seismic loading
μ = coefficient factor between soil and concrete face at abutment bottom

It is noted that the purpose of applying a factor of 0.75 to the design of shear keys is to reduce the possible damage to the abutment piles. For all transverse cases, if the design transverse earthquake force exceeds the sum of the capacities of the wingwalls and piles, the transverse stiffness for the analysis should equal zero ($EQ_T = 0$). Therefore, a released condition which usually results in larger lateral forces at adjacent bents should be studied.

Responding to seismic load, bridges usually accommodate a large displacement. To provide support at abutments for a bridge with large displacement, enough support width at the abutment must be designed. The minimum abutment support width, as shown in Figure 29.7, may be equal to the bridge displacement resulting from a seismic elastic analysis or be calculated as shown in Equation (29-2), whichever is larger:

$$N = (305 + 2.5L + 10H)(1 + 0.002\,S^2) \tag{29.2}$$

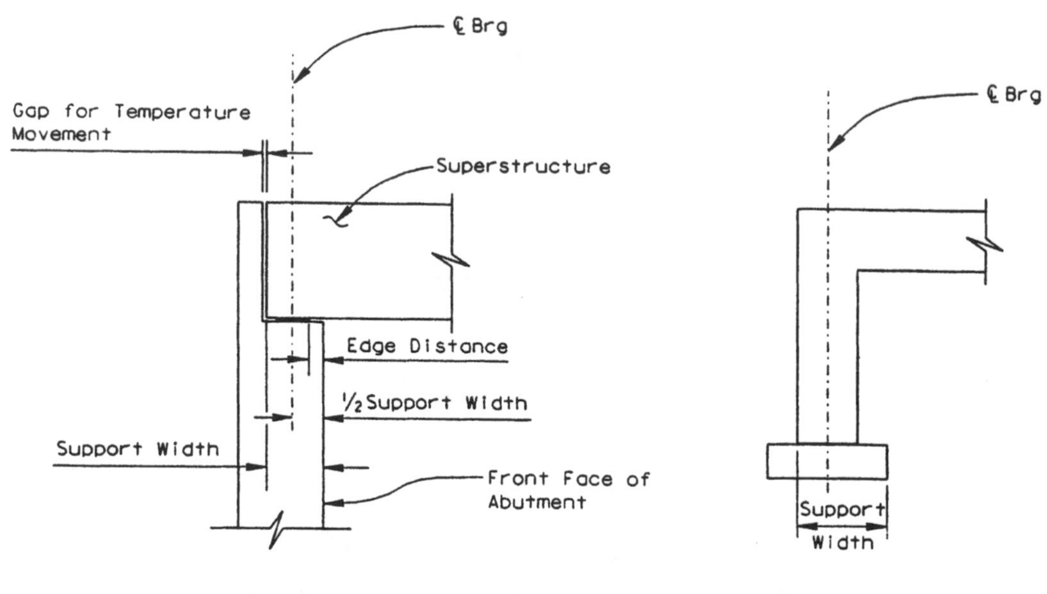

FIGURE 29.7 Abutment support width (seismic).

where

N = support width (mm)

L = length (m) of the bridge deck to the adjacent expansion joint, or to the end of bridge deck; for single-span bridges L equals the length of the bridge deck

S = angle of skew at abutment in degrees

H = average height (m) of columns or piers supporting the bridge deck from the abutment to the adjacent expansion joint, or to the end of the bridge deck; $H = 0$ for simple span bridges

29.2.4 Miscellaneous Design Considerations

Abutment Wingwall

Abutment wingwalls act as a retaining structure to prevent the abutment backfill soil and the roadway soil from sliding transversely. Several types of wingwall for highway bridges are shown in Figure 29.8. A wingwall design similar to the retaining wall design is presented in Section 29.3. However, live-load surcharge needs to be considered in wingwall design. Table 29.2 lists the live-load surcharge for different loading cases. Figure 29.9 shows the design loads for a conventional cantilever wingwall. For seismic design, the criteria in transverse direction discussed in Section 29.2.3 should be followed. Bridge wingwalls may be designed to sustain some damage in a major earthquake, as long as bridge collapse is not predicted.

Abutment Drainage

A drainage system is usually provided for the abutment construction. The drainage system embedded in the abutment backfill soil is designed to reduce the possible buildup of hydrostatic pressure, to control erosion of the roadway embankment, and to reduce the possibility of soil liquefaction during an earthquake. For a concrete-paved abutment slope, a drainage system also needs to be provided under the pavement. The drainage system may include pervious materials, PSP or PVC pipes, weep holes, etc. Figure 29.10 shows a typical drainage system for highway bridge construction.

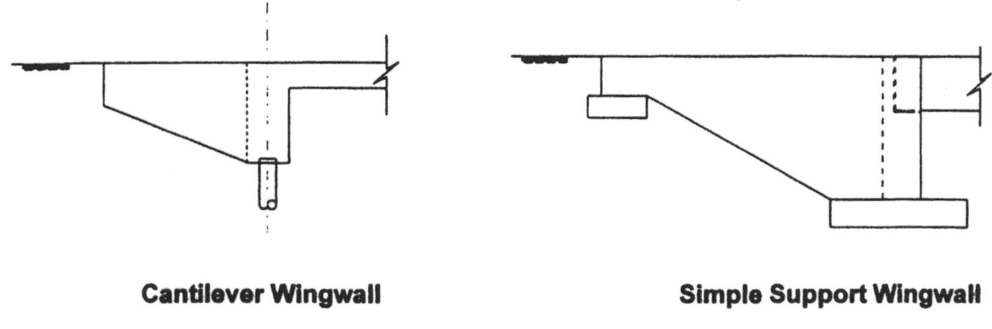

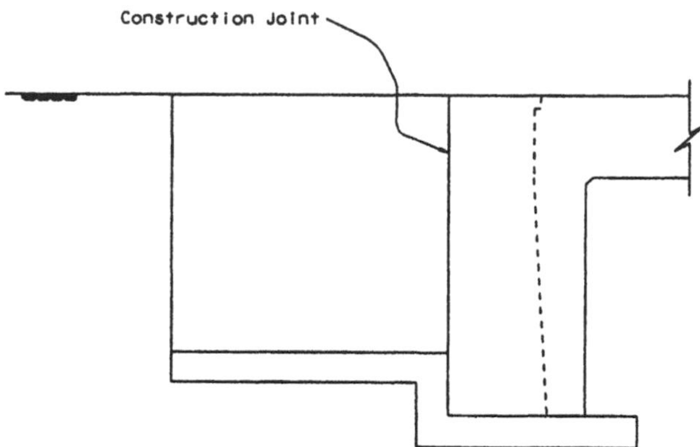

FIGURE 29.8 Typical wingwalls.

TABLE 29.2 Live Load Surcharges for Wingwall Design

Highway truck loading	2 ft 0 in. (610 mm) equivalent soil
Rail loading E-60	7 ft 6 in. (2290 mm) equivalent soil
Rail loading E-70	8 ft 9 in. (2670 mm) equivalent soil
Rail loading E-80	10 ft 0 in. (3050 mm) equivalent soil

Abutment Slope Protection

Flow water scoring may severely damage bridge structures by washing out the bridge abutment support soil. To reduce water scoring damage to the bridge abutment, pile support, rock slope protection, concrete slope paving, and gunite cement slope paving may be used. Figure 29.11 shows the actual design of rock slope protection and concrete slope paving protection for bridge abutments. The stability of the rock and concrete slope protection should be considered in the design. An enlarged block is usually designed at the toe of the protections.

Miscellaneous Details

Some details related to abutment design are given in Figure 29.12. Although they are only for regular bridge construction situations, those details present valuable references for bridge designers.

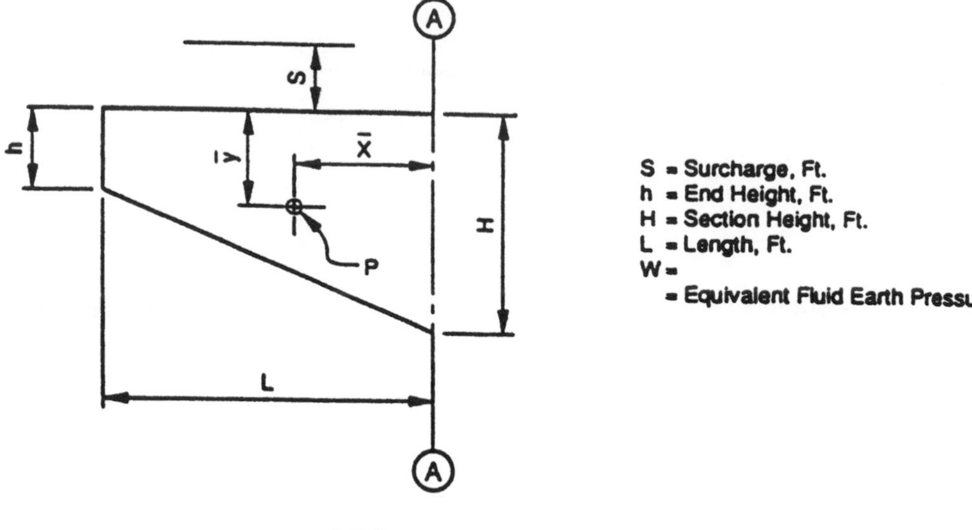

$$M_{AA} = \frac{WL^2}{24}\left[3h^2 + (H + 4S)(H + 2h)\right]$$

$$P = \frac{WL}{6}\left[H^2 + (h + H)(h + 3S)\right]$$

$$\overline{X} = \frac{M_{AA}}{P}$$

FIGURE 29.9 Design loading for cantilever wingwall.

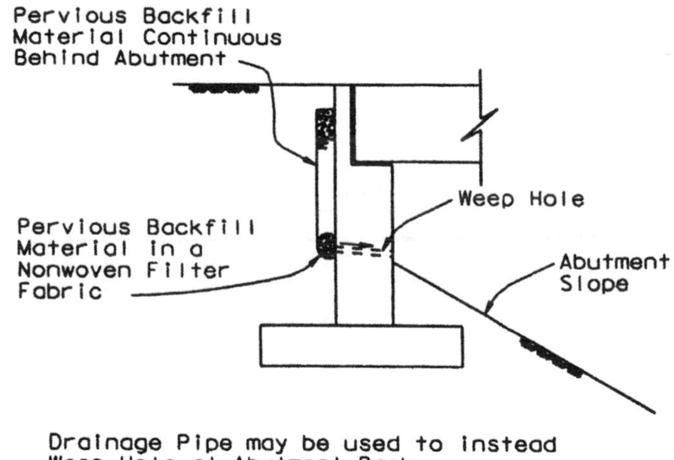

FIGURE 29.10 Typical abutment drainage system.

29.2.5 Design Example

A prestressed concrete box-girder bridge with 5° skew is proposed overcrossing a busy freeway as shown in Figure 29.13. Based on the roadway requirement, geotechnical information, and the details mentioned above, an open-end, seat-type abutment is selected. The abutment in transverse direction is 89 ft (27.13 m) wide. From the bridge analysis, the loads on abutment and bridge displacements are as listed bellow:

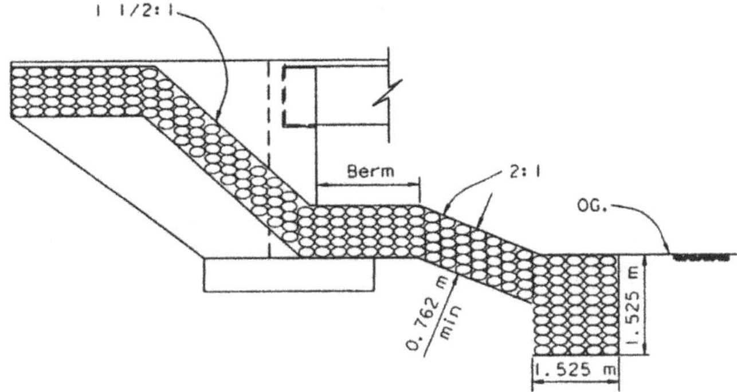

Rock Slope Protection

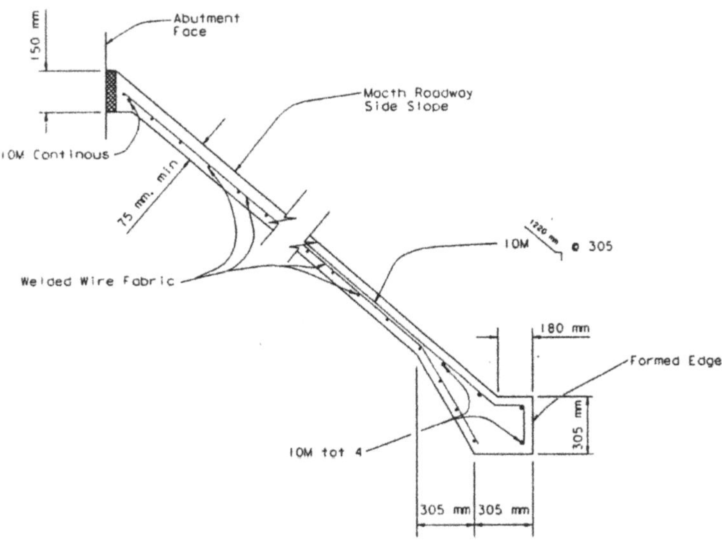

Concrete Slope Protection

FIGURE 29.11 Typical abutment slope protections.

Superstructure dead load	= 1630 kips (7251 kN)
HS20 live load	= 410 kips (1824 kN)
1.15 P-load + 1.0 HS load	= 280 kips (1245 kN)
Longitudinal live load	= 248 kips (1103 kN)
Longitudinal seismic load (bearing pad capacity)	= 326 kips (1450 kN)
Transverse seismic load	= 1241 kips (5520 kN)
Bridge temperature displacement	= 2.0 in. (75 mm)
Bridge seismic displacement	= 6.5 in. (165 mm)

Geotechnical Information

Live-load surcharge	= 2 ft (0.61 m)
Unit weight of backfill soil	= 120 pcf (1922 kg/m^3)

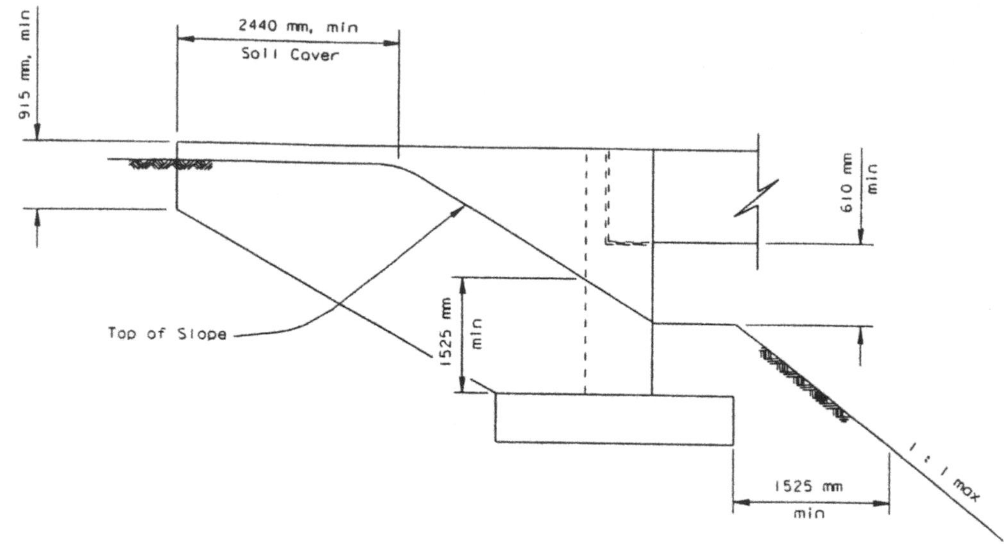

FIGURE 29.12 Abutment design miscellaneous details.

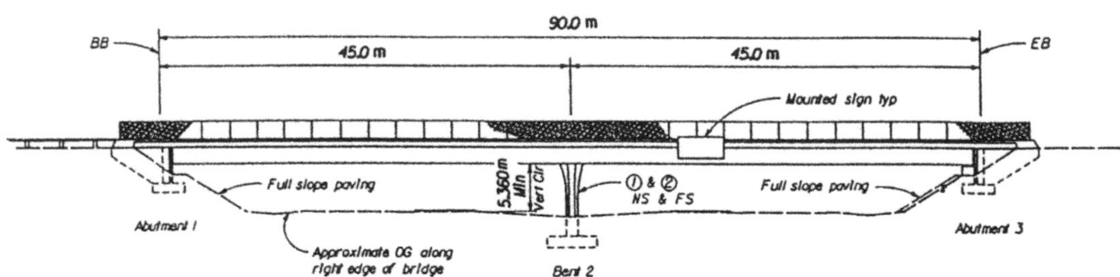

FIGURE 29.13 Bridge elevation (example).

Allowable soil bearing pressure	= 4.0 ksf (0.19 MPa)
Soil lateral pressure coefficient (Ka)	= 0.3
Friction coefficient	= tan 33°
Soil liquefaction potential	= very low
Ground acceleration	= 0.3 g

Design Criteria

Abutment design	Load factor method
Abutment stability	Service load method

Design Assumptions

1. Superstructure vertical loading acting on the center line of abutment footing;
2. The soil passive pressure by the soil at abutment toe is neglected;
3. 1.0 feet (0.305 m) wide of abutment is used in the design;
4. reinforcement yield stress, f_y = 60000 psi (414 MPa)
5. concrete strength, f_c' = 3250 psi (22.41 MPa)
6. abutment backwall allowed damage in the design earthquake

Solution

1. **Abutment Support Width Design**
   Applying Eq. (29.2) with

$$L = 6.5 \text{ m}$$
$$H = 90.0 \text{ m}$$
$$S = 5^\circ$$

the support width will be $N = 600$ mm. Add 75 mm required temperature movement, the total required support width equals 675 mm. The required minimum support width for seismic case equals the sum of the bridge seismic displacement, the bridge temperature displacement, and the reserved edge displacement (usually 75 mm). In this example, this requirement equals 315 mm, not in control. Based on the 675-mm minimum requirement, the design uses 760 mm, OK. A preliminary abutment configuration is shown in Figure 29.14 based on the given information and calculated support width.

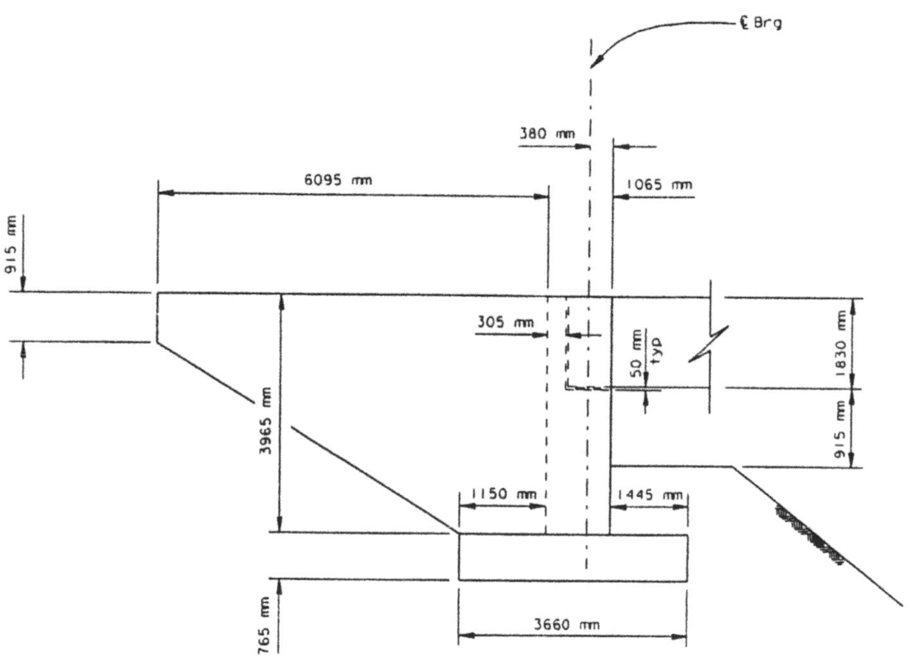

FIGURE 29.14 Abutment configuration (example).

2. **Abutment Stability Check**
   Figure 29.15 shows the abutment force diagram,

where

q_{sc} = soil lateral pressure by live-load surcharge
q_e = soil lateral pressure
q_{eq} = soil lateral pressure by seismic load
P_{DL} = superstructure dead load
P_{HS} = HS20 live load
P_P = permit live load

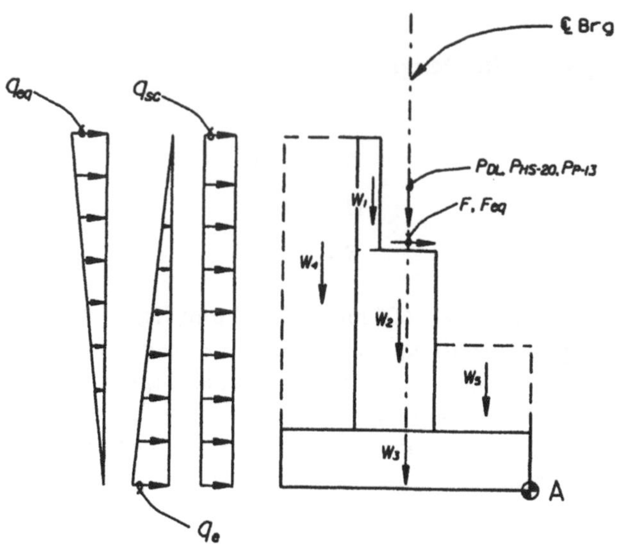

FIGURE 29.15 Abutment applying forces diagram (example).

F = longitudinal live load
F_{eq} = longitudinal bridge seismic load
P_{ac} = resultant of active seismic soil lateral pressure
h_{sc} = height of live-load surcharge
γ = unit weight of soil
W_i = weight of abutment component and soil block
$q_{sc} = k_a \times \gamma \times h_{sc} = 0.3 \times 0.12 \times 2 = 0.072$ ksf (0.0034 MPa)
$q_e = k_a \times \gamma \times H = 0.3 \times 0.12 \times 15.5 = 0.558$ ksf (0.0267 MPa)
$q_{eq} = k_{ae} \times \gamma \times H = 0.032 \times 0.12 \times 15.5 = 0.06$ ksf (0.003 MPa)

The calculated vertical loads, lateral loads, and moment about point A are listed in Table 29.3. The maximum and minimum soil pressure at abutment footing are calculated by

$$p = \frac{P}{B}\left(1 \pm \frac{6e}{B}\right) \tag{29.3}$$

where
p = soil bearing pressure
P = resultant of vertical forces
B = abutment footing width
e = eccentricity of resultant of forces and the center of footing

$$e = 2B - \frac{M}{P} \tag{29.4}$$

M = total moment to point A

Referring to the Table 29.1 and Eqs. (29.3) and (29.4) the maximum and minimum soil pressures under footing corresponding to different load cases are calculated as
Since the soil bearing pressures are less than the allowable soil bearing pressure, the soil bearing stability is OK.

Load Case	p_{max}	p_{min}	$p_{allowable}$ with Allowable % of Overstress	Evaluate
I	3.81	3.10	4.00	OK
II	3.42	2.72	4.00	OK
III	1.84	1.22	6.00	OK
IV	4.86	2.15	5.00	OK
V	2.79	1.93	6.00	OK
Seismic	6.73	0.54	8.00	OK

TABLE 29.3 Vertical Forces, Lateral Forces, and Moment about Point *A* (Example)

Load Description	Vertical Load (kips)	Lateral Load (kips)	Arm to *A* (ft)	Moment to *A* (k-ft)
Backwall W_1	0.94	—	7.75	7.28
Stem W_2	3.54	—	6.00	23.01
Footing W_3	4.50	—	6.00	27.00
Backfill soil	5.85	—	10.13	59.23
	—	4.33	5.17	–22.34
Soil surcharge	—	1.16	7.75	–8.65
Front soil W_4	1.71	—	2.38	4.06
Wingwalls	0.85	—	16.12	13.70
Keys	0.17	—	6.00	1.04
P_{DL}	18.31	—	6.00	110.00
P_{HS}	4.61	—	6.00	27.64
P_P	3.15	—	6.00	18.90
F	—	2.79	9.25	–25.80
F_{eq}	—	3.66	9.25	–33.90
Soil seismic load	—	0.47	9.30	–4.37

Check for the stability resisting the overturning (load case III and IV control):

Load Case	Driving Moment	Resist Moment	Factor of Safety	Evaluate
III	31	133.55	4.3	OK
IV	56.8	262.45	4.62	OK

Checking for the stability resisting the sliding (load case III and IV control)

Load Case	Driving Force	Resist Force	Factor of Safety	Evaluation
III	5.44	11.91	2.18	OK
IV	8.23	20.7	3.26	OK

Since the structure lateral dynamic force is only combined with dead load and static soil lateral pressures, and the factor of safety FS = 1.0 can be used, the seismic case is not in control.

3. **Abutment Backwall and Stem Design**

Referring to AASHTO guidelines for load combinations, the maximum factored loads for abutment backwall and stem are

Location	*V* (kips)	*M* (k-ft)
Backwall level	1.95	4.67
Bottom of stem	10.36	74.85

Abutment Backwall

Try #5 at 12 in. (305 mm) with 2 in. (50 mm) clearance

$d = 9.7$ in. (245 mm)

$$A_s \times f_y = 0.31 \times 60 \times \frac{12}{16} = 13.95 \text{ kips } (62.05 \text{ kN})$$

$$a = \frac{A_s \cdot f_y}{\phi \cdot f_c' \cdot b_w} = \frac{13.95}{(0.85)(3.25)(12)} = 0.42 \text{ in. } (10.67 \text{ mm})$$

$$M_u = \phi \cdot M_n = \phi \cdot A_s \cdot f_y\left(d - \frac{a}{2}\right) = 0.9 \times 13.95 \times \left(9.7 - \frac{0.42}{2}\right) = 9.33 \text{ k} \cdot \text{ft } (13.46 \text{ kN} \cdot \text{m})$$

$$> 4.67 \text{ k} \cdot \text{ft } (6.33 \text{ kN} \cdot \text{m}) \quad \text{OK}$$

$$V_c = 2\sqrt{f_c'} \cdot b_w \cdot d = 2 \times \sqrt{3250} \times 12 \times 9.7 = 13.27 \text{ kips } (59.03 \text{ kN})$$

$$V_u = \phi \cdot V_c = 0.85 \times 13.27 = 11.28 \text{ kip } (50.17 \text{ kN}) > 1.95 \text{ kips } (8.67 \text{ kN}) \quad \text{OK}$$

No shear reinforcement needed.

Abutment Stem

Abutment stem could be designed based on the applying moment variations along the abutment wall height. Here only the section at the bottom of stem is designed.

Try #6 at 12 in. (305 mm) with 2 in. (50 mm) clearance.

$$A_s \times f_y = 0.44 \times 60 = 26.40 \text{ kips } (117.43 \text{ kN})$$

$d = 39.4$ in. (1000 mm)

$$a = \frac{A_s \cdot f_y}{\phi \cdot f_c' \cdot b_w} = \frac{26.4}{(0.85)(3.25)(12)} = 0.796 \text{ in } (20.0 \text{ mm})$$

$$M_u = \phi \cdot A_s \cdot f_y\left(d - \frac{a}{2}\right) = 0.9 \times 26.4 \times \left(39.4 - \frac{0.8}{2}\right) = 77.22 \text{ k·ft } (104.7 \text{kN·m})$$

$$> 74.85 \text{ k·ft } (101.5 \text{kN·m}) \quad \text{OK}$$

$$V_c = 2\sqrt{f_c'}\, b_w\, d = 2 \times \sqrt{3250} \times 12 \times 39.4 = 53.91 \text{ kips } (238 \text{ kN})$$

$$V_u = \phi \cdot V_c = 0.85 \times 53.91 = 45.81 \text{ kips } (202.3 \text{ kN}) > 10.36 \text{ kips } (46.08 \text{ kN}) \quad \text{OK}$$

No shear reinforcement needed.

4. **Abutment Footing Design**

Considering all load combinations and seismic loading cases, the soil bearing pressure diagram under the abutment footing are shown in Figure 29.16.

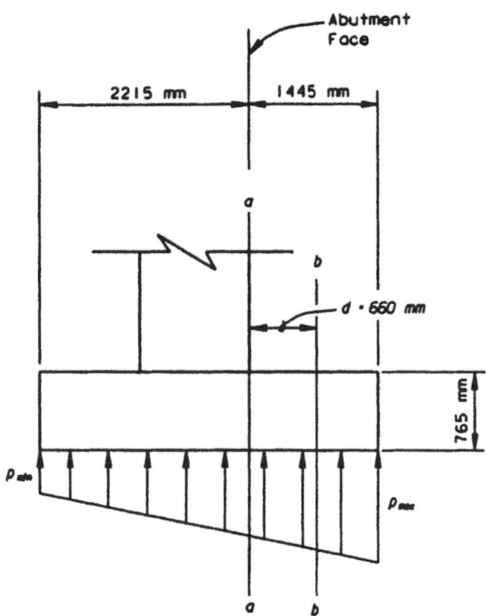

FIGURE 29.16 Bearing pressure under abutment footing (example).

a. *Design forces:*
Section at front face of abutment stem (design for flexural reinforcement):

$$q_{a\text{-}a} = 5.1263 \text{ ksf } (0.2454 \text{ MPa})$$

$$M_{a\text{-}a} = 69.4 \text{ k-ft } (94.1 \text{ kN·m})$$

Section at $d = 30 - 3 - 1 = 26$ in. (660 mm) from the front face of abutment stem (design for shear reinforcement):

$$q_{b\text{-}b} = 5.2341 \text{ ksf } (0.251 \text{ MPa})$$

$$V_{b\text{-}b} = 15.4 \text{ kips } (68.5 \text{ kN})$$

b. *Design flexural reinforcing (footing bottom):*
Try #8 at 12, with 3 in. (75 mm) clearance at bottom

$$d = 30 - 3 - 1 = 26 \text{ in. } (660 \text{ mm})$$

$$A_s \times f_y = 0.79 \times 60 = 47.4 \text{ kips} \quad (211 \text{ kN})$$

$$a = \frac{A_s \cdot f_y}{\phi \cdot f_c' \cdot b_w} = \frac{47.4}{(0.85)(3.25)(12)} = 1.43 \text{ in.} \quad (36 \text{ mm})$$

$$M_n = \phi \cdot A_s \cdot f_y\left(d - \frac{a}{2}\right) = 0.9 \times 47.4 \times \left(26 - \frac{1.43}{2}\right) = 89.9 \text{ k·ft} \quad (121.89 \text{ kN·m})$$

$$> 69.4 \text{ k·ft} \quad (94.1 \text{ kN·m}) \qquad \text{OK}$$

$$V_c = 2\sqrt{f_c'} \cdot b_w \cdot d = 2 \times \sqrt{3250} \times 12 \times 26 = 35.57 \text{ kips} \quad (158.24 \text{ kN})$$

$$V_u = \phi \cdot V_c = 0.85 \times 35.57 = 30.23 \text{ kips} \quad (134.5 \text{ kN}) \quad > \quad 15.5 \text{ kips} \quad (68.5 \text{ kN}) \qquad \text{OK}$$

No shear reinforcement needed.

Since the minimum soil bearing pressure under the footing is in compression, the tension at the footing top is not the case. However, the minimum temperature reinforcing, 0.308 in.²/ft (652 mm²/m) needs to be provided. Using #5 at 12 in. (305 mm) at the footing top yields

$$A_s = 0.31 \text{ in.}^2/\text{ft}, \ (656 \text{ mm}^2/\text{m}) \qquad \text{OK}$$

5. **Abutment Wingwall Design**

The geometry of wingwall is

$$h = 3.0 \text{ ft } (915 \text{ mm}); \qquad S = 2.0 \text{ ft } (610 \text{ mm});$$

$$H = 13.0 \text{ ft } (3960 \text{ mm}); \qquad L = 18.25 \text{ ft } (5565 \text{ mm})$$

Referring to the Figure 29.15, the design loads are

$$V_{A-A} = \frac{wL}{6}\left[H^2 + (h+H)(h+3S)\right]$$

$$= \frac{0.36 \times 18.25}{6}\left[13^2 + (3+13)(3+3\times 2)\right] = 34 \text{ kips} \quad (152.39 \text{ kN})$$

$$M_{A-A} = \frac{wL^2}{24}\left[3h^2 + (H+4S)(H+2h)\right]$$

$$= \frac{0.036 \times 18.25^2}{24}\left[3(3)^2 + (13+4\times 2)(12+2\times 3)\right] = 212.8 \text{ k·ft} \quad (3129 \text{ kN·m})$$

Design flexural reinforcing. Try using # 8 at 9 (225 mm).

$$A_s \times f_y = 13 \times (0.79) \times 60 \times \frac{12}{9} = 821.6 \text{ kips} \quad (3682 \text{ kN})$$

$$a = \frac{A_s \cdot f_y}{\phi \cdot f_c' \cdot b_w} = \frac{1280}{(0.85)(3.25)(13)(12)} = 2.97 \text{ in.} \quad (75 \text{ mm})$$

$$d = 12 - 2 - 0.5 = 9.5 \text{ in. } (240 \text{ mm})$$

$$M_n = \phi \cdot A_s \cdot f_y\left(d - \frac{a}{2}\right) = 0.9 \times (821.6) \times \left(9.5 - \frac{2.97}{2}\right) = 493.8 \text{ k·ft} \ (7261 \text{ kN}\cdot\text{m})$$

$$> 212.8 \text{ k·ft} \quad (3129 \text{ kN}\cdot\text{m})$$

Checking for shear

$$V_c = 2\sqrt{f_c'} \cdot b_w \cdot d = 2 \times \sqrt{3250} \times 13 \times 12 \times 9.5 = 168 \text{ kips} \quad (757.3 \text{ kN})$$

$$V_u = \varphi \cdot V_c = 0.85 \times 168 = 142 \text{ kips} \quad (636 \text{ kN}) > 34 \text{ kips} \ (152.3\text{kN}) \qquad \text{OK}$$

No shear reinforcing needed.

Since the wingwall is allowed to be broken off in a major earthquake, the adjacent bridge columns have to be designed to sustain the seismic loading with no wingwall resistance. The abutment section, footing, and wingwall reinforcing details are shown in Figures 29.17a and b.

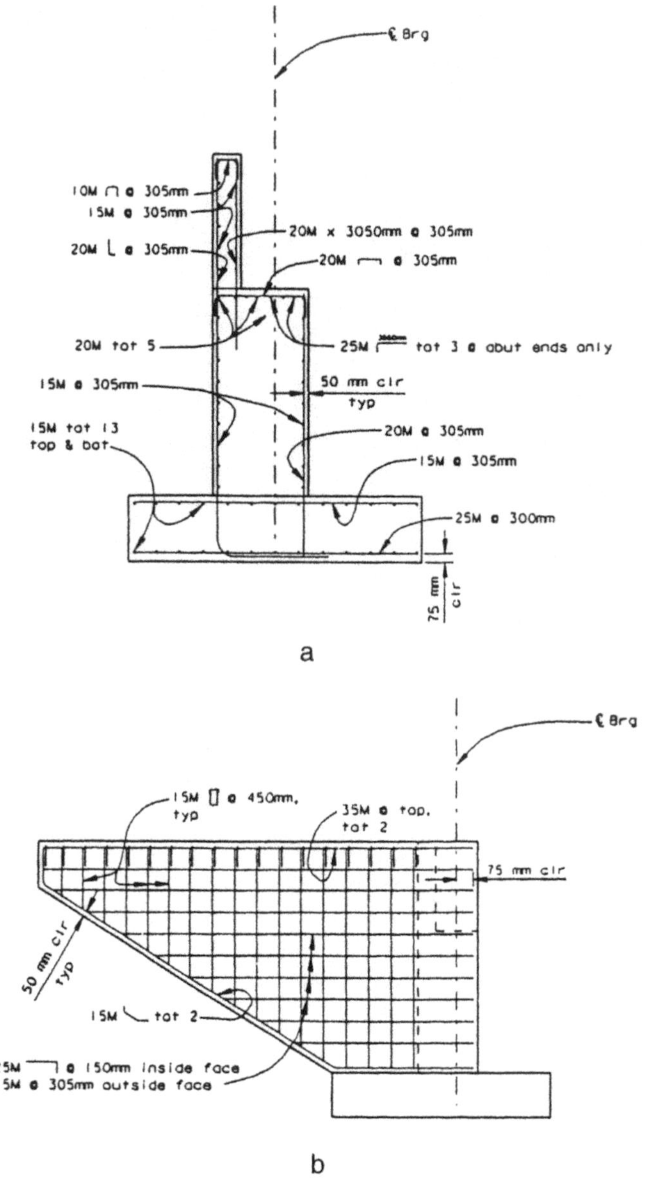

FIGURE 29.17 (a) Abutment typical section design (example). (b) Wingwall reinforcing (example).

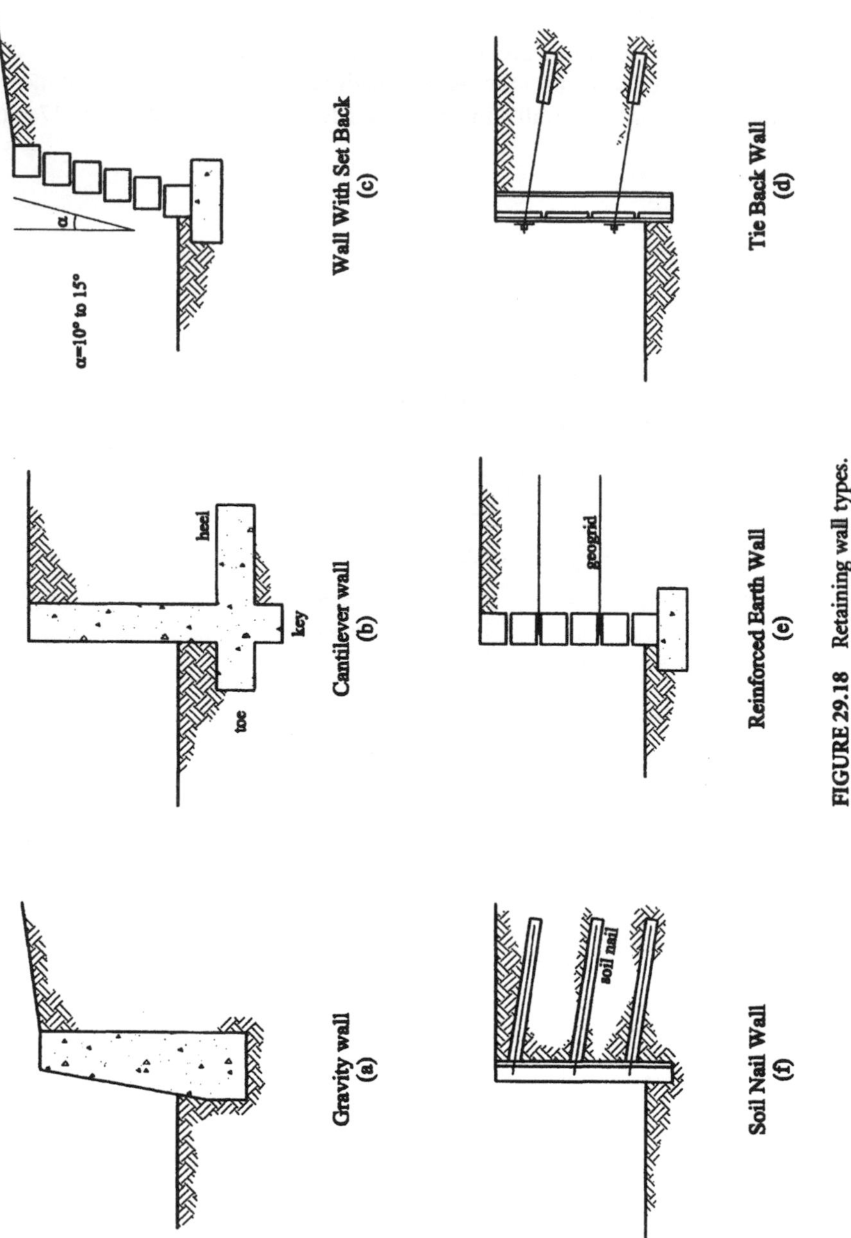

FIGURE 29.18 Retaining wall types.

29.3 Retaining Structures

29.3.1 Retaining Structure Types

The retaining structure, or, more specifically, the earth-retaining structure, is commonly required in a bridge design project. It is common practice that the bridge abutment itself is used as a retaining structure. The cantilever wall, tieback wall, soil nail wall and mechanically stabilized embankment (MSE) wall are the most frequently used retaining structure types. The major design function of a retaining structure is to resist lateral forces.

The cantilever retaining wall is a cantilever structure used to resist the active soil pressure in topography fill locations. Usually, the cantilever earth-retaining structure does not exceed 10 m in height. Some typical cantilever retaining wall sections are shown in Figure 29.18a.

The tieback wall can be used for topography cutting locations. High-strength tie strands are extended into the stable zone and act as anchors for the wall face elements. The tieback wall can be designed to have minimum lateral deflection. Figure 29.18d shows a tieback wall section.

The MSE wall is a kind of "reinforced earth-retaining" structure. By installing multiple layers of high-strength fibers inside of the fill section, the lateral deflection of filled soil will be restricted. There is no height limit for an MSE wall but the lateral deflection at the top of the wall needs to be considered. Figure 29.18e shows an example of an MSE wall.

The soil nail wall looks like a tieback wall but works like an MSE wall. It uses a series of soil nails built inside the soil body that resist the soil body lateral movement in the cut sections. Usually, the soil nails are constructed by pumping cement grout into predrilled holes. The nails bind the soil together and act as a gravity soil wall. A typical soil nail wall model is shown in Figure 29.18f.

29.3.2 Design Criteria

Minimum Requirements

All retaining structures must be safe from vertical settlement. They must have sufficient resistance against overturning and sliding. Retaining structures must also have adequate strength for all structural components.

1. *Bearing capacity*: Similar to any footing design, the bearing capacity factor of safety should be ≥1.0. Table 29.4 is a list of approximate bearing capacity values for some common materials. If a pile footing is used, the soil-bearing capacity between piles is not considered.
2. *Overturning resistance*: The overturning point of a typical retaining structure is located at the edge of the footing toe. The overturning factor of safety should be ≥1.50. If the retaining structure has a pile footing, the fixity of the footing will depend on the piles only.
3. *Sliding resistance*: The factor of safety for sliding should be ≥1.50. The typical retaining wall sliding capacity may include both the passive soil pressure at the toe face of the footing and the friction forces at the bottom of the footing. In most cases, friction factors of 0.3 and 0.4

TABLE 29.4 Bearing Capacity

	Bearing Capacity [N]	
Material	min, kPa	max, kPa
Alluvial soils	24	48
Clay	48	190
Sand, confined	48	190
Gravel	95	190
Cemented sand and gravel	240	480
Rock	240	—

can be used for clay and sand, respectively. If battered piles are used for sliding resistance, the friction force at the bottom of the footing should be neglected.

4. Structural strength: Structural section moment and shear capacities should be designed following common strength factors of safety design procedures.

Figure 29.19 shows typical loads for cantilever retaining structure design.

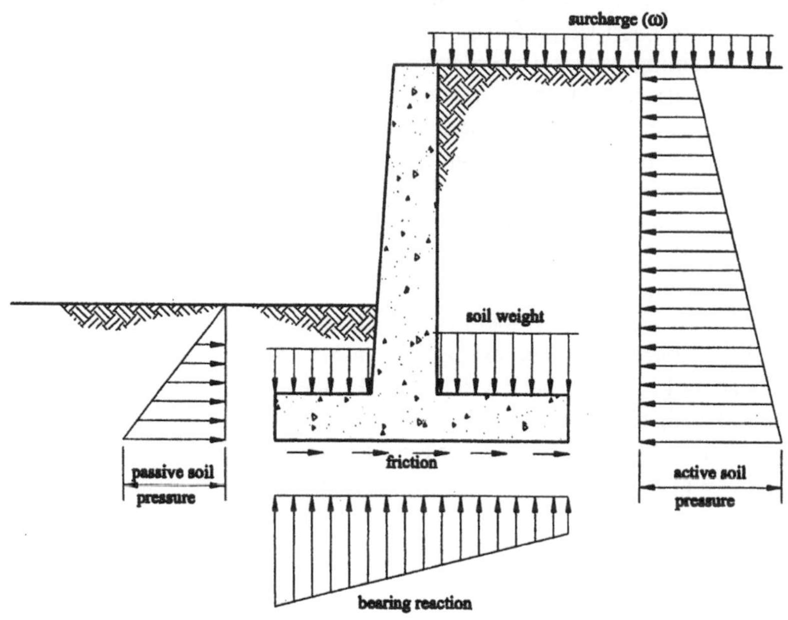

FIGURE 29 19 Typical loads on retaining wall.

Lateral Load

The unit weight of soil is typically in the range of 1.5 to 2.0 ton/m³. For flat backfill cases, if the backfill material is dry, cohesionless sand, the lateral earth pressure (Figure 29.20a) distribution on the wall will be as follows

The active force per unit length of wall (Pa) at bottom of wall can be determined as

$$p_a = k_a \gamma H \tag{29.5}$$

The passive force per unit length of wall (Pa) at bottom of wall can be determined as

$$p_p = k_p \gamma H \tag{29.6}$$

where
H = the height of the wall (from top of the wall to bottom of the footing)
γ = unit weight of the backfill material
k_a = active earth pressure coefficient
k_p = passive earth pressure coefficient

The coefficients k_a and k_p should be determined by a geologist using laboratory test data from a proper soil sample. The general formula is

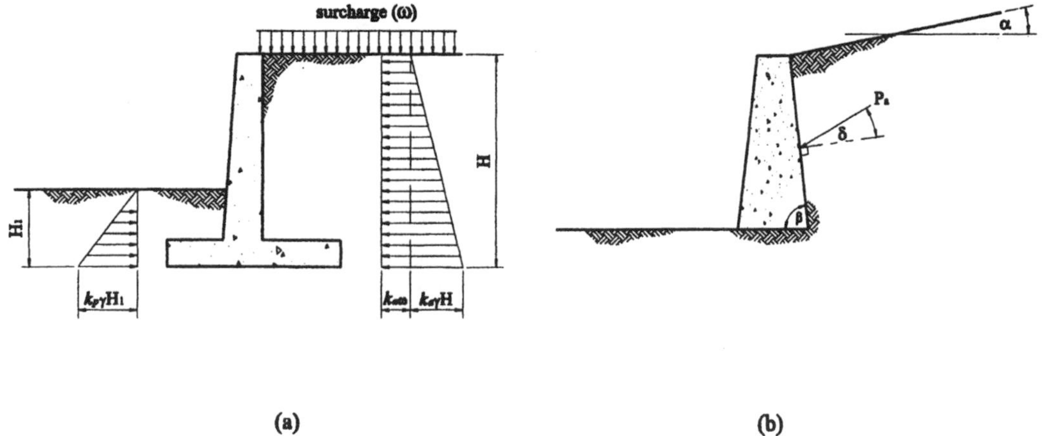

FIGURE 29.20 Lateral Earth pressure.

$$k_a = \frac{1-\sin\phi}{1+\sin\phi} \quad k_p = \frac{1}{k_a} = \frac{1+\sin\phi}{1-\sin\phi} \tag{29.7}$$

where ϕ is the internal friction angle of the soil sample.

Table 29.5 lists friction angles for some typical soil types which can be used if laboratory test data is not available. Generally, force coefficients of $k_a \geq 0.30$ and $k_p \leq 1.50$ should be used for preliminary design.

TABLE 29.5 Internal Friction Angle and Force Coefficients

Material	ϕ(*degrees*)	k_a	k_p
Earth, loam	30–45	0.33–0.17	3.00–5.83
Dry sand	25–35	0.41–0.27	2.46–3.69
Wet sand	30–45	0.33–0.17	3.00–5.83
Compact Earth	15–30	0.59–0.33	1.70–3.00
Gravel	35–40	0.27–0.22	3.69–4.60
Cinders	25–40	0.41–0.22	2.46–4.60
Coke	30–45	0.33–0.17	3.00–5.83
Coal	25–35	0.41–0.27	2.46–3.69

Based on the triangle distribution assumption, the total active lateral force per unit length of wall should be

$$P_a = \frac{1}{2} k_a \gamma H^2 \tag{29.8}$$

The resultant earth pressure always acts at distance of $H/3$ from the bottom of the wall.

When the top surface of backfill is sloped, the k_a coefficient can be determined by the Coulomb equation: (see Figure 29.20):

$$k_a = \frac{\sin^2(\phi+\beta)}{\sin^2\beta\,\sin(\beta-\delta)\left[1+\sqrt{\dfrac{\sin(\phi+\delta)\,\sin(\phi-\alpha)}{\sin(\beta-\delta)\,\sin(\alpha+\beta)}}\right]^2} \tag{29.9}$$

Note that the above lateral earth pressure calculation formulas do not include water pressure on the wall. A drainage system behind the retaining structures is necessary; otherwise the proper water pressure must be considered.

Table 29.6 gives values of k_a for the special case of zero wall friction.

TABLE 29.6 Active Stress Coefficient k_a Values from Coulomb Equation ($\delta = 0$)

		α					
		0.00°	18.43°	21.80°	26.57°	33.69°	45.00°
ϕ	β_o	Flat	1 to 3.0	1 to 2.5	1 to 2.0	1 to 1.5	1 to 1.0
20°	90°	0.490	0.731				
	85°	0.523	0.783				
	80°	0.559	0.842				
	75°	0.601	0.913				
	70°	0.648	0.996				
25°	90°	0.406	0.547	0.611			
	85°	0.440	0.597	0.667			
	80°	0.478	0.653	0.730			
	75°	0.521	0.718	0.804			
	70°	0.569	0.795	0.891			
30°	90°	0.333	0.427	0.460	0.536		
	85°	0.368	0.476	0.512	0.597		
	80°	0.407	0.530	0.571	0.666		
	75°	0.449	0.592	0.639	0.746		
	70°	0.498	0.664	0.718	0.841		
35°	90°	0.271	0.335	0.355	0.393	0.530	
	85°	0.306	0.381	0.404	0.448	0.602	
	80°	0.343	0.433	0.459	0.510	0.685	
	75°	0.386	0.492	0.522	0.581	0.781	
	70°	0.434	0.560	0.596	0.665	0.897	
40°	90°	0.217	0.261	0.273	0.296	0.352	
	85°	0.251	0.304	0.319	0.346	0.411	
	80°	0.287	0.353	0.370	0.402	0.479	
	75°	0.329	0.408	0.429	0.467	0.558	
	70°	0.375	0.472	0.498	0.543	0.651	
45°	90°	0.172	0.201	0.209	0.222	0.252	0.500
	85°	0.203	0.240	0.250	0.267	0.304	0.593
	80°	0.238	0.285	0.297	0.318	0.363	0.702
	75°	0.277	0.336	0.351	0.377	0.431	0.832
	70°	0.322	0.396	0.415	0.446	0.513	0.990

Any surface load near the retaining structure will generate additional lateral pressure on the wall. For highway-related design projects, the traffic load can be represented by an equivalent vertical surcharge pressure of 11.00 to 12.00 kPa. For point load and line load cases (Figure 29.21), the following formulas can be used to determine the additional pressure on the retaining wall:

For point load:

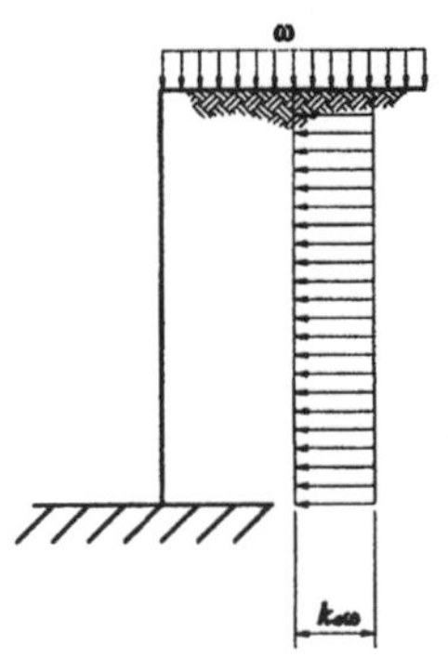

Uniform Surcharge
(a)

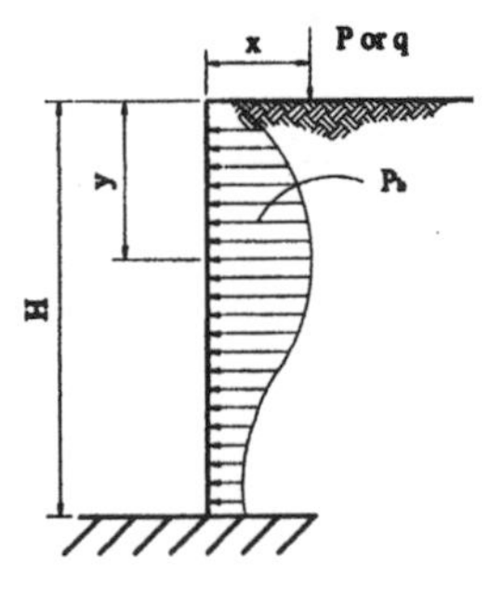

Point Load or Line Load
(b)

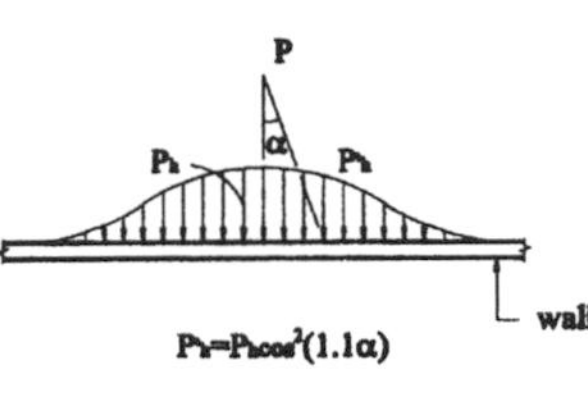

Horizontal Pressure Distribution
of Point Load
(c)

FIGURE 29.21 Additional lateral earth pressure. (a) Uniform surcharge; (b) point or line load; (c) horizontal pressure distribution of point load.

$$p_h = \frac{1.77V}{H^2}\frac{m^2n^2}{\left(m^2+n^2\right)^3} \quad (\text{m} \le 0.4) \quad p_h \frac{0.28V}{H^2}\frac{m^2n^2}{\left(0.16+n^3\right)} \quad (m > 0.4) \tag{29.10}$$

For line load:

$$p_h = \frac{\pi}{4}\frac{w}{H}\frac{m^2n}{\left(m^2+n^2\right)^2} \quad (m \le 0.4) \quad p_h = \frac{w}{H}\frac{0.203n}{\left(0.16+n^2\right)^2} \quad (m > 0.4) \tag{29.11}$$

where

$$m = \frac{x}{H}; \quad n = \frac{y}{H}$$

Table 29.7 gives lateral force factors and wall bottom moment factors which are calculated by above formulas.

29.3.3 Cantilever Retaining Wall Design Example

The cantilever wall is the most commonly used retaining structure. It has a good cost-efficiency record for walls less than 10 m in height. Figure 29.22a shows a typical cross section of a cantilever retaining wall and Table 29.8 gives the active lateral force and the active moment about bottom of the cantilever retaining wall.

For most cases, the following values can be used as the initial assumptions in the reinforced concrete retaining wall design process.

- $0.4 \le B/H \le 0.8$
- $1/12 \le t_{bot}/H \le 1/8$
- $L_{toe} \cong B/3$
- $t_{top} \ge 300$ mm
- $t_{foot} \ge t_{bot}$

TABLE 29.7 Line Load and Point Load Lateral Force Factors

Line Load Factors			Point Load Factors		
$m = x/H$	$(f)^a$	$(m)^b$	$m = x/H$	$(f)^c$	$(m)^d$
0.40	0.548	0.335	0.40	0.788	0.466
0.50	0.510	0.287	0.50	0.597	0.316
0.60	0.469	0.245	0.60	0.458	0.220
0.70	0.429	0.211	0.70	0.356	0.157
0.80	0.390	0.182	0.80	0.279	0.114
0.90	0.353	0.158	0.90	0.220	0.085
1.00	0.320	0.138	1.00	0.175	0.064
1.50	0.197	0.076	1.50	0.061	0.019
2.00	0.128	0.047	2.00	0.025	0.007

Notes:

[a] Total lateral force along the length of wall = factor(f) × ω (force)/(unit length).

[b] Total moment along the length of wall = factor(m) × ω × H (force × length)/(unit length) (at bottom of footing).

[c] Total lateral force along the length of wall = factor(f) × V/H (force)/(unit length).

[d] Total moment along the length of wall = factor(m) × V (force × length)/(unit length) (at bottom of footing).

Example

Given

A reinforced concrete retaining wall as shown in Figure 29.22b:

$H_o = 3.0$ m; surcharge $\omega = 11.00$ kPa
Earth internal friction angle $\phi = 30°$
Earth unit weight $\gamma = 1.8$ *ton/m³*
Bearing capacity $[\sigma] = 190$ kPa
Friction coefficient $f = 0.30$

Solution

1. **Select Control Dimensions**
   Try $h = 1.5$ m, therefore, $H = H_o + h = 3.0 + 1.5 = 4.5$ m.
   Use

$$t_{\text{bot}} = 1/10H = 0.45 \text{ m} \Rightarrow \; 500 \text{ mm}; \; t_{top} = t_{bot} = 500 \text{ mm}$$

$$t_{\text{foot}} = 600 \text{ mm}$$

   Use

$$B = 0.6H = 2.70 \text{ m} \Rightarrow 2700 \text{ mm};$$

$$L_{\text{toe}} = 900 \text{ mm; therefore, } L_{\text{heel}} = 2.7 - 0.9 - 0.5 = 1.3 \text{ m} = 1300 \text{ mm}$$

2. **Calculate Lateral Earth Pressure**
   From Table 29.4, $k_a = 0.33$ and $k_p = 3.00$.
   Active Earth pressure:

$$\text{Part 1 (surcharge) } P_1 = k_a\omega H = 0.33(11.0)(4.5) = 16.34 \text{ kN}$$

$$\text{Part 2 } P_2 = 0.5\, k_a\gamma\, H^2 = 0.5(0.33)(17.66)(4.5)^2 = 59.01 \text{ kN}$$

   Maximum possible passive Earth pressure:

$$P_p = 0.5k_p\gamma h^2 = 0.5(3.00)(17.66)(1.5)^2 = 59.60 \text{ kN}$$

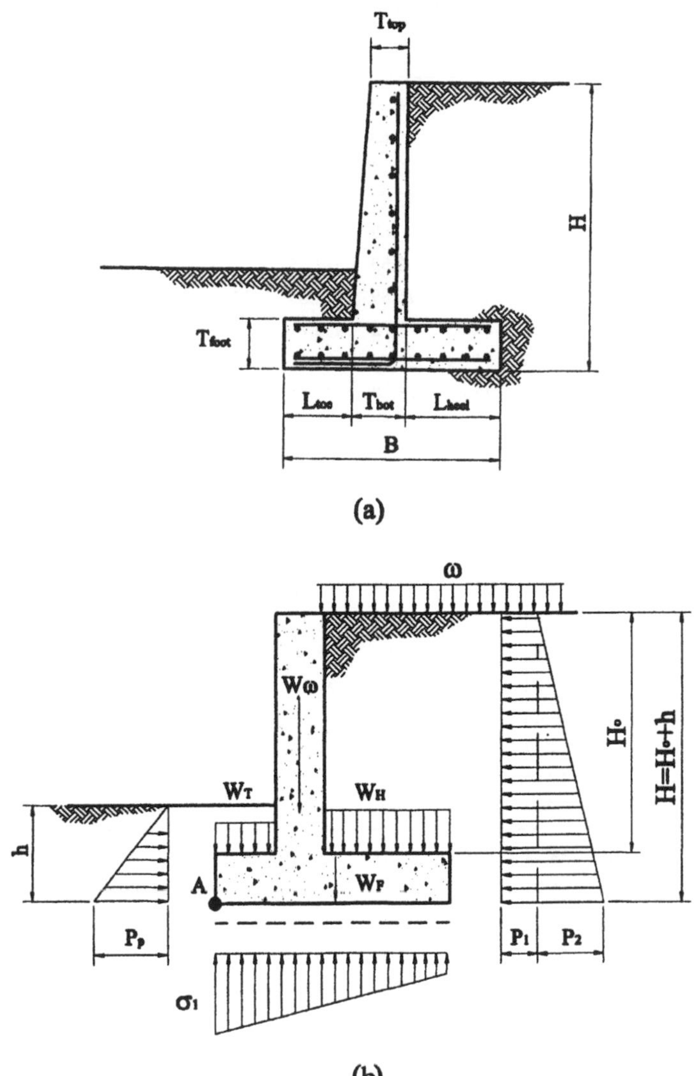

FIGURE 29.22 Design example.

3. Calculate Vertical Loads

Surcharge	W_s (11.00)(1.3)	= 14.30 kN

Use ρ = 2.50 ton/m³ as the unit weight of reinforced concrete

Wall	W_w 0.50 (4.5 – 0.6) (24.53)	= 47.83 kN
Footing	W_f 0.60 (2.70) (24.53)	= 39.74 kN
Soil cover at toe	W_t 17.66 (1.50 – 0.60) (0.90)	= 14.30 kN
Soil cover at heel	W_h 17.66 (4.50 – 0.60) (1.30)	= 89.54 kN
		Total 205.71 kN

TABLE 29.8 Cantilever Retaining Wall Design Data with Uniformly Distributed Surcharge Load

s	h	1.0	1.2	1.4	1.6	1.8	2.0	2.2	2.4	2.6	2.8	3.0
	p	2.94	4.24	5.77	7.53	9.53	11.77	14.24	16.94	19.89	23.06	26.48
0.00	y	0.33	0.40	0.47	0.53	0.60	0.67	0.73	0.80	0.87	0.93	1.00
	m	0.98	1.69	2.69	4.02	5.72	7.84	10.44	13.56	17.24	21.53	26.48
	p	5.30	7.06	9.06	11.30	13.77	16.47	19.42	22.59	26.01	29.65	33.54
0.40	y	0.41	0.48	0.55	0.62	0.69	0.76	0.83	0.90	0.97	1.04	1.11
	m	2.16	3.39	5.00	7.03	9.53	12.55	16.14	20.33	25.19	30.75	37.07
	p	6.47	8.47	10.71	13.18	15.89	18.83	22.00	25.42	29.06	32.95	37.07
0.60	y	0.42	0.50	0.57	0.65	0.72	0.79	0.86	0.93	1.00	1.07	1.14
	m	2.75	4.24	6.15	8.54	11.44	14.91	18.98	23.72	29.17	35.36	42.36
	p	7.65	9.88	12.36	15.06	18.00	21.18	24.59	28.24	32.12	36.24	40.60
0.80	y	0.44	0.51	0.59	0.67	0.74	0.81	0.89	0.96	1.03	1.10	1.17
	m	3.33	5.08	7.30	10.04	13.34	17.26	21.83	27.11	33.14	39.98	47.66
	p	8.83	11.30	14.00	16.94	20.12	23.53	27.18	31.07	35.18	39.54	44.13
1.00	y	0.44	0.53	0.60	0.68	0.76	0.83	0.91	0.98	1.06	1.13	1.20
	m	3.92	5.93	8.46	11.55	15.25	19.61	24.68	30.50	37.12	44.59	52.95
	p	11.77	14.83	18.12	21.65	25.42	29.42	33.65	38.13	42.83	47.77	52.95
1.50	y	0.46	0.54	0.63	0.71	0.79	0.87	0.94	1.02	1.10	1.17	1.25
	m	5.39	8.05	11.34	15.31	20.02	25.50	31.80	38.97	47.06	56.12	66.19
	p	14.71	18.36	22.24	26.36	30.71	35.30	40.13	45.19	50.48	56.01	61.78
2.00	y	0.47	0.55	0.64	0.72	0.81	0.89	0.97	1.05	1.13	1.21	1.29
	m	6.86	10.17	14.22	19.08	24.78	31.38	38.92	47.45	57.01	67.65	79.43

s	h	3.2	3.4	3.6	3.8	4.0	4.2	4.4	4.6	4.8	5.0	5.2
	p	30.12	34.01	38.13	42.48	47.07	51.89	56.95	62.25	67.78	73.55	79.55
0.00	y	1.07	1.13	1.20	1.27	1.33	1.40	1.47	1.53	1.60	1.67	1.73
	m	32.13	38.54	45.75	53.81	62.76	72.65	83.53	95.45	108.45	122.58	137.88
	p	37.66	42.01	46.60	51.42	56.48	61.78	67.31	73.07	79.08	85.31	91.78
0.40	y	1.17	1.24	1.31	1.38	1.44	1.51	1.58	1.65	1.71	1.78	1.85
	m	44.18	52.14	61.00	70.80	81.59	93.41	106.31	120.35	135.56	151.99	169.70
	p	41.42	46.01	50.83	55.89	61.19	66.72	72.49	78.49	84.72	91.20	97.90
0.60	y	1.21	1.28	1.35	1.42	1.49	1.56	1.62	1.69	1.76	1.83	1.90
	m	50.21	58.95	68.63	79.30	91.00	103.79	117.70	132.80	149.11	166.70	185.61
	p	45.19	50.01	55.07	60.37	65.90	71.66	77.66	83.90	90.37	97.08	104.02
0.80	y	1.24	1.31	1.38	1.45	1.52	1.59	1.66	1.73	1.80	1.87	1.94
	m	56.23	65.75	76.25	87.79	100.41	114.17	129.09	145.25	162.67	181.41	201.52
	p	48.95	54.01	59.31	64.84	70.60	76.60	82.84	89.31	96.02	102.96	110.14
1.00	y	1.27	1.34	1.41	1.49	1.56	1.63	1.70	1.77	1.84	1.90	1.97
	m	62.26	72.55	83.88	96.29	109.83	124.54	140.48	157.70	176.23	196.12	217.43
	p	58.37	64.01	69.90	76.02	82.37	88.96	95.79	102.85	110.14	117.67	125.44
1.50	y	1.32	1.40	1.47	1.55	1.62	1.69	1.76	1.84	1.91	1.98	2.05
	m	77.32	89.55	102.94	117.53	133.36	150.49	168.96	188.82	210.12	232.89	257.20
	p	67.78	74.02	80.49	87.19	94.14	101.32	108.73	116.38	124.26	132.38	140.74
2.00	y	1.36	1.44	1.52	1.59	1.67	1.74	1.82	1.89	1.96	2.04	2.11
	m	92.38	106.56	122.00	138.77	156.90	176.44	197.44	219.94	244.00	269.67	296.97

TABLE 29.8 Cantilever Retaining Wall Design Data with Uniformly Distributed Surcharge Load

s	h	5.4	5.6	5.8	6.0	6.2	6.4	6.6	6.8	7.0	7.2	7.4
	p	85.78	92.25	98.96	105.90	113.08	120.50	128.14	136.03	144.15	152.50	161.09
0.00	y	1.80	1.87	1.93	2.00	2.07	2.13	2.20	2.27	2.33	2.40	2.47
	m	154.41	172.21	191.33	211.81	233.70	257.06	281.92	308.33	336.35	366.01	397.36
	p	98.49	105.43	112.61	120.03	127.67	135.56	143.68	152.03	160.62	169.45	178.51
0.40	y	1.92	1.98	2.05	2.12	2.18	2.25	2.32	2.39	2.45	2.52	2.59
	m	188.72	209.11	230.91	254.17	278.94	305.26	333.18	362.74	394.01	427.01	461.80
	p	104.85	112.02	119.44	127.09	134.97	143.09	151.44	160.03	168.86	177.92	187.22
0.60	y	1.96	2.03	2.10	2.17	2.23	2.30	2.37	2.44	2.50	2.57	2.64
	m	205.88	227.56	250.70	275.35	301.55	329.36	358.81	389.95	422.83	457.51	494.02
	p	111.20	118.61	126.26	134.15	142.27	150.62	159.21	168.04	177.10	186.39	195.92
0.80	y	2.01	2.07	2.14	2.21	2.28	2.35	2.41	2.48	2.55	2.62	2.69
	m	223.04	246.01	270.50	296.53	324.17	353.46	384.43	417.16	451.66	488.01	526.24
	p	117.55	125.20	133.09	141.21	149.56	158.15	166.98	176.04	185.33	194.86	204.63
1.00	y	2.04	2.11	2.18	2.25	2.32	2.39	2.46	2.52	2.59	2.66	2.73
	m	240.19	264.46	290.29	317.71	346.79	377.55	410.06	444.36	480.49	518.51	558.46
	p	133.44	141.68	150.15	158.86	167.80	176.98	186.39	196.04	205.93	216.05	226.40
1.50	y	2.12	2.19	2.26	2.33	2.40	2.47	2.54	2.61	2.68	2.75	2.82
	m	283.08	310.59	339.77	370.67	403.33	437.80	474.14	512.38	552.57	594.76	639.00
	p	149.33	158.15	167.21	176.51	186.04	195.81	205.81	216.05	226.52	237.23	248.17
2.00	y	2.18	2.26	2.33	2.40	2.47	2.54	2.62	2.69	2.76	2.83	2.90
	m	325.97	356.72	389.25	423.62	459.87	498.05	538.21	580.39	624.64	671.01	719.55

s	h	7.6	7.8	7.0	8.2	8.4	8.6	8.8	9.0	9.2	9.5	10.0
	p	169.92	178.98	144.15	197.81	207.57	217.58	227.81	238.29	248.99	265.50	294.18
0.00	y	2.53	2.60	2.33	2.73	2.80	2.87	2.93	3.00	3.07	3.17	3.33
	m	430.46	465.35	336.35	540.67	581.21	623.72	668.25	714.86	763.58	840.74	980.60
	p	187.80	197.34	160.62	217.10	227.34	237.82	248.52	259.47	270.65	287.86	317.71
0.40	y	2.65	2.72	2.45	2.85	2.92	2.99	3.06	3.12	3.19	3.29	3.46
	m	498.43	536.94	394.01	619.79	664.23	710.75	759.38	810.17	863.18	946.94	1098.27
	p	196.75	206.51	168.86	226.75	237.23	247.93	258.88	270.06	281.47	299.03	329.48
0.60	y	2.71	2.77	2.50	2.91	2.98	3.04	3.11	3.18	3.24	3.34	3.51
	m	532.41	572.73	422.83	659.36	705.75	754.26	804.94	857.83	912.98	1000.04	1157.11
	p	205.69	215.69	177.10	236.40	247.11	258.05	269.23	280.65	292.30	310.21	341.25
0.80	y	2.75	2.82	2.55	2.96	3.02	3.09	3.16	3.23	3.29	3.39	3.56
	m	566.39	608.53	451.66	698.92	747.26	797.78	850.50	905.49	962.78	1053.14	1215.94
	p	214.63	224.87	185.33	246.05	257.00	268.17	279.59	291.24	303.12	321.39	353.02
1.00	y	2.80	2.87	2.59	3.00	3.07	3.14	3.20	3.27	3.34	3.44	3.61
	m	600.38	644.32	480.49	738.48	788.78	841.29	896.06	953.14	1012.58	1106.24	1274.78
	p	236.99	247.82	205.93	270.17	281.71	293.47	305.48	317.71	330.19	349.34	382.43
1.50	y	2.89	2.96	2.68	3.10	3.17	3.24	3.31	3.38	3.44	3.55	3.72
	m	685.34	733.81	552.57	837.38	892.57	950.08	1009.97	1072.29	1137.07	1238.99	1421.87
	p	259.35	270.76	226.52	294.30	306.42	318.77	331.36	344.19	357.25	377.29	411.85
2.00	y	2.97	3.04	2.76	3.18	3.25	3.32	3.39	3.46	3.53	3.64	3.81
	m	770.30	823.30	624.64	936.28	996.35	1058.87	1123.88	1191.43	1261.57	1371.74	1568.96

Notes:
1. s = equivalent soil thickness for uniformly distributed surcharge load (m).
2. h = wall height (m); the distance from bottom of the footing to top of the wall.
3. Assume soil density = 2.0 ton/m^3.
4. Active earth pressure factor k_a = 0.30.

Hence, the maximum possible friction force at bottom of footing

$$F = f\,N_{tot} = 0.30\ (205.71) = 61.71\ \text{kN}$$

4. **Check Sliding**
   Total lateral active force (include surcharge)

$$P_1 + P_2 = 16.34 + 59.01 = 75.35\ \text{kN}$$

Total maximum possible sliding resistant capacity

$$\text{Passive + friction} = 59.60 + 61.71 = 121.31\ \text{kN}$$

Sliding safety factor = 121.31/75.35 = 1.61 > 1.50 OK

5. **Check Overturning**
   Take point *A* as the reference point
   Resistant moment (do not include passive force for conservative)

Surcharge	14.30 (1.3/2 + 0.5 + 0.9)	= 29.32 kN·m
Soil cover at heel	89.54 (1.3/2 + 0.5 + 0.9)	= 183.56 kN·m
Wall	47.83 (0.5/2 + 0.9)	= 55.00 kN·m
Soil cover at toe	14.30 (0.9/2)	= 6.44 kN·m
Footing	39.74 (2.7/2)	= 53.65 kN·m
		Total 327.97 kN·m

Overturning moment

$$P_1(H/2) + P_2(H/3) = 16.34\ (4.5)/2 + 59.01\ (4.5)/3 = 125.28\ \text{kN·m}$$

Sliding safety factor = 327.97/125.28 = 2.62 > 1.50 OK

6. **Check Bearing**
   Total vertical load

$$N_{tot} = 205.71\ \text{kN}$$

Total moment about center line of footing:

- Clockwise (do not include passive force for conservative)

Surcharge	14.30 (2.70/2 – 1.30/2)	= 10.01 kN·m
Soil cover @ heel	89.54 (2.70/2 – 1.30/2)	= 62.68 kN·m
		72.69 kN·m

- Counterclockwise

Wall	47.83 (2.70/2 – 0.9 – 0.5/2)	= 9.57 kN·m
Soil cover at toe	14.30 (2.70/2 – 0.9/2)	= 12.87 kN·m
Active earth pressure		= 125.28 kN·m
		147.72 kN·m

Total moment at bottom of footing

$$M_{tot} = 147.72\text{-}72.69 = 75.03 \text{ kN·m (counterclockwise)}$$

Maximum bearing stress

$$\sigma = N_{tot}/A \pm M_{tot}/S$$

where
A = 2.70 (1.0) = 2.70 m^2
S = 1.0 $(2.7)^2$/6 = 1.22 m^3

Therefore:

$$\sigma_{max} = 205.71/2.70 + 75.03/1.22 = 137.69 \text{ kPa}$$

$$< [\sigma] = 190 \text{ kPa}$$

and

$$\sigma_{min} = 205.71/2.70 - 75.03/1.22 = 14.69 \text{ kPa}$$

$$> 0 \qquad \text{OK}$$

7. **Flexure and Shear Strength**
   Both wall and footing sections need to be designed to have enough flexure and shear capacity.

29.3.4 Tieback Wall

The tieback wall is the proper structure type for cut sections. The tiebacks are prestressed anchor cables that are used to resist the lateral soil pressure. Compared with other types of retaining structures, the tieback wall has the least lateral deflection. Figure 29.23 shows the typical components and the basic lateral soil pressure distribution on a tieback wall.

The vertical spacing of tiebacks should be between 1.5 and 2.0 m to satisfy the required clearance for construction equipment. The slope angle of drilled holes should be 10 to 15° for grouting convenience. To minimize group effects, the spacing between the tiebacks should be greater than three times the tieback hole diameter, or 1.5 m minimum.

The bond strength for tieback design depends on factors such as installation technique, hole diameter, etc. For preliminary estimates, an ultimate bound strength of 90 to 100 kPa may be assumed. Based on construction experience, most tieback hole diameters are between 150 and 300 mm, and the tieback design capacity is in the range of 150 to 250 kN. Therefore, the corresponding lateral spacing of the tieback will be 2.0 to 3.0 m. The final tieback capacity must be proof-tested by stressing the test tieback at the construction site.

A tieback wall is built from the top down in cut sections. The wall details consist of a base layer and face layer. The base layer may be constructed by using vertical soldier piles with timber or concrete lagging between piles acting as a temporary wall. Then, a final cast-in-place reinforced-concrete layer will be constructed as the finishing layer of the wall. Another type of base layer that has been used effectively is cast-in-place "shotcrete" walls.

29.3.5 Reinforced Earth-Retaining Structure

The reinforced earth-retaining structure can be used in fill sections only. There is no practical height limit for this retaining system, but there will be a certain amount of lateral movement. The essential

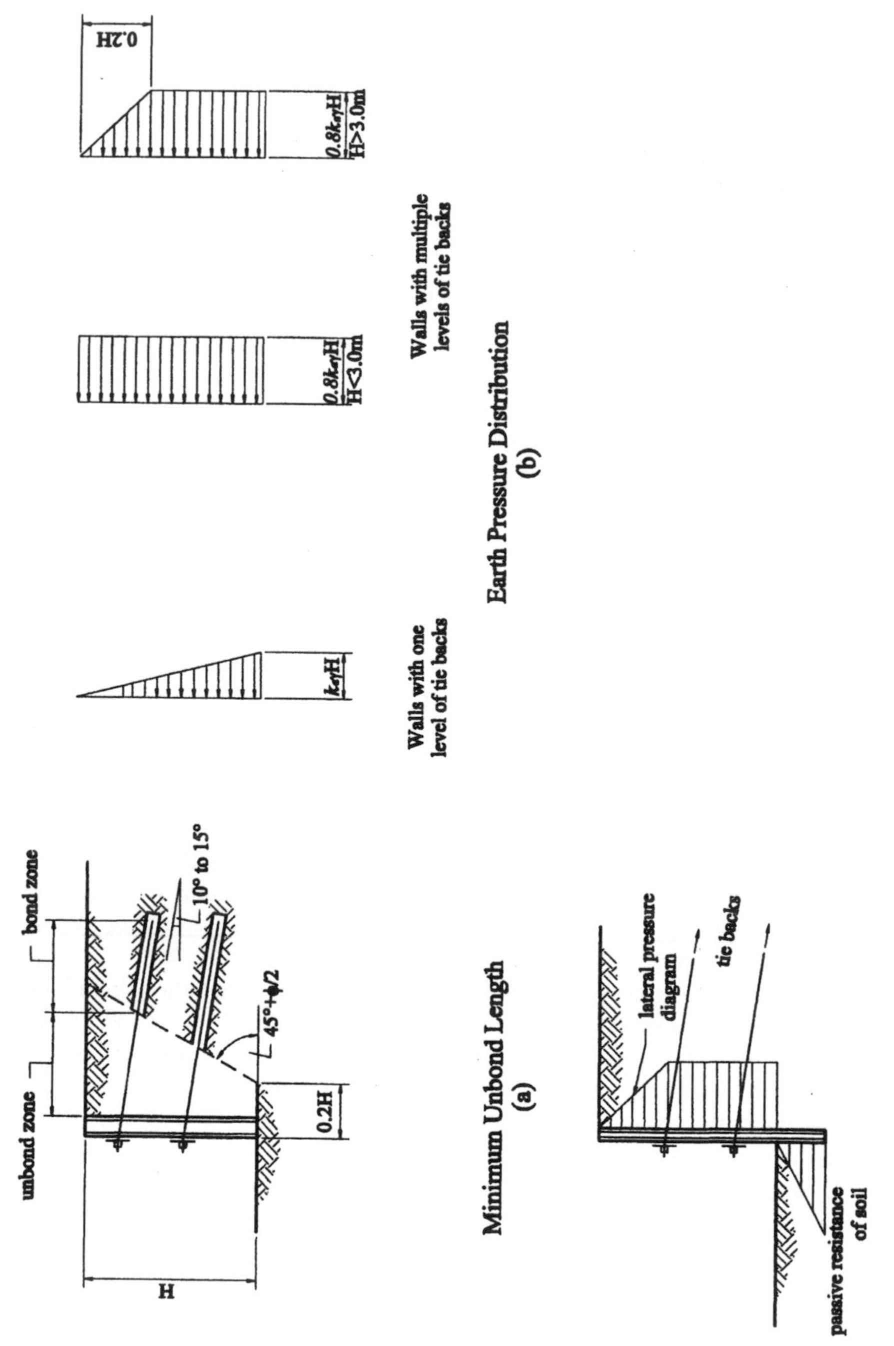

FIGURE 29.23 Tieback wall. (a) Minimum unbond length; (b) earth pressure distribution distribution; (c) typical load diagram.

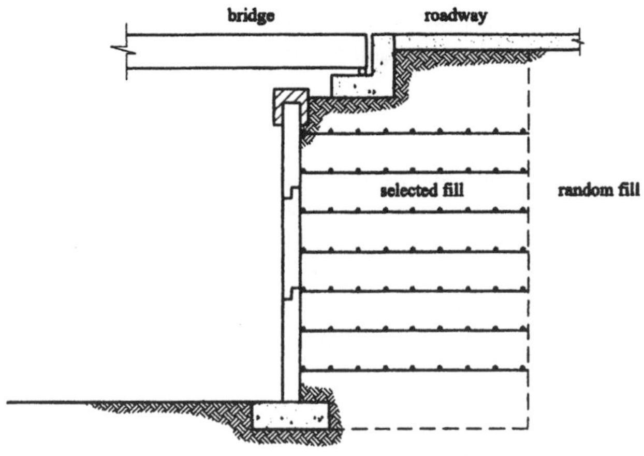

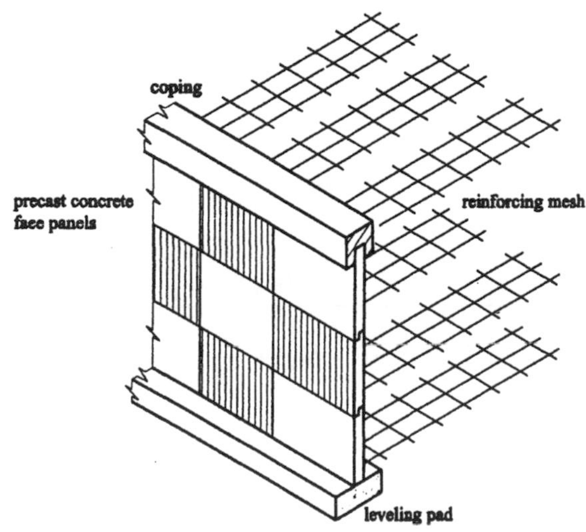

FIGURE 29.24 Mechanical Stabilized Earth (MSE).

concept is the use of multiple-layer strips or fibers to reinforce the fill material in the lateral direction so that the integrated fill material will act as a gravity retaining structure. Figure 29.24 shows the typical details of the MSE retaining structure.

Typically, the width of fill and the length of strips perpendicular to the wall face are on the order of 0.8 of the fill height. The effective life of the material used for the reinforcing must be considered. Metals or nondegradable fabrics are preferred.

Overturning and sliding need to be checked under the assumption that the reinforced soil body acts as a gravity retaining wall. The fiber strength and the friction effects between strip and fill material also need to be checked. Finally, the face panel needs to be designed as a slab which is anchored by the strips and subjected to lateral soil pressure.

29.3.6 Seismic Considerations for Retaining Structures

Seismic effects can be neglected in most retaining structure designs. For oversized retaining structures ($H > 10$ m), the seismic load on a retaining structure can be estimated by using the Mononobe–Okabe solution.

Soil Body ARS Factors

The factors k_v and k_h represent the maximum possible soil body acceleration values under seismic effects in the vertical and horizontal directions, respectively. Similar to other seismic load representations, the acceleration due to gravity will be used as the basic unit of k_v and k_h.

Unless a specific site study report is available, the maximum horizontal ARS value multiplied by 0.50 can be used as the k_h design value. Similarly, k_v will be equal to 0.5 times the maximum vertical ARS value. If the vertical ARS curve is not available, k_v can be assigned a value from $0.1k_h$ to $0.3k_h$.

Earth Pressure with Seismic Effects

Figure 29.25 shows the basic loading diagram for earth pressure with seismic effects. Similar to a static load calculation, the active force per unit length of wall (P_{ac}) can be determined as:

$$P_{ae} = \frac{1}{2} k_{ae} \gamma \left(1 - k_v\right) H^2 \tag{29.12}$$

where

$$\theta' = \tan^{-1}\left[\frac{k_h}{1 - k_v}\right] \tag{29.13}$$

$$k_{ae} = \frac{\sin^2(\phi + \beta - \theta')}{\cos\theta' \sin^2\beta \sin(\beta - \theta' - \delta)\left[1 + \sqrt{\dfrac{\sin(\phi + \delta)\sin(\phi - \theta' - \alpha)}{\sin(\beta - \theta' - \delta)\sin(\alpha + \beta)}}\right]^2} \tag{29.14}$$

Note that with no seismic load, $k_v = k_h = \theta' = 0$. Therefore, $K_{ac} = K_a$.

The resultant total lateral force calculated above does not act at a distance of $H/3$ from the bottom of the wall. The following simplified procedure is often used in design practice:

- Calculate P_{ae} (total active lateral earth pressure per unit length of wall)
- Calculate $P_a = ½\ k_a \gamma H^2$ (static active lateral earth pressure per unit length of wall)

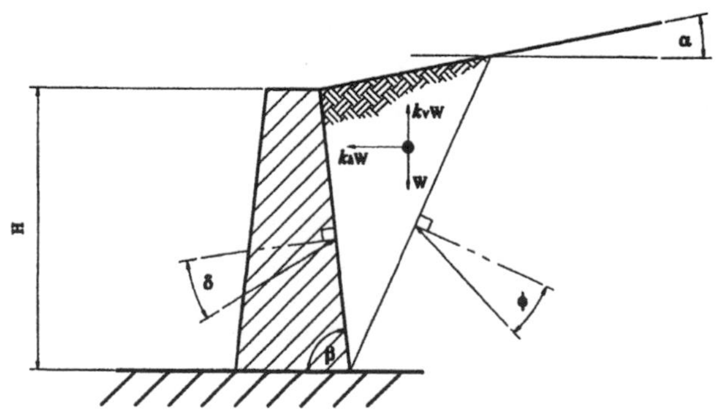

FIGURE 29.25 Load diagram for Earth pressure with seismic effects.

- Calculate $\Delta P = P_{ae} - P_a$
- Assume P_a acts at a distance of $H/3$ from the bottom of the wall
- Assume ΔP acts at a distance of $0.6H$ from the bottom of the wall

The total earth pressure, which includes seismic effects P_{ae}, should always be bigger than the static force P_a. If the calculation results indicate $\Delta P < 0$; use $k_v = 0$.

Using a procedure similar to the active Earth pressure calculation, the passive Earth pressure with seismic effects can be determined as follows:

$$P_{pe} = \frac{1}{2} k_{pe} \gamma (1 - k_v) H^2 \tag{29.15}$$

where

$$\theta' = \tan^{-1}\left[\frac{k_h}{1-k_v}\right]$$

$$k_{pe} = \frac{\sin^2(\beta + \theta' - \phi)}{\cos\theta' \sin^2\beta \sin(\beta + \theta' + \delta - 90)\left[1 - \sqrt{\dfrac{\sin(\phi+\delta)\sin(\phi - \theta' + \alpha)}{\sin(\beta + \theta' + \delta)\sin(\alpha+\beta)}}\right]^2} \tag{29.16}$$

Note that, with no seismic load, $k_{pc} = k_p$.

References

1. AASHTO, *Standard Specifications for Highway Bridges,* 16th ed., American Association of State Highway and Transportation Officials, Washington, D.C., 1996.
2. *Bridge Memo to Designers Manual,* Department of Transportation, State of California, Sacramento.
3. Brian H. Maroney, Matt Griggs, Eric Vanderbilt, Bruce Kutter, Yuk H. Chai and Karl Romstad, Experimental measurements of bridge abutment behavior, in *Proceeding of Second Annual Seismic Research Workshop,* Division of Structures, Department of Transportation, Sacramento, CA, March 1993.
4. Brian H. Maroney and Yuk H. Chai, Bridge abutment stiffness and strength under earthquake loadings, in *Proceedings of the Second International Workshop of Seismic Design and Retroffitting of Reinforced Concrete Bridges,* Queenstown, New Zealand, August 1994.
5. Rakesh K. Goel, Earthquake behavior of bridge with integral abutment, in *Proceedings of the National Seismic Conference on Bridges and Highways,* Sacramento, CA, July 1997.
6. E. C. Sorensen, Nonlinear soil-structure interaction analysis of a 2-span bridge on soft clay foundation, in *Proceedings of the National Seismic Conference on Bridges and Highways,* Sacramento, CA, July 1997.
7. AEAR, *Manual for Railway Engineering,* 1996.
8. Braja M. Das, *Principles of Foundation Engineering,* PWS-KENT Publishing Company, Boston, MA, 1990.
9. T. William Lambe and Robert V. Whitman, *Soil Mechanics,* John Wiley & Sons, New York, 1969.
10. Gregory P. Tschebotarioff, *Foundations, Retaining and Earth Structures,* 4th ed., McGraw-Hill, New York, 1973.
11. Joseph E. Bowles, *Foundation Analysis and Design,* McGraw-Hill, New York, 1988.
12. Whitney Clark Huntington, *Earth Pressure and Retaining Walls,* John Wiley & Sons, New York.

30

Geotechnical Considerations

Thomas W. McNeilan
Fugro West, Inc.

James Chai
California Department of Transportation

30.1 Introduction

A complete geotechnical study of a site will (1) determine the subsurface stratigraphy and stratigraphic relationships (and their variability), (2) define the physical properties of the earth materials, and (3) evaluate the data generated and formulate solutions to the project-specific and site-specific geotechnical issues. Geotechnical issues that can affect a project can be broadly grouped as follows:

- *Foundation Issues* — Including the determination of the strength, stability, and deformations of the subsurface materials under the loads imposed by the structure foundations, in and beneath slopes and cuts, or surrounding the subsurface elements of the structure.
- *Earth Pressure Issues* — Including the loads and pressures imposed by the earth materials on foundations and against supporting structures, or loads and pressures created by seismic (or other) external forces.

0-8493-7434-0/00/$0.00+$.50

- *Construction and Constructibility Considerations*— Including the extent and characteristics of materials to be excavated, and the conditions that affect deep foundation installation or ground improvement.
- *Groundwater Issues*— Including occurrence, hydrostatic pressures, seepage and flow, and erosion.

Site and subsurface characteristics directly affect the choice of foundation type, capacity of the foundation, foundation construction methods, and bridge cost. Subsurface and foundation conditions also frequently directly or indirectly affect the route alignment, bridge type selection, and/or foundation span lengths. Therefore, an appropriately scoped and executed foundation investigation and site characterization should:

1. Provide the required data for the design of safe, reliable, and economic foundations;
2. Provide data for contractors to use to develop appropriate construction cost estimates;
3. Reduce the potential for a "changed condition" claim during construction.

In addition, the site investigation objectives frequently may be to

1. Provide data for route selection and bridge type evaluation during planning and preliminary phase studies;
2. Provide data for as-built evaluation of foundation capacity, ground improvement, or other similar requirements.

For many projects, it is appropriate to conduct the geotechnical investigation in phases. For the first preliminary (or reconnaissance) phase, either a desktop study using only historical information or a desktop study and a limited field exploration program may be adequate. The results of the first-phase study can then be used to develop a preliminary geologic model of the site, which is used to determine the key foundation design issues and plan the design-phase site investigation.

Bridge projects may require site investigations to be conducted on land, over water, and/or on marginal land at the water's edge. Similarly, site investigations for bridge projects can range from conventional, limited-scope investigations for simple overpasses and grade separations to major state-of-the-practice investigations for large bridges over major bodies of water.

This chapter includes discussions of

- Field exploration techniques;
- Definition of the requirements for and extent of the site investigation program;
- Evaluation of the site investigation results and development/scoping of the laboratory testing program;
- Data presentation and site characterization.

The use of the site characterization results for foundation design is included in subsequent chapters.

30.2 Field Exploration Techniques

For the purpose of the following discussion, we have divided field exploration techniques into the following groupings:

- Borings (including drilling, soil sampling, and rock-coring techniques)
- Downhole geophysical logging
- *In situ* testing — including cone penetration testing (CPT) and vane shear, pressure meter and dilatometer testing)
- Test pits and trenches
- Geophysical survey techniques

FIGURE 30.1 Drilling methods. (a) On land; (b) over water; (c); on marginal land.

30.2.1 Borings and Drilling Methods

Drilled soil (or rock) borings are the most commonly used subsurface exploration technique. The drilled hole provides the opportunity to collect samples of the subsurface through the use of a variety of techniques and samplers. In addition to sample collection, drilling observations during the advancement of the borehole provide an important insight to the subsurface conditions. Drilling methods can be used for land, over water, and marginal land sites (Figure 30.1). It should be noted that the complexity introduced when working over water or on marginal land may require more-sophisticated and more-specialized equipment and techniques, and will significantly increase costs.

30.2.1.1 Wet (Mud) Rotary Borings

Wet rotary drilling is the most commonly used drilling method for the exploration of soil and rock, and also is used extensively for oil exploration and water well installation. It is generally the preferred method for (1) over water borings; (2) where groundwater is shallow; and (3) where the subsurface includes soft, squeezing, or flowing soils.

With this technique, the borehole is advanced by rapid rotation of the drill bit that cuts, chips, and grinds the material at the bottom of the borehole. The cuttings are removed from the borehole by circulating water or drilling fluid down through the drill string to flush the cuttings up through the annular space of the drill hole. The fluids then flow into a settling pit or solids separator. Drilling fluid is typically bentonite (a highly refined clay) and water, or one of a number of synthetic products. The drilling fluids are used to flush the cuttings from the hole, compensate the fluid pressure, and stabilize borehole sidewalls. In broken or fractured rock, coarse gravel and cobbles, or other formations with voids, it may be necessary to case the borehole to prevent loss of circulation. Wet rotary drilling is conducive to downhole geophysical testing, although the borehole must be thoroughly flushed before conducting some types of logging.

30.2.1.2 Air Rotary Borings

The air rotary drilling technology is similar to wet rotary except that the cuttings are removed with the circulation of high-pressure air rather than a fluid. Air rotary drilling techniques are typically used in hard bedrock or other conditions where drill hole stability is not an overriding issue. In very hard bedrock, a percussion hammer is often substituted for the bit. Air rotary drilling is conducive to downhole geophysical testing methods.

30.2.1.3 Bucket-Auger Borings

The rotary bucket is similar to a large- (typically 18- to 24-in.)-diameter posthole digger with a hinged bottom. The hole is advanced by rotating the bucket at the end of a kelly bar while pressing it into the soil. The bucket is removed from the hole to be emptied. Rotary-bucket-auger borings are used in alluvial soils and soft bedrock. This method is not always suitable in cobbly or rocky soils, but penetration of hard layers is sometimes possible with special coring buckets. Bucket-auger borings also may be unsuitable below the water table, although drilling fluids can be used to stabilize the borehole.

The rotary-bucket-auger drilling method allows an opportunity for continuous inspection and logging of the stratigraphic column of materials, by lowering the engineer or geologist on a platform attached to a drill rig winch. It is common in slope stability and fault hazards studies to downhole log 24-in.-diameter, rotary-bucket-auger boreholes advanced with this method.

30.2.1.4 Hollow-Stem-Auger Borings

The hollow-stem-auger drilling technique is frequently used for borings less than 20 to 30 m deep. The proliferation of the hollow-stem-auger technology in recent years occurred as the result of its use for contaminated soils and groundwater studies. The hollow-stem-auger consists of sections of steel pipe with welded helical flanges. The shoe end of the pipe has a hollow bit assembly that is plugged while rotating and advancing the auger. That plug is removed for advancement of the sampling device ahead of the bit.

Hollow-stem-auger borings are used in alluvial soils and soft bedrock. This method is not always suitable where groundwater is shallow or in cobbly and rocky soils. When attempting to sample loose, saturated sands, the sands may flow into the hollow auger and produce misleading data. The hollow-stem-auger drill hole is not conducive to downhole geophysical testing methods.

30.2.1.5 Continuous-Flight-Auger Borings

Continuous-flight-auger borings are similar to the hollow-stem-auger drilling method except that the auger must be removed for sampling. With the auger removed, the borehole is unconfined and hole instability often results. Continuous-flight-auger drill holes are used for shallow exploration above the groundwater level.

30.2.2 Soil-Sampling Methods

There are several widely used methods for recovering samples for visual classification and laboratory testing.

30.2.2.1 Driven Sampling

Driven sampling using standard penetration test (SPT) or other size samplers is the most widely used sampling method. Although this sampling method recovers a disturbed sample, the "blow count" measured with this type of procedure provides a useful index of soil density or strength. The most commonly used blow count is the SPT blow count (also referred to as the N-value). Although the N-value is an approximate and imprecise measurement (its value is affected by many operating factors that are part of the sampling process, as well as the presence of gravel or cementation), various empirical relationships have been developed to relate N-value to engineering and performance properties of the soils.

30.2.2.2 Pushed Samples

A thin-wall tube (or in some cases, other types of samplers) can be pushed into the soil using hydraulic pressure from the drill rig, the weight of the drill rod, or a fixed piston. Pushed sampling generally recovers samples that are less disturbed than those recovered using driven-sampling techniques. Thus, laboratory tests to determine strength and volume change characteristics should preferably be conducted on pushed samples rather than driven samples. Pushed sampling is the preferred sampling method in clay soils. Thin-wall samples recovered using push-sampling techniques can either be extruded in the field or sealed in the tubes.

30.2.2.3 Drilled or Cored Samplers

Drilled-in samplers also have application in some types of subsurface conditions, such as hard soil and soft rock. With these types of samplers (e.g., Denison barrel and pitcher barrel), the sample barrel is either cored into the sediment or rock or is advanced inside the drill rod while the rod is advanced.

30.2.3 Rock Coring

The two rock-coring systems most commonly used for engineering applications are the conventional core barrel and wireline (retrievable) system. At shallow depths above the water table, coring also sometimes can be performed with an air or a mist system.

Conventional core barrels consist of an inner and outer barrel with a bit assembly. To obtain a core at a discrete interval; (1) the borehole is advanced to the top of the desired interval, (2) the drill pipe is removed, (3) the core barrel/bit is placed on the bottom of the pipe, and (4) the assembly is run back to the desired depth. The selected interval is cored and the core barrel is removed to retrieve the core. Conventional systems typically are most effective at shallow depths or in cases where only discrete samples are required.

In contrast, wireline coring systems allow for continuous core retrieval without removal of the drill pipe/bit assembly. The wireline system has a retrievable inner core barrel that can be pulled to the surface on a wireline after each core run.

Variables in the coring process include the core bit type, fluid system, and drilling parameters. There are numerous bit types and compositions that are applicable to specific types of rock; however, commercial diamond or diamond-impregnated bits are usually the preferred bit from a core recovery and quality standpoint. Tungsten carbide core bits can sometimes be used in weak rock or in high-clay-content rocks. A thin bentonite mud is the typical drilling fluid used for coring. Thick mud can clog the small bit ports and is typically avoided. Drilling parameters include the revolutions per minute (RPM) and weight on bit (WOB). Typically, low RPM and WOB are used to start the core run and then both values are increased.

Rock engineering parameters include percent recovery, rock quality designation (RQD), coring rate, and rock strength. Percent recovery is a measure of the core recovery vs. the cored length, whereas RQD is a measure of the intact core pieces longer than 4 in. vs. the cored length. Both values typically increase as the rock mass becomes less weathered/fractured with depth; however, both values are highly dependent on the type of rock, amount of fracturing, etc. Rock strength (which is typically measured using unconfined triaxial compression test per ASTM guidelines) is used to evaluate bearing capacity, excavatability, etc.

30.2.4 *In Situ* Testing

There are a variety of techniques that use instrumented probes or testing devices to measure soil properties and conditions in the ground, the more widely used of which are described below. In contrast to sampling that removes a sample from its *in situ* stress conditions, *in situ* testing is used to measure soil and rock properties in the ground at their existing state of stress. The various *in*

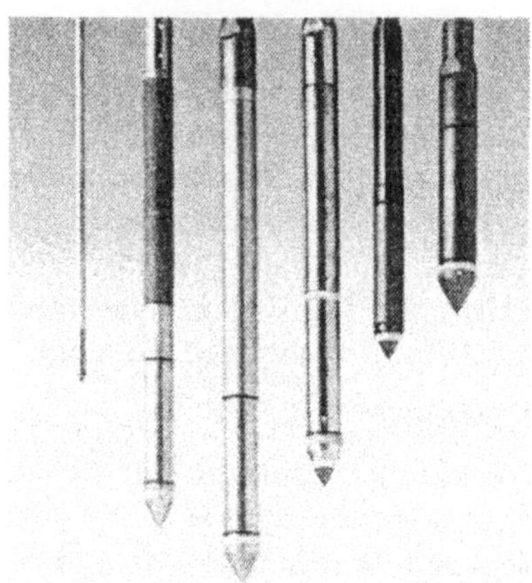

FIGURE 30.2 CPT cones.

situ tests can either be conducted in a borehole or as a continuous sounding from the ground surface. Except as noted, those techniques are not applicable to rock.

30.2.4.1 Cone Penetration Test Soundings

CPT sounding is one of the most versatile and widely used *in situ* test. The standard CPT cone consists of a 1.4-in.-diameter cone with an apex angle of 60°, although other cone sizes are available for special applications (Figure 30.2). The cone tip resistance beneath the 10-cm^2 cone tip and the friction along the 150 cm^2 friction sleeve are measured with strain gauges and recorded electronically at 1- or 2-cm intervals as the cone is advanced into the ground at a rate of about 2 cm/s. In addition to the tip and sleeve resistances, many cones also are instrumented to record pore water pressure or other parameters as the cone is advanced.

Because the CPT soundings provide continuous records of tip and sleeve resistances (and frequently pore pressure) vs. depth (Figure 30.3), they provide a continuous indicator of soil and subsurface conditions that are useful in defining soil stratification. Numerous correlations between the CPT measurements have been developed to define soil type and soil classification. In addition, empirical correlations have been published to relate the cone tip and sleeve friction resistances to engineering behavior, including undrained shear strength of clay soils and relative density and friction of granular soils.

Most land CPTs are performed as continuous soundings using large 20-ton cone trucks (Figure 30.4a), although smaller, more portable track-mounted equipment is also available. CPT soundings are commonly extended down to more than 20 to 50 m. CPT soundings also can be performed over water from a vessel using specialized equipment (Figure 30.4b) deployed by a crane or from a stern A-frame. In addition, downhole systems have been developed to conduct CPTs in boreholes during offshore site investigations. With a downhole system, CPT tests are interspersed with soil sampling to obtain CPT data to more than 100 m in depth.

30.2.4.2 *In Situ* Vane Shear Tests

The undrained shear strength of clay soils can be measured *in situ* using a vane shear test. This test is conducted by measuring the torque required to rotate a vane of known dimensions. The test can be conducted from the ground surface by attaching a vane blade onto a rod or downhole below the bottom of a borehole with a drop-in remote vane (Figure 30.5). The downhole vane is preferable, since the torque required to rotate the active rotating vane is not affected by the torque of the rod. The downhole vane is used both for land borings and over-water borings.

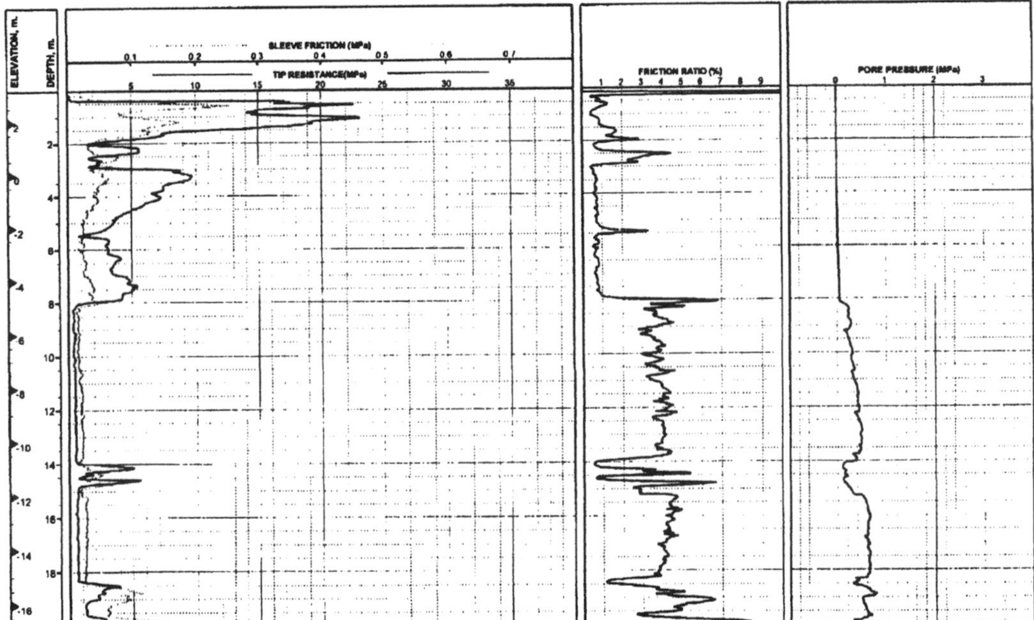

FIGURE 30.3 CPT data provide a continuous record of *in situ* conditions.

30.2.4.3 Pressure Meter and Dilatometer Tests

Pressure meter testing is used to measure the *in situ* maximum and average shear modulus of the soil or rock by inflating the pressure meter against the sidewalls of the borehole. The stresses, however, are measured in a horizontal direction, not in the vertical direction as would occur under most types of foundation loading. A test is performed by lowering the tool to the selected depth and expanding a flexible membrane through the use of hydraulic fluid. As the tool is inflated, the average displacement of the formation is measured with displacement sensors beneath the membrane, which is protected by stainless steel strips. A dilatometer is similar to a pressure meter, except that the dilatometer consists of a flat plate that is pushed into the soil below the bottom of the borehole. A dilatometer is not applicable to hard soils or rock.

30.2.5 Downhole Geophysical Logging

Geophysical logs are run to acquire data about the formation or fluid penetrated by the borehole. Each log provides a continuous record of a measured value at a specific depth in the boring, and is therefore useful for interpolating stratigraphy between sample intervals. Most downhole geophysical logs are presented as curves on grid paper or as electronic files (Figure 30.6). Some of the more prevalent geophysical tools, which are used for geotechnical investigations, are described below.

- *Electrical logs* (*E-logs*) include resistivity, induction, and spontaneous potential (SP) logs. Resistivity and induction logs are used to determine lithology and fluid type. A resistivity log is used when the borehole is filled with a conductive fluid, while an induction log is used when the borehole is filled with a non- or low-conductivity fluid. Resistivity tools typically require an open, uncased, fluid-filled borehole. Clay formations and sands with higher salinity will have low resistivity, while sands with fresh water will have higher resistivity values. Hard rock and dry formations have the highest resistivity values. An SP log is often used in suite with a resistivity or induction log to provide further information relative to formation permeability and lithology.

FIGURE 30.4 CPT sounding methods. (a) On land; (b) over water.

FIGURE 30.5 *In situ* vane shear device.

- *Suspension (velocity) logs* are used to measure the average primary, compression wave, and shear wave velocities of a 1-m-high segment of the soil and rock column surrounding the borehole. Those velocities are determined by measuring the elapsed time between arrivals of a wave propagating upward through the soil/rock column. The suspension probe includes both a shear wave source and a compression wave source, and two biaxial receivers that detect the source waves. This technique requires an open, fluid-filled hole.
- *Natural gamma logs* measure the natural radioactive decay occurring in the formation to infer soil or rock lithology. In general, clay soils will exhibit higher gamma counts than granular soils, although decomposed granitic sands are an exception to that generality. Gamma logs can be run in any salinity fluid as well as air, and also can be run in cased boreholes.
- *Caliper logs* are used to measure the diameter of a borehole to provide insight relative to caving and swelling. An accurate determination of borehole diameter also is important for the interpretation of other downhole logs.
- *Acoustic televiewer and digital borehole logs* are conducted in rock to image the rock surface within the borehole (Figure 30.7). These logs use sound in an uncased borehole to create an oriented image of the borehole surface. These logs are useful for determining rock layering, bedding, and fracture identification and orientation.
- *Crosshole, downhole, and uphole shear wave velocity measurements* are used to determine the primary and shear wave velocities either to determine the elastic soil properties of soil and rock or to calibrate seismic survey measurements. With the crosshole technique, the travel time is measured between a source in one borehole and a receiver in a second borehole. This technique can be used to measure directly the velocities of various strata. For downhole and uphole logs, the travel time is measured between the ground surface and a downhole source or receiver. Tests are conducted with the downhole source or receiver at different depths. These measurements should preferably be conducted in cased boreholes.

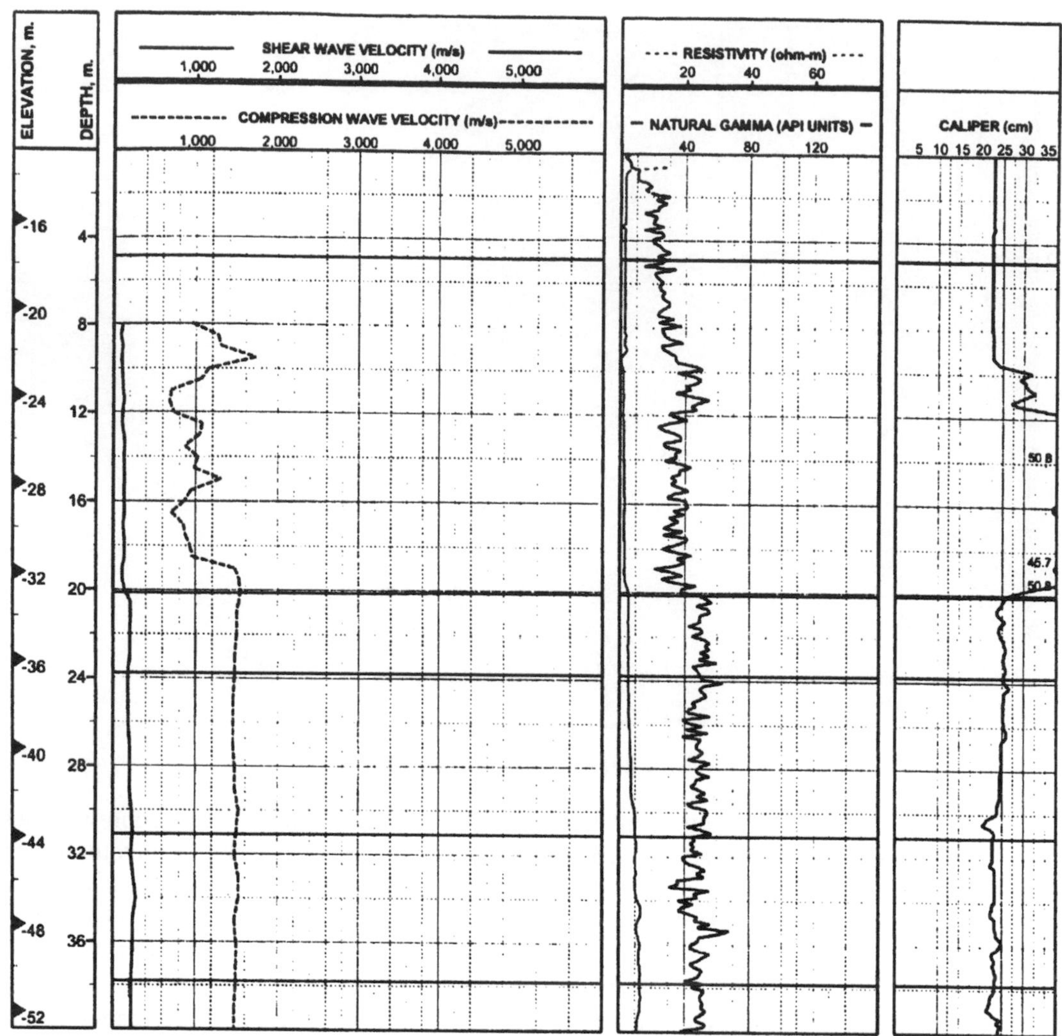

FIGURE 30.6 Example of downhole geophysical log.

30.2.6 Test Pits and Trenches

Where near-surface conditions are variable or problematic, the results of borings and *in situ* testing can be supplemented by backhoe-excavated or hand-excavated test pits or trenches. These techniques are particularly suitable for such purposes as: (1) collecting hand-cut, block samples of sensitive soils; (2) evaluating the variability of heterogeneous soils; (3) evaluating the extent of fill or rubble, (4) determining depth to groundwater, and (5) the investigation of faulting.

30.2.7 Geophysical Survey Techniques

Noninvasive (compared with drilling methods) geophysical survey techniques are available for remote sensing of the subsurface. In contrast to drilling and *in situ* testing methods, the geophysical survey methods explore large areas rapidly and economically. When integrated with boring data, these methods often are useful for extrapolating conditions between borings (Figure 30.8). Techniques are applicable either on land or below water. Some of the land techniques also are applicable for marginal land or in the shallow marine transition zone. Geophysical survey techniques can be used individually or as a group.

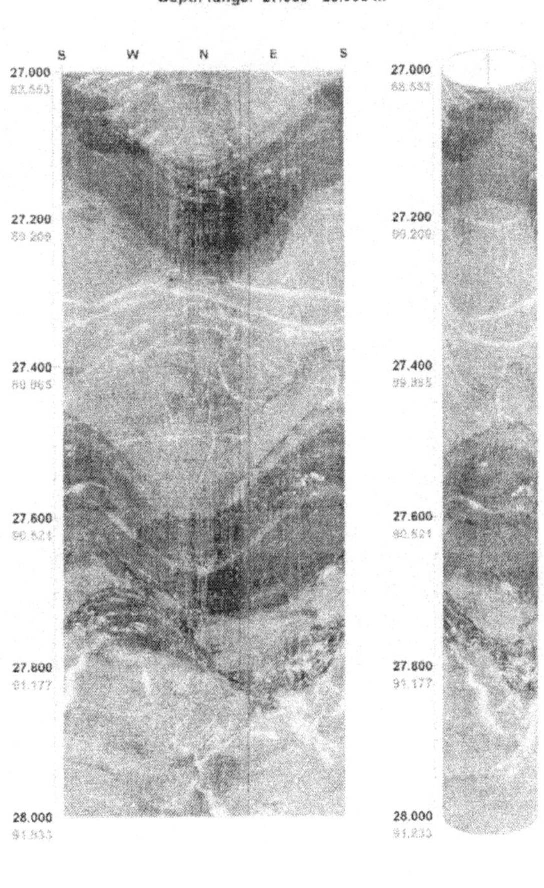

FIGURE 30.7 Example of digital borehole image in rock.

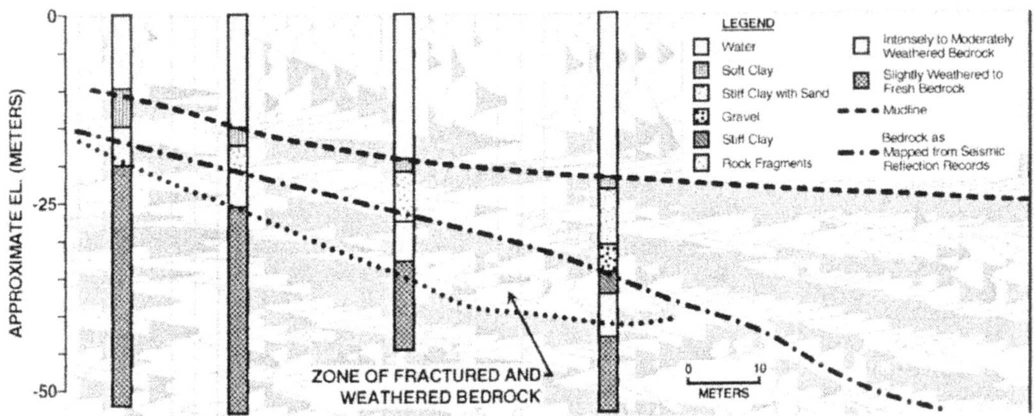

FIGURE 30.8 Example integration of seismic reflection and boring data.

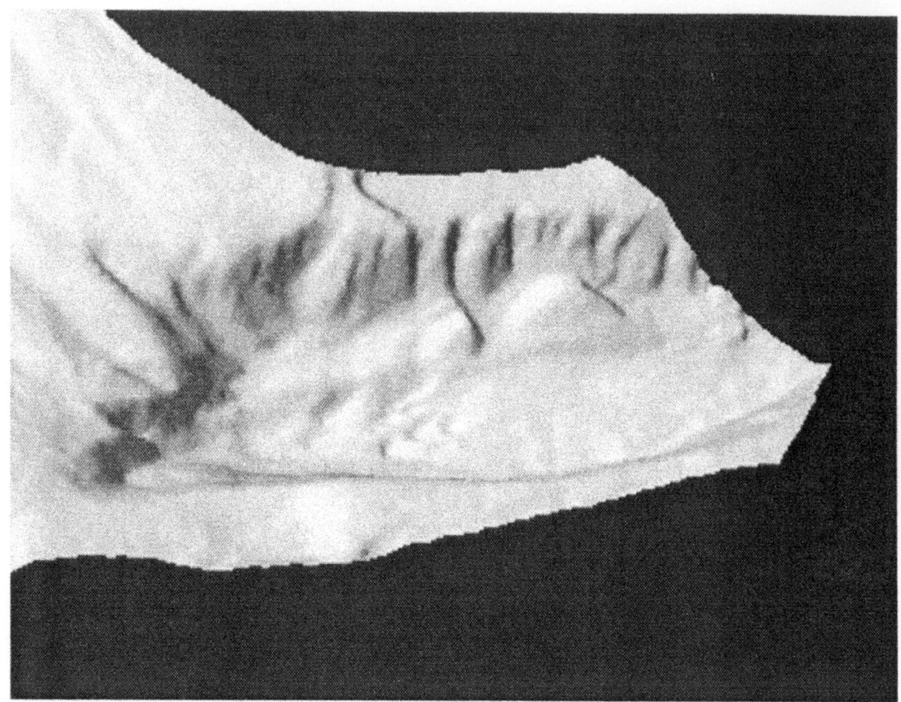

FIGURE 30.9 Multibeam image of river channel bathymetry.

30.2.7.1 Hydrographic Surveys

Hydrographic surveys provide bathymetric contour maps and/or profiles of the seafloor, lake bed, or river bottom. Water depth measurements are usually made using a high-frequency sonic pulse from a depth sounder transducer mounted on a survey vessel. The choice of depth sounder system (single-beam, multifrequency, multibeam, and swath) is dependent upon water depths, survey site conditions, and project accuracy and coverage requirements. The use and application of more-sophisticated multibeam systems (Figure 30.9) has increased dramatically within the last few years.

30.2.7.2 Side-Scan Sonar

Side-scan sonar is used to locate and identify man-made objects (shipwrecks, pipelines, cables, debris, etc.) on the seafloor and determine sediment and rock characteristics of the seafloor. The side-scan sonar provides a sonogram of the seafloor that appears similar to a continuous photographic strip (Figure 30.10). A mosaic of the seafloor can be provided by overlapping the coverage of adjacent survey lines.

30.2.7.3 Magnetometer

A magnetometer measures variations in the earth's magnetic field strength that result from metallic objects (surface or buried), variations in sediment and rock mineral content, and natural (diurnal) variations. Data are used to locate and identify buried objects for cultural, environmental, and archaeological site clearances.

30.2.7.4 High-Resolution Seismic Reflection and Subbottom Profilers

Seismic images of the subsurface beneath the seafloor can be developed by inducing sonic waves into the water column from a transducer, vibrating boomer plate, sparker, or small air or gas gun. Reflections of the sonic energy from the mudline and subsurface soils horizons are recorded to provide an image of the subsurface geologic structure and stratigraphy along the path of the survey

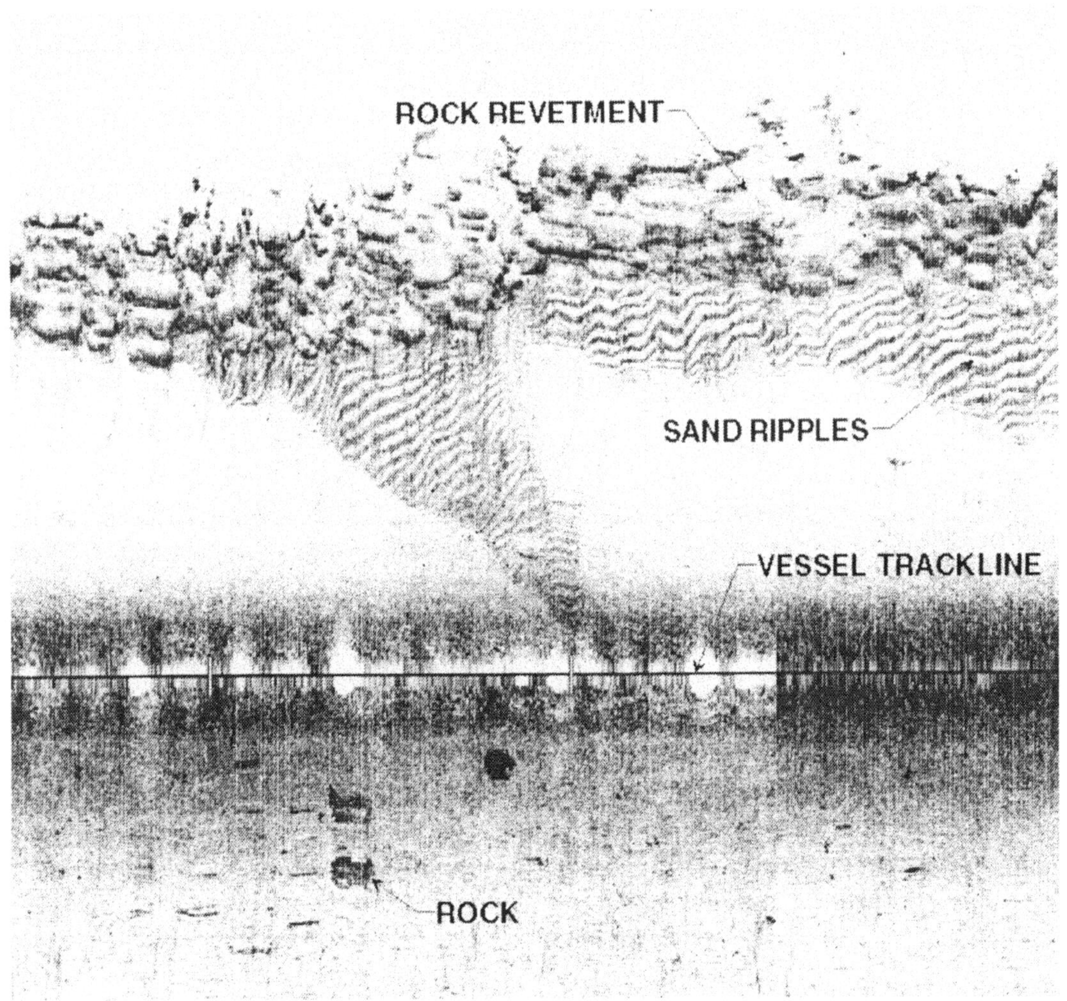

FIGURE 30.10 Side-scan sonar image of river bottom.

vessel. The effective depth of a system and resolution of subsurface horizons depend on a number of variables, including the system energy, output frequency spectrum, the nature of the seafloor, and the subsea sediments and rocks. Seismic reflection data are commonly used to determine the geologic structure (stratigraphy, depth to bedrock, folds, faults, subsea landslides, gas in sediments, seafloor seeps, etc.) and evaluate the horizon continuity between borings (Figure 30.11).

30.2.7.5 Seismic Refraction

Seismic refraction measurements are commonly used on land to estimate depth to bedrock and groundwater and to detect bedrock faulting. Measured velocities are also used for estimates of rippability and excavation characteristics. In the refraction technique, sonic energy is induced into the ground and energy refracted from subsurface soil and rock horizons is identified at a series of receivers laid out on the ground. The time–distance curves from a series of profiles are inverted to determine depths to various subsurface layers and the velocity of the layers. The data interpretation can be compromised where soft layers underlie hard layers and where the horizons are too thin to be detected by refraction arrivals at the surface. The technique also can be used in shallow water (surf zones, lakes, ponds, and river crossings) using bottom (bay) cables.

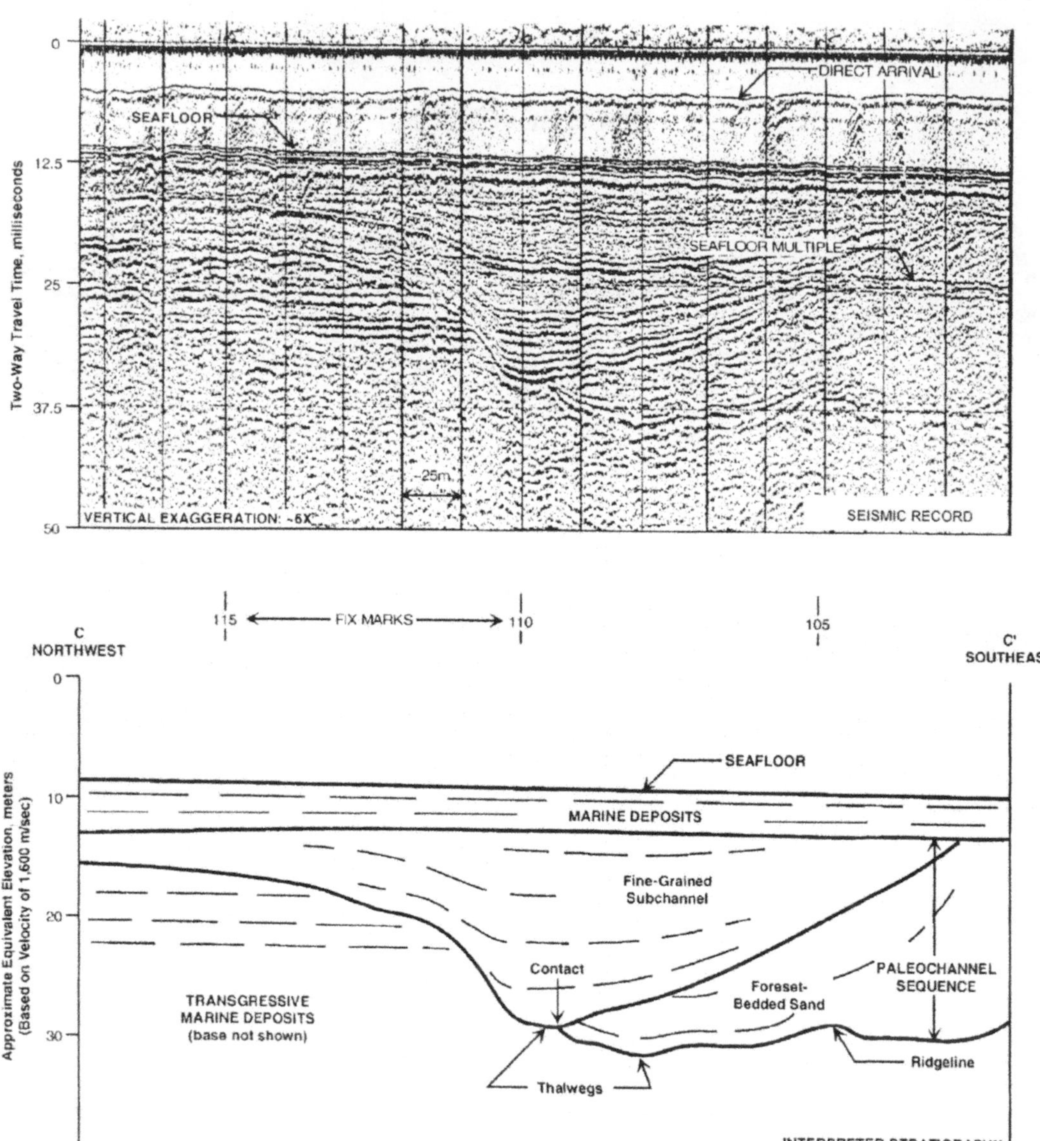

FIGURE 30.11 Interpreted stratigraphic relationships from seismic reflection data.

30.2.7.6 Ground Penetrating Radar Systems

Ground Penetrating Radar (GPR) systems measure the electromagnetic properties of the subsurface to locate buried utilities or rebar, estimate pavement thickness, interpret shallow subsurface stratigraphy, locate voids, and delineate bedrock and landslide surfaces. GPR also can be used in arctic conditions to estimate ice thickness and locate permafrost. Depths of investigation are usually limited to 50 ft or less. Where the surface soils are highly conductive, the effective depth of investigation may be limited to a few feet.

30.2.7.7 Resistivity Surveys

Resistivity surveys induce currents into the ground to locate buried objects and to investigate shallow groundwater. As electrodes are moved in specific patterns of separation, the resistivity is measured

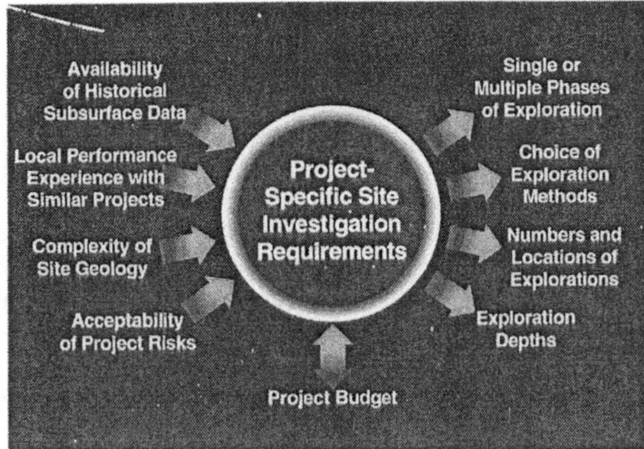

FIGURE 30.12 Key factors to consider when defining site investigation requirements.

and inverted to produce depth sections and contour maps of subsurface resistivity values. This method is used to identify and map subsurface fluids, including groundwater, surface and buried chemical plumes, and to predict corrosion potential.

30.2.8 Groundwater Measurement

Groundwater conditions have a profound effect on foundation design, construction, and performance. Thus, the measurement of groundwater depth (or depth of water when drilling over water) is one of the most fundamentally important elements of the site investigation. In addition to the measurement of the water level, the site investigation should consider and define the potential for artesian or perched groundwater. It is also important to recognize that groundwater levels may change with season, rainfall, or other temporal reasons. All groundwater and water depth measurements should document the time of measurement and, where practical, should determine variations in depth over some period of elapsed time. To determine the long-term changes in water level, it is necessary to install and monitor piezometers or monitoring wells.

30.3 Defining Site Investigation Requirements

Many factors should be considered when defining the requirements (including types, numbers, locations, and depths of explorations) for the site investigation (Figure 30.12). These factors include:

- Importance, uncertainty, or risk associated with bridge design, construction, and performance
- Geologic conditions and their potential variability
- Availability (or unavailability) of historical subsurface data
- Availability (or unavailability) of performance observations from similar nearby projects
- Investigation budget

The following factors should be considered when evaluating the project risk: (1) What are the risks? (2) How likely are the risks to be realized? (3) What are the consequences if the risks occur? Risks include:

- Certainty or uncertainty of subsurface conditions;
- Design risks (e.g., possibility that inadequate subsurface data will compromise design decisions or schedule);
- Construction risks (e.g., potential for changed conditions claims and construction delays);
- Performance risks (e.g., seismic performance).

Two additional requirements that should be considered when planning a subsurface investigation are (1) reliability of the data collected and (2) timeliness of the data generated. Unfortunately, these factors are too often ignored or underappreciated during the site investigation planning process or geotechnical consultant selection process. Because poor-quality or misleading subsurface data can lead to inappropriate selection of foundation locations, foundation types, and/or inadequate or inappropriate foundation capacities, selection of a project geotechnical consultant should be based on qualifications rather than cost. Similarly, the value of the data generated from the subsurface investigation is reduced if adequate data are not available when the design decisions, which are affected by subsurface conditions, are made. All too often, the execution of the subsurface exploration program is delayed, and major decisions relative to the general structure design and foundation locations have been cast in stone prior to the availability of the subsurface exploration results.

Frequently, the execution of the subsurface investigation is an iterative process that should be conducted in phases (i.e., desktop study, reconnaissance site investigation, detailed design-phase investigation). During each phase of site exploration, it is appropriate for data to be reviewed as they are generated so that appropriate modifications can be made as the investigation is ongoing. Appropriate adjustments in the investigation work scope can save significant expense, increase the quality and value of the investigation results, and/or reduce the potential for a remobilization of equipment to fill in missing information.

30.3.1 Choice of Exploration Methods and Consideration of Local Practice

Because many exploration techniques are suitable in some subsurface conditions, but not as suitable or economical in other conditions, the local practice for the methods of exploration vary from region to region. Therefore, the approach to the field exploration program should consider and be tailored to the local practice. Conversely, there are occasions where the requirements for a project may justify using exploration techniques that are not common in the project area. The need to use special techniques will increase with the size of the project and the uniqueness or complexity of the site conditions.

30.3.2 Exploration Depths

The depths to which subsurface exploration should be extended will depend on the structure, its size, and the subsurface conditions at the project location. The subsurface exploration for any project should extend down through unsuitable layers into materials that are competent relative to the design loads to be applied by the bridge foundations. Some of the exploration should be deep enough to verify that unsuitable materials do not exist beneath the bearing strata on which the foundations will be embedded. When the base of the foundation is underlain by layers of compressible material, the exploration should extend down through the compressible strata and into deeper strata whose compressibility will not influence foundation performance.

For lightly loaded structures, it may be adequate to terminate the exploration when rock is encountered, provided that the regional geology indicates that unsuitable strata do not underlie the rock surface. For heavily-loaded foundations or foundations bearing on rock, it is appropriate to verify that the explorations indeed have encountered rock and not a boulder. It is similarly appropriate to extend at least some of the explorations through the weathered rock into sound or fresh rock.

30.3.3 Numbers of Explorations

The basic intent of the site investigation is to determine the subsurface stratigraphy and its variations, and to define the representative soil (or rock) properties of the strata together with their lateral and vertical variations. The locations and spacing of explorations should be adequate to provide a reasonably accurate definition of the subsurface conditions, and should disclose the presence of any important irregularities in the subsurface conditions. Thus, the numbers of explorations will depend on both the project size and the geologic and depositional variability of the site location. When subsurface conditions are complex and variable, a greater number of more closely spaced explorations are warranted. Conversely, when subsurface conditions are relatively uniform, fewer and more widely spaced explorations may be adequate.

30.3.4 The Risk of Inadequate Site Characterization

When developing a site exploration program, it is often tempting to minimize the number of explorations or defer the use of specialized techniques due to their expense. The approach of minimizing the investment in site characterization is fraught with risk. Costs saved by the execution of an inadequate site investigation, whether in terms of the numbers of explorations or the exclusion of applicable site investigation techniques, rarely reduce the project cost. Conversely, the cost saved by an inadequate investigation frequently increases the cost of construction by many times the savings achieved during the site investigation.

30.4 Development of Laboratory Testing Program

30.4.1 Purpose of Testing Program

Laboratory tests are performed on samples for the following purposes:

- Classify soil samples;
- Evaluate basic index soil properties that are useful in evaluating the engineering properties of the soil samples;
- Measure the strength, compressibility, and hydraulic properties of the soils;
- Evaluate the suitability of on-site or borrow soils for use as fill;
- Define dynamic parameters for site response and soil–structure interaction analyses during earthquakes;
- Identify unusual subsurface conditions (e.g., presence of corrosive conditions, carbonate soils, expansive soils, or potentially liquefiable soils).

The extent of laboratory testing is generally defined by the risks associated with the project.

Soil classification, index property, and fill suitability tests generally can be performed on disturbed samples, whereas tests to determine engineering properties of the soils should preferably be performed on relatively undisturbed, intact specimen. The quality of the data obtained from the latter series of tests is significantly dependent on the magnitude of sample disturbance either during sampling or during subsequent processing and transportation.

30.4.2 Types and Uses of Tests

30.4.2.1 Soil Classification and Index Testing

Soil classification and index properties tests are generally performed for even low-risk projects. Engineering parameters often can be estimated from the available *in situ* data and basic index tests using published correlations. Site-specific correlations of these basic values may allow the results of a few relatively expensive advanced tests to be extrapolated. Index tests and their uses include the following:

- Unit weight and water content tests to evaluate the natural unit weight and water content.
- Atterberg (liquid and plastic) limit tests on cohesive soils for classification and correlation studies. Significant insight relative to strength and compressibility properties can be inferred from the natural water content and Atterberg limit test results.
- Sieve and hydrometer tests to define the grain size distribution of coarse- and fine-grained soils, respectively. Grain size data also are used for both classification and correlation studies.

Other index tests include tests for specific gravity, maximum and minimum density, expansion index, and sand equivalent.

30.4.2.2 Shear Strength Tests

Most bridge design projects require characterization of the undrained shear strength of cohesive soils and the drained strength of cohesionless soils. Strength determinations are necessary to evaluate the bearing capacity of foundations and to estimate the loads imposed on earth-retaining structures.

Undrained shear strength of cohesive soils can be estimated (often in the field) with calibrated tools such as a torvane, pocket penetrometer, fall cone, or miniature vane shear device. More definitive strength measurements are obtained in a laboratory by subjecting samples to triaxial compression (TX), direct simple shear (DSS), or torsional shear (TS) tests. Triaxial shear tests (including unconsolidated-undrained, UU, tests and consolidated-undrained, CU, tests) are the most common type of strength test. In this type of test, the sample is subject to stresses that mimic *in situ* states of stress prior to being tested to failure in compression or shear. Large and more high risk projects often warrant the performance of CU or DSS tests where samples are tested along stress paths which model the *in situ* conditions. In contrast, only less-sophisticated UU tests may be warranted for less important projects.

Drained strength parameters of cohesionless soils are generally measured in either relatively simple direct shear (DS) tests or in more-sophisticated consolidated-drained (CD) triaxial tests. In general, few laboratory strength tests are performed on *in situ* specimens of cohesionless soil because of the relative difficulty in obtaining undisturbed specimens.

30.4.2.3 Compaction Tests

Compaction tests are performed to evaluate the moisture–density relationship of potential fill material. Once the relationship has been evaluated and the minimum level of compaction of fill material to be used has been determined, strength tests may be performed on compacted specimens to evaluate design parameters for the project.

30.4.2.4 Subgrade Modulus

R-value and CBR tests are performed to determine subgrade modulus and evaluate the pavement support characteristics of the *in situ* or fill soils.

30.4.2.5 Consolidation Tests

Consolidation tests are commonly performed to (1) evaluate the compressibility of soil samples for the calculation of foundation settlement; (2) investigate the stress history of the soils at the boring locations to calculate settlement as well as to select stress paths to perform most advanced strength tests; (3) evaluate elastic properties from measured bulk modulus values; and (4) evaluate the time rate of settlement. Consolidation test procedures also can be modified to evaluate if foundation soils are susceptible to collapse or expansion, and to measure expansion pressures under various levels of confinement. Consolidation tests include incremental consolidation tests (which are performed at a number of discrete loads) and constant rate of strain (CRS) tests where load levels are constantly increased or decreased. CRS tests can generally be performed relatively quickly and provide a continuous stress–strain curve, but require more-sophisticated equipment.

30.4.2.6 Permeability Tests

In general, constant-head permeability tests are performed on relatively permeable cohesionless soils, while falling-head permeability tests are performed on relatively impermeable cohesive soils. Estimates of the permeability of cohesive soils also can be obtained from consolidation test data.

30.4.2.7 Dynamic Tests

A number of tests are possible to evaluate the behavior of soils under dynamic loads such as wave or earthquake loads. Dynamic tests generally are strength tests with the sample subjected to some sort of cyclic loading. Tests can be performed to evaluate variations of strength, modulus, and damping, with variations in rate and magnitude of cyclic stresses or strains. Small strain parameters for earthquake loading cases can be evaluated from resonant column tests.

For earthquake loading conditions, dynamic test data are often used to evaluate site response and soil–structure interaction. Cyclic testing also can provide insight into the behavior of potentially liquefiable soils, especially those which are not easily evaluated by empirical *in situ* test-based procedures.

30.4.2.8 Corrosion Tests

Corrosion tests are performed to evaluate potential impacts on steel or concrete structures due to chemical attack. Tests to evaluate corrosion potential include resistivity, pH, sulfate content, and chloride content.

30.5 Data Presentation and Site Characterization

30.5.1 Site Characterization Report

The site characterization report should contain a presentation of the site data and an interpretation and analysis of the foundation conditions at the project site. The site characterization report should:

- Present the factual data generated during the site investigation;
- Describe the procedures and equipment used to obtain the factual data;
- Describe the subsurface stratigraphic relationships at the project site;
- Define the soil and rock properties that are relevant to the planning, design, construction, and performance of the project structures;
- Formulate the solutions to the design and construction of the project.

The site data presented in the site characterization report may be developed from the current and/or past field investigations at or near the project site, as-built documents, maintenance records, and construction notes. When historic data are included or summarized, the original sources of the data should be cited.

30.5.2 Factual Data Presentation

The project report should include the accurate and appropriate documentation of the factual data collected and generated during the site investigation and testing program(s). The presentation and organization of the factual data, by necessity, will depend upon the size and complexity of the project and the types and extent of the subsurface data. Regardless of the project size or extent of exploration, all reports should include an accurate plan of exploration that includes appropriate graphical portrayal of surface features and ground surface elevation in the project area.

The boring log (Figure 30.13) is one of the most fundamental components of the data documentation. Although many styles of presentation are used, there are several basic elements that generally should be included on a boring log. Those typical components include:

- Documentation of location and ground surface elevation;
- Documentation of sampling and coring depths, types, and lengths — e.g., sample type, blow count (for driven samples), and sample length for soil samples; core run, recovery, and RQD for rock cores — as well as *in situ* test depths and lengths;
- Depths and elevations of groundwater and/or seepage encountered;
- Graphical representation of soil and rock lithology;
- Description of soil and rock types, characteristics, consistency/density, or hardness;
- Tabular or graphical representation of test data.

In addition to the boring logs, the factual data should include tabulated summaries of test types, depths, and results together with the appropriate graphical output of the tests conducted.

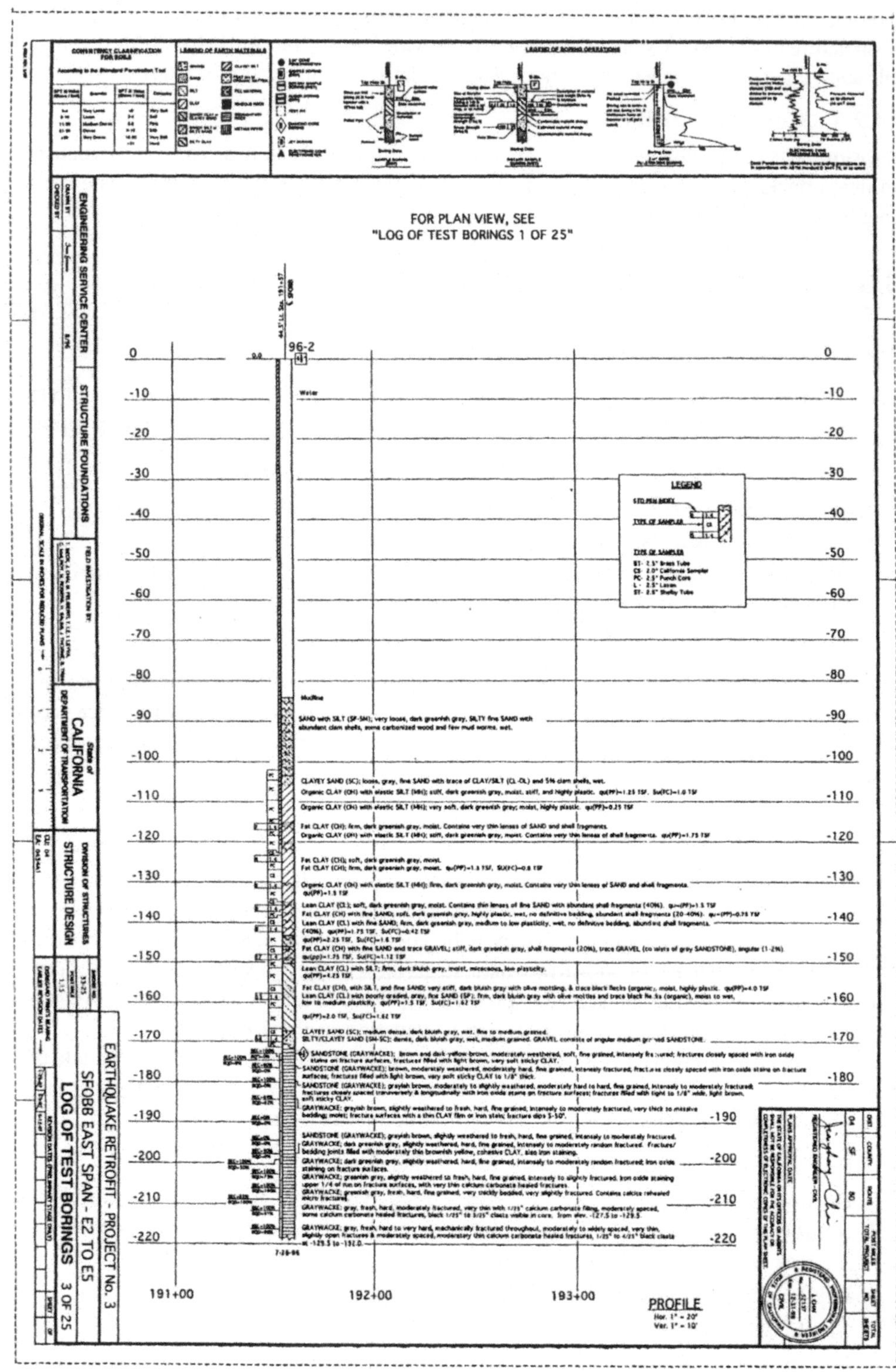

FIGURE 30.13 Typical log of test boring sheet for Caltrans project.

30.5.3 Description of Subsurface Conditions and Stratigraphy

A sound geologic interpretation of the exploration and testing data are required for any project to assess the subsurface conditions. The description of the subsurface conditions should provide users of the report with an understanding of the conditions, their possible variability, and the significance of the conditions relative to the project. The information should be presented in a useful format and terminology appropriate for the users, who usually will include design engineers and contractors who are not earth science professionals.

To achieve those objectives, the site characterization report should include descriptions of

1. Site topography and/or bathymetry,
2. Site geology,
3. Subsurface stratigraphy and stratigraphic relationships,
4. Continuity or lack of continuity of the various subsurface strata,
5. Groundwater depths and conditions, and
6. Assessment of the documented and possibie undocumented variability of the subsurface conditions.

Information relative to the subsurface conditions is usually provided in text, cross sections, and maps. Subsurface cross sections, or profiles, are commonly used to illustrate the stratigraphic sequence, subsurface strata and their relationships, geologic structure, and other subsurface features across a site. The cross section can range from simple line drawings to complex illustrations that include boring logs and plotted test data (Figure 30.14).

Maps are commonly used to illustrate and define the subsurface conditions at a site. The maps can include topographic and bathymetric contour maps, maps of the structural contours of a stratigraphic surface, groundwater depth or elevation maps, isopach thickness maps of an individual stratum (or sequence of strata), and interpreted maps of geologic features (e.g., faulting, bedrock outcrops, etc.). The locations of explorations should generally be included on the interpretive maps.

The interpretive report also should describe data relative to the depths and elevations of groundwater and/or seepage encountered in the field. The potential types of groundwater surface(s) and possible seasonal fluctuation of groundwater should be described. The description of the subsurface conditions also should discuss how the groundwater conditions can affect construction.

30.5.4 Definition of Soil Properties

Soil properties generally should be interpreted in terms of stratigraphic units or geologic deposits. The interpretation of representative soil properties for design should consider lateral and vertical variability of the different soil deposits. Representative soil properties should consider the potential for possible *in situ* variations that have not been disclosed by the exploration program and laboratory testing. For large or variable sites, it should be recognized that global averages of a particular soil property may not appropriately represent the representative value at all locations. For that condition, use of average soil properties may lead to unconservative design.

Soil properties and design recommendations are usually presented with a combination of narrative text, graphs, and data presented in tabular and/or bulleted list format. It is often convenient and helpful to reference generalized subsurface profiles and boring logs in those discussions. The narrative descriptions should include such factors as depth range, general consistency or density, plasticity or grain size, occurrence of groundwater, occurrence of layers or seams, degree of weathering, and structure. For each stratigraphic unit, ranges of typical measured field and laboratory data (e.g., strength, index parameters, and blow counts) should be described.

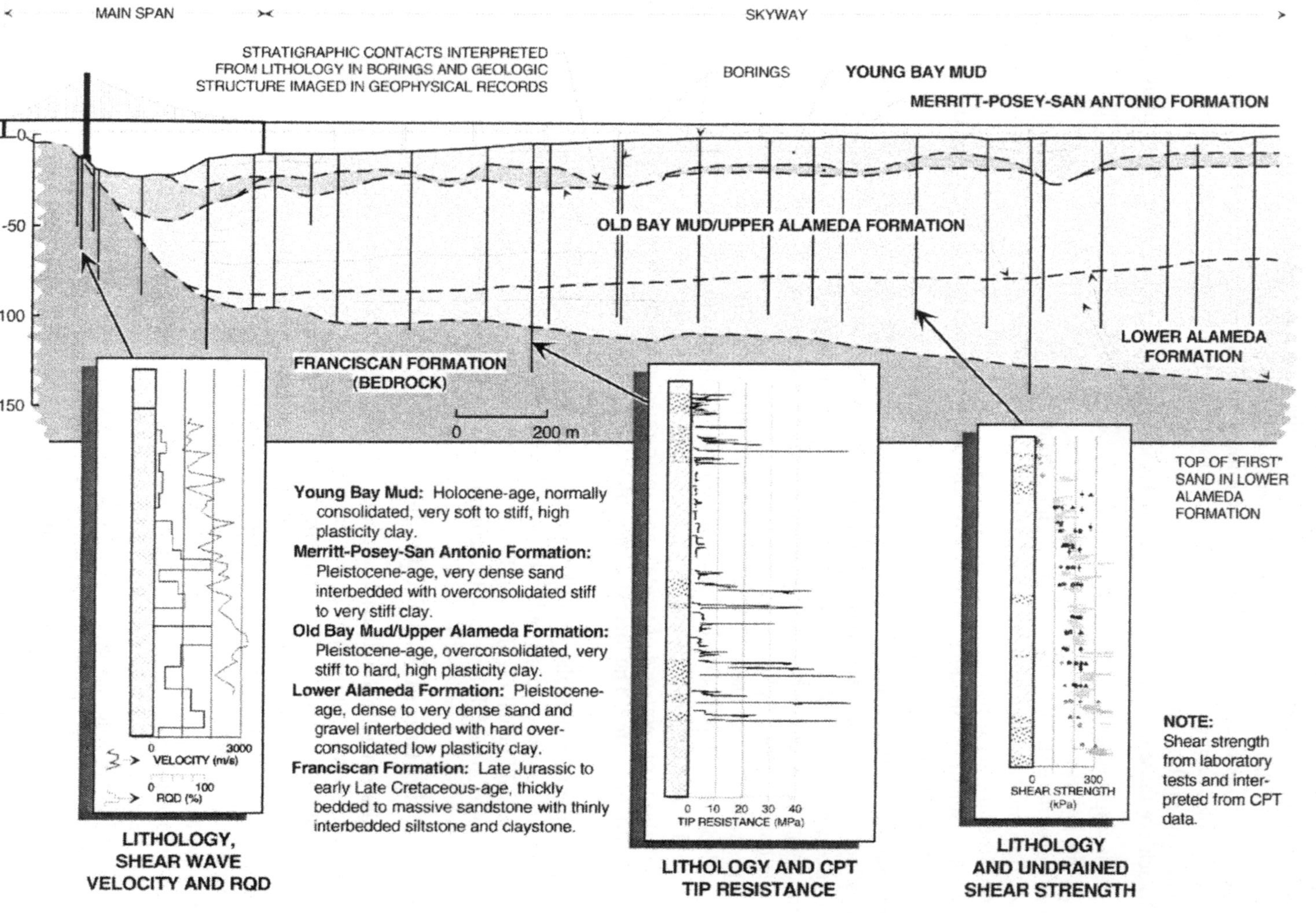

FIGURE 30.14 Subsurface cross section for San Francisco–Oakland Bay Bridge East Span alignment.

30.5.5 Geotechnical Recommendations

The site characterization report should provide solutions to the geotechnical issues and contain geotechnical recommendations that are complete, concise, and definitive. The recommended foundation and geotechnical systems should be cost-effective, performance-proven, and constructible. Where appropriate, alternative foundation types should be discussed and evaluated. When construction problems are anticipated, solutions to these problems should be described.

In addition to the standard consideration of axial and lateral foundation capacity, load–deflection characteristics, settlement, slope stability, and earth pressures, there are a number of subsurface conditions that can affect foundation design and performance:

- Liquefaction susceptibility of loose, granular soils;
- Expansive or collapsible soils;
- Mica-rich and carbonate soils;
- Corrosive soils;
- Permafrost or frozen soils;
- Perched or artesian groundwater.

When any of those conditions are present, they should be described and evaluated.

30.5.6 Application of Computerized Databases

Computerized databases provide the opportunity to compile, organize, integrate, and analyze geotechnical data efficiently. All collected data are thereby stored, in a standard format, in a central accessible location. Use of a computerized database has a number of advantages. Use of automated interactive routines allows the efficient production of boring logs, cross sections, maps, and parameter plots. Large volumes of data from multiple sources can be integrated and queried to evaluate or show trends and variability. New data from subsequent phases of study can be easily and rapidly incorporated into the existing database to update and revise the geologic model of the site.

31

Shallow Foundations

James Chai

California Department of Transportation

31.1 Introduction

A shallow foundation may be defined as one in which the foundation depth (D) is less than or on the order of its least width (B), as illustrated in Figure 31.1. Commonly used types of shallow foundations include spread footings, strap footings, combined footings, and mat or raft footings. Shallow foundations or footings provide their support entirely from their bases, whereas deep foundations derive the capacity from two parts, skin friction and base support, or one of these two. This chapter is primarily designated to the discussion of the bearing capacity and settlement of shallow foundations, although structural considerations for footing design are briefly addressed. Deep foundations for bridges are discussed in Chapter 32.

0-8493-7434-0/00/$0.00+$.50

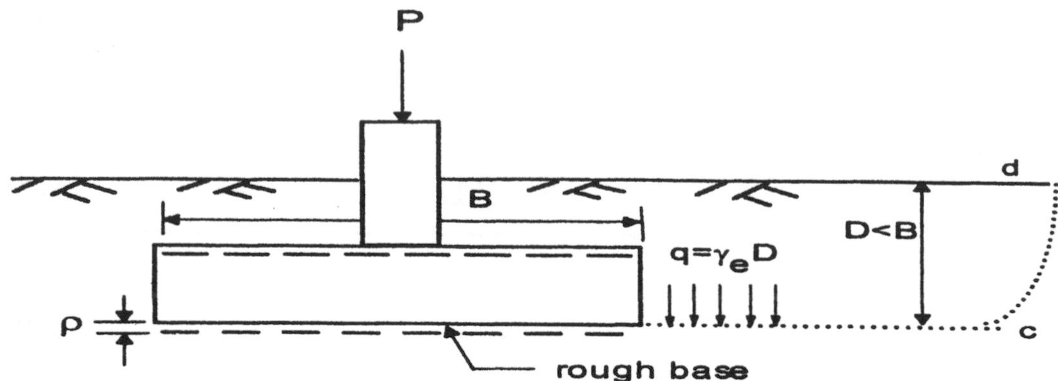

FIGURE 31.1 Definition sketch for shallow footings.

TABLE 31.1 Typical Values of Safety Factors Used in Foundation Design (after Barker et al. [9])

Failure Type	Failure Mode	Safety Factor	Remark
Shearing	Bearing capacity failure	2.0–3.0	The lower values are used when uncertainty in design is small and consequences of failure are minor; higher values are used when uncertainty in design is large and consequences of failure are major
	Overturning	2.0–2.5	
	Overall stability	1.5–2.0	
	Sliding	1.5–2.0	
Seepage	Uplift	1.5–2.0	
	Heave	1.5–2.0	
	Piping	2.0–3.0	

Source: Terzaghi, K. and Peck, R.B., *Soil Mechanics in Engineering Practice*, 2nd ed., John Wiley & Sons, New York, 1967. With permission.

31.2 Design Requirements

In general, any foundation design must meet three essential requirements: (1) providing adequate safety against structural failure of the foundation; (2) offering adequate bearing capacity of soil beneath the foundation with a specified safety against ultimate failure; and (3) achieving acceptable total or differential settlements under working loads. In addition, the overall stability of slopes in the vicinity of a footing must be regarded as part of the foundation design. For any project, it is usually necessary to investigate both the bearing capacity and the settlement of a footing. Whether footing design is controlled by the bearing capacity or the settlement limit rests on a number of factors such as soil condition, type of bridge, footing dimensions, and loads. Figure 31.2 illustrates the load–settlement relationship for a square footing subjected to a vertical load P. As indicated in the curve, the settlement p increases as load P increases. The ultimate load P_u is defined as a peak load (curves 1 and 2) or a load at which a constant rate of settlement (curve 3) is reached as shown in Figure 31.2. On the other hand, the ultimate load is the maximum load a foundation can support without shear failure and within an acceptable settlement. In practice, all foundations should be designed and built to ensure a certain safety against bearing capacity failure or excessive settlement. A safety factor (SF) can be defined as a ratio of the ultimate load P_u and allowable load P_a. Typical value of safety factors commonly used in shallow foundation design are given in Table 31.1.

31.3 Failure Modes of Shallow Foundations

Bearing capacity failure usually occurs in one of the three modes described as general shear, local shear, or punching shear failure. In general, which failure mode occurs for a shallow foundation depends on the relative compressibility of the soil, footing embedment, loading conditions, and drainage conditions. General shear failure has a well-defined rupture pattern consisting of three zones, I, II, and III, as shown in Figure 31.3a. Local shear failure generally consists of clearly defined rupture surfaces beneath the footing (zones I and II). However, the failure pattern on the sides of the footing (zone III) is not clearly defined. Punch shear failure has a poorly defined rupture pattern concentrated within zone I; it is usually associated with a large settlement and does not mobilize shear stresses in zones II and III as shown in Figure 31.3b and c. Ismael and Vesic [40] concluded that, with increasing overburden pressure (in cases of deep foundations), the failure mode changes from general shear to local or punch shear, regardless of soil compressibility. The further examination of load tests on footings by Vesic [68,69] and De Beer [29] suggested that the ultimate load occurs at the breakpoint of the load–settlement curve, as shown in Figure 31.2. Analyzing the modes of failure indicates that (1) it is possible to formulate a general bearing capacity equation for a loaded footing failing in the general shear mode, (2) it is very difficult to generalize the other two failure modes for shallow foundations because of their poorly defined rupture surfaces, and (3) it is of significance to know the magnitude of settlements of footings required to mobilize ultimate loads. In the following sections, theoretical and empirical methods for evaluating both bearing capacity and settlement for shallow foundations will be discussed.

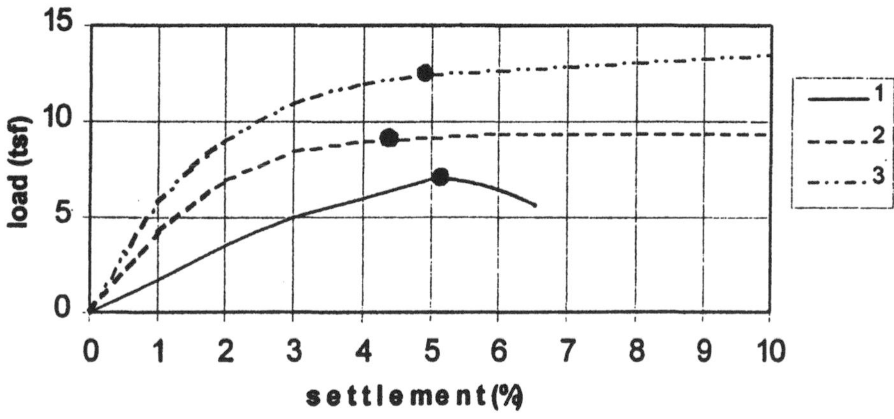

FIGURE 31.2 Load-settlement relationships of shallow footings.

31.4 Bearing Capacity for Shallow Foundations

31.4.1 Bearing Capacity Equation

The computation of ultimate bearing capacity for shallow foundations on soil can be considered as a solution to the problem of elastic–plastic equilibrium. However, what hinders us from finding closed analytical solutions rests on the difficulty in the selection of a mathematical model of soil constitutive relationships. Bearing capacity theory is still limited to solutions established for the rigid-plastic solid of the classic theory of plasticity [40,69]. Consequently, only approximate methods are currently available for the posed problem. One of them is the well-known Terzaghi's bearing capacity equation [19,63], which can be expressed as

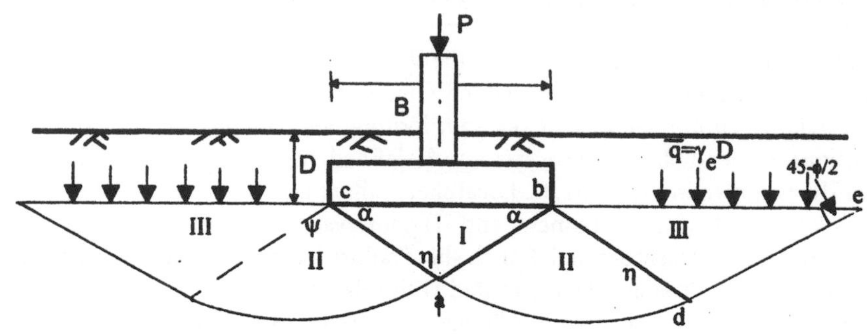

(a) General Shear Failure

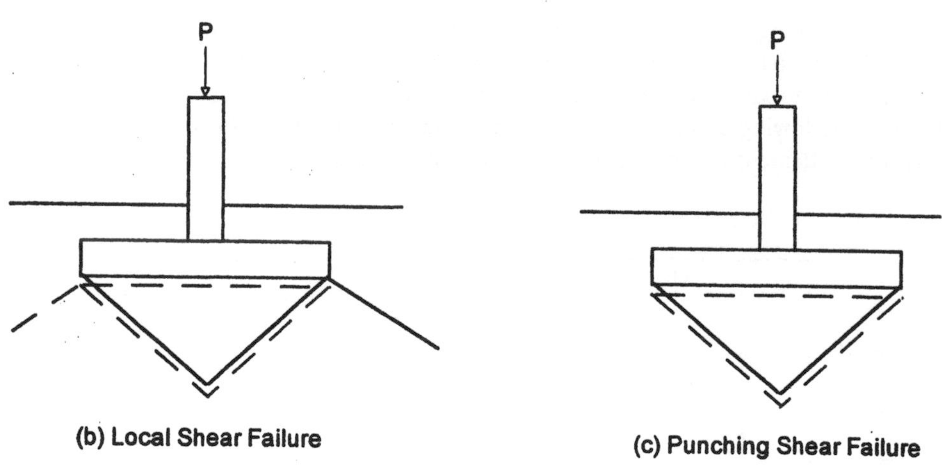

(b) Local Shear Failure

(c) Punching Shear Failure

FIGURE 31.3 Three failure modes of bearing capacity.

$$q_{\text{ult}} = cN_c s_c + \bar{q}N_q + 0.5\gamma BN_\gamma s_\gamma \tag{31.1}$$

where q_{ult} is ultimate bearing capacity, c is soil cohesion, $\bar{q}$ is effective overburden pressure at base of footing ($= \gamma_1 D$), γ is effective unit weight of soil or rock, and B is minimum plan dimension of footing. N_c, N_q, and N_γ are bearing capacity factors defined as functions of friction angle of soil and their values are listed in Table 31.2. s_c and s_r are shape factors as shown in Table 31.3.

These three N factors are used to represent the influence of the cohesion (N_c), unit weight (N_γ), and overburden pressure (N_q) of the soil on bearing capacity. As shown in Figures 31.1 and 31.3(a), the assumptions used for Eq. (31.1) include

1. The footing base is rough and the soil beneath the base is incompressible, which implies that the wedge *abc* (zone I) is no longer an active Rankine zone but is in an elastic state. Consequently, zone I must move together with the footing base.
2. Zone II is an immediate zone lying on a log spiral arc *ad*.

TABLE 31.2 Bearing Capacity Factors for the Terzaghi Equation

ϕ (°)	N_c	N_q	N_γ	$K_{p\gamma}$
0	5.7[a]	1.0	0	10.8
5	7.3	1.6	0.5	12.2
10	9.6	2.7	1.2	14.7
15	12.9	4.4	2.5	18.6
20	17.7	7.4	5.0	25.0
25	25.1	12.7	9.7	35.0
30	37.2	22.5	19.7	52.0
34	52.6	36.5	36.0	—
35	57.8	41.4	42.4	82.0
40	95.7	81.3	100.4	141.0
45	172.3	173.3	297.5	298.0
48	258.3	287.9	780.1	—
50	347.5	415.1	1153.2	800.0

[a] $N_c = 1.5\pi + 1$ (Terzaghi [63], p. 127); values of N_γ for ϕ of 0, 34, and 48° are original Terzaghi values and used to backcompute $K_{p\gamma}$

After Bowles, J.E., *Foundation Analysis and Design*, 5th ed., McGraw-Hill, New York, 1996. With permission.

TABLE 31.3 Shape Factors for the Terzaghi Equation

	Strip	Round	Square
s_c	1.0	1.3	1.3
s_γ	1.0	0.6	0.8

After Terzaghi [63].

3. Zone III is a passive Rankine zone in a plastic state bounded by a straight line *ed*.
4. The shear resistance along *bd* is neglected because the equation was intended for footings where $D < B$.

It is evident that Eq. (31.1) is only valid for the case of general shear failure because no soil compression is allowed before the failure occurs.

Meyerhof [45,48], Hansen [35], and Vesic [68,69] further extended Terzaghi's bearing capacity equation to account for footing shape (s_i), footing embedment depth (d_1), load inclination or eccentricity (i_i), sloping ground (g_i), and tilted base (b_i). Chen [26] reevaluated *N* factors in Terzaghi's equation using the limit analysis method. These efforts resulted in significant extensions of Terzaghi's bearing capacity equation. The general form of the bearing capacity equation [35,68,69] can be expressed as

$$q_{\text{ult}} = cN_c s_c d_c i_c g_c b_c + \bar{q} N_q s_q d_q b_q + 0.5\gamma B N_\gamma s_\gamma d_\gamma i_\gamma g_\gamma b_\gamma \tag{31.2}$$

when $\phi = 0$,

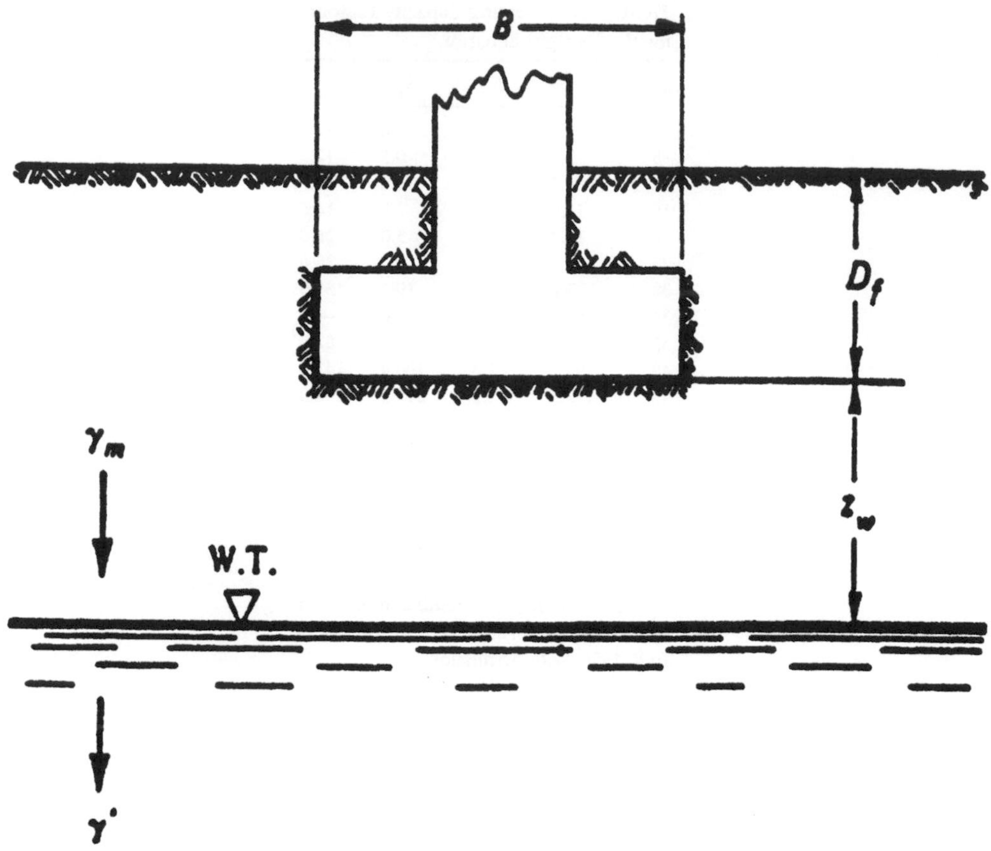

FIGURE 31.4 Influence of groundwater table on bearing capacity. (After AASHTO, 1997.)

$$q_{\text{ult}} = 5.14 s_u \left(1 + s_c' + d_c' - i_c' - b_c' - g_c'\right) + \bar{q} \tag{31.3}$$

where s_u is undrained shear strength of cohesionless. Values of bearing capacity factors N_c, N_q, and N_γ can be found in Table 31.4. Values of other factors are shown in Table 31.5. As shown in Table 31.4, N_c and N_q are the same as proposed by Meyerhof [48], Hansen [35], Vesic [68], or Chen [26]. Nevertheless, there is a wide range of values for N_γ as suggested by different authors. Meyerhof [48] and Hansen [35] use the plain-strain value of ϕ, which may be up to 10% higher than those from the conventional triaxial test. Vesic [69] argued that a shear failure in soil under the footing is a process of progressive rupture at variable stress levels and an average mean normal stress should be used for bearing capacity computations. Another reason causing the N_γ value to be unsettled is how to evaluate the impact of the soil compressibility on bearing capacity computations. The value of N_γ still remains controversial because rigorous theoretical solutions are not available. In addition, comparisons of predicted solutions against model footing test results are inconclusive.

Soil Density

Bearing capacity equations are established based on the failure mode of general shearing. In order to use the bearing capacity equation to consider the other two modes of failure, Terzaghi [63] proposed a method to reduce strength characteristics c and ϕ as follows:

$$c^* = 0.67c \quad \text{(for soft to firm clay)} \tag{31.4}$$

TABLE 31.4 Bearing Capacity Factors for Eqs. (31.2) and (31.3)

ϕ	N_c	N_q	$N_{\gamma(M)}$	$N_{\gamma(H)}$	$N_{\gamma(V)}$	$N_{\gamma(C)}$	N_q/N_c	$\tan\phi$
0	5.14	1.00	0.00	0.00	0.00	0.00	0.19	0.00
1	5.38	1.09	0.00	0.00	0.07	0.07	0.20	0.02
2	5.63	1.20	0.01	0.01	0.15	0.16	0.21	0.03
3	5.90	1.31	0.02	0.02	0.24	0.25	0.22	0.05
4	6.18	1.43	0.04	0.05	0.34	0.35	0.23	0.07
5	6.49	1.57	0.07	0.07	0.45	0.47	0.24	0.09
6	6.81	1.72	0.11	0.11	0.57	0.60	0.25	0.11
7	7.16	1.88	0.15	0.16	0.71	0.74	0.26	0.12
8	7.53	2.06	0.21	0.22	0.86	0.91	0.27	0.14
9	7.92	2.25	0.28	0.30	1.03	1.10	0.28	0.16
10	8.34	2.47	0.37	0.39	1.22	1.31	0.30	0.18
11	8.80	2.71	0.47	0.50	1.44	1.56	0.31	0.19
12	9.28	2.97	0.60	0.63	1.69	1.84	0.32	0.21
13	9.81	3.26	0.74	0.78	1.97	2.16	0.33	0.23
14	10.37	3.59	0.92	0.97	2.29	2.52	0.35	0.25
15	10.98	3.94	1.13	1.18	2.65	2.94	0.36	0.27
16	11.63	4.34	1.37	1.43	3.06	3.42	0.37	0.29
17	12.34	4.77	1.66	1.73	3.53	3.98	0.39	0.31
18	13.10	5.26	2.00	2.08	4.07	4.61	0.40	0.32
19	13.93	5.80	2.40	2.48	4.68	5.35	0.42	0.34
20	14.83	6.40	2.87	2.95	5.39	6.20	0.43	0.36
21	15.81	7.07	3.42	3.50	6.20	7.18	0.45	0.38
22	16.88	7.82	4.07	4.13	7.13	8.32	0.46	0.40
23	18.05	8.66	4.82	4.88	8.20	9.64	0.48	0.42
24	19.32	9.60	5.72	5.75	9.44	11.17	0.50	0.45
25	20.72	10.66	6.77	6.76	10.88	12.96	0.51	0.47
26	22.25	11.85	8.00	7.94	12.54	15.05	0.53	0.49
27	23.94	13.20	9.46	9.32	14.47	17.49	0.55	0.51
28	25.80	14.72	11.19	10.94	16.72	20.35	0.57	0.53
29	27.86	16.44	13.24	12.84	19.34	23.71	0.59	0.55
30	30.14	18.40	15.67	15.07	22.40	27.66	0.61	0.58
31	32.67	20.63	18.56	17.69	25.99	32.33	0.63	0.60
32	35.49	23.18	22.02	20.79	30.21	37.85	0.65	0.62
33	38.64	26.09	26.17	24.44	35.19	44.40	0.68	0.65
34	42.16	29.44	31.15	28.77	41.06	52.18	0.70	0.67
35	46.12	33.30	37.15	33.92	48.03	61.47	0.72	0.70
36	50.59	37.75	44.43	40.05	56.31	72.59	0.75	0.73
37	55.63	42.92	53.27	47.38	66.19	85.95	0.77	0.75
38	61.35	48.93	64.07	56.17	78.02	102.05	0.80	0.78
39	67.87	55.96	77.33	66.75	92.25	121.53	0.82	0.81
40	75.31	64.19	93.69	79.54	109.41	145.19	0.85	0.84
41	83.86	73.90	113.98	95.05	130.21	174.06	0.88	0.87
42	93.71	85.37	139.32	113.95	155.54	209.43	0.91	0.90
43	105.11	99.01	171.14	137.10	186.53	253.00	0.94	0.93
44	118.37	115.31	211.41	165.58	224.63	306.92	0.97	0.97
45	133.87	134.97	262.74	200.81	271.74	374.02	1.01	1.00
46	152.10	158.50	328.73	244.64	330.33	458.02	1.04	1.04
47	173.64	187.20	414.32	299.52	403.65	563.81	1.08	1.07
48	199.26	222.30	526.44	368.66	495.99	697.93	1.12	1.11
49	229.92	265.49	674.91	456.40	613.13	869.17	1.15	1.15
50	266.88	319.05	873.84	568.56	762.85	1089.46	1.20	1.19

Note: N_c and N_q are same for all four methods; subscripts identify author for N_γ: M = Meyerhof [48]; H = Hansen [35]; V = Vesic [69]; C = Chen [26].

TABLE 31.5 Shape, Depth, Inclination, Ground, and Base Factors for Eq. (31.3)

Shape Factors	Depth Factors
$s_c = 1.0 + \frac{N_q}{N_c} \quad \frac{B}{L}$ $s_c = 1.0 \quad$ (for strip footing)	$d_c = 1.0 + 0.4k \begin{cases} k = \frac{D_f}{B} & \text{for } \frac{D_f}{B} \le 1 \\ k = \tan^{-1}\left(\frac{D_f}{B}\right) & \text{for } \frac{D_f}{B} > 1 \end{cases}$
$s_q = 1.0 + \frac{B}{L} \tan\phi \quad$ (for all ϕ)	$d_q = 1 + 2\tan\phi(1-\sin\phi)^2 k$ (k defined above)
$s_\gamma = 1.0 - 0.4\frac{B}{L} \ge 0.6$	$d_\gamma = 1.00 \quad$ (for all ϕ)
Inclination Factors	**Ground Factors (base on slope)**
$i_c' = 1 - \frac{mHi}{A_f c_a N_c} \quad (\phi = 0)$	$g_c' = \frac{\beta}{5.14} \quad \beta$ in radius $\quad (\phi = 0)$
$i_c = i_q - \frac{1-i_q}{N_q - 1} \quad (\phi > 0)$	$g_c = i_q - \frac{1-i_q}{5.14\tan\phi} \quad (\phi > 0)$
$i_q = \left[1.0 - \frac{H_i}{V + A_f c_a \cot\phi}\right]^m$	$g_q = g_\gamma = (1.0 - \tan\beta)^2 \quad$ (for all ϕ)
	Base Factors (tilted base)
$i_\gamma = \left[1.0 - \frac{H_i}{V + A_f c_a \cot\phi}\right]^{m+1}$	$b_c' = g_c' \quad (\phi = 0)$
$m = m_B = \frac{2 + B/L}{1 + B/L} \quad$ or	$b_c = 1 - \frac{2\beta}{5.14\tan\phi} \quad (\phi > 0)$
$m = m_L = \frac{2 + L/B}{1 + L/B}$	$b_q = b_\gamma = (1.0 - \eta\tan\phi)^2 \quad$ (for all ϕ)

Notes:

1. When $\gamma = 0$ (and β 'ne 0) use $N_\gamma = 2\sin(\pm\beta)$ in N_γ term
2. Compute $m = m_B$ when $H_i = H_B$ (H parallel to B) and $m = m_L$ when $H_i = H_L$ (H parallel to L); for both H_B and H_L use

   $m = \sqrt{m_B^2 + m_L^2}$
3. $0 \le i_q, i_\gamma \le 1$
4. $\beta + \eta \le 90°; \beta \le \phi$

where

A_f = effective footing dimension as shown in Figure 31.6
D_f = depth from ground surface to base of footing
V = vertical load on footing
H_i = horizontal component of load on footing with $H_{max} \le V\tan\delta + c_a A_f$
c_a = adhesion to base ($0.6c \le c_a \le 1.0c$)
δ = friction angle between base and soil ($0.5\phi \le \delta \le \phi$)
β = slope of ground away from base with (+) downward
η = tilt angle of base from horizontal with (+) upward

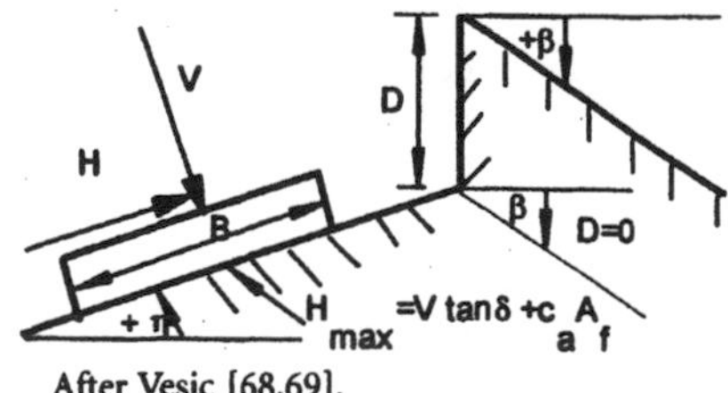

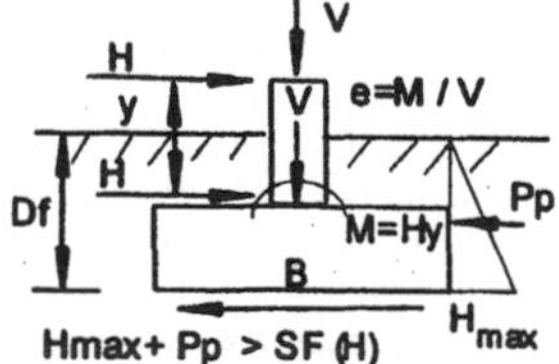

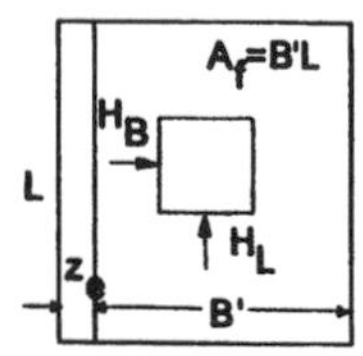

After Vesic [68,69].

$$\phi^* = \tan^{-1}(0.67 \tan\phi) \quad (\text{for loose sands with } \phi < 28°) \tag{31.5}$$

Vesic [69] suggested that a flat reduction of ϕ might be too conservative in the case of local and punching shear failure. He proposed the following equation for a reduction factor varying with relative density D_r:

$$\phi^* = \tan^{-1}\left(\left(0.67 + D_r - 0.75D_r^2\right) \tan\phi\right) \quad \left(\text{for } 0 < D_r < 0.67\right) \tag{31.6}$$

Groundwater Table

Ultimate bearing capacity should always be estimated by assuming the highest anticipated groundwater table. The effective unit weight γ_e shall be used in the qN_q and $0.5\gamma B$ terms. As illustrated in Figure 31.5, the weighted average unit weight for the $0.5\gamma B$ term can be determined as follows:

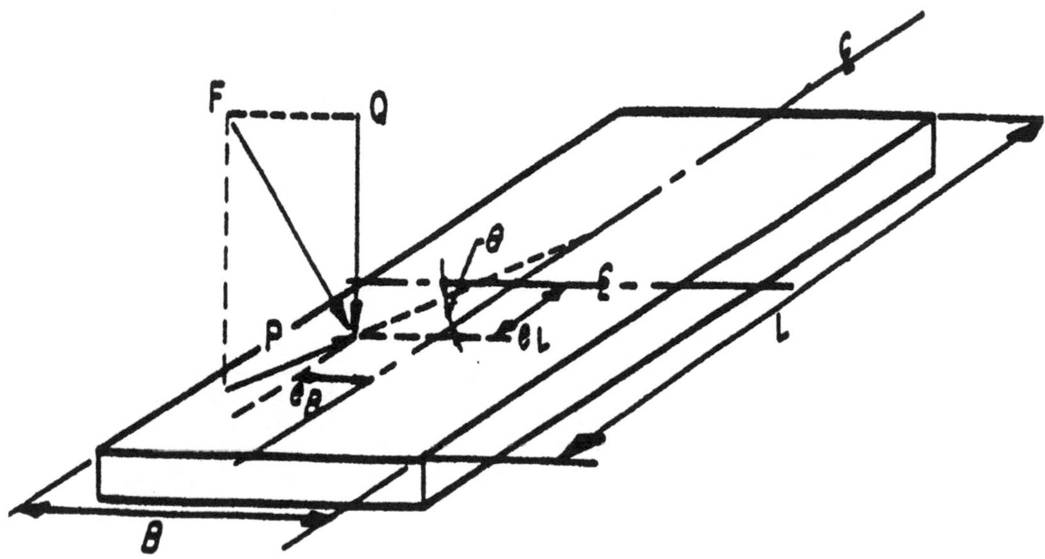

FIGURE 31.5 Definition sketch for loading and dimensions for footings subjected to eccentric or inclined loads. (After AASHTO, 1997.)

$$\gamma = \begin{cases} \gamma_{avg} & \text{for } d_w \geq B \\ \gamma' + (d_w/B)(\gamma_{avg} - \gamma') & \text{for } 0 < d_w < B \\ \gamma' & \text{for } d \leq 0 \end{cases} \tag{31.7}$$

Eccentric Load

For footings with eccentricity, effective footing dimensions can be determined as follows:

$$A_f = B'L' \tag{31.8}$$

where $L = L - 2e_L$ and $B = B - 2e_B$. Refer to Figure 31.5 for loading definitions and footing dimensions. For example, the actual distribution of contact pressure for a rigid footing with eccentric loading in the L direction (Figure 31.6) can be obtained as follows:

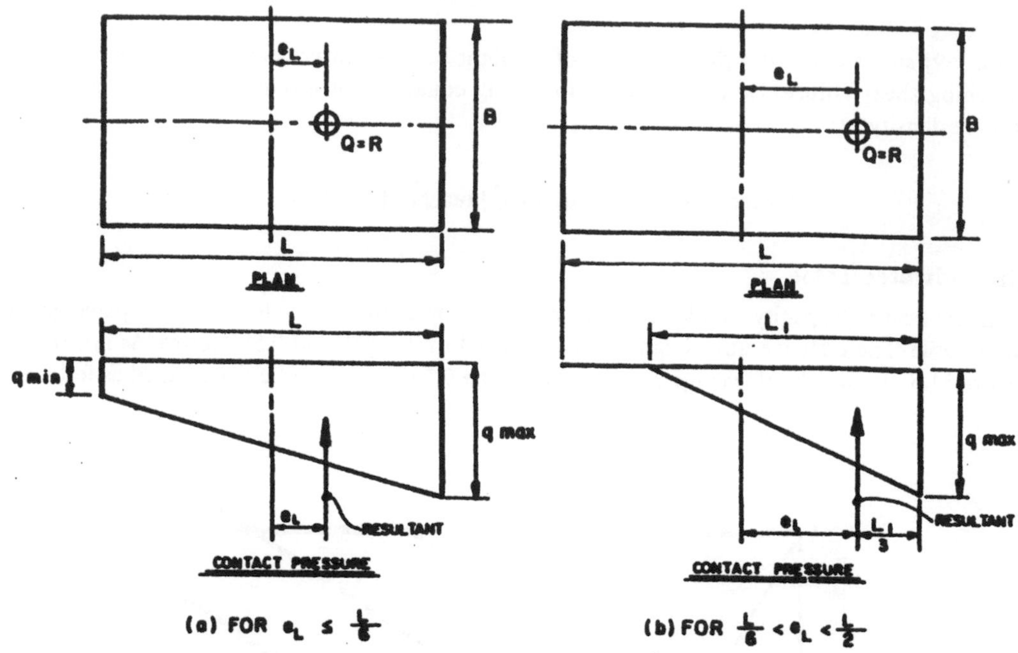

FIGURE 31.6 Contact pressure for footing loaded eccentrically about one axis. (After AASHTO 1997.)

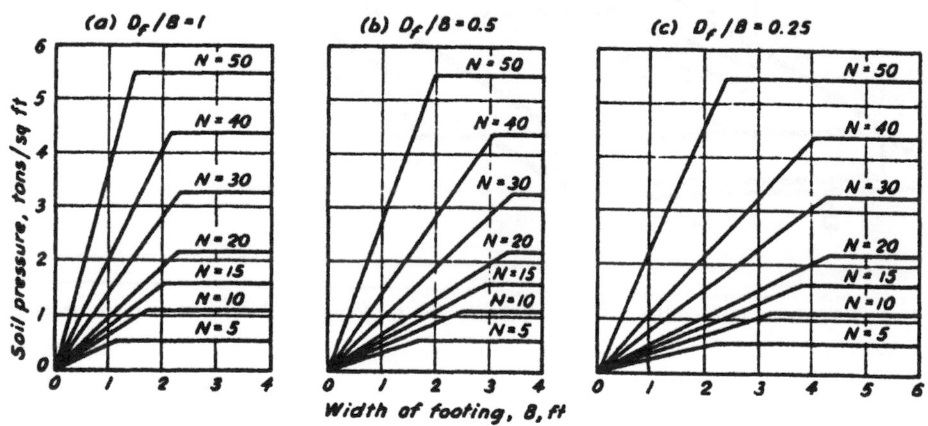

FIGURE 31.7 Design chart for proportioning shallow footings on sand. (a) Rectangular base; (b) round base. (After Peck et al. [53])

$$q_{\substack{max \\ min}} = P\left[1 \pm 6e_L/L\right]/BL \quad \left(\text{for } e_L < L/6\right) \tag{31.9}$$

$$q_{\substack{max \\ min}} = \begin{cases} 2P/\left[3B\left(L/2 - e_L\right)\right] \\ 0 \end{cases} \quad \left(\text{for } L/6 < e_L < L/2\right) \tag{31.10}$$

Contact pressure for footings with eccentric loading in the B direction may be determined using above equations by replacing terms L with B and terms B with L. For an eccentricity in both directions, reference is available in AASHTO [2,3].

31.4.2 Bearing Capacity on Sand from Standard Penetration Tests (SPT)

Terzaghi and Peck [64,65] proposed a method using SPT blow counts to estimate ultimate bearing capacity for footings on sand. Modified by Peck et al. [53], this method is presented in the form of the chart shown in Figure 31.7. For a given combination of footing width and SPT blow counts, the chart can be used to determine the ultimate bearing pressure associated with 25.4 mm (1.0 in.) settlement. The design chart applies to shallow footings ($D_f \le B$) sitting on sand with water table at great depth. Similarly, Meyerhof [46] published the following formula for estimating ultimate bearing capacity using SPT blow counts:

$$q_{\text{ult}} = N'_{\text{avg}} \frac{B}{10}\left(C_{w1} + C_{w2} \frac{D_f}{B} \right) R_I \tag{31.11}$$

where R_I is a load inclination factor shown in Table 31.6 (R_I = 1.0 for vertical loads). C_{w1} and C_{w2} are correction factors whose values depend on the position of the water table:

TABLE 31.6 Load Inclination Factor (R_I)

	For Square Footings		
	Load Inclination Factor (R_I)		
H/V	$D_f/B=0$	$D_f/B=1$	$D_f/B=3$
0.10	0.75	0.80	0.85
0.15	0.65	0.75	0.80
0.20	0.55	0.65	0.70
0.25	0.50	0.55	0.65
0.30	0.40	0.50	0.55
0.35	0.35	0.45	0.50
0.40	0.30	0.35	0.45
0.45	0.25	0.30	0.40
0.50	0.20	0.25	0.30
0.55	0.15	0.20	0.25
0.60	0.10	0.15	0.20

	For Rectangular Footings					
	Load Inclination Factor (R_I)					
H/H	$D_f/B=0$	$D_f/B=1$	$D_f/B=5$	$D_f/B=0$	$D_f/B=1$	$D_f/B=5$
0.10	0.70	0.75	0.80	0.80	0.85	0.90
0.15	0.60	0.65	0.70	0.70	0.80	0.85
0.20	0.50	0.60	0.65	0.65	0.70	0.75
0.25	0.40	0.50	0.55	0.55	0.65	0.70
0.30	0.35	0.40	0.50	0.50	0.60	0.65
0.35	0.30	0.35	0.40	0.40	0.55	0.60
0.40	0.25	0.30	0.35	0.35	0.50	0.55
0.45	0.20	0.25	0.30	0.30	0.45	0.50
0.50	0.15	0.20	0.25	0.25	0.35	0.45
0.55	0.10	0.15	0.20	0.20	0.30	0.40
0.60	0.05	0.10	0.15	0.15	0.25	0.35

After Barker et al. [9].

$$\begin{cases} C_{w1} = C_{w2} = 0.5 & \text{for } D_w = 0 \\ C_{w1} = C_{w2} = 1.0 & \text{for } D_w \geq D_f = 1.5B \\ C_{w1} = 0.5 \text{ and } C_{w2} = 1.0 & \text{for } D_w = D_f \end{cases} \tag{31.12}$$

N'_{avg} is an average value of the SPT blow counts, which is determined within the range of depths from footing base to 1.5B below the footing. In very fine or silty saturated sand, the measured SPT blow count (N) is corrected for submergence effect as follows:

$$N' = 15 + 0.5(N - 15) \quad \text{for } N > 15 \tag{31.13}$$

31.4.3 Bearing Capacity from Cone Penetration Tests (CPT)

Meyerhof [46] proposed a relationship between ultimate bearing capacity and cone penetration resistance in sands:

$$q_{ult} = q_c \frac{B}{40}\left(C_{w1} + C_{w2}\frac{D_f}{B}\right)R_I \tag{31.14}$$

where q_c is the average value of cone penetration resistance measured at depths from footing base to 1.5B below the footing base. C_{w1}, C_{w2}, and R_1 are the same as those as defined in Eq. (31.11).

Schmertmann [57] recommended correlated values of ultimate bearing capacity to cone penetration resistance in clays as shown in Table 31.7.

TABLE 31.7 Correlation between Ultimate Bearing Capacity (q^{ult}) and Cone Penetration Resistance (q_c)

	qult (ton/ft²)	
q_c (kg/cm² or ton/ft²)	Strip Footings	Square Footings
10	5	9
20	8	12
30	11	16
40	13	19
50	15	22

After Schmertmann [57] and Awkati, 1970.

31.4.4 Bearing Capacity from Pressure-Meter Tests (PMT)

Menard [44], Baguelin et al. [8], and Briaud [15,17] proposed using the limit pressure measured in PMT to estimate ultimate bearing capacity:

$$q_{ult} = r_0 + \kappa(p_1 - p_0) \tag{31.15}$$

where r_0 is the initial total vertical pressure at the foundation level, κ is the dimensionless bearing capacity coefficient from Figure 31.8, p_1 is limit pressure measured in PMT at depths from 1.5B above to 1.5B below foundation level, and p_0 is total horizontal pressure at the depth where the PMT is performed.

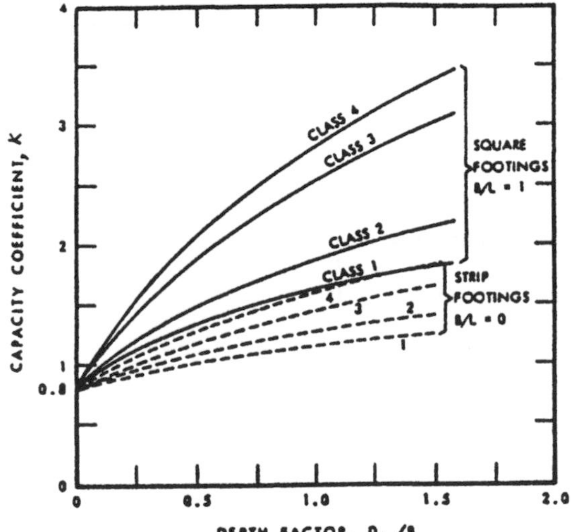

Soil Type	Consistency or Density	(P_l-P_0) (tsf)	Class
clay	soft to very firm	<12	1
	stiff	8 – 40	2
sand and gravel	loose	4-8	2
	medium to dense	10-20	3
	very dense	30-60	4
silt	loose to medium	<7	1
	dense	12-30	2
rock	very low strength	10-30	2
	low strength	30-60	3
	medium to high strength	60-100$^+$	4

FIGURE 31.8 Values of empirical capacity coefficient, κ. (After Canadian Geotechnical Society [24].)

31.4.5 Bearing Capacity According to Building Codes

Recommendations for bearing capacity of shallow foundations are available in most building codes. Presumptive value of allowable bearing capacity for spread footings are intended for preliminary design when site-specific investigation is not justified. Presumptive bearing capacities usually do not reflect the size, shape, and depth of footing, local water table, or potential settlement. Therefore, footing design using such a procedure could be either overly conservative in some cases or unsafe in others [9]. Recommended practice is to use presumptive bearing capacity as shown in Table 31.8 for preliminary footing design and to finalize the design using reliable methods in the preceding discussion.

31.4.6 Predicted Bearing Capacity vs. Load Test Results

Obviously, the most reliable method of obtaining the ultimate bearing capacity is to conduct a full-scale footing load test at the project site. Details of the test procedure have been standardized as ASTM D1194 [5]. The load test is not usually performed since it is very costly and not practical for routine design. However, using load test results to compare with predicted bearing capacity is a vital tool to verify the accuracy and reliability of various prediction procedures. A comparison between the predicted bearing capacity and results of eight load tests conducted by Milovic [49] is summarized in Table 31.9.

Recently, load testing of five large-scale square footings (1 to 3 m) on sand was conducted on the Texas A&M University National Geotechnical Experimental Site [94]. One of the main objects of the test is to evaluate the various procedures used for estimating bearing capacities and settlements of shallow foundations. An international prediction event was organized by ASCE Geotechnical Engineering Division, which received a total of 31 predictions (16 from academics and 15 from consultants) from Israel, Australia, Japan, Canada, the United States, Hong Kong, Brazil, France, and Italy. Comparisons of predicted and measured values of bearing capacity using various procedures were summarized in Tables 31.10 through 31.12. From those comparisons, it can be argued that the most accurate settlement prediction methods are the Schmertmann-DMT (1986) and the Peck and Bazarra (1967) although they are on the unconservative side. The most conservative

TABLE 31.8 Presumptive Values of Allowable Bearing Capacity for Spread Foundations

Type of Bearing Material	Consistency in Place	q_{all} (ton/ft²)	
		Range	Recommended Value for Use
Massive crystalline igneous and metamorphic rock: granite, diorite, basalt, gneiss, thoroughly cemented conglomerate (sound condition allows minor cracks)	Hard sound rock	60–100	80
Foliated metamorphic rock: slate, schist (sound condition allows minor cracks)	Medium-hard sound rock	30–40	35
Sedimentary rock: hard cemented shales, siltstone, sandstone, limestone without cavities	Medium-hard sound rock	15–25	20
Weathered or broken bedrock of any kind except highly argillaceous rock (shale); RQD less than 25	Soft rock	8–12	10
Compaction shale or other highly argillaceous rock in sound condition	Soft rock	8–12	10
Well-graded mixture of fine and coarse-grained soil: glacial till, hardpan, boulder clay (GW-GC, GC, SC)	Very compact	8–12	10
Gravel, gravel–sand mixtures, boulder gravel mixtures (SW, SP)	Very compact	6–10	7
	Medium to compact	4–7	5
	Loose	2–5	3
Coarse to medium sand, sand with little gravel (SW, SP)	Very compact	4–6	4
	Medium to compact	2–4	3
	Loose	1–3	1.5
Fine to medium sand, silty or clayey medium to coarse sand (SW, SM, SC)	Very compact	3–5	3
	Medium to compact	2–4	2.5
	Loose	1–2	1.5
Homogeneous inorganic clay, sandy or silty clay (CL, CH)	Very stiff to hard	3–6	4
	Medium to stiff	1–3	2
	Soft	0.5–1	0.5
Inorganic silt, sandy or clayey silt, varved silt-clay-fine sand	Very stiff to hard	2–4	3
	Medium to stiff	1–3	1.5
	Soft	0.5–1	0.5

Notes:

1. Variations of allowable bearing pressure for size, depth, and arrangement of footings are given in Table 2 of NAFVAC [52].
2. Compacted fill, placed with control of moisture, density, and lift thickness, has allowable bearing pressure of equivalent natural soil.
3. Allowable bearing pressure on compressible fine-grained soils is generally limited by considerations of overall settlement of structure.
4. Allowable bearing pressure on organic soils or uncompacted fills is determined by investigation of individual case.
5. If tabulated recommended value for rock exceeds unconfined compressive strength of intact specimen, allowable pressure equals unconfined compressive strength.

After NAVFAC [52].

methods are Briaud [15] and Burland and Burbidge [20]. The most accurate bearing capacity prediction method was the $0.2q_c$ (CPT) method [16].

TABLE 31.9 Comparison of Computed Theoretical Bearing Capacities and Milovic and Muh's Experimental Values

	Test							
Bearing Capacity Method	1	2	3	4	5	6	7	8
	D = 0.0 m	0.5	0.5	0.5	0.4	0.5	0.0	0.3
	B = 0.5 m	0.5	0.5	1.0	0.71	0.71	0.71	0.71
	L = 2.0 m	2.0	2.0	1.0	0.71	0.71	0.71	0.71
	γ = 15.69 kN/m³	16.38	17.06	17.06	17.65	17.65	17.06	17.06
	ϕ = 37°(38.5°)	35.5 (36.25)	38.5 (40.75)	38.5	22	25	20	20
	c = 6.37 kPa	3.92	7.8	7.8	12.75	14.7	9.8	9.8
Milovic (tests)					q_{ult} (kg/cm²) 4.1	5.5	2.2	2.6
Muhs (tests)	q_{ult} (kg/cm²) 10.8	12.2	24.2	33.0				
Terzaghi	9.4*	9.2	22.9	19.7	4.3*	6.5*	2.5	2.9*
Meyerhof	8.2*	10.3	26.4	28.4	4.8	7.6	2.3	3.0
Hansen	7.2	9.8	23.7*	23.4	5.0	8.0	2.2*	3.1
Vesic	8.1	10.4*	25.1	24.7	5.1	8.2	2.3	3.2
Balla	14.0	15.3	35.8	33.0*	6.0	9.2	2.6	3.8

[a] After Milovic (1965) but all methods recomputed by author and Vesic added.

Notes:
1. ϕ = triaxial value ϕ_{tr}; (plane strain value) = 1.5 ϕ_{tr} - 17.
2. * = best: Terzaghi = 4; Hansen = 2; Vesic = 1; and Balla = 1.

Source: Bowles, J.E., *Foundation Analysis and Design*, 5th ed., McGraw-Hill, New York, 1996. With permission.

TABLE 31.10 Comparison of Measured vs. Predicted Load Using Settlement Prediction Method

	Predicted Load (MN) @ s = 25 mm				
Prediction Methods	1.0 m Footing	1.5 m Footing	2.5 m Footing	3.0 m(n) Footing	3.0 m(s) Footing
Briaud [15]	0.904	1.314	2.413	2.817	2.817
Burland and Burbidge [20]	0.699	1.044	1.850	2.367	2.367
De Beer (1965)	1.140	0.803	0.617	0.597	0.597
Menard and Rousseau (1962)	0.247	0.394	0.644	1.017	1.017
Meyerhof CPT (1965)	0.288	0.446	0.738	0.918	0.918
Meyerhof — SPT (1965)	0.195	0.416	1.000	1.413	1.413
Peck and Bazarra (1967)	1.042	1.899	4.144	5.679	5.679
Peck, Hansen & Thornburn [53]	0.319	0.718	1.981	2.952	2.952
Schmertmann CPT (1970)	0.455	0.734	1.475	1.953	1.953
Schmertmann DMT (1970)	1.300	2.165	4.114	5.256	5.256
Schultze and Sherif (1973)	1.465	2.615	4.750	5.850	5.850
Terzaghi and Peck [65]	0.287	0.529	1.244	1.476	1.476
Measured Load @ s = 25mm	0.850	1.500	3.600	4.500	4.500

Source: FHWA, Publication No. FHWA-RD-97-068, 1997.

31.5 Stress Distribution Due to Footing Pressures

Elastic theory is often used to estimate the distribution of stress and settlement as well. Although soils are generally treated as elastic–plastic materials, the use of elastic theory for solving the problems is mainly due to the reasonable match between the boundary conditions for most footings and those of elastic solutions [37]. Another reason is the lack of availability of acceptable alternatives. Observation and experience have shown that this practice provides satisfactory solutions [14,37,54,59].

TABLE 31.11 Comparison of Measured vs. Predicted Load Using Bearing Capacity Prediction Method

	Predicted Bearing Capacity (MN)				
Prediction Methods	1.1 m Footing	1.5 m Footing	2.6 m Footing	3.0m(n) Footing	3.0m(s) Footing
Briaud — CPT [16]	1.394	1.287	1.389	1.513	1.513
Briaud — PMT [15]	0.872	0.779	0.781	0.783	0.783
Hansen [35]	0.772	0.814	0.769	0.730	0.730
Meyerhof [45,48]	0.832	0.991	1.058	1.034	1.034
Terzaghi [63]	0.619	0.740	0.829	0.826	0.826
Vesic [68,69]	0.825	0.896	0.885	0.855	0.855
Measured Load @ s = 150 mm					

Source: FHWA, Publication No. FHWA-RD-97-068, 1997.

TABLE 31.12 Best Prediction Method Determination

		Mean Predicted Load/ Mean Measured Load
	Settlement Prediction Method	
1	Briaud [15]	0.66
2	Burland & Burbidge [20]	0.62
3	De Beer [29]	0.24
4	Menard and Rousseau (1962)	0.21
5	Meyerhof CPT (1965)	0.21
6	Meyerhof SPT (1965)	0.28
7	Peck and Bazarra (1967)	1.19
8	Peck, et al. [53]	0.57
9	Schmertmann — CPT [56]	0.42
10	Schmertmann — DMT [56]	1.16
11	Shultze and Sherif (1973)	1.31
12	Terzaghi and Peck [65]	0.32
	Bearing Capacity Prediction Method	
1	Briaud — CPT [16]	1.08
2	Briaud — PMT [15]	0.61
3	Hansen [35]	0.58
4	Meyerhof [45,48]	0.76
5	Terzaghi [63]	0.59
6	Vesic [68,69]	0.66

Source: FHWA, Publication No. FHWA-RD-97-068, 1997.

31.5.1 Semi-infinite, Elastic Foundations

Bossinesq equations based on elastic theory are the most commonly used methods for obtaining subsurface stresses produced by surface loads on semi-infinite, elastic, isotropic, homogenous, weightless foundations. Formulas and plots of Bossinesq equations for common design problems are available in NAVFAC [52]. Figure 31.9 shows the isobars of pressure bulbs for square and continuous footings. For other geometry, refer to Poulos and Davis [55].

31.5.2 Layered Systems

Westergaard [70], Burmister [21-23], Sowers and Vesic [62], Poulos and Davis [55], and Perloff [54] discussed the solutions to stress distributions for layered soil strata. The reality of interlayer shear is very complicated due to *in situ* nonlinearity and material inhomogeneity [37,54]. Either zero (frictionless) or with perfect fixity is assumed for the interlayer shear to obtain possible

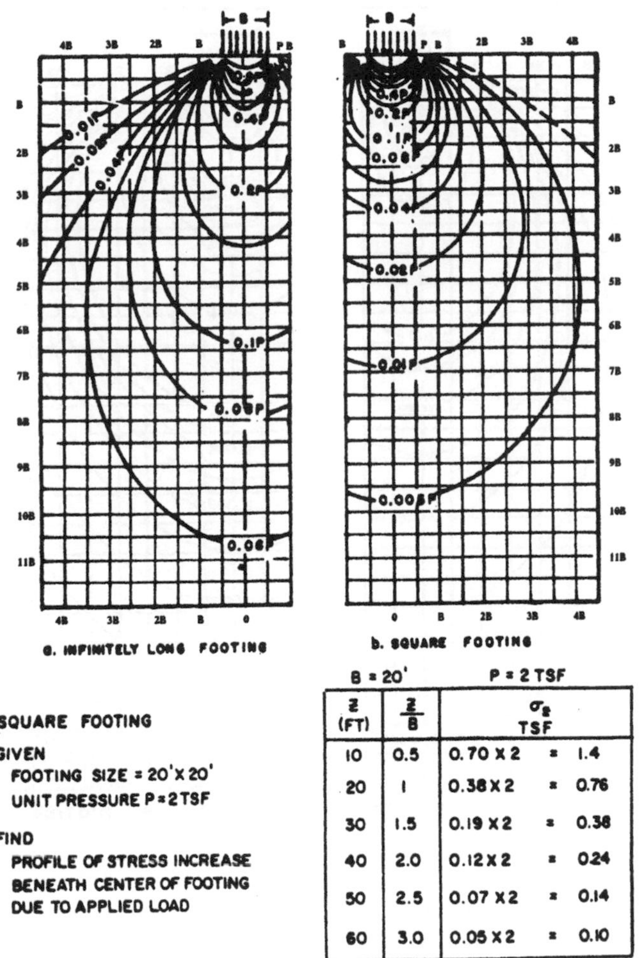

z (FT)	z/B	σ_z TSF
10	0.5	0.70 X 2 = 1.4
20	1	0.38 X 2 = 0.76
30	1.5	0.19 X 2 = 0.38
40	2.0	0.12 X 2 = 0.24
50	2.5	0.07 X 2 = 0.14
60	3.0	0.05 X 2 = 0.10

FIGURE 31.9 Pressure bulbs based on the Bossinesq equation for square and long footings. (After NAVFAC 7.01, 1986].)

solutions. The Westergaard method assumed that the soil being loaded is constrained by closed spaced horizontal layers that prevent horizontal displacement [52]. Figures 31.10 through 31.12 by the Westergaard method can be used for calculating vertical stresses in soils consisting of alternative layers of soft (loose) and stiff (dense) materials.

31.5.3 Simplified Method (2:1 Method)

Assuming a loaded area increasing systemically with depth, a commonly used approach for computing the stress distribution beneath a square or rectangle footing is to use the 2:1 slope method as shown in Figure 31.13. Sometimes a 60° distribution angle (1.73–to–1 slope) may be assumed. The pressure increase Δq at a depth z beneath the loaded area due to base load P is

$$\Delta q = \begin{cases} P/(B+z)(L+z) & \text{(for a rectangle footing)} \\ P/(B+z)^2 & \text{(for a square footing)} \end{cases} \tag{31.16}$$

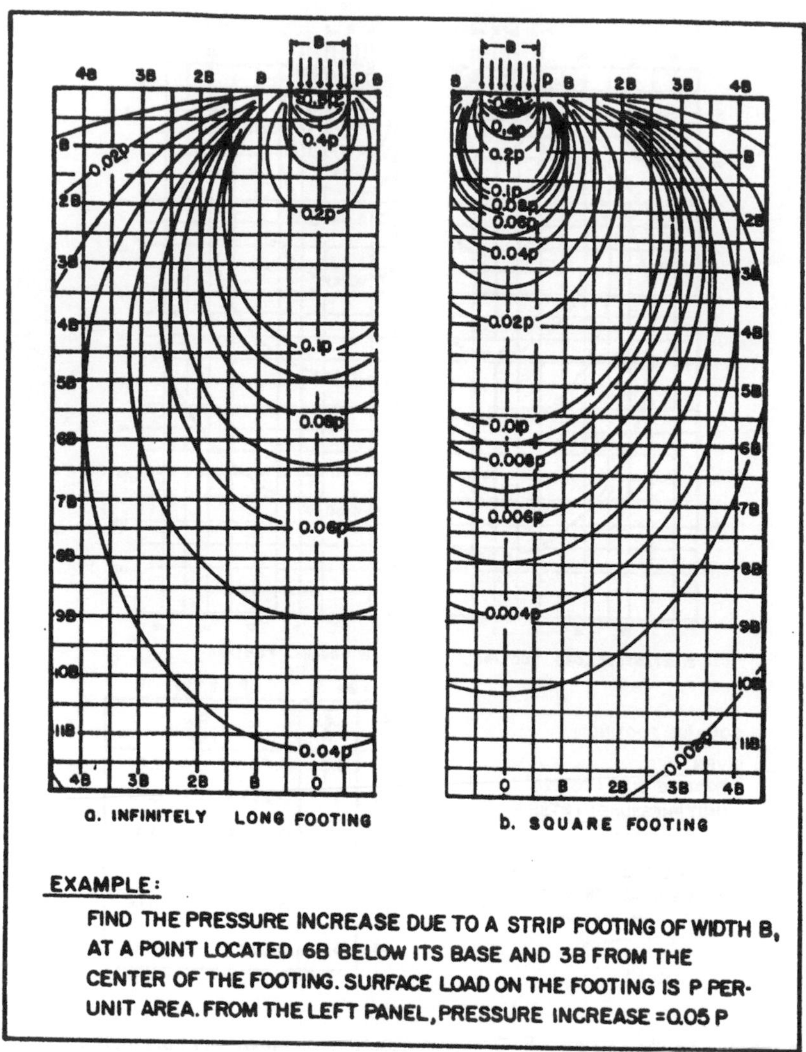

FIGURE 31.10 Vertical stress contours for square and strip footings [Westerqaard Case]. (After NAVFAC 7.01, 1986.)

where symbols are referred to Figure 31.14. The solutions by this method compare very well with those of more theoretical equations from depth z from B to about $4B$ but should not be used for depth z from 0 to B [14]. A comparison between the approximate distribution of stress calculated by a theoretical method and the 2:1 method is illustrated in Figure 31.15.

31.6 Settlement of Shallow Foundations

The load applied on a footing changes the stress state of the soil below the footing. This stress change may produce a time-dependent accumulation of elastic compression, distortion, or consolidation of the soil beneath the footing. This is often termed *foundation settlement*. True elastic deformation consists of a very small portion of the settlement while the major components of the settlement are due to a change of void ratio, particle rearrangement, or crushing. Therefore, very little of the settlement will be recovered even if the applied load is removed. The irrecoverable

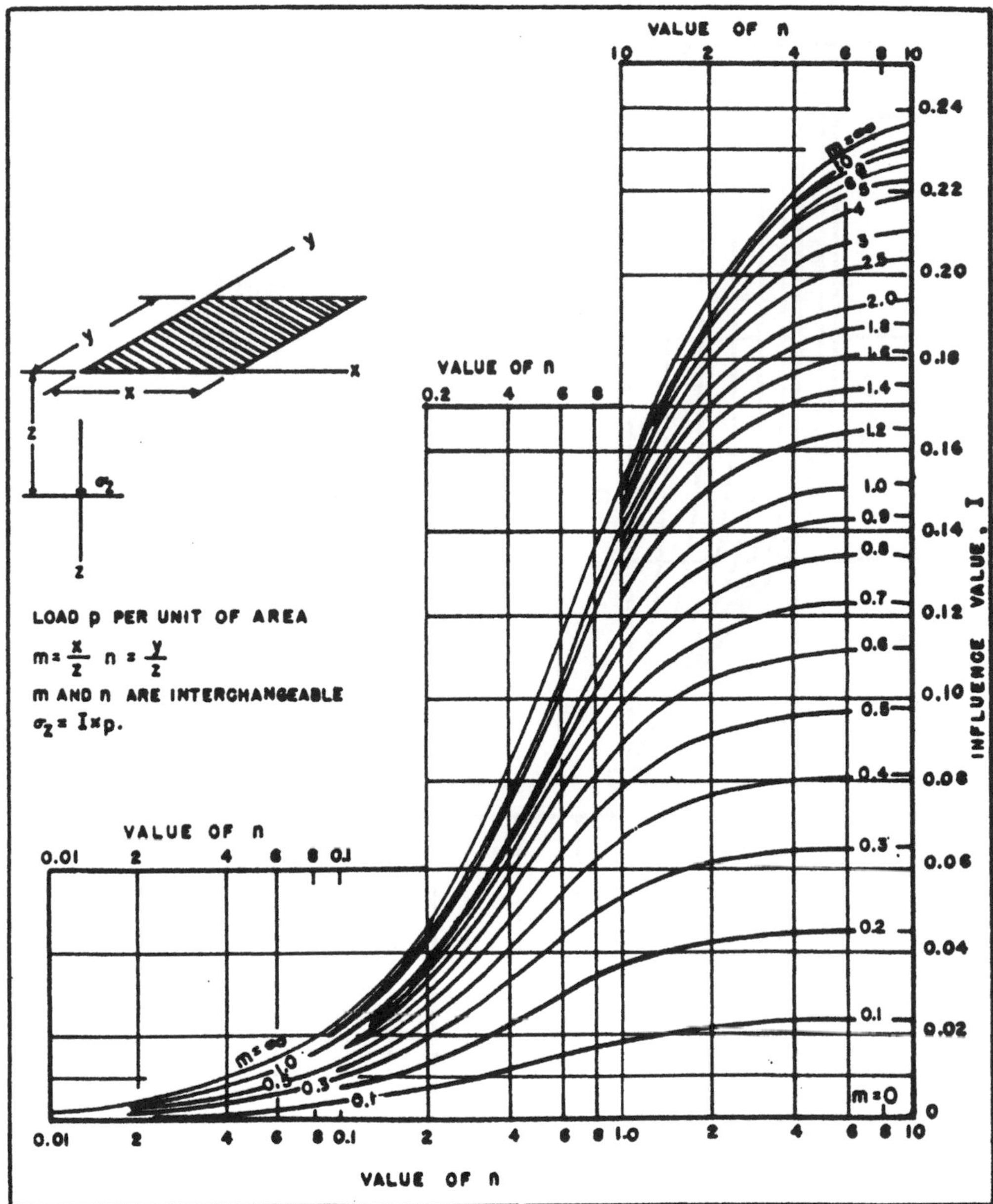

FIGURE 31.11 Influence value for vertical stress beneath a corner of a uniformly loaded rectangular area (Westerqaard Case). (After NAVFAC [52].)

deformation of soil reflects its inherent elastic–plastic stress–strain relationship. The reliability of settlement estimated is influenced principally by soil properties, layering, stress history, and the actual stress profile under the applied load [14,66]. The total settlement may be expressed as

$$s = s_i + s_c + s_s \tag{31.17}$$

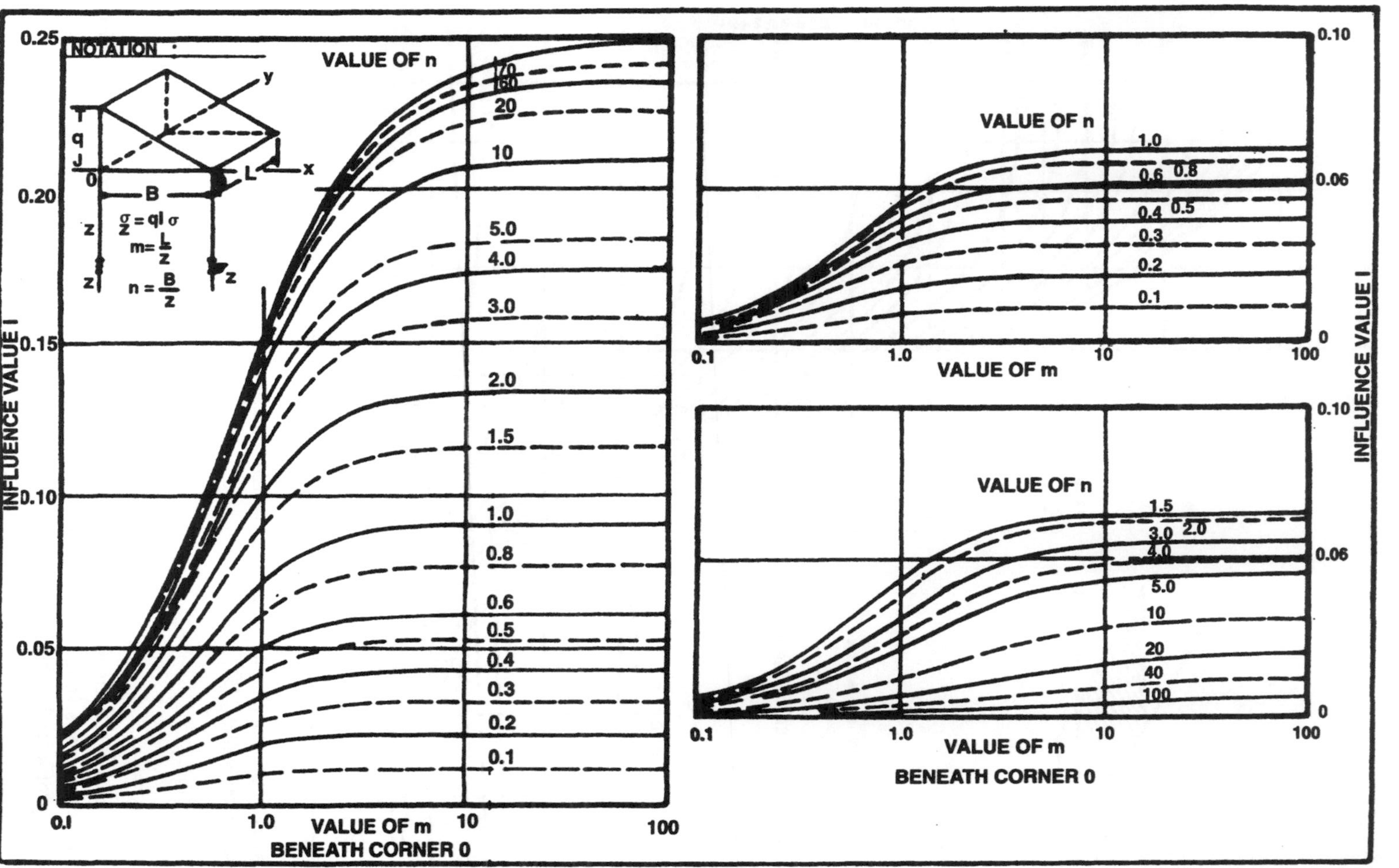

FIGURE 31.12 Influence value for vertical stress beneath triangular load (Westergaard Case). (After NAVFAC [52].)

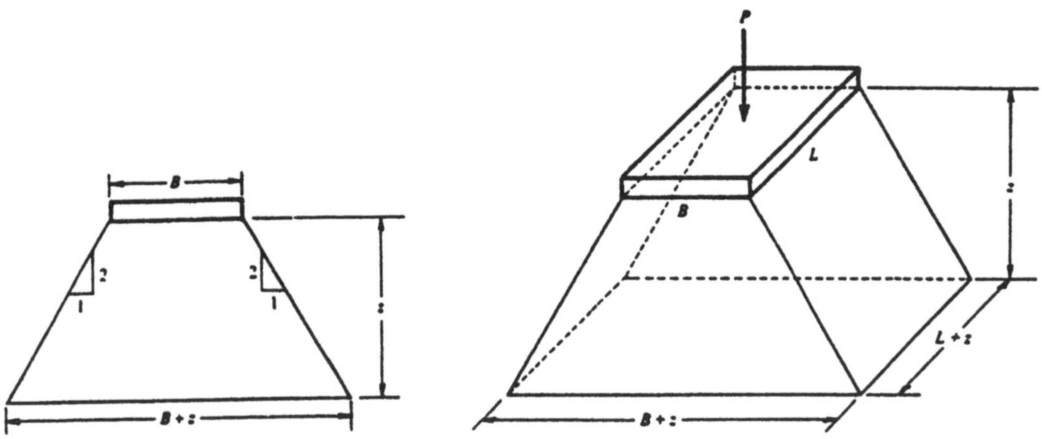

FIGURE 31.14 Approximate distribution of vertical stress due to surface load. (After Perloff [54].)

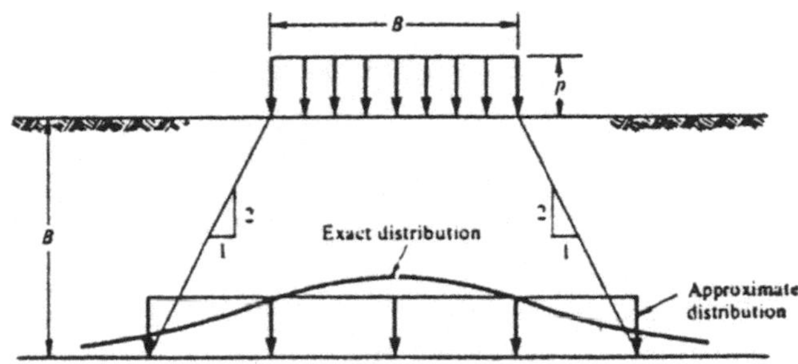

FIGURE 31.15 Relationship between vertical stress below a square uniformly loaded area as determined by approximate and exact methods. (After Perloff [54].)

where s is the total settlement, s_i is the immediate or distortion settlement, s_c is the primary consolidation settlement, and s_s is the secondary settlement. The time-settlement history of a shallow foundation is illustrated in Figure 31.15. Generally speaking, immediate settlement is not elastic. However, it is often referred to as elastic settlement because the elastic theory is usually used for computation. The immediate settlement component controls in cohesionless soils and unsaturated cohesive soils, while consolidation compression dictates in cohesive soils with a degree of saturation above 80% [3].

31.6.1 Immediate Settlement by Elastic Methods

Based on elastic theory, Steinbrenner [61] suggested that immediate settlements of footings on sands and clay could be estimated in terms of Young's modulus E of soils. A modified procedure developed by Bowles [14] may be used for computing settlements of footings with flexible bases on the half-space. The settlement equation can be expressed as follows

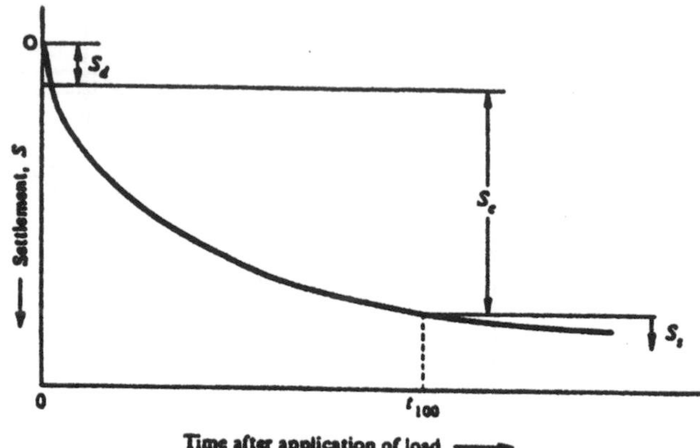

FIGURE 31.15 Schematic time–settlement history of typical point on a foundation. (After Perloff [54].)

$$s_i = q_0 B' \left(1-\mu^2\right) m I_s I_F / E_s \tag{31.18}$$

$$I_s = n\,(I_1 + (1-2\mu) I_2/(1-\mu)) \tag{31.19}$$

where q_0 is contact pressure, μ and E_s are weighted average values of Poisson's ratio and Young's modulus for compressive strata, B is the least-lateral dimension of contribution base area (convert round bases to equivalent square bases; $B = 0.5B$ for center and $B = B$ for corner I_i; $L' = 0.5L$ for center and $L' = L$ for corner I_i), I_i are influence factors depending on dimension of footings, base embedment depth, thickness of soil stratum, and Poisson's ratio (I_1 and I_2 are given in Table 31.13 and I_F is given in Figure 31.16; $M = L'/B'$ and $N = H/B'$), H is the stratum depth causing settlement (see discussion below), m is number of corners contributing to settlement ($m = 4$ at the footing center; $m = 2$ at a side; and $m = 1$ at a corner), and n equals 1.0 for flexible footings and 0.93 for rigid footings.

This equation applies to soil strata consisting of either cohesionless soils of any water content or unsaturated cohesive soils, which may be either organic or inorganic. Highly organic soils (both E_s and μ are subject to significant changes by high organic content) will be dictated by secondary or creep compression rather than immediate settlement; therefore, the applicability of the above equation is limited.

Suggestions were made by Bowles [14] to use the equations appropriately as follows: 1. Make the best estimate of base contact pressure q_0; 2. Identify the settlement point to be calculated and divide the base (as used in the Newmark stress method) so the point is at the corner or common corner of one or up to four contributing areas; 3. Determine the stratum depth causing settlement which does not approach to infinite rather at either the depth $z = 5B$ or depth to where a hard stratum is encountered (where F_s in the hard layer is about $10E_s$ of the adjacent upper layer); and 4. Calculate the weighted average E_s as follows:

$$E_{s,\text{avg}} = \sum_{n}^{1} H_i E_{si} \Big/ \sum_{n}^{1} H_i \tag{31.20}$$

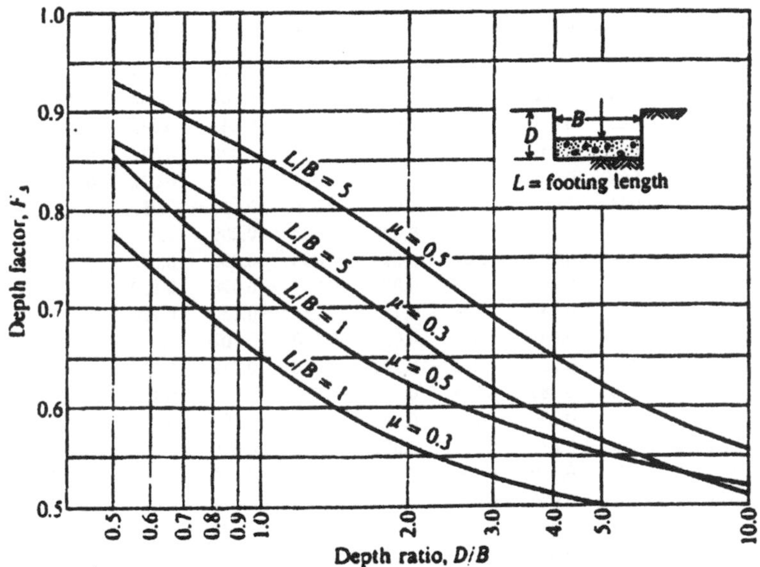

FIGURE 31.16 Influence factor I_F for footing at a depth D (use actual footing width and depth dimension for this D/B ratio). (After Bowles [14].)

31.6.2 Settlement of Shallow Foundations on Sand

SPT Method

D'Appolonio et al. [28] developed the following equation to estimate settlements of footings on sand using SPT data:

$$s = \mu_0 \mu_1 \, pB/M \tag{31.21}$$

where μ_0 and μ_1 are settlement influence factors dependent on footing geometry, depth of embedment, and depth to the relative incompressible layer (Figure 31.17), p is average applied pressure under service load and M is modulus of compressibility. The correlation between M and average SPT blow count is given in Figure 31.18.

Barker et al. [9] discussed the commonly used procedure for estimating settlement of footing on sand using SPT blow count developed by Terzaghi and Peck [64,65] and Bazaraa [10].

CPT Method

Schmertmann [56,57] developed a procedure for estimating footing settlements on sand using CPT data. This CPT method uses cone penetration resistance, q_c, as a measure of the *in situ* stiffness (compressibility) soils. Schmertmann's method is expressed as follows

$$s = C_1 C_2 \Delta p \Sigma \left(I_Z / E_s\right)_i \Delta z_i \tag{31.22}$$

$$C_1 = 1 - 0.5\left(\frac{\sigma'_{v0}}{\Delta p}\right) \geq 0.5 \tag{31.23}$$

$$C_2 = 1 + 0.2 \log\left(t_{yr}/0.1\right) \tag{31.24}$$

TABLE 31.13 Values of I_2 and I_2 to Compute Influence Factors as Used in Eq. (31.21)

N	*M* = 1.0	1.1	1.2	1.3	1.4	1.5	1.6	1.7	1.8	1.9	2.0
0.2	I_1 = 0.009	0.008	0.008	0.008	0.008	0.008	0.007	0.007	0.007	0.007	0.007
	I_2 = 0.041	0.042	0.042	0.042	0.042	0.042	0.043	0.043	0.043	0.043	0.043
0.4	0.033	0.032	0.031	0.030	0.029	0.028	0.028	0.027	0.027	0.027	0.027
	0.066	0.068	0.069	0.070	0.070	0.071	0.071	0.072	0.072	0.073	0.073
0.6	0.066	0.064	0.063	0.061	0.060	0.059	0.058	0.057	0.056	0.056	0.055
	0.079	0.081	0.083	0.085	0.087	0.088	0.089	0.090	0.091	0.091	0.092
0.8	0.104	0.102	0.100	0.098	0.096	0.095	0.093	0.092	0.091	0.090	0.089
	0.083	0.087	0.090	0.093	0.095	0.097	0.098	0.100	0.101	0.102	0.103
1.0	0.142	0.140	0.138	0.136	0.134	0.132	0.130	0.129	0.127	0.126	0.125
	0.083	0.088	0.091	0.095	0.098	0.100	0.102	0.104	0.106	0.108	0.109
1.5	0.224	0.224	0.224	0.223	0.222	0.220	0.219	0.217	0.216	0.214	0.213
	0.075	0.080	0.084	0.089	0.093	0.096	0.099	0.102	0.105	0.108	0.110
2.0	0.285	0.288	0.290	0.292	0.292	0.292	0.292	0.292	0.291	0.290	0.289
	0.064	0.069	0.074	0.078	0.083	0.086	0.090	0.094	0.097	0.100	0.102
3.0	0.363	0.372	0.379	0.384	0.389	0.393	0.396	0.398	0.400	0.401	0.402
	0.048	0.052	0.056	0.060	0.064	0.068	0.071	0.075	0.078	0.081	0.084
4.0	0.408	0.421	0.431	0.440	0.448	0.455	0.460	0.465	0.469	0.473	0.476
	0.037	0.041	0.044	0.048	0.051	0.054	0.057	0.060	0.063	0.066	0.069
5.0	0.437	0.452	0.465	0.477	0.487	0.496	0.503	0.510	0.516	0.522	0.526
	0.031	0.034	0.036	0.039	0.042	0.045	0.048	0.050	0.053	0.055	0.058
6.0	0.457	0.474	0.489	0.502	0.514	0.524	0.534	0.542	0.550	0.557	0.563
	0.026	0.028	0.031	0.033	0.036	0.038	0.040	0.043	0.045	0.047	0.050
7.0	0.471	0.490	0.506	0.520	0.533	0.545	0.556	0.566	0.575	0.583	0.590
	0.022	0.024	0.027	0.029	0.031	0.033	0.035	0.037	0.039	0.041	0.043
8.0	0.482	0.502	0.519	0.534	0.549	0.561	0.573	0.584	0.594	0.602	0.611
	0.020	0.022	0.023	0.025	0.027	0.029	0.031	0.033	0.035	0.036	0.038
9.0	0.491	0.511	0.529	0.545	0.560	0.574	0.587	0.598	0.609	0.618	0.627
	0.017	0.019	0.021	0.023	0.024	0.026	0.028	0.029	0.031	0.033	0.034
10.0	0.498	0.519	0.537	0.554	0.570	0.584	0.597	0.610	0.621	0.631	0.641
	0.016	0.017	0.019	0.020	0.022	0.023	0.025	0.027	0.028	0.030	0.031
20.0	0.529	0.553	0.575	0.595	0.614	0.631	0.647	0.662	0.677	0.690	0.702
	0.008	0.009	0.010	0.010	0.011	0.012	0.013	0.013	0.014	0.015	0.016
500	0.560	0.587	0.612	0.635	0.656	0.677	0.696	0.714	0.731	0.748	0.763
	0.000	0.000	0.000	0.000	0.000	0.000	0.001	0.001	0.001	0.001	0.001
0.2	I_1 = 0.007	0.006	0.006	0.006	0.006	0.006	0.006	0.006	0.006	0.006	0.006
	I_2 = 0.043	0.044	0.044	0.044	0.044	0.044	0.044	0.044	0.044	0.044	0.044
0.4	0.026	0.024	0.024	0.024	0.024	0.024	0.024	0.024	0.024	0.024	0.024
	0.074	0.075	0.075	0.075	0.076	0.076	0.076	0.076	0.076	0.076	0.076
0.6	0.053	0.051	0.050	0.050	0.050	0.049	0.049	0.049	0.049	0.049	0.049
	0.094	0.097	0.097	0.098	0.098	0.098	0.098	0.098	0.098	0.098	0.098
0.8	0.086	0.082	0.081	0.080	0.080	0.080	0.079	0.079	0.079	0.079	0.079
	0.107	0.111	0.112	0.113	0.113	0.113	0.113	0.114	0.114	0.014	0.014
1.0	0.121	0.115	0.113	0.112	0.112	0.112	0.111	0.111	0.110	0.110	0.110
	0.114	0.120	0.122	0.123	0.123	0.124	0.124	0.124	0.125	0.125	0.125
1.5	0.207	0.197	0.194	0.192	0.191	0.190	0.190	0.189	0.188	0.188	0.188
	0.118	0.130	0.134	0.136	0.137	0.138	0.138	0.139	0.140	0.140	0.140
2.0	0.284	0.271	0.267	0.264	0.262	0.261	0.260	0.259	0.257	0.256	0.256
	0.114	0.131	0.136	0.139	0.141	0.143	0.144	0.145	0.147	0.147	0.148
3.0	0.402	0.392	0.386	0.382	0.378	0.376	0.374	0.373	0.368	0.367	0.367
	0.097	0.122	0.131	0.137	0.141	0.144	0.145	0.147	0.152	0.153	0.154
4.0	0.484	0.484	0.479	0.474	0.470	0.466	0.464	0.462	0.453	0.451	0.451
	0.082	0.110	0.121	0.129	0.135	0.139	0.142	0.145	0.154	0.155	0.156
5.0	0.553	0.554	0.552	0.548	0.543	0.540	0.536	0.534	0.522	0.519	0.519
	0.070	0.098	0.111	0.120	0.128	0.133	0.137	0.140	0.154	0.156	0.157
6.0	0.585	0.609	0.610	0.608	0.604	0.601	0.598	0.595	0.579	0.576	0.575
	0.060	0.087	0.101	0.111	0.120	0.126	0.131	0.135	0.153	0.157	0.157

TABLE 31.13 (continued) Values of I_2 and I_2 to Compute Influence Factors as Used in Eq. (31.21)

N	M = 1.0	1.1	1.2	1.3	1.4	1.5	1.6	1.7	1.8	1.9	2.0
7.0	0.618	0.653	0.658	0.658	0.656	0.653	0.650	0.647	0.628	0.624	0.623
	0.053	0.078	0.092	0.103	0.112	0.119	0.125	0.129	0.152	0.157	0.158
8.0	0.643	0.688	0.697	0.700	0.700	0.698	0.695	0.692	0.672	0.666	0.665
	0.047	0.071	0.084	0.095	0.104	0.112	0.118	0.124	0.151	0.156	0.158
9.0	0.663	0.716	0.730	0.736	0.737	0.736	0.735	0.732	0.710	0.704	0.702
	0.042	0.064	0.077	0.088	0.097	0.105	0.112	0.118	0.149	0.156	0.158
10.0	0.679	0.740	0.758	0.766	0.770	0.770	0.770	0.768	0.745	0.738	0.735
	0.038	0.059	0.071	0.082	0.091	0.099	0.106	0.122	0.147	0.156	0.158
20.0	0.756	0.856	0.896	0.925	0.945	0.959	0.969	0.977	0.982	0.965	0.957
	0.020	0.031	0.039	0.046	0.053	0.059	0.065	0.071	0.124	0.148	0.156
500.0	0.832	0.977	1.046	1.102	1.150	1.191	1.227	1.259	2.532	1.721	1.879
	0.001	0.001	0.002	0.002	0.002	0.002	0.003	0.003	0.008	0.016	0.031

Source: Bowles, J.E., *Foundation Analysis and Design*, 5th ed., McGraw-Hill, New York, 1996. With permission.

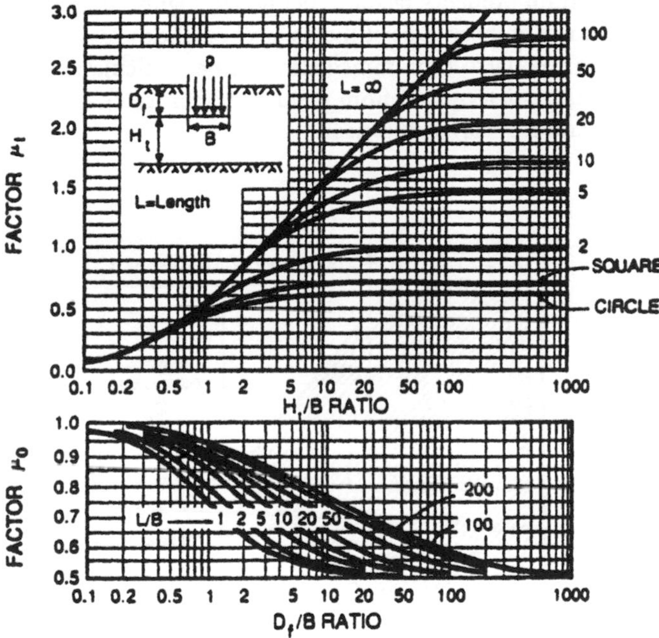

FIGURE 31.17 Settlement influence factors μ_0 and μ_1 for the D'Appolonia et al. procedure. (After D'Appolonia et al [28].)

$$E_s = \begin{cases} 2.5q_c & \text{for square footings (axisymmetric conditions)} \\ 3.5q_c & \text{for continuous footings with } L/B \geq 10 \text{ (plan strain conditions)} \\ \left[2.5+(L/B-1)/9\right]q_c & \text{for footings with } 1 \geq L/B \geq 10 \end{cases} \quad (31.25)$$

where $\Delta p = \sigma'_{vf} - \sigma'_{v0}$ is net load pressure at foundation level, σ'_{v0} is initial effective *in situ* overburden stress at the bottom of footings, σ'_{vf} is final effective *in situ* overburden stress at the

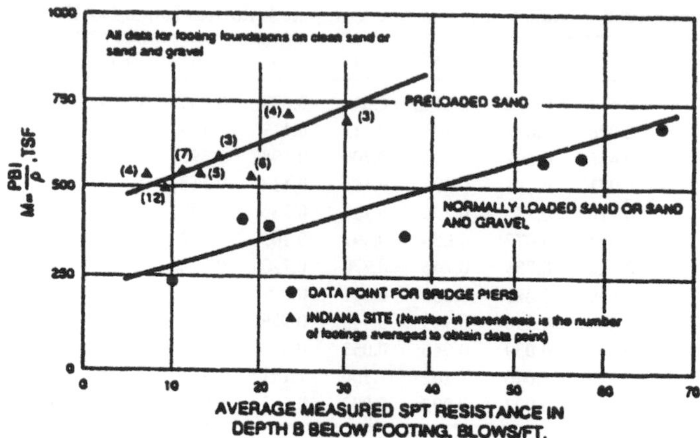

FIGURE 31.18 Correlation between modulus of compressibility and average value SPT blow count. (After D'Appolonia et al [28].)

bottom of footings, I_z is strain influence factor as defined in Figure 31.19 and Table 31.14, E_s is the appropriate Young's modulus at the middle of the *i*th layer of thickness Δz_1, C_1 is pressure correction factor, C_2 is time rate factor (equal to 1 for immediate settlement calculation or if the lateral pressure is less than the creep pressure determined from pressure-meter tests), q_c is cone penetration resistance, in pressure units, and Δz is layer thickness.

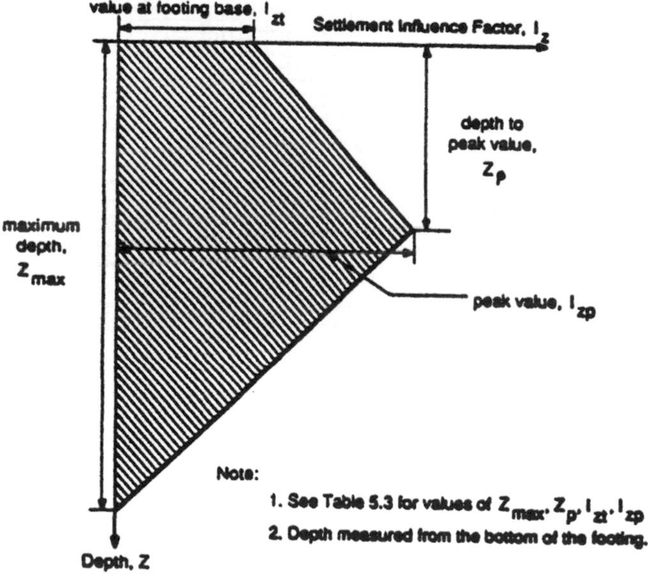

FIGURE 31.19 Variation of Schmertmann's improved settlement influence factors with depth. (After Schmertmann et al [58].)

Recent studies by Tan and Duncan [62] have compared measured settlements with settlements predicted using various procedures for footings on sand. These studies conclude that methods predicting settlements close to the average of measured settlement are likely to underestimate

TABLE 31.14 Coefficients to Define the Dimensions of Schmertmann's Improved Settlement Influence Factor Diagram in Figure 31.20

				Peak Value of Stress Influence Factor I_{zp}			
L/B	Max. Depth of Influence z_{max}/B	Depth to Peak Value z_p/B	Value of I_z at Top I_{zt}	$\frac{\Delta p}{\sigma'_{vp}}=1$	$\frac{\Delta p}{\sigma'_{vp}}=2$	$\frac{\Delta p}{\sigma'_{vp}}=4$	$\frac{\Delta p}{\sigma'_{vp}}=10$
1	2.00	0.50	0.10	0.60	0.64	0.70	0.82
2	2.20	0.55	0.11	0.60	0.64	0.70	0.82
4	2.65	0.65	0.13	0.60	0.64	0.70	0.82
8	3.55	0.90	0.18	0.60	0.64	0.70	0.82
≥10	4.00	1.00	0.20	0.60	0.64	0.70	0.82

Note: σ'_{vp} is the initial vertical pressure at depth of peak influence.

After Schmertmann et al. [57].

settlements half the time and to overestimate them half the time. The conservative methods (notably Terzaghi and Peck's) tend to overestimate settlements more than half the time and to underestimate them less often. On the other hand, there is a trade-off between accuracy and reliability. A relatively accurate method such as the D'Appolonia et al. method calculates settlements that are about equal to the average value of actual settlements, but it underestimates settlements half the time (a reliability of 50%). To ensure that the calculated settlements equal or exceed the measured settlements about 90% of the time (a reliability of 90%), an adjustment factor of two should be applied to the settlements predicted by the D'Appolonia et al. method. Table 31.15 shows values of the adjustment factor for 50 and 90% reliability in settlement predicted using Terzaghi and Peck, D'Appolonia et al., and Schmertmann methods.

TABLE 31.15 Value of Adjustment Factor for 50 and 90% Reliability in Displacement Estimates

		Adjustment Factor	
Method	Soil Type	For 50% Reliability	For 90% Reliability
Terzaghi and Peck [65]	Sand	0.45	1.05
Schmertmann	Sand	0.60	1.25
D'Appolonia et al. [28]	Sand	1.00	2.00

TABLE 31.16 Some Empirical Equations for C_c and C_α

Compression Index	Source	Comment
$C_c = 0.009(LL - 10)$	Terzaghi and Peck [65]	$S_t \leq 5$, $LL < 100$
$C_c = 0.2343e_0$	Nagaraj and Murthy [51]	
$C_c = 0.5G_s(PI/100)$	Worth and Wood [71]	Modified cam clay model
$C_c = 0PI/74$	EPRI (1990)	
$C_c = 0.37(e_0 + 0.003w_L + 0.0004w_N - 0.34)$	Azzouz et al. [7]	Statistical analysis
Recompression Index	**Source**	
$C_r = 0.0463w_LG_s$	Nagaraj and Murthy [50]	

31.6.3 Settlement of Shallow Foundations on Clay

Immediate Settlement

Immediate settlement of shallow foundations on clay can be estimated using the approach described in Section 31.6.1.

Consolidation Settlement

Consolidation settlement is time dependent and may be estimated using one-dimensional consolidation theory [43,53,66]. The consolidation settlement can be calculated as follows

$$s_c \begin{cases} \dfrac{H_c}{1+e_0}\left[C_r \log\left(\dfrac{\sigma'_p}{\sigma''_{vo}}\right) + C_c \log\left(\dfrac{\sigma'_{vf}}{\sigma'_p}\right)\right] & \left(\text{for OC soils, i.e., } \sigma'_p > \sigma'_{v0}\right) \\ \dfrac{H_c}{1+e_0} C_c \log\left(\dfrac{\sigma'_{vf}}{\sigma'_p}\right) & \left(\text{for NC soils, i.e., } \sigma'_p = \sigma'_{v0}\right) \end{cases} \tag{31.26}$$

where H_c is height of compressible layer, e_0 is void ratio at initial vertical effective stress, C_r is recompression index (see Table 31.16), C_c is compression index (see Table 31.16), σ'_p is maximum past vertical effective stress, σ'_{v0} is initial vertical effective stress, σ'_{vf} is final vertical effective stress. Highly compressible cohesive soils are rarely chosen to place footings for bridges where tolerable amount of settlement is relatively small. Preloading or surcharging to produce more rapid consolidation has been extensively used for foundations on compressible soils [54]. Alternative foundation systems would be appropriate if large consolidation settlement is expected to occur.

TABLE 31.17 Secondary Compression Index

C_α/C_c	Material
0.02 ± 0.01	Granular soils including rockfill
0.03 ± 0.01	Shale and mudstone
0.04 ± 0.01	Inorganic clays and silts
0.05 ± 0.01	Organic clays and silts
0.06 ± 0.01	Peat and muskeg

Source: Terzaghi, I. et al., *Soil Mechanics in Engineering Practice*, 3rd ed., John Wiley & Sons, New York, 1996. With permission.

Secondary Settlement

Settlements of footings on cohesive soils continuing beyond primary consolidation are called secondary settlement. Secondary settlement develops at a slow and continually decreasing rate and may be estimated as follows:

$$s_s = C_\alpha H_t \log \frac{t_{sc}}{t_p} \tag{31.27}$$

where C_α is coefficient of secondary settlement (Table 31.17), H_t is total thickness of layers undergoing secondary settlement, t_{sc} is time for which secondary settlement is calculated (in years), and t_p is time for primary settlement (>1 year).

31.6.4 Tolerable Settlement

Tolerable movement criteria for foundation settlement should be established consistent with the function and type of structure, anticipated service life, and consequences of unacceptable movements on structure performance as outlined by AASHTO [3]. The criteria adopted by AASHTO considering the angular distortion (δ/l) between adjacent footings is as follows:

$$\frac{\delta}{l} \leq \begin{cases} 0.008 & \text{for simple-span bridge} \\ 0.004 & \text{for continuous-span bridge} \end{cases} \tag{31.28}$$

where δ is differential settlement of adjacent footings and l is center–center spacing between adjacent footings. These (δ/l) limits are not applicable to rigid frame structures, which shall be designed for anticipated differential settlement using special analysis.

31.7 Shallow Foundations on Rock

Wyllie [72] outlines the following examinations which are necessary for designing shallow foundations on rock:

1. The bearing capacity of the rock to ensure that there will be no crushing or creep of material within the loaded zone;
2. Settlement of the foundation which will result from elastic strain of the rock, and possibly inelastic compression of weak seams within the volume of rock compressed by the applied load;
3. Sliding and shear failure of blocks of rock formed by intersecting fractures within the foundation.

This condition usually occurs where the foundation is located on a steep slope and the orientation of the fractures is such that the blocks can slide out of the free face.

31.7.1 Bearing Capacity According to Building Codes

It is common to use allowable bearing capacity for various rock types listed in building codes for footing design. As provided in Table 31.18, the bearing capacities have been developed based on rock strength from case histories and include a substantial factor of safety to minimize settlement.

31.7.2 Bearing Capacity of Fractured Rock

Various empirical procedures for estimating allowable bearing capacity of foundations on fractured rock are available in the literature. Peck et al. [53] suggested an empirical procedure for estimating allowable bearing pressures of foundations on jointed rock based on the RQD index. The predicted bearing capacities by this method shall be used with the assumption that the foundation settlement does not exceed 12.7 mm (0.5 in.) [53]. Carter and Kulhawy [25] proposed an empirical approach for estimating ultimate bearing capacity of fractured rock. Their method is based on the unconfined compressive strength of the intact rock core sample and rock mass quality.

Wyllie [72] detailed an analytical procedure for computing bearing capacity of fractured rock mass using Hoek–Brown strength criterion. Details of rational methods for the topic can also be found in Kulhawy and Goodman [42] and Goodman [32].

TABLE 31.18 Presumptive Bearing Pressures (tsf) for Foundations on Rock after Putnam, 1981

Code	Year[1]	Bedrock[2]	Sound Foliated Rock	Sound Sedimentary Rock	Soft Rock[3]	Soft Shale	Broken Shale
Baltimore	1962	100	35		10		
BOCA	1970	100	40	25	10	4	(4)
Boston	1970	100	50	10	10		1.5
Chicago	1970	100	100				(4)
Cleveland	1951/1969			25			
Dallas	1968	$0.2q_u$	$2q_u$	0.2qu	$0.2q_u$	$0.2q_u$	$0.2q_u$
Detroit	1956	100	100	9600	12	12	
Indiana	1967	$0.2q_u$	$2q_u$	0.2qu	$0.2q_u$	$0.2q_u$	$0.2q_u$
Kansas	1961/1969	$0.2q_u$	$2q_u$	0.2qu	$0.2q_u$	$0.2q_u$	$0.2q_u$
Los Angeles	1970	10	4	3	1	1	1
New York City	1970	60	60	60	8		
New York State		100	40	15			
Ohio	1970	100	40	15	10	4	
Philadelphia	1969	50	15	10–15	8		
Pittsburgh	1959/1969	25	25	25	8	8	
Richmond	1968	100	40	25	10	4	1.5
St. Louis	1960/1970	100	40	25	10	1.5	1.5
San Francisco	1969	3–5	3–5	3–5			
UBC	1970	$0.2q_u$	$2q_u$	$0.2q_u$	$0.2q_u$	$0.2q_u$	$0.2q_u$
NBC Canada	1970			100			
New South Wales, Australia	1974		33	13	4.5		

Notes:
1. Year of code or original year and date of revision.
2. Massive crystalline bedrock.
3. Soft and broken rock, not including shale.
4. Allowable bearing pressure to be determined by appropriate city official.
5. q_u = unconfined compressive strength.

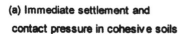

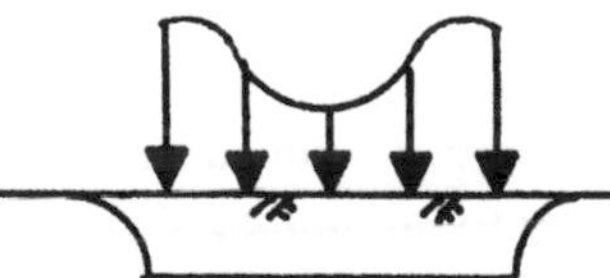

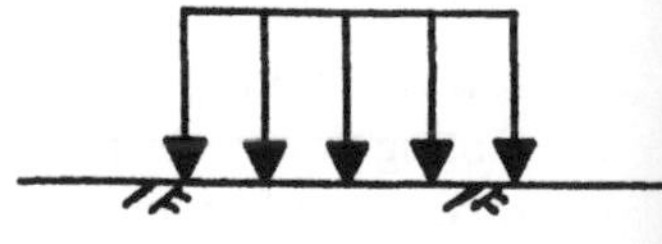

FIGURE 31.20 Contact pressure distribution for a rigid footing. (a) On cohesionless soils; (b) on cohesive soils; (c) usual assumed linear distribution.

31.7.3 Settlements of Foundations on Rock

Wyllie [72] summarizes settlements of foundations on rock as following three different types: 1. Elastic settlements result from a combination of strain of the intact rock, slight closure and movement of fractures and compression of any minor clay seams (less than a few millimeters). Elastic theory can be used to calculate this type of settlement. Detailed information can be found in Wyllie [72], Kulhawy, and AASHTO [3]. 2. Settlements result from the movement of blocks of rock due to shearing of fracture surfaces. This occurs when foundations are sitting at the top of a steep slope and unstable blocks of rocks are formed in the face. The stability of foundations on rock is influenced by the geologic characterization of rock blocks. The information required on structural geology consists of the orientation, length and spacing of fractures, and their surface and infilling materials. Procedures have been developed for identifying and analyzing the stability of sliding blocks [72], stability of wedge blocks [36], stability of toppling blocks [33], or three-dimensional stability of rock blocks [34]. 3. Time-dependent settlement occurs when foundations found on rock mass that consists of substantial seams of clay or other compressible materials. This type of settlement can be estimated using the procedures described in Section 31.6.3. Also time-dependent settlement can occur if foundations found on ductile rocks, such as salt where strains develop continuously at any stress level, or on brittle rocks when the applied stress exceeds the yield stress.

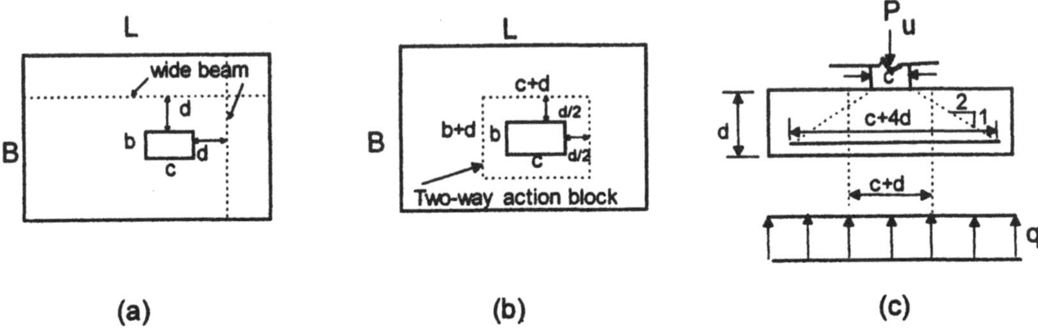

FIGURE 31.21 (a) Section for wide-beam shear; (b) section for diagonal-tension shear; (c) method of computing area for allowable column bearing stress.

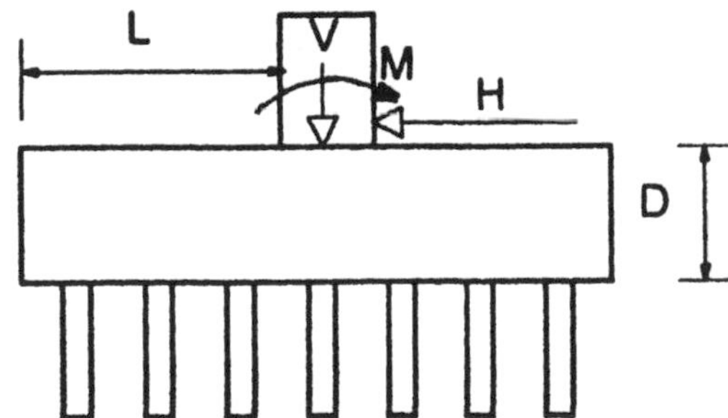

FIGURE 31.22 Illustration of the length-to-thickness ratio of cantilever of a footing or pile cap.

31.8 Structural Design of Spread Footings

The plan dimensions (*B* and *L*) of a spread footing are controlled by the allowable soil pressure beneath the footing. The pressure distribution beneath footings is influenced by the interaction of the footing rigidity with the soil type, stress–state, and time response to stress as shown in Figure 31.20 (a) (b). However, it is common practice to use the linear pressure distribution beneath rigid footings as shown in Figure 31.20 (c). The depth (*D*) for spread footings is usually controlled by shear stresses. Two-way action shear always controls the depth for centrally loaded square footings. However, wide-beam shear may control the depth for rectangular footings when the *L*/*B* ratio is greater than about 1.2 and may control for other *L*/*B* ratios when there is overturning or eccentric loading (Figure 31.21a). In addition, footing depth should be designed to satisfy diagonal (punching) shear requirement (Figure 31.21b). Recent studies by Duan and McBride [30] indicate that when the length-to-thickness ratio of cantilever (*L*/*D* as defined in Figure 31.22) of a footing (or pile-cap) is greater than 2.2, a nonlinear distribution of reaction should be used for footing or pile-cap design. The specifications and procedures for footing design can be found in AASHTO [2], ACI [4], or Bowles [12, 13].

Acknowledgment

I would like to take this opportunity to thank Bruce Kutter, who reviewed the early version of the chapter and provided many thoughtful suggestions. Advice and support from Prof. Kutter are greatly appreciated.

References

1. AASHTO, *LRFD Bridge Design Specifications*, American Association of State Highway and Transportation Officials, Washington, D.C., 1994.
2. AASHTO, *Standard Specifications for Highway Bridges (Interim Revisions)*, 16th ed., American Association of State Highway and Transportation Officials, Washington, D.C., 1997.
3. AASHTO, *LRFD Bridge System Design Specification (Interim Revisions)*, American Association of State Highway and Transportation Officials, Washington, D.C., 1997.
4. ACI, *Building Code Requirements for Reinforced Concrete* (ACI 318-89), American Concrete Institute, Detroit, MI, 1989, 353 pp. (with commentary).
5. ASTM, Section 4 Construction, 04.08 Soil and Rock (I): D420–D4914, American Society for Testing and Materials, Philadelphia, PA, 1997.
6. ATC-32, *Improved Seismic Design Criteria for California Bridges: Provisional Recommendations*, Applied Technology Council, Redwood City, CA, 1996.
7. Azzouz, A.S., Krizek, R.J., and Corotis, R.B., Regression of analysis of soil compressibility, *JSSMFE Soils and Foundations*, 16(2), 19–29, 1976.
8. Baguelin, F., Jezequel, J.F., and Shields, D.H., *The Pressuremeter and Foundation Engineering*, Transportation Technical Publications, Clausthal, 1978, 617 pp.
9. Barker, R.M., Duncan, J.M., Rojiani, K.B., Ooi, P.S.K., Tan, C.K., and Kim, S.G., Manuals for the Design of Bridge Foundations, National Cooperative Highway Research Program Report 343, Transportation Research Board, National Research Council, Washington, D.C., 1991.
10. Bazaraa, A.R.S.S., Use of Standard Penetration Test for Estimating Settlements of Shallow Foundations on Sands, Ph.D. dissertation, Department of Civil Engineering, University of Illinois, Urbana, 1967, 380 pp.
11. Bowles, J.E., *Analytical and Computer Methods in Foundation Engineering*, McGraw-Hill, New York, 1974.
12. Bowles, J.E., Spread footings, Chapter 15, in *Foundation Engineering Handbook*, Winterkorn, H.F. and Fang, H.Y., Eds., Van Nostrand Reinhold, New York, 1975.
13. Bowles, J.E., *Foundation Analysis and Design*, 5th ed., McGraw-Hill, New York, 1996.
14. Briaud, J.L., *The Pressuremeter*, A.A. Balkema Publishers, Brookfield, VT, 1992.
15. Briaud, J.L., Spread footing design and performance, FHWA Workshop at the Tenth Annual International Bridge Conference and Exhibition, 1993.
16. Briaud, J.L., Pressuremeter and foundation design, in *Proceedings of the Conference on Use of in situ tests in Geotechnical Engineering*, ASCE Geotechnical Publication No. 6, 74–116, 1986.
17. Briaud, J.L. and Gibben, R., Predicted and measured behavior of five spread footings on sand, Geotechnical Special Publication No. 41, ASCE Specialty Conference: Settlement 1994, ASCE, New York, 1994.
18. Buisman, A.S.K., *Grondmechanica*, Waltman, Delft, 190, 1940.
19. Burland, J.B. and Burbidge, M.C., Settlement of foundations on sand and gravel, *Proc. Inst. Civil Eng.*, Tokyo, 2, 517, 1984.
20. Burmister, D.M., The theory of stresses and displacements in layered systems and application to the design of airport runways, *Proc. Highway Res. Board*, 23, 126–148, 1943.

21. Burmister, D.M., Evaluation of pavement systems of WASHO road test layered system methods, Highway Research Board Bull. No. 177, 1958.
22. Burmister, D.M., Applications of dimensional analyses in the evaluation of asphalt pavement performances, paper presented at Fifth Paving Conference, Albuquerque, NM, 1967.
23. Canadian Geotechnical Society, *Canadian Foundation Engineering Manual*, 2nd ed., 1985, 456.
24. Carter. J.P. and Kulhawy, F.H., Analysis and Design of Drilled Shaft Foundations Socketed into Rock, Report No. EL-5918, Empire State Electric Engineering Research Corporation and Electric Power Research Institute, 1988.
25. Chen, W.F., *Limit Analysis and Soil Plasticity*, Elsevier, Amsterdam, 1975.
26. Chen, W.F. and Mccarron, W.O., Bearing capacity of shallow foundations, Chap. 4, in *Foundation Engineering Handbook*, 2nd ed., Fang, H.Y., Ed., Chapman & Hall, 1990.
27. D'Appolonia, D.J., D'Appolonia, E., and Brisette, R.F., Settlement of spread footings on sand (closure), *ASCE J. Soil Mech. Foundation Div.*, 96(SM2), 754–761, 1970.
28. De Beer, E.E., Bearing capacity and settlement of shallow foundations on sand, *Proc. Symposium on Bearing Capacity and Settlement of Foundations*, Duke University, Durham, NC, 315–355, 1965.
29. De Beer, E.E., Proefondervindelijke bijdrage tot de studie van het gransdraagvermogen van zand onder funderingen p staal, Bepaling von der vormfactor sb, *Ann. Trav. Publics Belg.*, 1967.
30. Duan, L. and McBride, S.B., The effects of cap stiffness on pile reactions, *Concrete International*, American Concrete Institue, 1995.
31. FHWA, Large-Scale Load Tests and Data Base of Spread Footings on Sand, Publication No. FHWA-RD-97-068, 1997.
32. Goodman, R.E. and Bray, J.W., Toppling of rock slopes, in *Proceedings of the Specialty Conference on Rock Engineering for Foundations and Slopes*, Vol. 2, ASCE, Boulder, CO, 1976, 201–234.
33. Goodman, R.E. and Shi, G., *Block Theory and Its Application to Rock Engineering*, Prentice-Hall, Englewood Cliffs, NJ, 1985.
34. Hansen, B.J., A Revised and Extended Formula for Bearing Capacity, Bull. No. 28, Danish Geotechnical Institute, Copenhagen, 1970, 5–11.
35. Hoek, E. and Bray, J., *Rock Slope Engineering*, 2nd ed., IMM, London, 1981.
36. Holtz, R.D., Stress distribution and settlement of shallow foundations, Chap. 5, in *Foundation Engineering Handbook*, 2nd ed., Fang, H.Y., Ed., Chapman & Hall, 1990.
37. Ismael, N.F. and Vesic, A.S., Compressibility and bearing capacity, *ASCE J. Geotech. Foundation Eng. Div.* 107(GT12), 1677–1691, 1981.
38. Kulhawy, F.H. and Mayne, P.W., Manual on Estimating Soil Properties for Foundation Design, Electric Power Research Institute, EPRI EL-6800, Project 1493-6, Final Report, August, 1990.
39. Kulhawy, F.H. and Goodman, R.E., Foundation in rock, Chap. 55, in *Ground Engineering Reference Manual*, F.G. Bell, Ed., Butterworths, 1987.
40. Lambe, T.W. and Whitman, R.V., *Soil Mechanics*, John Wiley & Sons, New York, 1969.
41. Menard, L., Regle pour le calcul de la force portante et du tassement des fondations en fonction des resultats pressionmetriques, in *Proceedings of the Sixth International Conference on Soil Mechanics and Foundation Engineering*, Vol. 2, Montreal, 1965, 295–299.
42. Meyerhof, G.G., The ultimate bearing capacity of foundations, *Geotechnique*, 2(4), 301–331, 1951.
43. Meyerhof, G.G., Penetration tests and bearing capacity of cohesionless soils, *ASCE J. Soil Mech. Foundation Div.*, 82(SM1), 1–19, 1956.
44. Meyerhof, G.G., Some recent research on the bearing capacity of foundations, *Can. Geotech. J.*, 1(1), 16–36, 1963.
45. Meyerhof, G.G., Shallow foundations, *ASCE J. Soil Mech. and Foundations Div.*, 91, No. SM2, 21–31, 1965.
46. Milovic, D.M., Comparison between the calculated and experimental values of the ultimate bearing capacity, in *Proceedings of the Sixth International Conference on Soil Mechanics and Foundation Engineering*, Vol. 2, Montreal, 142–144, 1965.

47. Nagaraj, T.S. and Srinivasa Murthy, B.R., Prediction of preconsolidation pressure and recompression index of soils, *ASTMA Geotech. Testing J.*, 8(4), 199–202, 1985.
48. Nagaraj. T.S. and Srinivasa Murthy, B.R., A critical reappraisal of compression index, *Geotechnique*, 36(1), 27–32, 1986.
49. NAVFAC, Design Manual 7.02, *Foundations & Earth Structures*, Naval Facilities Engineering Command, Department of the Navy, Washington, D.C., 1986.
50. NAVFAC, Design Manual 7.01, Soil Mechanics, Naval Facilities Engineering Command, Department of the Navy, Washington, D.C., 1986.
51. Peck, R.B., Hanson, W.E., and Thornburn, T.H., *Foundation Engineering*, 2nd ed., John Wiley & Sons, New York, 1974.
52. Perloff, W.H., Pressure distribution and settlement, Chap. 4, in *Foundation Engineering Handbook*, 2nd ed., Fang, H.Y., Ed., Chapman & Hall, 1975.
53. Poulos, H.G. and Davis, E.H., *Elastic Solutions for Soil and Rock Mechanics*, John Wiley & Sons, New York, 1974.
54. Schmertmann, J.H., Static cone to compute static settlement over sand, *ASCE J. Soil Mech. Foundation Div.*, 96(SM3), 1011–1043, 1970.
55. Schmertmann, J.H., Guidelines for cone penetration test performance, and design, Federal Highway Administration, Report FHWA-TS-78-209, 1978.
56. Schmertmann, J.H., Dilatometer to Computer Foundation Settlement, *Proc.* In Situ *'86, Specialty Conference on the Use of* In Situ *Tests and Geotechnical Engineering*, ASCE, New York, 303–321, 1986.
57. Schmertmann, J.H., Hartman, J.P., and Brown, P.R., Improved strain influence factor diagrams, *ASCE J. Geotech. Eng. Div.*, 104(GT8), 1131–1135, 1978.
58. Schultze, E. and Sherif, G. Prediction of settlements from evaluated settlement observations on sand, *Proc. 8th Int. Conference on Soil Mechanics and Foundation Engineering*, Moscow, 225–230, 1973.
59. Scott, R.F., *Foundation Analysis*, Prentice-Hall, Englewood Cliffs, NJ, 1981.
60. Sowers, G.F., and Vesic, A.B., Vertical stresses in subgrades beneath statically loaded flexible pavements, Highway Research Board Bulletin, No. 342, 1962.
61. Steinbrenner, W., *Tafeln zur Setzungberechnung*, Die Strasse, 1943, 121–124.
62. Tan, C.K. and Duncan, J.M., Settlement of footings on sand—accuracy and reliability, in *Proceedings of Geotechnical Congress*, Boulder, CO, 1991.
63. Terzaghi, J., *Theoretical Soil Mechanics*, John Wiley & Sons, New York, 1943.
64. Terzaghi, K. and Peck, R.B., *Soil Mechanics in Engineering Practice*, John Wiley & Sons, New York, 1948.
65. Terzaghi, K. and Peck, R.B., *Soil Mechanics in Engineering Practice*, 2nd ed., John Wiley & Sons, New York, 1967.
66. Terzaghi, K., Peck, R.B., and Mesri, G., *Soil Mechanics in Engineering Practice*, 3rd ed., John Wiley & Sons, New York, 1996.
67. Vesic, A.S., Bearing capacity of deep foundations in sand, National Academy of Sciences, National Research Council, Highway Research Record, 39, 112–153, 1963.
68. Vesic, A.S., Analysis of ultimate loads of shallow foundations, *ASCE J. Soil Mech. Foundation Eng. Div.*, 99(SM1), 45–73, 1973.
69. Vesic, A.S., Bearing capacity of shallow foundations, Chap. 3, in *Foundation Engineering Handbook*, Winterkorn, H.F. and Fang, H.Y., Ed., Van Nostrand Reinhold, New York, 1975.
70. Westergaard, H.M., A problem of elasticity suggested by a problem in soil mechanics: soft material reinforced by numerous strong horizontal sheets, in *Contributions to the Mechanics of Solids*, Stephen Timoshenko Sixtieth Anniversary Volume, Macmillan, New York, 1938.
71. Wroth, C.P. and Wood, D.M., The correlation of index properties with some basic engineering properties of soils, *Can. Geotech. J.*, 15(2), 137–145, 1978.
72. Wyllie, D.C., *Foundations on Rock*, E & FN SPON, 1992.

32

Deep Foundations

Youzhi Ma
Geomatrix Consultants, Inc.

Nan Deng
Bechtel Corporation

32.1 Introduction

A bridge foundation is part of the bridge substructure connecting the bridge to the ground. A foundation consists of man-made structural elements that are constructed either on top of or within existing geologic materials. The function of a foundation is to provide support for the bridge and to transfer loads or energy between the bridge structure and the ground.

A deep foundation is a type of foundation where the embedment is larger than its maximum plane dimension. The foundation is designed to be supported on deeper geologic materials because either the soil or rock near the ground surface is not competent enough to take the design loads or it is more economical to do so.

0-8493-7434-0/00/$0.00+$.50

The merit of a deep foundation over a shallow foundation is manifold. By involving deeper geologic materials, a deep foundation occupies a relatively smaller area of the ground surface. Deep foundations can usually take larger loads than shallow foundations that occupy the same area of the ground surface. Deep foundations can reach deeper competent layers of bearing soil or rock, whereas shallow foundations cannot. Deep foundations can also take large uplift and lateral loads, whereas shallow foundations usually cannot.

The purpose of this chapter is to give a brief but comprehensive review to the design procedure of deep foundations for structural engineers and other bridge design engineers. Considerations of selection of foundation types and various design issues are first discussed. Typical procedures to calculate the axial and lateral capacities of an individual pile are then presented. Typical procedures to analyze pile groups are also discussed. A brief discussion regarding seismic design is also presented for its uniqueness and importance in the foundation design.

32.2 Classification and Selection

32.2.1 Typical Foundations

Typical foundations are shown on Figure 32.1 and are listed as follows:

A *pile* usually represents a slender structural element that is driven into the ground. However, a pile is often used as a generic term to represent all types of deep foundations, including a (driven) pile, (drilled) shaft, caisson, or an anchor. A *pile group* is used to represent various grouped deep foundations.

A *shaft* is a type of foundation that is constructed with cast-in-place concrete after a hole is first drilled or excavated. A *rock socket* is a shaft foundation installed in rock. A shaft foundation also is called a *drilled pier* foundation.

A *caisson* is a type of large foundation that is constructed by lowering preconstructed foundation elements through excavation of soil or rock at the bottom of the foundation. The bottom of the caisson is usually sealed with concrete after the construction is completed.

An *anchor* is a type of foundation designed to take tensile loading. An anchor is a slender, small-diameter element consisting of a reinforcement bar that is fixed in a drilled hole by grout concrete. Multistrain high-strength cables are often used as reinforcement for large-capacity anchors. An *anchor for suspension bridge* is, however, a foundation that sustains the pulling loads located at the ends of a bridge; the foundation can be a deadman, a massive tunnel, or a composite foundation system including normal anchors, piles, and drilled shafts.

A *spread footing* is a type of foundation that the embedment is usually less than its smallest width. Normal spread footing foundation is discussed in detail in Chapter 31.

32.2.2 Typical Bridge Foundations

Bridge foundations can be individual, grouped, or combination foundations. Individual bridge foundations usually include individual footings, large-diameter drilled shafts, caissons, rock sockets, and deadman foundations. Grouped foundations include groups of caissons, driven piles, drilled shafts, and rock sockets. Combination foundations include caisson with driven piles, caisson with drilled shafts, large-diameter pipe piles with rock socket, spread footings with anchors, deadman with piles and anchors, etc.

For small bridges, small-scale foundations such as individual footings or drilled shaft foundations, or a small group of driven piles may be sufficient. For larger bridges, large-diameter shaft foundations, grouped foundations, caissons, or combination foundations may be required. Caissons, large-diameter steel pipe pile foundations, or other types of foundations constructed by using the cofferdam method may be necessary for foundations constructed over water.

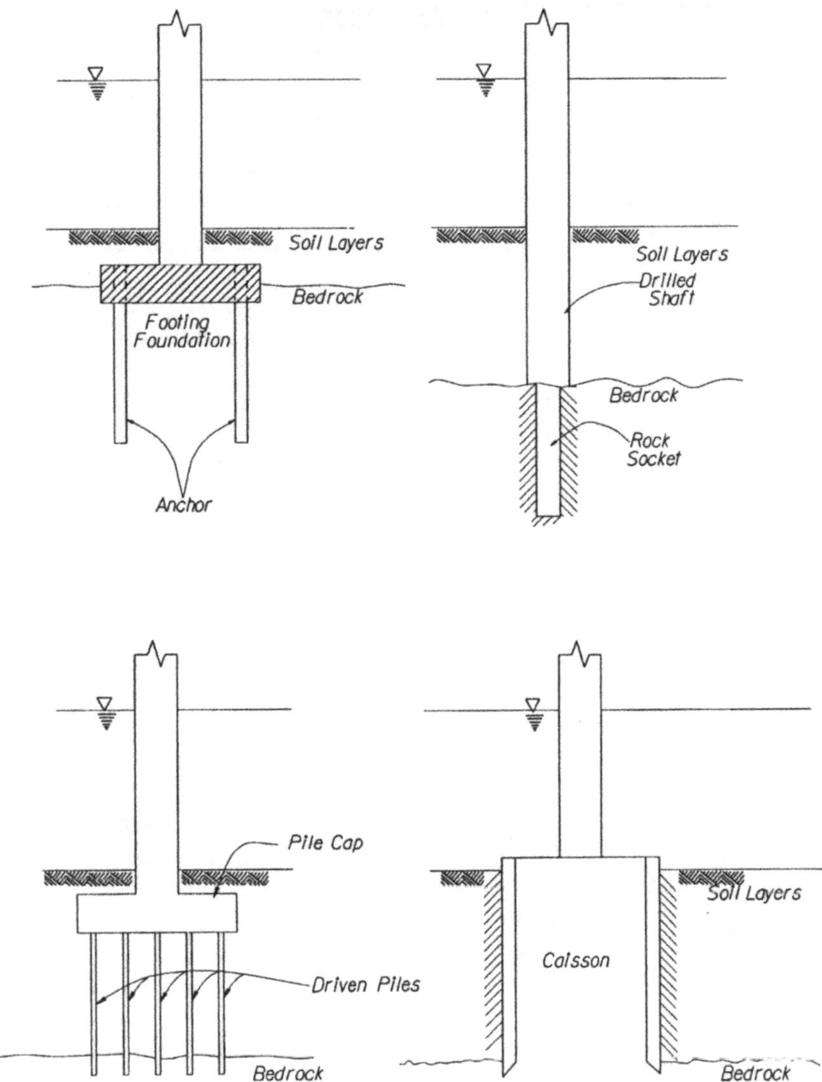

FIGURE 32.1 Typical foundations.

Bridge foundations are often constructed in difficult ground conditions such as landslide areas, liquefiable soil, collapsible soil, soft and highly compressible soil, swelling soil, coral deposits, and underground caves. Special foundation types and designs may be needed under these circumstances.

32.2.3 Classification

Deep foundations are of many different types and are classified according to different aspects of a foundation as listed below:

Geologic conditions— Geologic materials surrounding the foundations can be soil and rock. Soil can be fine grained or coarse grained; from soft to stiff and hard for fine-grained soil, or from loose to dense and very dense for coarse-grained soil. Rock can be sedimentary, igneous, or metamorphic; and from very soft to medium strong and hard. Soil and rock mass may possess predefined weaknesses and

TABLE 32.1 Range of Maximum Capacity of Individual Deep Foundations

Type of Foundation	Size of Cross Section	Maximum Compressive Working Capacity
Driven concrete piles	Up to 45 cm	100 to 250 tons (900 to 2200 kN)
Driven steel pipe piles	Up to 45 cm	50 to 250 tons (450 to 2200 kN)
Driven steel H-piles	Up to 45 cm	50 to 250 tons (450 to 2200 kN)
Drilled shafts	Up to 60 cm	Up to 400 tons (3500 kN)
Large steel pipe piles, concrete-filled; large-diameter drilled shafts; rock rocket	0.6 to 3 m	300 to 5,000 tons or more (2700 to 45000 kN)

discontinuities, such as rock joints, beddings, sliding planes, and faults. Water conditions can be different, including over river, lake, bay, ocean, or land with groundwater. Ice or wave action may be of concern in some regions.

Installation methods—Installation methods can be piles (driven, cast-in-place, vibrated, torqued, and jacked); shafts (excavated, drilled and cast-in-drilled-hole); anchor (drilled); caissons (Chicago, shored, benoto, open, pneumatic, floating, closed-box, Potomac, etc.); cofferdams (sheet pile, sand or gravel island, slurry wall, deep mixing wall, etc.); or combined.

Structural materials— Materials for foundations can be timber, precast concrete, cast-in-place concrete, compacted dry concrete, grouted concrete, post-tension steel, H-beam steel, steel pipe, composite, etc.

Ground effect— Depending on disturbance to the surrounding ground, piles can be displacement piles, low displacement, or nondisplacement piles. Driven precast concrete piles and steel pipes with end plugs are displacement piles; H-beam and unplugged steel pipes are low-displacement piles; and drilled shafts are nondisplacement piles.

Function— Depending on the portion of load carried by the side, toe, or a combination of the side and toe, piles are classified as frictional, end bearing, and combination piles, respectively.

Embedment and relative rigidity— Piles can be divided into long piles and short piles. A long pile, simply called a pile, is embedded deep enough that fixity at its bottom is established, and the pile is treated as a slender and flexible element. A short pile is a relatively rigid element that the bottom of the pile moves significantly. A caisson is often a short pile because of its large cross section and stiffness. An extreme case for short piles is a spread-footing foundation.

Cross section— The cross section of a pile can be square, rectangular, circular, hexagonal, octagonal, H-section; either hollow or solid. A pile cap is usually square, rectangular, circular, or bell-shaped. Piles can have different cross sections at different depths, such as uniform, uniform taper, step-taper, or enlarged end (either grouted or excavated).

Size— Depending on the diameter of a pile, piles are classified as pin piles and anchors (100 to 300 mm), normal-size piles and shafts (250 to 600 mm), large-diameter piles and shafts (600 to 3000 mm), caissons (600 mm and up to 3000 mm or larger), and cofferdams or other shoring construction method (very large).

Loading— Loads applied to foundations are compression, tension, moment, and lateral loads. Depending on time characteristics, loads are further classified as static, cyclic, and transient loads. The magnitude and type of loading also are major factors in determining the size and type of a foundation (Table 32.1).

Isolation— Piles can be isolated at a certain depth to avoid loading utility lines or other construction, or to avoid being loaded by them.

Inclination— Piles can be vertical or inclined. Inclined piles are often called battered or raked piles.

Multiple Piles— Foundation can be an individual pile, or a pile group. Within a pile group, piles can be of uniform or different sizes and types. The connection between the piles and the pile cap can be fixed, pinned, or restrained.

32.2.4 Advantages/Disadvantages of Different Types of Foundations

Different types of foundations have their unique features and are more applicable to certain conditions than others. The advantages and disadvantages for different types of foundations are listed as follows.

Driven Precast Concrete Pile Foundations

Driven concrete pile foundations are applicable under most ground conditions. Concrete piles are usually inexpensive compared with other types of deep foundations. The procedure of pile installation is straightforward; piles can be produced in mass production either on site or in a manufacture factory, and the cost for materials is usually much less than steel piles. Proxy coating can be applied to reduce negative skin friction along the pile. Pile driving can densify loose sand and reduce liquefaction potential within a range of up to three diameters surrounding the pile.

However, driven concrete piles are not suitable if boulders exist below the ground surface where piles may break easily and pile penetration may be terminated prematurely. Piles in dense sand, dense gravel, or bedrock usually have limited penetration; consequently, the uplift capacity of this type of piles is very small.

Pile driving produces noise pollution and causes disturbance to the adjacent structures. Driving of concrete piles also requires large overhead space. Piles may break during driving and impose a safety hazard. Piles that break underground cannot take their design loads, and will cause damage to the structures if the broken pile is not detected and replaced. Piles could often be driven out of their designed alignment and inclination and, as a result, additional piles may be needed. A special hardened steel shoe is often required to prevent pile tips from being smashed when encountering hard rock. End-bearing capacity of a pile is not reliable if the end of a pile is smashed.

Driven piles may not be a good option when subsurface conditions are unclear or vary considerably over the site. Splicing and cutting of piles are necessary when the estimated length is different from the manufactured length. Splicing is usually difficult and time-consuming for concrete piles. Cutting of a pile would change the pattern of reinforcement along the pile, especially where extra reinforcement is needed at the top of a pile for lateral capacity. A pilot program is usually needed to determine the length and capacity prior to mass production and installation of production piles.

The maximum pile length is usually up to 36 to 38 m because of restrictions during transportation on highways. Although longer piles can be produced on site, slender and long piles may buckle easily during handling and driving. Precast concrete piles with diameters greater than 45 cm are rarely used.

Driven Steel Piles

Driven steel piles, such as steel pipe and H-beam piles, are extensively used as bridge foundations, especially in seismic retrofit projects. Having the advantage and disadvantage of driven piles as discussed above, driven steel piles have their uniqueness.

Steel piles are usually more expensive than concrete piles. They are more ductile and flexible and can be spliced more conveniently. The required overhead is much smaller compared with driven concrete piles. Pipe piles with an open end can penetrate through layers of dense sand. If necessary, the soil inside the pipe can be taken out before further driving; small boulders may also be crushed and taken out. H-piles with a pointed tip can usually penetrate onto soft bedrock and establish enough end-bearing capacity.

Large-Diameter Driven, Vibrated, or Torqued Steel Pipe Piles

Large-diameter pipe piles are widely used as foundations for large bridges. The advantage of this type of foundation is manifold. Large-diameter pipe piles can be built over water from a barge, a trestle, or a temporary island. They can be used in almost all ground conditions and penetrate to a great depth to reach bedrock. Length of the pile can be adjusted by welding. Large-diameter pipe piles can also be used as casings to support soil above bedrock from caving in; rock sockets or rock anchors can then be constructed below the tip of the pipe. Concrete or reinforced concrete can be placed inside the pipe after it is cleaned. Another advantage is that no workers are required to work below water or the ground surface. Construction is usually safer and faster than other types of foundations, such as caissons or cofferdam construction.

Large-diameter pipe piles can be installed by methods of driving, vibrating, or torque. Driven piles usually have higher capacity than piles installed through vibration or torque. However, driven piles are hard to control in terms of location and inclination of the piles. Moreover, once a pile is out of location or installed with unwanted inclination, no corrective measures can be applied. Piles installed with vibration or torque, on the other hand, can be controlled more easily. If a pile is out of position or inclination, the pile can even be lifted up and reinstalled.

Drilled Shaft Foundations

Drilled shaft foundations are the most versatile types of foundations. The length and size of the foundations can be tailored easily. Disturbance to the nearby structures is small compared with other types of deep foundations. Drilled shafts can be constructed very close to existing structures and can be constructed under low overhead conditions. Therefore, drilled shafts are often used in many seismic retrofit projects. However, drilled shafts may be difficult to install under certain ground conditions such as soft soil, loose sand, sand under water, and soils with boulders. Drilled shafts will generate a large volume of soil cuttings and fluid and can be a mess. Disposal of the cuttings is usually a concern for sites with contaminated soils.

Drilled shaft foundations are usually comparable with or more expensive than driven piles. For large bridge foundations, their cost is at the same level of caisson foundations and spread footing foundations combined with cofferdam construction. Drilled shaft foundations can be constructed very rapidly under normal conditions compared with caisson and cofferdam construction.

Anchors

Anchors are special foundation elements that are designed to take uplift loads. Anchors can be added if an existing foundation lacks uplift capacity, and competent layers of soil or rock are shallow and easy to reach. Anchors, however, cannot take lateral loads and may be sheared off if combined lateral capacity of a foundation is not enough.

Anchors are in many cases pretensioned in order to limit the deformation to activate the anchor. The anchor system is therefore very stiff. Structural failure resulting from anchor rupture often occurs very quickly and catastrophically. Pretension may also be lost over time because of creep in some types of rock and soil. Anchors should be tested carefully for their design capacity and creep performance.

Caissons

Caissons are large structures that are mainly used for construction of large bridge foundations. Caisson foundations can take large compressive and lateral loads. They are used primarily for over-water construction and sometimes used in soft or loose soil conditions, with a purpose to sink or excavate down to a depth where bedrock or firm soil can be reached. During construction, large boulders can be removed.

Caisson construction requires special techniques and experience. Caisson foundations are usually very costly, and comparable to the cost of cofferdam construction. Therefore, caissons are usually not the first option unless other types of foundation are not favored.

Cofferdam and Shoring

Cofferdams or other types of shoring systems are a method of foundation construction to retain water and soil. A dry bottom deep into water or ground can be created as a working platform. Foundations of essentially any of the types discussed above can be built from the platform on top of firm soil or rock at a great depth, which otherwise can only be reached by deep foundations.

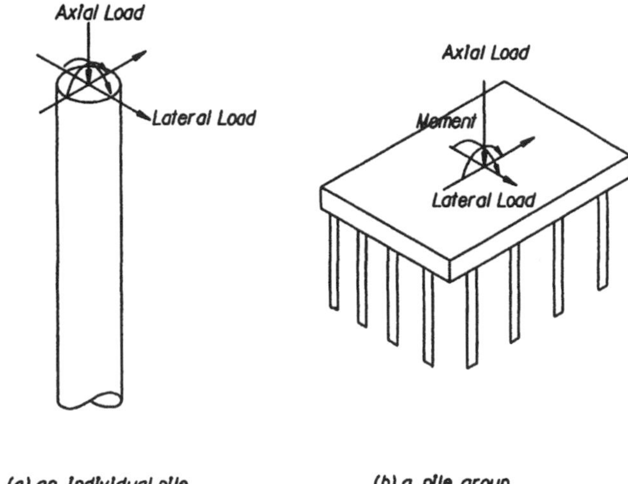

FIGURE 32.2 Acting loads on top of a pile or a pile group. (a) Individual pile; (b) pile group.

A spread footing type of foundation can be built from the platform. Pile foundations also can be constructed from the platform, and the pile length can be reduced substantially. Without cofferdam or shoring, a foundation may not be possible if constructed from the water or ground surface, or it may be too costly.

Cofferdam construction is often very expensive and should only be chosen if it is favorable compared with other foundation options in terms of cost and construction conditions.

32.2.5 Characteristics of Different Types of Foundations

In this section, the mechanisms of resistance of an individual foundation and a pile group are discussed. The function of different types of foundations is also addressed.

Complex loadings on top of a foundation from the bridge structures above can be simplified into forces and moments in the longitudinal, transverse, and vertical directions, respectively (Figure 32.2). Longitudinal and transverse loads are also called horizontal loads; longitudinal and transverse moments are called overturning moments, moment about the vertical axis is called torsional moment. The resistance provided by an individual foundation is categorized in the following (also see Figure 32.3).

End-bearing: Vertical compressive resistance at the base of a foundation; distributed end-bearing pressures can provide resistance to overturning moments;

Base shear: Horizontal resistance of friction and cohesion at the base of a foundation;

Side resistance: Shear resistance from friction and cohesion along the side of a foundation;

Earth pressure: Mainly horizontal resistance from lateral Earth pressures perpendicular to the side of the foundation;

Self-weight: Effective weight of the foundation.

Both base shear and lateral earth pressures provide lateral resistance of a foundation, and the contribution of lateral earth pressures decreases as the embedment of a pile increases. For long piles, lateral earth pressures are the main source of lateral resistance. For short piles, base shear and end-bearing pressures can also contribute part of the lateral resistance. Table 32.2 lists various types of resistance of an individual pile.

For a pile group, through the action of the pile cap, the coupled axial compressive and uplift resistance of individual piles provides the majority of the resistance to the overturning moment loading. Horizontal (or lateral) resistance can at the same time provide torsional moment resistance.

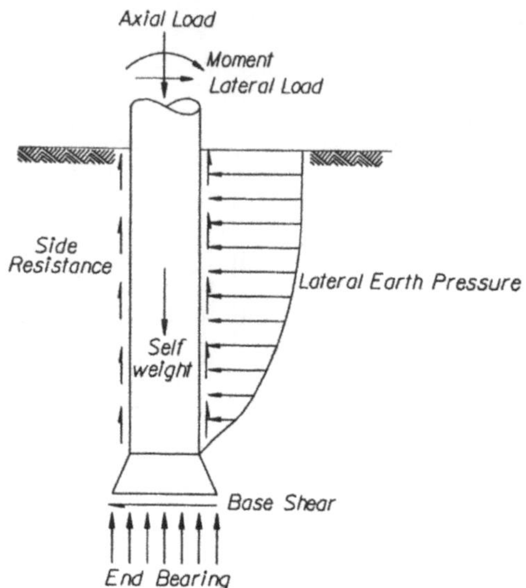

FIGURE 32.3 Resistances of an individual foundation.

TABLE 32.2 Resistance of an Individual Foundation

	Type of Resistance				
Type of Foundation	Vertical Compressive Load (Axial)	Vertical Uplift Load (Axial)	Horizontal Load (Lateral)	Overturning Moment (Lateral)	Torsional Moment (Torsional)
Spread footing (also see Chapter 31)	End bearing	—	Base shear, lateral earth pressure	End bearing, lateral earth pressure	Base shear, lateral earth pressure
Individual short pile foundation	End bearing; side friction	Side friction	Lateral earth pressure, base shear	Lateral earth pressure, end bearing	Side friction, lateral earth pressure, base shear
Individual end-bearing long pile foundation	End bearing	—	Lateral earth pressure	Lateral earth pressure	—
Individual frictional long pile foundation	Side friction	Side friction	Lateral earth pressure	Lateral earth pressure	Side friction
Individual long pile foundation	End bearing; side friction	Side friction	Lateral earth pressure	Lateral earth pressure	Side friction
Anchor	—	Side friction	—	—	—

TABLE 32.3 Additional Functions of Pile Group Foundations

	Type of Resistance	
Type of Foundation	Overturning moment (Lateral)	Torsional moment (Torsional)
Grouped spread footings	Vertical compressive resistance	Horizontal resistance
Grouped piles, foundations	Vertical compressive and uplift resistance	Horizontal resistance
Grouped anchors	Vertical uplift resistance	—

A pile group is more efficient in resisting overturning and torsional moment than an individual foundation. Table 32.3 summarizes functions of a pile group in addition to those of individual piles.

32.2.6 Selection of Foundations

The two predominant factors in determining the type of foundations are bridge types and ground conditions.

The bridge type, including dimensions, type of bridge, and construction materials, dictates the design magnitude of loads and the allowable displacements and other performance criteria for the foundations, and therefore determines the dimensions and type of its foundations. For example, a suspension bridge requires large lateral capacity for its end anchorage which can be a huge deadman, a high capacity soil or rock anchor system, a group of driven piles, or a group of large-diameter drilled shafts. Tower foundations of an over-water bridge require large compressive, uplift, lateral, and overturning moment capacities. The likely foundations are deep, large-size footings using cofferdam construction, caissons, groups of large-diameter drilled shafts, or groups of a large number of steel piles.

Surface and subsurface geologic and geotechnical conditions are another main factor in determining the type of bridge foundations. Subsurface conditions, especially the depths to the load-bearing soil layer or bedrock, are the most crucial factor. Seismicity over the region usually dictates the design level of seismic loads, which is often the critical and dominant loading condition. A bridge that crosses a deep valley or river certainly determines the minimum span required. Over-water bridges have limited options to chose in terms of the type of foundations.

The final choice of the type of foundation usually depends on cost after considering some other factors, such as construction conditions, space and overhead conditions, local practice, environmental conditions, schedule constraints, etc. In the process of selection, several types of foundations would be evaluated as candidates once the type of bridge and the preliminary ground conditions are known. Certain types of foundations are excluded in the early stage of study. For example, from the geotechnical point of view, shallow foundations are not an acceptable option if a thick layer of soft clay or liquefiable sand is near the ground surface. Deep foundations are used in cases where shallow foundations would be excessively large and costly. From a constructibility point of view, driven pile foundations are not suitable if boulders exist at depths above the intended firm bearing soil/rock layer.

For small bridges such as roadway overpasses, for example, foundations with driven concrete or steel piles, drilled shafts, or shallow spread footing foundations may be the suitable choices. For large over-water bridge foundations, single or grouped large-diameter pipe piles, large-diameter rock sockets, large-diameter drilled shafts, caissons, or foundations constructed with cofferdams are the most likely choice. Caissons or cofferdam construction with a large number of driven pile groups were widely used in the past. Large-diameter pipe piles or drilled shafts, in combination with rock sockets, have been preferred for bridge foundations recently.

Deformation compatibility of the foundations and bridge structure is an important consideration. Different types of foundation may behave differently; therefore, the same type of foundations should be used for one section of bridge structure. Diameters of the piles and inclined piles are two important factors to considere in terms of deformation compatibility and are discussed in the following.

Small-diameter piles are more "brittle" in the sense that the ultimate settlement and lateral deflection are relatively small compared with large-diameter piles. For example, 20 small piles can have the same ultimate load capacity as two large-diameter piles. However, the small piles reach the ultimate state at a lateral deflection of 50 mm, whereas the large piles do at 150 mm. The smaller piles would have failed before the larger piles are activated to a substantial degree. In other words, larger piles will be more flexible and ductile than smaller piles before reaching the ultimate state. Since ductility usually provides more seismic safety, larger-diameter piles are preferred from the point of view of seismic design.

Inclined or battered piles should not be used together with vertical piles unless the inclined piles alone have enough lateral capacity. Inclined piles provide partial lateral resistance from their axial capacity, and, since the stiffness in the axial direction of a pile is much larger than in the perpendicular directions, inclined piles tend to attract most of the lateral seismic loading. Inclined piles will fail or reach their ultimate axial capacity before the vertical piles are activated to take substantial lateral loads.

32.3 Design Considerations

32.3.1 Design Concept

The current practice of foundation design mainly employs two types of design concepts, i.e., the permissible stress approach and the limit state approach.

By using the permissible stress approach, both the demanded stresses from loading and the ultimate stress capacity of the foundation are evaluated. The foundation is considered to be safe as long as the demanded stresses are less than the ultimate stress capacity of the foundation. A factor of safety of 2 to 3 is usually applied to the ultimate capacity to obtain various allowable levels of loading in order to limit the displacements of a foundation. A separate displacement analysis is usually performed to determine the allowable displacements for a foundation, and for the bridge structures. Design based on the permissible concept is still the most popular practice in foundation design.

Starting to be adopted in the design of large critical bridges, the limit state approach requires that the foundation and its supported bridge should not fail to meet performance requirements when exceeding various limit states. Collapse of the bridge is the ultimate limit state, and design is aimed at applying various factors to loading and resistance to ensure that this state is highly improbable. A design needs to ensure the structural integrity of the critical foundations before reaching the ultimate limit state, such that the bridge can be repaired a relatively short time after a major loading incident without reconstruction of the time-consuming foundations.

32.3.2 Design Procedures

Under normal conditions, the design procedures of a bridge foundation should involve the following steps:

1. Evaluate the site and subsurface geologic and geotechnical conditions, perform borings or other field exploratory programs, and conduct field and laboratory tests to obtain design parameters for subsurface materials;
2. Review the foundation requirements including design loads and allowable displacements, regulatory provisions, space, or other constraints;
3. Evaluate the anticipated construction conditions and procedures;
4. Select appropriate foundation type(s);
5. Determine the allowable and ultimate axial and lateral foundation design capacity, load vs. deflection relationship, and load vs. settlement relationship;
6. Design various elements of the foundation structure; and
7. Specify requirements for construction inspection and/or load test procedures, and incorporate the requirements into construction specifications.

32.3.3 Design Capacities

Capacity in Long-Term and Short-Term Conditions

Depending on the loading types, foundations are designed for two different stress conditions. Capacity in total stress is used where loading is relatively quick and corresponds to an undrained condition. Capacity in effective stress is adopted where loading is slow and corresponds to a drained condition. For many types of granular soil, such as clean gravel and sand, drained capacity is very close to undrained capacity under most loading conditions. Pile capacity under seismic loading is usually taken 30% higher than capacity under static loading.

Axial, Lateral, and Moment Capacity

Deep foundations can provide lateral resistance to overturning moment and lateral loads and axial resistance to axial loads. Part or most of the moment capacity of a pile group are provided by the axial capacity of individual piles through pile cap action. The moment capacity depends on the axial capacity of the individual piles, the geometry arrangement of the piles, the rigidity of the pile cap, and the rigidity of the connection between the piles and the pile cap. Design and analysis is often concentrated on the axial and lateral capacity of individual piles. Axial capacity of an individual pile will be addressed in detail in Section 32.4 and lateral capacity in Section 32.5. Pile groups will be addressed in Section 32.6.

Structural Capacity

Deep foundations may fail because of structural failure of the foundation elements. These elements should be designed to take moment, shear, column action or buckling, corrosion, fatigue, etc. under various design loading and environmental conditions.

Determination of Capacities

In the previous sections, the general procedure and concept for the design of deep foundations are discussed. Detailed design includes the determination of axial and lateral capacity of individual foundations, and capacity of pile groups. Many methods are available to estimate these capacities, and they can be categorized into three types of methodology as listed in the following:

- Theoretical analysis utilizing soil or rock strength;
- Empirical methods including empirical analysis utilizing standard field tests, code requirements, and local experience; and
- Load tests, including full-scale load tests, and dynamic driving and restriking resistance analysis.

The choice of methods depends on the availability of data, economy, and other constraints. Usually, several methods are used; the capacity of the foundation is then obtained through a comprehensive evaluation and judgment.

In applying the above methods, the designers need to keep in mind that the capacity of a foundation is the sum of capacities of all elements. Deformation should be compatible in the foundation elements, in the surrounding soil, and in the soil–foundation interface. Settlement or other movements of a foundation should be restricted within an acceptable range and usually is a controlling factor for large foundations.

32.3.4 Summary of Design Methods

Table 32.4 presents a partial list of design methods available in the literature.

32.3.5 Other Design Issues

Proper foundation design should consider many factors regarding the environmental conditions, type of loading conditions, soil and rock conditions, construction, and engineering analyses, including:

- Various loading and loading combinations, including the impact loads of ships or vehicles
- Earthquake shaking
- Liquefaction

TABLE 32.4 Summary of Design Methods for Deep Foundations

Type	Design For	Soil Condition	Method and Author
Driven pile	End bearing	Clay	N_c method [67]
			N_c method [23]
			CPT methods [37,59,63]
			CPT [8,10]
		Sand	N_q method with critical depth concept [38]
			N_q method [3]
			N_q method [23]
			N_q by others [26,71,76]
			Limiting N_q values [1,13]
			Value of ϕ [27,30,39]
			SPT [37,38]
			CPT methods [37,59,63]
			CPT [8,10]
		Rock	[10]
	Side resistance	Clay	α-method [72,73]
			α-method [1]
			β-method [23]
			λ-method [28,80]
			CPT methods [37,59,63]
			CPT [8,10]
			SPT [14]
		Sand	α-method [72,73]
			β-method [7]
			β-method [23]
			CPT method [37,59,63]
			CPT [8,10]
			SPT [37,38]
	Side and end	All	Load test: ASTM D 1143, static axial compressive test
			Load test: ASTM D 3689, static axial tensile test
			Sanders' pile driving formula (1850) [50]
			Danish pile driving formula [68]
			Engineering News formula (Wellingotn, 1988)
			Dynamic formula — WEAP Analysis
			Strike and restrike dynamic analysis
			Interlayer influence [38]
			No critical depth [20,31]
	Load-settlement	Sand	[77]
			[41,81]
		All	Theory of elasticity, Mindlin's solutions [50]
			Finite-element method [15]
			Load test: ASTM D 1143, static axial compressive test
			Load test: ASTM D 3689, static axial tensile test
Drilled shaft	End bearing	Clay	N_c method [66]
			Large base [45,57]
			CPT [8,10]
		Sand	[74]
			[38]
			[55]
			[52]
			[37,38]
			[8,10]
		Rock	[10]
		Rock	Pressure meter [10]

TABLE 32.4 (continued) Summary of Design Methods for Deep Foundations

Type	Design For	Soil Condition	Method and Author
	Side resistance	Clay	α-method [52]
			α-method [67]
			α-method [83]
			CPT [8,10]
		Sand	[74]
			[38]
			[55]
			β-method [44,52]
			SPT [52]
			CPT [8,10]
		Rock	Coulombic [34]
			Coulombic [75]
			SPT [12]
			[24]
			[58]
			[11,32]
			[25]
	Side and end	Rock	[46]
			[84]
			[60]
			[48]
			[61,62]
			FHWA [57]
		All	Load test [47]
	Load-settlement	Sand	[57]
		Clay	[57]
			[85]
		All	Load test [47]
All	Lateral resistance	Clay	Broms' method [5]
		Sand	Broms' method [6]
		All	*p–y* method [56]
		Clay	*p–y* response [35]
		Clay (w/water)	*p–y* response [53]
		Clay (w/o water)	*p–y* response [82]
		Sand	*p–y* response [53]
		All	*p–y* response [1]
			p–y response for inclined piles [2,29]
			p–y response in layered soil
			p–y response [42]
		Rock	*p–y* response [86]
	Load-settlement	All	Theory of elasticity method [50]
			Finite-difference method [64]
			General finite-element method (FEM)
			FEM dynamic
	End bearing		Pressure meter method [36,78]
	Lateral resistance		Pressure meter method [36]
			Load test: ASTM D 3966
Group	Theory		Elasticity approach [50]
			Elasticity approach [21]
			Two-dimensional group [51]
			Three-dimensional group [52]
	Lateral g-factor		[10]
			[16]

- Rupture of active fault and shear zone
- Landslide or ground instability
- Difficult ground conditions such as underlying weak and compressible soils
- Debris flow
- Scour and erosion
- Chemical corrosion of foundation materials
- Weathering and strength reduction of foundation materials
- Freezing
- Water conditions including flooding, water table change, dewatering
- Environmental change due to construction of the bridge
- Site contamination condition of hazardous materials
- Effects of human or animal activities
- Influence upon and by nearby structures
- Governmental and community regulatory requirements
- Local practice

32.3.6 Uncertainty of Foundation Design

Foundation design is as much an art as a science. Although most foundation structures are man-made, the surrounding geomaterials are created, deposited, and altered in nature over the geologic times. The composition and engineering properties of engineering materials such as steel and concrete are well controlled within a variation of uncertainty of between 5 to 30%. However, the uncertainty of engineering properties for natural geomaterials can be up to several times, even within relatively uniform layers and formations. The introduction of faults and other discontinuities make generalization of material properties very hard, if not impossible.

Detailed geologic and geotechnical information is usually difficult and expensive to obtain. Foundation engineers constantly face the challenge of making engineering judgments based on limited and insufficient data of ground conditions and engineering properties of geomaterials.

It was reported that under almost identical conditions, variation of pile capacities of up to 50% could be expected within a pile cap footprint under normal circumstances. For example, piles within a nine-pile group had different restruck capacities of 110, 89, 87, 96, 86, 102, 103, 74, and 117 kips (1 kip = 4.45 kN) respectively [19].

Conservatism in foundation design, however, is not necessarily always the solution. Under seismic loading, heavier and stiffer foundations may tend to attract more seismic energy and produce larger loads; therefore, massive foundations may not guarantee a safe bridge performance.

It could be advantageous that piles, steel pipes, caisson segments, or reinforcement steel bars are tailored to exact lengths. However, variation of depth and length of foundations should always be expected. Indicator programs, such as indicator piles and pilot exploratory borings, are usually a good investment.

32.4 Axial Capacity and Settlement — Individual Foundation

32.4.1 General

The axial resistance of a deep foundation includes the tip resistance (Q_{end}), side or shaft resistance (Q_{side}), and the effective weight of the foundation (W_{pile}). Tip resistance, also called end bearing, is the compressive resistance of soil near or under the tip. Side resistance consists of friction, cohesion, and keyed bearing along the shaft of the foundation. Weight of the foundation is usually ignored under compression because it is nearly the same as the weight of the soil displaced, but is usually accounted for under uplift loading condition.

At any loading instance, the resistance of an individual deep foundation (or pile) can be expressed as follows:

$$Q = Q_{end} + \Sigma Q_{side} \pm W_{pile} \tag{32.1}$$

The contribution of each component in the above equation depends on the stress–strain behavior and stiffness of the pile and the surrounding soil and rock. The maximum capacity of a pile can be expressed as

$$Q^{c}{}_{max} \le Q^{c}{}_{end\_max} + \Sigma Q^{c}{}_{side\_max} - W_{pile} \quad \text{(in compression)} \tag{32.2}$$

$$Q^{t}{}_{max} \le Q^{t}{}_{end\_max} + \Sigma Q^{t}{}_{side\_max} + W_{pile} \quad \text{(in uplift)} \tag{32.3}$$

and is less than the sum of all the maximum values of resistance. The ultimate capacity of a pile undergoing a large settlement or upward movement can be expressed as

$$Q^{c}{}_{ult} = Q^{c}{}_{end\_ult} + \Sigma Q^{c}{}_{side\_ult} - W_{pile} \le Q^{c}_{max} \tag{32.4}$$

$$Q^{t}{}_{ult} = Q^{t}{}_{end\_ult} + \Sigma Q^{t}{}_{side\_ult} + W_{pile} \le Q^{t}_{max} \tag{32.5}$$

Side- and end-bearing resistances are related to displacement of a pile. Maximum end bearing capacity can be mobilized only after a substantial downward movement of the pile, whereas side resistance reaches its maximum capacity at a relatively smaller downward movement. Therefore, the components of the maximum capacities (Q_{max}) indicated in Eqs. (32.2) and (32.3) may not be realized at the same time at the tip and along the shaft. For a drilled shaft, the end bearing is usually ignored if the bottom of the borehole is not cleared and inspected during construction. Voids or compressible materials may exist at the bottom after concrete is poured; as a result, end bearing will be activated only after a substantial displacement.

Axial displacements along a pile are larger near the top than toward the tip. Side resistance depends on the amount of displacement and is usually not uniform along the pile. If a pile is very long, maximum side resistance may not occur at the same time along the entire length of the pile. Certain types of geomaterials, such as most rocks and some stiff clay and dense sand, exhibit strain softening behavior for their side resistance, where the side resistance first increases to reach its maximum, then drops to a much smaller residual value with further displacement. Consequently, only a fixed length of the pile segment may maintain high resistance values and this segment migrates downward to behave in a pattern of a progressive failure. Therefore, the capacity of a pile or drilled shaft may not increase infinitely with its length.

For design using the permissible stress approach, allowable capacity of a pile is the design capacity under service or routine loading. The allowable capacity (Q_{all}) is obtained by dividing ultimate capacity (Q_{ult}) by a factor of safety (FS) to limit the level of settlement of the pile and to account for uncertainties involving material, installation, loads calculation, and other aspects. In many cases, the ultimate capacity (Q_{ult}) is assumed to be the maximum capacity (Q_{max}). The factor of safety is usually between 2 to 3 for deep foundations depending on the reliability of the ultimate capacity estimated. With a field full-scale loading test program, the factor of safety is usually 2.

TABLE 32.5 Typical Values of Bearing Capacity Factor N_q

φ^a (degrees)	26	28	30	31	32	33	34	35	36	37	38	39	40
N_q (driven pile displacement)	10	15	21	24	29	35	42	50	62	77	86	120	145
N_q^b (drilled piers)	5	8	10	12	14	17	21	25	30	38	43	60	72

[a] Limit φ to 28° if jetting is used.

[b] 1. In case a bailer of grab bucket is used below the groundwater table, calculate end bearing based on φ not exceeding 28°.

2. For piers greater than 24-in. diameter, settlement rather than bearing capacity usually controls the design. For estimating settlement, take 50% of the settlement for an equivalent footing resting on the surface of comparable granular soils (Chapter 5, DM-7.01).

Source: NAVFAC [42].

32.4.2 End Bearing

End bearing is part of the axial compressive resistance provided at the bottom of a pile by the underlying soil or rock. The resistance depends on the type and strength of the soil or rock and on the stress conditions near the tip. Piles deriving their capacity mostly from end bearing are called end bearing piles. End bearing in rock and certain types of soil such as dense sand and gravel is usually large enough to support the designed loads. However, these types of soil or rock cannot be easily penetrated through driving. No or limited uplift resistance is provided from the pile tips; therefore, end-bearing piles have low resistance against uplift loading.

The end bearing of a pile can be expressed as:

$$Q_{\text{end\_max}} = \begin{cases} cN_c A_{\text{pile}} & \text{for clay} \\ \sigma'_v N_q A_{\text{pile}} & \text{for sand} \\ \dfrac{U_c}{2} N_k A_{\text{pile}} & \text{for rock} \end{cases} \tag{32.6}$$

where

$Q_{\text{end\_max}}$ = the maximum end bearing of a pile
A_{pile} = the area of the pile tip or base
N_c, N_q, N_k = the bearing capacity factors for clay, sand, and rock
c = the cohesion of clay
σ'_v = the effective overburden pressure
U_c = the unconfined compressive strength of rock and $\frac{U_c}{2} = S_u$, the equivalent shear strength of rock

Clay

The bearing capacity factor N_c for clay can be expressed as

$$N_c = 6.0\left(1 + 0.2\frac{L}{D}\right) \le 9 \tag{32.7}$$

where L is the embedment depth of the pile tip and D is the diameter of the pile.

Sand

The bearing capacity factor N_q generally depends on the friction angle ϕ of the sand and can be estimated by using Table 32.5 or the Meyerhof equation below.

$$N_q = e^{\pi \tan\varphi} \tan^2\left(45+\frac{\varphi}{2}\right) \tag{32.8}$$

The capacity of end bearing in sand reaches a maximum cutoff after a certain critical embedment depth. This critical depth is related to ϕ and D and for design purposes is listed as follows:

$$L_c = 7D, \quad \phi = 30^o \text{ for loose sand}$$

$$L_c = 10D, \quad \phi = 34^o \text{ for medium dense sand}$$

$$L_c = 14D, \quad \phi = 38^o \text{ for dense sand}$$

$$L_c = 22D, \quad \phi = 45^o \text{ for very dense sand}$$

The validity of the concept of critical depth has been challenged by some people; however, the practice to limit the maximum ultimate end bearing capacity in sand will result in conservative design and is often recommended.

Rock

The bearing capacity factor N_k depends on the quality of the rock mass, intact rock properties, fracture or joint properties, embedment, and other factors. Because of the complex nature of the rock mass and the usually high value for design bearing capacity, care should be taken to estimate N_k. For hard fresh massive rock without open or filled fractures, N_k can be taken as high as 6. N_k decreases with increasing presence and dominance of fractures or joints and can be as low as 1. Rock should be treated as soil when rock is highly fractured and weathered or in-fill weak materials control the behavior of the rock mass. Bearing capacity on rock also depends on the stability of the rock mass. Rock slope stability analysis should be performed where the foundation is based on a slope. A higher factor of safety, 3 to as high as 10 to 20, is usually applied in estimating allowable bearing capacity for rocks using the N_k approach.

The soil or rock parameters used in design should be taken from averaged properties of soil or rock below the pile tip within the influence zone. The influence zone is usually taken as deep as three to five diameters of the pile. Separate analyses should be conducted where weak layers exist below the tip and excessive settlement or punch failure might occur.

Empirical Methods

Empirical methods are based on information of the type of soil/rock and field tests or index properties. The standard penetration test (SPT) for sand and cone penetration test (CPT) for soil are often used.

Meyerhof [38] recommended a simple formula for piles driven into sand. The ultimate tip bearing pressure is expressed as

$$q_{\text{end\_max}} \le 4N_{\text{SPT}} \quad \text{in tsf (1 tsf = 8.9 kN)} \tag{32.9}$$

where N_{SPT} is the blow count of SPT just below the tip of the driven pile and $q_{\text{end\_max}} = Q_{\text{end\_max}} / A_{\text{pile}}$. Although the formula is developed for piles in sand, it also is used for piles in weathered rock for preliminary estimate of pile capacity.

Schmertmann [63] recommended a method to estimate pile capacity by using the CPT test:

$$q_{\text{end\_max}} = q_b = \frac{q_{c1} + q_{c2}}{2} \tag{32.10}$$

where

q_{c1} = averaged cone tip resistance over a depth of 0.7 to 4 diameters of the pile below tip of the pile
q_{c2} = the averaged cone tip resistance over a depth of 8 diameters of the pile above the tip of the pile

Chapter 31 presents recommended allowable bearing pressures for various soil and rock types for spread footing foundations and can be used as a conservative estimate of end-bearing capacity for end-bearing piles.

TABLE 32.6 Typical Values of α and f_s

Range of Shear Strength, S_u ksf	Formula to Estimate α	Range of α	Range of f_s ksf[a]	Description
0 to 0.600	$\alpha = 1.0$	1	0–0.6	Soft clay
0.600 to 3	$\alpha = 0.375\left(1+\frac{1}{S_u}\right)$,	1–0.5	0.6–1.5	Medium stiff clay to very stiff clay
3 to 11	$\alpha = 0.375\left(1+\frac{1}{S_u}\right)$,	0.5–0.41	1.5–4.5	Hard clay to very soft rock
11 to 576 (76 psi to 4000 psi)	$\alpha = \frac{5}{\sqrt{2S_u}}$, S_u in psi,	0.41–0.056	4.5–32 (31–220 psi)	Soft rock to hard rock

Note: 1 ksf = 1000 psf; 1 psi = 144 psf; 1 psf = 0.048 kPa; 1 psi = 6.9 kPa
[a] For concrete driven piles and for drilled piers without buildup of mud cakes along the shaft. (Verify if fs ≥ 3 ksf.)

32.4.3 Side Resistance

Side resistance usually consists of friction and cohesion between the pile and the surrounding soil or rock along the shaft of a pile. Piles that derive their resistance mainly from side resistance are termed *frictional piles*. Most piles in clayey soil are frictional piles, which can take substantial uplift loads.

The maximum side resistance of a pile Q_{side_max} can be expressed as

$$Q_{side\_max} = \sum f_s A_{side} \tag{32.11}$$

$$f_s = K_s \sigma'_v \tan\delta + c_a \tag{32.12}$$

$$c_a = \alpha S_u \tag{32.13}$$

where

$\sum$ = the sum for all layers of soil and rock along the pile
A_{side} = the shaft side area
f_s = the maximum frictional resistance on the side of the shaft
K_s = the lateral earth pressure factor along the shaft
σ'_v = the effective vertical stress along the side of the shaft
δ = the friction angle between the pile and the surrounding soil; for clayey soil under quick loading, δ is very small and usually omitted
c_a = the adhesion between pile and surrounding soil and rock
α = a strength factor, and
S_u = the cohesion of the soil or rock

TABLE 32.7 Typical Values Cohesion and Adhesion f_s

Pile Type	Consistency of Soil	Cohesion, S_u psf	Adhesion, f_s psf
Timber and concrete	Very soft	0–250	0–250
	Soft	250–500	250–480
	Medium stiff	500–1000	480–750
	Stiff	1000–2000	750–950
	Very stiff	2000–4000	950–1300
Steel	Very soft	0–250	0–250
	Soft	250–500	250–460
	Medium stiff	500–1000	480–700
	Stiff	1000–2000	700–720
	Very stiff	2000–4000	720–750

1 psf = 0.048 kPa.
Source: NAVFAC [42].

TABLE 32.8 Typical Values of Bond Stress of Rock Anchors for Selected Rock

Rock Type (Sound, Nondecayed)	Ultimate Bond Stresses between Rock and Anchor Plus (δ_{skin}), psi
Granite and basalt	250–450
Limestone (competent)	300–400
Dolomitic limestone	200–300
Soft limestone	150–220
Slates and hard shales	120–200
Soft shales	30–120
Sandstone	120–150
Chalk (variable properties)	30–150
Marl (stiff, friable, fissured)	25–36

Note: It is not generally recommended that design bond stresses exceed 200 psi even in the most competent rocks. 1 psi = 6.9 kPa.
Source: NAVFAC [42].

TABLE 32.9 Typical Values of earth Pressure Coefficient K_s

Pile Type	Earth Pressure Coefficients K_s		
	K_s [a] (compression)	K_s [a] (tension)	K_s [b]
Driven single H-pile	0.5–1.0	0.3–0.5	—
Driven single displacement pile	1.0–1.5	0.6–1.0	0.7–3.0
Driven single displacement tapered pile	1.5–2.0	1.0–1.3	—
Driven jetted pile	0.4–0.9	0.3–0.6	—
Drilled pile (less than 24-in. diameter)	0.7	0.4	—
Insert pile	—	—	0.7 (compression) 0.5 (tension)
Driven with predrilled hole	—	—	0.4–0.7
Drilled pier	—	—	0.1–0.4

[a] From NAVFAC [42].
[b] From Le Tirant (1979), K_s increases with OCR or D_R.

TABLE 32.10 Typical Value of Pile-Soil Friction Angles δ

Pile Type	δ, °	Alternate for δ
Concrete[a]	—	$\delta = \frac{3}{4}\varphi$
Concrete (rough, cast-in-place)[b]	33	$\delta = 0.85\varphi$
Concrete (smooth)[b]	30	$\delta = 0.70\varphi$
Steel[a]	20	—
Steel (corrugated)	33	$\delta = \varphi$
Steel (smooth)[c]	—	$\delta = \varphi - 5°$
Timber[a]	—	$\delta = \frac{3}{4}\varphi$

[a] NAVFAC [42].
[b] Woodward et al. [85]
[c] API [1] and de Ruiter and Beringen [13]

Typical values of α, f_s, K_s, δ are shown in Tables 32.6 through 32.10. For design purposes, side resistance f_s in sand is limited to a cutoff value at the critical depth, which is equal to about 10B for loose sand and 20B for dense sand.

Meyerhof [38] recommended a simple formula for driven piles in sand. The ultimate side adhesion is expressed as

$$f_s \leq \frac{N_{\text{SPT}}}{50} \quad \text{in tsf (1 tsf = 8.9 kN)} \tag{32.14}$$

where N_{SPT} is the averaged blow count of SPT along the pile.

Meyerhof [38] also recommended a formula to calculate the ultimate side adhesion based on CPT results as shown in the following.

For full displacement piles:

$$f_s = \frac{q_c}{200} \leq 1.0 \quad \text{in tsf} \tag{32.15}$$

or

$$f_s = 2f_c \leq 1.0 \tag{32.16}$$

For nondisplacement piles:

$$f_s = \frac{q_c}{400} \leq 0.5 \quad \text{in tsf} \tag{32.17}$$

or

$$f_s = f_c \leq 0.5 \tag{32.18}$$

in which

q_c, f_c = the cone tip and side resistance measured from CPT; averaged values should be used along the pile

Downdrag

For piles in soft soil, another deformation-related issue should be noted. When the soil surrounding the pile settles relative to a pile, the side friction, also called the negative skin friction, should be considered when there exists underlying compressible clayey soil layers and liquefiable loose sand layers. Downdrag can also happen when ground settles because of poor construction of caissons in sand. On the other hand, updrag should also be considered in cases where heave occurs around the piles for uplift loading condition, especially during installation of piles and in expansive soils.

32.4.4 Settlement of Individual Pile, *t–z*, *Q–z* Curves

Besides bearing capacity, the allowable settlement is another controlling factor in determining the allowable capacity of a pile foundation, especially if layers of highly compressible soil are close to or below the tip of a pile.

Settlement of a small pile (diameter less than 350 mm) is usually kept within an acceptable range (usually less than 10 mm) when a factor of safety of 2 to 3 is applied to the ultimate capacity to obtain the allowable capacity. However, in the design of large-diameter piles or caissons, a separate settlement analysis should always be performed.

The total settlement at the top of a pile consists of immediate settlement and long-term settlement. The immediate settlement occurs during or shortly after the loads are applied, which includes elastic compression of the pile and deformation of the soil surrounding the pile under undrained loading conditions. The long-term settlement takes place during the period after the loads are applied, which includes creep deformation and consolidation deformation of the soil under drained loading conditions.

Consolidation settlement is usually significant in soft to medium stiff clayey soils. Creep settlement occurs most significantly in overconsolidated (OC) clays under large sustained loads, and can be estimated by using the method developed by Booker and Poulos (1976). In principle, however, long-term settlement can be included in the calculation of ultimate settlement if the design parameters of soil used in the calculation reflect the long-term behavior.

Presented in the following sections are three methods that are often used:

- Method of solving ultimate settlement by using special solutions from the theory of elasticity [50,85]. Settlement is estimated based on equivalent elasticity in which all deformation of soil is assumed to be linear elastic.
- Empirical method [79].
- Method using localized springs, or the so called *t–z* and *Q–z* method [52a].

Method from Elasticity Solutions

The total elastic settlement S can be separated into three components:

$$S = S_b + S_s + S_{sh} \tag{32.19}$$

where S_b is part of the settlement at the tip or bottom of a pile caused by compression of soil layers below the pile under a point load at the pile tip, and is expressed as

$$S_b = \frac{p_b D_b I_{bb}}{E_s} \tag{32.20}$$

S_s is part of the settlement at the tip of a pile caused by compression of soil layers below the pile under the loading of the distributed side friction along the shaft of the pile, and can be expressed as

$$S_s = \sum_i \frac{(f_{si} l_i \Delta z_i) I_{bs}}{E_s} \tag{32.21}$$

and S_{sh} is the shortening of the pile itself, and can be expressed as

$$S_{sh} = \sum_i \frac{(f_{si} l_i \Delta z_i) + p_b A_b (\Delta z_i)}{E_c (A_i)} \tag{32.22}$$

where

p_b = averaged loading pressure at pile tip
A_b = cross section area of a pile at pile tip; $A_b p_b$ is the total load at the tip
D_b = diameter of pile at the pile tip
i = subscript for *i*th segment of the pile
l = perimeter of a segment of the pile
Δz = axial length of a segment of the pile; $L = \sum_i \Delta z_i$ is the total length of the pile.
f_s = unit friction along side of shaft; $f_{si} l_i \Delta z_i$ is the side frictional force for segment i of the pile
E_s = Young's modulus of uniform and isotropic soil
E_c = Young's modulus of the pile
I_{bb} = base settlement influence factor, from load at the pile tip (Figure 32.4)
I_{bs} = base settlement influence factor, from load along the pile shaft (Figure 32.4)

Because of the assumptions of linear elasticity, uniformity, and isotropy for soil, this method is usually used for preliminary estimate purposes.

Method by Vesic [79]

The settlement S at the top of a pile can be broken down into three components, i.e.,

$$S = S_b + S_s + S_{sh} \tag{32.23}$$

Settlement due to shortening of a pile is

$$S_{sh} = (Q_p + \alpha_s Q_s) \frac{L}{AE_c} \tag{32.24}$$

where

Q_p = point load transmitted to the pile tip in the working stress range
Q_s = shaft friction load transmitted by the pile in the working stress range (in force units)
α_s = 0.5 for parabolic or uniform distribution of shaft friction, 0.67 for triangular distribution of shaft friction starting from zero friction at pile head to a maximum value at pile tip, 0.33 for triangular distribution of shaft friction starting from a maximum at pile head to zero at the pile tip
L = pile length
A = pile cross-sectional area
E_c = modulus of elasticity of the pile

Settlement of the pile tip caused by load transmitted at the pile tip is

$$S_b = \frac{C_p Q_p}{D q_o} \tag{32.25}$$

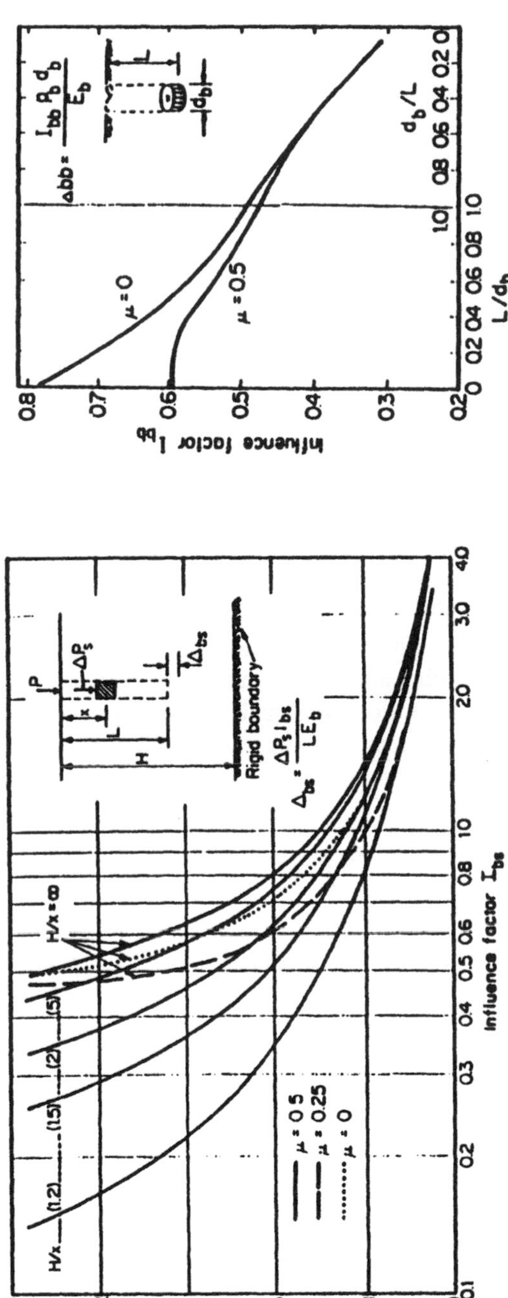

FIGURE 32.4 Influence factors I_{bb} and I_{bs}. [From Woodward, Gardner and Greer (1972)[85], used with permission of McGraw-Hill Book Company]

where

C_p = empirical coefficient depending on soil type and method of construction, see Table 32.11.
D = pile diameter
q_o = ultimate end bearing capacity

and settlement of the pile tip caused by load transmitted along the pile shaft is

$$S_s = \frac{C_s Q_s}{h q_o} \tag{32.26}$$

where

$C_s = (0.93 + 0.16 D/B) C_p$
h = embedded length

TABLE 32.11 Typical Values of C_p for Estimating Settlement of a Single Pile

Soil Type	Driven Piles	Bored Piles
Sand (dense to loose)	0.02–0.04	0.09–0.18
Clay (stiff to soft)	0.02–0.03	0.03–0.06
Silt (dense to loose)	0.03–0.05	0.09–0.12

Note: Bearing stratum under pile tip assumed to extend at least 10 pile diameters below tip and soil below tip is of comparable or higher stiffness.

Method Using Localized Springs: The *t–z* and *Q–z* method

In this method, the reaction of soil surrounding the pile is modeled as localized springs: a series of springs along the shaft (the *t–z* curves) and the spring attached to the tip or bottom of a pile (the *Q–z* curve). *t* is the load transfer or unit friction force along the shaft, *Q* is the tip resistance of the pile, and *z* is the settlement of soil at the location of a spring. The pile itself is also represented as a series of springs for each segment. A mechanical model is shown on Figure 32.5. The procedure to obtain the settlement of a pile is as follows:

- Assume a pile tip movement *zb_1*; obtain a corresponding tip resistance *Q_1* from the *Q–z* curve.
- Divide the pile into number of segments, and start calculation from the bottom segment. Iterations:
  1. Assume an averaged movement of the segment *zs_1*; obtain the averaged side friction along the bottom segment *ts_1* by using the *t–z* curve at that location.
  2. Calculate the movement at middle of the segment from elastic shortening of the pile under axial loading *zs_2*. The axial load is the tip resistance *Q_1* plus the added side friction *ts_1*.
  3. Iteration should continue until the difference between *zs_1* and *zs_2* is within an acceptable tolerance.

  Iteration continues for all the segments from bottom to top of the pile.
- A settlement at top of pile *zt_1* corresponding to a top axial load *Qt_1* is established.
- Select another pile tip movement *zb_2* and calculate *zt_2* and *Qt_2* until a relationship curve of load vs. pile top settlement is found.

The *t–z* and *Q–z* curves are established from test data by many authors. Figure 32.6 shows the *t–z* and *Q–z* curves for cohesive soil and cohesionless soil by Reese and O'Neil [57].

Although the method of *t–z* and *Q–z* curves employs localized springs, the calculated settlements are usually within a reasonable range since the curves are backfitted directly from the test results. Factors of nonlinear behavior of soil, complicated stress conditions around the pile, and partial

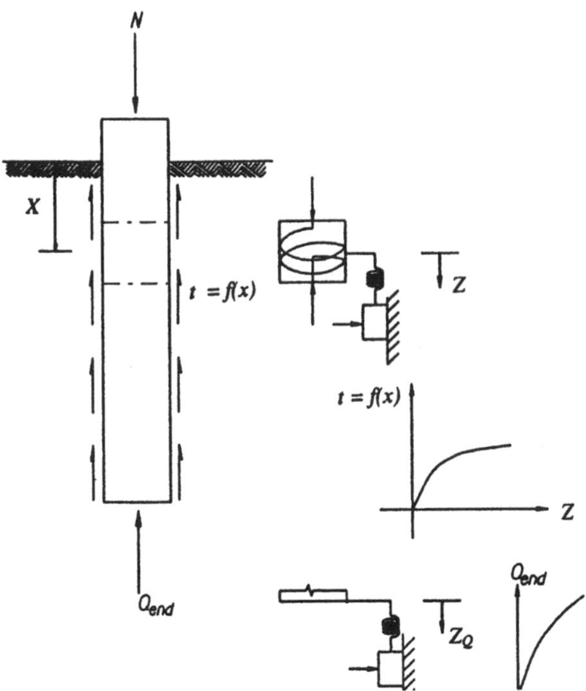

FIGURE 32.5 Analytical model for pile under axial loading with t–z and Q–z curves.

corrections to the Winkler's assumption are embedded in this methodology. Besides, settlement of a pile can be estimated for complicated conditions such as varying pile geometry, different pile materials, and different soil layers.

32.5 Lateral Capacity and Deflection — Individual Foundation

32.5.1 General

Lateral capacity of a foundation is the capacity to resist lateral deflection caused by horizontal forces and overturning moments acted on the top of the foundation. For an individual foundation, lateral resistance comes from three sources: lateral earth pressures, base shear, and nonuniformly distributed end-bearing pressures. Lateral earth pressure is the primary lateral resistance for long piles. Base shear and distributed end-bearing pressures are discussed in Chapter 31.

32.5.2 Broms' Method

Broms [5] developed a method to estimate the ultimate lateral capacity of a pile. The pile is assumed to be short and rigid. Only rigid translation and rotation movements are considered and only ultimate lateral capacity of a pile is calculated. The method assumes distributions of ultimate lateral pressures for cohesive and cohesionless soils; the lateral capacity of piles with different top fixity conditions are calculated based on the assumed lateral pressure as illustrated on Figures 32.7 and 32.8. Restricted by the assumptions, the Broms' method is usually used only for preliminary estimates of the ultimate lateral capacity of piles.

Ultimate Lateral Pressure

The ultimate lateral pressure $q_{h,u}$ along a pile is calculated as follows:

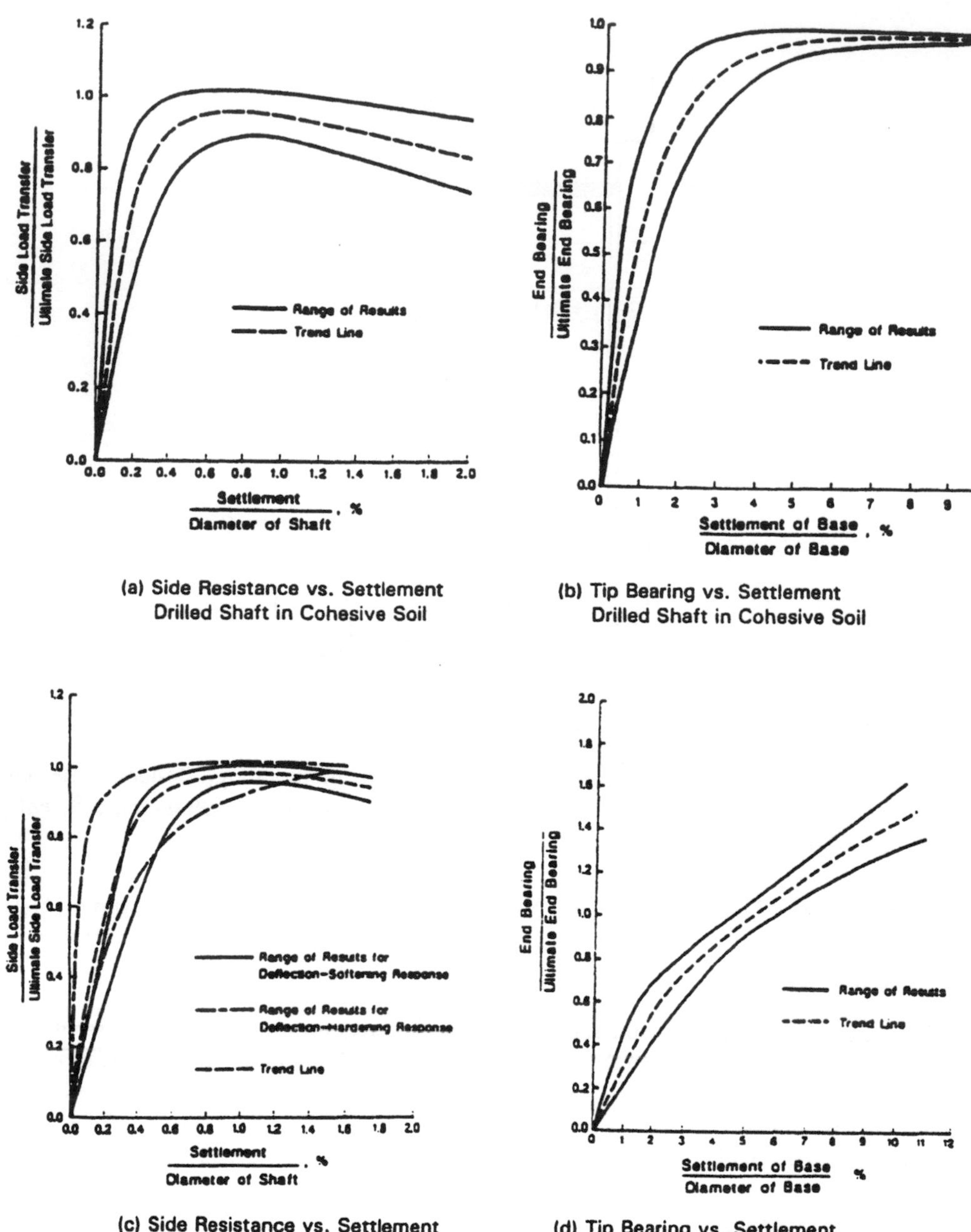

FIGURE 32.6 Load transfer for side resistance (*t–z*) and tip bearing (*Q–z*). (a) Side resistance vs. settlement, drilled shaft in cohesive soil; (b) tip bearing vs. settlement, drilled shaft in cohesive soil; (c) side resistance vs. settlement, drilled shaft in cohesionless soil; (d) tip bearing vs. settlement, drilled shaft in cohesionless soil. (From AASHTO LRFD Bridge Design Specifications, First Edition, coyyright 1996 by the American Association of State Highway and Transportation officials, Washington, D.C. Used by permission.)

$$q_{h,u} = \begin{cases} 9c_u & \text{for cohesive soil} \\ 3K_p p_0' & \text{for cohesionless soil} \end{cases} \tag{32.27}$$

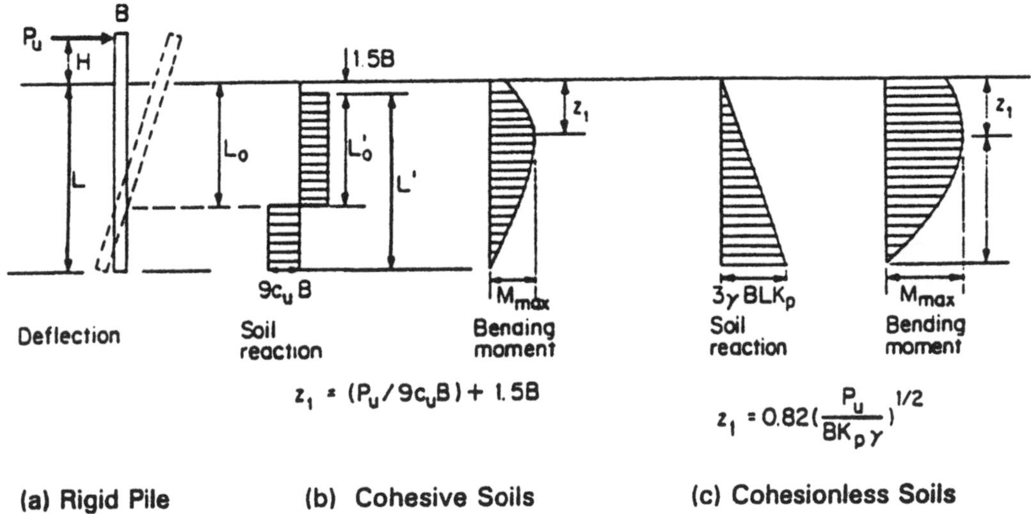

FIGURE 32.7 Free-head, short rigid piles — ultimate load conditions. (a) Rigid pile; (b) cohesive soils; (c) cohesionless soils. [After Broms (1964)[5,6]]

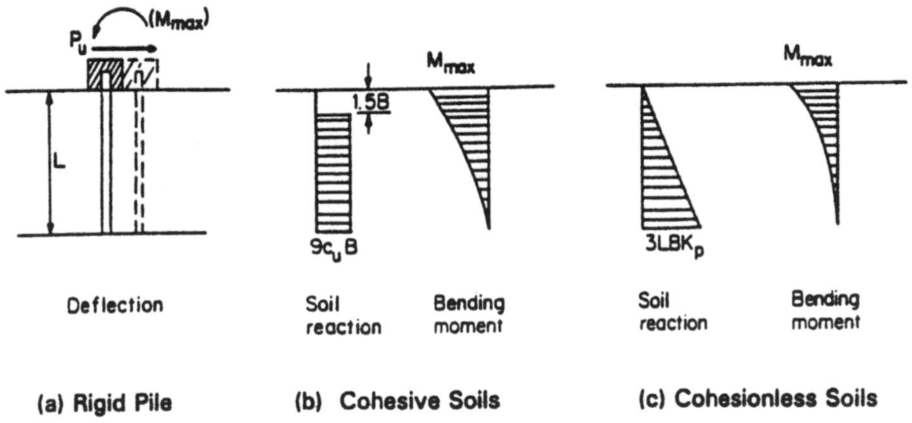

FIGURE 32.8 Fixed-head, short rigid piles — ultimate load conditions. (a) Rigid pile; (b) cohesive soils; (c) cohesionless soils. [After Broms (1964)[5,6]]

where

c_u = shear strength of the soil

K_p = coefficient of passive earth pressure, $K_p = \tan^2(45^o + \varphi/2)$ and φ is the friction angle of cohesionless soils (or sand and gravel)

p_0' = effective overburden pressure, $p_0' = \gamma_z'$ at a depth of z from the ground surface, where γ' is the effective unit weight of the soil

Ultimate Lateral Capacity for the Free-Head Condition

The ultimate lateral capacity P_u of a pile under the free-head condition is calculated by using the following formula:

$$P_u = \begin{cases} \left(\dfrac{L_0'^2 - 2L'L_0' + 0.5L'^2}{L' + H + 1.5B}\right)(9c_uB) & \text{for cohesive soil} \\ \dfrac{0.5BL^3K_p\gamma'}{H+L} & \text{for cohesionless soil} \end{cases} \tag{32.28}$$

where

L = embedded length of pile

H = distance of resultant lateral force above ground surface

B = pile diameter

L' = embedded pile length measured from a depth of $1.5B$ below the ground surface, or $L' = L - 1.5B$

L_0 = depth to center of rotation, and $L_0 = (H + 23L)/(2H + L)$

L_0' = depth to center of rotation measured from a depth of $1.5B$ below the ground surface, or $L_0' = L_0 - 1.5B$

Ultimate Lateral Capacity for the Fixed-Head Condition

The ultimate lateral capacity P_u of a pile under the fixed-head condition is calculated by using the following formula:

$$P_u = \begin{cases} 9c_uB(L - 1.5B) & \text{for cohesive soil} \\ 1.5\gamma'BL^2K_p & \text{for cohesionless soil} \end{cases} \tag{32.29}$$

32.5.3 Lateral Capacity and Deflection — *p–y* Method

One of the most commonly used methods for analyzing laterally loaded piles is the *p–y* method, in which soil reactions to the lateral deflections of a pile are treated as localized nonlinear springs based on the Winkler's assumption. The pile is modeled as an elastic beam that is supported on a deformable subgrade.

The *p–y* method is versatile and can be used to solve problems including different soil types, layered soils, nonlinear soil behavior; different pile materials, cross sections; and different pile head connection conditions.

Analytical Model and Basic Equation

An analytical model for pile under lateral loading with *p–y* curves is shown on Figure 32.9. The basic equation for the beam-on-a-deformable-subgrade problem can be expressed as

$$EI\frac{d^4y}{dx^4} - P_x\frac{d^2y}{dx^2} + p + q = 0 \tag{32.30}$$

where

y = lateral deflection at point x along the pile

EI = bending stiffness or flexural rigidity of the pile

P_x = axial force in beam column

p = soil reaction per unit length, and $p = -E_sy$; where E_s is the secant modulus of soil reaction.

q = lateral distributed loads

The following relationships are also used in developing boundary conditions:

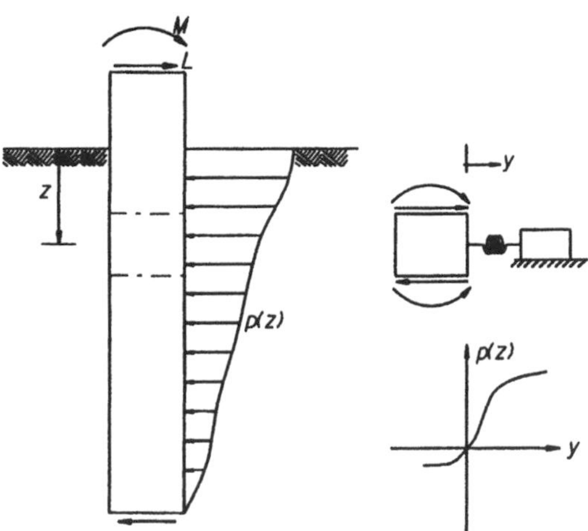

FIGURE 32.9 Analytical model for pile under lateral loading with *p–y* curves.

$$M = -EI\frac{d^4y}{dx^4} \tag{32.31}$$

$$Q = -\frac{dM}{dx} + P_x\frac{dy}{dx} \tag{32.32}$$

$$\theta = \frac{dy}{dx} \tag{32.33}$$

where M is the bending moment, Q is the shear force in the beam column, θ is the rotation of the pile.

The *p–y* method is a valuable tool in analyzing laterally loaded piles. Reasonable results are usually obtained. A computer program is usually required because of the complexity and iteration needed to solve the above equations using the finite-difference method or other methods. It should be noted that Winkler's assumption ignores the global effect of a continuum. Normally, if soil behaves like a continuum, the deflection at onc point will affect the deflections at other points under loading. There is no explicit expression in the *p–y* method since localized springs are assumed. Although *p–y* curves are developed directly from results of load tests and the influence of global interaction is included implicitly, there are cases where unexpected outcomes resulted. For example, excessively large shear forces will be predicted for large piles in rock by using the *p–y* method approach, where the effects of the continuum and the shear stiffness of the surrounding rock are ignored. The accuracy of the *p–y* method depends on the number of tests and the variety of tested parameters, such as geometry and stiffness of pile, layers of soil, strength and stiffness of soil, and loading conditions. One should be careful to extrapolate *p–y* curves to conditions where tests were not yet performed in similar situations.

Generation of *p–y* Curves

A *p–y* curve, or the lateral soil resistance p expressed as a function of lateral soil movement y, is based on backcalculations from test results of laterally loaded piles. The empirical formulations of *p–y* curves are different for different types of soil. *p–y* curves also depend on the diameter of the pile, the strength and stiffness of the soil, the confining overburden pressures, and the loading conditions. The effects of layered soil, battered piles, piles on a slope, and closely spaced piles are also usually considered. Formulation for soft clay, sand, and rock is provided in the following.

p–y Curves for Soft Clay

Matlock [35] proposed a method to calculate p–y curves for soft clays as shown on Figure 32.10. The lateral soil resistance p is expressed as

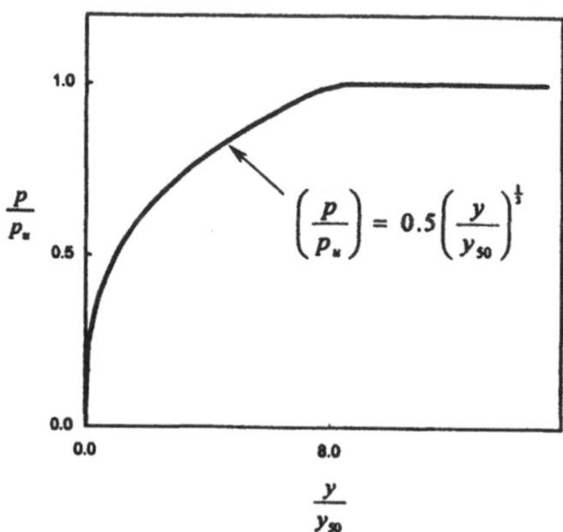

FIGURE 32.10 Characteristic shape of p–y curve for soft clay. [After Matlock, (1970)[35]]

$$p = \begin{cases} 0.5\left(\dfrac{y}{y_{50}}\right)^{1/3} p_u & y < y_p = 8y_{50} \\ p_u & y \geq y_p \end{cases} \tag{32.34}$$

in which
p_u = ultimate lateral soil resistance corresponding to ultimate shear stress of soil
y_{50} = lateral movement of soil corresponding to 50% of ultimate lateral soil resistance
y = lateral movement of soil

The ultimate lateral soil resistance p_u is calculated as

$$p_u = \begin{cases} \left(3 + \dfrac{\gamma' x}{c} + J\dfrac{x}{B}\right)cB & x < x_r = (6B)/\left(\dfrac{\gamma' B}{c} + J\right) \\ 9cB & x \geq x_r \end{cases} \tag{32.35}$$

where γ' is the effective unit weight, x is the depth from ground surface, c is the undrained shear strength of the clay, and J is a constant frequently taken as 0.5.

The lateral movement of soil corresponding to 50% of ultimate lateral soil resistance y_{50} is calculated as

$$y_{50} = 2.5\varepsilon_{50}B \tag{32.36}$$

where ε_{50} is the strain of soil corresponding to half of the maximum deviator stress. Table 32.12 shows the representative values of ε_{50}.

p–y Curves for Sands

Reese et al. [53] proposed a method for developing p–y curves for sandy materials. As shown on Figure 32.11, a typical p–y curve usually consists of the following four segments:

Segment	Curve type	Range of y	Range of p	p–y curve
1	Linear	0 to y_k	0 to p_k	$p=(kx)y$
2	Parabolic	y_k to y_m	p_k to p_m	$p=p_m\left(\frac{y}{y_m}\right)^n$
3	Linear	y_m to y_u	p_m to p_u	$p=p_m+\frac{p_u-p_m}{y_u-y_m}(y-y_m)$
4	Linear	$\geq y_u$	p_u	$p=p_u$

TABLE 32.12 Representative Values of ε_{50}

Consistency of Clay	Undrained Shear Strength, psf	ε_{50}
Soft	0–400	0.020
Medium stiff	400–1000	0.010
Stiff	1000–2000	0.007
Very stiff	2000–4000	0.005
Hard	4000–8000	0.004

1 psf = 0.048 kPa.

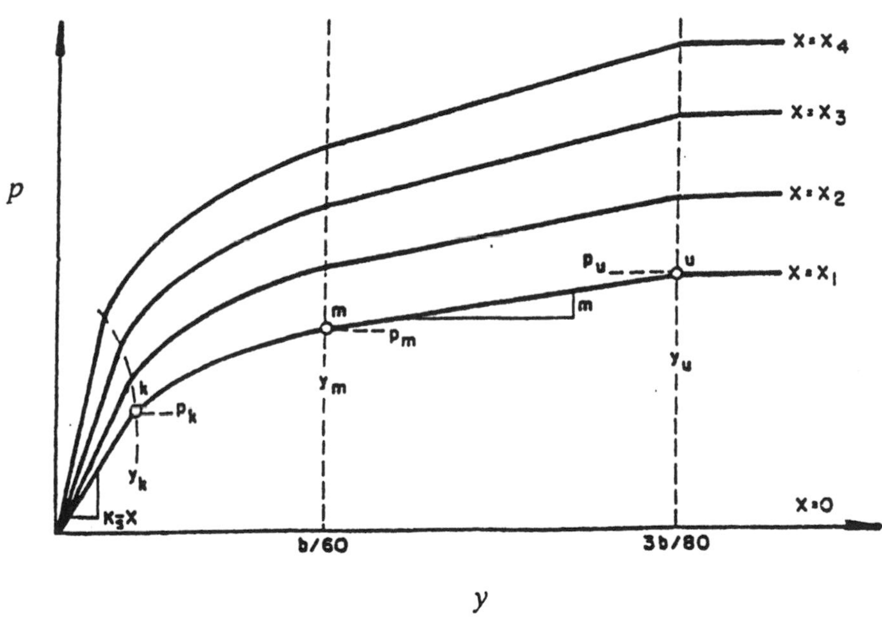

FIGURE 32.11 Characteristic shape of p–y curves for sand. [After Reese, et al. (1974)[53]]

where y_m, y_u, p_m, and p_u can be determined directly from soil parameters. The parabolic form of Segment 2, and the intersection with Segment 1 (y_k and p_k) can be determined based on y_m, y_u, p_m, and p_u as shown below.

Segment 1 starts with a straight line with an initial slope of kx, where x is the depth from the ground surface to the point where the p–y curve is calculated. k is a parameter to be determined based on relative density and is different whether above or below water table. Representative values of k are shown in Table 32.13.

TABLE 32.13 Friction Angle and Consistency

Relative to Water Table	Friction Angle and Consistency 29°–30° (Loose)	30°–36° (Medium Dense)	36°–40° (Dense)
Above	20 pci	60 pci	125 pci
Below	25 pci	90 pci	225 pci

1 pci = 272 kPa/m.

Segment 2 is parabolic and starts from end of Segment 1 at

$$y_k = \left[\frac{p_m / y_m}{(kx)^n}\right]^{1/(n-1)}$$

and $p_k = (kx)y_k$, the power of the parabolic

$$n = \frac{y_m}{p_m}\left(\frac{p_u - p_m}{y_u - y_m}\right)$$

Segments 3 and 4 are straight lines. y_m, y_u, p_m, and p_u are expressed as

$$y_m = \frac{b}{60} \tag{32.37}$$

$$y_u = \frac{3b}{80} \tag{32.38}$$

$$p_m = B_s p_s \tag{32.39}$$

$$p_u = A_s p_s \tag{32.40}$$

where b is the diameter of a pile; A_s and B_s are coefficients that can be determined from Figures 32.12 and 32.13, depending on either static or cyclic loading conditions; p_s is equal to the minimum of p_{st} and p_{sd}, as

$$p_{st} = \gamma x\left[\begin{array}{l}\dfrac{K_o x \tan\varphi \sin\beta}{\tan(\beta-\varphi)\cos\alpha} + \dfrac{\tan\beta}{\tan(\beta-\varphi)}(b + x\tan\beta\tan\alpha) \\ + K_o x \tan\beta(\tan\phi\tan\varphi - \tan\alpha) - K_a b\end{array}\right] \tag{32.41}$$

$$p_{sd} = K_a b x\gamma[\tan^8\beta - 1] + K_o b\gamma x\tan\phi\tan^4\beta \tag{32.42}$$

$$p = \min(p_{st}, p_{sd}) \tag{32.43}$$

in which φ is the friction angle of soil; α is taken as $\varphi/2$; β is equal to $45^o + \varphi/2$; K_o is the coefficient of the earth pressure at rest and is usually assumed to be 0.4; and K_a is the coefficient of the active earth pressure and equals to $\tan^2(45^o - \varphi/2)$.

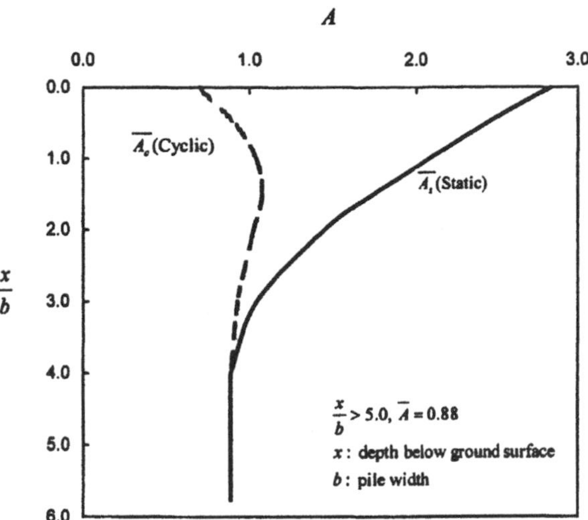

FIGURE 32.12 Variation of A_s with depth for sand. [After Reese, et al. (1974)[53]]

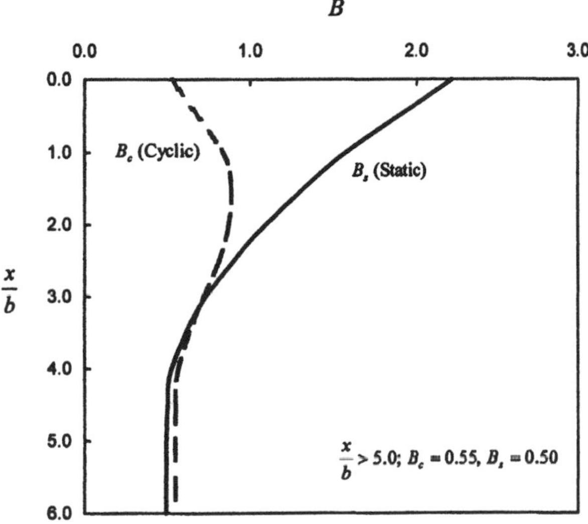

FIGURE 32.13 Variation of B_s with depth for sand. [After Reese, et al. (1974)[53]]

32.5.4 Lateral Spring: *p–y* Curves for Rock

Reese[86] proposed a procedure to calculate *p–y* curves for rock using basic rock and rock mass properties such as compressive strength of intact rock q_{ur}, rock quality designation (RQD), and initial modulus of rock E_{ir}. A description of the procedure is presented in the following.

A *p–y* curve consists of three segments:

$$\begin{aligned}
&\text{Segment 1: } p = K_{ir} y && \text{for } y \le y_a \\
&\text{Segment 2: } p = \frac{p_{ur}}{2}\left(\frac{y}{y_{rm}}\right)^{0.25} && \text{for } y_a < y < 16y_{rm} \qquad (32.44) \\
&\text{Segment 3: } p = p_{ur} && \text{for } y \ge 16y_{rm}
\end{aligned}$$

where p is the lateral force per unit pile length and y is the lateral deflection.

K_{ir} is the initial slope and is expressed as

$$K_{ir} = k_{ir} E_{ir} \tag{32.45}$$

k_{ir} is a dimensionless constant and is determined by

$$k_{ir} = \begin{cases} \left(100 + \dfrac{400 x_r}{3b}\right) & \text{for} \quad 0 \le x_r \le 3b \\ 500 & \text{for} \quad x_r > 3b \end{cases} \tag{32.46}$$

x_r = depth below bedrock surface, b is the width of the rock socket
E_{ir} = initial modulus of rock.

y_a is the lateral deflection separating Segment 1 and 2, and

$$y_a = \left(\frac{p_{ur}}{2 y_{rm}^{0.25} K_{ir}}\right)^{1.333} \tag{32.47}$$

where

$$y_{rm} = k_{rm} b \tag{32.48}$$

k_{rm} is a constant, ranging from 0.0005 to 0.00005.

p_{ur} is the ultimate resistance and can be determined by

$$P_{ur} = \begin{cases} a_r q_{ur} b \left(1 + 1.4 \dfrac{x_r}{b}\right) & \text{for} \quad 0 \le x_r \le 3b \\ 5.2 a_r q_{ur} b & \text{for} \quad x_r > 3b \end{cases} \tag{32.49}$$

where

q_{ur} = compressive strength of rock and α_r is a strength reduction factor determined by

$$\alpha_r = 1 - \frac{\text{RQD}}{150} \qquad 0 \le \text{RQD} \le 100 \tag{32.50}$$

RQD = rock quality designation for rock.

32.6 Grouped Foundations

32.6.1 General

Although a pile group is composed of a number of individual piles, the behavior of a pile group is not equivalent to the sum of all the piles as if they were separate individual piles. The behavior of a pile group is more complex than an individual pile because of the effect of the combination of piles, interactions between the piles in the group, and the effect of the pile cap. For example, stresses in soil from the loading of an individual pile will be insignificant at a certain depth below the pile tip. However, the stresses superimposed from all neighboring piles may increase the level of stress at that depth and result in considerable settlements or a bearing capacity failure, especially if there exists an underlying weak soil layer. The interaction and influence between piles usually diminish for piles spaced at approximately 7 to 8 diameters.

The axial and lateral capacity and the corresponding settlement and lateral deflection of a pile group will be discussed in the following sections.

32.6.2 Axial Capacity of Pile Group

The axial capacity of a pile group is the combination of piles in the group, with consideration of interaction between the piles. One way to account for the interaction is to use the group efficiency factor η_a, which is expressed as:

$$\eta_a = \frac{P_{Group}}{\sum_i P_{Single\_Pile,i}} \tag{32.51}$$

where P_{Group} is the axial capacity of a pile group. $\sum_i P_{Single\_Pile,i}$ is the sum of the axial capacity of all the individual piles. Individual piles are discussed in detail in Section 32.4. The group efficiency for axial capacity depends on many factors, such as the installation method, ground conditions, and the function of piles, which are presented in Table 32.14.

TABLE 32.14 Group Efficiency Factor for Axial Capacity

Pile Installation Method	Function	Ground Conditions	Expected Group Efficiency	Design Group Efficiency (with minimum spacing equal to 2.5 pile diameter)
Driven Pile	End bearing	Sand	1.0	1.0
	Side friction	Loose to medium dense sand	>1.0, up to 2.0	1.0, or increase with load test
	Side friction	Dense sand	May be ≧ 1.0	1.0
Drilled shaft	All	Sand	<1.0	0.67–1.0
Driven pile and drilled shaft	Side friction	Soft to medium stiff clay	<1.0	0.67–1.0
	End bearing	Soft to medium stiff clay	<1.0	0.67–1.0
	Side friction	Stiff clay	1.0	1.0
	End bearing	Stiff clay	1.0	1.0
	Side friction	Clay	<1.0	Also use "Group Block"
	End bearing	Clay, or underlying clay layers	<1.0	Also use "Group Block"

At close spacings, driven piles in loose to medium dense sand may densify the sand and consequently increase the lateral stresses and frictions along the piles. However, driven piles in dense sand may cause dilation of the sand and consequently cause heave and damage to other piles. The influence of spacing to the end bearing for sand is usually limited and the group efficiency factor η_a is taken as 1.0, under normal conditions.

For drilled piers in loose to medium dense sand, no densification of sand is made. The group efficiency factor η_a is usually less than 1.0 because of the influence of other close piles.

For driven piles in stiff to very stiff clay, the piles in a pile group tend to form a "group block" that behaves like a giant, short pile. The size of the group block is the extent of soil enclosed by the piles, including the perimeter piles as shown on Figure 32.14. The group efficiency factor η_a is usually equal to 1.0. For piles in soft to medium stiff clay, the group efficiency factor η_a is usually less than 1.0 because the shear stress levels are increased by loading from adjacent piles.

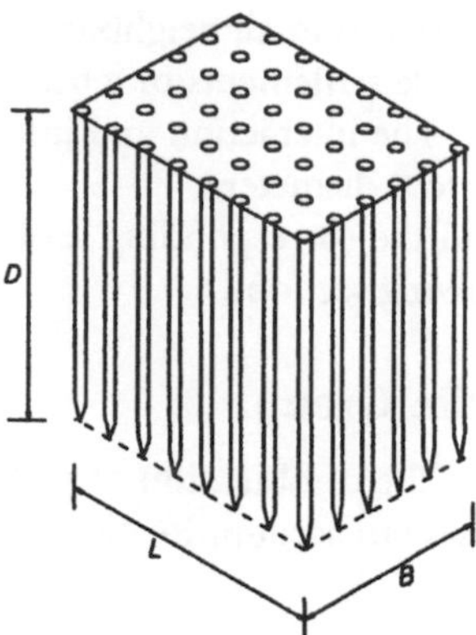

FIGURE 32.14 Block failure model for pile group in clay.

The group block method is also often used to check the bearing capacity of a pile group. The group block is treated as a large deep spread footing foundation and the assumed bottom level of the footing is different depending on whether the pile is end bearing or frictional. For end-bearing piles, the capacity of the group block is examined by assuming the bottom of the footing is at the tip of the piles. For frictional piles, the capacity of the group pile is checked by assuming that the bottom of the footing is located at ⅓ of the total embedded length above the tip. The bearing capacity of the underlying weaker layers is then estimated by using methods discussed in Chapter 31. The smaller capacity, by using the group efficiency approach, the group block approach, and the group block approach with underlying weaker layers, is selected as the capacity of the pile group.

32.6.3 Settlement of a Pile Group

The superimposed stresses from neighboring piles will raise the stress level below the tip of a pile substantially, whereas the stress level is much smaller for an individual pile. The raised stress level has two effects on the settlement of a pile group. The magnitude of the settlement will be larger for a pile group and the influence zone of a pile group will be much greater. The settlement of a pile group will be much larger in the presence of underlying highly compressible layers that would not be stressed under the loading of an individual pile.

The group block method is often used to estimate the settlement of a group. The pile group is simplified to an equivalent massive spread footing foundation except that the bottom of the footing is much deeper. The plane dimensions of the equivalent footing are outlined by the perimeter piles of the pile group. The method to calculate settlement of spread footings is discussed in Chapter 31. The assumed bottom level of the footing block is different depending on either end bearing or frictional piles. For end-bearing piles, the bottom of the footing is at the tip of the piles. For frictional piles, the bottom of the footing is located at ⅓ of total embedded length above the tip. In many cases, settlement requirement also is an important factor in the design of a pile group.

Vesic [79] introduced a method to calculate settlement of a pile group in sand which is expressed as

$$S_g = S_s\sqrt{\frac{B_g}{B_s}} \tag{32.52}$$

where

S_g = the settlement of a pile group

S_s = the settlement of an individual pile

B_g = the smallest dimension of the group block
B_s = the diameter of an individual pile

32.6.4 Lateral Capacity and Deflection of a Pile Group

The behavior of a pile group under lateral loading is not well defined. As discussed in the sections above, the lateral moment capacity is greater than the sum of all the piles in a group because piles would form couples resulting from their axial resistance through the action of the pile cap. However, the capacity of a pile group to resist lateral loads is usually smaller than the sum of separate, individual piles because of the interaction between piles.

The approach used by the University of Texas at Austin (Reese, O'Neil, and co-workers) provides a comprehensive and practical method to analyze a pile group under lateral loading. The finite-difference method is used to model the structural behavior of the foundation elements. Piles are connected through a rigid pile cap. Deformations of all the piles, in axial and lateral directions, and force and moment equilibrium are established. The reactions of soil are represented by a series of localized nonlinear axial and lateral springs. The theory and procedures to calculate axial and lateral capacity of individual piles are discussed in detail in Sections 32.4 and 32.5. A computer program is usually required to analyze a pile group because of the complexity and iteration procedure involving nonlinear soil springs.

The interaction of piles is represented by the lateral group efficiency factors, which is multiplied to the *p–y* curves for individual piles to reduce the lateral soil resistance and stiffness. Dunnavant and O'Neil [16] proposed a procedure to calculate the lateral group factors. For a particular pile *i*, the group factor is the product of influence factors from all neighboring piles *j*, as

$$\beta_i = \beta_0 \prod_{\substack{j=1 \\ j\neq i}}^{n} \beta_{ij} \tag{32.53}$$

where β_i is the group factor for pile *i*, β_0 is a total reduction factor and equals 0.85, β_{ij} is the influence factor from a neighboring pile *j*, and *n* is the total number of piles. Depending on the location of the piles *i* and *j* in relation to the direction of loading, β_{ij} is calculated as follows:

$$i \text{ is leading, or directly ahead of } j\ (\theta = 0°)\ \beta_l = \beta_{ij} = 0.69 + 0.5\log_{10}\left(\frac{S_{ij}}{B}\right) \le 1 \tag{32.54}$$

$$i \text{ is trailing, or directly behead of } j\ (\theta = 180°)\ \beta_t = \beta_{ij} = 0.48 + 0.6\log_{10}\left(\frac{S_{ij}}{B}\right) \le 1 \tag{32.55}$$

$$i \text{ and } j \text{ are abreast, or side-by-side } (\theta = 90°)\ \beta_s = \beta_{ij} = 0.78 + 0.36\log_{10}\left(\frac{S_{ij}}{B}\right) \le 1 \tag{32.56}$$

where S_{ij} is the center-to-center distance between *i* and *j*, *B* is the diameter of the piles *i* and *j*, and θ is the angle between the loading direction and the connection vector from *i* to *j*. When the piles *i* and *j* are at other angles to the direction of loading, β_{ij} is computed by interpolation, as

$$0° < \theta < 90° \qquad \beta_{\theta 1} = \beta_{ij} = \beta_l + (\beta_s - \beta_l)\frac{\theta}{90} \tag{32.57}$$

$$90° < \theta < 180° \qquad \beta_{\theta 2} = \beta_{ij} = \beta_t + (\beta_s - \beta_t)\frac{\theta - 90}{90} \tag{32.58}$$

In cases that the diameters of the piles i and j are different, we propose to use the diameter of pile j. To avoid an abrupt change of β_0 from 0.85 to 1.0, we propose to use:

$$\beta_o = \begin{cases} 0.85 & \text{for} \quad \frac{S_{ij}}{B_j} \le 3 \\ 0.85 + 0.0375\left(\frac{S_{ij}}{B_j} - 3\right) & \text{for} \quad 3 < \frac{S_{ij}}{B_j} < 7 \\ 1.0 & \text{for} \quad \frac{S_{ij}}{B_j} \ge 7 \end{cases}$$

32.7 Seismic Design

Seismic design of deep bridge foundations is a broad issue. Design procedures and emphases vary with different types of foundations. Since pile groups, including driven piles and drilled cast-in-place shafts, are the most popular types of deep bridge foundations, following discussion will concentrate on the design issues for pile group foundations only.

In most circumstances, seismic design of pile groups is performed to satisfy one or more of the following objectives:

- Determine the capacity and deflection of the foundation under the action of the seismic lateral load;
- Provide the foundation stiffness parameters for dynamic analysis of the overall bridge structures; and
- Ensure integrity of the pile group against liquefaction and slope instability induced ground movement.

32.7.1 Seismic Lateral Capacity Design of Pile Groups

In current practice, seismic lateral capacity design of pile groups is often taken as the same as conventional lateral capacity design (see Section 32.5). The seismic lateral force and the seismic moment from the upper structure are first evaluated for each pile group foundation based on the tributary mass of the bridge structure above the foundation level, the location of the center of gravity, and the intensity of the ground surface acceleration. The seismic force and moment are then applied on the pile cap as if they were static forces, and the deflections of the piles and the maximum stresses in each pile are calculated and checked against the allowable design values. Since seismic forces are of transient nature, the factor of safety required for resistance of seismic load can be less than those required for static load. For example, in the Caltrans specification, it is stipulated that the design seismic capacity can be 33% higher than the static capacity [9].

It should be noted that in essence the above procedure is pseudostatic, only the seismic forces from the upper structure are considered, and the effect of seismic ground motion on the behavior of pile group is ignored. The response of a pile group during an earthquake is different from its response to a static lateral loading. As seismic waves pass through the soil layers and cause the soil layers to move laterally, the piles are forced to move along with the surrounding media. Except for the case of very short piles, the pile cap and the pile tip at any moment may move in different directions. This movement induces additional bending moments and stresses in the piles.

Depending on the intensity of the seismic ground motion and the characteristics of the soil strata, this effect can be more critical to the structural integrity of the pile than the lateral load from the upper structure.

Field measurements (e.g., Tazoh et al. [70]), post-earthquake investigation (e.g., Seismic Advisory Committee, [65]), and laboratory model tests (e.g., Nomura et al., [43]) all confirm that seismic ground movements dictate the maximum responses of the piles. The more critical situation is when the soil profile consists of soft layer(s) sandwiched by stiff layers, and the modulus contrast among the layers is large. In this case, local seismic moments and stresses in the pile section close to the soft layer/hard layer interface may very well be much higher than the moments and stresses caused by the lateral seismic loads from the upper structure. If the site investigation reveals that the underground soil profile is of this type and the bridge is of critical importance, it is desirable that a comprehensive dynamic analysis be performed using one of more sophisticated computer programs capable of modeling the dynamic interaction between the soil and the pile system, e.g., SASSI [33]. Results of such dynamic analysis can provide a better understanding of the seismic responses of a pile group.

32.7.2 Determination of Pile Group Spring Constants

An important aspect in bridge seismic design is to determine, through dynamic analysis, the magnitude and distribution of seismic forces and moments in the bridge structure. To accomplish this goal, the characteristics of the bridge foundation must be considered appropriately in an analytical model.

At the current design practice, the force–displacement relationships of a pile foundation are commonly simplified in an analytical model as a stiffness matrix, or a set of translational and rotational springs. The characteristics of the springs depend on the stiffness at pile head for individual piles and the geometric configuration of piles in the group. For a pile group consisting of vertical piles, the spring constants can be determined by the following steps:

- The vertical and lateral stiffnesses at the pile head of a single pile, K_{vv} and K_{hh}, are first evaluated based on the pile geometry and the soil profile. These values are determined by calculating the displacement at the pile head corresponding to a unit force. For many bridge foundations, a rigid pile cap can be assumed. Design charts are available for uniform soil profiles (e.g., NAVFAC [42]). For most practical soil profiles, however, it is convenient to use computer programs, such as APILE [18] and LPILE [17], to determine the single pile stiffness values. It should be noted that the force–deformation behavior of a pile is highly nonlinear. In evaluating the stiffness values, it is desirable to use the secant modulus in the calculated pile-head force–displacement relationship compatible to the level of pile-head displacement to be developed in the foundation. This is often an iterative process.

  In calculating the lateral stiffness values, it is common practice to introduce a group factor η, $\eta \leq 1.0$, to account for the effect of the other piles in the same group. The group factor depends on the relative spacing *S/D* in the pile group, where *S* is the spacing between two piles and *D* is the diameter of the individual pile. There are studies reported in the literature about the dynamic group factors for pile groups of different configurations. However, in the current design practice, static group factors are used in calculation of the spring constants. Two different approaches exist in determining the group factor: one is based on reduction of the subgrade reaction moduli; the other is based on the measurement of plastic deformation of the pile group. Since the foundation deformations in the analysis cases involving the spring constants are mostly in the small-strain range, the group factors based on subgrade reaction reduction should be used (e.g., NAVFAC [42]).
- The spring constants of the pile group can be calculated using the following formulas:

$$K_{G,x} = \sum_{i=1}^{N} K_{hh,i} \tag{32.59}$$

$$K_{G,y} = \sum_{i=1}^{N} K_{hh,i} \tag{32.60}$$

$$K_{G,z} = \sum_{i=1}^{N} K_{vv,i} \tag{32.61}$$

$$K_{G,yy} = \sum_{i=1}^{N} K_{vv,i} \cdot x_i^2 \tag{32.62}$$

$$K_{G,xx} = \sum_{i=1}^{N} K_{vv,i} \cdot y_i^2 \tag{32.63}$$

where $K_{G,x}$, $K_{G,y}$, $K_{G,z}$ are the group translational spring constants, $K_{G,yy}$, $K_{G,xx}$ are the group rotational spring constants with respect to the center of the pile cap. All springs are calculated at the center of the pile cap; $K_{vv,i}$ and $K_{hh,i}$ are the lateral and vertical stiffness values at pile head of the ith pile; x_i, y_i are the coordinates of the ith pile in the group; and N is the total number of piles in the group.

In the above formulas, the bending stiffness of a single pile at the pile top and the off-diagonal stiffness terms are ignored. For most bridge pile foundations, these ignored items have only minor significance. Reasonable results can be obtained using the above simplified formulas.

It should be emphasized that the behavior of the soil–pile system is greatly simplified in the concept of "spring constant." The responses of a soil–pile structure system are complicated and highly nonlinear, frequency dependent, and are affected by the inertia/stiffness distribution of the structure above ground. Therefore, for critical structures, it is advisable that analytical models including the entire soil–pile structure system should be used in the design analysis.

32.7.3 Design of Pile Foundations against Soil Liquefaction

Liquefaction of loose soil layers during an earthquake poses a serious hazard to pile group foundations. Field observations and experimental studies (e.g., Nomura et al. [43], Miyamoto et al. [40], Tazoh and Gazetas [69], Boulanger et al. [4]) indicate that soil liquefaction during an earthquake has significant impacts on the behavior of pile groups and superstructures. The impacts are largely affected by the intensity of liquefaction-inducing earthquakes and the relative locations of the liquefiable loose soil layers. If a loose layer is close to the ground surface and the earthquake intensity is moderate, the major effect of liquefaction of the loose layer is to increase the fundamental period of the foundation–structure system, causing significant lateral deflection of the pile group and superstructure. For high-intensity earthquakes, and especially if the loose soil layer is sandwiched in hard soil layers, liquefaction of the loose layer often causes cracking and breakage of the piles and complete loss of capacity of the foundation, thus the collapse of the superstructure.

There are several approaches proposed in the literature for calculation of the dynamic responses of a pile or a pile group in a liquefied soil deposit. In current engineering practice, however, more emphasis is on taking proper countermeasures to mitigate the adverse effect of the liquefaction hazard. These mitigation methods include

- Densify the loose, liquefiable soil layer. A stone column is often satisfactory if the loose layer is mostly sand. Other approaches, such as jet grouting, deep soil mixing with cementing agents, and *in situ* vibratory densification, can all be used. If the liquefiable soil layer is close to the ground surface, a complete excavation and replacement with compacted engineering fill is sometimes also feasible.
- Isolate the pile group from the surrounding soil layers. This is often accomplished by installing some types of isolation structures, such as sheet piles, diaphragm walls, soil-mixing piles, etc., around the foundation to form an enclosure. In essence, this approach creates a huge block surrounding the piles with increased lateral stiffness and resistance to shear deformation while limiting the lateral movement of the soil close to the piles.
- Increase the number and dimension of the piles in a foundation and therefore increase the lateral resistance to withstand the forces induced by liquefied soil layers. An example is 10 ft (3.3 m) diameter cast-in-steel shell piles used in bridge seismic retrofit projects in the San Francisco Bay Area following the 1989 Loma Prieta earthquake.

References

1. API, *API Recommended Practice for Planning, Designing and Constructing Fixed Offshore Platforms*, 15th ed., API RP2A, American Petroleum Institute, 115 pp, 1984.
2. Awoshika, K. and L. C. Reese, Analysis of Foundation with Widely-Spaced Batter Piles, Research Report 117-3F, Center for Highway Research, The University of Texas at Austin, February, 1971.
3. Berezantzev, V. G., V. S. Khristoforov, and V. N. Golubkov, Load bearing capacity and deformation of piled foundations, *Proc. 5th Int. Conf. Soil Mech.*, Paris, 2, 11–15, 1961.
4. Boulanger, R. W., D. W. Wilson, B. L. Kutter, and A. Abghari, Soil–pile-structure interaction in liquefiable sand, *Transp. Res. Rec.*, 1569, April, 1997.
5. Broms, B. B., Lateral resistance of piles in cohesive soils, *Proc. ASCE, J. Soil Mech. Found. Eng. Div.*, 90(SM2), 27–64, 1964.
6. Broms, B. B., Lateral resistance of piles in cohesionless soils, *Proc. ASCE J. Soil Mech. Found. Eng. Div.*, 90(SM3), 123–156, 1964.
7. Burland, J. B., Shaft friction of piles in clay — a simple fundamental approach, *Ground Eng.*, 6(3), 30–42, 1973.
8. Bustamente, M. and L. Gianeselli, Pile bearing capacity prediction by means of static penetrometer CPT, *Proc. of Second European Symposium on Penetration Testing* (ESOPT II), Vol. 2, A. A. Balkema, Amsterdam, 493–500, 1982.
9. Caltrans, Bridge Design Specifications, California Department of Transportation, Sacramento, 1990.
10. CGS, *Canadian Foundation Engineering Manual*, 3rd ed., Canadian Geotechnical Society, BiTech Publishers, Vancouver, 512 pp, 1992.
11. Carter, J. P. and F. H. Kulhawy, Analysis and Design of Drilled Shaft Foundations Socketed into Rock, EPRI Report EI-5918, Electric Power Research Institute, Palo Alto, CA, 1988.
12. Crapps, D. K., Design, construction and inspection of drilled shafts in limerock and limestone, paper presented at the Annual Meeting of the Florida Section of ASCE, 1986.
13. De Ruiter, J. and F. L., Beringen, Pile foundations for large North Sea structures, *Marine Geotechnol.*, 3(2), 1978.
14. Dennis, N. D., Development of Correlations to Improve the Prediction of Axial Pile Capacity, Ph.D. dissertation, University of Texas at Austin, 1982.
15. Desai, C. S. and J. T. Christian, *Numerical Methods in Geotechnical Engineering*, McGraw-Hill Book, New York, 1977.

16. Dunnavant, T. W. and M. W. O'Neil, Evaluation of design-oriented methods for analysis of vertical pile groups subjected to lateral load, *Numerical Methods in Offshore Piling*, Institut Francais du Petrole, Labortoire Central des Ponts et Chausses, 303–316, 1986.
17. Ensoft Inc., Lpile Plus for Windows. Version 3.0. A Computer Program for Analysis of Laterally Loaded Piles, Austin, TX, 1997.
18. Ensoft Inc., Apile Plus. Version 3.0, Austin, TX, 1998.
19. Fellenius, B. H., in an ASCE meeting in Boston as quoted by R. E. Olson in 1991, Capacity of Individual Piles in Clay, internal report, 1986.
20. Fellenius, B. H., The critical depth — how it came into being and why it does not exist, *Proc. of the Inst. Civil Eng., Geotech. Eng.*, 108(1), 1994.
21. Focht, J. A. and K. J. Koch, Rational analysis of the lateral performance of offshore pile groups, in *Proceedings, Fifth Offshore Technology Conference*, Vol. 2, Houston, 701–708, 1973.
22. Geordiadis, M., Development of *p–y* curves for layered soils, in *Proceedings, Geotechnical Practice in Offshore Engineering*, ASCE, April, 1983, 536–545.
23. Goudreault, P. A. and B. H. Fellenius, A Program for the Design of Piles and Piles Groups Considering Capacity, Settlement, and Dragload Due to Negative Skin Friction, 1994.
24. Gupton, C. and T. Logan, Design Guidelines for Drilled Shafts in Weak Rocks in South Florida, Preprint, Annual Meeting of South Florida Branch of ASCE, 1984.
25. Horvath, R. G. and T. C. Kenney, Shaft resistance of rock-socketed drilled piers, in *Symposium on Deep Foundations*, ASCE National Convention, Atlanta, GA, 1979, 182–214.
26. Janbu, N., Static bearing capacity of friction piles, in *Proc. 6th European Conference on Soil Mech. & Found. Eng.*, Vol. 1.2, 1976, 479–488.
27. Kishida, H., Ultimate bearing capacity of piles driven into loose sand, *Soil Foundation*, 7(3), 20–29, 1967.
28. Kraft L. M., J. A. Focht, and S. F. Amerasinghe, Friction capacity of piles driven into clay, *J. Geot. Eng. Div. ASCE*, 107(GT 11), 1521–1541, 1981.
29. Kubo, K., Experimental Study of Behavior of Laterally Loaded Piles, Report, Transportation Technology Research Institute, Vol. 12, No. 2, 1962.
30. Kulhawy, F. H., Transmission Line Structures Foundations for Uplift-Compression Loading, Report No. EL-2870, Report to the Electrical Power Research Institute, Geotechnical Group, Cornell University, Ithaca, NY, 1983.
31. Kulhawy, F. H., Limiting tip and side resistance: fact or fallacy? in *Proc. of the American Society of Civil Engineers, ASCE, Symposium on Analysis and Design of Pile Foundations*, R. J. Meyer, Ed., San Francisco, 1984, 80–89.
32. Kulhawy, F. H. and K. K. Phoon, Drilled shaft side resistance in clay soil or rock, in *Geotechnical Special Publication No. 38, Design and Performance of Deep Foundations: Piles and Piers in Soil to Soft Rock*, Ed. P. P. Nelson, T. D. Smith, and E. C. Clukey, Eds., ASCE, 172–183, 1993.
33. Lysmer, J., M. Tabatabaie-Raissi, F. Tajirian, S. Vahdani, and F. Ostadan, SASSI — A System for Analysis of Soil-Structure Interaction, Report No. UCB/GT/81-02, Department of Civil Engineering, University of California, Berkeley, April, 1981.
34. McVay, M. C., F. C. Townsend, and R. C. Williams, Design of socketed drilled shafts in limestone, *J. Geotech. Eng.*, 118-GT10, 1626–1637, 1992.
35. Matlock, H., Correlations for design of laterally-loaded piles in soft clay, Paper No. OTC 1204, *Proc. 2nd Annual Offshore Tech. Conf.*, Vol. 1, Houston, TX, 1970, 577–594.
36. Menard, L. F., Interpretation and application of pressuremeter test results, *Sols-Soils*, Paris, 26, 1–23, 1975.
37. Meyerhof, G. G., Penetration tests and bearing capacity of cohesionless soils, *J. Soil Mech. Found. Div. ASCE*, 82(SM1), 1–19, 1956.
38. Meyerhof, G. G., Bearing capacity and settlement of pile foundations, *J. Geotech. Eng. Div. ASCE*, 102(GT3), 195–228, 1976.

39. Mitchell, J. K. and T. A. Lunne, Cone resistance as a measure of sand strength, *Proc. ASCE J. Geotech. Eng. Div.*, 104(GT7), 995–1012, 1978.
40. Miyamoto, Y., Y. Sako, K. Miura, R. F. Scott, and B. Hushmand, Dynamic behavior of pile group in liquefied sand deposit, *Proceedings, 10th World Conference on Earthquake Engineering*, 1992, 1749–1754.
41. Mosher, R. L., Load Transfer Criteria for Numerical Analysis of Axially Loaded Piles in Sand, U.S. Army Engineering Waterways Experimental Station, Automatic Data Processing Center, Vicksburg, MI, January, 1984.
42. NAVFAC, Design Manual DM7.02: Foundations and Earth Structures, Department of the Navy, Naval Facilities Engineering Command, Alexandra, VA, September, 1986.
43. Nomura, S., K. Tokimatsu, and Y. Shamoto, Behavior of soil-pile-structure system during liquefaction, in *Proceedings, 8th Japanese Conference on Earthquake Engineering*, Tokyo, December 12–14, Vol. 2, 1990, 1185–1190.
44. O'Neil M. W. and L. C. Reese, Load transfer in a slender drilled pier in sand, ASCE, ASCE Spring Convention and Exposition, Pittsburgh, PA, Preprint 3141, April, 1978, 30 pp.
45. O'Neil, M. W. and S. A. Sheikh, Geotechnical behavior of underrems in pleistocene clay, in *Drilled Piers and Caissons, II*, C. N. Baker, Jr., Ed., ASCE, May, 57–75, 1985.
46. O'Neil, M. W., F. C. Townsend, K. M. Hassan, A. Buller, and P. S. Chan, Load Transfer for Drilled Shafts in Intermediate Geo'materials, FHWA-RD-95-172, November, 184 pp, 1996.
47. Osterberg, J. O., New load cell testing device, in *Proc. 14th Annual Conf.*, Vol. 1, Deep Foundations Institute, 1989, 17–28.
48. Pells, P. J. N. and R. M. Turner, Elastic solutions for the design and analysis of rock-socketed piles, *Can. Geotech. J.*, 16(3), 481–487, 1979.
49. Pells, P. J. N. and R. M. Turner, End bearing on rock with particular reference to sandstone, in Structural Foundations on Rock, *Proc. Intn. Conf. on Structural Found. on Rock*, Vol. 1, Sydney, May 7–9, 1980, 181–190.
50. Poulos, H. G. and E. H. Davis, *Pile Foundation Analysis and Design*, John Wiley & Sons, New York, 1980.
51. Reese, L. C. and H. Matlock, Behavior of a two-dimensional pile group under inclined and eccentric loading, in *Proc. Offshore Exploration Conf.*, Long Beach, CA, February, 1966.
52. Reese, L. C. and M. W. O'Neil, The analysis of three-dimensional pile foundations subjected to inclined and eccentric loads, *Proc. ASCE Conf.*, September, 1967, 245–276.
53. Reese, L. C., W. R. Cox, and F. D. Koop, Analysis of laterally loaded piles in sand, paper OTC 2080, *Proc. Fifth Offshore Tech. Conf.*, Houston, TX, 1974.
54. Reese, L. C., W. R. Cox, and F. D. Koop, Field testing and analysis of laterally loaded piles in stiff clay, paper OTC 2313, in *Proc. Seventh Offshore Tech. Conf.*, Houston, TX, 1975.
55. Reese, L. C. and S. J. Wright, Drilled Shafts: Design and Construction, Guideline Manual, Vol. 1; Construction Procedures and Design for Axial Load, U.S. Department of Transportation, Federal Highway Administration, July, 1977.
56. Reese, L. C., Behavior of Piles and Pile Groups under Lateral Load, a report submitted to the Federal Highway Administration, Washington, D.C., July, 1983, 404 pp.
57. Reese, L. C. and M. W. O'Neil, Drilled Shafts: Construction Procedures and Design Methods, U.S. Department of Transportation, Federal Highway Administration, McLean, VA, 1988.
58. Reynolds, R. T. and T. J. Kaderabek, Miami Limestone Foundation Design and Construction, Preprint No. 80-546, South Florida Convention, ASCE, 1980.
59. Robertson, P. K., R. G. Campanella, et al., Axial Capacity of Driven Piles in Deltaic Soils Using CPT, Penetration Testing 1988, ISOPT-1, De Ruite, Ed., 1988.
60. Rosenberg, P. and N. L. Journeaux, Friction and end bearing tests on bedrock for high capacity socket design, *Can. Geotech. J.*, 13(3), 324–333, 1976.
61. Rowe, R. K. and H. H. Armitage, Theoretical solutions for axial deformation of drilled shafts in rock, *Can. Geotech. J.*, Vol. 24(1), 114–125, 1987.

62. Rowe, R. K. and H. H. Armitage, A design method for drilled piers in soft rock, *Can. Geotech. J.*, 24(1), 126–142, 1987.
63. Schmertmann, J. H., Guidelines for Cone Penetration Test: Performance and Design, FHWA-TS-78-209, Federal Highway Administration, Office of Research and Development, Washington, D.C., 1978.
64. Seed, H. B. and L. C. Reese, The action of soft clay along friction piles, *Trans. Am. Soc. Civil Eng.*, Paper No. 2882, 122, 731–754, 1957.
65. Seismic Advisory Committee on Bridge Damage, Investigation Report on Highway Bridge Damage Caused by the Hyogo-ken Nanbu Earthquake, Japan Ministry of Construction, 1995.
66. Skempton, A. W., The bearing capacity of clay, *Proc. Building Research Congress*, Vol. 1, 1951, 180–189.
67. Skempton, A. W., Cast-*in situ* bored piles in London clay, *Geotechnique*, 9, 153–173, 1959.
68. Sörensen, T. and B. Hansen, Pile driving formulae — an investigation based on dimensional considerations and a statistical analysis, *Proc. 4th Int. Conf. Soil Mech.*, London, 2, 61–65, 1957.
69. Tazoh, T. and G. Gazetas, Pile foundations subjected to large ground deformations: lessons from kobe and research needs, *Proceedings, 11th World Conference on Earthquake Engineering*, Paper No. 2081, 1996.
70. Tazoh, T., K. Shimizu, and T. Wakahara, Seismic Observations and Analysis of Grouped Piles, Dynamic Response of Pile Foundations — Experiment, Analysis and Observation, ASCE Geotechnical Special Publication No. 11, 1987.
71. Terzaghi, K., *Theoretical Soil Mechanics*, John Wiley & Sons, New York, 510 pp, 1943.
72. Tomlinson, M. J., The adhesion of piles in clay soils, *Proc., Fourth Int. Conf. Soil Mech. Found. Eng.*, 2, 66–71, 1957.
73. Tomlinson, M. J., Some effects of pile driving on skin friction, in *Behavior of Piles*, Institution of Civil Engineers, London, 107–114, and response to discussion, 149–152, 1971.
74. Touma, F. T. and L. C. Reese, Load Tests of Instrumented Drilled Shafts Constructed by the Slurry Displacement Method. Research report conducted under Interagency contract 108 for the Texas Highway Department, Center for Highway Research, the University of Texas at Austin, January, 1972, 79 pp.
75. Townsend, F. C., Comparison of deep foundation load test method, in *FHWA 25th Annual Southeastern Transportation Geotechnical Engineering Conference*, Natchez, MS, October 4–8, 1993.
76. Vesic, A. S., Ultimate loads and settlements of deep foundations in sand, in *Proc. Symp. on Bearing Capacity and Settlement of Foundations*, Duke University, Durham, NC, 1967.
77. Vesic, A. S., Load transfer in pile-soil system, in *Design and Installation of Pile Foundations and Cellar Structures*, Fang and Dismuke, Eds., Envo, Lehigh, PA, 47–74, 1970.
78. Vesic, A. S., Expansion of cavities in infinite soil mass, *Proc. ASCE J. Soil Mech. Found. Eng. Div.*, 98(SM3), 1972.
79. Vesic, A. S., Design of Pile Foundations, National Cooperative Highway Research Program Synthesis 42, Transportation Research Board, 1977.
80. Vijayvergiya, V. N. and J. A. Focht, A new way to predict the capacity of piles in clay, *Offshore Technology Conference*, Vol. 2, Houston, TX, 1972, 965–874.
81. Vijayvergiya, V. N., Load-movement characteristics of piles, *Proc. of Ports '77 Conf.*, Long Beach, CA, 1977.
82. Welsh, R. C. and Reese, L. C., Laterally Loaded Behavior of Drilled Shafts, Research Report No. 3-5-65-89, conducted for Texas Highway Department and U.S. Department of Transportation, Federal Highway Administration, Bureau of Public Roads, by Center for Highway Research, the University of Texas at Austin, May, 1972.
83. Weltman, A. J. and P. R. Healy, Piling in boulder clay and other glacial tills, *Construction Industry Research and Information Association*, Report PG5, 1978.

84. Williams, A. F., I. W. Johnson, and I. B. Donald, The Design of Socketed Piles in Weak Rock, *Structural Foundations on Rock, Proc. Int. Conf. on Structural Found. on Rock*, Vol. 1, Sydney, May 7–9, 1980, 327–347.
85. Woodward, R. J., W. S. Gardner, and D. M. Greer, *Drilled Pier Foundations,* McGraw-Hill, New York, 1972.
86. Reese, L. C., Analysis of Laterally Loaded Piles in Weak Rock, *J. of Geotech & Geoenvironmental Engr.*, Vol. 123, No., 11, 1010–1017, 1997.

Section IV

Seismic Design

33

Geotechnical Earthquake Considerations

Steven Kramer
University of Washington

Charles Scawthorn
EQE International

33.1 Introduction

Earthquakes are naturally occurring broad-banded vibratory ground motions, that are due to a number of causes including tectonic ground motions, volcanism, landslides, rockbursts, and man-made explosions, the most important of which are caused by the fracture and sliding of rock along tectonic **faults** within the Earth's crust. For most earthquakes, shaking and ground failure are the dominant and most widespread agents of damage. Shaking near the actual earthquake rupture lasts only during the time when the fault ruptures, a process which takes seconds or at most a few minutes. The seismic waves generated by the rupture propagate long after the movement on the fault has stopped, however, spanning the globe in about 20 min. Typically, earthquake ground

0-8493-7434-0/00/$0.00+$.50

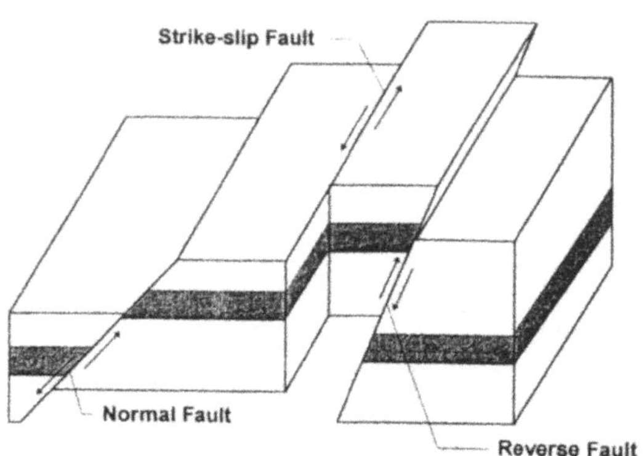

FIGURE 33.1 Fault types.

motions are powerful enough to cause damage only in the near field (i.e., within a few tens of kilometers from the causative fault) — in a few instances, long-period motions have caused significant damage at great distances, to selected lightly damped structures, such as in the 1985 Mexico City earthquake, where numerous collapses of mid- and high-rise buildings were due to a magnitude 8.1 earthquake occurring at a distance of approximately 400 km from Mexico City.

33.2 Seismology

Plate Tectonics: In a global sense, tectonic earthquakes result from motion between a number of large plates comprising the Earth's crust or lithosphere (about 15 in total). These plates are driven by the convective motion of the material in the Earth's mantle, which in turn is driven by heat generated at the Earth's core. Relative plate motion at the fault interface is constrained by friction and/or **asperities** (areas of interlocking due to protrusions in the fault surfaces). However, strain energy accumulates in the plates, eventually overcomes any resistance, and causes slip between the two sides of the fault. This sudden slip, termed **elastic rebound** by Reid [49] based on his studies of regional deformation following the 1906 San Francisco earthquake, releases large amounts of energy, which constitute the earthquake. The location of initial radiation of seismic waves (i.e., the first location of dynamic rupture) is termed the **hypocenter**, while the projection on the surface of the Earth directly above the hypocenter is termed the **epicenter**. Other terminology includes **near-field** (within one source dimension of the epicenter, where source dimension refers to the length of faulting), **far-field** (beyond near-field) and **meizoseismal** (the area of strong shaking and damage). Energy is radiated over a broad spectrum of frequencies through the Earth, in **body waves** and **surface waves** [4]. Body waves are of two types: P waves (transmitting energy via push–pull motion) and slower S waves (transmitting energy via shear action at right angles to the direction of motion). Surface waves are also of two types: horizontally oscillating **Love waves** (analogous to S body waves) and vertically oscillating **Rayleigh waves.**

Faults are typically classified according to their sense of motion, Figure 33.1. Basic terms include **transform** or **strike slip** (relative fault motion occurs in the horizontal plane, parallel to the strike of the fault), **dip-slip** (motion at right angles to the strike, up- or down-slip), **normal** (dip-slip motion, two sides in tension, move away from each other), **reverse** (dip-slip, two sides in compression, move toward each other), and **thrust** (low-angle reverse faulting).

Generally, earthquakes will be concentrated in the vicinity of faults; faults that are moving more rapidly than others will tend to have higher rates of seismicity, and larger faults are more likely than others to produce a large event. Many faults are identified on regional geologic maps, and useful information on fault location and displacement history is available from local and national geologic surveys in areas of high seismicity. An important development has been the growing recognition of **blind thrust faults**, which emerged as a result of the several earthquakes in the 1980s, none of which was accompanied by surface faulting [61].

33.3 Measurement of Earthquakes

Magnitude

An individual earthquake is a unique release of strain energy — quantification of this energy has formed the basis for measuring the earthquake event. C.F. Richter [51] was the first to define earthquake **magnitude**, as

$$M_L = \log A - \log A_0 \tag{33.1}$$

where M_L is **local magnitude** (which Richter only defined for Southern California), A is the maximum trace amplitude in microns recorded on a standard Wood–Anderson short-period torsion seismometer at a site 100 km from the epicenter, and $\log A_0$ is a standard value as a function of distance, for instruments located at distances other than 100 km and less than 600 km. A number of other magnitudes have since been defined, the most important of which are **surface wave** magnitude M_S, body wave magnitude m_b, and **moment magnitude** M_W. Magnitude can be related to the total energy in the expanding wave front generated by an earthquake, and thus to the total energy release — an empirical relation by Richter is

$$\log_{10} E_s = 11.8 + 1.5 M_s \tag{33.2}$$

where E_S is the total energy in ergs. Due to the observation that deep-focus earthquakes commonly do not register measurable surface waves with periods near 20 s, a body wave magnitude m_b was defined [25], which can be related to M_S [16]:

$$m_b = 2.5 + 0.63 M_S \tag{33.3}$$

Body wave magnitudes are more commonly used in eastern North America, due to the deeper earthquakes there. More recently, **seismic moment** has been employed to define a moment magnitude M_W [26] (also denoted as bold-face **M**) which is finding increased and widespread use:

$$\text{Log } M_0 = 1.5 M_W + 16.0 \tag{33.4}$$

where seismic moment M_0 (dyne-cm) is defined as [33]

$$M_0 = \mu A \bar{u} \tag{33.5}$$

where μ is the material shear modulus, A is the area of fault plane rupture, and $\bar{u}$ is the mean relative displacement between the two sides of the fault (the averaged fault slip). Comparatively, M_W and M_S are numerically almost identical up to magnitude 7.5. Figure 33.2 indicates the relationship between moment magnitude and various magnitude scales.

From the foregoing discussion, it can be seen that magnitude and energy are related to fault rupture length and slip. Slemmons [60] and Bonilla et al. [5] have determined statistical relations between these parameters, for worldwide and regional data sets, aggregated and segregated by type of faulting (normal, reverse, strike-slip). Bonilla et al.'s worldwide results for all types of faults are

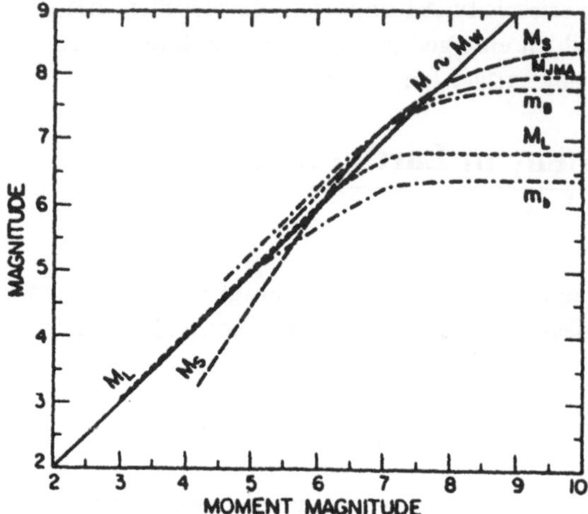

FIGURE 33.2 Relationship between moment magnitude and various magnitude scales. (*Source*: Campbell, K. W., *Earthquake Spectra*, 1(4), 759–804, 1985. With permission.)

$$M_s = 6.04 + 0.708 \log_{10} L \qquad s = 0.306 \tag{33.6}$$

$$\log_{10} L = -2.77 + 0.619 M_s \qquad s = 0.286 \tag{33.7}$$

$$M_s = 6.95 + 0.723 \log_{10} d \qquad s = 0.323 \tag{33.8}$$

$$\log_{10} d = -3.58 + 0.550 M_s \qquad s = 0.282 \tag{33.9}$$

which indicates, for example, that, for $M_S = 7$, the average fault rupture length is about 36 km (and the average displacement is about 1.86 m). Conversely, a fault of 100 km length is capable of about an $M_S = 7.5$* event (see also Wells and Coppersmith [66] for alternative relations).

Intensity

In general, seismic intensity is a metric of the effect, or the strength, of an earthquake hazard at a specific location. While the term can be generically applied to engineering measures such as peak ground acceleration, it is usually reserved for qualitative measures of location-specific earthquake effects, based on observed human behavior and structural damage. Numerous intensity scales were developed in preinstrumental times — the most common in use today are the Modified Mercalli (MMI) [68] (Table 33.1), the Rossi–Forel (R-F), the Medvedev-Sponheur-Karnik (MSK-64, 1981), and the Japan Meteorological Agency (JMA) scales.

Time History

Sensitive strong motion seismometers have been available since the 1930s, and they record actual ground motions specific to their location, Figure 33.3. Typically, the ground motion records, termed **seismographs** or **time histories**, have recorded acceleration (these records are termed **accelerograms**), for

*Note that $L = g(M_S)$ should not be inverted to solve for $M_S = f(L)$, as a regression for $y = f(x)$ is different from a regression for $x = g(y)$.

TABLE 33.1 Modified Mercalli Intensity Scale of 1931

I	Not felt except by a very few under especially favorable circumstances
II	Felt only by a few persons at rest, especially on upper floors of buildings. Delicately suspended objects may swing.
III	Felt quite noticeably indoors, especially on upper floors of buildings, but many people do not recognize it as an earthquake; standing automobiles may rock slightly; vibration like passing truck; duration estimated
IV	During the day felt indoors by many, outdoors by few; at night some awakened; dishes, windows, and doors disturbed; walls make creaking sound; sensation like heavy truck striking building; standing automobiles rock noticeably
V	Felt by nearly everyone; many awakened; some dishes, windows, etc., broken; a few instances of cracked plaster; unstable objects overturned; disturbance of trees, poles, and other tall objects sometimes noticed; pendulum clocks may stop
VI	Felt by all; many frightened and run outdoors; some heavy furniture moved; a few instances of fallen plaster or damaged chimneys; damage slight
VII	Everybody runs outdoors; damage negligible in buildings of good design and construction, slight to moderate in well-built ordinary structures; considerable in poorly built or badly designed structures; some chimneys broken; noticed by persons driving automobiles
VIII	Damage slight in specially designed structures, considerable in ordinary substantial buildings, with partial collapse, great in poorly built structures; panel walls thrown out of frame structures; fall of chimneys, factory stacks, columns, monuments, walls; heavy furniture overturned; sand and mud ejected in small amounts; changes in well water; persons driving automobiles disturbed
IX	Damage considerable in specially designed structures; well-designed frame structures thrown out of plumb; great in substantial buildings, with partial collapse; buildings shifted off foundations; ground cracked conspicuously; underground pipes broken
X	Some well-built wooden structures destroyed; most masonry and frame structures destroyed with foundations; ground badly cracked; rails bent; landslides considerable from river banks and steep slopes; shifted sand and mud; water splashed over banks
XI	Few, if any (masonry) structures remain standing; bridges destroyed; broad fissures in ground; underground pipelines completely out of service; earth slumps and land slips in soft ground; rails bent greatly
XII	Damage total; waves seen on ground surfaces; lines of sight and level distorted; objects thrown upward into the air

After Wood and Neumann [68].

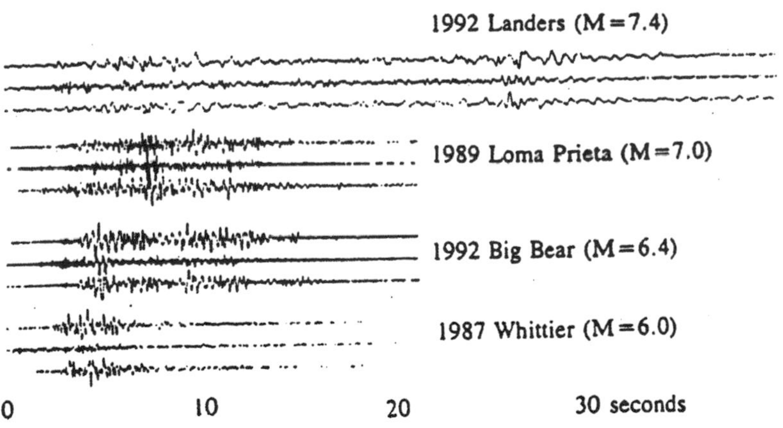

FIGURE 33.3 Typical earthquake accelerograms. (Courtesy of Darragh et al., 1994.)

many years in analog form on photographic film and, more recently, digitally. Analog records required considerable effort for correction due to instrumental drift, before they could be used.

Time histories theoretically contain complete information about the motion at the instrumental location, recording three *traces* or orthogonal records (two horizontal and one vertical). Time histories (i.e., the earthquake motion at the site) can differ dramatically in duration, frequency, content, and amplitude. The maximum amplitude of recorded acceleration is termed the **peak ground acceleration,** PGA (also termed the ZPA, or **zero period acceleration**); peak ground velocity

(PGV) and peak ground displacement (PGD) are the maximum respective amplitudes of velocity and displacement. Acceleration is normally recorded, with velocity and displacement being determined by integration; however, velocity and displacement meters are deployed to a lesser extent. Acceleration can be expressed in units of cm/s^2 (termed *gals*), but is often also expressed in terms of the fraction or percent of the acceleration of gravity (980.66 gals, termed 1 *g*). Velocity is expressed in cm/s (termed *kine*). Recent earthquakes — 1994 Northridge, M_W 6.7 and 1995 Hanshin (Kobe) M_W 6.9 — have recorded PGAs of about 0.8 *g* and PGVs of about 100 kine, while almost 2 g was recorded in the 1992 Cape Mendocino earthquake.

Elastic Response Spectra

If a single-degree-of-freedom (SDOF) mass is subjected to a time history of ground (i.e., base) motion similar to that shown in Figure 33.3, the mass or elastic **structural response** can be readily calculated as a function of time, generating a **structural response time history**, as shown in Figure 33.4 for several oscillators with differing natural periods. The response time history can be calculated by direct integration of Eq. (33.1) in the **time domain**, or by solution of the **Duhamel** integral. However, this is time-consuming, and the elastic response is more typically calculated in the **frequency domain** [12].

For design purposes, it is often sufficient to know only the maximum amplitude of the response time history. If the natural period of the SDOF is varied across a spectrum of engineering interest (typically, for natural periods from 0.03 to 3 or more seconds, or frequencies of 0.3 to 30+ Hz), then the plot of these maximum amplitudes is termed a **response spectrum**. Figure 33.4 illustrates this process, resulting in S_d, the *displacement response spectrum*, while Figure 33.5 shows (a) the S_d, displacement response spectrum, (b) S_v, the *velocity response spectrum* (also denoted PSV, the pseudo-spectral velocity, "pseudo" to emphasize that this spectrum is not exactly the same as the relative velocity response spectrum), and (c) S_a, the *acceleration response spectrum*. Note that

$$S_v = \frac{2\pi}{T} S_d = \varpi S_d \tag{33.10}$$

and

$$S_a = \frac{2\pi}{T} S_v = \varpi S_v = \left(\frac{2\pi}{T}\right)^2 S_d = \varpi^2 S_d \tag{33.11}$$

Response spectra form the basis for much modern earthquake engineering structural analysis and design. They are readily calculated *if* the ground motion is known. For design purposes, however, response spectra must be estimated — this process is discussed below. Response spectra may be plotted in any of several ways, as shown in Figure 33.5 with arithmetic axes, and in Figure 33.6, where the velocity response spectrum is plotted on tripartite logarithmic axes, which equally enables reading of displacement and acceleration response. Response spectra are most normally presented for 5% of critical **damping**.

Inelastic Response Spectra

While the foregoing discussion has been for elastic response spectra, most structures are not expected, or even designed, to remain elastic under strong ground motions. Rather, structures are expected to enter the *inelastic* region — the extent to which they behave inelastically can be defined by the **ductility factor**, μ:

$$\mu = \frac{u_m}{u_y} \tag{33.12}$$

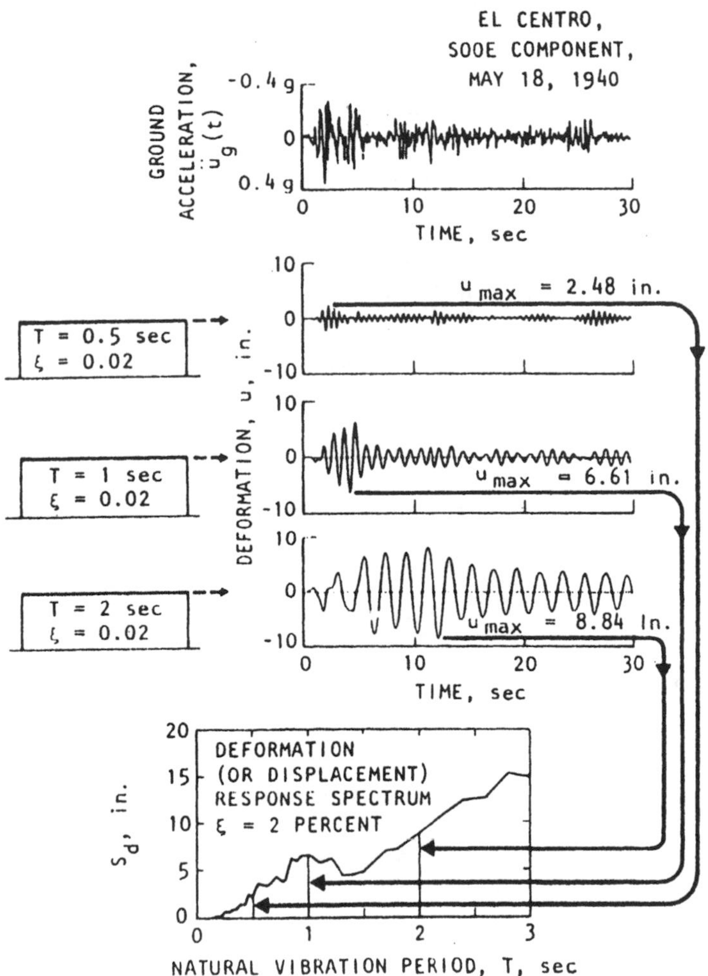

FIGURE 33.4 Computation of deformation (or displacement) response spectrum. (*Source*: Chopra, A. K., *Dynamics of Structures, A Primer*, Earthquake Engineering Research Institute, Oakland, CA, 1981. With permission.)

where u_m is the actual displacement of the mass under actual ground motions, and u_y is the displacement at yield (i.e., that displacement which defines the extreme of elastic behavior). Inelastic response spectra can be calculated in the time domain by direct integration, analogous to elastic response spectra but with the structural stiffness as a nonlinear function of displacement, $k = k(u)$. If elastoplastic behavior is assumed, then elastic response spectra can be readily modified to reflect inelastic behavior, on the basis that (1) at low frequencies (<0.3 Hz) displacements are the same, (2) at high frequencies (>33 Hz), accelerations are equal, and (3) at intermediate frequencies, the absorbed energy is preserved. Actual construction of inelastic response spectra on this basis is shown in Figure 33.9, where $DVAA_o$ is the elastic spectrum, which is reduced to D' and V' by the ratio of $1/\mu$ for frequencies less than 2 Hz, and by the ratio of $1/(2\mu - 1)^{1/2}$ between 2 and 8 Hz. Above 33 Hz, there is no reduction. The result is the inelastic acceleration spectrum ($D'V'A'A_o$), while $A''A_o'$ is the inelastic displacement spectrum. A specific example, for ZPA = 0.16 *g*, damping = 5% of critical and $\mu = 3$ is shown in Figure 33.10.

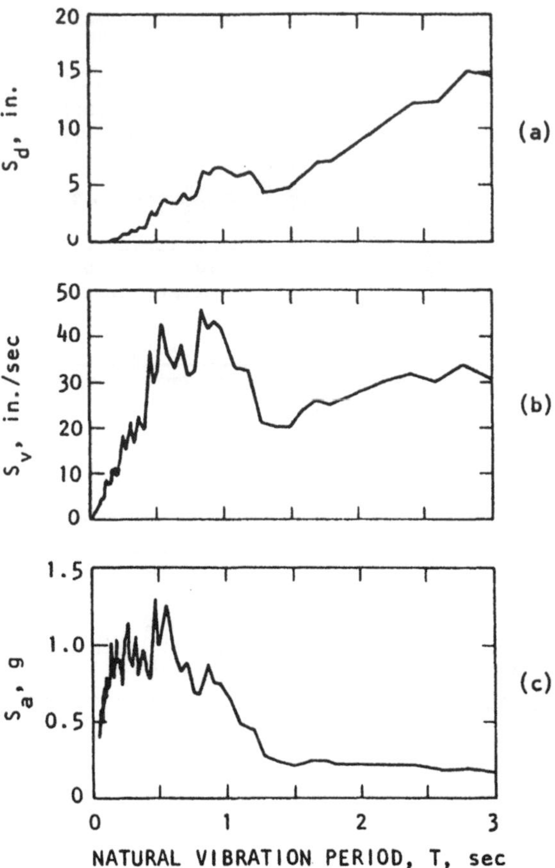

FIGURE 33.5 Response spectra. (*Source*: Chopra, A. K., *Dynamics of Structures, A Primer*, Earthquake Engineering Research Institute, Oakland, CA, 1981. With permission.)

33.4 Strong Motion Attenuation and Duration

The rate at which earthquake ground motion decreases with distance, termed **attenuation**, is a function of the regional geology and inherent characteristics of the earthquake and its source. Campbell [10] offers an excellent review of North American relations up to 1985. Initial relationships were for PGA, but regression of the amplitudes of response spectra at various periods is now common, including consideration of fault type and effects of soil. A currently favored relationship is

Campbell and Bozorgnia [11] (PGA — Worldwide Data)

$$\begin{aligned}\ln(\text{PGA}) = {} & -3.512 + 0.904M - 1.328\ln\sqrt{\left\{R_s^2 + \left[0.149\exp(0.647M)\right]^2\right\}} \\ & + \left[1.125 - 0.112\ln\left(R_s\right) - 0.0957M\right]F \\ & + \left[0.440 - 0.171\ln\left(R_s\right)\right]S_{sr} + \left[0.405 - 0.222\ln\left(R_s\right)\right]S_{hr} + \varepsilon\end{aligned} \tag{33.13}$$

RESPONSE SPECTRUM

IMPERIAL VALLEY EARTHQUAKE

MAY 18, 1940 — 2037 PST

IIIA001 40.001.0 EL CENTRO SITE

IMPERIAL VALLEY IRRIGATION DISTRICT COMP SOOE

DAMPING VALUES ARE 0, 2, 5, 10, AND 20 PERCENT OF CRITICAL

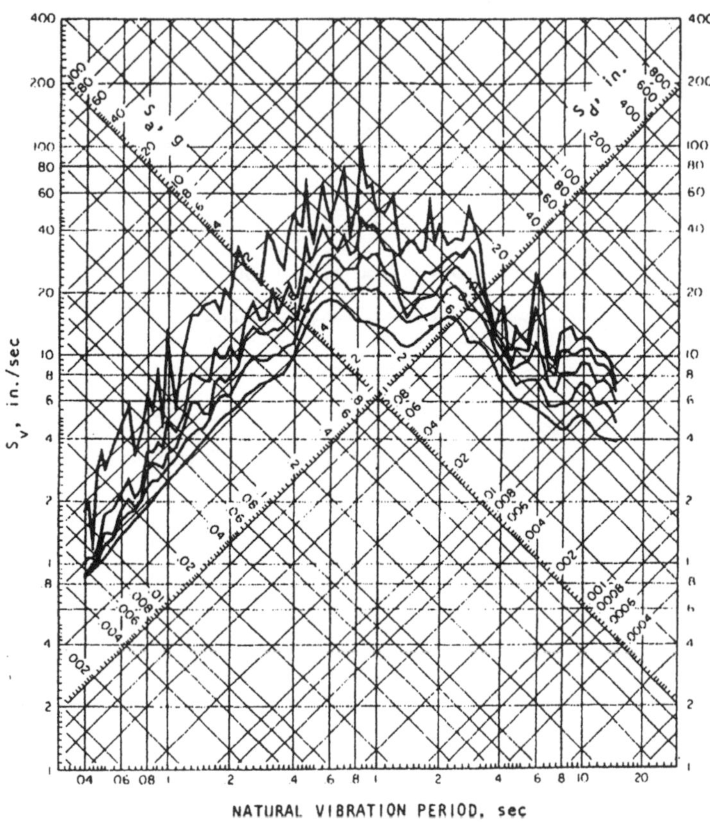

FIGURE 33.6 Response spectra, tripartite plot (El Centro S 0° E component). (*Source*: Chopra, A. K., *Dynamics of Structures, A Primer*, Earthquake Engineering Research Institute, Oakland, CA, 1981. With permission.)

where

PGA = the geometric mean of the two horizontal components of peak ground acceleration (g)

M = moment magnitude (M_w)

R_s = the closest distance to seismogenic rupture on the fault (km)

F = 0 for strike-slip and normal faulting earthquakes, and 1 for reverse, reverse-oblique, and thrust faulting earthquakes

S_{sr} = 1 for soft-rock sites

S_{hr} = 1 for hard-rock sites

$S_{sr} = S_{hr} =$ 0 for alluvium sites

ε = is a random error term with zero mean and standard deviation equal to $\sigma_{\ln}$(PGA), the standard error of estimate of ln(PGA)

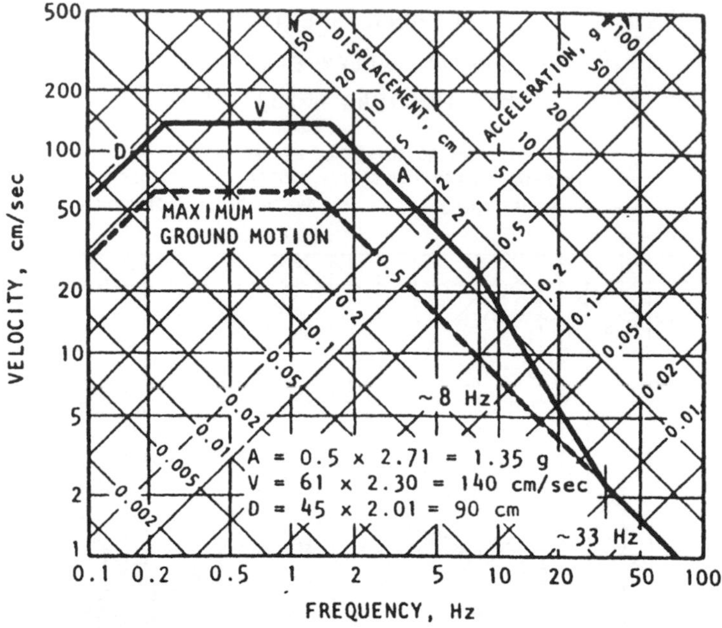

FIGURE 33.7 Idealized elastic design spectrum, horizontal motion (ZPA = 0.5 *g*, 5% damping, one sigma cumulative probability. (*Source*: Newmark, N. M. and Hall, W. J., *Earthquake Spectra and Design*, Earthquake Engineering Research Institute, Oakland, CA, 1982. With permission.)

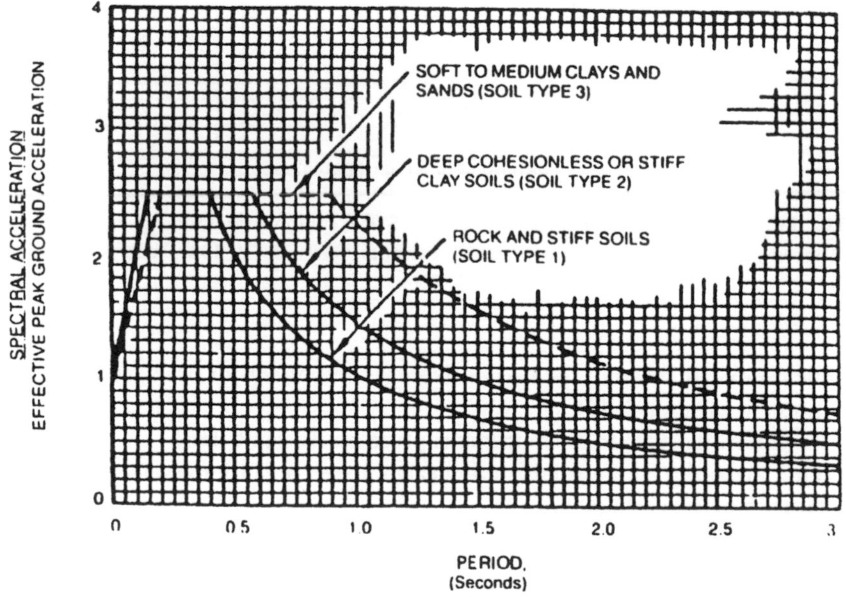

FIGURE 33.8 Normalized response spectra shapes. (*Source*: Uniform Building Code, Structural Engineering Design Provisions, Vol. 2, Intl. Conf. Building Officials, Whittier, 1994. With permission.)

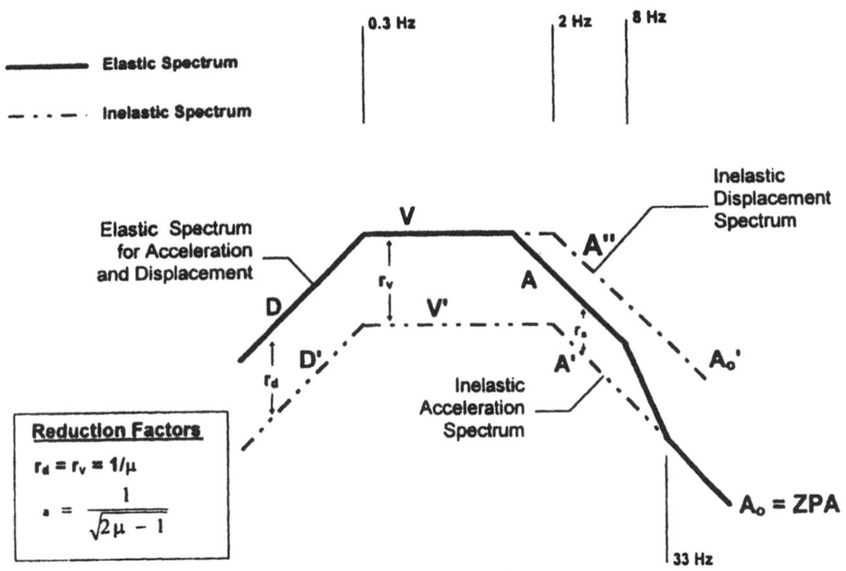

FIGURE 33.9 Inelastic response spectra for earthquakes. (*Source*: Newmark, N. M. and Hall, W. J., *Earthquake Spectra and Design*, Earthquake Engineering Research Institute, Oakland, CA, 1982.)

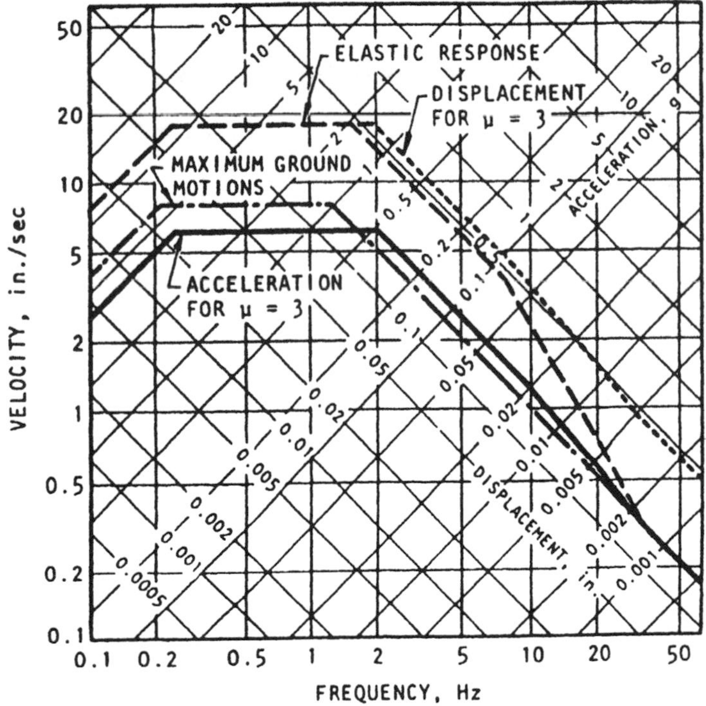

FIGURE 33.10 Example inelastic response spectra. (*Source*: Newmark, N. M. and Hall, W. J., *Earthquake Spectra and Design*, Earthquake Engineering Research Institute, Oakland, CA, 1982.)

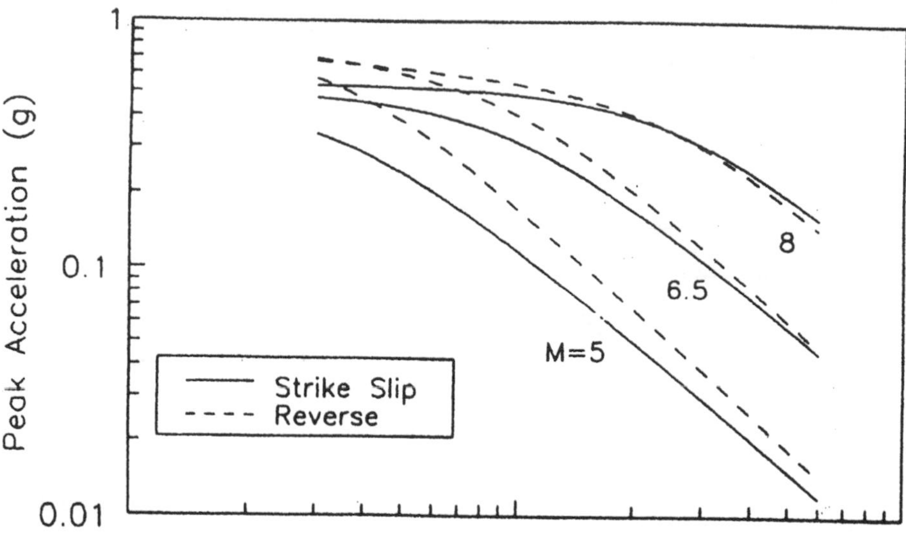

FIGURE 33.11 Campbell and Bozorgnia worldwide attenuation relationship showing (for alluvium) the scaling of peak horizontal acceleration with magnitude and style of faulting. (*Source:* Campbell, K. W. and Bozorgnia, Y., in *Proc. Fifth U.S. National Conference on Earthquake Engineering*, Earthquake Engineering Research Institute, Oakland, CA, 1994. With permission.)

Regarding the uncertainty, ε was estimated as

$$\sigma_{\ln}(\text{PGA}) = \begin{vmatrix} 0.55 & \text{if PGA} < 0.068 \\ 0.173 - 0.140\ln(\text{PGA}) & \text{if } 0.068 \leq \text{PGA} \leq 0.21 \\ 0.39 & \text{if PGA} > 0.21 \end{vmatrix}$$

Figure 33.11 indicates, for alluvium, median values of the attenuation of peak horizontal acceleration with magnitude and style of faulting. Many other relationships are also employed (e.g., Boore et al.[6]).

33.5 Probabilistic Seismic Hazard Analysis

The probabilistic seismic hazard analysis (PSHA) approach entered general practice with Cornell's [13] seminal paper, and basically employs the theorem of total probability to formulate:

$$P(Y) = \sum_F \sum_M \sum_R p(Y|M,R)p(M)p(R) \tag{33.14}$$

where

Y = a measure of intensity, such as PGA, response spectral parameters PSV, etc.
$p(Y|M,R)$ = the probability of Y given earthquake magnitude M and distance R (i.e., attenuation)
$p(M)$ = the probability of a given earthquake magnitude M
$p(R)$ = the probability of a given distance R, and
F = seismic sources, whether discrete such as faults, or distributed

This process is illustrated in Figure 33.12, where various seismic sources (faults modeled as line sources and dipping planes, and various distributed or area sources, including a background source to account for miscellaneous seismicity) are identified, and their seismicity characterized on the basis of historic seismicity and/or geologic data. The effects at a specific site are quantified on the

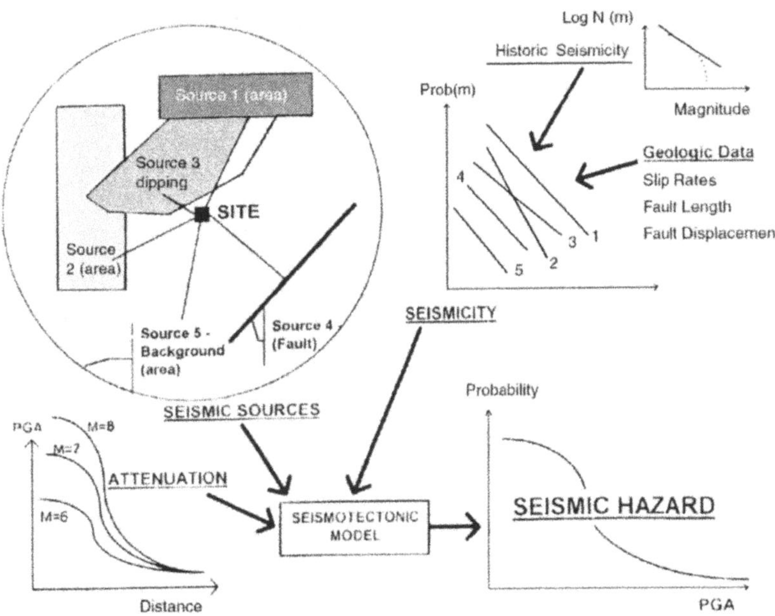

FIGURE 33.12 Elements of seismic hazard analysis — seismotectonic model is composed of seismic sources, whose seismicity is characterized on the basis of historic seismicity and geologic data, and whose effects are quantified at the site via strong motion attenuation models.

basis of strong ground motion modeling, also termed attenuation. These elements collectively are the **seismotectonic model** — their integration results in the **seismic hazard.**

There is an extensive literature on this subject [42,50] so that only key points will be discussed here. Summation is indicated, as integration requires closed-form solutions, which are usually precluded by the empirical form of the attenuation relations. The *p*(*YÔM,R*) term represents the full probabilistic distribution of the attenuation relation — summation must occur over the full distribution, due to the significant uncertainty in attenuation. The *p*(*M*) term is referred to as the **magnitude–frequency relation**, which was first characterized by Gutenberg and Richter [24] as

$$\log \mathbf{N}(m) = a_N - b_N m \tag{33.15}$$

where $\mathbf{N}(m)$ = the number of earthquake events equal to or greater than magnitude m occurring on a seismic source per unit time, and a_N and b_N are regional constants (10^{a_N} = the total number of earthquakes with magnitude >0, and b_N is the rate of seismicity; b_N is typically 1 ± 0.3). The Gutenberg–Richter relation can be normalized to

$$\mathbf{F}(m) = 1. - \exp\,[-\,\mathbf{B}_M\,(m - \mathbf{M}_{o)}] \tag{33.16}$$

where $F(m)$ is the cumulative distribution function (CDF) of magnitude, $\mathbf{B}_M$ is a regional constant and $\mathbf{M}_o$ is a small enough magnitude such that lesser events can be ignored. Combining this with a Poisson distribution to model large earthquake occurrence [20] leads to the CDF of earthquake magnitude per unit time

$$F(m) = \exp\,[-\exp\,\{-\,a_M\,(m - \mu_M)\}] \tag{33.17}$$

which has the form of a Gumbel [23] extreme value type I (largest values) distribution (denoted $EX_{I,L}$), which is an unbounded distribution (i.e., the variate can assume any value). The parameters

a_M and μ_M can be evaluated by a least-squares regression on historical seismicity data, although the probability of very large earthquakes tends to be overestimated. Several attempts have been made to account for this (e.g., Cornell and Merz [14]). Yegulalp and Kuo [70] have used Gumbel's Type III (largest value, denoted $EX_{III,L}$) to successfully account for this deficiency. This distribution

$$F(m) = \exp\left[-\left(\frac{w-m}{w-u}\right)^k\right] \tag{33.18}$$

has the advantage that w is the largest possible value of the variate (i.e., earthquake magnitude), thus permitting (when w, u, and k are estimated by regression on historical data) an estimate of the source's largest possible magnitude. It can be shown (Yegulalp and Kuo [70]) that estimators of w, u, and k can be obtained by satisfying Kuhn–Tucker conditions although, if the data is too incomplete, the $EX_{III,L}$ parameters approach those of the $EX_{I,L}$. Determination of these parameters requires careful analysis of historical seismicity data (which is highly complex and something of an art [17], and the merging of the resulting statistics with estimates of maximum magnitude and seismicity made on the basis of geologic evidence (i.e., as discussed above, maximum magnitude can be estimated from fault length, fault displacement data, time since last event, and other evidence, and seismicity can be estimated from fault slippage rates combined with time since the last event, see Schwartz [55] for an excellent discussion of these aspects). In a full probabilistic seismic hazard analysis, many of these aspects are treated fully or partially probabilistically, including the attenuation, magnitude–frequency relation, upper- and lower-bound magnitudes for each source zone, geographic bounds of source zones, fault rupture length, and many other aspects. The full treatment requires complex specialized computer codes, which incorporate uncertainty via use of multiple alternative source zonations, attenuation relations, and other parameters [3,19] often using a logic tree format. A number of codes have been developed using the public domain FRISK (Fault RISK) code first developed by McGuire [37].

33.6 Site Response

When seismic waves reach a site, the ground motions they produce are affected by the geometry and properties of the geologic materials at that site. At most bridge sites, rock will be covered by some thickness of soil which can markedly influence the nature of the motions transmitted to the bridge structure as well as the loading on the bridge foundation. The influence of local site conditions on ground response has been observed in many past earthquakes, but specific provisions for site effects were not incorporated in codes until 1976.

The manner in which a site responds during an earthquake depends on the near-surface stiffness gradient and on how the incoming waves are reflected and refracted by the near-surface materials. The interaction between seismic waves and near-surface materials can be complex, particularly when surface topography and/or subsurface stratigraphy is complex. Quantification of site response has generally been accomplished by analytical or empirical methods.

Basic Concepts

The simplest possible case of site response would consist of a uniform layer of viscoelastic soil of density, ρ, shear modulus, G, viscosity, η, and thickness, H, resting on rigid bedrock and subjected to vertically propagating shear waves (Figure 33.13a). The response of the layer would be governed by the wave equation

$$\rho\frac{\partial^2 u}{\partial t^2} = G\frac{\partial^2 u}{\partial z^2} + \eta\frac{\partial^3 u}{\partial z^2 \partial t} \tag{33.19}$$

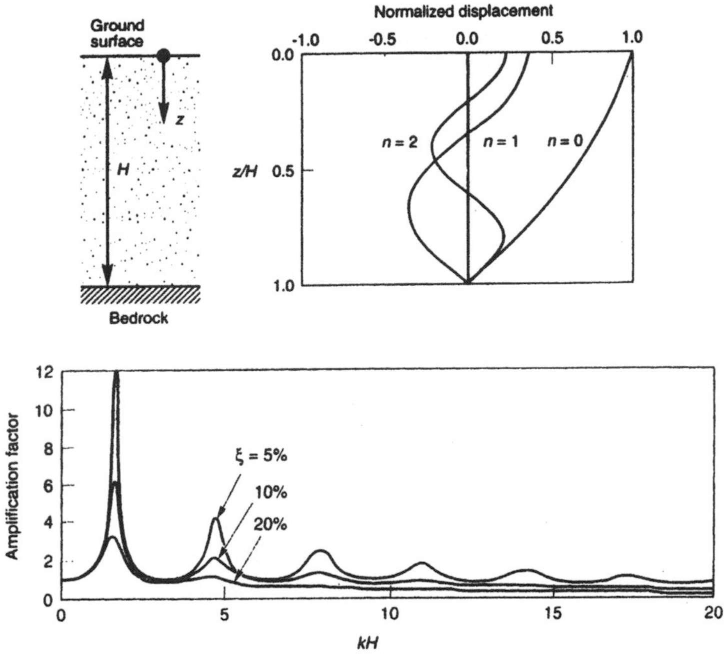

FIGURE 33.13 Illustration of (top) mode shapes and (bottom) amplification function for uniform elastic layer underlain by rigid boundary. (*Source*: Kramer, S.L., *Geotechnical Earthquake Engineering*, Prentice-Hall, Upper Saddle River, NJ, 1996.)

which has a solution that can be expressed in the form of upward and downward traveling waves. At certain frequencies, these waves interfere constructively to produce increased amplitudes; at other frequencies, the upward and downward traveling waves tend to cancel each other and produce lower amplitudes. Such a system can easily be shown to have an infinite number of natural frequencies and mode shapes (Figure 33.13 (top)) given by

$$\omega_n = \frac{v_s}{H}\left(\frac{\pi}{2} + n\pi\right) \quad \text{and} \quad \phi_n = \cos\left[\frac{z}{H}\left(\frac{\pi}{2} + n\pi\right)\right] \tag{33.20}$$

Note that the fundamental, or characteristic site period, is given by $T_s = 2\pi/\omega_0 = 4H/v_s$. The ratio of ground surface to bedrock amplitude can be expressed in the form of an **amplification function** as

$$A(\omega) = \frac{1}{\sqrt{\cos^2\left(\omega H / v_s + \left[\xi(\omega H / v_S)\right]^2\right)}} \tag{33.21}$$

Figure 33.13(b) shows the amplification function which illustrates the frequency-dependent nature of site amplification. The amplification factor reaches its highest value when the period of the input motion is equal to the characteristic site period. More realistic site conditions produce more-complicated amplification functions, but all amplification functions are frequency-dependent. In a sense, the surficial soil layers act as a filter that amplifies certain frequencies and deamplifies others. The overall effect on site response depends on how these frequencies match up with the dominant frequencies in the input motion.

The example illustrated above is mathematically convenient, but unrealistically simple for application to actual sites. First, the assumption of rigid bedrock implies that all downward-traveling waves are perfectly reflected back up into the overlying layer. While generally quite stiff, bedrock is not perfectly rigid and therefore a portion of the energy in a downward-traveling wave is transmitted into the bedrock to continue traveling downward — as a result, the energy carried by the reflected wave that travels back up is diminished. The relative proportions of the transmitted and reflected waves depends on the ratio of the **specific impedance** of the two materials on either side of the boundary. At any rate, the amount of wave energy that remains within the surficial layer is decreased by waves radiating into the underlying rock. The resulting reduction in wave amplitudes is often referred to as **radiation damping**. Second, subsurface stratigraphy is generally more complicated than that assumed in the example. Most sites have multiple layers of different materials with different specific impedances. The boundaries between the layers may be horizontal or may be inclined, but all will reflect and refract seismic waves to produce wave fields that are much more complicated than described above. This is often particularly true in the vicinity of bridges located in fluvial geologic environments where soil stratigraphy may be the result of an episodic series of erosional and depositional events. Third, site topography is generally not flat, particularly in the vicinity of bridges which may be supported in sloping natural or man-made materials, or on man-made embankments. Topographic conditions can strongly influence the amplitude and frequency content of ground motions. Finally, subsurface conditions can be highly variable, particularly in the geologic environments in which many bridges are constructed. Conditions may be different at each end of a bridge, and even at the locations of intermediate supports — this effect is particularly true for long bridges. These factors, combined with the fact that seismic waves may reach one end of the bridge before the other, can reduce the **coherence** of ground motions. Different motions transmitted to a bridge at different support points can produce loads and displacements that would not occur in the case of perfectly coherent motions.

Evidence for Local Site Effects

Theoretical evidence for the existence of local site effects has been supplemented by instrumental and observational evidence in numerous earthquakes. Nearly 200 years ago [35], variations in damage patterns were correlated to variations in subsurface conditions; such observations have been repeated on a regular basis since that time. With the advent of modern seismographs and strong motion instruments, quantitative evidence for local site effects is now available. In the Loma Prieta earthquake, for example, strong motion instruments at Yerba Buena Island and Treasure Island were at virtually identical distances and azimuths from the hypocenter. However, the Yerba Buena Island instrument was located on a rock outcrop and the Treasure Island instrument on about 14 m of loose hydraulically placed sandy fill underlain by nearly 17 m of soft San Francisco Bay Mud. The measured motions, which differed significantly (Figure 33.14), illustrate the effects of local site effects. At a small but increasing number of locations, strong motion instruments have been placed in a boring directly below a surface instrument (Figure 33.15a). Because such vertical arrays can measure motions at the surface and at bedrock level, they allow direct computation of measured amplification functions. Such an empirical amplification function is shown in Figure 33.15b. The general similarity of the measured amplification function, particularly the strong frequency dependence, to even the simple theoretical amplification (Figure 33.13) is notable.

Methods of Analysis

Development of suitable design ground motions, and estimation of appropriate foundations loading, generally requires prediction of anticipated site response. This is usually accomplished using empirical or analytical methods. For small bridges, or for projects in which detailed subsurface information is not available, the empirical approach is more common. For larger and more important structures, a subsurface exploration program is generally undertaken to provide information for site-specific analytical prediction of site response.

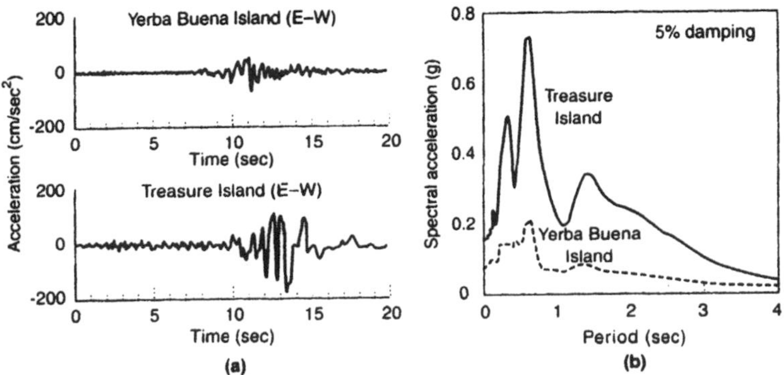

FIGURE 33.14 Ground surface motions at Yerba Buena Island and Treasure Island in the Loma Prieta earthquake. sources Kramer, S.L., *Geotechnical Earthquake Engineering*, Prentice-Hall, Upper Saddle River, NJ, 1996.)

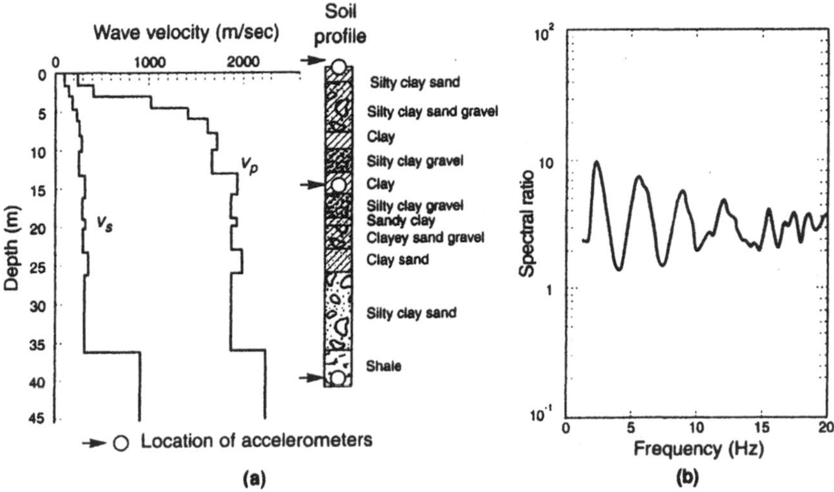

FIGURE 33.15 (a) Subsurface profile at location of Richmond Field Station downhole array, and (b) measured surface/bedrock amplification function in Briones Hills (M_L = 4.3) earthquake. sources Kramer, S.L., *Geotechnical Earthquake Engineering*, Prentice-Hall, Upper Saddle River, NJ, 1996.)

Empirical Methods

In the absence of site-specific information, local site effects can be estimated on the basis of empirical correlation to measured site response from past earthquakes. The database of strong ground motion records has increased tremendously over the past 30 years. Division of records within this database according to general site conditions has allowed the development of empirical correlations for different site conditions.

The earliest empirical approach involved estimation of the effects of local soil conditions on peak ground surface acceleration and spectral shape. Seed et al. [59] divided the subsurface conditions at the sites of 104 strong motion records into four categories — rock, stiff soils (<61 m), deep cohesionless soils (>76 m), and soft to medium clay and sand. Comparing average peak ground surface accelerations measured at the soil sites with those anticipated at equivalent rock sites allowed development of curves such as those shown in Figure 33.16. These curves show that soft profiles amplify peak acceleration over a wide range of rock accelerations, that even stiff soil profiles amplify peak acceleration when peak accelerations are relatively low, and that peak accelerations are deamplified at very high

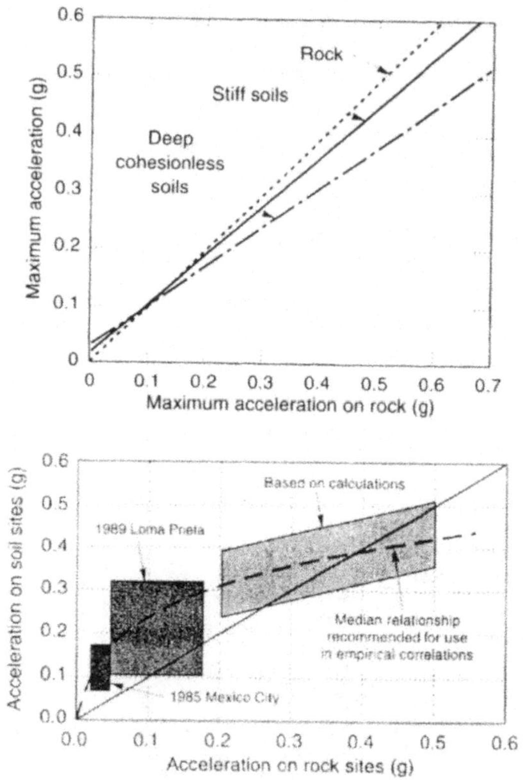

FIGURE 33.16 Approximate relationship between beak accelerations on rock and soil sites (after Seed et al. [59]; Idriss, 1990).

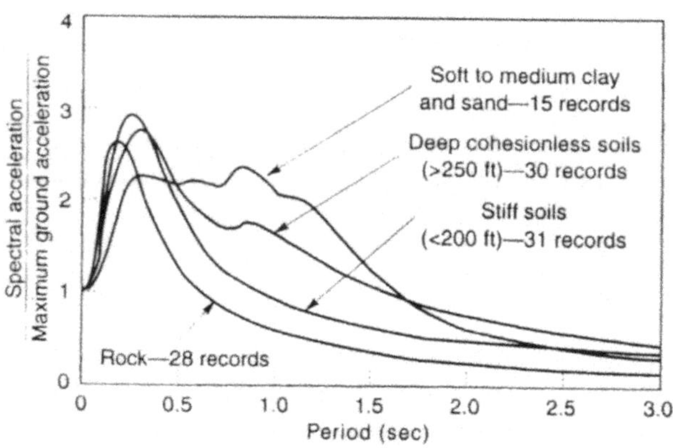

FIGURE 33.17 Average normalized response spectra (5% damping) for different local site conditions (after Seed et al. [59]).

input acceleration levels. Computation of average response spectra, when normalized by peak acceleration (Figure 33.17), showed the significant effect of local soil conditions on spectral shape, a finding that has strongly influenced the development of seismic codes and standards.

A more recent empirical approach has been to include local site conditions directly in attenuation relationships. By developing a site parameter to characterize the soil conditions at the locations of strong motion instruments and incorporating that parameter into the basic form of an attenuation

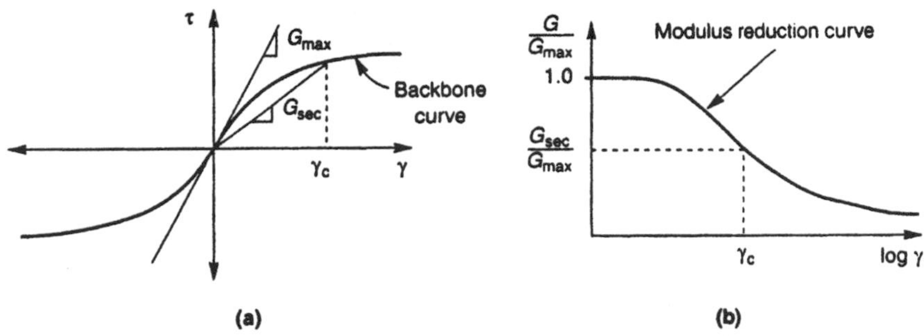

FIGURE 33.18 Relationship between backbone curve and modulus reduction curve.

relationship, regression analyses can produce attenuation relationships that include the effects of local site conditions. In such relationships, site conditions are typically grouped into different site classes on the basis of such characteristics as surficial soil/rock conditions [see the factors S_{sr} and S_{hr} in Eq. (33.13)] or average shear wave velocity within the upper 30 m of the ground surface (e.g., Boore et al. [6]). Such relationships can be used for empirical prediction of peak acceleration and response spectra, and incorporated into probabilistic seismic hazard analyses to produce uniform risk spectra for the desired class of subsurface conditions.

The reasonableness of empirically based methods for estimation of site response effects depends on the extent to which site conditions match the site conditions in the databases from which the empirical relationships were derived. It is important to recognize the empirical nature of such methods and the significant uncertainty inherent in the results they produce.

Analytical Methods

When sufficient information to characterize the geometry and dynamic properties of subsurface soil layers is available, local site effects may be computed by site-specific ground response analyses. Such analyses may be conducted in one, two, or three dimensions; one-dimensional analyses are most common, but the topography of many bridge sites may require two-dimensional analyses.

Unlike most structural materials, soils are highly nonlinear, even at very low strain levels. This nonlinearity causes soil stiffness to decrease and material damping to increase with increasing strain amplitude. The variation of stiffness with strain can be represented in two ways — by nonlinear **backbone** (stress–strain) **curves** or by **modulus reduction curves**, both of which are related as illustrated in Figure 33.18. The modulus reduction curve shows how the secant shear modulus of the soil decreases with increasing strain amplitude. To account for the effects of nonlinear soil behavior, ground response analyses are generally performed using one of two basic approaches: the **equivalent linear approach** or the **nonlinear approach**.

In the equivalent linear approach, a linear analysis is performed using shear moduli and damping ratios that are based on an initial estimate of strain amplitude. The strain level computed using these properties is then compared with the estimated strain amplitude and the properties adjusted until the computed strain levels are very close to those corresponding to the soil properties. Using this iterative approach, the effects of nonlinearity are approximated in a linear analysis by the use of *strain-compatible* soil properties. Modulus reduction and damping behavior has been shown to be influenced by soil plasticity, with highly plastic soils exhibiting higher linearity and lower damping than low-plasticity soils (Figure 33.19). The equivalent linear approach has been incorporated into such computer programs as SHAKE [53] and ProShake [18] for one-dimensional analyses, FLUSH [34] for two-dimensional analyses, and TLUSH [29] for three-dimensional analyses.

In the nonlinear approach, the equations of motion are assumed to be linear over each of a series of small time increments. This allows the response at the end of a time increment to be computed from the conditions at the beginning of the time increment and the loading applied during the time

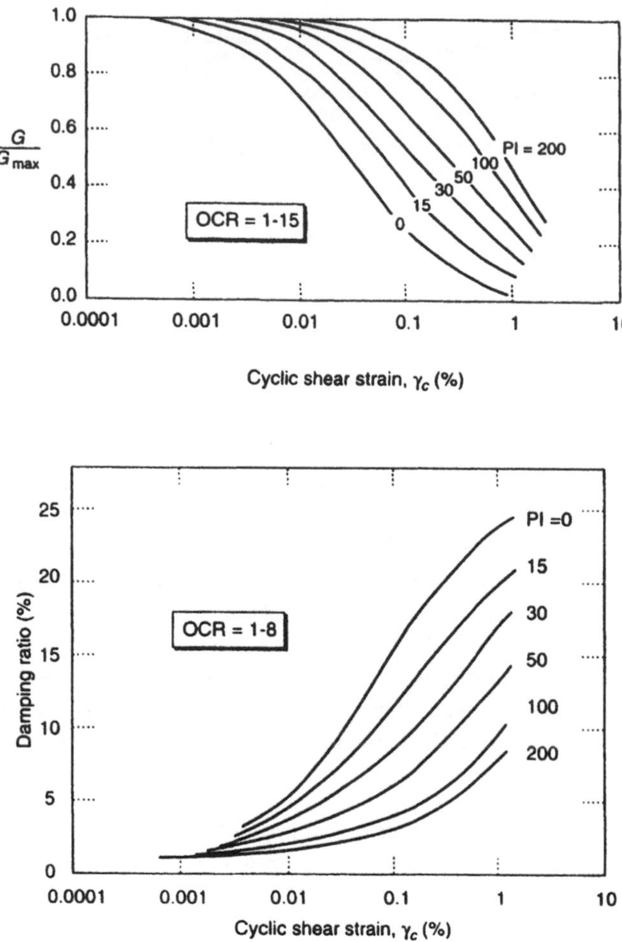

FIGURE 33.19 Equivalent linear soil behavior: (a) modulus reduction curves and (b) damping curves. (*Source*: Vucetic and Dobry, 1991.)

increment. At the end of the time increment, the properties are updated for the next time increment. In this way, the stiffness of each element of soil can be changed depending on the current and past stress conditions and hysteretic damping can be modeled directly. For seismic analysis, the nonlinear approach requires a constitutive (stress–strain) model that is capable of representing soil behavior under dynamic loading conditions. Such models can be complicated and can require calibration of a large number of soil parameters by extensive laboratory testing. With a properly calibrated constitutive model, however, nonlinear analyses can provide reasonable predictions of site response and have two significant advantages over equivalent linear analyses. First, nonlinear analyses are able to predict permanent deformations such as those associated with ground failure (Section 33.7). Second, nonlinear analyses are able to account for the generation, redistribution, and eventual dissipation of porewater pressures which makes them particularly useful for sites that may be subject to liquefaction and/or lateral spreading. The nonlinear approach has been incorporated into such computer programs as DESRA [31], TESS [48], and SUMDES for one-dimensional analysis, and TARA [21] for two-dimensional analyses. General-purpose programs such as FLAC can also be used for nonlinear two-dimensional analyses. In practice, however, the use of nonlinear analyses has lagged behind the use of equivalent linear analyses, principally because of the difficulty in characterizing nonlinear constitutive model parameters.

TABLE 33.2 Site Coefficient

Soil Type	Description	S
I	Rock of any characteristic, either shalelike or crystalline in nature (such material may be characterized by a shear wave velocity greater than 760 m/s, or by other appropriate means of classification; or Stiff soil conditions where the soil depth is less than 60 m and the soil types overlying rock are stable deposits of sands, gravels, or stiff clays.	1.0
II	Stiff clay or deep cohesionless conditions where the soil depth exceeds 60 m and the soil types overlying rock are stable deposits of sands, gravels, or stiff clays	1.2
III	Soft to medium-stiff clays and sands, characterized by 9 m or more of soft to medium-stiff clays with or without intervening layers of sand or other cohesionless soils	1.5
IV	Soft clays or silts greater than 12 m in depth; these materials may be characterized by a shear wave velocity less than 150 m/s and might include loose natural deposits or synthetic nonengineered fill	2.0

Site Effects for Different Soil Conditions

As indicated previously, soil deposits act as filters, amplifying response at some frequencies and deamplifying it at others. The greatest degree of amplification occurs at frequencies corresponding to the characteristic site period, $T_s = 4H/v_s$. Because the characteristic site period is proportional to shear wave velocity and inversely proportional to thickness, it is clear that the response of a given soil deposit will be influenced by the stiffness and thickness of the deposit. Thin and/or stiff soil deposits will amplify the short-period (high-frequency) components, and thick and/or soft soil deposits will amplify the long-period (low-frequency) components of an input motion. As a result, generalizations about site effects for different soil conditions are generally based on the average stiffness and thickness of the soil profile.

These observations of site response are reflected in bridge design codes. For example, the 1997 Interim Revision of the 1996 Standard Specifications for Highway Bridges (AASHTO, 1997) require the use of an elastic seismic response coefficient for an SDOF structure of natural period, T, taken as

$$C_s = \frac{1.2AS}{T^{2/3}} \tag{33.22}$$

where A is an acceleration coefficient that depends on the location of the bridge and S is a dimensionless site coefficient obtained from Table 33.2. In accordance with the behavior illustrated in Figure 33.17, the site coefficient prescribes increased design requirements at long periods for bridges underlain by thick deposits of soft soil (Figure 33.20).

33.7 Earthquake-Induced Settlement

Settlement is an important consideration in the design of bridge foundations. In most cases, settlement results from *consolidation*, a process that takes place relatively slowly as porewater is squeezed from the soil as it seeks equilibrium under a new set of stresses. Consolidation settlements are most significant in fine-grained soils such as silts and clays. However, the tendency of coarse-grained soils (sands and gravels) to densify due to vibration is well known; in fact, it is frequently relied upon for efficient compaction of sandy soils. Densification due to the cyclic stresses imposed by earthquake shaking can produce significant settlements during earthquakes. Whether caused by consolidation or earthquakes, bridge designers are concerned with **total settlement** and, because settlements rarely occur uniformly, also with **differential settlement.** Differential settlement can induce very large loads in bridge structures.

While bridge foundations may settle due to shearing failure in the vicinity of abutments (Chapter 30), shallow foundations (Chapter 31), and deep foundations (Chapter 32), this section

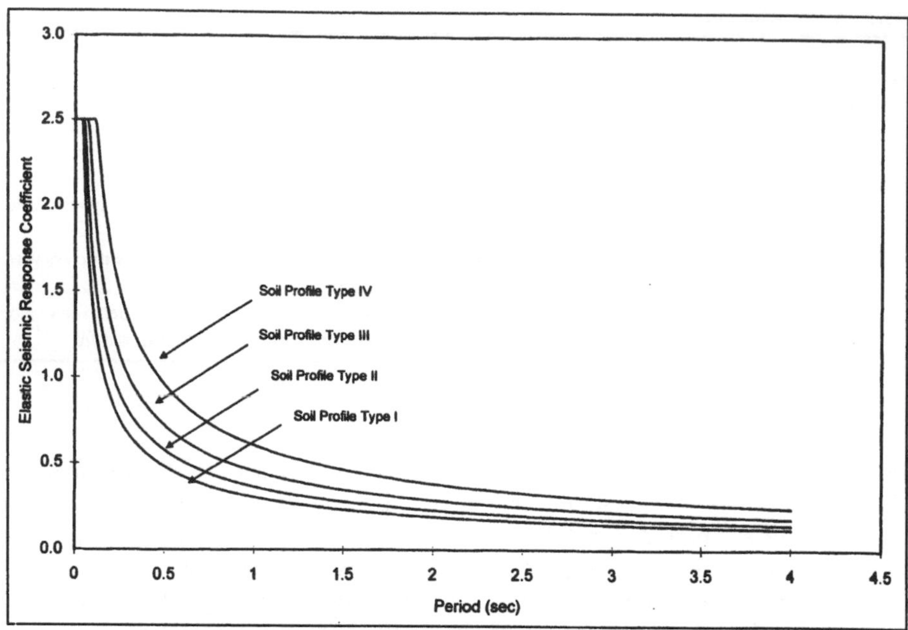

FIGURE 33.20 Variation of elastic seismic response coefficient with period for $A = 0.25$.

deals with settlement due to earthquake-induced soil densification. Densification of soils beneath shallow bridge foundations can cause settlement of the foundation. Densification of soils adjacent to deep foundations can cause downdrag loading on the foundations (and bending loading if the foundations are battered). Densification of soils beneath approach fills can lead to differential settlements at the ends of the bridge that can be so abrupt as to render the bridge useless.

Accurate prediction of earthquake-induced settlements is difficult. Errors of 25 to 50% are common in estimates of consolidation settlement, so even less accuracy should be expected in the more-complicated case of earthquake-induced settlement. Nevertheless, procedures have been developed that account for the major factors known to influence earthquake-induced settlement and that have been shown to produce reasonable agreement with many cases of observed field performance. Such procedures are generally divided into cases of dry sands and saturated sands.

Settlement of Dry Sands

Dry sandy soils are often found above the water table in the vicinity of bridges. The amount of densification experienced by dry sands depends on the density of the sand, the amplitude of cyclic shear strain induced in the sand, and on the number of cycles of shear strain applied during the earthquake. Settlements can be estimated using cyclic strain amplitudes from site response analyses with corrections for the effects of multidirectional shaking [47,58] or by simplified procedures [63]. Because of the high air permeability of sands, settlement of dry sands occurs almost instantaneously.

In the simplified procedure, the effective cyclic strain amplitude is estimated as

$$\gamma_{cyc} = 0.65 \frac{a_{max}}{g} \frac{\sigma_v r_d}{G} \tag{33.23}$$

Because the shear modulus, G, is a function of γ_{cyc}, several iterations may be required to calculate a value of γ_{cyc} that is consistent with the shear modulus. When the low strain stiffness, G_{max} ($= \rho v_s^2$), is known, the effective cyclic strain amplitude can be estimated using Figures 33.21 and 33.22.

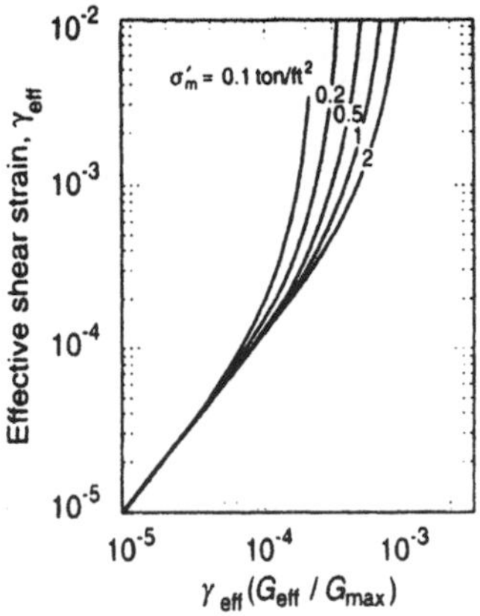

FIGURE 33.21 Plot for determination of effective cyclic shear strain in sand deposits. (Tokimatsu and Seed [63]).

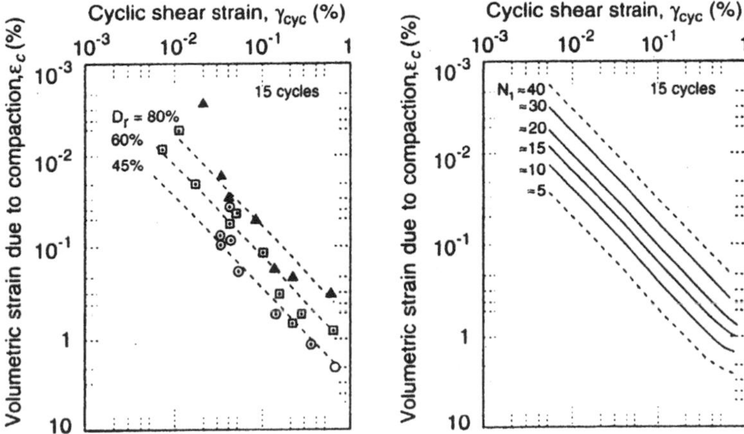

FIGURE 33.22 Relationship between volumetric strain and cyclic shear strain in dry sands as function of (a) relative density and (b) SPT resistance. (Tokimatsu and Seed [63]).

Figure 33.22 then allows the effective cyclic strain amplitude, along with the relative density or SPT resistance of the sand, to be used to estimate the volumetric strain due to densification. These volumetric strains are based on durations associated with a M = 7.5 earthquake; corrections for other magnitudes can be made with the aid of Table 33.3. The effects of multidirectional shaking are generally accounted for by doubling the computed volumetric strain. Because the stiffness, density, and cyclic shear strain amplitude generally vary with depth, a given soil deposit is usually divided into sublayers with the volumetric strain for each sublayer computed independently. The resulting settlement of each sublayer can then be computed as the product of the volumetric strain and thickness. The total settlement is obtained by summing the settlements of the individual sublayers.

Settlement of Saturated Sands

The dissipation of high excess porewater pressures generated in saturated sands (*reconsolidation*) can lead to settlement following earthquakes. Settlements of 50 to 70 cm occurred in a 5-m-thick

TABLE 33.3 Correction of Cyclic Stress Ratio for Earthquake Magnitude

Magnitude, M	5¼	6	6¾	7½	8½
$\varepsilon_{v,M}/\varepsilon_{v,M=7.5}$	0.4	0.6	0.85	1.0	1.25

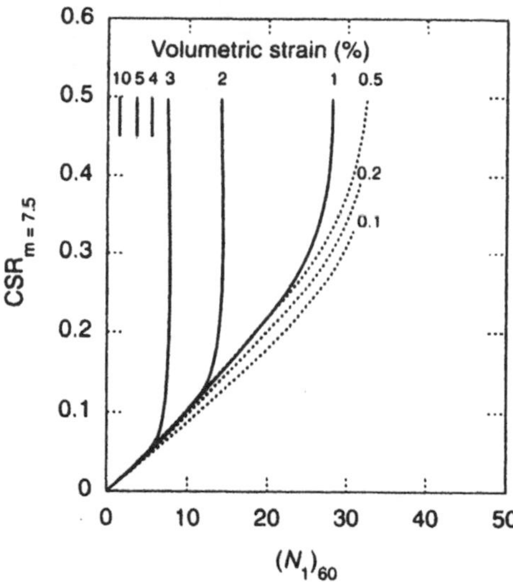

FIGURE 33.23 Plot for estimation of postliquefaction volumetric strain in saturated sands. (Tokimatsu and Seed [63]).

layer of very loose sand in the Tokachioki earthquake [44] and settlements of 50 to 100 cm were observed on Port Island and Rokko Island in Kobe, Japan following the 1995 Hyogo-ken Nambu earthquake. Because water flows much more slowly through soil than air, settlements of saturated sands occur much more slowly than earthquake-induced settlements of dry sands. Nevertheless, the main factors that influence the magnitude of saturated soil settlements are basically the same as those that influence that of dry sands.

Tokimatsu and Seed [63] developed charts to estimate the volumetric strains that develop in saturated soils. In this approach, the volumetric strain resulting from reconsolidation can be estimated from the corrected standard penetration resistance, $(N_1)_{60}$, and the cyclic stress ratio (Figure 33.23). The value of $(N_1)_{60}$ is obtained by correcting the measured standard penetration resistance, N_m, to a standard overburden pressure of 95.8 kPa (1 ton/ft^2) and to an energy of 60% of the theoretical free-fall energy of an SPT hammer using the equation:

$$(N_1)_{60} = N_m C_N \frac{E_m}{0.60\, E_{ff}} \tag{33.24}$$

where C_N is an overburden correction factor that can be estimated as $C_N = (\sigma'_{vo})^{-0.5}$, E_m is the measured hammer energy and E_{ff} is the theoretical free-fall energy. In Figure 33.23, the cyclic stress ratio, defined as $CSR_{M=7.5} = \tau_{cyc}/\sigma'_{vo}$, corresponds to a magnitude 7.5 earthquake. For other magnitudes, the corresponding value of the cyclic stress ratio can be obtained using Table 33.4. As in the case of dry sands, the soil layer is typically divided into sublayers with the total settlement taken as the sum of the products of the thickness and volumetric strain of all sublayers. In some cases, earthquake-induced porewater pressures may be insufficient to cause liquefaction but still may produce post-earthquake settlement. The volumetric strain produced by reconsolidation in such cases may be estimated from Figure 33.24.

TABLE 33.4 Correction of Cyclic Stress Ratio for Earthquake Magnitude

Magnitude, *M*	5¼	6	6¾	7½	8½
$CSR_M/CSR_{M=7.5}$	1.50	1.32	1.13	1.00	0.89

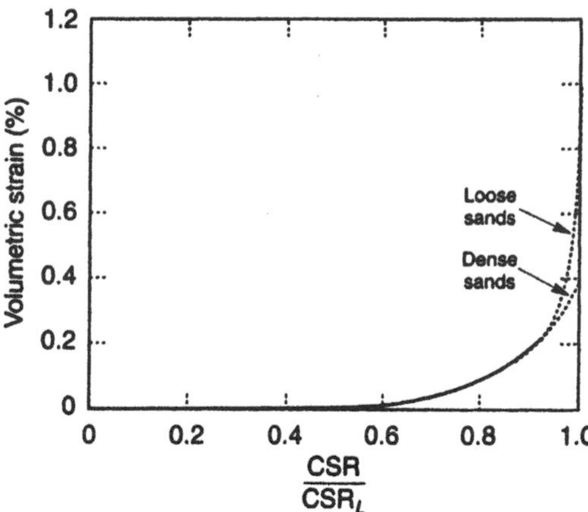

FIGURE 33.24 Plot for estimation of volumetric strain in saturated sands that do not liquefy. (Tokimatsu and Seed [63]).

33.8 Ground Failure

Strong earthquake shaking can produce a dynamic response of soils that is so energetic that the stress waves exceed the strength of the soil. In such cases, ground failure characterized by permanent soil deformations may occur. Ground failure may be caused by weakening of the soil or by temporary exceedance of the strength of the soil by transient inertial stresses. The former case results in phenomena such as liquefaction and lateral spreading; the latter in inertial failures of slopes and retaining wall backfills.

Liquefaction

The term *liquefaction* has been widely used to describe a range of phenomena in which the strength and stiffness of a soil deposit are reduced due to the generation of porewater pressure. It occurs most commonly in loose, saturated sands, although it has also been observed in gravels and non-plastic silts. The effects of liquefaction can range from massive landslides with displacements measured in tens of meters to relatively small slumps or spreads with small displacements. Many bridges, particularly those that cross bodies of water, are located in areas with geologic and hydrologic conditions that tend to produce liquefaction.

The mechanisms that produce liquefaction-related phenomena can be divided into two categories. The first, **flow liquefaction,** can occur when the shear stresses required for static equilibrium of a soil mass is greater than the shear strength of the soil in its liquefied state. While not common, flow liquefaction can produce tremendous instabilities known as **flow failures.** In such cases, the earthquake serves to trigger liquefaction, but the large deformations that result are actually driven by the preexisting static stresses. The second phenomenon, **cyclic mobility,** occurs when the initial static stresses are less than the strength of the liquefied soil. The effects of cyclic mobility lead to deformations that develop incrementally during the period of earthquake shaking, and are commonly called **lateral spreading.**

Lateral spreading can occur on very gentle slopes, in the vicinity of free surfaces such as riverbanks, and beneath and adjacent to embankments. Lateral spreading occurs much more frequently than flow failure, and can cause significant distress to bridges and their foundations.

Liquefaction Susceptibility

The first step in an evaluation of liquefaction hazards is the determination of whether or not the soil is susceptible to liquefaction. If the soils at a particular site are not susceptible to liquefaction, liquefaction hazards do not exist and the liquefaction hazard evaluation can be terminated. If the soil is susceptible, however, the issues of initiation and effects of liquefaction must be considered.

Liquefaction occurs most readily in loose, clean, uniformly graded, saturated soils. Therefore, geologic processes that sort soils into uniform grain size distributions and deposit them in loose states produce soil deposits with high liquefaction susceptibility. As a result, fluvial deposits, and colluvial and aeolian deposits when saturated, are likely to be susceptible to liquefaction. Liquefaction also occurs in alluvial, beach, and estuarine deposits, but not as frequently as in those previously listed. Because bridges are commonly constructed in such geologic environments, liquefaction is a frequent and important consideration in their design.

Liquefaction susceptibility also depends on the stress and density characteristics of the soil. Very dense soils, even if they have the other characteristics listed in the previous paragraph, will not generate high porewater pressures during earthquake shaking and hence are not susceptible to liquefaction. The minimum density at which soils are not susceptible to liquefaction increases with increasing effective confining pressure. This characteristic indicates that, for a soil deposit of constant density, the deeper soils are more susceptible to liquefaction than the shallower soils. For the general range of soil conditions encountered in the field, cohesionless soils with $(N_1)_{60}$ values greater than 30 or normalized cone penetration test (CPT) tip resistances (q_{c1N}, see next section) greater than about 175 are generally not susceptible to liquefaction.

Initiation of Liquefaction

The fact that a soil deposit is susceptible to liquefaction does not mean that liquefaction will occur in a given earthquake. Liquefaction must be triggered by some disturbance, such as earthquake shaking with sufficient strength to exceed the liquefaction resistance of the soil. Even a liquefaction-susceptible soil will have some liquefaction resistance. Evaluating the potential for the occurrence of liquefaction (liquefaction potential) involves comparison of the loading imposed by the anticipated earthquake with the liquefaction resistance of the soil. Liquefaction potential is most commonly evaluated using the cyclic stress approach in which both earthquake loading and liquefaction resistance are expressed in terms of cyclic stresses, thereby allowing direct and consistent comparison.

Characterization of Earthquake Loading

The level of porewater pressure generated by an earthquake is related to the amplitude and duration of earthquake-induced shear stresses. Such shear stresses can be predicted in a site response analysis using either the equivalent linear method or nonlinear methods. Alternatively, they can be estimated using a simplified approach that does not require site response analyses.

Early methods of liquefaction evaluation were based on the results of cyclic triaxial tests performed with harmonic (constant-amplitude) loading, and it remains customary to characterize loading in terms of an equivalent shear stress amplitude,

$$\tau_{cyc} = 0.65\tau_{max} \tag{33.25}$$

When sufficient information is available to perform site response analyses, it is advisable to compute τ_{max} in a site response analysis and use Eq. (33.6) to compute τ_{cyc}. When such information is not available, τ_{cyc} at a particular depth can be estimated as

$$\tau_{cyc} = 0.65 \frac{a_{max}}{g} \sigma_v r_d \tag{33.26}$$

where a_{max} is the peak ground surface acceleration, g is the acceleration of gravity, σ_v is the total vertical stress at the depth of interest, and r_d is the value of a site response reduction factor which can be estimated from

$$r_d = \frac{1.0 - 0.4113z^{0.5} + 0.04052z + 0.001753z^{1.5}}{1.0 - 0.4177z^{0.5} + 0.05729z - 0.006205z^{1.5} + 0.001210z^2} \tag{33.27}$$

where z is the depth of interest in meters. For evaluation of liquefaction potential, it is common to normalize τ_{cyc} by the initial (pre-earthquake) vertical effective stress, thereby producing the **cyclic stress ratio** (CSR)

$$\mathrm{CSR} = \frac{\tau_{cyc}}{\sigma'_{vo}} \tag{33.28}$$

Characterization of Liquefaction Resistance

While early liquefaction potential evaluations relied on laboratory tests to measure liquefaction resistance, increasing recognition of the deleterious effects of sampling disturbance on laboratory test results has led to the use of field tests for measurement of liquefaction resistance. Although the use of new soil freezing and sampling techniques offers considerable promise for acquisition of undisturbed samples, liquefaction resistance is currently evaluated using *in situ* tests such as the standard penetration test (SPT) and the CPT *and* observations of liquefaction behavior in past earthquakes.

Case histories in which liquefaction was and was not observed can be analyzed to obtain empirical estimates of liquefaction resistance. By characterizing each of a series of case histories in terms of a loading parameter, $\mathcal{L}$, and a resistance parameter, $\mathcal{R}$, all combinations of $\mathcal{L}$ and $\mathcal{R}$ can be plotted with symbols that indicate whether liquefaction was observed or was not observed (Figure 33.25).

In this approach, the cyclic stress ratio induced in the soil for each case history is used as the loading parameter and an *in situ* test measurement is used as the resistance parameter. Two *in situ* tests are commonly used — the SPT which produces the resistance parameter $(N_1)_{60}$, and the CPT which produces the resistance parameter, q_{c1N}. Because the value of the cyclic stress ratio given by the curve represents the minimum cyclic stress ratio required to produce liquefaction, it is commonly referred to as the **cyclic resistance ratio,** CRR.

Because liquefaction involves the cumulative buildup of porewater pressure, the ultimate porewater pressure level is a function of the duration of ground shaking. In the development of procedures for evaluation of liquefaction potential, duration was implicitly correlated to earthquake magnitude. As a result, the procedures have been keyed to magnitude 7.5 earthquakes with corrections developed that can be applied for other magnitudes. The procedures have also been keyed to clean sands (<5% fines), again with corrections developed for application to silty sands.

Recent review of SPT-based procedures for characterization of CRR resulted in recommendation of the curve shown in Figure 33.26. This CRR curve is for clean sand and magnitude 7.5 earthquakes. For a silty sand with fines content, FC, an equivalent clean sand SPT resistance can be computed from

$$(N_1)_{60\text{-}cs} = \alpha + \beta\,(N_1)_{60} \tag{33.29}$$

where

$\alpha = 0$	and	$\beta = 1.0$	for FC < 5%
$\alpha = \exp[1.76 - 190/FC^2]$	and	$\beta = 0.99 + FC^{1.5}/1000$	for 5% < FC < 35%
$\alpha = 5.0$	and	$\beta = 1.2$	for FC > 35%

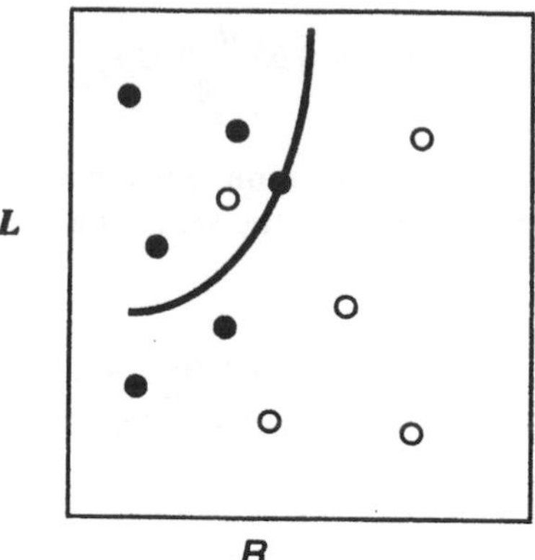

FIGURE 33.25 Discrimination between case histories in which liquefaction was observed (solid circles) and was not observed (open circles). Curve represents conservative estimate of resistance, $\mathcal{R}$, for given level of loading, $\mathcal{L}$.

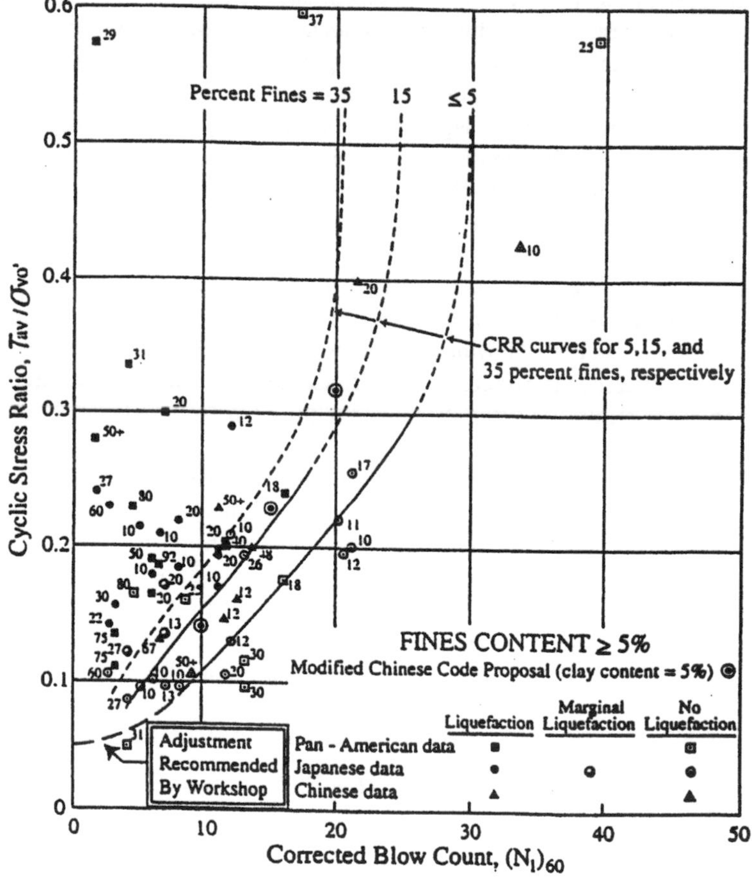

FIGURE 33.26 Relationship between cyclic stress ratios causing liquefaction and $(N_1)_{60}$ values for clean sand (after Youd and Idriss, 1998).

TABLE 33.5 Magnitude Scaling Factor

Magnitude, *M*	MSF
5.5	2.20–2.80
6.0	1.76–2.10
6.5	1.44–1.60
7.0	1.19–1.25
7.5	1.00
8.0	0.84
8.5	0.72

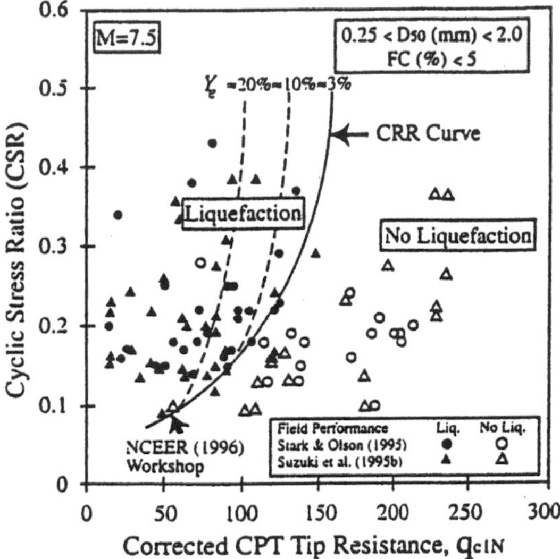

FIGURE 33.27 Relationship between cyclic stress ratios causing liquefaction and $(q_c)_1$ values for clean sand (after Youd and Idriss, 1998).

Correction for magnitudes other than 7.5 is accomplished by correcting the CRR according to

$$CRR_M = CRR_{7.5} \times MSF \tag{33.30}$$

where MSF is a magnitude scaling factor obtained from Table 33.5.

The CPT offers two distinct advantages over the SPT for evaluation of liquefaction resistance. First, the CPT provides a nearly continuous profile of penetration resistance, a characteristic that allows it to identify thin layers that can easily be missed in an SPT-based investigation. Second, the CPT shows greater consistency and repeatability than the SPT. However, the CPT is a more recent development and there is less professional experience with it than with the SPT, particularly in the United States. As more data correlating CPT resistance to liquefaction resistance become available, the CPT is likely to be come the primary *in situ* test for evaluation of liquefaction potential. At present, however, a general consensus on the most appropriate technique for CPT-based evaluation of liquefaction potential has not emerged. One of the most well-developed procedures for CPT-based evaluation of liquefaction potential was described by Robertson and Wride. In this procedure, the measured CPT resistance, q_c, is normalized to a dimensionless resistance

$$q_{c1N} = C_Q \frac{q_c}{p_a} \tag{33.31}$$

where $C_Q = (p_a/\sigma'_{vo})^n$, p_a is atmospheric pressure, and n is an exponent that ranges from 0.5 (clean sand) to 1.0 (clay). A maximum C_Q value of 2.0 is generally applied to CPT data at shallow depths. Soil type can be inferred from CPT tip resistance, q_c, and sleeve resistance, f_s, with the aid of a soil behavior type index

$$I_c = \sqrt{(3.47 - \log Q)^2 + (1.22 + \log F)^2} \tag{33.32}$$

where

$$Q = C_Q \frac{q_c - \sigma_{vo}}{p_a}$$

$$F = \frac{f_s}{q_c - \sigma_{vo}} \times 100\%$$

If I_c (computed with $n = 1.0$) is greater than 2.6, the soil is considered too clayey to liquefy. If I_c (computed with $n = 0.5$ and $Q = q_{c1N}$) is less than 2.6, the soil is most likely granular and nonplastic and capable of liquefying. If I_c (computed with $n = 0.5$ and $Q = q_{c1N}$) is greater than 2.6, however, the soil is likely to be very silty and possibly plastic; in this case, I_c should be recalculated with $n = 0.7$ and $Q = q_{c1N}$. Once I_c has been determined, the effects of fines and plasticity can be considered by computing the clean sand normalized tip resistance

$$q_{c1N\text{-}cs} = K_c\, q_{c1N} \tag{33.33}$$

where

$K_c = 1.0$ for $I_c < 1.64$

$K_c = -0.403 I_c^4 + 5.581 I_c^3 - 21.63 I_c^2 + 33.75 I_c - 17.88$ for $I_c > 1.64$

With the clean sand normalized tip resistance, $CRR_{7.5}$ can be determined using Figure 33.27. For other magnitudes, the appropriate value of CRR can be obtained using the same magnitude scaling factor used for the SPT-based procedure, (Eq. 30). Other procedures for CPT-based evaluation of liquefaction potential include those of Seed and De Alba (57), Mitchell and Tseng [39], and Olson [46].

Liquefaction resistance has also been correlated to other *in situ* test measurements such as shear wave velocity [62,64], dilatometer index, and Becker penetration tests. In addition, probabilistic approaches that yield a probability of liquefaction have also been developed [32].

Lateral Spreading

Lateral spreading has often caused damage to bridges and bridge foundations in earthquakes. Lateral spreading generally involves the lateral movement of soil at and below the ground surface, often in the form of relatively intact surficial blocks riding on a mass of softened and weakened soil. The lateral soil movement can impose large lateral loads on abutments and wingwalls, and can induce large bending moments in pile foundations. The damage produced by lateral spreading is closely related to the magnitude of the lateral soil displacements.

Because cyclic mobility, the fundamental phenomenon that produces lateral spreading, is so complex, analytical procedures for prediction of lateral spreading displacements have not yet reached the point at which they can be used for design. As a result, currently accepted procedures for prediction of lateral spreading displacements are empirically based.

Bartlett and Youd [1] used multiple regression on a large database of lateral spreading case histories to develop empirical expressions for lateral spreading ground surface displacements. Two expressions were developed — a ground slope expression for sites with gentle, uniformly sloping

TABLE 33.6 Range of Verified Values for Eq. 33.32

Input Parameter	Range of Values
Magnitude	$6.0 < M_w < 8.0$
Free-face ratio	$1.0\% < W < 20\%$
Thickness of loose layer	$0.3\text{ m} < T_{15} < 12\text{ m}$
Fines content	$0\% < F_{15} < 50\%$
Mean grain size	$0.1\text{ mm} < (D_{50})_{15} < 1.0\text{ mm}$
Ground slope	$0.1\% < S < 6\%$
Depth to bottom of section	Depth to bottom of liquefied zone <15 m

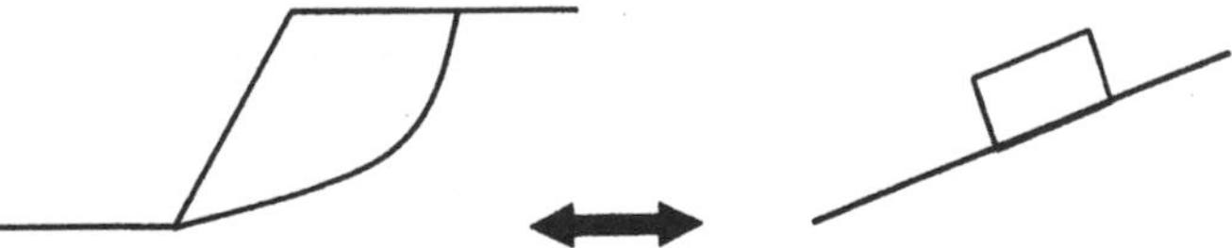

FIGURE 33.28 Illustration of sliding block analogy for evaluation of permanent slope displacements.

surfaces, and a free-face expression for sites near steep banks. For the former, displacements can be estimated from

$$\begin{aligned}\log D_H = &-16.3658 + 1.1782 M_w - 0.9275 \log R - 0.0133 R + 0.4293 \log S \\ &+ 0.3483 \log T_{15} + 4.5720 \log(100 - F_{15}) - 0.9224 (D_{50})_{15}\end{aligned} \tag{33.34a}$$

where D_H is the estimated lateral ground displacement in meters, M_w is the moment magnitude, R is the horizontal distance from the seismic energy source in km, S is the ground slope in percent, T_{15} is the cumulative thickness of saturated granular layers with $(N_1)_{60} < 15$ in meters, F_{15} is the average fines content for the granular layers comprising T_{15} in percent, and $(D_{50})_{15}$ is the average mean grain size for the granular layers comprising T_{15} in millimeters. For free-face sites, displacements can be estimated from

$$\begin{aligned}\log D_H = &-16.3658 + 1.1782 M_w - 0.9275 \log R - 0.0133 R + 0.6572 \log W \\ &+ 0.3483 \log T_{15} + 4.5720 \log(100 - F_{15}) - 0.9224 (D_{50})_{15}\end{aligned} \tag{33.34b}$$

where W is the ratio of the height of the bank to the horizontal distance between the toe of the bank and the point of interest. With these equations, 90% of the predicted displacements were within a factor of two of those observed in the corresponding case histories. The range of parameters for which the predicted results have been verified by case histories is presented in Table 33.6.

Global Instability

Ground failure may also occur due to the temporary exceedance of the shear strength of the soil by earthquake-induced shear stresses. These failures may take the form of large, deep-seated soil failures that can encompass an entire bridge abutment or foundation as illustrated in Figure 33.28. The potential for such failures, often referred to as global instabilities, must be evaluated during design.

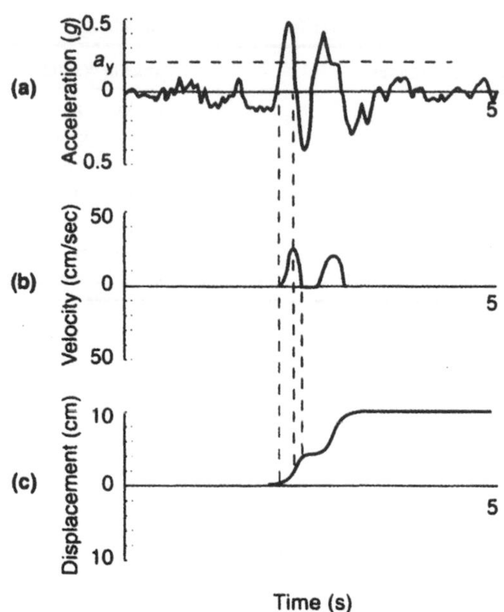

FIGURE 33.29 Illustration of computation of permanent slope displacements using sliding block method.

Historically, inertial failures were evaluated using **pseudo-static methods** in which the transient, dynamic effects of earthquake shaking were represented by constant, pseudo-static accelerations. The resulting destabilizing pseudo-static forces were included in a limit equilibrium analysis to compute a pseudo-static factor of safety. A pseudo-static factor of safety greater than one was considered indicative of stability. However, difficulty in selection of the pseudo-static acceleration, interpretation of the significance of computed factors of safety less than 1, and increasing recognition that serviceability is closely related to permanent deformations led to the development of alternative approaches.

The most common current procedure uses pseudo-static principles to establish the point at which permanent displacements would begin, but then uses a simple slope analogy to estimate the magnitude of the resulting permanent displacements. This procedure is commonly known as the sliding block procedure [43]. By using the common assumptions of rigid, perfectly plastic behavior embedded in limit equilibrium analyses, a potentially unstable slope is considered to be analogous to a block resting on an inclined plane (Figure 33.28) in the sliding block procedure. In both cases, base accelerations above a certain level will result in permanent relative displacements of the potentially unstable mass.

In the sliding block procedure, a pseudo-static analysis is performed to determine the horizontal pseudo-static acceleration that produces a factor of safety of 1.0. This pseudo-static acceleration, referred to as the **yield acceleration**, represents the level of acceleration above which permanent slope displacements are expected to occur. When the input acceleration exceeds the yield acceleration, the shear stress between the sliding block and the plane exceeds the available shear resistance and the block is unable to accelerate as quickly as the underlying plane. As a result, there is a relative acceleration between the block and the plane that lasts until the shear stress drops below the strength long enough to decelerate the block to zero relative acceleration. Integration of the relative acceleration over time yields a relative velocity, and integration of the relative velocity produces the relative displacement between the block and the plane. By this process, illustrated in Figure 33.29, the sliding block procedure allows estimation of the permanent displacement of a slope.

For embankments subjected to ground motions perpendicular to their axes, Makdisi and Seed [36] developed a simplified procedure for estimation of earthquake-induced displacements based

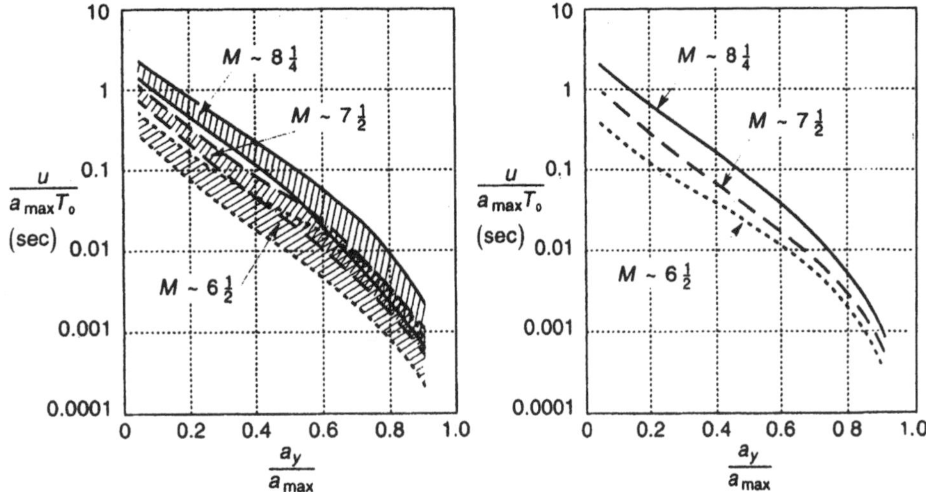

FIGURE 33.30 Variation of normalized permanent slope displacement with yield acceleration for earthquakes of different magnitudes: (a) summary for several different earthquakes and embankments and (b) average values (after Makdisi and Seed [36]).

on sliding block analyses of dams and embankments subjected to several recorded and synthetic input motions. By knowing the yield acceleration of the slope in addition to the peak acceleration and fundamental period of the embankment, Figure 33.30 can be used to estimate permanent slope displacements.

Retaining Structures

Earth-retaining structures are commonly constructed as parts of bridge construction projects and, in the form of abutment walls and wingwalls, as parts of bridge structures themselves. However, there are many different types of retaining structures, several of which have been developed in recent years. Historically, rigid retaining structures have been most commonly used; their static design is based on classical earth pressure theories. However, newer types of retaining structures, such as flexible anchored walls, soil nailed walls, and reinforced walls, have required the development of new approaches, even for static conditions. Under seismic conditions, classical earth pressure theories can be extended in a logical way to account for the effects of earthquake shaking, but seismic design procedures for the newer types of retaining structures remain under development.

Free-standing rigid retaining structures typically maintain equilibrium through the development of active and passive earth pressures that develop as the wall translates and rotates under the action of the imposed stresses. By assuming that static stresses develop through mobilization of the shear strength of the backfill soil on a planar potential failure surface, Coulomb earth pressure theory predicts a static active thrust of

$$P_A = \frac{1}{2} K_A \gamma H^2 \tag{33.35}$$

where

$$K_A = \frac{\cos^2(\phi - \theta)}{\cos^2\theta \cos(\delta + \theta)\left[\dfrac{\sin(\delta + \phi)\sin(\phi - \beta)}{\cos(\delta + \theta)\cos(\beta - \theta)}\right]^2}$$

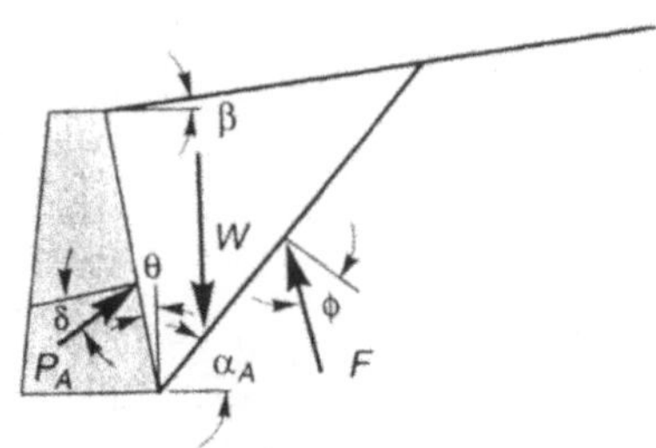

FIGURE 33.31 Illustration of variables for computation of active Earth thrust.

δ is the angle of interface friction between the wall and the soil, and β and θ are as shown in Figure 33.31.

Under earthquake shaking, active earth pressures tend to increase above static levels. In one of the first geotechnical earthquake engineering analyses, Okabe [45] and Mononobe and Matsuo [40] developed a pseudo-static extension of Coulomb theory to predict the active earth thrust under seismic conditions. Assuming pseudo-static accelerations of $a_h = k_h g$ and $a_v = k_v g$ in the horizontal and vertical directions, respectively, the Mononobe–Okabe total thrust is given by

$$P_{AE} = \frac{1}{2} K_{AE} \gamma H^2 (1 - k_v) \tag{33.36}$$

where

$$K_{AE} = \frac{\cos^2(\phi - \theta - \psi)}{\cos\psi \cos^2\theta \cos(\delta + \theta + \psi)\left[1 + \sqrt{\dfrac{\sin(\delta + \phi)\sin(\phi - \beta - \psi)}{\cos(\delta + \theta + \psi)\cos(\beta - \theta)}}\right]^2}$$

where $\phi - \beta \geq \psi$ and $\Psi = \tan^{-1}[k_h/(1 - k_v)]$. Although the assumptions used in the Mononobe-Okabe analysis imply that the total active thrust should act at a height of $H/3$ above the base of the wall, experimental results indicate that it acts at a higher point. The total active thrust of Eq. (33.36) can be divided into a static component, P_A, given by Eq. (33.35), and a dynamic component,

$$\Delta P_{AE} = P_{AE} - P_A \tag{33.37}$$

which acts at a height of approximately $0.6H$ above the base of the wall. On this basis, the total active thrust can be taken to act at a height

$$h = \frac{P_A H/3 + \Delta P_{AE}(0.6H)}{P_{AE}} \tag{33.38}$$

above the base of the wall.

When retaining walls are braced against lateral movement at top and bottom, as can occur with abutment walls, the shear strength of the soil will not be fully mobilized under static or seismic conditions. As a result, the limiting conditions of minimum active or maximum passive conditions cannot be developed. In such cases, it is common to estimate lateral Earth pressures using the elastic solution of Wood [69] for a linear elastic material of height, H, trapped between rigid walls separated by a horizontal distance, L. For motions at less than half the fundamental frequency of the unrestrained

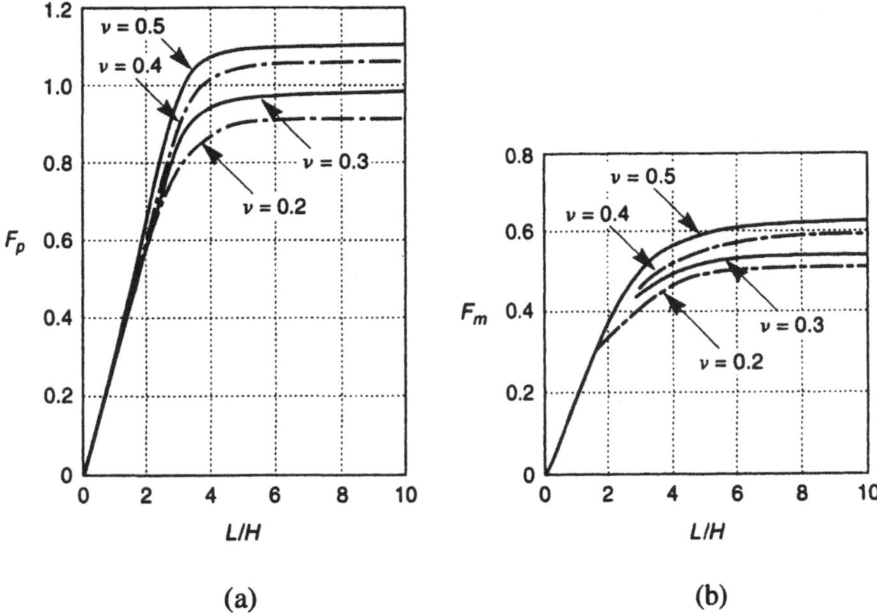

FIGURE 33.32 Charts for determination of (a) dimensionless thrust factor and (b) dimensionless moment factor for various geometries and Poisson ratios (after Wood [69]).

backfill ($f_o = v_s/4H$) the dynamic thrust and dynamic overturning moment (about the base of the wall) can be expressed as

$$\Delta P_{eq} = \gamma H^2 \frac{a_h}{g} F_p \tag{33.39}$$

$$\Delta M_{eq} = \gamma H^3 \frac{a_h}{g} F_m \tag{33.40}$$

where a_h is the amplitude of the harmonic base acceleration and F_p and F_m are dimensionless factors that can be obtained from Figure 33.32. It should be noted that Eqs. 33.39 and 33.40 refer to dynamic thrusts and moments; static thrusts and moments must be added to obtain total thrusts and moments.

33.9 Soil Improvement

When existing subsurface conditions introduce significant seismic hazards that adversely affect safety or impact construction costs, improved performance may be achieved through a program of soil improvement. A variety of techniques are available for soil improvement and may be divided into four main categories: densification, drainage, reinforcement, and grouting/mixing. Each soil improvement technique has advantages and disadvantages that influence the cost and effectiveness under different circumstances. Soil improvement techniques for both seismic and nonseismic areas are described in detail in such references as Welsh [67], Van Impe [65], Hausmann [27], Broms [8], Bell [2], and Mosely [41].

Densification Techniques

Virtually all mechanical properties of soil (e.g., strength, stiffness, etc.) improve with increasing soil density. This is particularly true when earthquake problems are considered — the tendency of loose soils to densify under dynamic loading is responsible for such hazards as liquefaction, lateral spreading, and earthquake-induced settlement. This tendency can be used to advantage, however, as most densification techniques rely on vibrations to densify granular soil efficiently. Because fines inhibit densification for much the same reason as they inhibit liquefaction, densification techniques are most efficient in clean sands and gravels.

Vibratory densification of large volumes of soil can be accomplished most economically by dynamic compaction. In this procedure, a site is densified by repeatedly lifting and dropping a heavy weight in a grid pattern across the surface of the site. By using weights that can range from 53 to 267 kN and drop heights of 10 to 30 m, densification can be achieved to depths of up to 12 m. The process is rather intrusive in terms of ground surface disturbance, noise, dust, and vibration of surrounding areas, so it is used primarily in undeveloped areas. Vibrations from probes that penetrate below the ground surface have also proved to be effective for densification. Vibroflotation, for example, is accomplished by lowering a vibrating probe into the ground (with the aid of water jets, in some cases). By vibrating the probe as it is pulled back toward the surface, a column of densified soil surrounding the vibroflot is produced. Gravel or crushed stone may be introduced into the soil at the surface or, using a bottom-feed vibroflot, at the tip of the probe to form stone columns. Blasting can also be used to densify cohesionless soils. Blast densification is usually accomplished by detonating multiple explosive charges spaced vertically at distances of 3 to 6 m in borings spaced horizontally at distances of 5 to 15 m. The charges at different elevations are often detonated at small time delays to enhance the amplitude, and therefore the densification capacity, of the blast waves. Two or three rounds of blasting, with later rounds detonated at locations between those of the earlier rounds, are often used to achieve the desired degree of densification. Finally, densification may be achieved using static means using compaction grouting. Compaction grouting involved the injection of very low slump (usually less than 25 mm) cementitious grout into the soil under high pressure. The grout forms an intact bulb or column that densifies the surrounding soil by displacement. Compaction grouting may be performed at a series of points in a grid or along a line. Grout points are typically spaced at distances of about 1 to 4 m, and have extended to depths of 30 m.

Drainage Techniques

Excessive soil and foundation movements can often be eliminated by lowering the groundwater table, and construction techniques for dewatering are well developed. The buildup of high porewater pressures in liquefiable soils can also be suppressed using drainage techniques, although drainage alone is rarely relied upon for mitigation of liquefaction hazards. Stone columns provide means for rapid drainage by horizontal flow, but also improve the soil by densification (during installation) and reinforcement.

Reinforcement Techniques

The strength and stiffness of some soil deposits can be improved by installing discrete inclusions that reinforce the soil. Stone columns are columns of dense angular gravel or crushed stone (stone columns) that reinforce the soil in which they are installed. Stone columns also improve the soil due to their drainage capabilities and the densification and lateral stress increase that generally occurs during their installation. Granular soils can also be improved by the installation of compaction piles, usually prestressed concrete or timber, driven in a grid pattern and left in place. Compaction piles can often increase relative densities to 75 to 80% within a distance of 7 to 12 pile diameters. Drilled inclusions such as drilled shafts or drilled piers have been used to stabilize many slopes, although the difficulty in drilling through loose granular soils limits their usefulness for slopes with liquefiable soils. Soil nails, tiebacks, micropiles, and root piles have also been used.

Grouting/Mixing Techniques

The characteristics of many soils can be improved by the addition of cementitious materials. Introduced by injection or mixing, these materials both strengthen the contacts between soil grains and fill the space between the grains. Grouting involves injection of cementitious materials into the voids of the soil or into fractures in the soil; in both cases, the particle structure of the majority of the soil remains intact. In mixing, the cementitious materials are mechanically or hydraulically mixed into the soil, completely destroying the initial particle structure.

Permeation grouting involves the injection of low-viscosity grouts into the voids of the soil without disturbing the particle structure. Both particulate grouts (aqueous suspensions of cement, fly ash, bentonite, microfine cement, etc.) and chemical grouts (silica and lignin gels, or phenolic and acrylic resins) may be used. The more viscous particulate grouts are generally used in coarser-grained soils with large voids such as gravels and coarse sands; chemical grouts can be used in fine sands. The presence of fines can significantly reduce the effectiveness of permeation grouting. Grout pipes are usually arranged in a grid pattern at spacings of 1.2 to 2.4 m and can produce grouted soil strengths of 350 to 2100 kPa. Intrusion grouting involves the injection of more viscous (and hence stronger) cementitious grouts under pressure to cause controlled fracturing of the ground. The first fractures generally follow weak bedding planes or minor principal stress planes; after allowing the initially placed grout to cure, repeated grouting fractures the soil along additional planes, eventually producing a three-dimensional network of intersecting grout lenses.

Using a mechanical system consisting of hollow stem augers and rotating paddles, soil mixing produces an amorphous mixture of soil and cementitious material. The soil-mixing process produces columns of soil–cement that can be arranged in a grid pattern or in a linear series of overlapping columns to produce subsurface walls and/or cellular structures. Soil mixing, which can be used in virtually all inorganic soils, has produced strengths of 1400 kPa and improvement to depths of 60 m. In jet grouting, cement grout is injected horizontally under high pressure through ports in the sides of a hollow rod lowered into a previously drilled borehole. Jet grouting begins at the bottom of the borehole and proceeds to the top. Rotation of the injection nozzle as the process occurs allows the jet to cut through and hydraulically mix columns of soil up to 2.4 m in diameter. Air or air and water may also be injected to aid in the mixing process. Jet grouting can be performed in any type of inorganic soil to depths limited only by the range of the drilling equipment.

Defining Terms

Selected terms used in this section are compiled below:

Amplification function: A function that describes the ratio of ground surface motion to bedrock motion as a function of frequency.

Attenuation: The rate at which earthquake ground motion decreases with distance.

Backbone curve: The nonlinear stress–strain curve of a monotonically loaded soil.

Blind thrust faults: Faults at depth occurring under anticlinal folds — since they have only subtle surface expression, their seismogenic potential can only be evaluated by indirect means [22]. Blind thrust faults are particularly worrisome because they are hidden, are associated with folded topography in general, including areas of lower and infrequent seismicity, and therefore result in a situation where the potential for an earthquake exists in any area of anticlinal geology, even if there are few or no earthquakes in the historic record. Recent major earthquakes of this type have included the 1980 M_w 7.3 El Asnam (Algeria), 1988 M_w 6.8 Spitak (Armenia), and 1994 M_w 6.7 Northridge (California) events.

Body waves: Vibrational waves transmitted through the body of the Earth, and are of two types: p waves (transmitting energy via dilatational or push–pull motion), and slower s waves (transmitting energy via shear action at right angles to the direction of motion).

Characteristic earthquake: A relatively narrow range of magnitudes at or near the maximum that can be produced by the geometry, mechanical properties, and state of stress of a fault. [56].

Coherence: The similarity of ground motions at different locations. The coherence of ground motions at closely spaced locations is higher than at greater spacings. At a given spacing, the coherence of low-frequency (long wavelength) components is greater than that of high-frequency (short-wavelength) components.

Cyclic mobility: A phenomenon involving accumulation of porewater pressure during cyclic loading in soils for which the residual shear strength is greater than the shear stress required to maintain static equilibrium.

Cyclic resistance ratio (CRR): The ratio of equivalent shear stress amplitude required to trigger liquefaction to the initial vertical effective stress acting on the soil.

Cyclic stress ratio (CSR): The ratio of equivalent shear stress amplitude of an earthquake ground motion to the initial vertical effective stress acting on the soil.

Damping: The force or energy lost in the process of material deformation (damping coefficient c = force per velocity).

Damping curve: A plot of equivalent viscous damping ratio as a function of shear strain amplitude.

Differential settlement: The relative amplitudes of settlement at different locations. Differential settlement may be particularly damaging to bridges and other structures.

Dip: The angle between a plane, such as a fault, and the Earth's surface.

Dip-slip: Motion at right angles to the strike, up- or down-slip.

Ductility factor: The ratio of the total displacement (elastic plus inelastic) to the elastic(i.e., yield) displacement.

Epicenter: The projection on the surface of the Earth directly above the hypocenter.

Equivalent linear analysis: An analysis in which the stress–strain behavior of the soil is characterized by a secant shear modulus and damping ratio that, through a process of iteration, are compatible with the level of shear strain induced in the soil.

Far-field: Beyond near-field, also termed teleseismic.

Fault: A zone of the Earth's crust within which the two sides have moved — faults may be hundreds of miles long, from 1 to over 100 miles deep, and may not be readily apparent on the ground surface.

Flow failure: A soil failure resulting from flow liquefaction. Flow failures can involve very large deformations.

Flow liquefaction: A phenomenon that can occur when liquefaction is triggered in a soil with a residual shear strength lower than the shear stress required to maintain static equilibrium.

Hypocenter: The location of initial radiation of seismic waves (i.e., the first location of dynamic rupture).

Intensity: A metric of the effect, or the strength, of an earthquake hazard at a specific location, commonly measured on qualitative scales such as MMI, MSK, and JMA.

Lateral spreading: A phenomenon resulting from cyclic mobility in soils with some nonzero initial shear stress. Lateral spreading is characterized by the incremental development of permanent lateral soil deformations.

Magnitude: A unique measure of an individual earthquake's release of strain energy, measured on a variety of scales, of which the moment magnitude M_w (derived from seismic moment) is preferred.

Magnitude-frequency relation: The probability of occurrence of a selected magnitude — the commonest is $\log_{10} n(m) = a - bm$ [25]

Meizoseismal: The area of strong shaking and damage.

Modulus reduction curve: The ratio of secant shear modulus at a particular shear strain to maximum shear modulus (corresponding to very low strains) plotted as a function of shear strain amplitude.

Near-field: Within one source dimension of the epicenter, where source dimension refers to the length or width of faulting, whichever is less.

Nonlinear approach: An analysis in which the nonlinear, inelastic stress–strain behavior of the soil is explicitly modeled.

Normal fault: A fault that exhibits dip-slip motion, where the two sides are in tension and move away from each other.

Peak ground acceleration (PGA): The maximum amplitude of recorded acceleration (also termed the ZPA, or zero-period acceleration)

Pseudo-static approach: A method of analysis in which the complex, transient effects of earthquake shaking are represented by constant accelerations. The inertial forces produced by these accelerations are considered, along with the static forces, in limit equilibrium stability analyses.

Radiation damping: A reduction in wave amplitude due to geometric spreading of traveling waves, or radiation into adjacent or underlying materials.

Response spectrum: A plot of maximum amplitudes (acceleration, velocity or displacement) of an sdof oscillator, as the natural period of the SDOF is varied across a spectrum of engineering interest (typically, for natural periods from 0.03 to 3 or more seconds or frequencies of 0.3 to 30+ Hz).

Reverse fault: A fault that exhibits dip-slip motion, where the two sides are in compression and move away toward each other.

Seismic hazards: The phenomena and/or expectation of an earthquake-related agent of damage, such as fault rupture, vibratory ground motion (i.e., shaking), inundation (e.g., tsunami, seiche, dam failure), various kinds of permanent ground failure (e.g., liquefaction), fire or hazardous materials release.

Seismic moment: The moment generated by the forces generated on an earthquake fault during slip.

Seismotectonic model: A mathematical model representing the seismicity, attenuation, and related environment.

Specific impedance: Product of density and wave propagation velocity.

Spectrum amplification factor: The ratio of a response spectral parameter to the ground motion parameter (where parameter indicates acceleration, velocity, or displacement).

Strike: The intersection of a fault and the surface of the Earth, usually measured from the north (e.g., the fault strike is N 60° W).

Strike slip fault: See Transform or Strike slip fault.

Subduction: The plunging of a tectonic plate (e.g., the Pacific) beneath another (e.g., the North American) down into the mantle, due to convergent motion.

Surface waves: Vibrational waves transmitted within the surficial layer of the Earth, of two types: horizontally oscillating Love waves (analogous to S body waves) and vertically oscillating Rayleigh waves.

Thrust fault: Low-angle reverse faulting (blind thrust faults are faults at depth occurring under anticlinal folds — they have only subtle surface expression).

Total settlement: The total amplitude of settlement at a particular location.

Transform or strike slip fault: A fault where relative fault motion occurs in the horizontal plane, parallel to the strike of the fault.

Uniform hazard spectra: Response spectra with the attribute that the probability of exceedance is independent of frequency.

Yield acceleration: The horizontal acceleration that produces a pseudo-static factor of safety of 1. Accelerations greater than the yield acceleration are expected to produce permanent deformations.

References

1. Bartlett, S.F. and Youd, T.L., Empirical analysis of horizontal ground displacement generated by liquefaction-induced lateral spread, *Technical Report NCEER-92-0021*, National Center for Earthquake Engineering Research, Buffalo, NY, 1992.
2. Bell, F.G., *Engineering Treatment of Soils*, E & FN Spon, London, 1993, 302 pp.

3. Bernreuter, D.L. et al., Seismic Hazard Characterization of 69 Nuclear Power Plant Sites East of the Rocky Mountains, U.S. Nuclear Regulatory Commission, NUREG/CR-5250, 1989.
4. Bolt, B.A., *Earthquakes,* W.H. Freeman, New York, 1993.
5. Bonilla, M.G. et al., Statistical relations among earthquake magnitude, surface rupture length, and surface fault displacement, *Bull. Seismol. Soc. Am.*, 74(6), 2379–2411, 1984.
6. Boore, D.M., Joyner, W.B., and Fumal, T.E., Estimation of Response Spectra and Peak Acceleration from Western North American Earthquakes: An Interim Report, U.S.G.S Open-File Report 93-509, Menlo Park, CA, 1993.
7. Bozorgnia, Y. and Campbell, K.W., Spectral characteristics of vertical ground motion in the Northridge and other earthquakes, *Proc. 4th U.S. Conf. On Lifeline Earthquake Eng.*, American Society of Civil Engineers, New York, 660–667, 1995.
8. Broms, B., Deep compaction of granular soil, in H.-Y. Fang, Ed., *Foundation Engineering Handbook,* 2nd ed., Van Nostrand Reinhold, New York, 1991, 814–832.
9. BSSC, NEHRP Recommended Provisions for Seismic Regulations for New Buildings, 1994.
10. Campbell, K.W., Strong ground motion attenuation relations: a ten-year perspective, *Earthquake Spectra,* 1(4), 759–804, 1985.
11. Campbell, K.W. and Bozorgnia, Y., Near-source attenuation of peak horizontal acceleration from worldwide accelerograms recorded from 1957 to 1993, *Proc. Fifth U.S. National Conference on Earthquake Engineering,* Earthquake Engineering Research Institute, Oakland, CA, 1994.
12. Clough, R.W. and Penzien, J., *Dynamics of Structures,* McGraw-Hill, New York, 1975.
13. Cornell, C.A., Engineering seismic risk analysis, *Bull. Seismol Soc. Am.*, 58(5), 1583–1606, 1968.
14. Cornell, C.A. and Merz, H.A., Seismic risk analysis based on a quadratic magnitude frequency law, *Bull. Seis. Soc. Am.*, 63(6), 1992–2006, 1973.
15. Crouse, C. B., Ground-motion attenuation equations for earthquakes on the cascadia subduction zone, *Earthquake Spectra,* 7(2), 201–236, 1991.
16. Darragh, R.B., Huang, M.J., and Shakal, A.F., Earthquake engineering aspects of strong motion data from recent California earthquakes, *Proc. Fifth U.S. National Conf. Earthquake Engineering,* V. III, Earthquake Engineering Research Institute, Oakland, CA, 99–108, 1994.
17. Donovan, N.C. and Bornstein, A.E., Uncertainties in seismic risk procedures, *J. Geotech. Div.*, 104(GT7), 869–887, 1978.
18. EduPro Civil Systems, Inc., *ProShake User's Manual,* EduPro Civil Systems, Inc., Redmond, WA, 1998, 52 pp.
19. Electric Power Research Institute, Seismic Hazard Methodology for the Central and Eastern United States, EPRI NP-4726, Menlo Park, 1986.
20. Esteva, L., Seismicity, in *Seismic Risk and Engineering Decisions,* Lomnitz, C. and Rosenblueth, E., Eds., Elsevier, New York, 1976.
21. Finn, W.D.L., Yogendrakumar, M., Yoshida, M., and Yoshida, N., TARA-3: A program to compute the response of 2-D embankments and soil-structure interaction systems to seismic loadings, Department of Civil Engineering, University of Brirish Columbia, Vancouver, Canada, 1986.
22. Greenwood, R.B., Characterizing blind thrust fault sources — an overview, in Woods, M.C. and Seiple. W.R., Eds., The Northridge California Earthquake of 17 January 1994, California Department Conservation, Division of Mines and Geology, Special Publ. 116, 1995, 279–287.
23. Gumbel, E.J., *Statistics of Extremes,* Columbia University Press, New York, 1958.
24. Gutenberg, B. and Richter, C.F., *Seismicity of the Earth and Associated Phenomena,* Princeton University Press, Princeton, NJ, 1954.
25. Gutenberg, B. and Richter, C. F., Magnitude and energy of earthquakes, *Ann. Geof.*, 9(1), p 1–15, 1956.
26. Hanks, T.C. and Kanamori, H., A moment magnitude scale, *J. Geophys. Res.*, 84, 2348–2350, 1979.
27. Hausmann, M.R., *Engineering Principles of Ground Modification,* McGraw-Hill, New York, 1990, 632 pp.

28. Housner, G., Historical view of earthquake engineering, in *Proc. Post-Conf. Volume, Eighth World Conf. on Earthquake Engineering,* Earthquake Engineering Research Institute, Oakland, CA, 1984, 25–39, as quoted by S. Otani [1995].
29. Kagawa, T., Mejia, L., Seed, H.B., and Lysmer, J., TLUSH — a computer program for three-dimensional dynamic analysis of Earth dams, Report No. UCB/EERC-81/14, University of California, Berkeley, 1981.
30. Lawson, A. C. and Reid, H.F., *The California Earthquake of April 18, 1906. Report of the State Earthquake Investigation Commission,* California. State Earthquake Investigation Commission, Carnegie Institution of Washington, Washington, D.C., 1908–1910.
31. Lee, M.K.W and Finn, W.D.L., DESRA-2, Dynamic effective stress response analysis of soil deposits with energy transmitting boundary including assessment of liquefaction potential, Soil Mechanics Series No. 38, University of British Columbia, Vancouver, 1978.
32. Liao, S.S.C., Veneziano, D., and Whitman, R.V., Regression models for evaluating liquefaction probability, *J. Geotech. Eng. ASCE,* 114(4), 389–411, 1988.
33. Lomnitz, C., *Global Tectonics and Earthquake Risk,* Elsevier, New York, 1974.
34. Lysmer, J., Udaka, T., Tsai, C.F., and Seed, H.B., FLUSH — A computer program for approximate 3-D analysis of soil-structure interaction problems, Report No. EERC 75-30, Earthquake Engineering Research Center, University of California, Berkeley, 1975, 83 pp.
35. MacMurdo, J., Papers relating to the earthquake which occurred in India in 1819, *Philos. Mag.,* 63, 105–177, 1824.
36. Makdisi, F.I. and Seed, H.B., Simplified procedure for estimating dam and embankment earthquake-induced deformations, *J. Geotech. Eng. Div. ASCE,* 104(GT7), 849–867, 1978.
37. McGuire, R. K., FRISK: Computer Program for Seismic Risk Analysis Using Faults as Earthquake Sources. U.S. Geological Survey, Reports, U.S. Geological Survey Open file 78-1007, 1978, 71 pp.
38. Meeting on Updating of MSK-64, Report on the ad hoc Panel Meeting of Experts on Updating of the MSK-64 Seismic Intensity Scale, Jene, 10–14 March 1980, *Gerlands Beitr. Geophys.,* Leipzeig 90(3), 261–268, 1981.
39. Mitchell, J.K. and Tseng, D.-J., Assessment of liquefaction potential by cone penetration resistance, *Proceedings, H. Bolton Seed Memorial Symposium,* Berkeley, CA, Vol. 2, J.M. Duncan, Ed., 1990, 335–350.
40. Mononobe, N. and Matsuo, H., On the determination of Earth pressures during earthquakes, *Proceedings, World Engineering Congress,* 9, 1929.
41. Mosely, M.P., ed., *Ground Improvement,* Blackie Academic & Professional, London, 1993, 218 pp.
42. National Academy Press, *Probabilistic Seismic Hazard Analysis,* National Academy of Sciences, Washington, D.C., 1988.
43. Newmark, N., Effects of earthquakes on dams and embankments, *Geotechnique,* 15(2), 139–160, 1965.
44. Ohsaki, Y., Effects of sand compaction on liquefaction during Tokachioki earthquake, *Soils Found.,* 10(2), 112–128, 1970.
45. Okabe, S., General theory of Earth pressures, *J. Jpn. Soc. Civil Eng.,* 12(1), 1926.
46. Olson, R.S., Cyclic liquefaction based on the cone penetrometer test, Proceedings of the NCEER Workshop on Evaluation of Liquefaction Resistance of Soils, Technical Report NCEER-97-0022, T.L. Youd and I.M. Idriss, Eds., National Center for Earthquake Engineering Research, Buffalo, NY, 1997, 280 pp.
47. Pyke, R., Seed, H.B., and Chan, C.K., Settlement of sands under multi-directional loading, *J. Geotech. Eng. Div. ASCE,* 101(GT4), 379–398, 1975.
48. Pyke, R.L., TESS1 User's Guide, TAGA Engineering Software Services, Berkeley, CA, 1985.
49. Reid, H.F., The Mechanics of the Earthquake, The California Earthquake of April 18, 1906, Report of the State Investigation Committee, vol. 2, Carnegie Institution of Washington, Washington, D.C., 1910.

50. Reiter, L., *Earthquake Hazard Analysis, Issues and Insights*, Columbia University Press, New York, 1990.
51. Richter, C. F., An instrumental earthquake scale, *Bull. Seismol. Soc. Am.*, 25 1–32, 1935.
52. Richter, C.F., *Elementary Seismology*, W.H. Freeman, San Francisco, 1958.
53. Schnabel, P.B., Lysmer, J., and Seed, H.B., SHAKE: A computer program for earthquake response analysis of horizontally layered sites, Report No. EERC 72-12, Earthquake Engineering Research Center, University of California, Berkeley, 1972.
54. Scholz, C.H., *The Mechanics of Earthquakes and Faulting*, Cambridge University Press, New York, 1990.
55. Schwartz, D.P., Geologic characterization of seismic sources: moving into the 1990s, in *Earthquake Engineering and Soil Dynamics II — Recent Advances in Ground-Motion Evaluation*, J.L. v. Thun, Ed., Geotechnical Spec. Publ. No. 20., American Society of Civil Engineers, New York, 1988.
56. Schwartz, D.P. and Coppersmith, K.J., Fault behavior and characteristic earthquakes: examples from the Wasatch and San Andreas faults, *J. Geophys. Res.*, 89, 5681–5698, 1984.
57. Seed, H.B. and De Alba, P., Use of SPT and CPT tests for evaluating the liquefaction resistance of soils, *Proceedings, Insitu '86*, ASCE, 1986.
58. Seed, H.B. and Silver, M.L., Settlement of dry sands during earthquakes, *J. Soil Mech. Found. Div.*, 98(SM4), 381–397, 1972.
59. Seed, H.B., Ugas, C., and Lysmer, J., Site-dependent spectra for earthquake-resistant design, *Bull. Seismol. Soc. Am.*, 66, 221–243, 1976.
60. Slemmons, D.B., State-of-the-art for assessing earthquake hazards in the United States, Report 6: Faults and earthquake magnitude, U.S. Army Corps of Engineers, Waterways Experiment Station, Misc. Paper s-73-1, 1977, 129 pp.
61. Stein, R.S. and Yeats, R.S., Hidden earthquakes, *Sci. Am.*, June, 1989.
62. Stokoe, K.H. II, Roesset, J.M., Bierschwale, J.G., and Aouad, M., Liquefaction potential of sands from shear wave velocity, *Proceedings, 9th World Conference on Earthquake Engineering*, Tokyo, Japan, Vol. 3, 1988, 213–218.
63. Tokimatsu, K. and Seed, H.B., Evaluation of settlements in sand due to earthquake shaking, *J. Geotechn. Eng. ASCE*, 113(8), 861–878, 1987.
64. Tokimatsu, K., Kuwayama, S., and Tamura, S., Liquefaction potential evaluation based on Rayleigh wave investigation and its comparison with field behavior, *Proceedings, 2nd International Conference on Recent Advances in Geotechnical Earthquake Engineering and Soil Dynamics*, St. Louis, MO, Vol. 1, 1991, 357–364.
65. Van Impe, W.F., *Soil Improvement Techniques and Their Evolution*, A.A. Balkema, Rotterdam, 1989, 125 pp.
66. Wells, D.L. and Coppersmith, K.J., Empirical relationships among magnitude, rupture length, rupture width, rupture area and surface displacement, *Bull. Seisinol. Soc. Am.*, 84(4), 974–1002, 1994.
67. Welsh, J.P., Soil Improvement: A Ten Year Update, Geotechnical Special Publication No. 12, ASCE, New York, 1987, 331 pp.
68. Wood, H.O. and Neumann, Fr., Modified Mercalli intensity scale of 1931, *Bull. Seismol. Soc. Am.*, 21, 277–283, 1931.
69. Wood, J., Earthquake-Induced Soil Pressures on Structures, Report No. EERL 73-05, California Institute of Technology, Pasadena, 1973, 311 pp.
70. Yegulalp, T.M. and Kuo, J.T., Statistical prediction of the occurrence of maximum magnitude earthquakes, *Bull. Seis. Soc. Am.*, 64(2), 393–414, 1974.
71. Youngs, R R. and Coppersmith, K J., Attenuation relationships for evaluation of seismic hazards from large subduction zone earthquakes, *Proceedings of Conference XLVIII: 3rd Annual Workshop on Earthquake Hazards in the Puget Sound, Portland Area*, March 28–30, 1989, Portland, OR; Hays-Walter-W., Ed., U.S. Geological Survey, Reston, VA, 1989, 42–49.

72. Youngs, R.R. and Coppersmith, K.J., Implication of fault slip rates and earthquake recurrence models to probabilistic seismic hazard estimates, *Bull. Seismol. Soc. Am.*, 75, 939–964, 1987.

34

Earthquake Damage to Bridges

Jack P. Moehle
University of California, Berkeley

Marc O. Eberhard
University of Washington

34.1 Introduction

Earthquake damage to a bridge can have severe consequences. Clearly, the collapse of a bridge places people on or below the bridge at risk, and it must be replaced after the earthquake unless alternative transportation paths are identified. The consequences of less severe damage are less obvious and dramatic, but they are nonetheless important. A bridge closure, even if it is temporary, can have tremendous consequences, because bridges often provide vital links in a transportation system. In the immediate aftermath of an earthquake, closure of a bridge can impair emergency response operations. Later, the economic impact of a bridge closure increases with the length of time the bridge is closed, the economic importance of the traffic using the route, the traffic delay caused by following alternate routes, and the replacement cost for the bridge.

The purpose of this chapter is to identify and classify types of damage to bridges that earthquakes commonly induce and, where possible, to identify the causes of the damage. This task is not straightforward. Damage usually results from a complex and interacting set of contributing variables. The details of damage often are obscured by the damage itself, so that some speculation is required in reconstructing the event. In many cases, the cause of damage can be understood only after detailed analysis, and, even then, the actual causes and effects may be elusive.

Even when the cause of a particular collapse is well understood, it is difficult to generalize about the causes of bridge damage. In past earthquakes, the nature and extent of damage that each bridge

0-8493-7434-0/00/$0.00+$.50

suffered have varied with the characteristics of the ground motion at the particular site and the construction details of the particular bridge. No two earthquakes or bridge sites are identical. Design and construction practices vary extensively throughout the world and even within the United States. These practices have evolved with time, and, in particular, seismic design practice improved significantly in the western United States during the 1970s as a result of experience gained from the 1971 San Fernando earthquake.

Despite these uncertainties and variations, one can learn from past earthquake damage, because many types of damage occur repeatedly. By being aware of typical vulnerabilities that bridges have experienced, it is possible to gain insight into structural behavior and to identify potential weaknesses in existing and new bridges. Historically, observed damage has provided the impetus for many improvements in earthquake engineering codes and practice.

An effort is made to distinguish damage according to two classes, as follows:

Primary damage— Damage caused by earthquake ground shaking or deformation that was the primary cause of damage to the bridge, and that may have triggered other damage or collapse.

Secondary damage— Damage caused by earthquake ground shaking or deformation that was the result of structural failures elsewhere in the bridge, and was caused by redistribution of internal actions for which the structure was not designed.

The emphasis in this chapter is on primary damage. It must be accepted, however, that in many cases the distinction between primary and secondary damage is obscure because the bridge geometry is complex or, in the case of collapse, because it is difficult to reconstruct the failure sequence.

The following sections are organized according to which element in the overall set of contributing factors appears to be the primary cause of the bridge damage. The first three sections address general issues related to the site conditions, construction era, and current condition of the bridge. The next section focuses on the effects of structural configuration, including curved layout, skew, and redundancy. Unseating of superstructures at expansion joints is discussed in the subsequent section. Then, the chapter describes typical types of damage to the superstructure, followed by discussion of damage related to bearings and restrainers supporting or interconnecting segments of the superstructure. The final section describes damage associated with the substructure, including the foundation.

34.2 Effects of Site Conditions

Performance of a bridge structure during an earthquake is likely to be influenced by proximity of the bridge to the fault and site conditions. Both of these factors affect the intensity of ground shaking and ground deformations, as well as the variability of those effects along the length of the bridge.

The influence of site conditions on bridge response became widely recognized following the 1989 Loma Prieta earthquake. Figure 34.1 plots the locations of minor and major bridge damage from the Loma Prieta earthquake [16]. With some exceptions, the most significant damage occurred around the perimeter or within San Francisco Bay where relatively deep and soft soil deposits amplified the bedrock ground motion. In the same earthquake, the locations of collapse of the Cypress Street Viaduct nearly coincided with zones of natural and artificial fill where ground shaking was likely to have been the strongest (Figure 34.2) [10]. A major conclusion to be drawn from this and other earthquakes is the significant impact that local site conditions have on amplifying strong ground motion, and the subsequent increased vulnerability of bridges on soft soil sites. This observation is important because many bridges and elevated roadways traverse bodies of water where soft soil deposits are common.

During the 1995 Hyogo-Ken Nanbu (Kobe) earthquake, significant damage and collapse likewise occurred in elevated roadways and bridges founded adjacent to or within Osaka Bay [2]. Several types of site conditions contributed to the failures. First, many of the bridges were founded on sand–gravel terraces (alluvial deposits) overlying gravel–sand–mud deposits at depths of less than 33 ft (10 m), a condition which is believed to have led to site amplification of the bedrock motions.

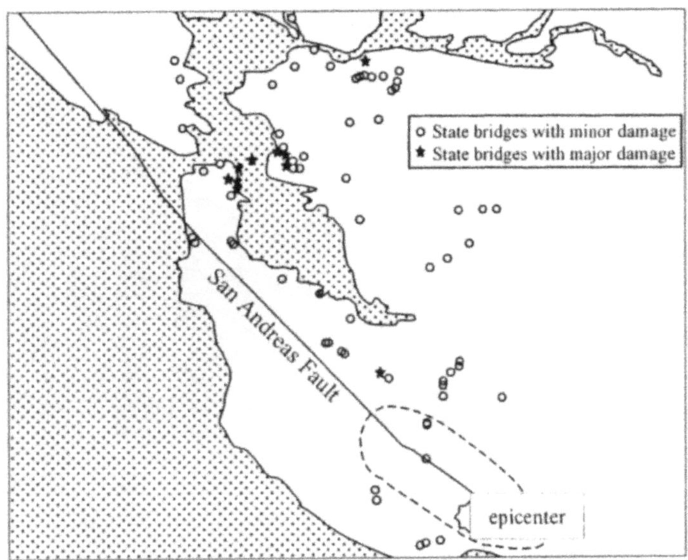

FIGURE 34.1 Incidence of minor and major damage in the 1989 Loma Prieta earthquake [modified from Zelinski, 16].

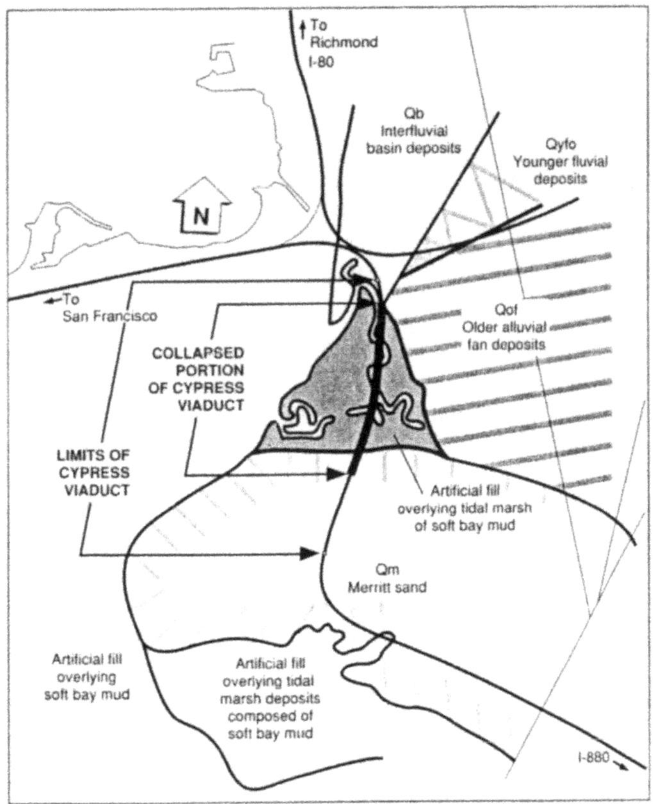

FIGURE 34.2 Geologic map of Cypress Street Viaduct site. (*Source:* Housner, G., Report to the Governor, Office of Planning and Research, State of California, 1990.)

FIGURE 34.3 Nishinomiya-ko Bridge approach span collapse in the 1995 Hyogo-Ken Nanbu earthquake [Kobe Collection, EERC Library, University of California, Berkeley].

Furthermore, many of the sites were subject to liquefaction and lateral spreading, resulting in permanent substructure deformations and loss of superstructure support (Figure 34.3). Finally, the site was directly above the fault rupture, resulting in ground motions having high horizontal and vertical ground accelerations as well as large velocity pulses. Near-fault ground motions can impose large deformation demands on yielding structures, as was evident in the overturning collapse of all 17 bents of the Higashi-Nada Viaduct of the Hanshin Expressway, Route 3, in Kobe (Figure 34.4). Other factors contributed to the behavior of structures in Kobe; several of these will be discussed in subsequent portions of this chapter.

34.3 Correlation of Damage with Construction Era

Bridge seismic design practices have changed over the years, largely reflecting lessons learned from performance in past earthquakes. Several examples in the literature demonstrate that the construction

FIGURE 34.4 Higashi-Nada Viaduct collapse in the 1995 Hyogo-Ken Nanbu earthquake. (*Source:* EERI, The Hyogo-Ren Nambu Earthquake, January 17, 1995, Preliminary Reconnaissance Report, Feb. 1995.)

era of a bridge is a good indicator of likely performance, with higher damage levels expected in older construction than in newer construction.

An excellent example of the effect of construction era is provided by observing the relative performances of bridges on Routes 3 and 5 of the Hanshin Expressway in Kobe. Route 3 was constructed from 1965 through 1970, while Route 5 was completed in the early to mid-1990s [2]. The two routes are parallel to one another, with Route 3 being farther inland and Route 5 being built largely on reclaimed land. Despite the potentially worse soil conditions for Route 5, it performed far better than Route 3, losing only a single span owing apparently to permanent ground deformation and span unseating (Figure 34.3). In contrast, Route 3 has been estimated to have sustained moderate-to-large-scale damage in 637 piers, with damage in over 1300 spans, and approximately 50 spans requiring replacement (see, for example, Figure 34.4).

The superior performance of newer construction in the Hyogo-Ken Nanbu earthquake and other earthquakes [2,8,10] has led to the use of benchmark years as a crude but effective method for rapidly assessing the likely performance of bridge construction. This method has been an effective tool for bridge assessment in California. The reason for its success there is the rapid change in bridge construction practice following the 1971 San Fernando earthquake [8]. Before that time, California design and construction practice was based on significantly lower design forces and less stringent detailing requirements compared with current requirements. In the period following that earthquake, the California Department of Transportation (Caltrans) developed new design approaches requiring increased strength and improved detailing for ductile response.

The 1994 Northridge earthquake provides an insightful study on the use of benchmarking. Over 2500 bridges existed in the metropolitan Los Angeles freeway system at that time. Table 34.1 summarizes cases of major damage and collapse [8]. All these cases correspond to bridges designed before or around the time of the major change in the Caltrans specifications. It is interesting to note that some bridges constructed as late as 1976 appear in this table. This reflects the fact that the new design provisions did not take full effect until a few years after the earthquake and that these did not govern construction of some bridges that were at an advanced design stage at that time. Some caution is therefore required in establishing and interpreting the concept of benchmark years.

34.4 Effects of Changes in Condition

Changes in the condition of a bridge can greatly affect its seismic performance. In many regions of North America, extensive deterioration of bridge superstructures, bearings, and substructures has

TABLE 34.1 Summary of Bridges with Major Damage — Northridge Earthquake

Bridge Name	Route	Construction Year	Prominent Damage
		Collapse	
La Cienega-Venice Undercrossing	I-10	1964	Column failures
Gavin Canyon Undercrossing	I-5	1967	Unseating at skewed expansion hinges
Route 14/5 Separation and Overhead	I-5/SR14	1971/1974	Column failure
North Connector	I-5/SR14	1975	Column failure
Mission-Gothic Undercrossing	SR118	1976	Column failures
		Major Damage	
Fairfax-Washington Undercrossing	I-10	1964	Column failures
South Connector Overcrossing	I-5/SR14	1971/1972	Pounding at expansion hinges
Route 14/5 Separation and Overhead	I-5/SR14	1971/1974	Pounding at expansion hinges
Bull Creek Canyon Channel Bridge	SR118	1976	Column failures

accumulated. It is evident that the current conditions will lead to reduced seismic performance in future earthquakes, although hard evidence is lacking because of a paucity of earthquakes in these regions in modern times.

Construction modifications, either during the original construction or during the service life, can also have a major effect on bridge performance. Several graphic examples were provided by the Northridge earthquake [8]. Figure 34.5 shows a bridge column that was unintentionally restrained by a reinforced concrete channel wall. The wall shortened the effective length of the column, increased the column shear force, and shifted nonlinear response from a zone of heavy confinement upward to a zone of light transverse reinforcement, where the ductility capacity was inadequate. Failures of this type illustrate the importance of careful inspection during construction and during the service life of a bridge.

34.5 Effects of Structural Configuration

Ideally, earthquake-resistant construction should be designed to have a regular configuration so that the behavior is simple to conceptualize and analyze, and so that inelastic energy dissipation is promoted in a large number of readily identified yielding components. This ideal often is not achievable in bridge construction because of irregularities imposed by site conditions and traffic flow requirements. In theory, any member or joint can be configured to resist the induced force and deformation demands. However, in practice, bridges with certain configurations are more vulnerable to earthquakes than others.

Experience indicates that a bridge is most likely to be vulnerable if (1) excessive deformation demands occur in a few brittle elements, (2) the structural configuration is complex, or (3) a bridge lacks redundancy. The bridge designer needs to recognize the potential consequences of these irregularities and to design accordingly either to reduce the irregularity or to toughen the structure to compensate for it.

A common form of irregularity arises when a bridge traverses a basin requiring columns of nonuniform length. Although the response of the superstructure may be relatively uniform, the deformation demands on the individual substructure piers are highly irregular; the largest strains are imposed on the shortest columns. In some cases, the deformation demands on the short columns can induce their failure before longer, more flexible adjacent columns can fully participate. The Route 14/5 Separation and Overhead structure provides an example of these phenomena. The structure comprised a box-girder monolithic with single-column bents that varied in height depending on the road and grade elevations (Figure 34.6a). Apparently, the short column at Bent 2 failed in shear because of large deformation demands in that column, resulting in the collapse of the adjacent spans (Figure 34.6b).

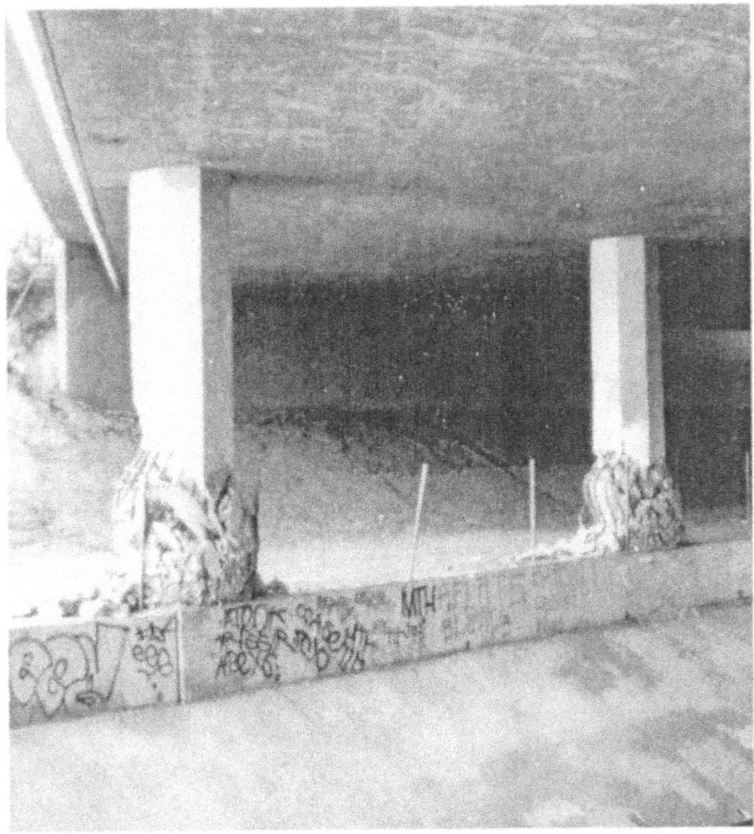

FIGURE 34.5 Bull Creek Canyon Channel Bridge damage in the 1994 Northridge earthquake.

The effects identified above can be exacerbated in long-span bridges. In addition to changes in subgrade and structural irregularities that may be required to resolve complex foundation and transportation requirements, long bridges can be affected by spatial and temporal variations in the ground motions. Expressed in simple terms, different piers are subjected to different ground motions at any one time, because seismic waves take time to travel from one bridge pier to another. This effect can result in one pier being pulled in one direction while the other is being pushed in the opposite direction. This complex behavior is not accounted for directly in conventional bridge design. An example where this behavior may have resulted in increased damage and collapse is the eastern portion of the San Francisco–Oakland Bay Bridge (Figure 34.7a). This bridge includes a variety of different superstructure and substructure configurations, traverses variable subsoils, and is long enough for spatial and temporal variations in ground motions to induce large relative displacements between adjacent bridge segments. The bridge lost two spans, one upper and one lower, at a location where the superstructure was required to accommodate differential movements of adjacent bridge segments (Figure 34.7b).

34.6 Unseating at Expansion Joints

Expansion joints introduce a structural irregularity that can have catastrophic consequences. Such joints are commonly provided in bridges to alleviate stresses associated with volume changes that occur as a bridge ages and as the temperature changes. These joints can occur within a span (in-span hinges), or they can occur at the supports, as is the case for simply supported bridges.

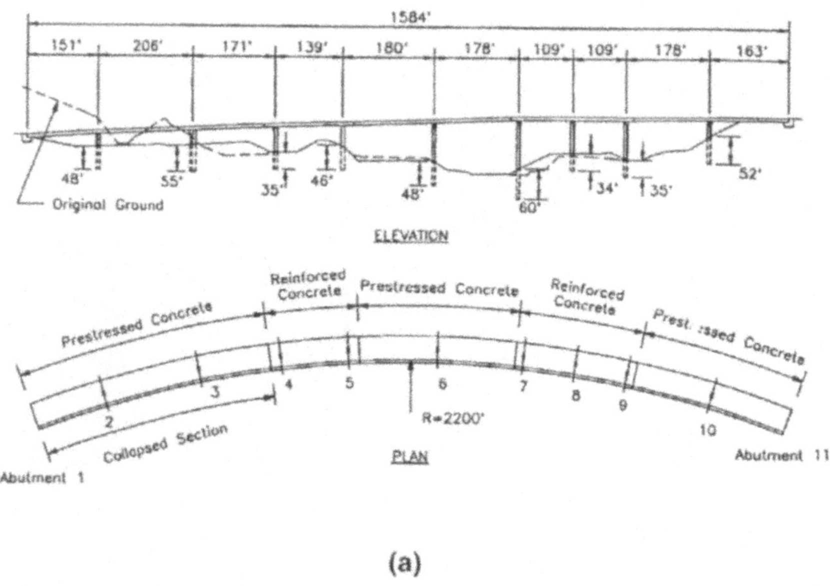

FIGURE 34.6 Geometry and collapse of the Route 14/5 Separation and Overhead in the 1994 Northridge earthquake. (a) Configuration [8]; (b) photograph of collapse.

Earthquake ground shaking, or transient or permanent ground deformations resulting from the earthquake, can induce superstructure movements that cause the supported span to unseat. Unseating is especially a problem with the shorter seats that were common in older construction (e.g., References [2,6–8,12]).

Bridges with Short Seats and Simple Spans

In much of the United States and in many other areas of the world, bridges often comprise a series of simple spans supported on bents. These spans are prone to being toppled from their supporting substructures either due to shaking or differential support movement associated with ground

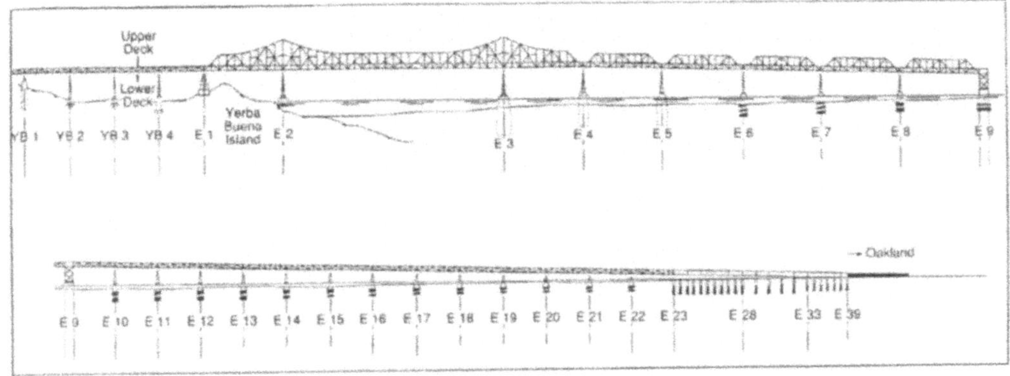

(a)

(b)

FIGURE 34.7 San Francisco–Oakland Bay Bridge, east crossing; geometry and collapse in the 1989 Loma Prieta earthquake. (a) Configuration [10]; (b) photograph of collapse.

deformation. Unseating of simple spans was observed in California in earlier earthquakes, leading in recent decades to development of bridge construction practices based on monolithic box-girder-substructure construction. Problems of unseating still occur with older bridge construction and with new bridges in regions where simple spans are still common. For example, during the 1991 Costa Rica earthquake, widespread liquefaction led to abutment and internal bent rotations, resulting in the collapse of no fewer than four bridges with simple supports [7]. The collapse of the Showa Bridge in the 1964 Niigata earthquake demonstrates one result of the unseating of simple spans (Figure 34.8).

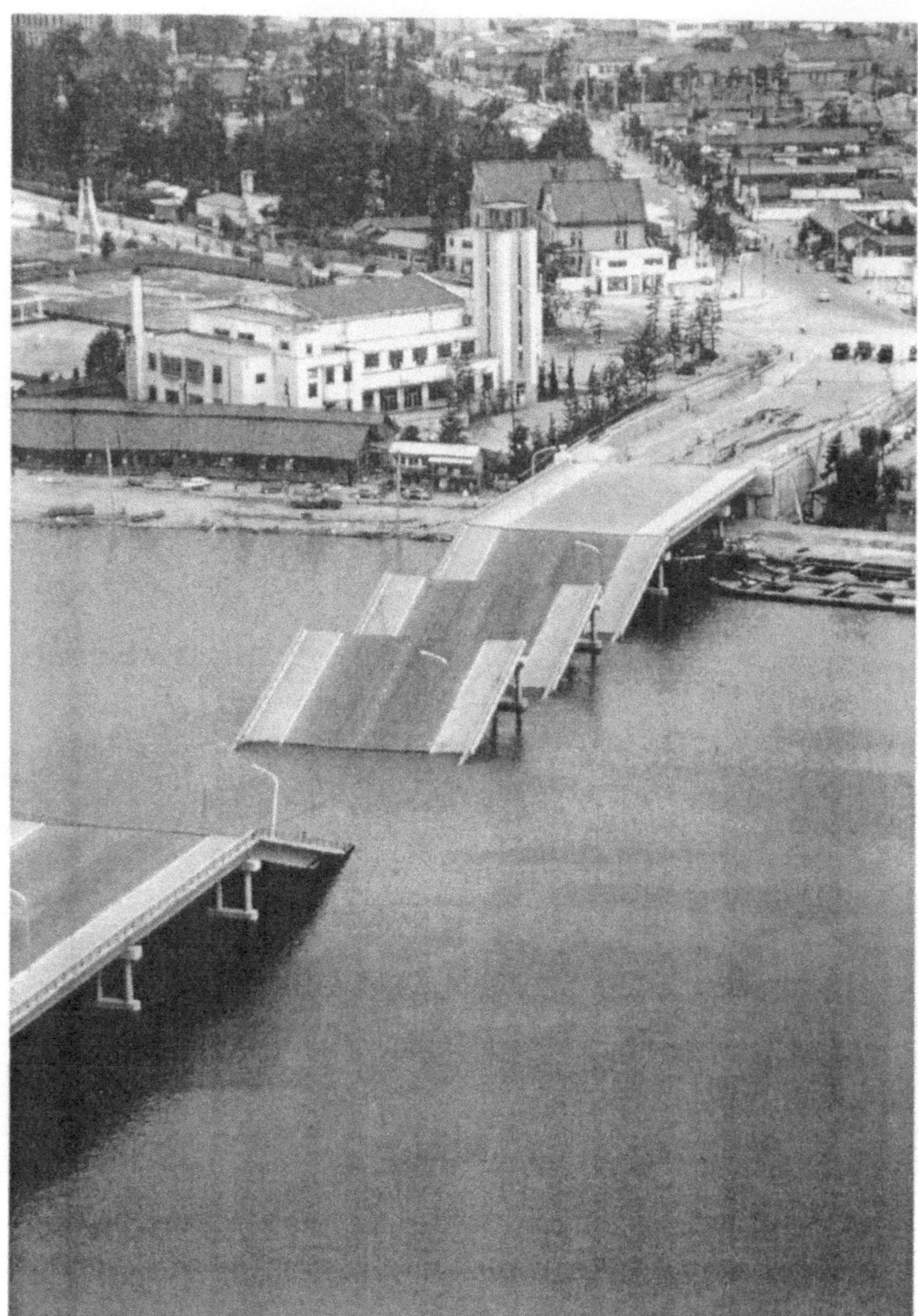

FIGURE 34.8 Showa Bridge collapse in 1964 Niigata earthquake.

Skewed Bridges

Skewed bridges are defined as those having supports that are not perpendicular to the alignment of the bridge. Collisions between a skewed bridge and its abutments (or adjacent frames) can cause a bridge to rotate about a vertical axis. Because the abutments resist compression but not tension, the sense of this rotation is the same (for a given bridge configuration) regardless of whether the

FIGURE 34.9 Rio Bananito Bridge collapse in the 1991 Costa Rica earthquake. (*Source*: EERI, *Earthquake Spectra*, Special Suppl. to Vol. 7, 1991.)

bridge collides with one abutment or the other. If the rotations are large and the seat lengths small, a bridge can come unseated at the acute corners of the decks.

Several examples of skewed bridge damage and collapse can be found in the literature [7,8,12]. A typical example is the Rio Bananito Bridge, in which the bridge and central slab pier were skewed at 30°, which lost both spans off the central pier in the direction of the skew during the 1991 Costa Rica earthquake (Figure 34.9) [7]. Another example of skewed bridge failure is the Gavin Canyon Undercrossing, which failed during the 1994 Northridge earthquake [8]. Both skewed hinges became unseated during the earthquake, resulting in collapse of the unseated spans (Figure 34.10).

Curved Bridges

Curved bridges can have asymmetrical response similar to that of skewed bridges. For loading in one direction, an in-span hinge tends to close, while for loading in the other direction, the hinge opens. An example in which the curved alignment may have contributed to bridge collapse is the curved ramps of the I-5/SR14 interchange, which sustained collapses in both the 1971 San Fernando earthquake [13] and the 1994 Northridge earthquake (see Figure 34.6) [8]. Other factors that may have contributed to the failures include inadequate hinge seats and column deformability.

Hinge Restrainers

Hinge restrainers appear to have been effective in preventing unseating in both the Loma Prieta [10] and Northridge earthquakes [8]. In some other cases, hinge restrainers were not fully effective in preventing unseating. For example, the hinge restrainers in the Gavin Canyon Undercrossing, which were aligned parallel the bridge alignment, did not prevent unseating (see Figure 34.10).

FIGURE 34.10 Gavin Canyon Undercrossing collapse in the 1994 Northridge earthquake.

34.7 Damage to Superstructures

Superstructures are designed to support service gravity loads elastically, and, for seismic applications, they are usually designed to be a strong link in the earthquake-resisting system. As a result, superstructures tend to be sufficiently strong to remain essentially elastic during earthquakes. In general, superstructure damage is unlikely to be the primary cause of collapse of a span.

Instead, damage typically is focused in bearings and substructures. The superstructure may rest on elastomeric pads, pin supports, or rocker bearings, or may be monolithic with the substructure. As bearings and substructures are damaged and in some cases collapse, a wide range of damage and failure of superstructures may result, but these failures are often secondary; that is, they result from failures elsewhere in the bridge. There are, however, some cases of primary superstructure damage as well. Some examples are highlighted below.

With the exception of bridge superstructures that come unseated and collapse, the most common form of damage to superstructures is due to pounding of adjacent segments at the expansion hinges. This type of damage occurs in bridges of all construction materials. Figure 34.11a shows pounding damage at an in-span expansion joint of the Santa Clara River Bridge during the 1994 Northridge earthquake, and Figure 34.11b shows pounding damage at an abutment of the same structure.

Following the 1971 San Fernando earthquake, Caltrans initiated the first phase of its retrofit program, which involved installation of hinge and joint restrainers to prevent deck joints from separating. Both cable restrainers and pipe restrainers (the former intended only to restrain longitudinal movement and the latter intended also to restrain transverse motions) were installed in bridge superstructures. The restrainers extended through end diaphragms that had not been designed originally for the forces associated with restraint. Some punching shear damage to end diaphragms retrofitted with cable restrainers was observed in the I-580/I-980/SR24 connectors following the 1989 Loma Prieta earthquake [15].

FIGURE 34.11 Santa Clara River Bridge pounding damage in 1994 Northridge earthquake. (a) Barrier rail pounding damage; (b) abutment pounding damage.

FIGURE 34.12 Buckling of braces near pier 209 of the Hanshin Expressway in the 1995 Hyogo-Ken Nanbu earthquake.

Steel superstructures commonly comprise lighter framing elements, especially for transverse bracing. These have been found to be susceptible to damage due to transverse loading, especially following failure of bearings [1,8]. Several cases of steel superstructure damage occurred in the Hyogo-Ken Nanbu earthquake. Figure 34.12 shows buckling of cross braces beneath the roadway of a typical steel girder bridge span of the Hanshin Expressway. Figure 34.13 shows girder damage in the same expressway due to excessive lateral movement at the support. Figure 34.14 shows buckled cross-members between the upper chords of the Rokko Island Bridge. That single-span, 710-ft (217-m) tied-arch span bridge slipped from its expansion bearings, allowing the bridge to move laterally about 10 ft (3 m). The movement was sufficient for one end of one arch to drop off the cap beam, twisting the superstructure and apparently resulting in the buckling of the top chord bracing members [2,3].

A spectacular example of steel superstructure failure and collapse is that of the eastern portion of the San Francisco–Oakland Bay Bridge during the 1989 Loma Prieta earthquake (see Figure 34.7) [10]. In this bridge, a 50-ft (15-m) span over tower E9 was a transition point between 506-ft (154-m) truss spans to the west and 290-ft (88-m) truss spans to the east, serving to transmit longitudinal forces among the adjacent spans and the massive steel tower at E9. Failure of a bolted connection between the 290-ft (88-m) span truss and the tower resulted in sliding of the span and unseating of the transition span over tower E9. This collapse resulted in closure for 1 month of this critical link between San Francisco and the East Bay.

34.8 Damage to Bearings

In some regions of the world, the prevalent bridge construction consists of steel superstructures supported on bearings, which, in turn, rest on a substructure. In the United States, this form of

FIGURE 34.13 Girder damage at Bent 351 of the Hanshin Expressway apparently due to transverse movement during the 1995 Hyogo-Ken Nanbu earthquake.

construction is common in new bridges east of the Sierra Nevada Mountains as well as throughout the country in older existing bridges. In such bridges, the bearings commonly consist of steel components designed to provide restraint in one or more directions and, in some cases, to permit movement in one or more directions. Failure of these bearings in an earthquake can cause redistribution of internal forces, which may overload either the superstructure or substructure, or both. Collapse is also possible when bearing support is lost.

The predominant type of bridge construction in Japan involves steel superstructures supported on bearings, which, in turn, are supported on concrete substructures. The Hyogo-Ken Nanbu earthquake provides several examples of bearing failures in these types of bridges [2,3]. One example is provided by the Hamate Bypass, which was a double-deck elevated roadway comprising steel box girders on either fixed or expansion steel bearings. Bearing failure at several locations led to large superstructure rotations that can be seen in Figure 34.15. Another example is provided by the Nishinomiya-ko Bridge, a 830-ft (252-m) span-arch bridge supported on two fixed bearings at one end and two expansion bearings at the other end. The fixed-end bearings, which apparently were designed to have a capacity of approximately 70% of the bridge weight [2], failed, apparently leading to unseating of the adjacent approach span (see Figure 34.3). The failed bearing is shown in Figure 34.16.

FIGURE 34.14 Buckling of cross-members in the upper chord of the Rokko Island Bridge in the 1995 Hyogo-Ken Nanbu earthquake.

FIGURE 34.15 Hamate Bypass superstructure rotations as a result of bearing failures in the 1995 Hyogo-Ken Nanbu earthquake.

FIGURE 34.16 Nishinomiya-ko Bridge bearing failure in the 1995 Hyogo-Ken Nanbu earthquake.

34.9 Damage to Substructures

Columns

Unlike building design, current practice in bridge design is to proportion members of a frame (bent) such that its lateral-load capacity is limited by the flexural strength of its columns. For this strategy to be successful, the connecting elements (e.g., footings, joints, cross-beams) need to be strong enough to force yielding into the columns, and the columns need to be sufficiently ductile (or tough) to sustain the imposed deformations. Even in older bridges, where the "weak column" design approach may not have been adopted explicitly, columns tend to be weaker than the beam–diaphragm–slab assembly to which they connect. Consequently, columns can be subjected to large inelastic demands during strong earthquakes. Failure of a column can result in loss of vertical load-carrying capacity; column failure is often the primary cause of bridge collapse.

Most damage to columns can be attributed to inadequate detailing, which limits the ability of the column to deform inelastically. In concrete columns, the detailing inadequacies can produce flexural, shear, splice, or anchorage failures, or as is often the case, a failure that combines several mechanisms. In steel columns, local buckling has been observed to lead progressively to collapse.

Ideally, a concrete column should be designed such that the lateral load strength is controlled by flexure. However, even if most of the inelastic action is flexural, a column may not be sufficiently tough to sustain the imposed flexural deformations without failure. Such failures are particularly common in older bridges. In the United States, the transverse reinforcement of reinforced concrete columns designed before 1971 commonly consists of #4 hoops (ϕ = 13 mm) or ties at 12-in. (305-mm) spacing. Moreover, the ends of the transverse reinforcement rarely are anchored into the

FIGURE 34.17 San Fernando Road Overhead damage in the 1971 San Fernando earthquake.

core of these columns. This amount and type of reinforcement provides negligible confinement to the concrete, particularly in large columns. Figure 34.17 shows bridge columns that had insufficient flexural ductility to withstand the 1971 San Fernando earthquake. Figure 34.18 shows similar damage in a circular cross section column in the 1995 Hyogo-Ken Nanbu earthquake.

Other detailing practices (in addition to providing little confinement) may lead to flexural failure in reinforced concrete columns. A common practice in Japan has been to terminate some of the longitudinal reinforcement within the column height. The resulting development length of the terminated reinforcement can be inadequate, and may lead to splitting failure along the terminated bars or to flexural and shear distress near the cutoff point. Figure 34.19 illustrates failure of a column with bars terminated near the column midheight. In the case of the Hanshin Expressway Route 3, which collapsed during the 1995 Hyogo-Ken Nanbu earthquake (see Figure 34.4), the curtailment of one third of the main reinforcement was accompanied by the use of gas-pressure butt welding of the continuing longitudinal reinforcement. In tests following the earthquake, approximately half of the undamaged, butt-welded bars failed at the welds [14].

Shear failures of concrete bridge columns have occurred in many earthquakes (e.g., [5,8]). Such failures can occur at relatively low structural displacements, at which point the longitudinal reinforcement may not yet have yielded. Alternatively, because shear strength degrades with inelastic loading cycles, shear failures can occur after flexural yielding. Examples of shear failure can be found in several of the references provided at the end of this chapter. Figure 34.20 illustrates shear failure of a column having relatively light transverse reinforcement typical of bridges constructed in the western United States prior to the mid-1970s. The failure features a steeply inclined diagonal crack and dilation of the core into discrete blocks of concrete. Under the action of several deformation

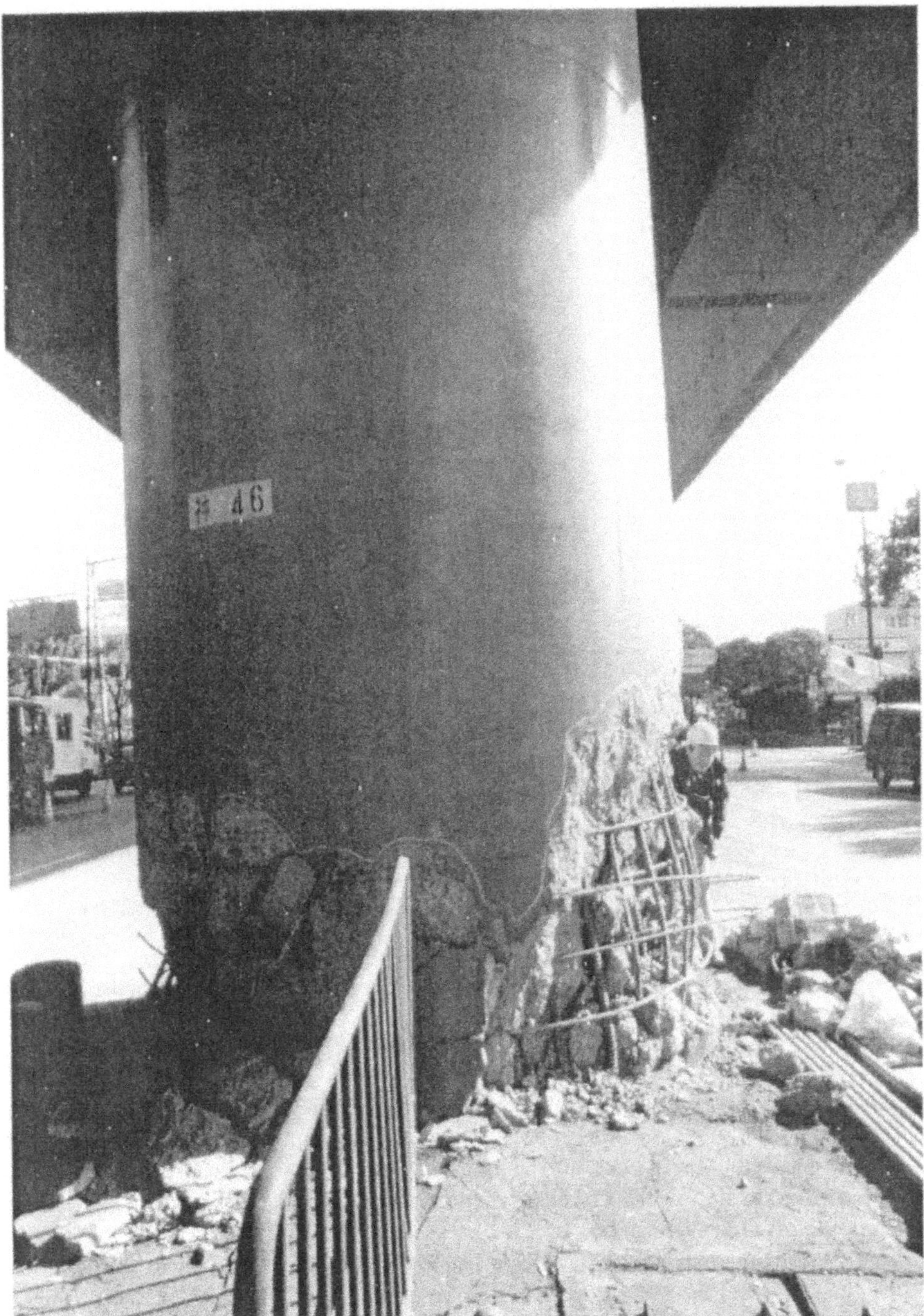

FIGURE 34.18 Hanshin Expressway, Pier 46, damage in the 1995 Hyogo-Ken Nanbu earthquake.

cycles combined with vertical loads, a column can degrade to nearly complete loss of load-carrying capacity, as suggested by the heavily damaged column in Figure 34.21. Provision of closely spaced transverse reinforcement as required in some modern codes is required to prevent this type of failure.

Shear failures in reinforced concrete columns can be induced by interactions with "nonstructural" elements. These elements can decrease the distance between locations of flexural yielding, and therefore increase the shear demand for a column. Figure 34.5, discussed previously, shows a case

FIGURE 34.19 Failure of column with longitudinal reinforcement cutoffs near midheight in the 1995 Hyogo-Ken Nanbu earthquake.

in which a channel wall restrained the column at the base and forced the location of yielding to occur higher in the column than was anticipated in design [8]. Figure 34.22 shows a case in which an architectural flare strengthened the upper portion of the column, forcing yielding to occur lower than was intended [8]. In both cases, an element that was not considered in designing the column forced failure to occur in a lightly confined portion of the column that was incapable of resisting the force and deformation demands.

Figure 34.23 illustrates the failure of a stout, two-column bent on a spur just to the north of the Hanshin Expressway in the 1995 Hyogo-Ken Nanbu earthquake. The failure involves shattering of the columns, bent cap, and joints, and shatters the notion that strength alone is an adequate provision for bridge seismic design.

Lap splices of longitudinal reinforcement in older reinforced-concrete bridges may be vulnerable because, typically, the splices are short (on the order of 20 to 30 bar diameters), poorly confined, and are located in regions of high flexural demand. In particular, for construction convenience, splices are often located directly above a footing. With these details, the splices may be unable to develop the flexural capacity of the column, and they may be more vulnerable to shear failure. Despite these vulnerabilities, there is little field evidence of lap splice failures at the bases of bridge columns. However, failures associated with welded splices and terminated longitudinal reinforcement were identified in the 1995 Hyogo-Ken Nanbu earthquake (Figures 34.4 and 34.19), as discussed previously.

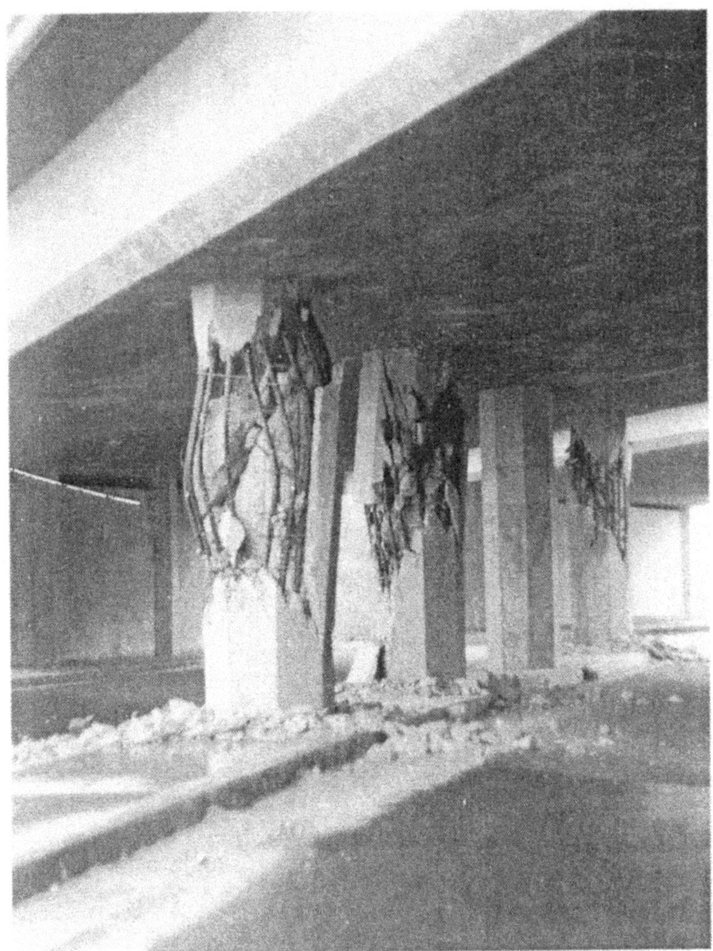

FIGURE 34.20 Failure of columns of the Route 5/210 interchange during the 1971 San Fernando earthquake.

Concrete columns also can fail if the anchorage of the longitudinal reinforcement is inadequate. Such failures can occur both at the top of a column at the connection with the bent cap and at the bottom of a column at the connection with the foundation. Figure 34.24 shows a column that failed at its base during the 1971 San Fernando earthquake (12). The column had been supported on a single 6-ft (1.8-m)-diameter, cast-in-drilled-hole pile. Other columns having hooked longitudinal reinforcement anchored in footings also failed with similar results in that earthquake. The consequences of foundation anchorage failures perhaps are larger in single-column bents than multicolumn bents, because the lateral-force resistance of single-column bents depends on the column developing its flexural strength at the base.

The record of steel column failures is sparse, because few bridges with steel columns have been subjected to strong earthquakes elsewhere than Japan. In the 1995 Hyogo-Ken Nanbu earthquake, failures apparently were associated with local buckling and subsequent splitting at welds or tearing of steel near the buckle. In columns with circular cross sections, the local buckling sometimes occurred at locations where section thicknesses changed. Figure 34.25 illustrates the formation of a local buckle in a circular cross section column accompanied by visible plastic deformation. In rectangular columns, local buckling of web and flange plates was insufficiently restrained by small web stiffeners [14]. Figure 34.26 illustrates the collapse of a rectangular column. A nearby column sustained local buckling at the base and tearing of the vertical welded seam between the two steel plates forming a corner of the column, suggesting the nature of failure that resulted in the collapse shown in Figure 34.26.

FIGURE 34.21 Failure of columns of Interstate 10, La Cienega-Venice Undercrossing in the 1994 Northridge earthquake. (Masonry walls of storage units are supporting the collapsed frame.)

Beams

Beams traditionally have received much less attention than columns in seismic design and evaluation. In many bridges, the transverse beams are stronger than the columns because of gravity load requirements and composite action with the superstructure. Also, in many bridges, the consequences of beam failures are less severe than the consequences of column failures. In bridges with outriggers, however, the beams can be critical components of the bent and can be subjected to loadings that may result in failure. An example illustrating possible damage to an outrigger beam is in Figure 34.27. This outrigger beam was monolithically framed with the superstructure and supporting column such that, under longitudinal load, significant torsion was required to be resisted by the outrigger portion of the beam. In some modern designs, torsion is reduced by providing nominal "pinned" connections between the beams and columns.

Joints

As with beams, joints traditionally have received little attention in seismic design, and they similarly may be exposed to critically damaging actions when the joints lie outside of the superstructure. Although joint failures occurred in previous earthquakes (e.g., Jennings [13]), significant attention was not paid to joints until several spectacular failures were observed following the 1989 Loma Prieta earthquake [6,10]. Figure 34.28 shows joint damage to the Embarcadero Viaduct in San Francisco during the 1989 Loma Prieta earthquake. The occurrence of damage at the relatively large epicentral distance of approximately 60 miles (100 km) is attributed in part to site amplification and focusing of seismic waves as well as the vulnerability of the framing.

The collapse of the Cypress Street Viaduct during the Loma Prieta earthquake had more severe consequences (Figure 34.29). Failure of a concrete pedestal located just above the first-level joint

FIGURE 34.22 Failure of flared column in the Route 118, Mission-Gothic Undercrossing, in the 1994 Northridge earthquake.

led to the collapse of the upper deck on the lower deck, at a cost of 42 lives. Such pedestals are not common, but this collapse demonstrates that each earthquake has the potential to reveal a mode of failure that has not yet been considered routinely.

The Loma Prieta earthquake also identified an apparent weakness of a modern design. For example, damage occurred to the outrigger knee joints of the Route 980/880 connector, which had been constructed just a few years before the earthquake. This damage identified the need for special details in bridge construction, which has been the subject of important studies identified elsewhere in this book.

Abutments

The types of failures that can occur at abutments vary from one bridge to the next. The foundation type varies greatly (e.g., spread footing, pile-supported footing, drilled shafts), and the properties of the soil can be important, particularly if the soil liquefies during an earthquake. The situation is further complicated by the interaction of the backwalls, wingwalls, footings and piles with the surrounding soil. A common practice has been to treat abutments or abutment components as sacrificial elements, acting as fuses to relieve large seismic forces arriving at the stiff abutment. The occurrence of widespread and extensive damage in the 1994 Northridge earthquake [8] suggests that an alternative approach might be economical.

In most seat-type abutments, longitudinal motion is unrestrained, because there is a joint at the interface of the superstructure and abutment backwall. This configuration is attractive, because it

FIGURE 34.23 Failure of a two-column bent in the 1995 Hyogo-Ken Nanbu earthquake.

reduces the superstructure forces induced by temperature and shrinkage-induced displacements. The most important vulnerability of such abutments is span unseating, which can occur when there are large relative displacements between the superstructure and abutment seat. Abutment unseating failures are often attributable to displacement or rotation of the abutment, usually the result of liquefaction or lateral spreading [7].

Shear keys can be damaged also. Shear keys are components that restrain relative displacements (usually in the transverse direction) between the superstructure and the abutments. External shear keys are located outside of the superstructure cross section, while internal shear keys are located within the superstructure cross section. Since these elements are stocky, it is nearly impossible to make them ductile, and they will fail if their strength is exceeded.

Shear key failures were widespread during the 1994 Northridge earthquake [8]. Figure 34.30 shows a typical failure in which the external shear keys failed. Figure 34.31 shows a failed internal shear key. It appears that these failures can occur with small transverse displacement and little energy dissipation. Damage to internal shear keys usually is accompanied by damage to the interlocking backwall. In seat-type abutments, damage has also occurred in seat abutments due to pounding of backwalls by the superstructure. This type of damage is similar to that shown in Figure 34.11.

In monolithic abutments, the superstructure is cast monolithically with the abutments. This configuration is attractive, because it reduces the likelihood of span unseating. However, the abutment can be damaged as the superstructure displaces in the longitudinal direction away from the abutment. Also, depending on the geometry and details of the abutment, the wingwall may serve as an external shear key. In such cases, the wingwall can fail in the same manner as an external shear key.

FIGURE 34.24 Failure at the base of a column supported on a single cast-in-place pile in the 1971 San Fernando earthquake [Steinbrugge Collection, EERC Library, University of California, Berkeley].

Foundations

Reports of foundation failures during earthquakes are relatively rare, with the notable exception of situations in which liquefaction occurred. It is not clear whether failures are indeed that rare or whether many foundation failures are undetected because they remain underground. There are many reasons why older foundations might be vulnerable. Piles might have little confinement reinforcement, yet be subjected to large deformation demands. Older spread and pile-supported footings rarely have top flexural reinforcement or any shear reinforcement.

The 1995 Hyogo-Ken Nanbu earthquake resulted in extensive damage to superstructures and substructures above the ground, as reported elsewhere in this chapter. The occurrence of that damage provided impetus to conduct extensive investigations of the conditions of foundation components [11]. Along the older inland Route 3, an investigation of 109 foundations identified only cases of "small" flexural cracks in piles. Along the newer coastal Route 5, more extensive liquefaction occurred, resulting in lateral spreading in several cases. An investigation of 153 foundations for this route found cases of flexural cracks in piles where large residual displacements occurred, but the investigators found no spalling or reinforcement buckling. The absence of extensive damage was attributed to the spread of deformations along a significant length of the piles.

Foundation damage associated with liquefaction-induced lateral spreading has probably been the single greatest cause of extreme distress and collapse of bridges [12]. The problem is especially critical for bridges with simple spans (see Figure 34.8). The 1991 Costa Rica earthquake provides many examples of foundation damage [7]. For example, Figure 34.32 shows an abutment that

FIGURE 34.25 Local buckling of a circular cross-section column of the Hanshin Expressway in the 1995 Hyogo-Ken Nanbu earthquake.

rotated due to liquefaction and lateral spreading. Figure 34.33 shows a situation in which soil movements have led to extensive damage to the batter piles. Use of batter piles should be considered carefully in design in light of the extensive damage observed in these piles in this and other earthquakes [6].

Approaches

Even if the bridge structure remains intact, a bridge may be placed out of service if the roadway leading to it settles significantly. For example, during the 1971 San Fernando earthquake [12] and the 1985 Chile earthquake [4], settlement of the backfill abutments led to abrupt differential settlements in many locations. Such settlements can be large enough to pose a hazard to the traveling public. Approach or settlement slabs can be effective means of spanning across backfills, as shown in Figure 34.34.

34.10 Summary

This chapter has reviewed various types of damage that can occur in bridges during earthquakes. Damage to a bridge can have severe consequences for a local economy, because bridges provide vital links in the transportation system of a region. In general, the likelihood of damage increases

(a)

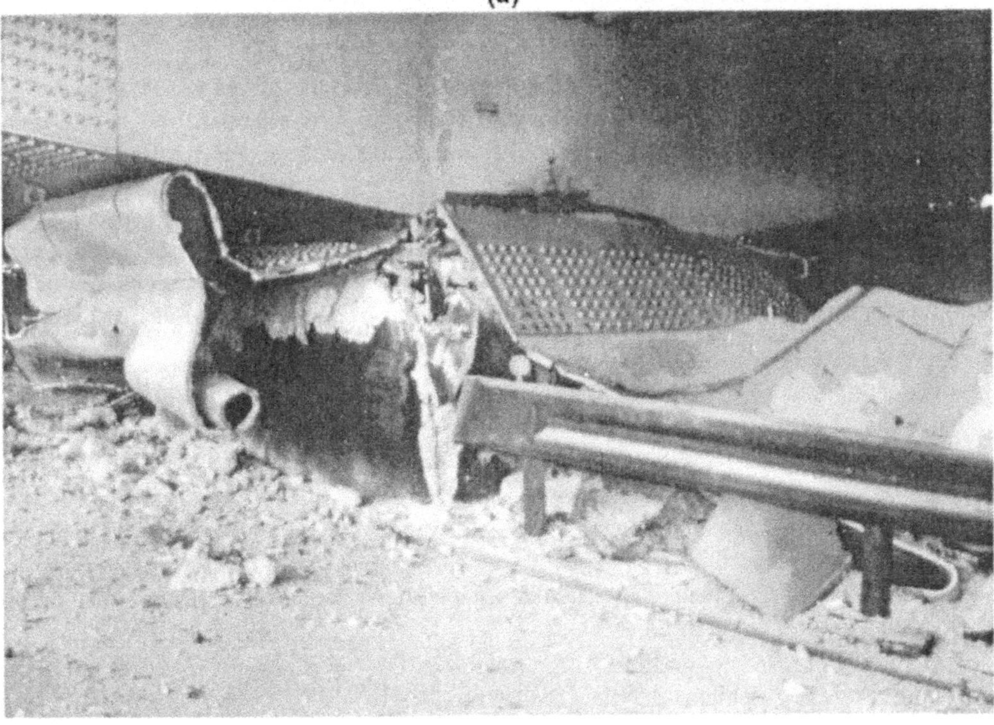

(b)

FIGURE 34.26 Collapse of a rectangular cross section steel column in the 1995 Hyogo-Ken Nanbu earthquake. (a) Collapsed bent and superstructure; (b) close-up of collapsed column.

FIGURE 34.27 Outrigger damage in the 1989 Loma Prieta earthquake.

if the ground motion is particularly intense, the soils are soft, the bridge was constructed before modern codes were implemented, or the bridge configuration is irregular. Even a well-designed bridge can suffer damage if nonstructural modifications and structural deterioration have increased the vulnerability of the bridge.

Depending on the ground motion, site conditions, overall configuration, and specific details of the bridge, the damage induced in a particular bridge can take many forms. Despite these complexities, the record is clear. Damage within the superstructure is rarely the primary cause of collapse. Though exceptions abound, most of the severe damage to bridges has taken one of the following forms:

- Unseating of superstructure at in-span hinges or simple supports attributable to inadequate seat lengths or restraint. The presence of a skewed or curved configuration further exacerbates the vulnerability. For simply supported bridges, these failures are most likely when ground failure induces relative motion between the spans and their supports.
- Column failure attributable to inadequate ductility (toughness). In reinforced-concrete columns, the inadequate ductility usually stems from inadequate confinement reinforcement. In steel columns, the inadequate ductility usually stems from local buckling, which progresses to collapse.
- Damage to shear keys at abutments. Because of their geometry, it is nearly impossible to make these stiff elements ductile.
- Unique failures in complex structures. In the Cypress Street Viaduct, the unique vulnerability was the inadequately reinforced pedestal above the first level. In outrigger column bents, the vulnerability may be in the cross-beam or the beam–column joint.

FIGURE 34.28 Embarcadero Viaduct damage during the 1989 Loma Prieta earthquake.

Acknowledgments

This work was made possible through the reconnaissance work of many individuals identified in the cited references, as well as many not directly cited. The writers acknowledge the significant effort and risk made by those experts, as well as the funding agencies, in particular the U.S. National Science Foundation. Many of the photographs were provided generously by the Earthquake Engineering Research Institute, as cited in the figure captions. Where no citation is given in the caption, the photograph was from the extensive collection of the Earthquake Engineering Research Center Library, Pacific Earthquake Engineering Research Center, University of California, Berkeley. That collection is made possible through funding from the U.S. National Science Foundation and the Federal Emergency Management Agency, and from generous donations from earthquake experts. Photographs in this chapter were made possible by donations from K. Steinbrugge, W. Godden, M. Nakashima, and M. Yashinski.

FIGURE 34.29 Cypress Street Viaduct collapse in the 1989 Loma Prieta earthquake.

FIGURE 34.30 Damage to external shear key in an abutment in the 1994 Northridge earthquake. (*Source*: EERI, *Earthquake Spectra*, Special Suppl. to Vol. II, 1995.)

FIGURE 34.31 Damage to internal shear key in an abutment in the 1994 Northridge earthquake. (*Source:* EERI, *Earthquake Spectra,* Special Suppl. to Vol. II, 1995.)

FIGURE 34.32 Rotation of abutment due to liquefaction and lateral spreading during the 1991 Costa Rica earthquake. (*Source:* EERI, *Earthquake Spectra,* Special Suppl. to Vol. 7, 1991.)

FIGURE 34.33 Abutment piles damaged during the 1991 Costa Rica earthquake. (*Source:* EERI, *Earthquake Spectra*, Special Suppl. to Vol. 7, 1991.)

FIGURE 34.34 Settlement slab spanning across slumped abutment fill material at the Rio Quebrada Calderon bridge in the 1991 Costa Rica earthquake. (*Source:* EERI, *Earthquake Spectra*, Special Suppl. to Vol. 7, 1991.)

References

1. Astaneh-Asl, A. et al., Seismic Performance of Steel Bridges during the 1994 Northridge Earthquake: A Preliminary Report, Report No. UCB/CE-STEEL-94/01, Department of Civil Engineering, University of California, Berkeley, April 1994, 296 pp.
2. Chung, R. et al., The January 17, 1995 Hyogoken-Nanbu (Kobe) Earthquake, *NIST Special Publication 901*, National Institute of Standards and Technology, July 1996, 544 pp.
3. EERC, Earthquake Engineering Research Center, Seismological and Engineering Aspects of the 1995 Hyogoken-Nanbu Earthquake, Report No. UCB/EERC-95/10, Nov. 1995, 250 pp.
4. EERI, Earthquake Engineering Research Institute, The Chile earthquake of March 3, 1985, *Earthquake Spectra*, Special Supplement to 2(2), Feb. 1986, 513 pp.
5. EERI, Earthquake Engineering Research Institute, The Whittier Narrows earthquake of October 1, 1987, *Earthquake Spectra*, 4(2), May 1988, 409 pp.
6. EERI, Earthquake Engineering Research Institute, Loma Prieta earthquake reconnaissance report, *Earthquake Spectra*, Special Suppl. to Vol. 6, May 1990, 448 pp.
7. EERI, Earthquake Engineering Research Institute, Costa Rica earthquake reconnaissance report, *Earthquake Spectra*, Special Suppl. to Vol. 7, Oct. 1991, 127 pp.
8. EERI, Earthquake Engineering Research Institute, Northridge earthquake reconnaissance report, *Earthquake Spectra*, Special Suppl. to Vol. 11, April 1995, 523 pp.
9. EERI, Earthquake Engineering Research Institute, The Hyogo-Ken Nanbu Earthquake, January 17, 1995, Preliminary Reconnaissance Report, Feb. 1995, 116 pp.
10. Housner, G., Competing against time, Report to Governor George Deukmejian from the Governor's Board of Inquiry on the 1989 Loma Prieta Earthquake, Office of Planning and Research, State of California, May 1990, 264 pp.
11. Ishizaki, H. et al., Inspection and restoration of damaged foundations due to the Great Hanshin Earthquake 1995, *Proceedings, Third U.S.–Japan Workshop on Seismic Retrofit of Bridges*, Tsukuba, Japan, 1996, 327-341.
12. Iwasaki, T., Penzien, J., and Clough, R. Literature Survey-Seismic Effects on Highway Bridges, Earthquake Engineering Research Report No. 72-11, University of California, Berkeley, November 1972, 397 pp.
13. Jennings, P. C., Ed., Engineering Features of the San Fernando Earthquake of February 9, 1971, Report ERL 71-02, California Institute of Technology, June 1971.
14. Kawashima, K. and Unjoh, S., The damage of highway bridges in the 1995 Hyogo-Ken Nanbu earthquake and its impact on Japanese seismic design, *J. Earthquake Eng.*, 1(3), 505–541, 1997.
15. Saiidi, M., Maragakis, E., and Feng, S., Field performance and design issues for bridge hinge restrainers, *Proceedings, Fifth U.S. National Conference on Earthquake Engineering*, Earthquake Engineering Research Institute, Oakland, CA, Vol. I, 1994, 439–448.
16. Zelinski, R., Post Earthquake Investigation Team Report for the Loma Prieta Earthquake, California Department of Transportation, Division of Structures, Sacramento, 1994.

35

Dynamic Analysis

Rambabu Bavirisetty
California Department of Transportation

Murugesu Vinayagamoorthy
California Department of Transportation

Lian Duan
California Department of Transportation

35.1 Introduction

The primary purpose of this chapter is to present dynamic methods for analyzing bridge structures when subjected to earthquake loads. Basic concepts and assumptions used in typical dynamic analysis are presented first. Various approaches to bridge dynamics are then discussed. A few examples are presented to illustrate their practical applications.

35.1.1 Static vs. Dynamic Analysis

The main objectives of a structural analysis are to evaluate structural behavior under various loads and to provide the information necessary for design, such as forces, moments, and deformations. Structural analysis can be classified as *static* or *dynamic*: while *statics* deals with time-independent loading, *dynamics* considers any load where the magnitude, direction, and position vary with time. Typical dynamic loads for a bridge structure include vehicular motions and wave actions such as winds, stream flow, and earthquakes.

0-8493-7434-0/00/$0.00+$.50

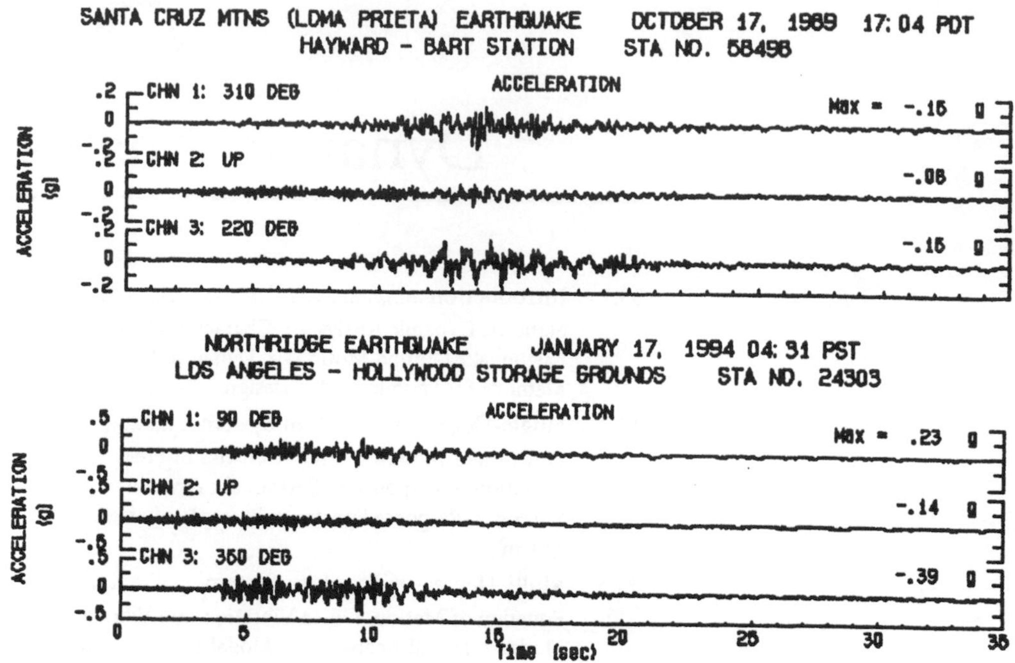

FIGURE 35.1 Ground motions recorded during recent earthquakes.

35.1.2 Characteristics of Earthquake Ground Motions

An earthquake is a natural ground movement caused by various phenomena including global tectonic processes, volcanism, landslides, rock-bursts, and explosions. The global tectonic processes are continually producing mountain ranges and ocean trenches at the Earth's surface and causing earthquakes. This section briefly discusses the earthquake input for seismic bridge analysis. Detailed discussions of ground motions are presented in Chapter 33.

Ground motion is represented by the time history or seismograph in terms of acceleration, velocity, and displacement for a specific location during an earthquake. Time history plots contain complete information about the earthquake motions in the three orthogonal directions (two horizontal and one vertical) at the strong-motion instrument location. Acceleration is usually recorded by strong-motion accelerograph and the velocities and displacements are determined by numerical integration. The accelerations recorded at locations that are approximately the same distance away from the epicenter may differ significantly in duration, frequency content, and amplitude due to different local soil conditions. Figure 35.1 shows several time histories of recent earthquakes.

From a structural engineering view, the most important characteristics of an earthquake are the peak ground acceleration (PGA), duration, and frequency content. The PGA is the maximum acceleration and represents the intensity of a ground motion. Although the ground velocity may be a more significant measure of intensity than the acceleration, it is not often measured directly, but determined using supplementary calculations [1]. The duration is the length of time between the first and the last peak exceeding a specified strong motion level. The longer the duration of a strong motion, the more energy is imparted to a structure. Since the elastic strain energy absorbed by a structure is very limited, a longer strong earthquake has a greater possibility to enforce a structure into the inelastic range. The frequency content can be represented by the number of zero crossings per second in the accelerogram. It is well understood that when the frequency of a regular disturbing force is the same as the natural vibration frequency of a structure (resonance), the oscillation of structure can be greatly magnified and effects of damping become minimal. Although earthquake motions are never as regular as a sinusoidal waveform, there is usually a period that dominates the response.

Since it is impossible to measure detailed ground motions for all structure sites, the rock motions or ground motions are estimated at a fault and then propagated to the Earth surface using a computer program considering the local soil conditions. Two guidelines [2, 3] recently developed by the California Department of Transportation provide the methods to develop seismic ground motions for bridges.

35.1.3 Dynamic Analysis Methods for Seismic Bridge Design

Depending on the seismic zone, geometry, and importance of the bridge, the following analysis methods may be used for seismic bridge design:

- The single-mode method (single-mode spectral and uniform load analysis) [4,5] assumes that seismic load can be considered as an equivalent static horizontal force applied to an individual frame in either the longitudinal or transverse direction. The equivalent static force is based on the natural period of a single degree of freedom (SDOF) and code-specified response spectra. Engineers should recognize that the single-mode method (sometimes referred to as equivalent static analysis) is best suited for structures with well-balanced spans with equally distributed stiffness.
- Multimode spectral analysis assumes that member forces, moments, and displacements due to seismic load can be estimated by combining the responses of individual modes using the methods such as complete quadratic combination (CQC) method and the square root of the sum of the squares (SRSS) method. The CQC method is adequate for most bridge systems [6], and the SRSS method is best suited for combining responses of well-separated modes.
- The multiple support response spectrum (MSRS) method provides response spectra and the peak displacements at individual support degrees of freedom by accurately accounting for the spatial variability of ground motions including the effects of incoherence, wave passage, and spatially varying site response. This method can be used for multiply supported long structures [7].
- The time history method is a numerical step-by-step integration of equations of motion. It is usually required for critical/important or geometrically complex bridges. Inelastic analysis provides a more realistic measure of structural behavior when compared with an elastic analysis.

Selection of the analysis method for a specific bridge structure should not be purely based on performing structural analysis, but be based on the effective design decisions [8]. Detailed discussions of the above methods are presented in the following sections.

35.2 Single-Degree-of-Freedom System

The familiar spring–mass system represents the simplest dynamic model and is shown in Figure 35.2a. When the *idealized, undamped* structures are excited by either moving the support or by displacing the mass in one direction, the mass oscillates about the equilibrium state forever without coming to rest. But, real structures do come to rest after a period of time due to a phenomenon called *damping*. To incorporate the effect of the damping, a massless viscous damper is always included in the dynamic model, as shown in Figure 35.2b.

In a dynamic analysis, the number of displacements required to define the displaced positions of all the masses relative to their original positions is called the number of degrees of freedom (DOF). When a structural system can be idealized with a single mass concentrated at one location and moved only in one direction, this dynamic system is called an SDOF system. Some structures,

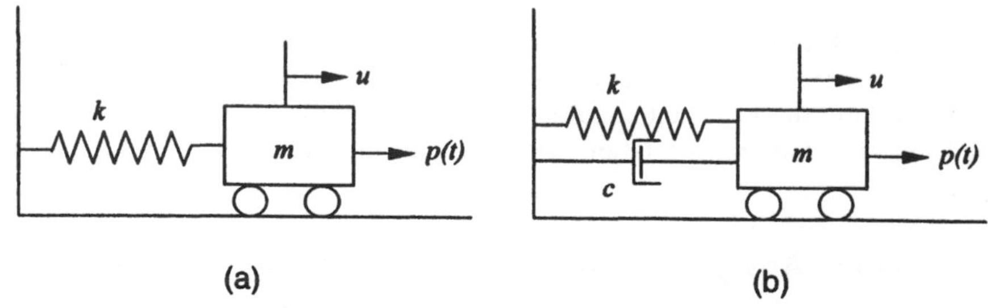

FIGURE 35.2 Idealized dynamic model. (a) Undamped SDOF system; (b) damped SDOF system.

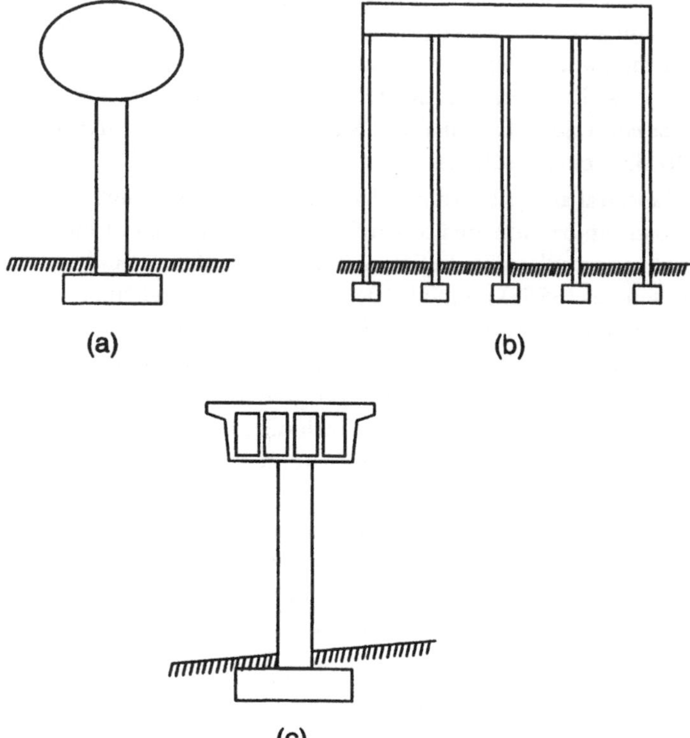

FIGURE 35.3 Examples of SDOF structures. (a) Water tank supported by single column; (b) one-story frame building; (c) two-span bridge supported by single column.

such as a water tank supported by a single-column, one-story frame structure and a two-span bridge supported by a single column, could be idealized as SDOF models (Figure 35.3).

In the SDOF system shown in Figure 35.3c, the mass of the bridge superstructure is the mass of the dynamic system. The stiffness of the dynamic system is the stiffness of the column against side sway and the viscous damper of the system is the internal energy absorption of the bridge structure.

35.2.1 Equation of Motion

The response of a structure depends on its mass, stiffness, damping, and applied load or displacement. The structure could be excited by applying an external force $p(t)$ on its mass or by a ground motion $u(t)$ at its supports. In this chapter, since the seismic loading is induced by exciting the support, we focus mainly on the equations of motion of an SDOF system subjected to ground excitation.

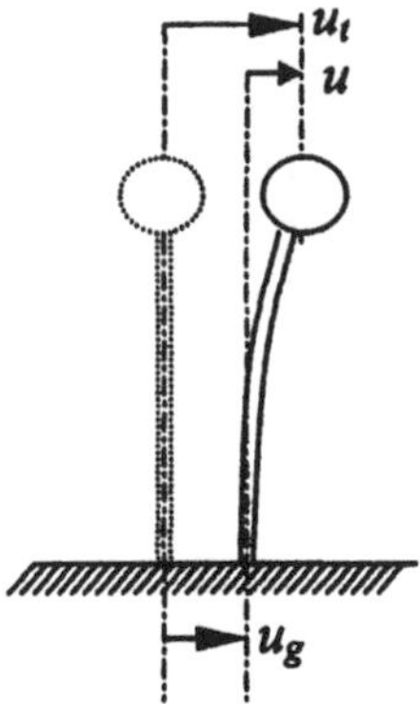

FIGURE 35.4 Earthquake–induced motion of an SDOF system.

The displacement of the ground motion u_g, the total displacement of the single mass u_t, and the relative displacement between the mass and ground u (Figure 35.4) are related by

$$u_t = u + u_g \tag{35.1}$$

By applying Newton's law and D'Alembert's principle of dynamic equilibrium, it can be shown that

$$f_I + f_D + f_S = 0 \tag{35.2}$$

where f_I is the inertial force of the single mass and is related to the acceleration of the mass by $f_I = m\ddot{u}_t$; f_D is the damping force on the mass and related to the velocity across the viscous damper by $f_D = c\dot{u}$; f_S is the elastic force exerted on the mass and related to the relative displacement between the mass and the ground by $f_S = ku$, where k is the spring constant; c is the damping ratio; and m is the mass of the dynamic system.

Substituting these expressions for f_I, f_D, and f_S into Eq. (35.2) gives

$$m\ddot{u}_t + c\dot{u} + ku = 0 \tag{35.3}$$

The equation of motion for an SDOF system subjected to a ground motion can then be obtained by substituting the Eq. (35.1) into Eq. (35.3), and is given by

$$m\ddot{u} + c\dot{u} + ku = -m\ddot{u}_g \tag{35.4}$$

35.2.2 Characteristics of Free Vibration

To determine the characteristics of the oscillations such as the time to complete one cycle of oscillation (T_n) and number of oscillation cycles per second (ω_n), we first look at the *free* vibration of a dynamic system. Free vibration is typically initiated by disturbing the structure from its equilibrium state by an external force or displacement. Once the system is disturbed, the system vibrates without any external input. Thus, the equation of motion for free vibration can be obtained by setting $\ddot{u}_g$ to zero in Eq. (35.4) and is given by

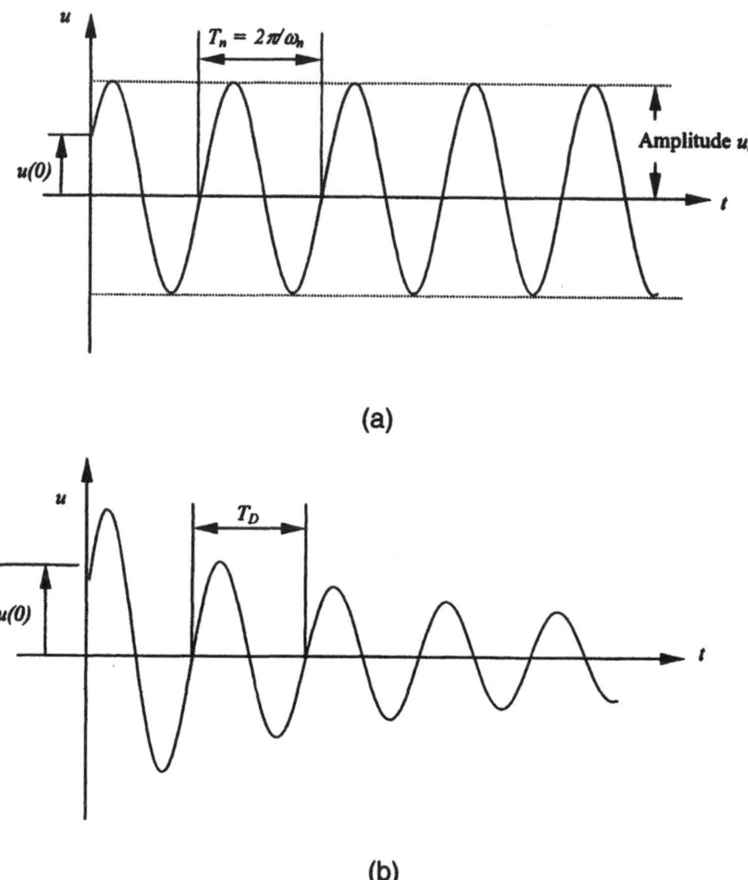

FIGURE 35.5 Typical response of an SDOF system. (a) Undamped; (b) damped.

$$m\ddot{u} + c\dot{u} + ku = 0 \tag{35.5}$$

Dividing the Equation (35.5) by its mass m will result in

$$\ddot{u} + \left(\frac{c}{m}\right)\dot{u} + \left(\frac{k}{m}\right)u = 0 \tag{35.6}$$

$$\ddot{u} + 2\xi\omega_n + \omega_n^2 u = 0 \tag{35.7}$$

where $\omega_n = \sqrt{k/m}$ the natural circular frequency of vibration or the undamped frequency; $\xi = c/c_{cr}$ the damping ratio; $c_{cr} = 2m\omega_n = 2\sqrt{km} = 2k/\omega_n$ the critical damping coefficient.

Figure 35.5a shows the response of a typical idealized, *undamped* SDOF system. The time required for the SDOF system to complete one cycle of vibration is called the natural period of vibration (T_n) of the system and is given by

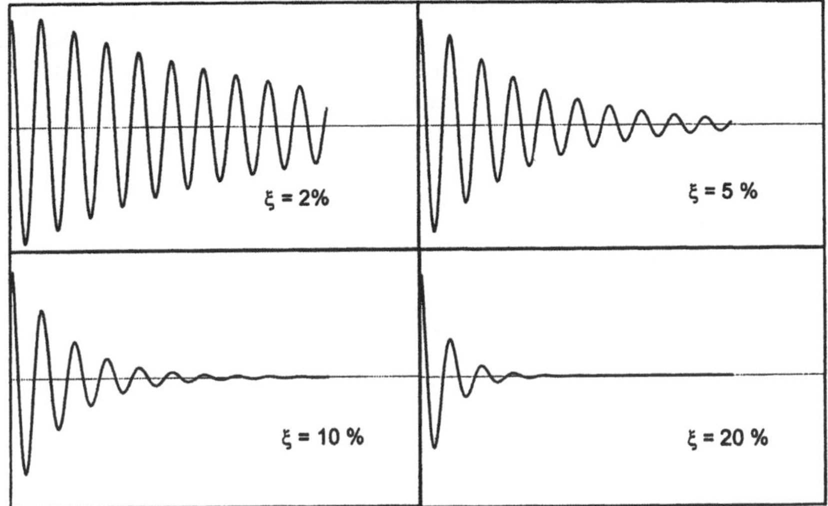

FIGURE 35.6 Response of an SDOF system for various damping ratios.

$$T_n = \frac{2\pi}{\omega_n} = 2\pi\sqrt{\frac{m}{k}} \tag{35.8}$$

Furthermore, the natural cyclic frequency of vibration f_n is given by

$$f_n = \frac{\omega_n}{2\pi} = \frac{1}{2\pi}\sqrt{\frac{k}{m}} \tag{35.9}$$

Figure 35.5b shows the response of a typical *damped* SDOF structure. The circular frequency of the vibration or damped vibration frequency of the SDOF structure, ω_d, is given by $\omega_d = \omega_n\sqrt{1-\xi^2}$.

The damped period of vibration (T_d) of the system is given by

$$T_d = \frac{2\pi}{\omega_d} = \frac{2\pi}{\sqrt{1-\xi^2}}\sqrt{\frac{m}{k}} \tag{35.10}$$

When $\xi = 1$ or $c = c_{cr}$ the structure returns to its equilibrium position without oscillating and is referred to as a critically damped structure. When $\xi > 1$ or $c > c_{cr}$, the structure is *overdamped* and comes to rest without oscillating, but at a slower rate. When $\xi < 1$ or $c < c_{cr}$, the structure is *underdamped* and oscillates about its equilibrium state with progressively decreasing amplitude. Figure 35.6 shows the response of SDOF structures with different damping ratios.

For structures such as buildings, bridges, dams, and offshore structures, the damping ratio is less than 0.15 and thus can be categorized as *underdamped* structures. The basic dynamic properties estimated using damped or undamped assumptions are approximately the same. For example, when $\xi = 0.10$, $\omega_d = 0.995\omega_n$, and $T_d = 1.01T_n$.

Damping dissipates the energy out of a structure in opening and closing of microcracks in concrete, stressing of nonstructural elements, and friction at the connection of steel members. Thus, the damping coefficient accounts for all energy-dissipating mechanisms of the structure and can only be estimated by experimental methods. Two seemingly identical structures may have slightly different material properties and may dissipate energy at different rates. Since damping does not play an important quantitative role except for resonant responses in structural responses, it is common to use average damping ratios based on the types of construction materials. Relative damping ratios for common types of structures, such as welded metal of 2 to 4%, bolted metal structures of 4 to 7%, prestressed concrete structures of 2 to 5%, reinforced-concrete structures of 4 to 7% and wooden structures of 5 to 10%, are recommended by Chmielewski et al. [9].

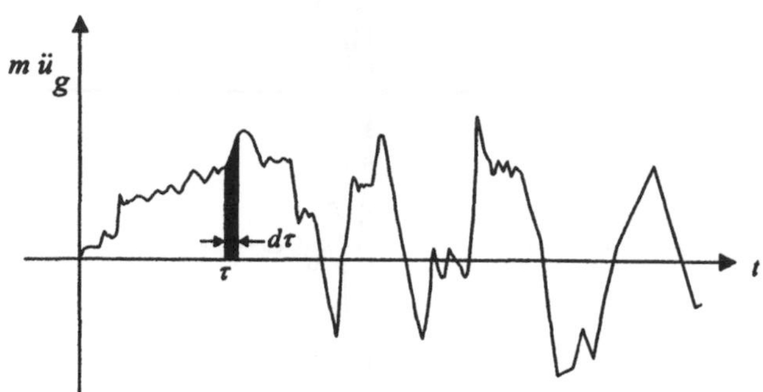

FIGURE 35.7 Induced earthquake force vs. time on an SDOF system.

35.2.3 Response to Earthquake Ground Motion

A typical excitation of an earth movement is shown in Figure 35.7. The basic equation of motion of an SDOF system is expressed in Eq. (35.4). Since the excitation force $m\ddot{u}_g$ cannot be described by simple mathematical expression, closed-form solutions for Eq. (35.4) are not available. Thus, the entire ground excitation needs to be treated as a superposition of short-duration impulses to evaluate the response of the structure to the ground excitation. An impulse is defined as the product of the force times duration. For example, the impulse of the force at time τ during the time interval $d\tau$ equals $-m\ddot{u}_g(\tau)d\tau$ and is represented by the shaded area in Figure 35.7. The total response of the structure for the earthquake motion can then be obtained by integrating all responses of the increment impulses. This approach is sometimes referred to as "time history analysis." Various solution techniques are available in the technical literature on structural dynamics [1,10].

In seismic structural design, designers are interested in the maximum or extreme values of the response of a structure as discussed in the following sections. Once the dynamic characteristics (T_n and ω_n) of the structure are determined, the maximum displacement, moment, and shear on the SDOF system can easily be estimated using basic principles of mechanics.

35.2.4 Response Spectra

The response spectrum is a relationship of the peak values of a response quantity (acceleration, velocity, or displacement) with a structural dynamic characteristic (natural period or frequency). Its core concept in earthquake engineering provides a much more convenient and meaningful measure of earthquake effects than any other quantity. It represents the peak response of all possible SDOF systems to a particular ground motion.

Elastic Response Spectrum

This, the response spectrum of an elastic structural system, can be obtained by the following steps [10]:

1. Define the ground acceleration time history (typically at a 0.02-second interval).
2. Select the natural period T_n and damping ratio ξ of an elastic SDOF system.
3. Compute the deformation response $u(t)$ using any numerical method.
4. Determine u_o, the peak value of $u(t)$.
5. Calculate the spectral ordinates by $D=u_o$, $V=2\pi D/T_n$, and $A=(2\pi/T_n)^2 D$.
6. Repeat Steps 2 and 5 for a range of T_n and ξ values for all possible cases.
7. Construct results graphically to produce three separate spectra as shown in Figure 35.8 or a combined tripartite plot as shown in Figure 35.9.

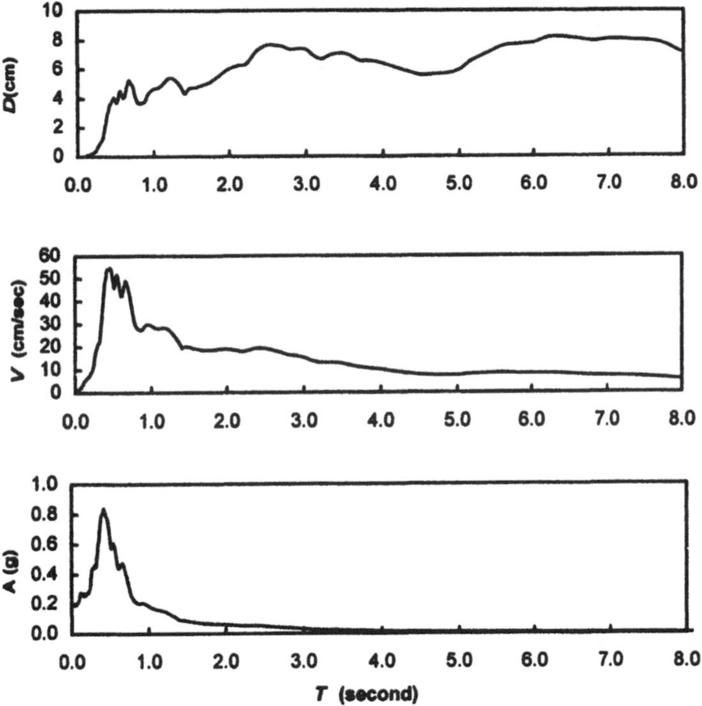

FIGURE 35.8 Example of response spectra (5% critical damping) for Loma Prieta 1989 motion.

It is noted that although three spectra (displacement, velocity, and acceleration) for a specific ground motion contain the same information, each provides a physically meaningful quantity. The displacement spectrum presents the peak displacement. The velocity spectrum is related directly to the peak strain energy stored in the system. The acceleration spectrum is related directly to the peak value of the equivalent static force and base shear.

A response spectrum (Figure 35.9) can be divided into three ranges of periods [10]:

- Acceleration-sensitive region (very short period region): A structure with a very short period is extremely stiff and expected to deform very little. Its mass moves rigidly with the ground and its peak acceleration approximately equals the ground acceleration.
- Velocity-sensitive region (intermediate-period region): A structure with an intermediate period responds greatly to the ground velocity than other ground motion parameters.
- Displacement-sensitive region (very long period region): A structure with a very long period is extremely flexible and expected to remain stationary while the ground moves. Its peak deformation is closer to the ground displacement. The structural response is most directly related to ground displacement.

Elastic Design Spectrum

Since seismic bridge design is intended to resist future earthquakes, use of a response spectrum obtained from a particular past earthquake motion is inappropriate. In addition, jagged spectrum values over small ranges would require an unreasonable accuracy in the determination of the structure period [11]. It is also impossible to predict a jagged response spectrum in all its details for a ground motion that may occur in the future. To overcome these shortcomings, the elastic design spectrum, a smoothened idealized response spectrum, is usually developed to represent the envelopes of ground motions recorded at the site during past earthquakes. The development of an elastic design spectrum is based on statistical analysis of the response spectra for the ensemble of ground motions. Figure 35.10 shows a set of elastic design spectra in Caltrans Bridge Design Specifications [12]. Figure 35.11 shows project-specific acceleration response spectra for the California Sonoma Creek Bridge.

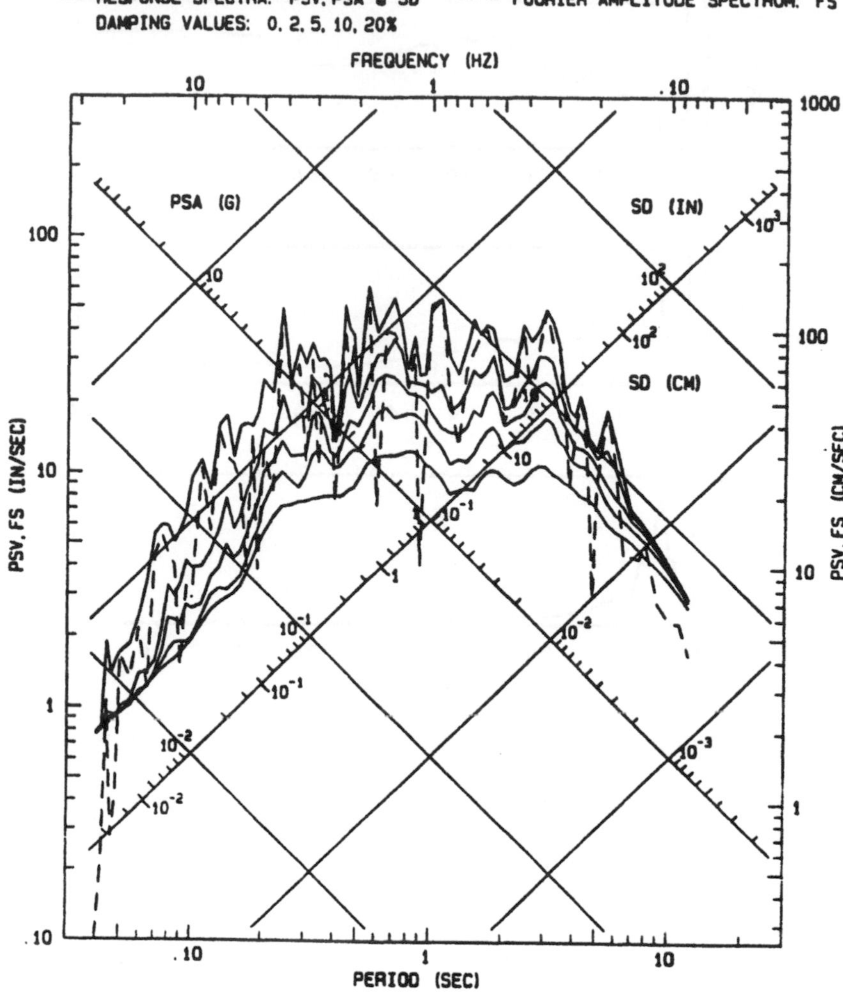

FIGURE 35.9 Tripartite plot–response spectra (1994 Northridge Earthquake, Arleta–Rordhoff Ave. Fire Station).

Engineers should recognize the conceptual differences between a response spectrum and a design spectrum [10]. A response spectrum is only the peak response of all possible SDOF systems due to a particular ground motion, whereas a design spectrum is a specified level of seismic design forces or deformations and is the envelope of two different elastic design spectra. The elastic design spectrum provides a basis for determining the design force and deformation for elastic SDOF systems.

Inelastic Response Spectrum

A bridge structure may experience inelastic behavior during a major earthquake. The typical elastic and elastic–plastic responses of an idealized SDOF to severe earthquake motions are shown in Figure 35.12. The input seismic energy received by a bridge structure is dissipated by both viscous damping and yielding (localized inelastic deformation converting into heat and other irrecoverable forms of energy). Both viscous damping and yielding reduce the response of inelastic structures compared with elastic structures. Viscous damping represents the internal friction loss of a structure when deformed and is approximately a constant because it depends mainly on structural materials. Yielding, on the other hand, varies depending on structural materials, structural configurations, and loading patterns and histories. Damping has negligible effects on the response of structures for

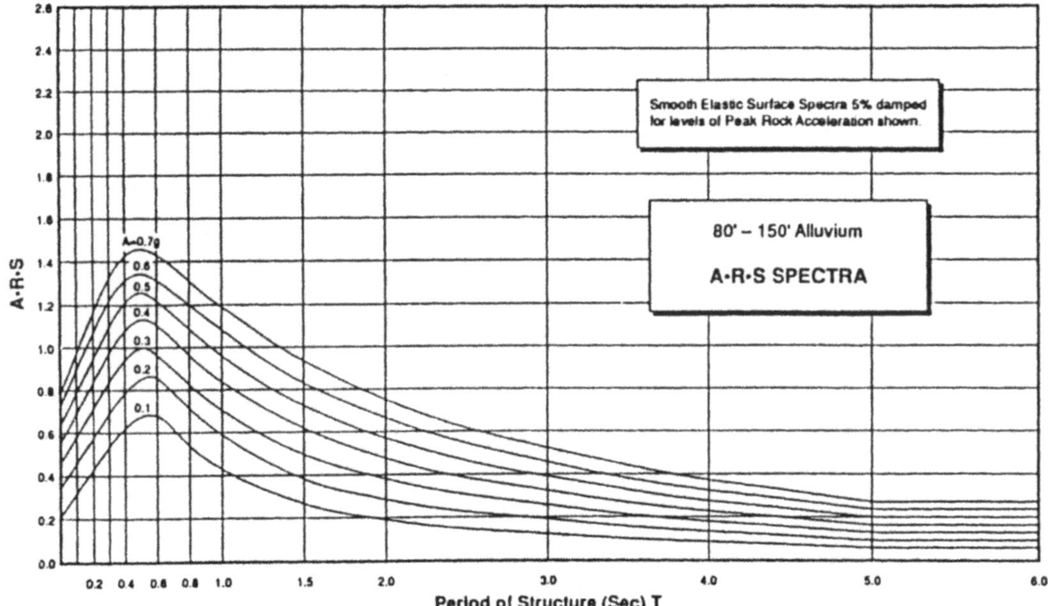

FIGURE 35.10 Typical Caltrans elastic design response spectra.

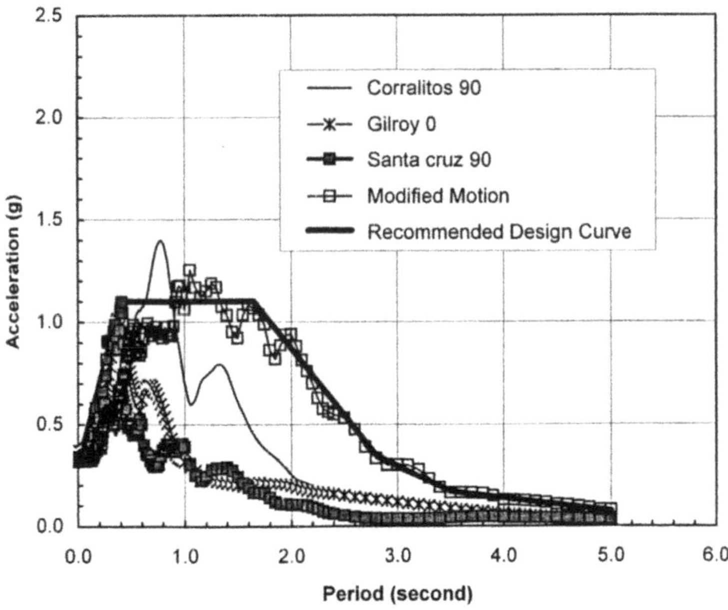

FIGURE 35.11 Acceleration response spectra for Sonoma Creek Bridge.

the long-period and short-period systems and is most effective in reducing response of structures for intermediate-period systems.

In seismic bridge design, a main objective is to ensure that a structure is capable of deforming in a ductile manner when subjected to a larger earthquake loading. It is desirable to consider the inelastic response of a bridge system to a major earthquake. Although a nonlinear inelastic dynamic analysis is not difficult in concept, it requires careful structural modeling and intensive computing

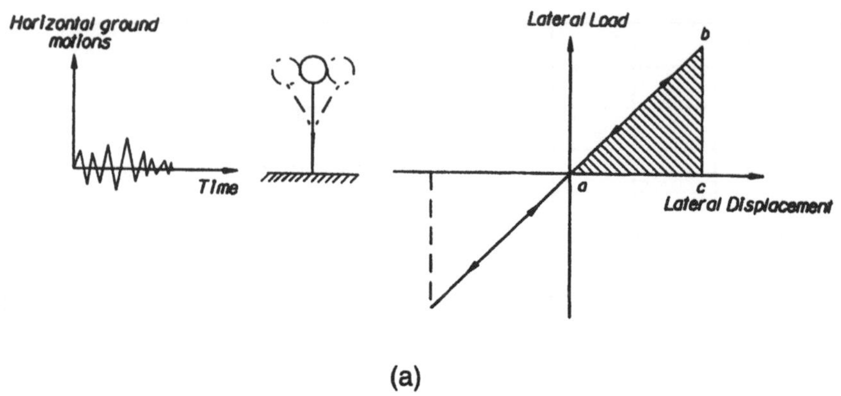

(a)

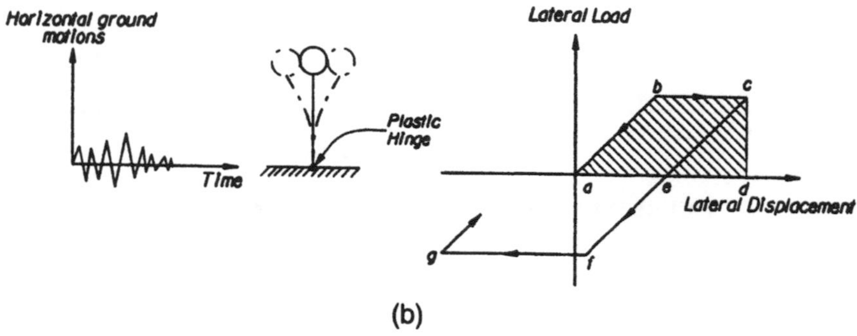

(b)

FIGURE 35.12 Response of an SDOF to earthquake ground motions. (a) Elastic system; (b) inelastic system.

effort [8]. To consider inelastic seismic behavior of a structure without performing a true nonlinear inelastic analysis, the ductility-factor method can be used to obtain the inelastic response spectra from the elastic response spectra. The ductility of a structure is usually referred as the displacement ductility factor μ defined by (Figure 35.13):

$$\mu = \frac{\Delta_u}{\Delta_y} \tag{35.11}$$

where Δ_u is ultimate displacement capacity and Δ_y is yield displacement.

The simplest approach to developing the inelastic design spectrum is to scale the elastic design spectrum down by some function of the available ductility of a structural system:

$$ARS_{\text{inelastic}} = \frac{ARS_{\text{elastic}}}{f(\mu)} \tag{35.12}$$

$$f(\mu) = \begin{cases} 1 & \text{for } T_n \leq 0.03 \text{ sec.} \\ 2\mu - 1 & \text{for } 0.03 \text{ sec.} < T_n \leq 0.5 \text{ sec.} \\ \mu & \text{for } T_n \geq 0.5 \text{ sec.} \end{cases} \tag{35.13}$$

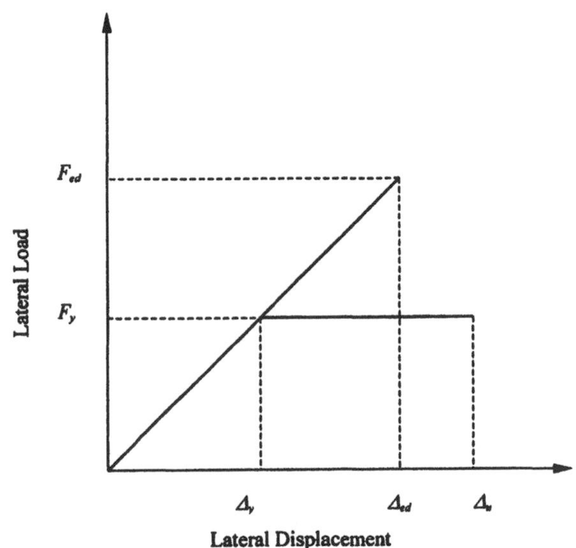

FIGURE 35.13 Lateral load–displacement relations.

For very short period (T_n ≤ 0.03 sec) in the acceleration-sensitive region, the elastic displacement demand Δ_{ed} is less than displacement capacity Δ_u (see Figure 35.13). The reduction factor $f(\mu) = 1$ implies that the structure should be designed and remained at elastic to avoid excessive inelastic deformation. For intermediate period (0.03 sec < T_n ≤ 0.5 sec) in the velocity-sensitive region, elastic displacement demand Δ_{ed} may be greater or less than displacement capacity Δ_u and the reduction factor is based on the equal-energy concept. For the very long period (T_n > 0.5 sec) in the displacement-sensitive region, the reduction factor is based on the equal displacement concept.

35.2.5 Example of an SDOF system

Given

An SDOF bridge structure is shown in Figure 35.14. To simplify the problem, the bridge is assumed to move only in the longitudinal direction. The total resistance against the longitudinal motion comes in the form of friction at bearings and this could be considered a damper. Assume the following properties for the structure: damping ratio $\xi = 0.05$, area of superstructure $A = 3.57$ m^2, moment of column $I_c = 0.1036$ m^4, E_c of column = 20,700 MPa, material density $\rho = 2400$ kg/m^3, length of column $L_c = 9.14$ m, and length of the superstructure $L_s = 36.6$ m. The acceleration response curve of the structure is given in the Figure 35.11. Determine (1) natural period of the structure, (2) damped period of the structure, (3) maximum displacement of the superstructure, and (4) maximum moment in the column.

Solution

$$\text{Stiffness: } k = \frac{12E_c I_c}{L_c^3} = \frac{12(20700\times10^6)(0.1036)}{9.14^3} = 33690301 \text{ N/m}$$

$$\text{Mass: } m = A L_s \rho = (3.57)(36.6)(2400) = 313{,}588.8 \text{ kg}$$

$$\text{Natural circular frequency: } \omega_n = \sqrt{\frac{k}{m}} = \sqrt{\frac{33{,}690{,}301}{313{,}588.8}} = 10.36 \text{ rad/s}$$

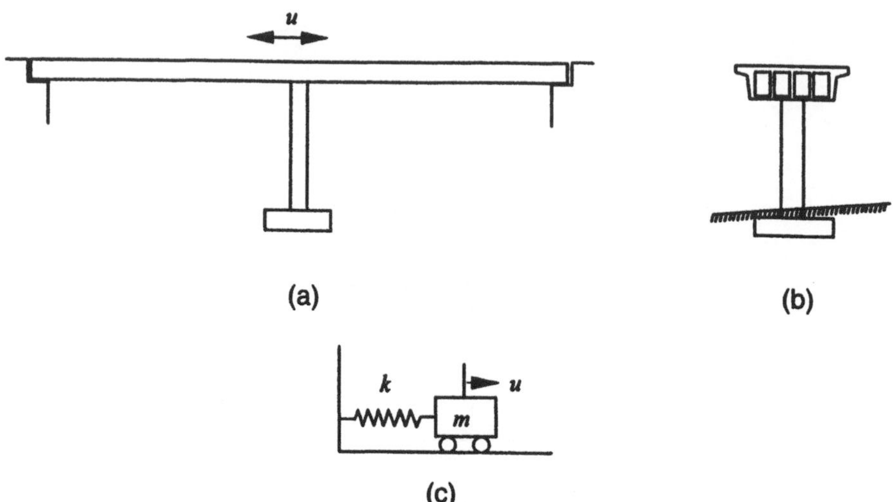

FIGURE 35.14 SDOF bridge example. (a) Two-span bridge schematic diagram; (b) single column bent; (c) idealized equivalent model for longitudinal response.

$$\text{Natural cyclic frequency: } f_n = \frac{\omega_n}{2\pi} = \frac{10.36}{2\pi} = 1.65 \text{ cycles/s}$$

$$\text{Natural period of the structure: } T_n = \frac{1}{f_n} = \frac{1}{1.65} = 0.606 \text{ s}$$

The *damped* circular frequency is given by

$$\omega_d = \omega_n \sqrt{1-\xi^2} = 10.36\sqrt{1-0.05^2} = 10.33 \text{ rad/s}$$

The *damped* period of the structure is given by

$$T_d = \frac{2\pi}{\omega_d} = \frac{2\pi}{10.33} = 0.608 \text{ s}$$

From the ARS curve, for a period of 0.606 s, the maximum acceleration of the structure will be 0.9 g = 1.13 × 9.82 = 11.10 m/s. Then,

$$\text{The force acting on the mass} = m \times 11.10 = 313588.8 \times 11.10 = 3.48 \text{MN}$$

$$\text{The maximum displacement} = \frac{FL_c^3}{12EI_c} = \frac{3.48 \times 9.14^3}{12 \times 20700 \times 0.1036} = 0.103 \text{m}$$

$$\text{The maximum moment in the column} = \frac{FL_c}{2} = \frac{3.48 \times 9.14}{2} = 15.90 \text{ MN-m}$$

35.3 Multidegree-of-Freedom System

The SDOF approach may not be applicable for complex structures such as multilevel frame structure and bridges with several supports. To predict the response of a complex structure, the structure is discretized with several members of lumped masses. As the number of lumped masses increases, the number of displacements required to define the displaced positions of all masses increases. The response of a multidegree of freedom (MDOF) system is discussed in this section.

35.3.1 Equation of Motion

The equation of motion of an MDOF system is similar to the SDOF system, but the stiffness **k**, mass **m**, and damping **c** are matrices. The equation of motion to an MDOF system under ground motion can be written as

$$[\mathbf{M}]\{\ddot{u}\}+[\mathbf{C}]\{\dot{u}\}+[\mathbf{K}]\{u\}=-[\mathbf{M}]\{\mathbf{B}\}\ddot{u}_g \tag{35.14}$$

The stiffness matrix $[\mathbf{K}]$ can be obtained from standard static displacement-based analysis models and may have off-diagonal terms. The mass matrix $[\mathbf{M}]$ due to the negligible effect of mass coupling can best be expressed in the form of tributary lumped masses to the corresponding displacement degree of freedoms, resulting in a diagonal or uncoupled mass matrix. The damping matrix $[\mathbf{C}]$ accounts for all the energy-dissipating mechanisms in the structure and may have off-diagonal terms. The vector $\{\mathbf{B}\}$ is a displacement transformation vector that has values 0 and 1 to define degrees of freedoms to which the earthquake loads are applied.

35.3.2 Free Vibration and Vibration Modes

To understand the response of MDOF systems better, we look at the *undamped, free* vibration of an *N* degrees of freedom (*N*-DOF) system first.

Undamped Free Vibration

By setting $[\mathbf{C}]$ and $\ddot{u}_g$ to zero in the Eq. (35.14), the equation of motion of undamped, free vibration of an *N*-DOF system can be shown as:

$$[\mathbf{M}]\{\ddot{u}\}+[\mathbf{K}]\{u\}=0 \tag{35.15}$$

where $[\mathbf{M}]$ and $[\mathbf{K}]$ are $n \times n$ square matrices.

Equation (35.15) could then be rearranged to

$$\left[[\mathbf{K}]-\omega_n^{\,2}[\mathbf{M}]\right]\{\phi_n\}=0 \tag{35.16}$$

where $\{\phi_n\}$ is the deflected shape matrix. Solution to this equation can be obtained by setting

$$\left|[\mathbf{K}]-\omega_n^{\,2}[\mathbf{M}]\right|=0 \tag{35.17}$$

The roots or eigenvalues of Eq. (35.17) will be the *N* natural frequencies of the dynamic system. Once the natural frequencies (ω_n) are estimated, Eq. (35.16) can be solved for the corresponding *N* independent, deflected shape matrices (or eigenvectors), $\{\phi_n\}$. In other words, a vibrating system with N-DOFs will have N natural frequencies (usually arranged in sequence from smallest to largest), corresponding N natural periods T_n, and N natural mode shapes $\{\phi_n\}$. These eigenvectors are sometimes referred to as natural modes of vibration or natural mode shapes of vibration. It is important to recognize that the eigenvectors or mode shapes represent only the deflected shape corresponding to the natural frequency, not the actual deflection magnitude.

The N eigenvectors can be assembled in a single $n \times n$ square matrix $[\Phi]$, modal matrix, where each column represents the coefficients associated with the natural mode. One of the important aspects of these mode shapes is that they are orthogonal to each other. Stated mathematically,

$$\text{If} \quad \omega_n \neq \omega_r, \qquad \{\phi_n\}^T [K] \{\phi_r\} = 0 \quad \text{and} \quad \{\phi_n\}^T [M] \{\phi_r\} = 0 \tag{35.18}$$

$$[K^*] = [\Phi]^T [K] [\Phi] \tag{35.19}$$

$$[M^*] = [\Phi]^T [M] [\Phi] \tag{35.20}$$

where $[\mathbf{K}]$ and $[\mathbf{M}]$ have off-diagonal elements, whereas $[\mathbf{K}^*]$ and $[\mathbf{M}^*]$ are diagonal matrices.

Damped Free Vibration

When damping of the MDOF system is included, the free vibration response of the damped system will be given by

$$[\mathbf{M}]\{\ddot{u}\} + [\mathbf{C}]\{\dot{u}\} + [\mathbf{K}]\{u\} = 0 \tag{35.21}$$

The displacements are first expressed in terms of natural mode shapes, and later they are multiplied by the transformed natural mode matrix to obtain the following expression:

$$[\mathbf{M}^*]\{\ddot{Y}\} + [\mathbf{C}^*]\{\dot{Y}\} + [\mathbf{K}^*]\{Y\} = 0 \tag{35.22}$$

where, $[\mathbf{M}^*]$ and $[\mathbf{K}^*]$ are diagonal matrices given by Eqs. (35.19) and (35.20) and

$$[\mathbf{C}^*] = [\mathbf{\Phi}]^T [\mathbf{C}] [\mathbf{\Phi}] \tag{35.23}$$

While $[\mathbf{M}^*]$ and $[\mathbf{K}^*]$ are diagonal matrices, $[\mathbf{C}^*]$ may have off diagonal terms. When $[\mathbf{C}^*]$ has off diagonal terms, the damping matrix is referred to as a *nonclassical* or *nonproportional* damping matrix. When $[\mathbf{C}^*]$ is diagonal, it is referred to as a *classical* or *proportional* damping matrix. Classical damping is an appropriate idealization when similar damping mechanisms are distributed throughout the structure. Nonclassical damping idealization is appropriate for the analysis when the damping mechanisms differ considerably within a structural system.

Since most bridge structures have predominantly one type of construction material, bridge structures could be idealized as a classical damping structural system. Thus, the damping matrix of Eq. (35.22) will be a diagonal matrix for most bridge structures. And, the equation of nth mode shape or generalized nth modal equation is given by

$$\ddot{Y}_n + 2\xi_n \omega_n \dot{Y}_n + \omega^2 Y_n = 0 \tag{35.24}$$

Equation (35.24) is similar to the Eq. (35.7) of an SDOF system. Also, the vibration properties of each mode can be determined by solving the Eq. (35.24).

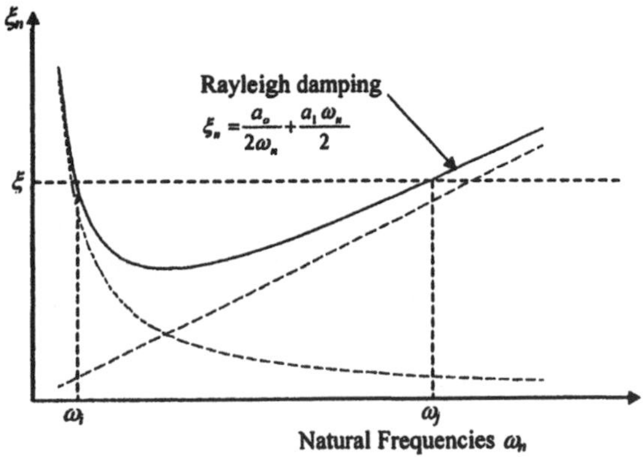

FIGURE 35.15 Rayleigh damping variation with natural frequency.

Rayleigh Damping

The damping of a structure is related to the amount of energy dissipated during its motion. It could be assumed that a portion of the energy is lost due to the deformations, and thus damping could be idealized as proportional to the stiffness of the structure. Another mechanism of energy dissipation could be attributed to the mass of the structure, and thus damping idealized as proportional to the mass of the structure. In Rayleigh damping, it is assumed that the damping is proportional to the mass and stiffness of the structure.

$$[\mathbf{C}] = a_o[\mathbf{M}] + a_1[\mathbf{K}] \tag{35.25}$$

The generalized damping of the nth mode is then given by

$$C_n = a_o M_n + a_1 K_n \tag{35.26}$$

$$C_n = a_o M_n + a_1 \omega_n^2 M_n \tag{35.27}$$

$$\xi_n = \frac{C_n}{2M_n\omega_n} \tag{35.28}$$

$$\xi_n = \frac{a_o}{2}\frac{1}{\omega_n} + \frac{a_1}{2}\omega_n \tag{35.29}$$

Figure 35.15 shows the Rayleigh damping variation with natural frequency. The coefficients a_o and a_1 can be determined from specified damping ratios at two independent dominant modes (say, i^{th} and j^{th} modes). Expressing Eq. (35.29) for these two modes will lead to the following equations:

$$\xi_i = \frac{a_o}{2}\frac{1}{\omega_i} + \frac{a_1}{2}\omega_i \tag{35.30}$$

$$\xi_j = \frac{a_o}{2}\frac{1}{\omega_j} + \frac{a_1}{2}\omega_j \tag{35.31}$$

When the damping ratio at both the ith and j^{th} modes is the same and equals ξ, it can be shown that

$$a_o = \xi \frac{2\omega_i \omega_j}{\omega_i + \omega_j} \qquad a_1 = \xi \frac{2}{\omega_i + \omega_j} \tag{35.32}$$

It is important to note that the damping ratio at a mode between the ith and j^{th} mode is less than ξ. And, in practical problems the specified damping ratios should be chosen to ensure reasonable values in all the mode shapes that lie between the ith and jth mode shapes.

35.3.3 Modal Analysis and Modal Participation Factor

In previous sections, we have discussed the basic vibration properties of an MDOF system. Now, we will look at the response of an MDOF system to earthquake ground motion. The basic equation of motion of the MDOF for an earthquake ground motion given by Eq. (35.14) is repeated here:

$$[\mathbf{M}]\{\ddot{u}\} + [\mathbf{C}]\{\dot{u}\} + [\mathbf{K}]\{u\} = -[\mathbf{M}]\{B\}\ddot{u}_g$$

The displacement is first expressed in terms of natural mode shapes, and later it is multiplied by the transformed natural mode matrix to obtain the following expression:

$$[\mathbf{M}^*]\{\ddot{Y}\} + [\mathbf{C}^*]\{\dot{Y}\} + [\mathbf{K}^*]\{Y\} = -[\mathbf{\Phi}]^T[\mathbf{M}]\{B\}\ddot{u}_g \tag{35.33}$$

And, the equation of the nth mode shape is given by

$$M_n^* \ddot{Y}_n + 2\xi_n \omega_n M_n^* \dot{Y}_n + \omega^2 M_n^* Y_n = L_n \ddot{u}_g \tag{35.34}$$

where

$$M_n^* = \{\phi_n\}^T [\mathbf{M}] \{\phi_n\} \tag{35.35}$$

$$L_n = -\{\phi_n\}^T [\mathbf{M}] [\mathbf{B}] \tag{35.36}$$

The L_n is referred to as the *modal participation factor* of the nth mode.

By dividing the Eq. (35.34) by M_n^*, the generalized modal equation of the nth mode becomes

$$\ddot{Y}_n + 2\xi_n \omega_n \dot{Y}_n + \omega^2 Y_n = \left(\frac{L_n}{M_n^*}\right)\ddot{u}_g \tag{35.37}$$

Equation (35.34) is similar to the equation motion of an SDOF system, and thus Y_n can be determined by using methods similar to those described for SDOF systems. Once Y_n is established, the displacement due to the nth mode will be given by $u_n(t) = \phi_n Y_n(t)$. The total displacement due to combination of all mode shapes can then be determined by summing up all displacements for each mode and is given by

$$u(t) = \sum \phi_n Y_n(t) \tag{35.38}$$

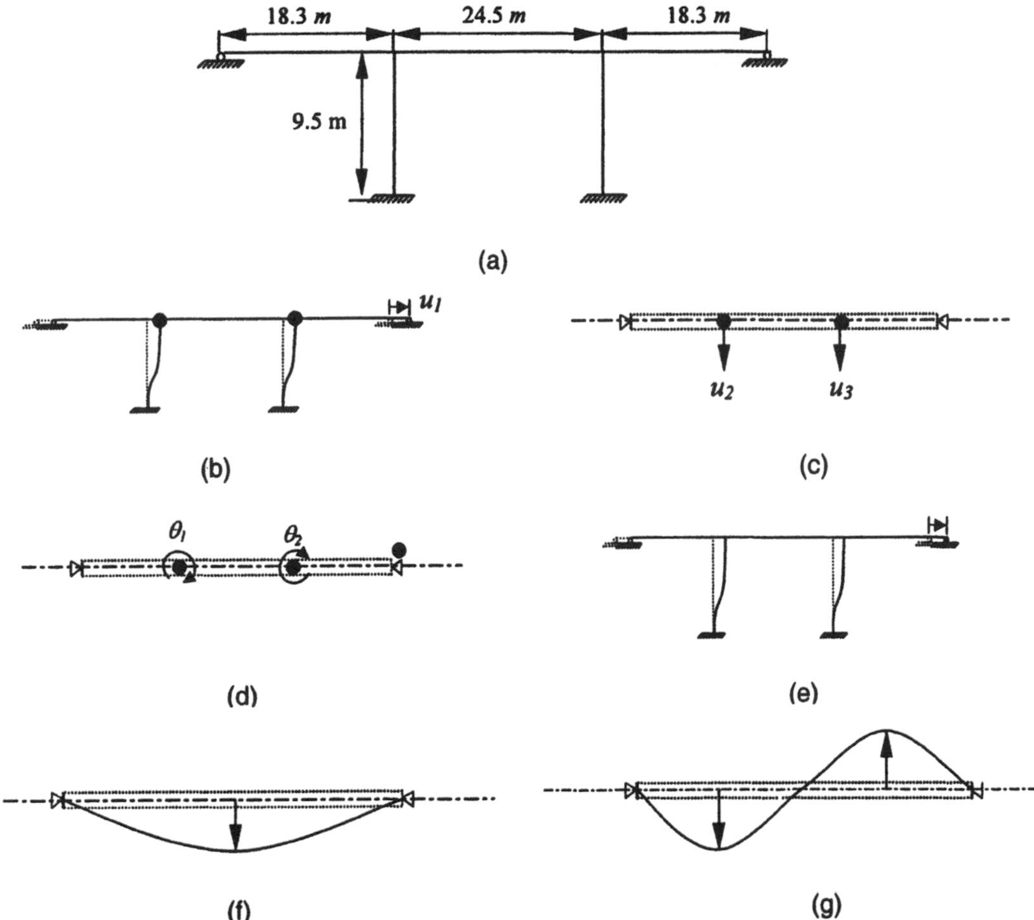

FIGURE 35.16 Three-span continuous framed bridge structure of MDOF example. (a) Schematic diagram; (b) longitudinal degree of freedom; (c) transverse degree of freedom; (d) rotational degree of freedom; (e) mode shape 1; (f) mode shape 2; (g) mode shape 3.

This approach is sometimes referred to as the classical mode superposition method. Similar to the estimation of the total displacement, the element forces can also be estimated by adding the element forces for each mode shape.

35.3.4 Example of an MDOF System

Given

The bridge shown in Figure 35.16 is a three-span continuous frame structure. Details of the bridge are as follows: span lengths are 18.3, 24.5, and 18.3 m.; column length is 9.5 m; area of superstructure is 5.58 m^2; moment of inertia of superstructure is 70.77 m^4; moment of inertia of column is 0.218 m^4; modulus of elasticity of concrete is 20,700 MPa. Determine the vibration modes and frequencies of the bridge.

Solution

As shown in Figures 35.16b, c, and d, five degrees of freedom are available for this structure. Stiffness and mass matrices are estimated separately and the results are given here.

$$[\mathbf{K}]=\begin{bmatrix} 126318588 & 0 & 0 & 0 & 0 \\ 0 & 1975642681 & -1194370500 & -1520122814 & -14643288630 \\ 0 & -1194370500 & 1975642681 & 14643288630 & 1520122814 \\ 0 & -1520122814 & 14643288630 & 479327648712 & 119586857143 \\ 0 & -14643288630 & 1520122814 & 119586857143 & 479327648712 \end{bmatrix}$$

$$[\mathbf{M}]=\begin{bmatrix} 81872 & 0 & 0 & 0 & 0 \\ 0 & 286827 & 0 & 0 & 0 \\ 0 & 0 & 286827 & 0 & 0 \\ 0 & 0 & 0 & 0 & 0 \\ 0 & 0 & 0 & 0 & 0 \end{bmatrix}$$

Condensation procedure will eliminate the rotational degrees of freedom and will result in three degrees of freedom. (The condensation procedure is performed separately and the result is given here.) The equation of motion of free vibration of the structure is

$$[M]\{\ddot{u}\} + [K]\{u\} = \{0\}$$

Substituting condensed stiffness and mass matrices into the above equation gives

$$\begin{bmatrix} 81872 & 0 & 0 \\ 0 & 286827 & 0 \\ 0 & 0 & 286827 \end{bmatrix}\begin{Bmatrix} \ddot{u}_1 \\ \ddot{u}_2 \\ \ddot{u}_3 \end{Bmatrix} + \begin{bmatrix} 126318588 & 0 & 0 \\ 0 & 1975642681 & -1194370500 \\ 0 & -1194370500 & 1975642681 \end{bmatrix}\begin{Bmatrix} u_1 \\ u_2 \\ u_3 \end{Bmatrix} = \begin{Bmatrix} 0 \\ 0 \\ 0 \end{Bmatrix}$$

The above equation can be rearranged in the following form:

$$\frac{1}{\omega^2}[M]^{-1}[K]\{\phi\} = \{\phi\}$$

Substitution of appropriate values in the above expression gives the following

$$\frac{1}{\omega_n^2}\begin{bmatrix} \frac{1}{818172} & 0 & 0 \\ 0 & \frac{1}{286827} & 0 \\ 0 & 0 & \frac{1}{286827} \end{bmatrix}\begin{bmatrix} 126318588 & 0 & 0 \\ 0 & 1518171572 & -1215625977 \\ 0 & -1215625977 & 1518171572 \end{bmatrix}\begin{Bmatrix} \phi_{1n} \\ \phi_{2n} \\ \phi_{3n} \end{Bmatrix} = \begin{Bmatrix} \phi_{1n} \\ \phi_{2n} \\ \phi_{3n} \end{Bmatrix}$$

$$\frac{1}{\omega_n^2}\begin{bmatrix} 154.39 & 0 & 0 \\ 0 & 5292.9 & -4238.2 \\ 0 & -4238.2 & 5292.9 \end{bmatrix}\begin{Bmatrix} \phi_{1n} \\ \phi_{2n} \\ \phi_{3n} \end{Bmatrix} = \begin{Bmatrix} \phi_{1n} \\ \phi_{2n} \\ \phi_{3n} \end{Bmatrix}$$

By assuming different vibration modes, natural frequencies of the structure can be estimated.

Substitution of vibration mode $\{1\ 0\ 0\}^T$ will result in the first natural frequency.

$$\frac{1}{\omega_n^2}\begin{bmatrix}154.39 & 0 & 0\\ 0 & 5292.9 & -4238.2\\ 0 & -4238.2 & 5292.9\end{bmatrix}\begin{Bmatrix}1\\0\\0\end{Bmatrix}=\frac{1}{\omega_n^2}\begin{bmatrix}154.39\\0\\0\end{bmatrix}=\begin{Bmatrix}1\\0\\0\end{Bmatrix}$$

Thus, $\omega_n^2 = 154.39$ and $\omega_n = 12.43\ \text{rad/s}$

By substituting the vibration modes of $\{0\ 1\ 1\}^T$ and $\{0\ 1\ -1\}^T$ in the above expression, the other two natural frequencies are estimated as 32.48 and 97.63 rad/s.

35.3.5 Multiple-Support Excitation

So far we have assumed that all supports of a structural system undergo the same ground motion. This assumption is valid for structures with foundation supports close to each other. However, for long-span bridge structures, supports may be widely spaced. As described in Section 35.1.2, earth motion at a location depends on the localized soil layer and the distance from the epicenter. Thus, bridge structures with supports that lie far from each other may experience different earth excitation. For example, Figure 35.17c, d, and e shows the predicted earthquake motions at Pier W3 and Pier W6 of the San Francisco–Oakland Bay Bridge (SFOBB) in California. The distance between Pier W3 and Pier W6 of the SFOBB is approximately 1411 m. These excitations are predicted by the California Department of Transportation by considering the soil and rock properties in the vicinity of the SFOBB and expected Earth movements at the San Andreas and Hayward faults. Note that the Earth motion at Pier W3 and Pier W6 are very different. Furthermore, Figures 35.17c, d, and e indicates that the Earth motion not only varies with the location, but also varies with direction. Thus, to evaluate the response of long, multiply supported, and complicated bridge structures, use of the actual earthquake excitation at each support is recommended.

The equation of motion of a multisupport excitation would be similar to Eq. (35.14), but the only difference is now that $\{B\}\ddot{u}_g$ is replaced by an displacement array $\{\ddot{u}_g\}$. And, the equation of motion for the multisupport system becomes

$$[\mathbf{M}]\{\ddot{u}\}+[\mathbf{C}]\{\dot{u}\}+[\mathbf{K}]\{u\}=-[\mathbf{M}]\{\ddot{u}_g\} \tag{35.39}$$

where $\{\ddot{u}_g\}$ has the acceleration at each support locations and has zero value at nonsupport locations. By using the uncoupling procedure described in the previous sections, the modal equation of the nth mode can be written as

$$\ddot{Y}_n + 2\xi_n\omega_n\dot{Y}_n + \omega^2 Y_n = -\sum_{l=1}^{N_g}\frac{L_n}{M_n^*}\ddot{u}_g \tag{35.40}$$

where N_g is the total number of externally excited supports.

The deformation response of the nth mode can then be determined as described in previous sections. Once the displacement responses of the structure for all the mode shapes are estimated, the total dynamic response can be obtained by combining the displacements.

35.3.6 Time History Analysis

When the structure enters the nonlinear range, or has nonclassical damping properties, modal analysis cannot be used. A numerical integration method, sometimes referred to as time history analysis, is required to get more accurate responses of the structure.

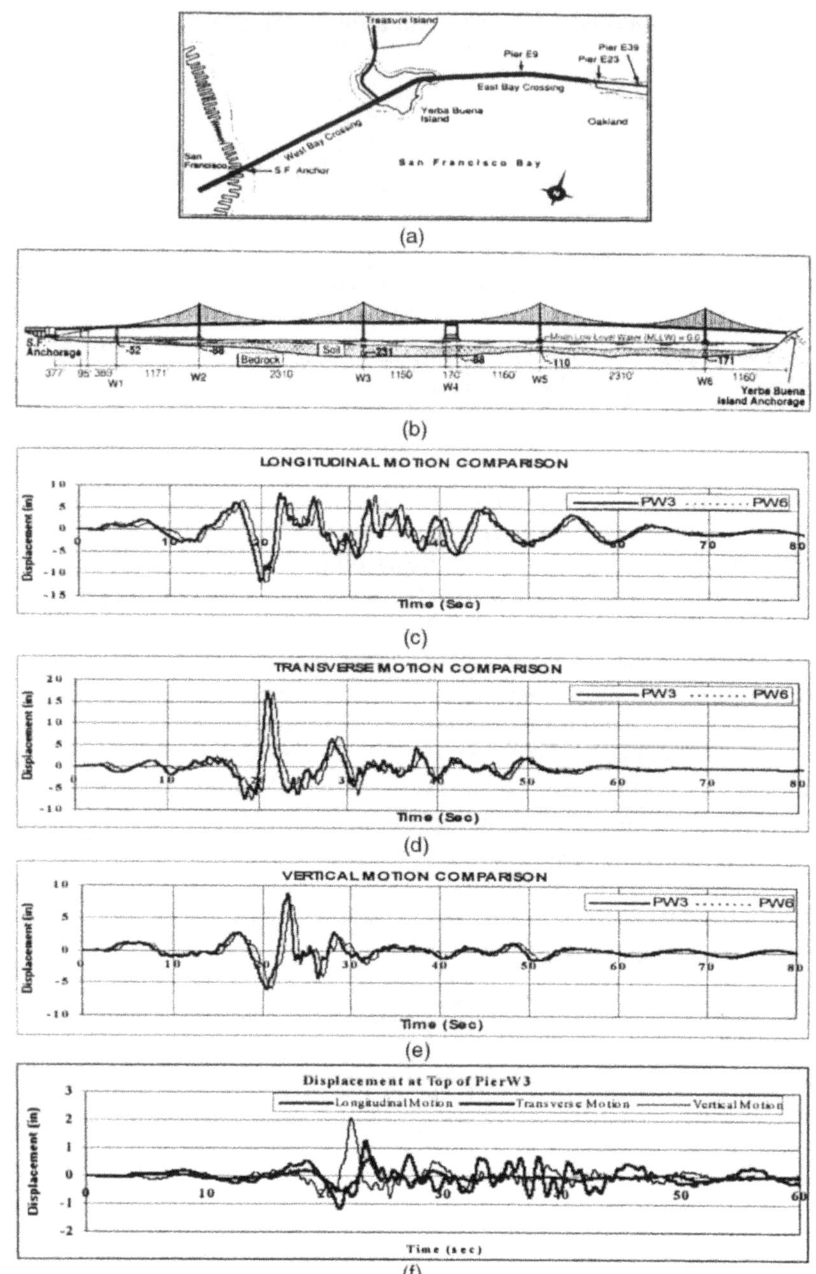

FIGURE 35.17 San Francisco–Oakland Bay Bridge. (a) Vicinity map; (b) general plan elevation; (c) longitudinal motion at rock level; (d) transverse motion at rock level; (e) vertical motion at rock level; (f) displacement response at top of Pier W3.

In a time history analysis, the timescale is divided into a series of smaller steps, d_τ. Let us say the response at i^{th} time interval has already determined and is denoted by $u_i, \dot{u}_i, \ddot{u}_i$. Then, the response of the system at i^{th} time interval will satisfy the equation of motion (Eq. 35.39).

$$[\mathbf{M}]\{\ddot{u}_i\}+[\mathbf{C}]\{\dot{u}_i\}+[\mathbf{K}]\{u_i\}=-[\mathbf{M}]\{\ddot{u}_{gi}\} \tag{35.41}$$

The time-stepping method enables us to step ahead and determine the responses $u_{i+1}, \dot{u}_{i+1}, \ddot{u}_{i+1}$ at the $i+1^{th}$ time interval by satisfying Eq. (35.39). Thus, the equation of motion at $i+1^{th}$ time interval will be

$$[\mathbf{M}]\{\ddot{u}_{i+1}\}+[\mathbf{C}]\{\dot{u}_{i+1}\}+[\mathbf{K}]\{u_{i+1}\}=-[\mathbf{M}]\{\ddot{u}_{gi+1}\} \tag{35.42}$$

Equation (35.42) needs to be solved prior to proceeding to the next time step. By stepping through all the time steps, the actual response of the structure can be determined at all time instants.

Example of Time History Analysis

The Pier W3 of the SFOBB was modeled using the ADINA [13] program and nonlinear analysis was performed using the displacement time histories. The displacement time histories in three directions are applied at the bottom of the Pier W3 and the response of the Pier W3 was studied to estimate the demand on Pier W3. One of the results, the displacement response at top of Pier W3, is shown in Figure 35.17f.

35.4 Response Spectrum Analysis

Response spectrum analysis is an approximate method of dynamic analysis that gives the maximum response (acceleration, velocity, or displacement) of an SDOF system with the same damping ratio, but with different natural frequencies, respond to a specified seismic excitation. Structural models with n degrees of freedom can be transformed to n single-degree systems and response spectra principles can be applied to systems with many degrees of freedom. For most ordinary bridges, a complete time history is not required. Because the design is generally based on the maximum earthquake response, response spectrum analysis is probably the most common method used in design offices to determine the maximum structural response due to transient loading. In this section, we will discuss basic procedures of response spectrum analysis for bridge structures.

35.4.1 Single-Mode Spectral Analysis

Single-mode spectral analysis is based on the assumption that earthquake design forces for structures respond predominantly in the first mode of vibration. This method is most suitable to regular linear elastic bridges to compute the forces and deformations, but is not applicable to irregular bridges (unbalanced spans, unequal stiffness in the columns, etc.) because higher modes of vibration affect the distribution of the forces and resulting displacements significantly. This method can be applied to both continuous and noncontinuous bridge superstructures in either the longitudinal or transverse direction. Foundation flexibility at the abutments can be included in the analysis.

Single-mode analysis is based on Rayleigh's energy method— an approximate method which assumes a vibration shape for a structure. The natural period of the structure is then calculated by equating the maximum potential and kinetic energies associated with the assumed shape. The inertial forces $p_e(x)$ are calculated using the natural period, and the design forces and displacements are then computed using static analysis. The detailed procedure can be described in the following steps:

1. Apply uniform loading p_o over the length of the structure and compute the corresponding static displacements $u_s(x)$. The structure deflection under earthquake loading, $u_s(x,t)$ is then approximated by the shape function, $u_s(x)$, multiplied by the generalized amplitude function, $u(t)$, which satisfies the geometric boundary conditions of the structural system. This dynamic deflection is shown as

$$u(x,t) = u_s(x)\ u(t) \tag{35.43}$$

2. Calculate the generalized parameters $\alpha, \beta,$ and γ using the following equations:

$$\alpha = \int u_s(x)\,dx \tag{35.44}$$

$$\beta = \int w(x)\,u_s(x)\;dx \tag{35.45}$$

$$\gamma = \int w(x)\left[u_s(x)\right]^2 dx \tag{35.46}$$

where $w(x)$ is the weight of the dead load of the bridge superstructure and tributary substructure.

3. Calculate the period T_n

$$T_n = 2\pi\sqrt{\frac{\gamma}{p_o g \alpha}} \tag{35.47}$$

where g is acceleration of gravity (mm/s²).

4. Calculate the static loading $p_e(x)$ which approximates the inertial effects associated with the displacement $u_s(x)$ using the ARS curve or the following equation [4]:

$$p_e(x) = \frac{\beta\, C_{sm}}{\gamma}\; w(x)\; u_s(x) \tag{35.48}$$

$$C_{sm} = \frac{1.2AS}{T_m^{\;2/3}} \tag{35.49}$$

where C_{sm} is the dimensionless elastic seismic response coefficient; A is the acceleration coefficient from the acceleration coefficient map; S is the dimensionless soil coefficient based on the soil profile type; T_n is the period of the structure as determined above; $p_e(x)$ is the intensity of the equivalent static seismic loading applied to represent the primary mode of vibration (N/mm).

5. Apply the calculated loading $p_e(x)$ to the structure as shown in the Figure 35.18 and compute the structure deflections and member forces.

This method is an iterative procedure, and the previous calculations are used as input parameters for the new iteration leading to a new period and deflected shape. The process is continued until the assumed shape matches the fundamental mode shape.

35.4.2 Uniform-Load Method

The uniform-load method is essentially an equivalent static method that uses the uniform lateral load to compute the effect of seismic loads. For simple bridge structures with relatively straight alignment, small skew, balanced stiffness, relatively light substructure, and with no hinges, the uniform-load method may be applied to analyze the structure for seismic loads. This method is not suitable for bridges with stiff substructures such as pier walls. This method assumes continuity of the structure and distributes earthquake force to all elements of the bridge and is based on the fundamental mode of vibration in either a longitudinal or transverse direction [5]. The period of

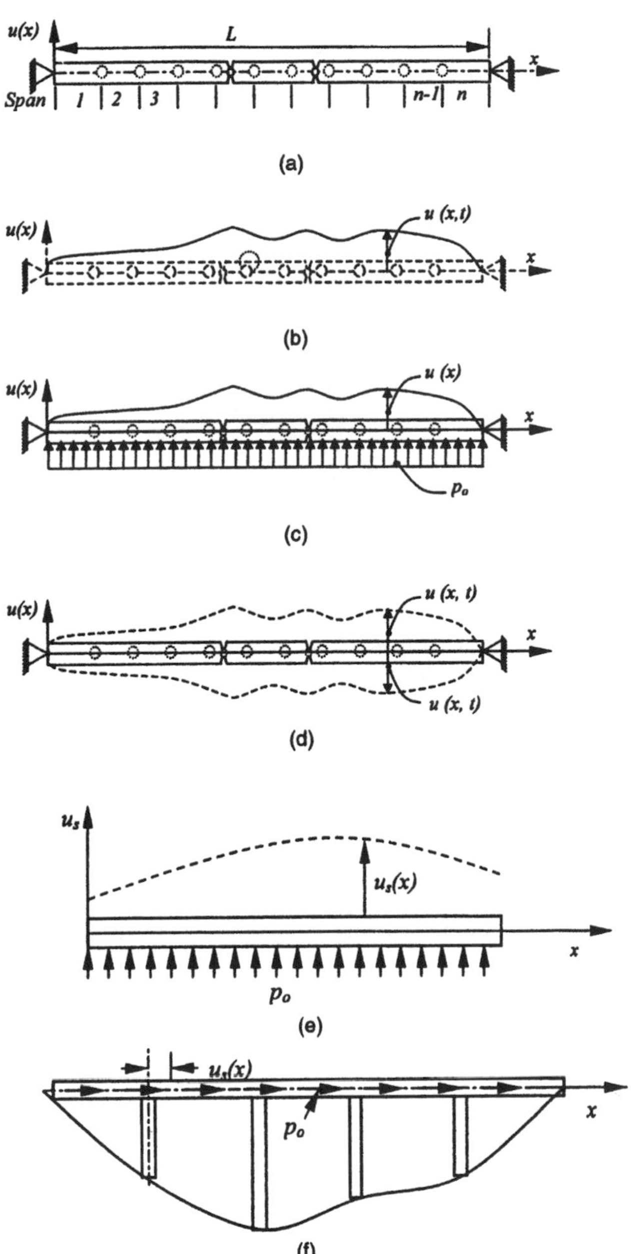

FIGURE 35.18 Single-mode spectral analysis method. (a) Plan view of a bridge subjected to transverse earthquake motion. (b) Displacement function describing the transverse position of the bridge deck. (c) Deflected shape due to uniform static loading. (d) Transverse free vibration of the bridge in assumed mode shape. (e) Transverse loading (f) longitudinal loading.

vibration is taken as that of an equivalent single mass–spring oscillator. The maximum displacement that occurs under the arbitrary uniform load is used to calculate the stiffness of the equivalent spring. The seismic elastic response coefficient C_{sm} or the ARS curve is then used to calculate the equivalent uniform seismic load, using which the displacements and forces are calculated. The following steps outline the uniform load method:

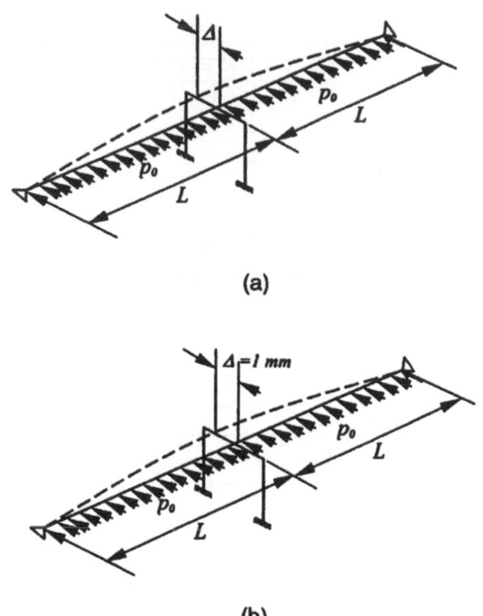

FIGURE 35.19 Structure idealization and deflected shape for uniform load method. (a) Structure idealization; (b) deflected shape with maximum displacement of 1 mm.

1. Idealize the structure into a simplified model and apply a uniform horizontal load (p_o) over the length of the bridge as shown in Figure 35.19. It has units of force/unit length and may be arbitrarily set equal to 1 N/mm.
2. Calculate the static displacements $u_s(x)$ under the uniform load p_o using static analysis.
3. Calculate the maximum displacement $u_{s,\max}$ and adjust it to 1 mm by adjusting the uniform load p_o.
4. Calculate bridge lateral stiffness K using the following equation:

$$K = \frac{p_o L}{u_{s,\max}} \tag{35.50}$$

where L is total length of the bridge (mm); and $u_{s,\max}$ is maximum displacement (mm).

5. Calculate the total weight W of the structure including structural elements and other relevant loads such as pier walls, abutments, columns, and footings, by

$$W = \int w(x)dx \tag{35.51}$$

where $w(x)$ is the nominal, unfactored dead load of the bridge superstructure and tributary substructure.

6. Calculate the period of the structure T_n using the following equation:

$$T_n = \frac{2\pi}{31.623}\sqrt{\frac{W}{gK}} \tag{35.52}$$

where g is acceleration of gravity (m/s^2).

7. Calculate the equivalent static earthquake force p_e using the ARS curve or using the following equation:

$$p_e = \frac{C_{sm}W}{L} \tag{35.53}$$

8. Calculate the structure deflections and member forces by applying p_e to the structure.

35.4.4 Multimode Spectral Analysis

The multimode spectral analysis method is more sophisticated than single-mode spectral analysis and is very effective in analyzing the response of more complex linear elastic structures to an earthquake excitation. This method is appropriate for structures with irregular geometry, mass, or stiffness. These irregularities induce coupling in three orthogonal directions within each mode of vibration. Also, for these bridges, several modes of vibration contribute to the complete response of the structure. A multimode spectral analysis is usually done by modeling the bridge structure consisting of three-dimensional frame elements with structural mass lumped at various locations to represent the vibration modes of the components. Usually, five elements per span are sufficient to represent the first three modes of vibration. A general rule of thumb is, to capture the ith mode of vibration, the span should have at least $(2i-1)$ elements. For long-span structures many more elements should be used to capture all the contributing modes of vibration. To obtain a reasonable response, the number of modes should be equal to at least three times the number of spans. This analysis is usually performed with a dynamic analysis computer program such as ADINA [13], GTSTRUDL [14], SAP2000 [15], ANSYS [16], and NASTRAN [17]. For bridges with outrigger bents, C-bents, and single column bents, rotational moment of inertia of the superstructure should be included. Discontinuities at the hinges and abutments should be included in the model. The columns and piers should have intermediate nodes at quarter points in addition to the nodes at the ends of the columns.

By using the programs mentioned above, frequencies, mode shapes, member forces, and joint displacements can be computed. The following steps summarize the equations used in the multimode spectral analysis [5].

1. Calculate the dimensionless mode shapes $\{\phi_i\}$ and corresponding frequencies ω_i by

$$\left[[\mathbf{K}] - \omega^2 [\mathbf{M}]\right]\{u\} = 0 \tag{35.54}$$

where

$$u_i = \sum_{j=1}^{n} \phi_j y_j = \Phi y_i \tag{35.55}$$

y_j = modal amplitude of jth mode; ϕ_j = shape factor of jth mode; Φ = mode-shape matrix. The periods for i^{th} mode can then be calculated by

$$T_i = \frac{2\pi}{\omega_i} \quad (i = 1, 2, \ldots, n) \tag{35.56}$$

2. Determine the maximum absolute mode amplitude for the entire time history is given by

$$Y_i(t)_{max} = \frac{T_i^2 S_a(\xi_i, T_i)}{4\pi^2} \frac{\{\phi_i\}^T [\mathbf{M}]\{B\}\ddot{u}_g}{\{\phi_i\}^T \quad [\mathbf{M}]\{\phi_i\}} \tag{35.57}$$

where $S_a(\xi_i, T_i) = gC_{sm}$ is the acceleration response spectral value; C_{sm} is the elastic seismic response coefficient for mode $m = 1.2AS/T_n^{2/3}$; A is the acceleration coefficient from the acceleration coefficient map; S is the dimensionless soil coefficient based on the soil profile type; T_n is the period of the nth mode of vibration.

3. Calculate the value of any response quantity $Z(t)$ (shear, moment, displacement) using the following equation:

$$Z(t) = \sum_{i=1}^{n} A_i Y_i(t) \tag{35.58}$$

where coefficients A_i are functions of mode shape matrix (Φ) and force displacement relationships.

4. Compute the maximum value of $Z(t)$ during an earthquake using the mode combination methods described in the next section.

Modal Combination Rules

The mode combination method is a very useful tool for analyzing bridges with a large number of degrees of freedom. In a linear structural system, maximum response can be estimated by mode combination after calculating natural frequencies and mode shapes of the structure using free vibration analysis. The maximum response cannot be computed by adding the maximum response of each mode because different modes attain their maximum values at different times. The absolute sum of the individual modal contributions provides an upper bound which is generally very conservative and not recommended for design. There are several different empirical or statistical methods available to estimate the maximum response of a structure by combining the contributions of different modes of vibrations in a spectral analysis. Two commonly used methods are the square root of sum of squares (SRSS) and the complete quadratic combination (CQC).

For an undamped structure, the results computed using the CQC method are identical to those using the SRSS method. For structures with closely spaced dominant mode shapes, the CQC method is precise whereas SRSS estimates inaccurate results. Closely spaced modes are those within 10% of each other in terms of natural frequency. The SRSS method is suitable for estimating the total maximum response for structures with well-spaced modes. Theoretically, all mode shapes must be included to calculate the response, but fewer mode shapes can be used when the corresponding mass participation is over 85% of the total structure mass. In general, the factors considered to determine the number of modes required for the mode combination are dependent on the structural characteristics of the bridge, the spatial distribution, and the frequency content of the earthquake loading. The following list [14] summarizes several commonly used mode combination methods to compute the maximum total response. The variable Z represents the maximum value of some response quantity (displacement, shear, etc.), Z_i is the peak value of that quantity in the ith mode, and N is the total number of contributing modes.

1. *Absolute Sum*: The absolute sum is sum of the modal contributions:

$$Z = \sum_{i=1}^{N} |Z_i| \tag{35.59}$$

2. *SRSS or Root Mean Square (RMS) Method*: This method computes the maximum by taking the square root of sum of squares of the modal contributions:

$$Z = \left[\sum_{i=1}^{N} Z_i^2 \right]^{1/2} \tag{35.60}$$

3. *Peak Root Mean Square (PRMS)*: Absolute value of the largest modal contribution is added to the root mean square of the remaining modal contributions:

$$Z_j = |\max\ Z_i| \tag{35.61}$$

$$Z = \left[\sum_{i=1}^{N} Z_i^2 \right]^{1/2} + Z_j \quad \text{with} \quad i \neq j \tag{35.62}$$

4. *CQC*: Cross correlations between all modes are considered:

$$Z = \left[\sum_{i=1}^{N} \sum_{j=1}^{N} Z_i \ \rho_{ij} \ Z_j \right]^{1/2} \tag{35.63}$$

$$\rho_{ij} = \frac{8\sqrt{\xi_i \xi_j}\left(\xi_i + r\xi_j\right) r^{3/2}}{\left(1-r^2\right)^2 + 4\xi_i \xi_j r\left(1+r^2\right) + 4\left(\xi_i^2 + \xi_j^2\right) r^2} \tag{35.64}$$

where

$$r = \frac{\omega_j}{\omega_i} \tag{35.65}$$

5. *Nuclear Regulatory Commission Grouping Method*: This method is similar to RMS method with additional accounting for groups of modes whose frequencies are within 10%.

$$Z = \left[\sum_{i=1}^{N} Z_i^2 + \sum_{g=1}^{G} \sum_{n=s}^{e} \sum_{m=s}^{e} \left| Z_n^g \times Z_m^g \right| \right]^{1/2} \quad n \neq m \tag{35.66}$$

where G is number of groups; s is mode shape number where the gth group starts; e is mode shape number where the gth group ends; and Z_i^g is the ith modal contribution in the gth group.

6. *Nuclear Regulatory Commission Ten Percent Method*: This method is similar to the RMS method with additional accounting for all modes whose frequencies are within 10%.

$$Z = \left[\sum_{i=1}^{N} Z_i^2 + 2 \sum \left| Z_n Z_m \right| \right]^{1/2} \tag{35.67}$$

The additional terms must satisfy

$$\frac{\omega_n - \omega_m}{\omega_m} \leq 0.1 \quad \text{for} \quad 0.1 \leq m \leq n \leq N \tag{35.68}$$

7. *Nuclear Regulatory Commission Double Sum Method*: This method is similar to the CQC method.

$$Z = \left[\sum_{i=1}^{N} \sum_{j=1}^{N} \left| Z_i Z_j \right| \ \varepsilon_{ij} \right]^{1/2} \tag{35.69}$$

$$\varepsilon_{ij} = \left[1 + \left\{ \frac{\left(\omega'_i - \omega'_j \right)}{\left(\xi'_i \omega_i + \xi'_j \omega_j \right)} \right\} \right]^{-1} \tag{35.70}$$

$$\omega'_i = \omega \left[1 - \xi_i^2 \right]^{1/2} \tag{35.71}$$

$$\xi'_i = \xi_i + \frac{2}{t_d \omega_i} \tag{35.72}$$

where t_d is the duration of support motion.

Combination Effects

Effects of ground motions in two orthogonal horizontal directions should be combined while designing bridges with simple geometric configurations. For bridges with long spans, outrigger bents, and with cantilever spans, or where effects due to vertical input are significant, vertical input should be included in the design along with two orthogonal horizontal inputs. When bridge structures are analyzed independently along each direction using response spectra analysis, then responses are combined either using methods, such as the SRSS combination rule as mentioned in the previous section, or using the alternative method described below. For structures designed using equivalent static analysis or modal analysis, seismic effects should be determined using the following alternative method for the following load cases:

1. *Seismic load case* 1: 100% Transverse + 30% Longitudinal + 30% Vertical
2. *Seismic load case* 2: 30% Transverse + 100% Longitudinal + 30% Vertical
3. *Seismic load case* 3: 30% Transverse + 30% Longitudinal + 100% Vertical

For structures designed using time-history analysis, the structure response is calculated using the input motions applied in orthogonal directions simultaneously. Where this is not feasible, the above alternative procedure can be used to combine the independent responses.

35.4.4 Multiple-Support Response Spectrum Method

Records from recent earthquakes indicate that seismic ground motions can significantly vary at different support locations for multiply supported long structures. When different ground motions are applied at various support points of a bridge structure, the total response can be calculated by superposition of responses due to independent support input. This analysis involves combination of dynamic response from single-input and pseudo-static response resulting from the motion of the supports relative to each other. The combination effects of dynamic and pseudo-static forces due to multiple support excitation on a bridge depend on the structural configuration of the bridge and the ground motion characteristics. Recently, Kiureghian et al. [7] presented a comprehensive study on the multiple-support response spectrum (MSRS) method based on fundamental principles of stationary random vibration theory for seismic analysis of multiply supported structures which accounts for the effects of variability between the support motions. Using the MSRS combination rule, the response of a linear structural system subjected to multiple support excitation can be computed directly in terms of conventional response spectra at the support degrees of freedom and a coherency function describing the spatial variability of the ground motion. This method accounts for the three important effects of ground motion spatial variability, namely, the incoherence effects, the wave passage effect, and the site response effect. These three components of ground motion spatial variability can strongly influence the response of multiply supported bridges and may amplify or deamplify the response by one order of magnitude. Two important limitations of this method are nonlinearities in the bridge structural components and/or connections and the effects of soil–structure interaction. This method is an efficient, accurate, and versatile solution and requires less computational time than a true time history analysis. Following are the steps that describe the MSRS analysis procedure.

1. *Determine the necessity of variable support motion analysis*: Three factors that influence the response of the structure under multiple support excitation are the distance between the supports of the structure, the rate of variability of the local soil conditions, and the stiffness of the structure. The first factor, the distance between the supports, influences the incoherence and wave passage effects. The second factor, the rate of variability of the local conditions, influences the site response. The third factor, the stiffness of the superstructure, plays an important role in determining the necessity of variable-support motion analysis. Stiff structures such as box-girder bridges may generate large internal forces under variable support motion, whereas flexible structures such as suspension bridges easily conform to the variable support motion.
2. *Determine the frequency response function for each support location.* Programs such as SHAKE [18] can be used to develop these functions using borehole data and time-domain site response analysis. Response spectra plots, peak ground displacements in three orthogonal directions for each support location, and a coherency function for each pair of degrees of freedom are required to perform the MSRS analysis. The comprehensive report by Kiureghian [7] provides all the formulas required to account for the effect of nonlinearity in the soil behavior and the site frequency involving the depth of the bedrock.
3. *Calculate the Structural Properties*: such as effective modal frequencies, damping ratios, influence coefficients and effective modal participation factors (ω_i, ξ_i, a_k, and b_{ki}) are to be computed externally and provided as input.
4. *Determine the response spectra plots, peak ground displacements in three directions, and a coherency function for each pair of support degrees of freedom required to perform MSRS analysis*: Three components of the coherency function are incoherence, wave passage effect, and site response effect. Analysis by an array of recordings is used to determine the incoherence component. The models for this empirical method are widely available [19]. Parameters such as shear wave velocity, the direction of propagation of seismic waves, and the angle of incidence are used to calculate the wave passage effect. The frequency response function determined in the previous steps is used to calculate the site response component.

35.5 Inelastic Dynamic Analysis

35.5.1 Equations of Motion

Inelastic dynamic analysis is usually performed for the safety evaluation of important bridges to determine the inelastic response of bridges when subjected to design earthquake ground motions.

Inelastic dynamic analysis provides a realistic measure of response because the inelastic model accounts for the redistribution of internal actions due to the nonlinear force displacement behavior of the components [20–25]. Inelastic dynamic analysis considers nonlinear damping, stiffness, load deformation behavior of members including soil, and mass properties. A step-by-step integration procedure is the most powerful method used for nonlinear dynamic analysis. One important assumption of this procedure is that acceleration varies linearly while the properties of the system such as damping and stiffness remain constant during the time interval. By using this procedure, a nonlinear system is approximated as a series of linear systems and the response is calculated for a series of small equal intervals of time Δt and equilibrium is established at the beginning and end of each interval.

The accuracy of this procedure depends on the length of the time increment Δt. This time increment should be small enough to consider the rate of change of loading $p(t)$, nonlinear damping and stiffness properties, and the natural period of the vibration. An SDOF system and its characteristics are shown in the Figure 35.20. The characteristics include spring and damping forces, forces acting on mass of the system, and arbitrary applied loading. The force equilibrium can be shown as

$$f_i(t) + f_d(t) + f_s(t) = p(t) \tag{35.73}$$

and the incremental equations of motion for time t can be shown as

$$m\,\Delta\ddot{u}(t) + c(t)\,\Delta\dot{u}(t) + k(t)\Delta u(t) = \Delta p(t) \tag{35.74}$$

Current damping $f_d(t)$, elastic forces $f_s(t)$ are then computed using the initial velocity $\dot{u}(t)$, displacement values $u(t)$, nonlinear properties of the system, damping $c(t)$, and stiffness $k(t)$ for that interval. New structural properties are calculated at the beginning of each time increment based on the current deformed state. The complete response is then calculated by using the displacement and velocity values computed at the end of each time step as the initial conditions for the next time interval and repeating until the desired time.

35.5.2 Modeling Considerations

A bridge structural model should have sufficient degrees of freedom and proper selection of linear/nonlinear elements such that a realistic response can be obtained. Nonlinear analysis is usually preceded by a linear analysis as a part of a complete analysis procedure to capture the physical and mechanical interactions of seismic input and structure response. Output from the linear response solution is then used to predict which nonlinearities will affect the response significantly and to model them appropriately. In other words, engineers can justify the effect of each nonlinear element introduced at the appropriate locations and establish the confidence in the nonlinear analysis. While discretizing the model, engineers should be aware of the trade-offs between the accuracy, computational time, and use of the information such as the regions of significant geometric and material nonlinearities. Nonlinear elements should have material behavior to simulate the hysteresis relations under reverse cyclic loading observed in the experiments.

The general issues in modeling of bridge structures include geometry, stiffness, mass distribution, and boundary conditions. In general, abutments, superstructure, bent caps, columns and pier walls, expansion joints, and foundation springs are the elements included in the structural model. The mass distribution in a structural model depends on the number of elements used to represent the bridge components. The model must be able to simulate the vibration modes of all components contributing to the seismic response of the structure.

Superstructure: Superstructure and bent caps are usually modeled using linear elastic three-dimensional beam elements. Detailed models may require nonlinear beam elements.

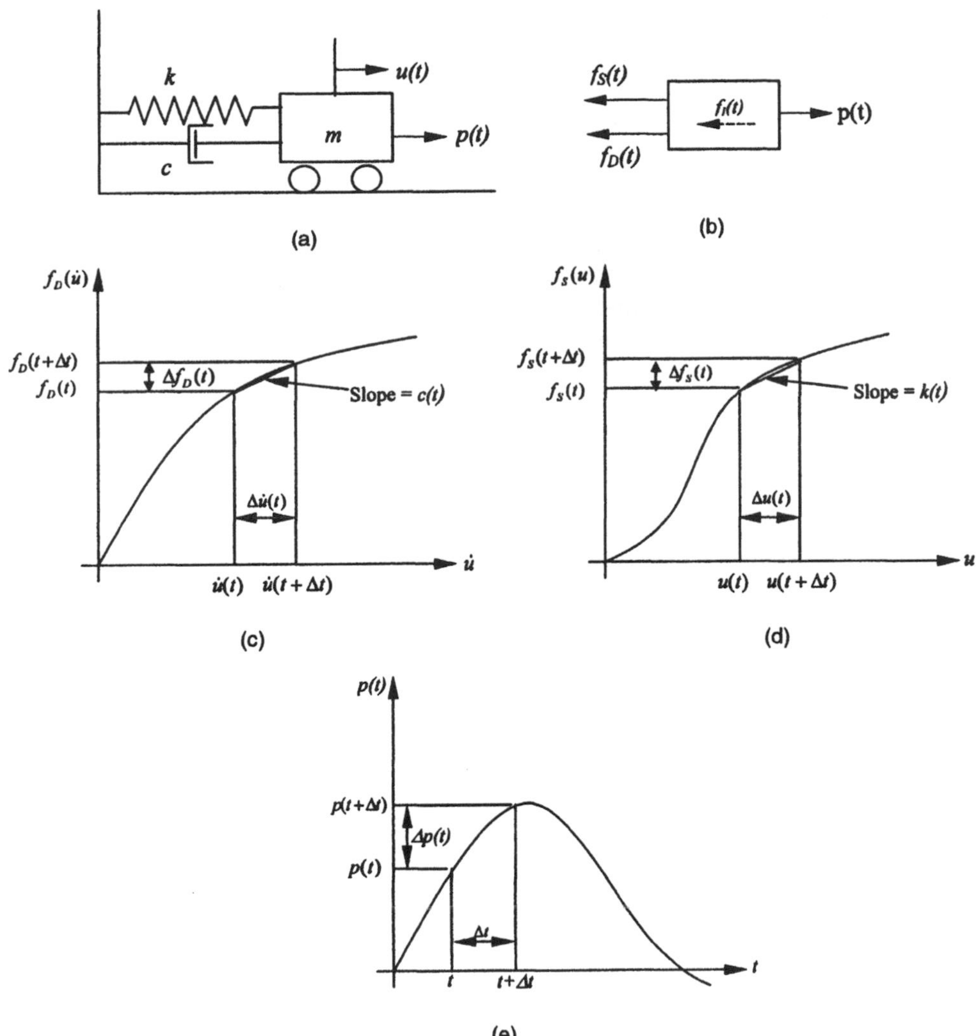

FIGURE 35.20 Definition of a nonlinear dynamic system. (a) Basis SDOF structure; (b) force equilibrium; (c) nonlinear damping; (d) nonlinear stiffness; (e) applied load.

Columns and pier walls: Columns and pier walls are usually modeled using nonlinear beam elements having response properties with a yield surface described by the axial load and biaxial bending. Some characteristics of the column behavior include initial stiffness degradation due to concrete cracking, flexural yielding at the fixed end of the column, strain hardening, pinching at the point of load reversal. Shear actions can be modeled using either linear or nonlinear load deformation relationships for columns. For both columns and pier walls, torsion can be modeled with linear elastic properties. For out-of-plane loading, flexural response of a pier wall is similar to that of columns, whereas for in-plane loading the nonlinear behavior is usually shear action.

Expansion joints: Expansion joints can be modeled using gap elements that simulate the nonlinear behavior of the joint. The variables include initial gap, shear capacity of the joint, and nonlinear load deformation characteristics of the gap.

Foundations and abutments: Foundations are typically modeled using nonlinear spring elements to represent the translational and rotational stiffness of the foundations to represent the expected behavior during a design earthquake. Abutments are modeled using nonlinear spring and gap elements to represent the soil action, stiffness of the pile groups, and gaps at the seat.

35.6 Summary

This chapter has presented the basic principles and methods of dynamic analysis for the seismic design of bridges. Response spectrum analysis — the SDOF or equivalent SDOF-based equivalent static analysis — is efficient, convenient, and most frequently used for ordinary bridges with simple configurations. Elastic dynamic analysis is required for bridges with complex configurations. A multisupport response spectrum analysis recently developed by Kiureghian et al. [7] using a lumped-mass beam element mode may be used in lieu of an elastic time history analysis.

Inelastic response spectrum analysis is a useful concept, but the current approaches apply only to SDOF structures. An actual nonlinear dynamic time history analysis may be necessary for some important and complex bridges, but linearized dynamic analysis (dynamic secant stiffness analysis) and inelastic static analysis (static push-over analysis) (Chapter 36) are the best possible alternatives [8] for the most bridges.

References

1. Clough, R.W. and Penzien, J., *Dynamics of Structures,* 2nd ed., McGraw-Hill, New York, 1993.
2. Caltrans, Guidelines for Generation of Response — Spectrum-Compatible Rock Motion Time History for Application to Caltrans Toll Bridge Seismic Retrofit Projects, the Caltrans Seismic Advisor Board ad hoc Committee on Soil–Foundation-Structure Interaction, California Department of Transportation, Sacramento, 1996.
3. Caltrans, Guidelines for Performing Site Response Analysis to Develop Seismic Ground Motions for Application to Caltrans Toll Bridge Seismic Retrofit Projects, The Caltrans Seismic Advisor Board ad hoc Committee on Soil–Foundation-Structure Interaction, California Department of Transportation, Sacramento, 1996.
4. AASHTO, *LRFD Bridge Design Specifications,* American Association of State Highway and Transportation Officials, Washington, D.C., 1994.
5. AASHTO, *LRFD Bridge Design Specifications,* 1996 Interim Version, American Association of State Highway and Transportation Officials, Washington, D.C., 1996.
6. Wilson, E. L, der Kiureghian, A., and Bayom, E. P., A replacement for SSRS method in seismic analysis, *J. Earthquake Eng. Struct. Dyn.*, 9, 187, 1981.
7. Kiureghian, A. E., Keshishian, P., and Hakobian, A., Multiple Support Response Spectrum Analysis of Bridges Including the Site-Response Effect and the MSRS Code, Report No. UCB/EERC-97/02, University of California, Berkeley, 1997.
8. Powell, G. H., Concepts and Principles for the Application of Nonlinear Structural Analysis in Bridge Design, Report No.UCB/SEMM-97/08, Department of Civil Engineering, University of California, Berkeley, 1997.
9. Chmielewski, T., Kratzig, W. B., Link, M., Meskouris, K., and Wunderlich, W., Phenomena and evaluation of dynamic structural responses, in *Dynamics of Civil Engineering Structures,* W. B. Kratzig and H.-J. Niemann, Eds., A.A. Balkema, Rotterdams, 1996.
10. Chopra, A.K., *Dynamics of Structures,* Prentice-Hall, Englewood Cliffs, NJ, 1995.
11. Lindeburg, M., *Seismic Design of Building Structures: A Professional's Introduction to Earthquake Forces and Design Details,* Professional Publications, Belmont, CA, 1998.
12. Caltrans, Bridge Design Specifications, California Department of Transportation, Sacramento, CA, 1991.
13. ADINA, *User's Guide,* Adina R&D, Inc., Watertown, MA, 1995.
14. GTSTRUDL, *User's Manual,* Georgia Institute of Technology, Atlanta, 1996.
15. SAP2000, *User's Manual,* Computers and Structures Inc., Berkeley, CA, 1998.
16. ANSYS, *User's Manual,* Vols. 1 and 2, Version 4.4, Swanson Analysis Systems, Inc., Houston, TX, 1989.
17. NASTRAN, *User's Manual,* MacNeil Schwindler Corporation, Los Angeles, CA.

18. Idriss, I. M., Sun J. I., and Schnabel, P. B., User's manual for SHAKE91: a computer program for conducting equivalent linear seismic response analyses of horizontally layered soil deposits, *Report of Center for Geotechnical Modeling*, Department of Civil and Environmental Engineering, University of California at Davis, 1991.
19. Abrahamson, N. A., Schneider, J. F., and Stepp, J. C., Empirical spatial coherency functions for application to soil-structure interaction analysis, *Earthquake Spectra*, 7, 1991.
20. Imbsen & Associates, *Seismic Design of Highway Bridges*, Sacramento, CA, 1992.
21. Priestly, M. J. N., Seible, F., and Calvi, G. M., *Seismic Design and Retrofit of Bridges*, John Wiley & Sons, New York, 1996.
22. Bathe, K.-J., *Finite Element Procedures in Engineering Analysis*, 2nd ed., Prentice-Hall, Englewood Cliffs, NJ, 1996.
23. ATC 32, Improved Seismic Design Criteria for California Bridges: Provisional Recommendations, Applied Technology Council, 1996.
24. Buchholdt, H. A., *Structural Dynamics for Engineers*, Thomas Telford, London, 1997.
25. Paz, M., *Structural Dynamics — Theory and Computation*, 3rd ed., Van Nostrand Reinhold, New York, 1991.

18. [illegible], [illegible] and [illegible], [illegible] manual for SHAKE[illegible] [illegible] conducting equivalent linear seismic response analyses of horizontally [illegible] of [illegible] Center for Geotechnical [illegible] Univ. of California, [illegible] 1991.
19. [illegible] applications [illegible]
20. [illegible]
21. [illegible] New York, [illegible]
22. [illegible] Cliffs, NJ, [illegible]
23. [illegible] and [illegible]
[illegible]
[illegible]

36

Nonlinear Analysis of Bridge Structures

Mohammed Akkari
California Department of Transportation

Lian Duan
California Department of Transportation

36.1 Introduction

In recent years, nonlinear bridge analysis has gained a greater momentum because of the need to assess inelastic structural behavior under seismic loads. Common seismic design philosophies for ordinary bridges allow some degree of damage without collapse. To control and evaluate damage, a postelastic nonlinear analysis is required. A nonlinear analysis is complex and involves many simplifying assumptions. Engineers must be familiar with those complexities and assumptions to design bridges that are safe and economical.

Many factors contribute to the nonlinear behavior of a bridge. These include factors such as material inelasticity, geometric or second-order effects, nonlinear soil–foundation–structure interaction, gap opening and closing at hinges and abutment locations, time-dependent effects due to concrete creep and shrinkage, etc. The subject of nonlinear analysis is extremely broad and cannot be covered in detail in this single chapter. Only material and geometric nonlinearities as well as

0-8493-7434-0/00/$0.00+$.50

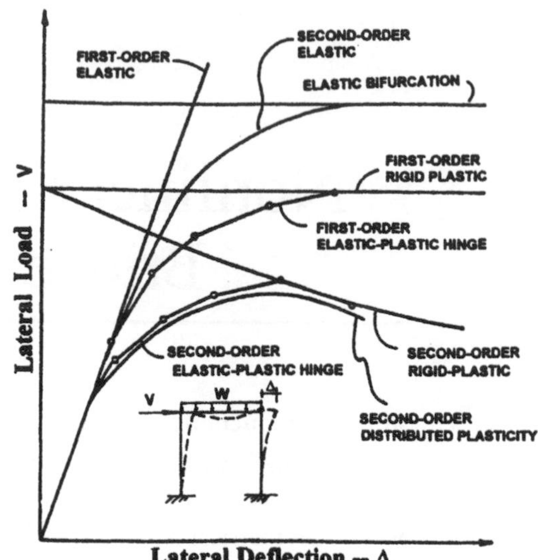

FIGURE 36.1 Lateral load–displacement curves of a frame.

some of the basic formulations of nonlinear static analysis with their practical applications to seismic bridge design will be presented here. The reader is referred to the many excellent papers, reports, and books [1-8] that cover this type of analysis in more detail.

In this chapter, some general guidelines for nonlinear static analysis are presented. These are followed by discussion of the formulations of geometric and material nonlinearities for section and frame analysis. Two examples are given to illustrate the applications of static nonlinear push-over analysis in bridge seismic design.

36.2 Analysis Classification and General Guidelines

Engineers use structural analysis as a fundamental tool to make design decisions. It is important that engineers have access to several different analysis tools and understand their development assumptions and limitations. Such an understanding is essential to select the proper analysis tool to achieve the design objectives.

Figure 36.1 shows lateral load vs. displacement curves of a frame using several structural analysis methods. Table 36.1 summarizes basic assumptions of those methods. It can be seen from Figure 36.1 that the first-order elastic analysis gives a straight line and no failure load. A first-order inelastic analysis predicts the maximum plastic load-carrying capacity on the basis of the undeformed geometry. A second-order elastic analysis follows an elastic buckling process. A second-order inelastic analysis traces load–deflection curves more accurately.

36.2.1 Classifications

Structural analysis methods can be classified on the basis of different formulations of equilibrium, the constitutive and compatibility equations as discussed below.

Classification Based on Equilibrium and Compatibility Formulations

First-order analysis: An analysis in which equilibrium is formulated with respect to the undeformed (or original) geometry of the structure. It is based on small strain and small displacement theory.

TABLE 36.1 Structural Analysis Methods

Methods		Features: Constitutive Relationship	Features: Equilibrium Formulation	Features: Geometric Compatibility
First-order	Elastic Rigid–plastic Elastic–plastic hinge Distributed plasticity	Elastic Rigid plastic Elastic perfectly plastic Inelastic	Original undeformed geometry	Small strain and small displacement
Second-order	Elastic Rigid–plastic Elastic–plastic hinge Distributed plasticity	Elastic Rigid plastic Elastic perfectly plastic Inelastic	Deformed structural geometry (P-Δ and P-δ)	Small strain and moderate rotation (displacement may be large)
True large displacement	Elastic Inelastic	Elastic Inelastic	Deformed structural geometry	Large strain and large deformation

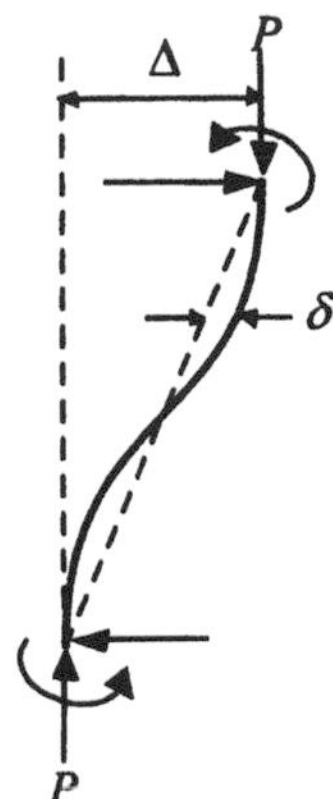

FIGURE 36.2 Second–order effects.

Second-order analysis: An analysis in which equilibrium is formulated with respect to the deformed geometry of the structure. A second-order analysis usually accounts for the P-Δ effect (influence of axial force acting through displacement associated with member chord rotation) and the P-δ effect (influence of axial force acting through displacement associated with member flexural curvature) (see Figure 36.2). It is based on small strain and small member deformation, but moderate rotations and large displacement theory.

True large deformation analysis: An analysis for which large strain and large deformations are taken into account.

Classification Based on Constitutive Formulation

Elastic analysis: An analysis in which elastic constitutive equations are formulated.

Inelastic analysis: An analysis in which inelastic constitutive equations are formulated.

Rigid–plastic analysis: An analysis in which elastic rigid–plastic constitutive equations are formulated.

Elastic–plastic hinge analysis: An analysis in which material inelasticity is taken into account by using concentrated "zero-length" plastic hinges.

Distributed plasticity analysis: An analysis in which the spread of plasticity through the cross sections and along the length of the members are modeled explicitly.

Classification Based on Mathematical Formulation

Linear analysis: An analysis in which equilibrium, compatibility, and constitutive equations are linear.

Nonlinear analysis: An analysis in which some or all of the equilibrium, compatibility, and constitutive equations are nonlinear.

36.2.4 General Guidelines

The following guidelines may be useful in analysis type selection:

- A first-order analysis may be adequate for short- to medium-span bridges. A second-order analysis should always be encouraged for long-span, tall, and slender bridges. A true large displacement analysis is generally unnecessary for bridge structures.
- An elastic analysis is sufficient for strength-based design. Inelastic analyses should be used for displacement-based design.
- The bowing effect (effect of flexural bending on member's axial deformation), the Wagner effect (effect of bending moments and axial forces acting through displacements associated with the member twisting), and shear effects on solid-webbed members can be ignored for most of bridge structures.
- For steel nonlinearity, yielding must be taken into account. Strain hardening and fracture may be considered. For concrete nonlinearity, a complete strain–stress relationship (in compression up to the ultimate strain) should be used. Concrete tension strength can be neglected.
- Other nonlinearities, most importantly, soil–foundation–structural interaction, seismic response modification devices (dampers and seismic isolations), connection flexibility, gap close and opening should be carefully considered.

36.3 Geometric Nonlinearity Formulation

Geometric nonlinearities can be considered in the formulation of member stiffness matrices. The general force–displacement relationship for the prismatic member as shown in Figure 36.3 can be expressed as follows:

$$\{F\} = [K]\{D\} \tag{36.1}$$

where $\{F\}$ and $\{D\}$ are force and displacement vectors and $[K]$ is stiffness matrix.

For a two-dimensional member as shown in Figure 36.3a

$$\{F\} = \{P_{1a}, F_{2a}, M_{3a}, P_{1b}, F_{2b}, M_{3b}\}^T \tag{36.2}$$

$$\{D\} = \{u_{1a}, u_{2a}, \theta_{3a}, u_{1b}, u_{2b}, \theta_{3b}\}^T \tag{36.3}$$

For a three-dimensional member as shown in Figure 36.3b

$$\{F\} = \{P_{1a}, F_{2a}, F_{3a}, M_{1a}, M_{2a}, M_{3a}, P_{1b}, F_{2b}, F_{3b}, M_{1b}, M_{2b}, M_{3b}\}^T \tag{36.4}$$

$$\{D\} = \{u_{1a}, u_{2a}, u_{3a}, \theta_{1a}, \theta_{2a}, \theta_{3a}, u_{1b}, u_{2b}, u_{3b}, \theta_{1b}, \theta_{2b}, \theta_{3b}\}^T \tag{36.5}$$

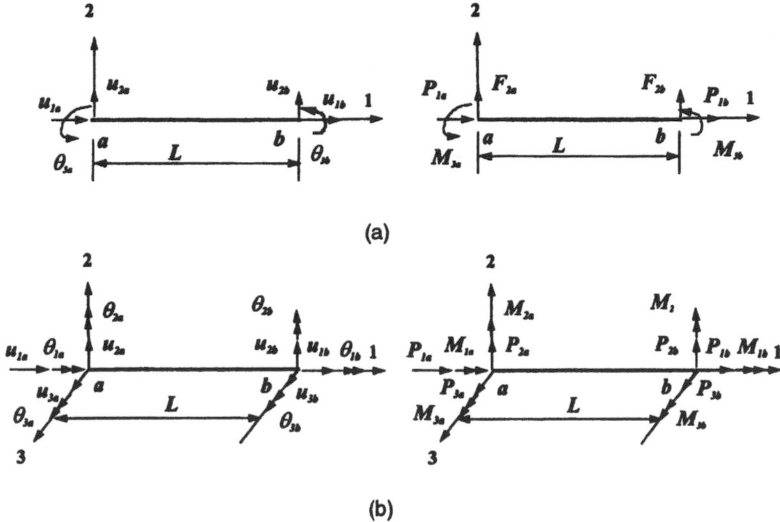

FIGURE 36.3 Degrees of freedom and nodal forces for a framed member. (a) Two-dimensional and (b) three-dimensional members.

Two sets of formulations of stability function-based and finite-element-based stiffness matrices are presented in the following section.

36.3.1 Two-Dimensional Members

For a two-dimensional prismatic member as shown in Figure 36.3a, the stability function-based stiffness matrix [9] is as follows:

$$[K] = \begin{bmatrix} \frac{AE}{L} & 0 & 0 & -\frac{AE}{L} & 0 & 0 \\ & \frac{12EI}{L^3}\phi_1 & \frac{-6EI}{L^2}\phi_2 & 0 & \frac{-12EI}{L^3}\phi & \frac{-6EI}{L^2}\phi_2 \\ & & 4\phi_3 & 0 & \frac{6EI}{L^2}\phi_2 & 2\phi_4 \\ & & & \frac{AE}{L} & 0 & 0 \\ & & & & \frac{12EI}{L^3}\phi & \frac{6EI}{L^2}\phi_2 \\ & & & & & 4\phi_3 \end{bmatrix} \tag{36.6}$$

where A is cross section area; E is the material modulus of elasticity; L is the member length; ϕ_1, ϕ_2, ϕ_3, and ϕ_4 can be expressed by stability equations and are listed in Table 36.2. Alternatively, ϕ_i functions can also be expressed in the power series derived from the analytical solutions [10] as listed in Table 36.3.

Assuming polynomial displacement functions, the finite-element-based stiffness matrix [11,12] has the following form:

$$[K] = [K_e] + [K_g] \tag{36.7}$$

where $[K_e]$ is the first-order conventional linear elastic stiffness matrix and $[K_g]$ is the geometric stiffness matrix which considers the effects of axial load on the bending stiffness of a member.

equilibrium state by an external force or displacement. Once the system is disturbed, the system vibrates without any external input. Thus, the equation of motion for free vibration can be obtained by setting $\ddot{u}_g$ to zero in Eq. (35.4) and is given by

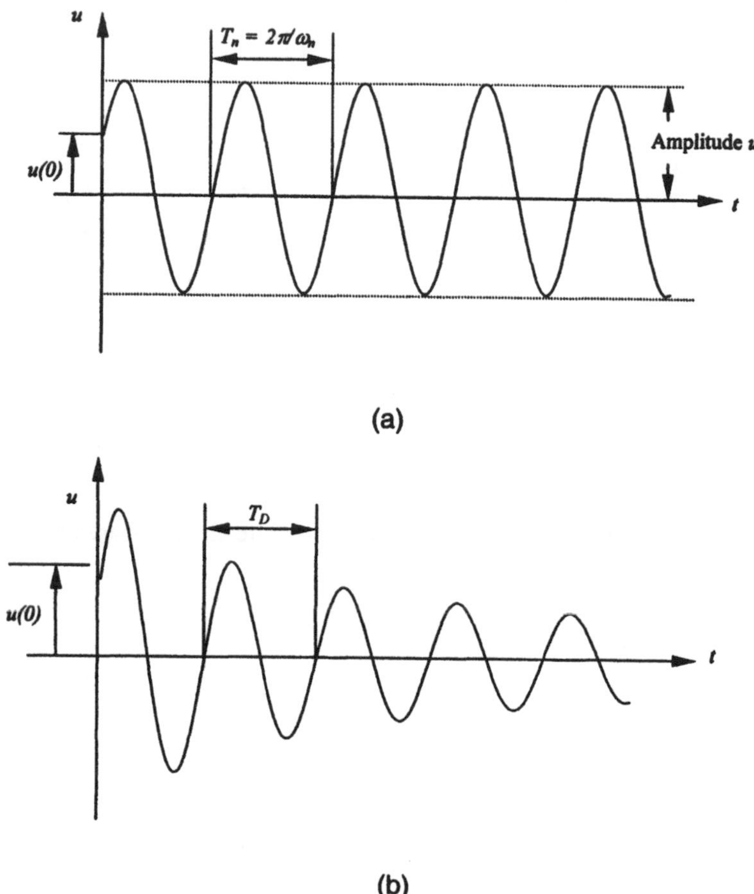

FIGURE 35.5 Typical response of an SDOF system. (a) Undamped; (b) damped.

$$m\ddot{u} + c\dot{u} + ku = 0 \tag{35.5}$$

Dividing the Equation (35.5) by its mass m will result in

$$\ddot{u} + \left(\frac{c}{m}\right)\dot{u} + \left(\frac{k}{m}\right)u = 0 \tag{35.6}$$

$$\ddot{u} + 2\xi\omega_n + \omega_n^2 u = 0 \tag{35.7}$$

where $\omega_n = \sqrt{k/m}$ the natural circular frequency of vibration or the undamped frequency; $\xi = c/c_{cr}$ the damping ratio; $c_{cr} = 2m\omega_n = 2\sqrt{km} = 2k/\omega_n$ the critical damping coefficient.

Figure 35.5a shows the response of a typical idealized, *undamped* SDOF system. The time required for the SDOF system to complete one cycle of vibration is called the natural period of vibration (T_n) of the system and is given by

$$
\left[K_g\right] = \mp\frac{P}{L}\begin{bmatrix}
0 & 0 & 0 & 0 & 0 & 0 \\
 & \frac{6}{5} & \frac{-L}{10} & 0 & \frac{-6}{5} & \frac{-L}{10} \\
 & & \frac{2L^2}{15} & 0 & \frac{L}{10} & -\frac{L^2}{30} \\
 & & & 0 & 0 & 0 \\
 & \text{sym.} & & & \frac{6}{5} & \frac{L}{10} \\
 & & & & & \frac{2L^2}{15}
\end{bmatrix} \tag{36.9}
$$

It is noted [13] that Eqs. (36.8) and (36.9) exactly coincide with the stability function-based stiffness matrix when taken only the first two terms of the Taylor series expansion in Eq. (36.6).

36.3.2 Three-Dimensional Members

For a three-dimensional frame member as shown in Figure 36.3b, the stability function-based stiffness matrix has the following form [14]:

$$
[K] = \begin{bmatrix}
\phi_{s1} & 0 & 0 & 0 & 0 & 0 & -\phi_{s1} & 0 & 0 & 0 & 0 & 0 \\
 & \phi_{s7} & 0 & 0 & 0 & \phi_{s6} & 0 & -\phi_{s7} & 0 & 0 & 0 & \phi_{s6} \\
 & & \phi_{s9} & 0 & -\phi_{s8} & 0 & 0 & 0 & -\phi_{s9} & 0 & -\phi_{s8} & 0 \\
 & & & \frac{GJ}{L} & 0 & 0 & 0 & 0 & 0 & -\frac{GJ}{L} & 0 & 0 \\
 & & & & \phi_{s4} & 0 & 0 & 0 & \phi_{s8} & 0 & \phi_{s5} & 0 \\
 & & & & & \phi_{s2} & 0 & -\phi_{s6} & 0 & 0 & 0 & \phi_{s3} \\
 & & & & & & \phi_{s1} & 0 & 0 & 0 & 0 & 0 \\
 & & & & & & & \phi_{s7} & 0 & 0 & 0 & -\phi_{s6} \\
 & & \text{Sym.} & & & & & & \phi_{s9} & 0 & \phi_{s8} & 0 \\
 & & & & & & & & & \frac{GJ}{L} & 0 & 0 \\
 & & & & & & & & & & \phi_{s4} & 0 \\
 & & & & & & & & & & & \phi_{s2}
\end{bmatrix} \tag{36.10}
$$

where G is shear modulus of elasticity; J is torsional constant; ϕ_{s1} to ϕ_{s9} are expressed by stability equations and listed in Table 36.4.

Finite-element-based stiffness matrix has the form [15]:

$$
[K_e] = \begin{bmatrix}
\phi_{e1} & 0 & 0 & 0 & 0 & 0 & -\phi_{e1} & 0 & 0 & 0 & 0 & 0 \\
 & \phi_{e7} & 0 & 0 & 0 & \phi_{e6} & 0 & -\phi_{e7} & 0 & 0 & 0 & \phi_{e6} \\
 & & \phi_{e9} & 0 & -\phi_{e8} & 0 & 0 & 0 & -\phi_{e9} & 0 & -\phi_{e8} & 0 \\
 & & & \frac{GJ}{L} & 0 & 0 & 0 & 0 & 0 & -\frac{GJ}{L} & 0 & 0 \\
 & & & & \phi_{e4} & 0 & 0 & 0 & -\phi_{e8} & 0 & \phi_{e5} & 0 \\
 & & & & & \phi_{e2} & 0 & -\phi_{e6} & 0 & 0 & 0 & \phi_{e3} \\
 & & & & & & \phi_{e1} & 0 & 0 & 0 & 0 & 0 \\
 & & & & & & & \phi_{e7} & 0 & 0 & 0 & -\phi_{e6} \\
 & & \text{Sym.} & & & & & & \phi_{e9} & 0 & \phi_{e8} & 0 \\
 & & & & & & & & & \frac{GJ}{L} & 0 & 0 \\
 & & & & & & & & & & \phi_{e4} & 0 \\
 & & & & & & & & & & & \phi_{e2}
\end{bmatrix} \tag{36.11}
$$

TABLE 36.4 Stability Function-Based ϕ_{si} for Three-Dimensional Member

ϕ_{si}		Stability Functions S_i	
		Compression	Tension
$\phi_{s1}=S_1\dfrac{EA}{L}$	S_1	$\dfrac{1}{1-\dfrac{EA}{4P^3L^2}\left[H_y+H_z\right]}$	$\dfrac{1}{1-\dfrac{EA}{4P^3L^2}\left[H'_y+H'_z\right]}$
$\phi_{s2}=S_2\dfrac{(4+\phi_y)EI_Z}{(1+\phi_y)L}$	S_2	$\dfrac{(\alpha L)(\sin\alpha L-\alpha L\cos\alpha L)}{4\phi_\alpha}$	$\dfrac{(\alpha L)(\alpha L\cosh\alpha L-\sinh\alpha L)}{4\phi_\alpha}$
$\phi_{s3}=S_2\dfrac{(2-\phi_y)EI_Z}{(1+\phi_y)L}$	S_3	$\dfrac{(\alpha L)(\alpha L-\sin\alpha L)}{2\phi_\alpha}$	$\dfrac{(\alpha L)(\sinh\alpha L-\alpha L)}{2\phi_\alpha}$
$\phi_{s4}=S_4\dfrac{(4+\phi_z)EI_y}{(1+\phi_z)L}$	S_4	$\dfrac{(\beta L)(\sin\beta L-\beta L\cos\beta L)}{4\phi_\beta}$	$\dfrac{(\beta L)(\beta L\cosh\beta L-\sinh\beta L)}{4\phi_\beta}$
$\phi_{s5}=S_2\dfrac{(2-\phi_z)EI_y}{(1+\phi_z)L}$	S_5	$\dfrac{(\beta L)(\beta L-\sin\beta L)}{2\phi_\beta}$	$\dfrac{(\beta L)(\sinh\beta L-\beta L)}{2\phi_\beta}$
$\phi_{s6}=S_6\dfrac{6EI_Z}{(1+\phi_y)L^2}$	S_6	$\dfrac{(\alpha L)^2(1-\cos\alpha L)}{6\phi_\alpha}$	$\dfrac{(\alpha L)^2(\cosh\alpha L-1)}{6\phi_\alpha}$
$\phi_{s7}=S_7\dfrac{12EI_Z}{(1+\phi_y)L^3}$	S_7	$\dfrac{(\alpha L)^3\sin\alpha L}{12\phi_\alpha}$	$\dfrac{(\alpha L)^3\sinh\alpha L}{12\phi_\alpha}$
$\phi_{s8}=S_8\dfrac{6EI_y}{(1+\phi_z)L^2}$	S_8	$\dfrac{(\beta L)^2(1-\cos\beta L)}{6\phi_\beta}$	$\dfrac{(\beta L)^2(\cosh\beta L-1)}{6\phi_\beta}$
$\phi_{s9}=S_9\dfrac{12EI_y}{(1+\phi_z)L^3}$	S_9	$\dfrac{(\beta L)^3\sin\beta L}{12\phi_\beta}$	$\dfrac{(\beta L)^3\sinh\beta L}{12\phi_\beta}$
$\alpha=\sqrt{P/EI_Z}$	ϕ_α	$2-2\cos\alpha L-\alpha L\sin\alpha L$	$2-2\cosh\alpha L+\alpha L\sinh\alpha L$
$\beta=\sqrt{P/EI_y}$	ϕ_β	$2-2\cos\beta L-\beta L\sin\beta L$	$2-2\cosh\beta L+\beta L\sinh\beta L$

$$H_y=\beta L(M_{ya}^2+M_{yb}^2)(\cot\beta L+\beta L\cos ec^2\beta L)-2(M_{ya}+M_{yb})^2+2\beta LM_{ya}M_{yb}(\cos ec\beta L)(1+\beta L\cot\beta L)$$

$$H_z=\alpha L(M_{za}^2+M_{zb}^2)(\cot\alpha L+\alpha L\cos ec^2\alpha L)-2(M_{za}+M_{zb})^2+2\alpha LM_{za}M_{zb}(\cos ec\alpha L)(1+\alpha L\cot\alpha L)$$

$$H'_y=\beta L(M_{ya}^2+M_{yb}^2)(\coth\beta L+\beta L\cos ech^2\beta L)-2(M_{ya}+M_{yb})^2+2\beta LM_{ya}M_{yb}(\cos ech\beta L)(1+\beta L\coth\beta L)$$

$$H'_z=\alpha L(M_{za}^2+M_{zb}^2)(\coth\alpha L+\alpha L\cos ech^2\alpha L)-2(M_{za}+M_{zb})^2+2\alpha LM_{za}M_{zbb}(\cos ech\alpha L)(1+\alpha L\coth\alpha L)$$

$$[K_g]=\begin{bmatrix}
\phi_{g1} & \phi_{g10} & -\phi_{g11} & 0 & 0 & 0 & 0 & -\phi_{g10} & \phi_{g11} & 0 & 0 & 0\\
 & \phi_{g7} & 0 & \phi_{g12} & \phi_{g13} & \phi_{g6} & -\phi_{g10} & -\phi_{g7} & 0 & \phi_{g14} & -\phi_{g13} & \phi_{g6}\\
 & & \phi_{g9} & \phi_{g15} & -\phi_{g6} & \phi_{g13} & \phi_{g11} & 0 & -\phi_{g9} & \phi_{g16} & -\phi_{g6} & -\phi_{g13}\\
 & & & \phi_{g17} & \phi_{g18} & \phi_{g19} & & -\phi_{g12} & -\phi_{g15} & -\phi_{g17} & -\phi_{g20} & \phi_{g21}\\
 & & & & \phi_{g4} & 0 & 0 & -\phi_{g13} & \phi_{g6} & -\phi_{g20} & -\phi_{g5} & \phi_{g13}\\
 & & & & & \phi_{g2} & 0 & -\phi_{g6} & -\phi_{g13} & \phi_{g21} & -\phi_{g13} & -\phi_{g3}\\
 & & & & & & \phi_{g1} & \phi_{g10} & -\phi_{g11} & 0 & 0 & 0\\
 & & & & & & & \phi_{g7} & 0 & -\phi_{g14} & \phi_{g13} & -\phi_{g6}\\
 & & \text{Sym.} & & & & & & \phi_{g9} & -\phi_{g16} & \phi_{g6} & \phi_{g13}\\
 & & & & & & & & & \phi_{g17} & \phi_{g18} & \phi_{g19}\\
 & & & & & & & & & & \phi_{g4} & 0\\
 & & & & & & & & & & & \phi_{g2}
\end{bmatrix} \tag{36.12}$$

where ϕ_{ei} and ϕ_{gi} are given in Table 36.5.

TABLE 36.5 Elements of Finite-Element-Based Stiffness Matrix

Linear Elastic Matrix	Geometric Nonlinear Matrix
$\phi_{e1} = \frac{AE}{L}$; $\phi_{e2} = \frac{4EI_Z}{L}$	$\phi_{g1} = 0$; $\phi_{g2} = \phi_{g4} = \frac{2F_{xb}L}{15}$; $\phi_{g3} = \phi_{g5} = \frac{F_{xb}L}{30}$
$\phi_{e3} = \frac{2EI_z}{L}$ $\phi_{e4} = \frac{4EI_y}{L}$	$\phi_{g7} = \phi_{g9} = \frac{6F_{xb}}{5L}$; $\phi_{g6} = \phi_{g8} = \frac{F_{xb}}{10}$; $\phi_{g10} = \frac{M_{za} + M_{zb}}{L^2}$
$\phi_{e5} = \frac{2EI_y}{L}$; $\phi_{e6} = \frac{6EI_Z}{L^2}$	$\phi_{g11} = \frac{M_{ya} + M_{yb}}{L^2}$; $\phi_{g12} = \frac{M_{ya}}{L}$; $\phi_{g13} = \frac{M_{xb}}{L}$
$\phi_{e7} = \frac{12EI_Z}{L^3}$; $\phi_{e8} = \frac{6EI_y}{L^2}$	$\phi_{g14} = \frac{M_{yb}}{L}$; $\phi_{g15} = \frac{M_{za}}{L}$; $\phi_{g16} = \frac{M_{zb}}{L}$
$\phi_{e9} = \frac{12EI_y}{L^3}$	$\phi_{g17} = \frac{F_{xb}I_p}{AL}$; $\phi_{g18} = \frac{M_{zb}}{6} - \frac{M_{za}}{3}$; $\phi_{g19} = \frac{M_{ya}}{3} - \frac{M_{yb}}{6}$
	$\phi_{g20} = \frac{M_{za} + M_{zyb}}{6}$; $\phi_{g21} = \frac{M_{ya} + M_{yb}}{6}$

I_z and I_y are moments of inertia about *z–z* and *y–y* axis, respectively; I_p is the polar moment of inertia.

Stiffness matrices considering warping degree of freedom and finite rotations for a thin-walled member were derived by Yang and McGuire [16,17].

In conclusion, both sets of the stiffness matrices have been used successfully when considering geometric nonlinearities (*P*-Δ and *P*-δ effects). The stability function-based formulation gives an accurate solution using fewer degrees of freedom when compared with the finite-element method. Its power series expansion (Table 36.3) can be implemented easily without truncation to avoid numerical difficulty.

The finite-element-based formulation produces an approximate solution. It has a simpler form and may require dividing the member into a large number of elements in order to keep the (P/L) term a small quantity to obtain accurate results.

36.4 Material Nonlinearity Formulations

36.4.1 Structural Concrete

Concrete material nonlinearity is incorporated into analysis using a nonlinear stress–strain relationship. Figure 36.4 shows idealized stress–strain curves for unconfined and confined concrete in uniaxial compression. Tests have shown that the confinement provided by closely spaced transverse reinforcement can substantially increase the ultimate concrete compressive stress and strain. The confining steel prevents premature buckling of the longitudinal compression reinforcement and increases the concrete ductility. Extensive research has been made to develop concrete stress–strain relationships [18-25].

36.4.1.1 Compression Stress–Strain Relationship

Unconfined Concrete

A general stress–strain relationship proposed by Hognestad [18] is widely used for plain concrete or reinforced concrete with a small amount of transverse reinforcement. The relation has the following simple form:

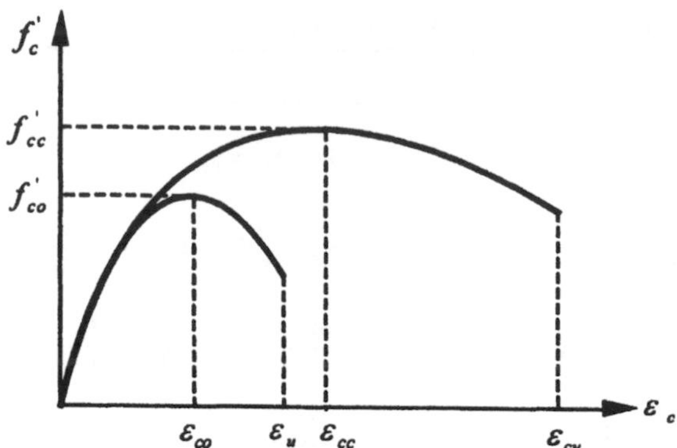

FIGURE 36.4 Idealized stress-strain curves for concrete in uniaxial compression.

$$f_c = \begin{cases} f_{co}'\left[\dfrac{2\varepsilon_c}{\varepsilon_{co}} - \left(\dfrac{\varepsilon_c}{\varepsilon_{co}}\right)^2\right] & \varepsilon_c \le \varepsilon_{co} \\ f_{co}'\left[1-\beta\left(\dfrac{\varepsilon_c - \varepsilon_o}{\varepsilon_u - \varepsilon_{co}}\right)\right] & \varepsilon_{co} < \varepsilon_c \le \varepsilon_u \end{cases} \tag{36.13}$$

$$\varepsilon_{co} = \frac{2f_{co}'}{E_c} \tag{36.14}$$

where f_c and ε_c are the concrete stress and strain; f_{co}' is the peak stress for unconfined concrete usually taken as the cylindrical compression strength f_c'; ε_{co} is strain at peak stress for unconfined concrete usually taken as 0.002; ε_u is the ultimate compression strain for unconfined concrete taken as 0.003; E_c is the modulus of elasticity of concrete; β is a reduction factor for the descending branch usually taken as 0.15. Note that the format of Eq. (36.13) can be also used for confined concrete if the concrete-confined peak stress f_{cc}' and strain ε_{cu} are known or assumed and substituted for f_{co}' and ε_u, respectively.

Confined Concrete — Mander's Model

Analytical models describing the stress–strain relationship for confined concrete depend on the confining transverse reinforcement type (such as hoops, spiral, or ties) and shape (such as circular, square, or rectangular). Some of those analytical models are more general than others in their applicability to various confinement types and shapes. A general stress–strain model (Figure 36.5) for confined concrete applicable (in theory) to a wide range of cross sections and confinements was proposed by Mander et al. [23,24] and has the following form:

$$f_c = \frac{f_{cc}'(\varepsilon_c / \varepsilon_{cc})r}{r-1+(\varepsilon_c / \varepsilon_{cc})^r} \tag{36.15}$$

$$\varepsilon_{cc} = \varepsilon_{co}\left[1+5\left(\frac{f_{cc}'}{f_{co}'}-1\right)\right] \tag{36.16}$$

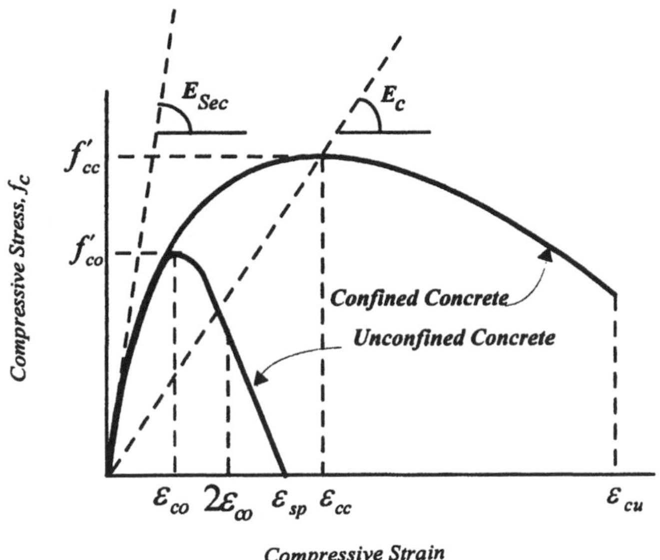

FIGURE 36.5 Stress–strain curves — mander model.

$$r = \frac{E_c}{E_c - E_{sec}} \tag{36.17}$$

$$E_{sec} = \frac{f'_{cc}}{\varepsilon_{cc}} \tag{36.18}$$

where f'_{cc} and ε_{cc} are peak compressive stress and corresponding strain for confined concrete. f'_{cc} and ε_{cu} which depend on the confinement type and shape, are calculated as follows:

Confined Peak Stress

1. **For concrete circular section confined by circular hoops or spiral** (Figure 36.6a):

$$f'_{cc} = f'_{co}\left(2.254\sqrt{1 + \frac{7.94 f'_l}{f'_{co}}} - \frac{2 f'_l}{f'_{co}} - 1.254\right) \tag{36.19}$$

$$f'_l = \frac{1}{2} K_e \rho_s f_{yh} \tag{36.20}$$

$$K_e = \begin{cases} \left(1 - s'/2d_s\right)^2 / \left(1 - \rho_{cc}\right) & \text{for circular hoops} \\ \left(1 - s'/2d_s\right) / \left(1 - \rho_{cc}\right) & \text{for circular spirals} \end{cases} \tag{36.21}$$

$$\rho_s = \frac{4 A_{sp}}{d_s s} \tag{36.22}$$

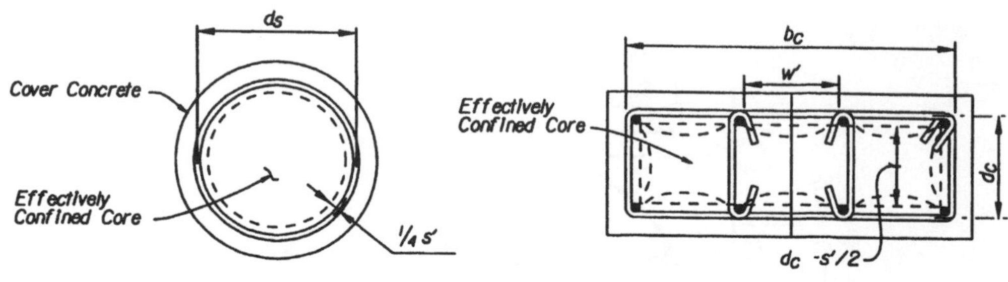

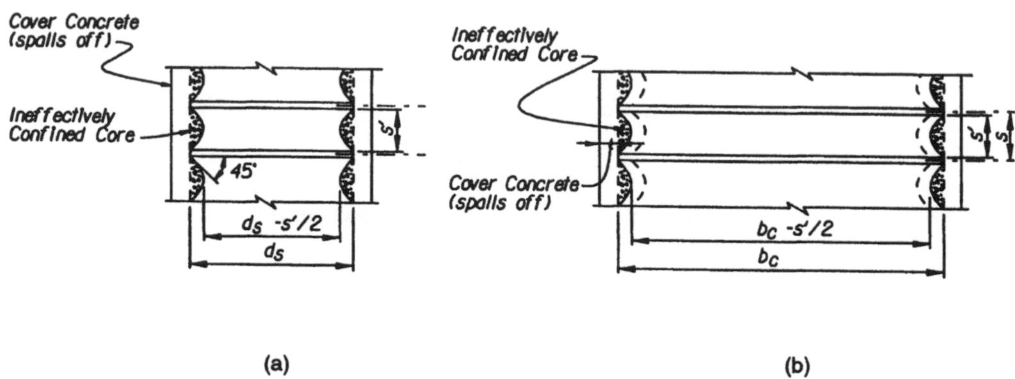

(a) (b)

FIGURE 36.6 Confined core for hoop reinforcement. (a) Circular hoop and (b) rectangular hoop reinforcement.

where f_l' is the effective lateral confining pressure; K_e is confinement effectiveness coefficient, f_{yh} is the yield stress of the transverse reinforcement, s' is the clear vertical spacing between hoops or spiral; s is the center-to-center spacing of the spiral or circular hoops; d_s is the centerline diameter of the spiral or hoops circle; ρ_{cc} is the ratio of the longitudinal reinforcement area to section core area; ρ_s is the ratio of the transverse confining steel volume to the confined concrete core volume; and A_{sp} is the bar area of transverse reinforcement.

2. **For rectangular concrete section confined by rectangular hoops** (Figure 36.6b)

The rectangular hoops may produce two unequal effective confining pressures f_{lx}' and f_{ly}' in the principal x and y direction defined as follows:

$$f_{lx}' = K_e\ \rho_x f_{yh} \tag{36.23}$$

$$f_{ly}' = K_e\ \rho_y f_{yh} \tag{36.24}$$

$$K_e = \frac{\left[1 - \sum_{i=1}^{n} \frac{(w_i')^2}{6b_c d_c}\right]\left(1 - \frac{s'}{2b_c}\right)\left(1 - \frac{s'}{2d_c}\right)}{\left(1 - \rho_{cc}\right)} \tag{36.25}$$

$$\rho_x = \frac{A_{sx}}{s d_c} \tag{36.26}$$

$$\rho_y = \frac{A_{sy}}{s b_c} \tag{36.27}$$

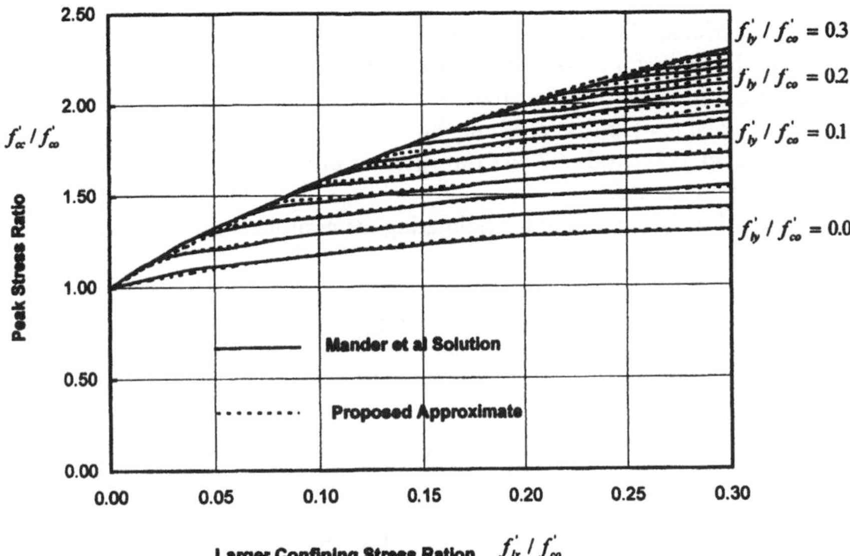

FIGURE 36.7 Peak stress of confined concrete.

where f_{yh} is the yield strength of transverse reinforcement; w'_i is the ith clear distance between adjacent longitudinal bars; b_c and d_c are core dimensions to centerlines of hoop in x and y direction (where $b \geq d$), respectively; A_{sx} and A_{sy} are the total area of transverse bars in x and y direction, respectively.

Once f'_{lx} and f'_{ly} are determined, the confined concrete strength f'_{cc} can be found using the chart shown in Figure 36.7 with f'_{lx} being greater or equal to f'_{ly}. The chart depicts the general solution of the "five-parameter" multiaxial failure surface described by William and Warnke [26].

As an alternative to the chart, the authors derived the following equations for estimating f'_{cc}:

$$f_{cc} = \begin{cases} Af'^2_{lx} + Bf'_{lx} + C & f'_{ly} < f'_{lx} \text{ and } f'_{ly} \leq 0.15 \\ \dfrac{f'_{lx} - f'_{lx}}{0.3 - f'_{ly}} D + C & f'_{ly} < f'_{lx} \text{ and } f'_{ly} > 0.15 \\ C & f'_{ly} = f'_{lx} \end{cases} \tag{36.28}$$

$$A = 196.5f'^2_{lx} + 29.1f'_{lx} - 4 \tag{36.29}$$

$$B = -69.5f'^2_{lx} - 8.9f'_{lx} + 2.2 \tag{36.30}$$

$$C = -6.83f'^2_{lx} + 6.38f'_{lx} + 1 \tag{36.31}$$

$$D = -1.5f'^2_{lx} - 0.55f'_{lx} + 0.3 \tag{36.32}$$

Note that by setting $f'_l = 0.0$ in Eqs. (36.19), Eqs. (36.16) and (36.15) will produce to Mander's expression for unconfined concrete. In this case and for concrete strain $\varepsilon_c > 2\,\varepsilon_{co}$, a straight line which reaches zero stress at the spalling strain ε_{sp} is assumed.

Confined Concrete Ultimate Compressive Strain

Experiments have shown that a sudden drop in the confined concrete stress–strain curve takes place when the confining transverse steel first fractures. Defining the ultimate compressive strain as the longitudinal strain at which the first confining hoop fracture occurs, and using the energy balance approach, Mander et al. [27] produced an expression for predicting the ultimate compressive strain which can be solved numerically.

A conservative and simple equation for estimating the confined concrete ultimate strain is given by Priestley et al. [7]:

$$\varepsilon_{cu} = 0.004 + \frac{1.4\rho_s f_{yh}\varepsilon_{su}}{f'_{cc}} \tag{36.33}$$

where ε_{su} is the steel strain at maximum tensile stress. For rectangular section $\rho_s = \rho_x + \rho_y$ as defined previously. Typical values for ε_{cu} range from 0.012 to 0.05.

Equation (36.33) is formulated for confined sections subjected to axial compression. It is noted that when Eq. (36.33) is used for a section in bending or in combined bending and axial compression, then it tends to be conservative by a least 50%.

Chai et al. [28] used an energy balance approach to derive the following expression for calculating the concrete ultimate confined strain as

$$\varepsilon_{cu} = \varepsilon_{sp} + \begin{cases} \rho_s \varepsilon_{su} \dfrac{\gamma_2 f_{yh}}{\gamma_1 f'_{cc}} & \text{confined by reinconcement} \\ \rho_{sj} \varepsilon_{suj} \dfrac{\gamma_2 f_{yj}}{\gamma_1 f'_{cc}} & \text{confined by circular steel jackets} \end{cases} \tag{36.34}$$

where ε_{sp} is the spalling strain of the unconfined concrete (usually = 0.003 to 0.005), γ_1 is an integration coefficient of the area between the confined and unconfined stress–strain curves; and γ_2 is an integration coefficient of the area under the transverse steel stress–strain curve. The confining ratio for steel jackets $\rho_{sj} = 4t_j/(D_j - 2t_j)$; D_i and t_j are outside diameter and thickness of the jacket, respectively; f_{yj} is yield stress of the steel jacket. For high- and mild-strength steels and concrete compressive strengths of 4 to 6 ksi (27.58 to 41.37 MPa), Chai et al. [28] proposed the following expressions

$$\frac{\gamma_2}{\gamma_1} = \begin{cases} \dfrac{2000\rho_s}{\left(1+(1428\rho_s)^4\right)^{0.25}} & \text{for Grade 40 Steel} \\ \dfrac{2000\rho_s}{\left(1+(1480\rho_s)^{0.25}\right)^{0.4}} & \text{for Grade 60 Steel} \end{cases} \tag{36.35}$$

Confined Concrete — Hoshikuma's Model

In additional to Mander's model, Table 36.6 lists a stress–strain relationship for confined concrete proposed by Hoshikuma et al. [25]. The Hoshikuma model was based on the results of a series of experimental tests covering circular, square, and wall-type cross sections with various transverse reinforcement arrangement in bridge piers design practice in Japan.

36.4.1.2 Tension Stress-Strain Relationship

Two idealized stress–strain curves for concrete in tension is shown in Figure 36.8. For plain concrete, the curve is linear up to cracking stress f_r. For reinforced concrete, there is a descending branch

TABLE 36.6 Hoshikuma et al. [25] Stress–Strain Relationship of Confined Concrete

$$f_c = \begin{cases} E_c \varepsilon_c \left[1 - \frac{1}{n} \left(\frac{\varepsilon_c}{\varepsilon_{cc}} \right)^{n-1} \right] & \varepsilon_c \le \varepsilon_{cc} \\ f'_{cc} - E_{des}(\varepsilon_c - \varepsilon_{cc}) & \varepsilon_{cc} < \varepsilon_c \le \varepsilon_{cu} \end{cases}$$

$$n = \frac{E_c \varepsilon_{cc}}{E_c \varepsilon_{cc} - f'_{cc}}; \quad \varepsilon_{cu} = \varepsilon_{cc} + \frac{f'_{cc}}{2E_{des}}; \quad E_{des} = 11.2 \frac{f'^2_{co}}{\rho_s f_{yh}}$$

$$\frac{f'_{cc}}{f'_{co}} = \begin{cases} 1.0 + 3.8 \frac{\rho_s f_{yh}}{f'_{co}} & \text{for circular section} \\ 1.0 + 0.76 \frac{\rho_s f_{yh}}{f'_{co}} & \text{for square section} \end{cases}$$

$$\varepsilon_{cc} = \begin{cases} 0.002 + 0.033 \frac{\rho_s f_{sh}}{f'_{co}} & \text{for circular section} \\ 0.002 + 0.013 \frac{\rho_s f_{sh}}{f'_{co}} & \text{for square section} \end{cases}$$

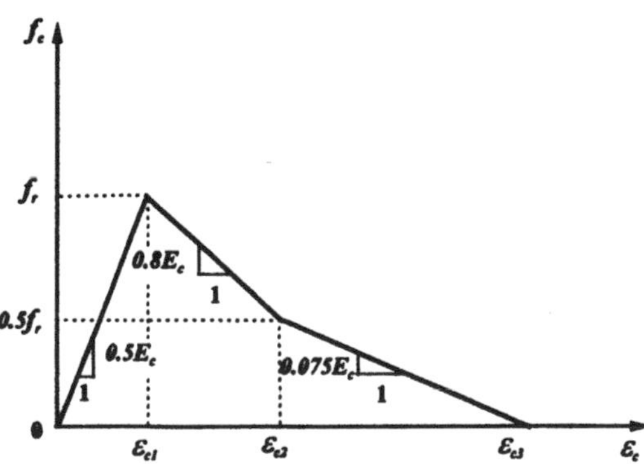

FIGURE 36.8 Idealized stress–strain curve of concrete in uniaxial tension.

because of bond characteristics of reinforcement. A trilinear expression proposed by Vebe et al. [29] is as follows:

$$f_c = \begin{cases} 0.5E_c\varepsilon_c & \varepsilon_c \le \varepsilon_{c1} = 2f_r/E_c \\ f_r[1 - 0.8E_c(\varepsilon_c - 2f_r/E_c)] & \varepsilon_{c1} < \varepsilon_c \le \varepsilon_{c2} = 2.625f_r/E_c \\ f_r[0.5 - 0.075E_c(\varepsilon_c - 2.625f_r/4E_c)] & \varepsilon_c < \varepsilon_{c3} = 9.292f_r/E_c \end{cases} \tag{36.36}$$

where f_r is modulus of rupture of concrete.

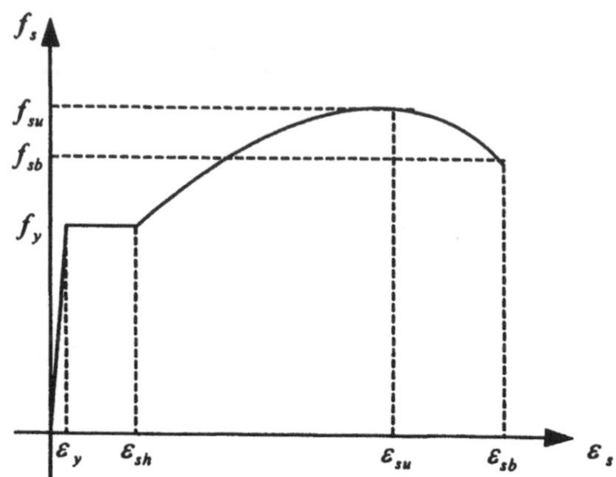

FIGURE 36.9 Idealized stress–strain curve of structural steel and reinforcement.

36.4.2 Structural and Reinforcement Steel

For structural steel and nonprestressed steel reinforcement, its stress–strain relationship can be idealized as four parts: elastic, plastic, strain hardening, and softening, as shown in Figure 36.9. The relationship if commonly expressed as follows:

$$f_s = \begin{cases} E_s\varepsilon_s & 0 \le \varepsilon_s \le \varepsilon_y \\ f_y & \varepsilon_{sy} < \varepsilon_s \le \varepsilon_{sh} \\ f_y + \dfrac{\varepsilon_s - \varepsilon_{sh}}{\varepsilon_{su} - \varepsilon_{sh}}(f_{su} - f_y) & \varepsilon_{sh} < \varepsilon_s \le \varepsilon_{su} \\ f_u\left[1 - \dfrac{\varepsilon_s - \varepsilon_{su}}{\varepsilon_{sb} - \varepsilon_{su}}(f_{su} - f_{sb})\right] & \varepsilon_{cu} < \varepsilon_s \le \varepsilon_{sb} \end{cases} \tag{36.37}$$

where f_s and ε_s is stress of strain in steel; E_s is the modulus of elasticity of steel; f_y and ε_y is yield stress and strain; ε_{sh} is hardening strain; f_{su} and ε_{su} is maximum stress and corresponding strain; f_{sb} and ε_{sb} are rupture stress and corresponding strain.

$$\varepsilon_{sh} = \begin{cases} 14\varepsilon_y & \text{for Grade 40} \\ 5\,\varepsilon_y & \text{for Grade 60} \end{cases} \tag{36.38}$$

$$\varepsilon_{su} = \begin{cases} 0.14 + \varepsilon_{sh} & \text{for Grade 40} \\ 0.12 & \text{for Grade 60} \end{cases} \tag{36.39}$$

For the reinforcing steel, the following nonlinear form can also be used for the strain-hardening portion [28]:

$$f_s = f_y\left[\frac{m(\varepsilon_s - \varepsilon_{sh}) + 2}{60(\varepsilon_s - \varepsilon_{sh}) + 2} + \frac{(\varepsilon_s - \varepsilon_{sh})(60 - m)}{2(30r + 1)^2}\right] \quad \text{for } \varepsilon_{sh} < \varepsilon_s \le \varepsilon_{su} \tag{36.40}$$

$$m = \frac{(f_{su}/f_y)(30r+1)^2 - 60r - 1}{15r^2} \tag{36.41}$$

$$r = \varepsilon_{su} - \varepsilon_{sh} \tag{36.42}$$

$$f_{su} = 1.5 f_y \tag{36.43}$$

For both strain-hardening and -softening portions, Holzer et al. [30] proposed the following expression

$$f_s = f_y\left[1 + \frac{\varepsilon_s - \varepsilon_{sh}}{\varepsilon_{su} - \varepsilon_{sh}}\left(\frac{f_{su}}{f_y} - 1\right)\exp\left(1 - \frac{\varepsilon_s - \varepsilon_{sh}}{\varepsilon_{su} - \varepsilon_{sh}}\right)\right] \quad \text{for } \varepsilon_{sh} < \varepsilon_s \le \varepsilon_{sb} \tag{36.44}$$

For prestressing steel, its stress–strain behavior is different from the nonprestressed steel. There is no obvious yield flow plateau in its response. The stress-stress expressions presented in Chapter 10 can be used in an analysis.

36.5 Nonlinear Section Analysis

36.5.1 Basic Assumptions and Formulations

The main purpose of section analysis is to study the moment–thrust–curvature behavior. In a nonlinear section analysis, the following assumptions are usually made:

- Plane sections before bending remain plane after bending;
- Shear and torsional deformation is negligible;
- Stress-strain relationships for concrete and steel are given;
- For reinforced concrete, a prefect bond between concrete and steel rebar exists.

The mathematical formulas used in the section analysis are (Figure 36.10):

Compatibility equations

$$\phi_x = \varepsilon / y \tag{36.45}$$

$$\phi_y = \varepsilon / x \tag{36.46}$$

Equilibrium equations

$$P = \int_A \sigma \, dA = \sum_{i=1}^{n} \sigma_i A_i \tag{36.47}$$

$$M_x = \int_A \sigma y \, dA = \sum_{i=1}^{n} \sigma_i y_i A_i \tag{36.48}$$

$$M_y = \int_A \sigma x \, dA = \sum_{i=1}^{n} \sigma_i x_i A_i \tag{36.49}$$

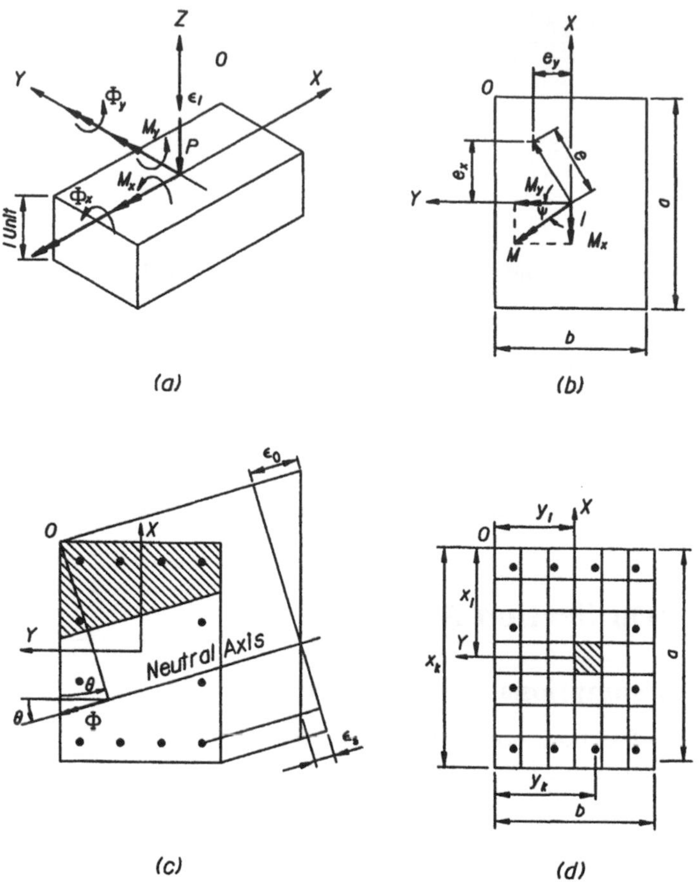

FIGURE 36.10 Moment–curvature–strain of cross section.

36.5.2 Modeling and Solution Procedures

For a reinforced-concrete member, the cross section is divided into a proper number of concrete and steel filaments representing the concrete and reinforcing steel as shown in Figure 36.10d. Each concrete and steel l filament is assigned its corresponding stress–strain relationships. Confined and unconfined stress–strain relationships are used for the core concrete and for the cover concrete, respectively.

For a structural steel member, the section is divided into steel filaments and a typical steel stress–strain relationship is used for tension and compact compression elements, and an equivalent stress–strain relationship with reduced yield stress and strain can be used for a noncompact compression element.

The analysis process starts by selecting a strain for the extreme concrete (or steel) fiber. By using this selected strain and assuming a section neutral axis (NA) location, a linear strain profile is constructed and the corresponding section stresses and forces are computed. Section force equilibrium is then checked for the given axial load. By changing the location of the NA, the process is repeated until equilibrium is satisfied. Once equilibrium is satisfied, for the assumed strain and the given axial load, the corresponding section moment and curvature are computed by Eqs. (36.48) and (36.49).

A moment–curvature (M–Φ) diagram for a given axial load is constructed by incrementing the extreme fiber strain and finding the corresponding moment and the associated curvature. An interaction diagram (M–P) relating axial load and the ultimate moment is constructed by incrementing the axial load and finding the corresponding ultimate moment using the above procedure.

For a reinforced-concrete section, the yield moment is usually defined as the section moment at onset of yielding of the tension reinforcing steel. The ultimate moment is defined as the moment at peak moment capacity. The ultimate curvature is usually defined as the curvature when the extreme concrete fiber strain reaches ultimate strain or when the reinforcing rebar reaches its ultimate (rupture) strain (whichever takes place first). Figure 36.11a shows typical M–P–Φ curves for a reinforced-concrete section.

For a simple steel section, such as rectangular, circular-solid, and thin-walled circular section, a closed-form of M–P–Φ can be obtained using the elastic-perfectly plastic stress–strain relations [4, 31]. For all other commonly used steel section, numerical iteration techniques are used to obtain M–P–Φ curves. Figure 36.11b shows typical M–P–Φ curves for a wide-flange section.

36.5.3 Yield Surface Equations

The yield or failure surface concept has been conveniently used in inelastic analysis to describe the full plastification of steel and concrete sections under the action of axial force combined with biaxial bending. This section will present several yield surface expressions for steel and concrete sections suitable for use in a nonlinear analysis.

36.5.3.1 Yield Surface Equations for Concrete Sections

The general interaction failure surface for a reinforced-concrete section with biaxial bending, as shown in Figure 36.12a can be approximated by a nondimensional interaction equation [32]:

$$\left(\frac{M_x}{M_{xo}}\right)^m + \left(\frac{M_y}{M_{yo}}\right)^n = 1.0 \tag{36.50}$$

where M_x and M_y are bending moments about x–x and y–y principal axes, respectively; M_{xo} and M_{yo} are the uniaxial bending capacity about the x–x and y–y axes under axial load P; the exponents m and n depend on the reinforced-concrete section properties and axial force. They can be determined by a numerical analysis or experiments. In general, the values of m and n usually range from 1.1 to 1.4 for low and moderate axial compression.

36.5.3.2 Yield Surface Equation for Doubly Symmetrical Steel Sections

The general shape of yield surface for a doubly symmetrical steel section as shown in Figure 36.12b can be described approximately by the following general equation [33]

$$\left(\frac{M_x}{M_{pcx}}\right)^{\alpha_x} + \left(\frac{M_y}{M_{pcy}}\right)^{\alpha_y} = 1.0 \tag{36.51}$$

where M_{pcx} and M_{pcy} are the moment capacities about respective axes, reduced for the presence of axial load; they can be obtained by the following formulas:

$$M_{pcx} = M_{px}\left[1 - \left(\frac{P}{P_y}\right)^{\beta_x}\right] \tag{36.52}$$

$$M_{pcy} = M_{py}\left[1 - \left(\frac{P}{P_y}\right)^{\beta_y}\right] \tag{36.53}$$

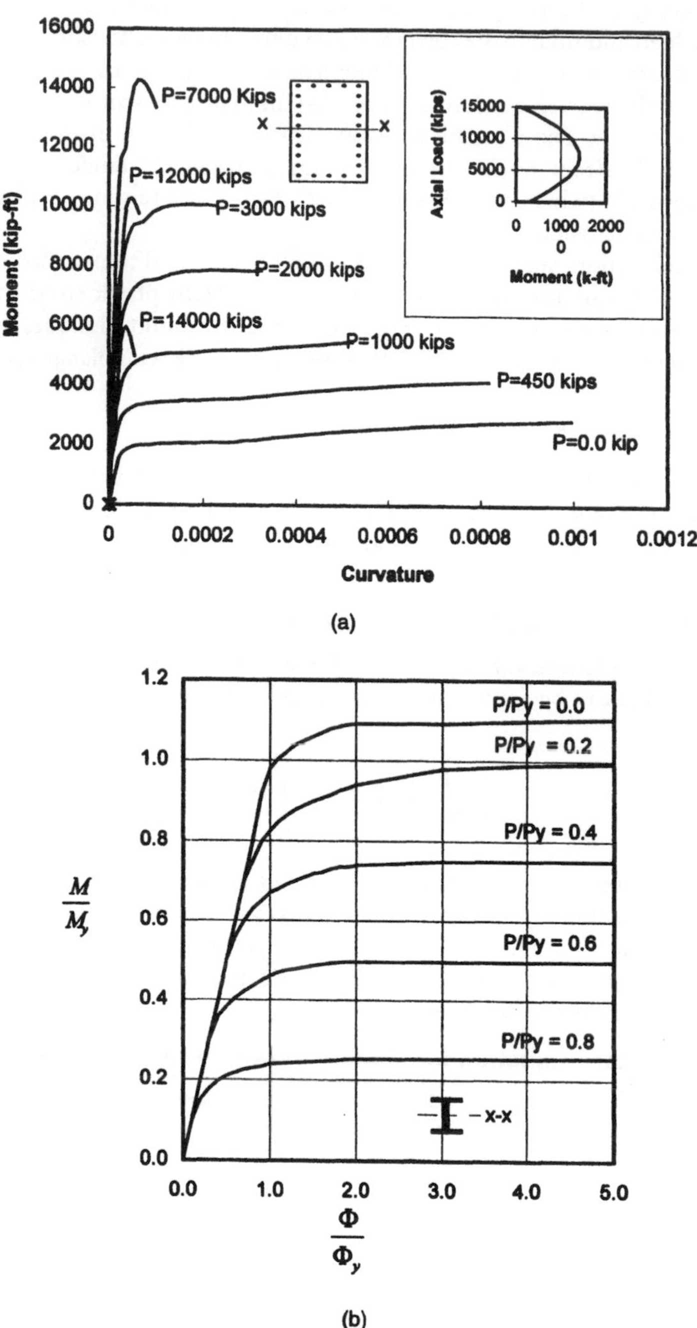

FIGURE 36.11 Moment–thrust–curvature curve. (a) Reinforced concrete section (b) steel I-section.

where P is the axial load; M_{px} and M_{py} are the plastic moments about x–x and y–y principal axes, respectively; α_x, α_y, β_x, and β_y are parameters that depend on cross-sectional shapes and area distribution and are listed in Table 36.7.

Equation (36.51) represents a smooth and convex surface in the three-dimensional stress-resultant space. It is easy to implement in a computer-based structural analysis.

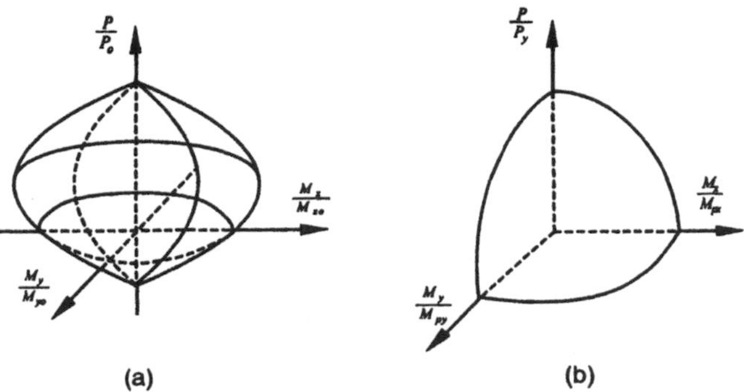

FIGURE 36.12 General yield surfaces. (a) Reinforced concrete section; (b) steel section.

TABLE 36.7 Parameters for Doubly Symmetrical Steel Sections

Section Types	α_x	α_y	β_x	β_y
Solid rectangular	1.7 +1.3 (P/P_y)	1.7 +1.3 (P/P_y)	2.0	2.0
Solid circular	2.0	2.0	2.1	2.1
I-shape	2.0	1.2 + 2 (P/P_y)	1.3	2 + 1.2 (A_w/A_f)
Thin-walled box	1.7 + 1.5 (P/P_y)	1.7 + 1.5 (P/P_y)	$2-0.5\bar{B}\geq 1.3$	$2-0.5\bar{B}\geq 1.3$
Thin-walled circular	2.0	2.0	1.75	1.75

Where $\bar{B}$ is the ratio of width to depth of the box section with respect to the bending axis.

Orbison [15] developed the following equation for a wide-flange section by trial and error and curve fitting:

$$1.15\left(\frac{P}{P_y}\right)^2+\left(\frac{M_x}{M_{px}}\right)^2+\left(\frac{M_y}{M_{py}}\right)^4+3.67\left(\frac{P}{P_y}\right)\left(\frac{M_x}{M_{px}}\right)^2 +3.0\left(\frac{P}{P_y}\right)^2\left(\frac{M_y}{M_{py}}\right)^2+4.65\left(\frac{M_x}{M_{px}}\right)^4\left(\frac{M_y}{M_{py}}\right)^2=1.0 \tag{36.54}$$

36.6 Nonlinear Frame Analysis

Both the first-order and second-order inelastic frame analyses can be categorized into three types of analysis: (1) elastic–plastic hinge, (2) refined plastic hinge, and (3) distributed plasticity. This section will discuss the basic assumptions and applications of those analyses.

36.6.1 Elastic–Plastic Hinge Analysis

In an elastic-plastic hinge (lumped plasticity) analysis, material inelasticity is taken into account using concentrated "zero-length" plastic hinges. The traditional plastic hinge is defined as a zero-length point along the structure member which can maintain plastic moment capacity and rotate freely. When the section reaches its plastic capacity (for example, the yield surface as shown in Figures 36.12 or 36.13), a plastic hinge is formed and the element stiffness is adjusted [34, 35] to reflect the hinge formation. For regions in a framed member away from the plastic hinge, elastic behavior is assumed.

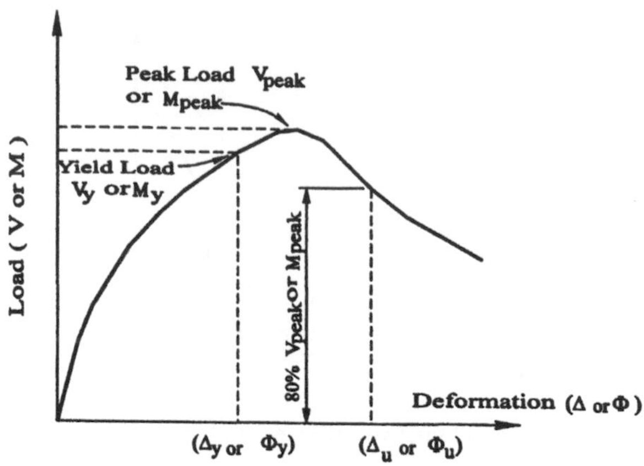

FIGURE 36.13 Load-deformation curves.

For a framed member subjected to end forces only, the elastic–plastic hinge method usually requires only one element per member making the method computationally efficient. It does not, however, accurately represent the distributed plasticity and associated P-δ effects. This analysis predicts an upper-bound solution (see Figure 36.1).

36.6.2 Refined Plastic Hinge Analysis

In the refined plastic hinge analysis [36], a two-surface yield model considers the reduction of plastic moment capacity at the plastic hinge (due to the presence of axial force) and an effective tangent modulus accounts for the stiffness degradation (due to distributed plasticity along a frame member). This analysis is similar to the elastic–plastic hinge analysis in efficiency and simplicity and, to some extent, also accounts for distributed plasticity. The approach has been developed for advanced design of steel frames, but detailed considerations for concrete structures still need to be developed.

36.6.3 Distributed Plasticity Analysis

Distributed plasticity analysis models the spread of inelasticity through the cross sections and along the length of the members. This is also referred to as plastic zone analysis, spread-of-plasticity analysis, and elastoplastic analysis by various researchers. In this analysis, a member needs to be subdivided into several elements along its length to model the inelastic behavior more accurately. There are two main approaches which have been successfully used to model plastification of members in a second-order distributed plasticity analysis:

1. Cross-sectional behavior is described as an input for the analysis by means of moment–thrust–curvature (M–P–Φ) and moment–trust–axial strain (M–P–ε) relations which may be obtained separately from section analysis as discussed in Section 36.5 or approximated by closed-form expressions [31].
2. Cross sections are subdivided into elemental areas and the state of stresses and strains are traced explicitly using the proper stress–strain relations for all elements during the analysis.

In summary, the elastic–plastic hinge analysis is the simplest one, but provides an upper-bound solution. Distributed plasticity analysis is considered the most accurate and is generally computationally intensive for larger and complex structures. Refined plastic hinge analysis seems to be an alternative that can reasonably achieve both computational efficiency and accuracy.

36.7 Practical Applications

In this section, the concept and procedures of displacement-based design and the bases of the static push-over analysis are discussed briefly. Two real bridges are analyzed as examples to illustrate practical application of the nonlinear static push-over analysis approach for bridge seismic design. Additional examples and detailed discussions of nonlinear bridge analysis can be found in the literature [7, 37].

36.7.1 Displacement-Based Seismic Design

36.7.1.1 Basic Concept

In recent years, displacement-based design has been used in the bridge seismic design practice as a viable alternative approach to strength-based design. Using displacements rather than forces as a measurement of earthquake damage allows a structure to fulfill the required function (damage-control limit state) under specified earthquake loads.

In a common design procedure, one starts by proportioning the structure for strength and stiffness, performs the appropriate analysis, and then checks the displacement ductility demand against available capacity. This procedure has been widely used in bridge seismic design in California since 1994. Alternatively, one could start with the selection of a target displacement, perform the analysis, and then determine strength and stiffness to achieve the design level displacement. Strength and stiffness do not enter this process as variables; they are the end results [38, 39].

In displacement-based design, the designer needs to define a criterion clearly for acceptable structural deformation (damage) based on postearthquake performance requirements and the available deformation capacity. Such criteria are based on many factors including structural type and importance.

36.7.1.2 Available Ultimate Deformation Capacity

Because structural survival without collapse is commonly adopted as a seismic design criterion for ordinary bridges, inelastic structural response and some degradation in strength can be expected under seismic loads. Figure 36.13 shows a typical load–deformation curve. A gradual degrading response as shown in Figure 36.13 can be due to factors such as P-Δ effects and/or plastic hinge formulation. The available ultimate deformation capacity should be based on how great a reduction (degradation) in structure load-carrying capacity response can be tolerated [21].

In general, the available ultimate deformation capacity can be referred to as the deformation that a structure can undergo without losing significant load-carrying capacity [40]. It is, therefore, reasonable to define available ultimate deformation as that deformation when the load-carrying capacity has been reduced by an acceptable amount after the peak load, say, 20%, as shown in Figure 36.13. This acceptable reduction amount may vary depending on required performance criteria of the particular case.

The available deformation capacity based on the design criteria requirements needs not correspond to the ultimate member or system deformation capacity. For a particular member cross section, the ultimate deformation in terms of the curvature depends on the shape, material properties, and loading conditions of the section (i.e., axial load, biaxial bending) and corresponds to the condition when the section extreme fiber reaches its ultimate strain (ε_{cu} for concrete and ε_{sp} for steel). The available ultimate curvature capacity ϕ_u can be chosen as the curvature that corresponds to the condition when section moment capacity response reduces by, say, 20%, from the peak moment.

For a framed structure system, the ultimate deformation in terms of the lateral displacement depends on structural configurations, section behavior, and loading conditions and corresponds to a failure state of the frame system when a collapse mechanism forms. The available lateral displacement capacity Δ_u

can be chosen as the displacement that corresponds to the condition when lateral load-carrying capacity reduces by some amount, say, 20%, from its peak load. In current seismic design practice in California, the available frame lateral displacement capacity commonly corresponds to the first plastic hinge reaching its ultimate rotational capacity.

36.7.1.3 Analysis Procedures

Seismic analysis procedures used in displacement-based design can be divided into three groups:

Group I: Seismic displacement and force demands are estimated from an elastic dynamic time history or a response spectrum analysis with effective section properties. For concrete structures, cracked section properties are usually used to determine displacement demands, and gross section properties are used to determine force demands. Strength capacity is evaluated from nonlinear section analysis or other code-specified methods, and displacement capacity is obtained from a static nonlinear push-over analysis.

Group II: Seismic displacement demand is obtained from a specified response spectrum and initial effective stiffness or a substitute structural model [38] considering both the effective stiffness and the effective damping. Effective stiffness and displacement capacity are estimated from a nonlinear static push-over analysis.

Group III: A nonlinear inelastic dynamic time history analysis is performed. Bridge assessment is based on displacement (damage) comparisons between analysis results and the given acceptance criteria. This group of analyses is complex and time-consuming and used only for important structures.

36.7.2 Static Push-Over Analysis

In lieu of a nonlinear time history dynamic analysis, bridge engineers in recent years have used static push-over analyses as an effective and simple alternative when assessing the performance of existing or new bridge structures under seismic loads. Given the proper conditions, this approximate alternative can be as reliable as the more accurate and complex ones. The primary goal of such an analysis is to determine the displacement or ductility capacity which is then compared with displacement or ductility demand obtained for most cases from linear dynamic analysis with effective section properties. However, under certain conditions, the analysis can also be used in the assessment of the displacement demand, as will be illustrated in the examples to follow.

In this analysis, a stand-alone portion from a bridge structure (such as bent-frame with single or multicolumns) is isolated and statically analyzed taking into account whatever nonlinear behavior deemed necessary (most importantly and commonly, material and geometric nonlinear behavior). The analysis can utilize any of the modeling methods discussed in Section 36.6, but plastic hinges or distributed plasticity models are commonly used. The analytical frame model is first subjected to the applied tributary gravity load and then is pushed laterally in several load (or displacement) increments until a collapse mechanism or a given failure criterion is reached. Figure 36.14 shows a flowchart outlining a procedure using static push-over analysis in seismic design and retrofit evaluation.

When applying static push-over analysis in seismic design, it is assumed that such analysis can predict with reasonable accuracy the dynamic lateral load–displacement behavior envelope, and that an elastic acceleration response spectrum can provide the best means for establishing required structural performance.

36.7.3 Example 36.1 — Reinforced Concrete Multicolumn Bent Frame with P-Δ Effects

Problem Statement

The as-built details of a reinforced concrete bridge bent frame consisting of a bent cap beam and two circular columns supported on pile foundations are shown in Figure 36.15. An as-built unconfined

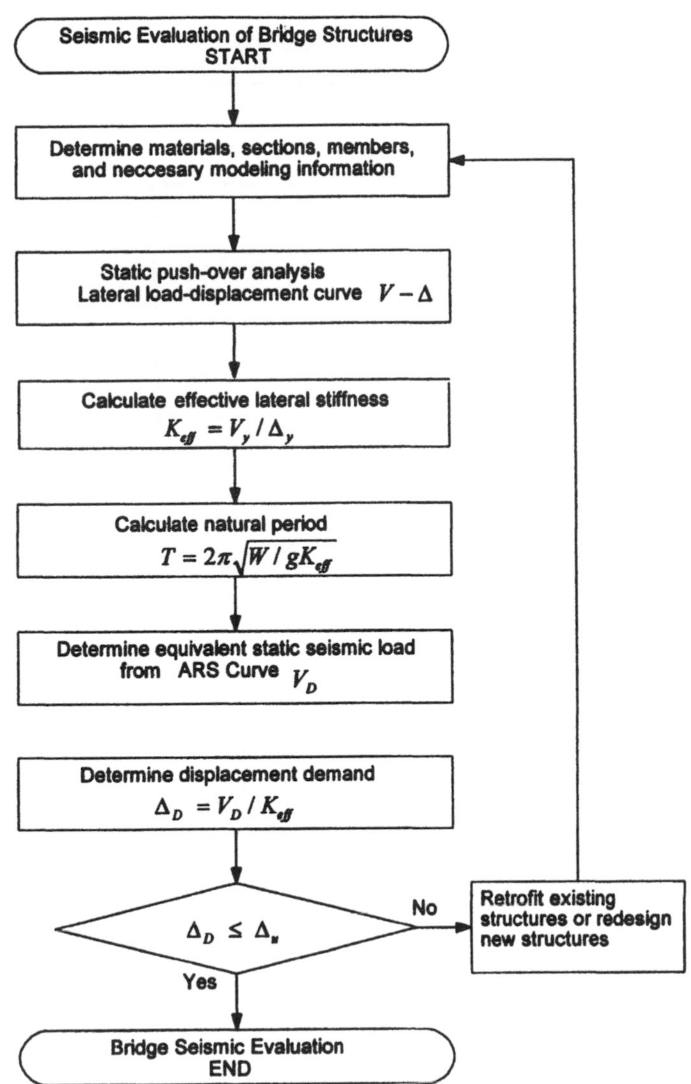

FIGURE 36.14 An alternative procedure for bridge seismic evaluation.

concrete strength of 5 ksi (34.5 MPa) and steel strength of 40 ksi (275.8 MPa) are assumed. Due to lack of adequate column transverse reinforcement, the columns are retrofitted with 0.5-in. (12.7-mm)-thick steel jacket. The bottom of the column is assumed to be fixed, however, since the footing lacks top mat and shear reinforcement, the bottom with a pinned connection is also to be considered. The frame is supported on a stiff pile–foundation and the soil–foundation–structure interaction is to be ignored.

Use static nonlinear push-over analysis to study the extent of the P-Δ effect on the lateral response of the bent frame when the columns are assumed fixed at the base in one case and pinned in another case. Assume the columns are retrofitted with steel jacket in both cases and determine if the footing retrofit is also required. Use 0.7 g ground acceleration and the ARS spectrum with 5% damping shown in Figure 36.16.

Analysis Procedure

The idealized bent frame, consisting of the cap beam and the two retrofitted column members, is discretized into a finite number of beam elements connected at joints, as shown in Figure 36.17.

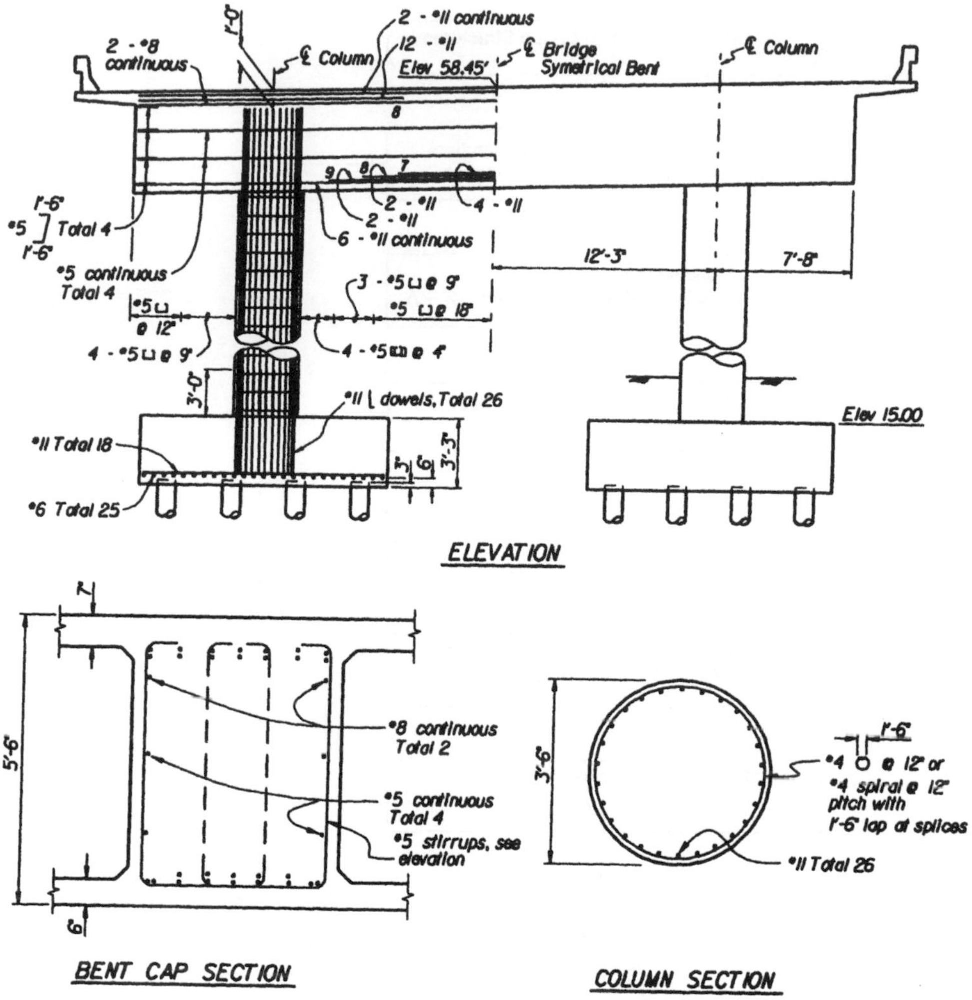

ELEVATION

BENT CAP SECTION

COLUMN SECTION

FIGURE 36.15 As-built plan — Example 36.1.

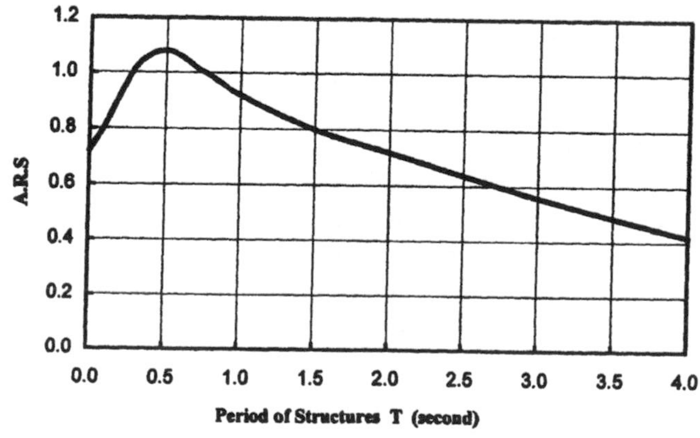

FIGURE 36.16 Specific ars curve — Example 36.1.

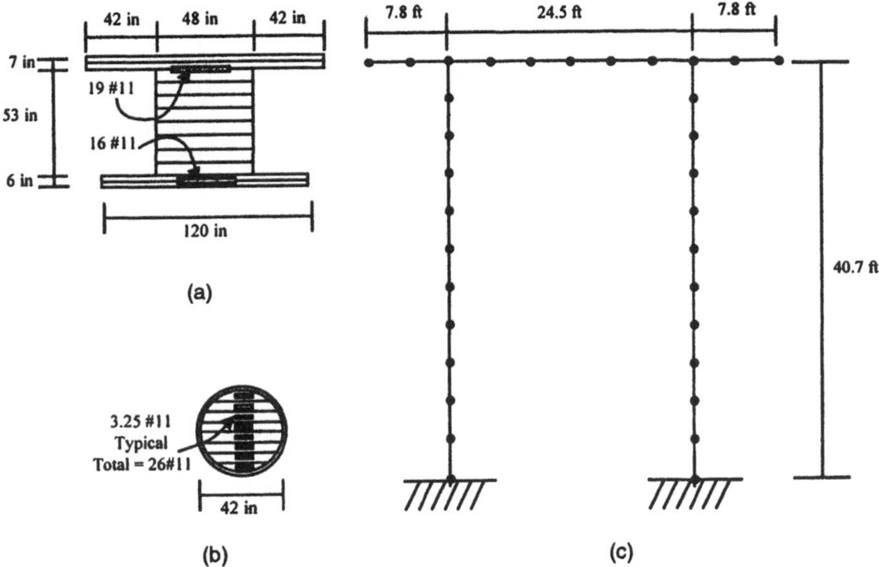

FIGURE 36.17 Analytical model — Example 36.1. (a) Local layered cab beam section: 12 concrete and 2 steel layers; (b) layered column section: 8 concrete and 8 steel layers; (c) discretized frame model.

The idealized column and cap beam cross sections are divided into several concrete layers and reinforcing steel layers as shown. Two different concrete material properties are used for the column and cap beam cross sections. The column concrete properties incorporated the increase in concrete ultimate stress and strain due to the confinement provided by the steel jacket. In this study the column confined ultimate concrete compressive stress and strain of 7.5 ksi (51.7 MPa) and 0.085 are used respectively. The total tributary superstructure dead load of 1160 kips (5160 kN) is applied uniformly along the length of the cap beam. The frame is pushed laterally in several load increments until failure is reached.

For this study, failure is defined as the limit state when one of the following conditions first take place:

1. A concrete layer strain reaches the ultimate compressive strain at any member section;
2. A steel layer strain reaches the rupture strain at any member section;
3. A 20% reduction from peak lateral load of the lateral load response curve (this condition is particularly useful when considering *P*-Δ).

The lateral displacement corresponding to this limit state at the top of the column defines the frame failure (available) displacement capacity.

A nonlinear analysis computer program NTFrame [41, 42] is used for the push-over analysis. The program is based on distributed plasticity model and the P-Δ effect is incorporated in the model second-order member stiffness formulation.

Discussion of the Results

The resulting frame lateral load vs. displacement responses are shown in Figures 36.18 for the cases when the bottom of the column is fixed and pinned. Both cases will be discussed next, followed by concluding remarks.

Column Fixed at Bottom Case

In this case the column base is modeled with a fixed connection. The lateral response with and without the *P*-Δ effect is shown in Figure 36.18a. The sharp drop in the response curve is due to several extreme concrete layers reaching their ultimate compressive strain at the top of the column.

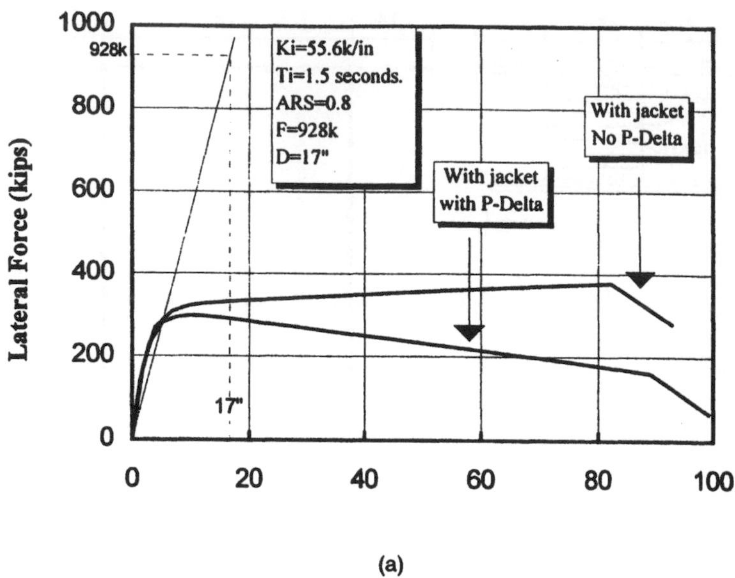

(a)

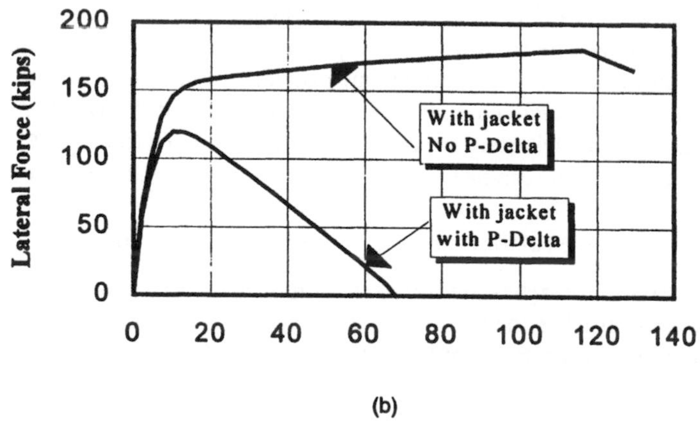

(b)

FIGURE 36.18 Lateral load vs. displacement responses — Example 36.1. Lateral response (a) fixed column (b) with penned column.

The effect of *P*-Δ at failure can be seen to be considerable but not as severe as shown in Figure 36.18b with the pinned connection. Comparing Figures 36.19a and b, one can observe that fixing the bottom of the column resulted in stiffer structural response.

Using the curve shown in Figure 36.18a, the displacement demand for the fixed column case with *P*-Δ effect is calculated as follows:

Step 1: Calculate the Initial Effective Stiffness K_{eff}

The computer results showed that the first column extreme longitudinal rebar reached yield at lateral force of 928 kips (4128 kN) at a corresponding lateral yield displacement of 17 in. (431.8 mm), therefore

$$K_{eff} = 928/17 = 55 \text{ kips/in. } (9.63 \text{ kN/mm})$$

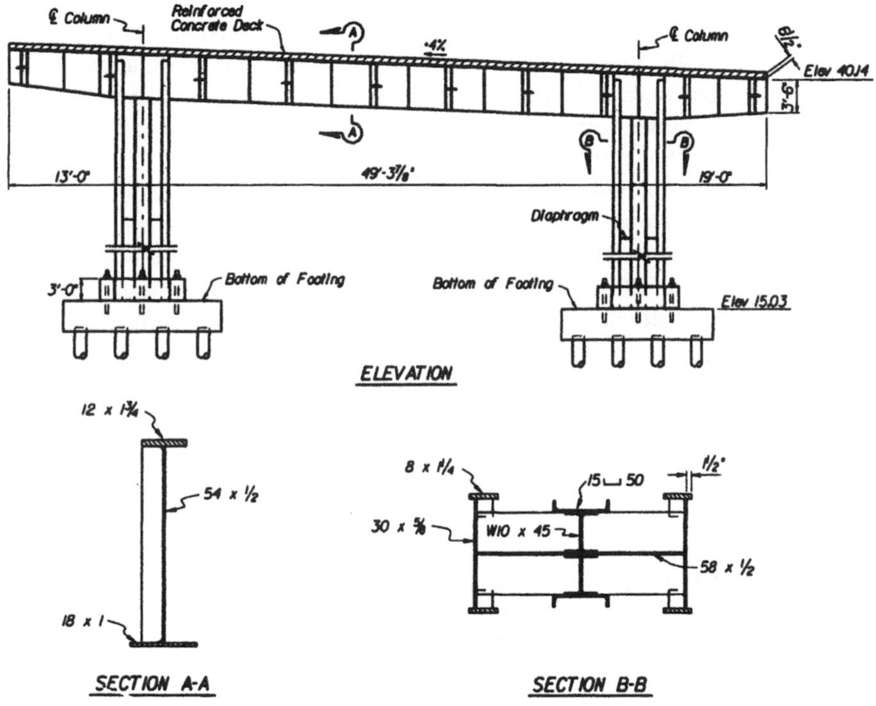

FIGURE 36.19 As-built plane — Example 36.2.

Step 2: Calculate an Approximate Fundamental Period T_f

$$T_f = 0.32\sqrt{\frac{W}{K_{\text{eff}}}} = 0.32\sqrt{\frac{1160}{55}} = 1.5 \text{ s}$$

Step 3: Determine the Damped Elastic Acceleration Response Spectrum (ARS) at the Site in g's
By using the given site spectrum shown in Figure 36.16 and the above calculated period, the corresponding ARS for 5% damping is 0.8.

Step 4: Calculate the Displacement Demand D_d

$$D_d = \frac{ARS(W)}{K_{eff}} = \frac{0.8(1160)}{55} = 16.9 \text{ in. } (429.3 \text{ m})$$

(in this case the yield and demand displacements are found to be practically equal).

In much of the seismic design practice in California, the effect of P-Δ is usually ignored if the P-Δ moment is less than 20% of the design maximum moment capacity. Adopting this practice and assuming the reduction in the moment is directly proportional to the reduction in the lateral force, one may conclude that at displacement demand of 16.9 in. (429.3 mm), the reduction in strength (lateral force) is less then 20%, and as a result the effects of P-Δ are negligible.

The displacement demand of 16.9 in. (429.3 mm) is less than the failure state displacement capacity of about 40 in. (1016 mm) (based on a 20% lateral load reduction from the peak). Note that for the fixed bottom case with P-Δ, the displacement when the extreme concrete layer at the top of the column reached its ultimate compressive strain is about 90 in. (2286 mm).

Column Pinned at Bottom Case

In this case the column bottom is modeled with a pinned connection. Note that the pinned condition assumption is based on the belief that in the event of a maximum credible earthquake the column/footing connection would quickly degenerate (degrade) and behave like a pinned connection. The resulting lateral responses with and without the *P*-Δ effect are shown in Figure 36.18b. In this case the effects of *P*-Δ is shown to be quite substantial.

When considering the response without the *P*-Δ one obtains a displacement demand of 38 in. (965.2 mm) (based on a calculated initial stiffness of 18.5 kips/in. (3.24 kN/mm) and a corresponding structure period of 2.5 s). This displacement demand is well below the ultimate (at failure) displacement capacity of about 115 in. (2921 mm). As a result, one would conclude that the retrofit measure of placing a steel jacket around the column with no footing retrofit is adequate.

The actual response, however, is the one that includes the *P*-Δ effect. In this case the effect of *P*-Δ resulted in a slight change in initial stiffness and frame period — 15.8 kips/in. (2.77 kN/mm) and 2.7 s, respectively. However, beyond the initial stages, the effects are quite severe on the load–displacement response. The failure mode in this case will most likely be controlled by dynamic instability of the frame. MacRae et al. [43] performed analytical studies of the effect of *P*-Δ on single-degree-of-freedom bilinear oscillators (i.e., single-column frame) and proposed some procedures to obtain a limiting value at which the structure becomes dynamically unstable. The process requires the generation of the proper hysteresis loops and the determination of what is termed the *effective bilinear stiffness factor.* Setting aside the frame dynamic instability issue, the calculated initial stiffness displacement demand is about 38 in. (965.2 mm) and the displacement capacity at 20% reduction from peak load is 24 in. (609.6 mm).

Referring to the curves with *P*-Δ in Figure 36.18, it is of interest to mention, as pointed out by Mahin and Boroschek [44], that continued pushing of the frame will eventually lead to a stage when the frame structure becomes statically unstable. At that stage the forces induced by the *P*-Δ effect overcome the mechanical resistance of the structure. Note that the point when the curve with *P*-Δ effect intersects the displacement-axis in (as shown Figure 36.19b) will determine the lateral displacement at which the structure becomes statically unstable. Dynamic instability limits can be 20 to 70% less than the static instability depending on the ground motion and structural characteristic [44]. Note that dynamic instability is assumed not to be a controlling factor in the previous case with fixed column.

In conclusion, if the as-built column–footing connection can support the expected column moment obtained from the fixed condition case (which is unlikely), then retrofitting the column with steel jacket without footing retrofit is adequate. Otherwise, the footing should also be retrofitted to reduce (limit) the effect of *P*-Δ.

It should be pointed out that in this example the analysis is terminated at the completion of the first plastic hinge (conservative), whereas in other types of push-over analysis such as event-to-event analysis, the engineer may chose to push the frame farther until it forms a collapse mechanism. Also, unlike the substitute structure procedure described by Priestly et al. [7] in which both the effective system stiffness and damping ratio are adjusted (iterated) several times before final displacement demand is calculated, here only the initial effective stiffness and a constant specified structure damping are used.

As a final remark, the *P*-Δ effect in bridge analysis is normally assumed small and is usually ignored. This assumption is justified in most cases under normal loading conditions. However, as this example illustrated, under seismic loading, the *P*-Δ effect should be incorporated in the analysis, when large lateral displacements are expected before the structure reaches its assumed failure state. In the design of a new bridge, the lateral displacement and the effect of *P*-Δ can be controlled. When assessing an existing bridge for possible seismic retrofit, accurate prediction of the lateral displacement with *P*-Δ effects can be an essential factor in determining the retrofit measures required.

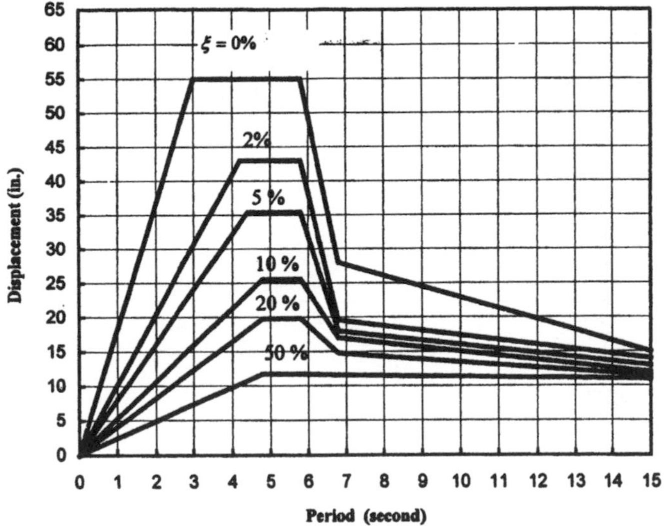

FIGURE 36.20 Displacement Response spectra — Example 36.2.

36.7.4 Example 36.2 — Steel Multicolumn Bent Frame Seismic Evaluation

Problem Statement

The as-built details of a steel bridge bent frame consisting of a bent cap plate girder and two builtup columns supported on a stiff pile–foundation, as shown in Figure 36.19. Steel is Grade 36. Site-specific displacement response spectra are given in Figure 36.20. For simplicity and illustration purposes, fixed bases of columns are assumed and the soil–foundation–structure interaction is ignored.

Evaluate lateral displacement capacity by using static nonlinear push-over analysis. Estimate seismic lateral displacement demands by using the substitute structure approach considering both the effective stiffness and the effective damping. The effective damping ξ can be calculated by Takeda's formula [45]:

$$\xi = 0.05 + \frac{\left(1 - \frac{0.95}{\sqrt{\mu_{\Delta d}}} - 0.05\sqrt{\mu_{\Delta d}}\right)}{\pi} \tag{36.55}$$

$$\mu_{\Delta d} = \frac{\Delta_{ud}}{\Delta_y} \tag{36.56}$$

where $\mu_{\Delta d}$ is displacement ductility demand; Δ_{ud} and Δ_y are displacement demand and yield displacement, respectively.

Analysis Modeling

The bent frame members are divided into several beam elements as shown in Figure 36.21. The properties of beam elements are defined by two sets of relationships for moment–curvature, axial force–strain, and torsion–twist for the cap beam and columns, respectively. The available ultimate curvature is assumed as 20 times yield curvature. The total tributary superstructure dead load of 800 kips (3558 kN) is applied at longitudinal girder locations. A lateral displacement is applied incrementally at the top of the bent column until a collapse mechanism of the bent frame is formed.

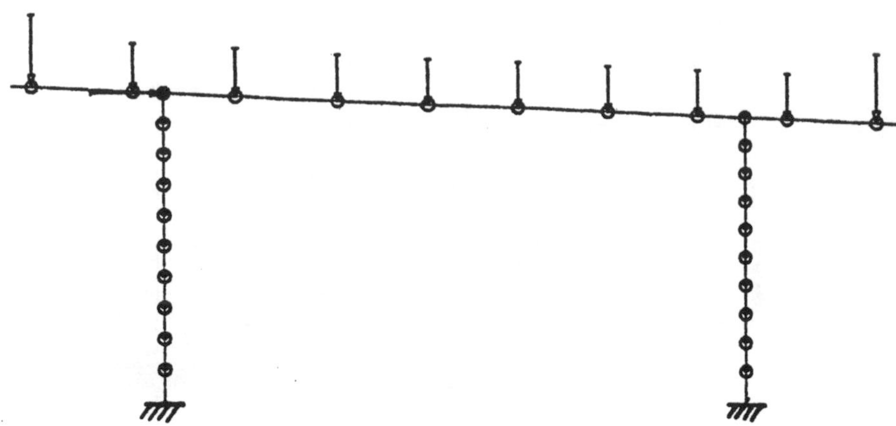

FIGURE 36.21 Analytical model — Example 36.2.

Displacement Capacity Evaluation

The displacement capacity evaluation is performed by push-over analysis using the ADINA [46] analysis program. Large displacements are considered in the analysis. The resulting lateral load vs. displacement response at the top of columns is shown in Figures 36.22. The sudden drops in the response curve are due to the several beam elements reaching their available ultimate curvatures. The yield displacement $\Delta_y = 1.25$ in. (31.8 mm) and the available ultimate displacement capacity (corresponding to a 20% reduction from the peak lateral load) $\Delta_u = 2.61$ in. (66.3 mm) are obtained.

Displacement Demand Estimation

A substitute structure approach with the effective stiffness and effective damping will be used to evaluate displacement demand.

1. Try $\Delta_{ud} = 3$ in. (76.2 mm); from Figure 36.20, Eqs. (36.55) and (36.56), we obtain

$$K_{\text{eff}} = \frac{600}{3} = 200 \text{ kips/in. } (35.04 \text{ kN/mm})$$

$$T_{\text{eff}} = 0.32\sqrt{\frac{W(\text{kips})}{k_{\text{eff}}(\text{kips/in.})}} = 0.32\sqrt{\frac{800}{200}} = 0.64 \text{ (s)}$$

$$\mu_{\Delta d} = \frac{\Delta_{ud}}{\Delta_y} = \frac{3}{1.25} = 2.4$$

$$\xi = 0.05 + \frac{\left(1 - \frac{0.95}{\sqrt{2.4}} - 0.05\sqrt{2.4}\right)}{\pi} = 0.15$$

From Figure 36.20, find $\Delta_d = 2.5$ in. < $\Delta_{ud} = 3$ in. (76.2 mm).

2. Try $\Delta_{ud} = 2.5$ in. (63.5 mm); from Figures 36.20 and 36.22 Eqs. (36.55), and (36.56), we obtain

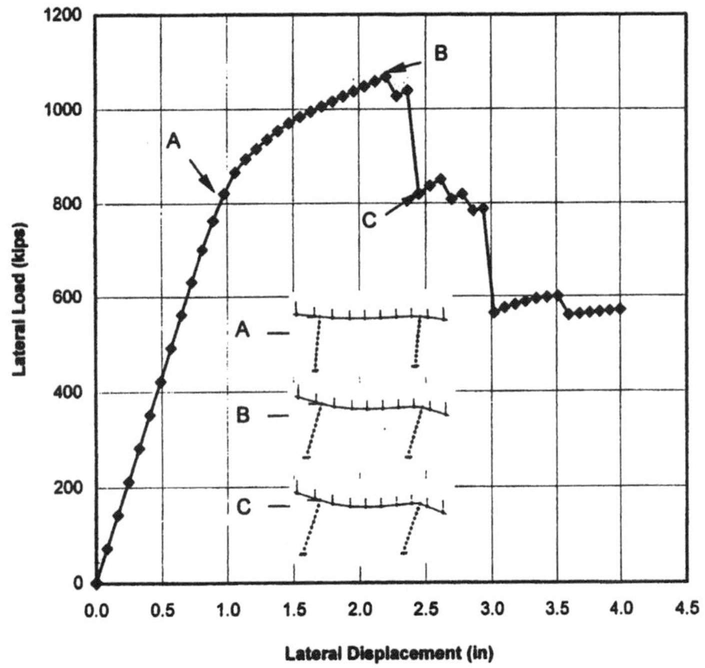

FIGURE 36.22 Lateral load vs. displacement — Example 36.2.

$$K_{\text{eff}} = \frac{830}{2.5} = 332 \text{ kips/in. (58.14 kN/mm)}$$

$$T_{\text{eff}} = 0.32\sqrt{\frac{W(\text{kips})}{K_{eff}(\text{kips/in.})}} = 0.32\sqrt{\frac{800}{332}} = 0.50$$

$$\mu_{\Delta d} = \frac{\Delta_{ud}}{\Delta_y} = \frac{2.5}{1.25} = 2$$

$$\xi = 0.05 + \frac{\left(1 - \frac{0.95}{\sqrt{2}} - 0.05\sqrt{2}\right)}{\pi} = 0.13$$

From Figure 36.20, find $\Delta_d = 2.45$ in. (62.2 mm) close to $\Delta_{ud} = 2.5$ in. (63.5 mm) OK
Displacement demand $\Delta_d = 2.45$ in. (62.2 mm).

Discussion

It can be seen that the displacement demand Δ_d of 2.45 in. (62.2 mm) is less than the available ultimate displacement capacity of $\Delta_u = 2.61$ in. (66.3 mm). It should be pointed out that in the actual seismic evaluation of this frame, the flexibility of the steel column to the footing bolted connection should be considered.

References

1. Chen, W. F., *Plasticity in Reinforced Concrete*, McGraw-Hill, New York, NY, 1982.
2. Clough, R. W. and Penzien, J., *Dynamics of Structures*, 2nd ed., McGraw-Hill, New York, 1993.
3. Fung, Y. C., *First Course in Continuum Mechanics*, 3rd ed., Prentice-Hall Engineering, Science & Math, Englewood Cliffs, NJ, 1994.

4. Chen, W. F. and Han, D. J., *Plasticity for Structural Engineers*, Gau Lih Book Co., Ltd., Taipei, Taiwan, 1995.
5. Chopra, A. K., *Dynamics of Structures: Theory and Applications to Earthquake Engineering*, Prentice-Hall, Englewood Cliffs, NJ, 1995.
6. Bathe, K. J., *Finite Element Procedures*, Prentice-Hall Engineering, Science & Math, Englewood Cliffs, NJ, 1996.
7. Priestley, M. J. N., Seible, F., and Calvi, G. M., *Seismic Design and Retrofit of Bridges*, John Wiley & Sons, New York, 1996.
8. Powell, G. H., Concepts and Principles for the Applications of Nonlinear Structural Analysis in Bridge Design, Report No. UCB/SEMM-97/08, Department of Civil Engineering, University of California, Berkeley, 1997.
9. Chen, W. F. and Lui, E. M., *Structural Stability: Theory and Implementation*, Elsevier, New York, 1987.
10. Goto, Y. and Chen, W. F., Second-order elastic analysis for frame design, *J. Struct. Eng. ASCE*, 113(7), 1501, 1987.
11. Allen, H. G. and Bulson, P. S., *Background of Buckling*, McGraw-Hill, London, 1980.
12. White, D. W. and McGuire, W., Method of Analysis in LRFD, Reprints of ASCE Structure Engineering Congress '85, Chicago, 1985.
13. Schilling, C. G. Buckling of one story frames, *AISC Eng. J.*, 2, 49, 1983.
14. Ekhande, S. G., Selvappalam, M., and Madugula, M. K. S., Stability functions for three-dimensional beam-column, *J. Struct. Eng. ASCE*, 115(2), 467, 1989.
15. Orbison, J. G., Nonlinear Static Analysis of Three-dimensional Steel Frames, Department of Structural Engineering, Cornell University, Ithaca, NY, 1982.
16. Yang, Y. B. and McGuire, W., Stiffness matrix for geometric nonlinear analysis, *J. Struct. Eng. ASCE*, 112(4), 853, 1986.
17. Yang, Y. B. and McGuire, W., Joint rotation and geometric nonlinear analysis, *J. Struct. Eng. ASCE*, 112(4), 879, 1986.
18. Hognestad, E., A Study of Combined Bending and Axial Load in Reinforced Concrete Members, University of Illinois Engineering Experimental Station, Bulletin Series No. 399, Nov., Urbana, IL, 1951.
19. Kent, D. C. and Park, R., Flexural members with confined concrete, *J. Struct. Div.* ASCE, 97(ST7), 1969, 1971.
20. Popovics, S. A., Review of stress-strain relationship for concrete, *J. ACI*, 67(3), 234, 1970.
21. Park, R. and Paulay, T., *Reinforced Concrete Structures*, John Wiley & Sons, New York, 1975.
22. Wang, W. C. and Duan, L., The stress-strain relationship for concrete, *J. Taiyuan Inst. Technol.*, 1, 125, 1981.
23. Mander, J. B., Priestley, M. J. N., and Park, R., Theoretical stress-strain model for confined concrete, *J. Struct. Eng. ASCE*, 114(8), 1804, 1988.
24. Mander, J. B., Priestley, M. J. N., and Park, R., Observed stress-strain behavior of confined concrete, *J. Struct. Eng. ASCE*, 114(8), 1827, 1988.
25. Hoshikuma, J., et al., Stress-strain model for confined reinforced concrete in bridge piers, *J. Struct. Eng. ASCE*, 123(5), 624, 1997.
26. William, K. J. and Warnke, E. P., Constitutive model for triaxial behavior of concrete, *Proc. IABSE*, 19, 1, 1975.
27. Mander, J. B., Priestley, M. J. N., and Park, R., Seismic Design of Bridge Piers, Research Report No. 84-2, University of Canterbury, New Zealand, 1984.
28. Chai, Y. H., Priestley, M. J. N., and Seible, F., Flexural Retrofit of Circular Reinforced Bridge Columns by Steel Jacketing, Report No. SSRP-91/05, University of California, San Diego, 1990.
29. Vebe, A. et al., Moment-curvature relations of reinforced concrete slab, *J. Struct. Div. ASCE*, 103(ST3), 515, 1977.

30. Holzer, S. M. et al., SINDER, A Computer Code for General Analysis of Two-Dimensional Reinforced Concrete Structures, AFWL-TR-74-228 Vol. 1, Air Force Weapons Laboratory, Kirtland AFB, NM, 1975.
31. Chen, W. F. and Atsuta, T., *Theory of Beam-Columns*, Vol. 1 and 2, McGraw-Hill, New York, 1977.
32. Bresler, B., Design criteria for reinforced concrete columns under axial load and biaxial bending, *J. ACI*, 32(5), 481, 1960
33. Duan, L. and Chen, W. F., A yield surface equation for doubly symmetrical section, *Struct. Eng.*, 12(2), 114, 1990.
34. King, W. S., White, D. W., and Chen, W. F., Second-order inelastic analysis methods for steel-frame design, *J. Struct. Eng. ASCE*, 118(2), 408, 1992.
35. Levy, R., Joseph, F., and Spillers, W. R., Member stiffness with offset hinges, *J. Struct. Eng. ASCE*, 123(4), 527, 1997.
36. Chen, W. F. and S. Toma, *Advanced Analysis of Steel Frames*, CRC Press, Boca Raton, FL, 1994.
37. Aschheim, M., Moehle, J. P., and Mahin, S. A., Design and Evaluation of Reinforced Concrete Bridges for Seismic Resistance, Report, UCB/EERC-97/04, University of California, Berkeley, 1997.
38. Priestley, N., Myths and Fallacies in Earthquake Engineering — Conflicts between Design and Reality, in *Proceedings of Tom Paulay Symposium — Recent Development in Lateral Force Transfer in Buildings*, University of California, San Diego, 1993.
39. Kowalsky, M. J., Priestley, M. J. N., and MacRae, G. A., Displacement-Based Design, Report No. SSRP-94/16, University of California, San Diego, 1994.
40. Duan, L. and Cooper, T. R., Displacement ductility capacity of reinforced concrete columns, *ACI Concrete Int.*, 17(11). 61, 1995.
41. Akkari, M. M., Nonlinear push-over analysis of reinforced and prestressed concrete frames, *Structure Notes*, State of California, Department of Transportation, Sacramento, July 1993.
42. Akkari, M. M., Nonlinear push-over analysis with p-delta effects, *Structure Notes*, State of California, Department of Transportation, Sacramento, November 1993.
43. MacRae, G. A., Priestly, M. J. N., and Tao, J., P-delta design in seismic regions, Structure System Research Project Report No. SSRP-93/05, University of California, San Diego, 1993.
44. Mahin, S. and Boroschek, R., Influence of geometric nonlinearities on the seismic response and design of bridge structures, *Background Report*, California Department of Transportation, Division of Structures, Sacramento, 1991.
45. Takeda, T., Sozen, M. A., and Nielsen, N. N., Reinforced concrete response to simulated earthquakes, *J. Struct. Div. ASCE*, 96(ST12), 2557, 1970.
46. ADINA, *ADINA-IN for ADINA User's Manual*, ADINA R & D, Inc., Watertown, MA, 1994.

37

Seismic Design Philosophies and Performance-Based Design Criteria

Lian Duan
California Department of Transportation

Fang Li
California Department of Transportation

37.1 Introduction

Seismic design criteria for highway bridges have been improving and advancing based on research findings and lessons learned from past earthquakes. In the United States, prior to the 1971 San Fernando earthquake, the seismic design of highway bridges was partially based on lateral force requirements for buildings. Lateral loads were considered as levels of 2 to 6% of dead loads. In 1973, the California Department of Transportation (Caltrans) developed new seismic design criteria related to site, seismic response of the soils at the site, and the dynamic characteristics of bridges. The American Association of State Highway and Transportation Officials (AASHTO) modified the Caltrans 1973 Provisions slightly, and adopted Interim Specifications. The Applied Technology Council (ATC) developed guidelines ATC-6 [1] for seismic design of bridges in 1981. AASHTO adopted ATC-6 [1] as the Guide Specifications in 1983 and later incorporated it into the Standard Specifications for Highway Bridges in 1991.

0-8493-7434-0/00/$0.00+$.50

Since the 1989 Loma Prieta earthquake in California [2], extensive research [3-15] has been conducted on seismic design and retrofit of bridges in the United States, especially in California. The performance-based project-specific design criteria [16,17] were developed for important bridges. Recently, ATC published improved seismic design criteria recommendations for California bridges [18] in 1996, and for U.S. bridges and highway structures [19] in 1997, respectively. Caltrans published the new seismic Design Methodology in 1999. [20] The new Caltrans Seismic Design Criteria [43] is under development. Great advances in earthquake engineering have been made during this last decade of the 20th century.

This chapter first presents the bridge seismic design philosophy and the current practice in the United States. It is followed by an introduction to the newly developed performance-based criteria [17] as a reference guide.

37.2 Design Philosophies

37.2.1 No-Collapse-Based Design

For seismic design of ordinary bridges, the basic philosophy is to prevent collapse during severe earthquakes [21-26]. To prevent collapse, two alternative approaches are commonly used in design. The first is a conventional force-based approach where the adjustment factor Z for ductility and risk assessment [26], or the response modification factor R [23], is applied to elastic member forces obtained from a response spectra analysis or an equivalent static analysis. The second approach is a more recent displacement-based approach [20] where displacements are a major consideration in design. For more-detailed information, reference can be made to a comprehensive discussion in *Seismic Design and Retrofit of Bridges* by Priestley, Seible, and Calvi [15].

37.2.2 Performance-Based Design

Following the 1989 Loma Prieta earthquake, bridge engineers [2] have faced three essential challenges:

- Ensure that earthquake risks posed by new construction are acceptable.
- Identify and correct unacceptable seismic safety conditions in existing structures.
- Develop and implement a rapid, effective, and economic response mechanism for recovering structural integrity after damaging earthquakes.

In the California, although the Caltrans Bridge Design Specifications [26] have not been formally revised since 1989, project-specific criteria and design memoranda have been developed and implemented for the design of new bridges and the retrofitting of existing bridges. These revised or supplementary criteria included guidelines for development of site-specific ground motion estimates, capacity design to preclude brittle failure modes, rational procedures for joint shear design, and definition of limit states for various performance objectives [14]. As shown in Figure 37.1, the performance requirements for a specific project must be established first. Loads, materials, analysis methods, and detailed acceptance criteria are then developed to achieve the expected performance.

37.3 No-Collapse-Based Design Approaches

37.3.1 AASHTO-LRFD Specifications

Currently, AASHTO has issued two design specifications for highway bridges: the second edition of AASHTO-LRFD [23] and the 16th edition of the Standard Specifications [24]. This section mainly discusses the design provisions of the AASHTO-LRFD Specifications.

The principles used for the development of AASHTO-LRFD [23] seismic design specifications are as follows:

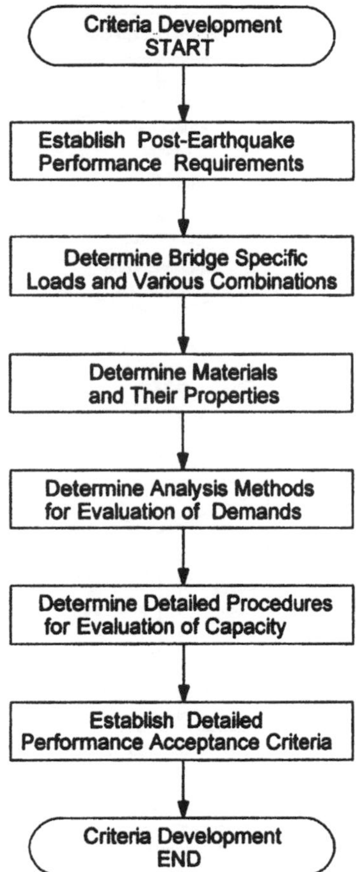

FIGURE 37.1 Development of performance-based seismic design criteria.

- Small to moderate earthquakes should be resisted within the elastic range of the structural components without significant damage.
- Realistic seismic ground motion intensities and forces should be used in the design procedures.
- Exposure to shaking from a large earthquake should not cause collapse of all or part of bridges where possible; damage that does occur should be readily detectable and accessible for inspection and repair.

Seismic force effects on each component are obtained from the elastic seismic response coefficient C_{sm} and divided by the elastic response modification factor *R*. Specific detailing requirements are provided to maintain structural integrity and to ensure ductile behavior. The AASHTO-LRFD seismic design procedure is shown in Figure 37.2.

Seismic Loads

Seismic loads are specified as the horizontal force effects and are obtained by production of C_{sm} and the equivalent weight of the superstructures. The seismic response coefficient is given as:

$$C_{sm} = \begin{cases} \dfrac{1.25AS}{T_m^{2/3}} \le 2.5A & \\ A(0.8+4T_m) & \text{for Soil III, IV, and nonfundamental } T_m < 0.3s \\ 3AST_m^{0.75} & \text{for Soil III, IV and } T_m > 0.4s \end{cases} \tag{37.1}$$

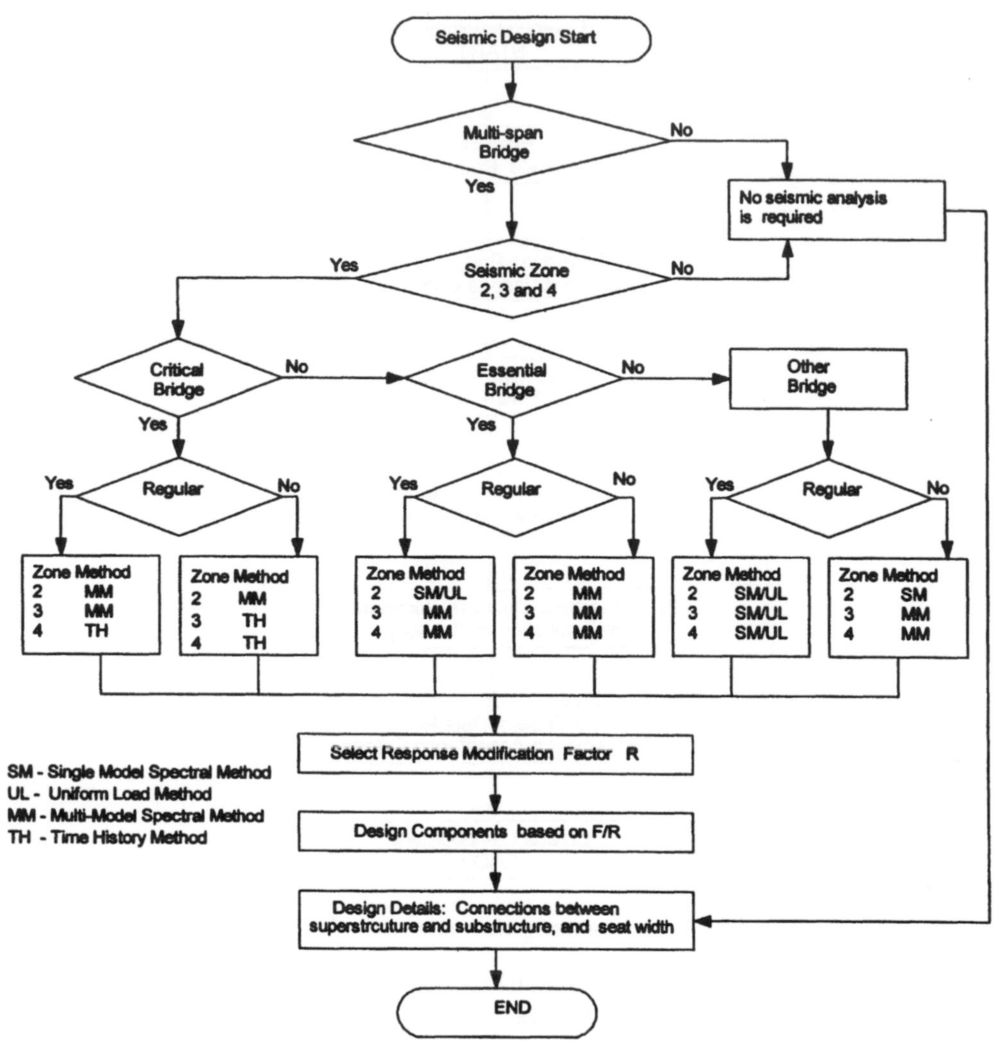

FIGURE 37.2 AASHO-LRFD seismic design procedure.

where A is the acceleration coefficient obtained from a contour map (Figure 37.3) which represents the 10% probability of an earthquake of this size being exceeded within a design life of 50 years; S is the site coefficient and is dependent on the soil profile types as shown in Table 37.1; T_m is the structural period of the mth mode in second.

Analysis Methods

Four seismic analysis methods specified in AASHTO-LRFD [23] are the uniform-load method, the single-mode spectral method, the multimode spectral method, and the time history method. Depending on the importance, site, and regularity of a bridge structure, the minimum complexity analysis methods required are shown in Figure 37.2. For single-span bridges and bridges located seismic Zone 1, no seismic analysis is required.

The importance of bridges is classified as critical, essential, and other in Table 37.2 [23], which also shows the definitions of a regular bridge. All other bridges not satisfying the requirements of Table 37.2 are considered irregular.

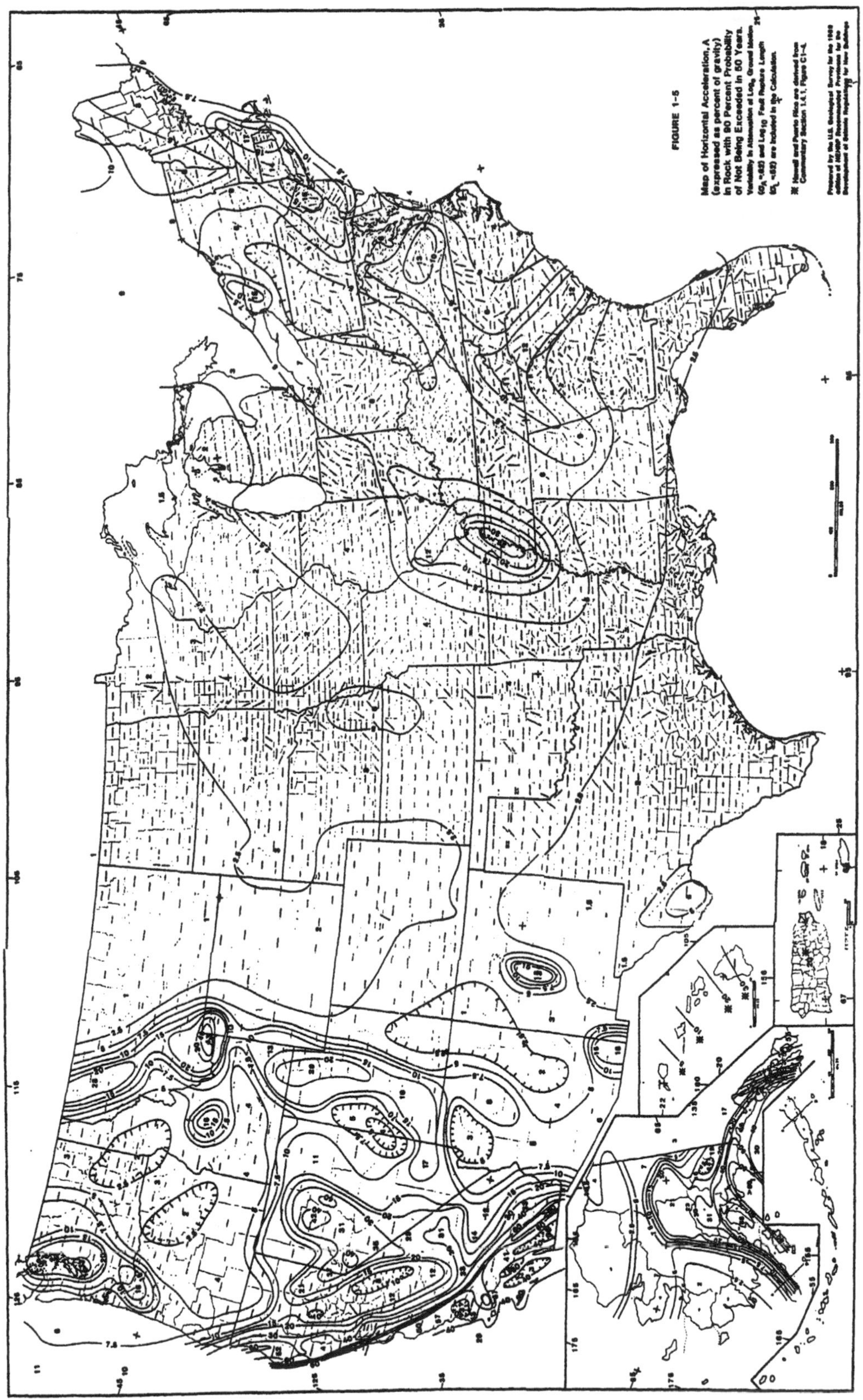

FIGURE 37.3 AASHTO-LRFD seismic contour map.

TABLE 37.1 AASHTO-LRFD Site Coefficient — *S*

Soil Profile Type	Descriptions	Site Coefficient, *S*
I	• Rock characterized by a shear wave velocity > 765 m/s • Stiff soil where the soil depth < 60 m and overlying soil are stable deposits of sands, gravel, or stiff clays	1.0
II	Stiff cohesive or deep cohesionless soil where the soil depth > 60 m and the overlying soil are stable deposits of sands, gravel, or stiff clays	1.2
III	Stiff to medium-stiff clays and sands, characterized by 9 m or more soft to medium-stiff clays without intervening layers of sands or other cohesionless soils	1.5
IV	Soft clays of silts > 12 m in depth characterized by a shear wave velocity < 153 m/s	2.0

TABLE 37.2 AASHTO-LRFD Bridge Classifications for Seismic Analysis

Importance	Critical	• Remain open to all traffic after design earthquake • Usable by emergency vehicles and for security/defense purposes immediately after a large earthquake (2500-year return period event)					
	Essential	Remain open emergency vehicles and for security/defense purposes immediately after the design earthquake (475-year return period event)					
	Others	Not required as critical and essential bridges					
Regularity	Regular	Structural Features Number of Span	2	3	4	5	6
		Maximum subtended angles for a curved bridge			90°		
		Maximum span length ratio from span to span	3	2	2	1.5	1.5
		Maximum bent/pier stiffness ratio from span to span excluding abutments	—	4	4	3	2
	Irregular	Multispan not meet requirement of regular bridges					

TABLE 37.3 Response Modification Factor, *R*

	Structural Component	Important Category: Critical	Essential	Others
Substructure	Wall-type pier — Large dimension	1.5	1.5	2.0
	Reinforced concrete pile bent			
	• Vertical pile only	1.5	2.0	3.0
	• With batter piles	1.5	1.5	2.0
	Single column	1.5	2.0	3.0
	Steel or composite steel and concrete pile bents			
	• Vertical pile only	1.5	3.5	5.0
	• With batter piles	1.5	2.0	3.0
	Multiple column bents	1.5	3.5	5.0
	Foundations		1.0	
Connection	Substructure to abutment		0.8	
	Expansion joints with a span of the superstructure		0.8	
	Column, piers, or pile bents to cap beam or superstructure		1.0	
	Columns or piers to foundations		1.0	

Component Design Force Effects

Design seismic force demands for a structural component are determined by dividing the forces calculated using an elastic dynamic analysis by appropriate response modification factor *R*

(Table 37.3) to account for inelastic behavior. As an alternative to the use of *R* factor for connection, the maximum force developed from the inelastic hinging of structures may be used for designing monolithic connections.

To account for uncertainty of earthquake motions, the elastic forces obtained from analysis in each of two perpendicular principal axes shall be combined using 30% rule, i.e., 100% of the absolute response in one principal direction plus 30% of the absolute response in the other.

The design force demands for a component should be obtained by combining the reduced seismic forces with the other force effects caused by the permanent and live loads, etc. Design resistance (strength) are discussed in Chapter 38 for concrete structures and Chapter 39 for steel structures.

37.3.2 Caltrans Bridge Design Specifications

The current Caltrans Bridge Design Specifications [26] adopts a single-level force-based design approach based on the no-collapse design philosophy and includes:

- Seismic force levels defined as elastic acceleration response spectrum (ARS);
- Multimodal response spectrum analysis considering abutment stiffness effects;
- Ductility and risk *Z* factors used for component design to account for inelastic effects;
- Properly designed details.

Seismic Loads

A set of elastic design spectra ARS curves are recommended to consider peak rock accelerations (*A*), normalized 5% damped rock spectra (*R*), and soil amplification factor (*S*). Figure 37.4 shows typical ARS curves.

Analysis Methods

For ordinary bridges with well-balanced span and bent/column stiffness, an equivalent static analysis with the ARS times the weight of the structure applied at the center of gravity of total structures can be used. This method is used mostly for hinge restrainer design. For ordinary bridges with significantly irregular geometry configurations, a dynamic multimodal response spectrum analysis is recommended. The following are major considerations in seismic design practice:

- A beam-element model with three or more lumped masses in each span is usually used [25-27].
- A larger cap stiffness is often used to simulate a stiff deck.
- Gross section properties of columns are commonly used to determine force demands, and cracked concrete section properties of columns are used for displacement demands.
- Soil–spring elements are used to simulate the soil–foundation–structure–interaction. Adjustments are often made to meet force–displacement compatibility, particularly for abutments. The maximum capacity of the soil behind abutments with heights larger than 8 ft (2.44 m) is 7.7 ksf (369 kPa) and lateral pile capacity of 49 kips (218 kN) per pile.
- Compression and tension models are used to simulate the behavior of expansion joints.

Component Design Force Effects

Seismic design force demands are determined using elastic forces from the elastic response analysis divided by the appropriate component- and period-based (stiffness) adjustment factor *Z*, as shown in Figure 38.4a to consider ductility and risk. In order to account for directional uncertainty of earthquake motions, elastic forces obtained from analysis of two perpendicular seismic loadings are combined as the 30% rule, the same as the AASHTO-LRFD [23].

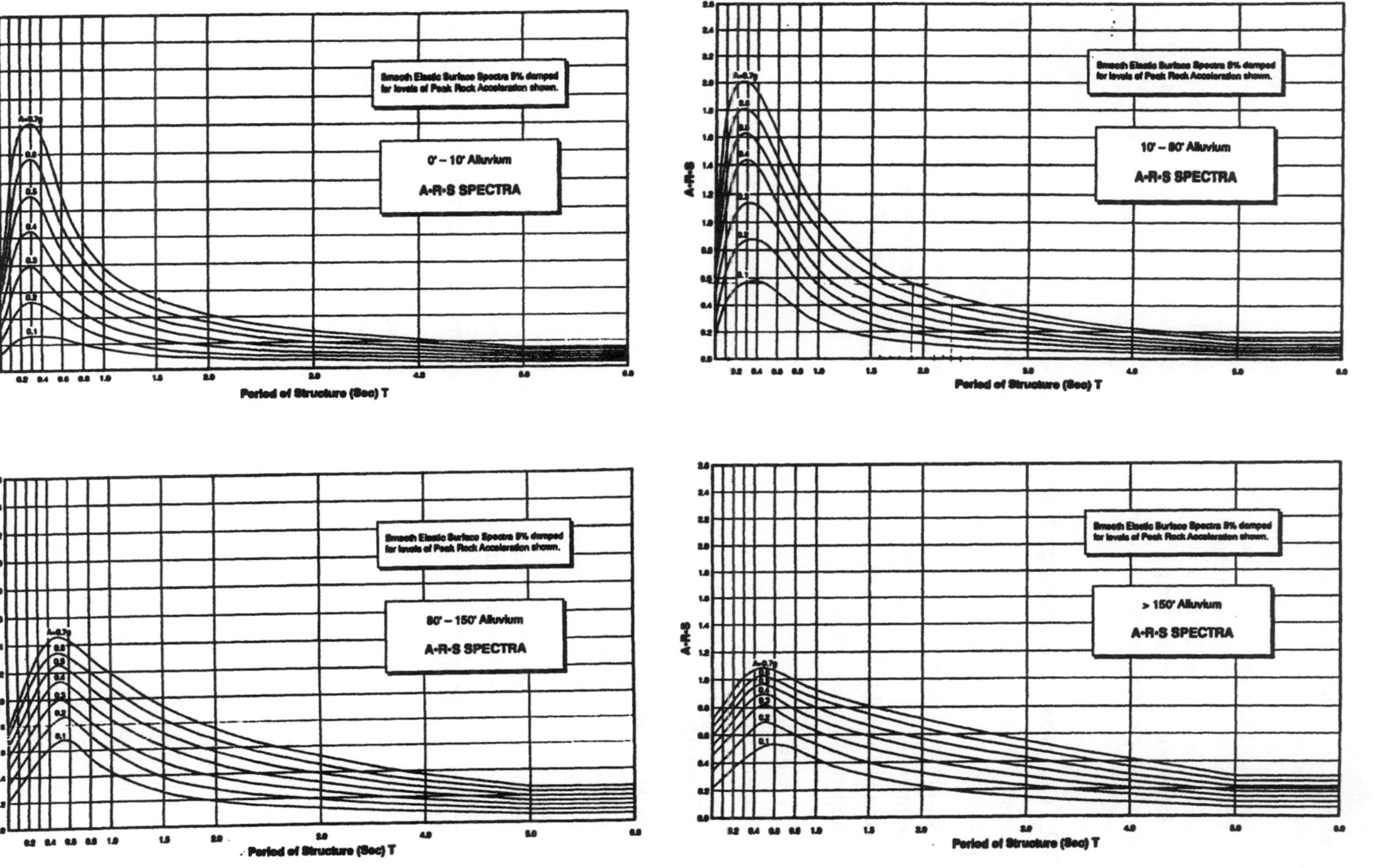

FIGURE 37.4 Caltrans ARS curves.

TABLE 37.4 Caltrans Seismic Performance Criteria

Ground motions at the site	Minimum (ordinary bridge) performance level	Important bridge performance level
Functional evaluation	Immediate service; repairable damage	Immediate service level; minimum damage
Safety evaluation	Limited service level; significant damage	Immediate service level; repairable damage

Definitions:
Important Bridge (one of more of following items present):
- Bridge required to provide secondary life safety
- Time for restoration of functionality after closure creates a major economic impact
- Bridge formally designed as critical by a local emergency plan

(*Ordinary Bridge*: Any bridge not classified as an important bridge.)
Functional Evaluation Ground Motion (*FEGM*): Probabilistic assessed ground motions that have a 40% probability of occurring during the useful lifetime of the bridge. The determination of this event shall be reviewed by a Caltrans-approved consensus group. A separate functionality evaluation is required for important bridges. All other bridges are only required to meet the specified design requirement to assure minimum functionality performance level compliance.

Safety Evaluation Ground Motion (*SEGM*): Up to two methods of defining ground motion may be used:
- Deterministically assessed ground motions from the maximum earthquake as defined by the Division of Mines and Geology Open-File Report 92-1 [1992].
- Probabilistically assessed ground motions with a long return period (approximately 1000–2000 years).

For important bridges both methods should be given consideration; however, the probabilistic evaluation should be reviewed by a Caltrans-approved consensus group. For all other bridges, the motions should be based only on the deterministic evaluation. In the future, the role of the two methods for other bridges should be reviewed by a Caltrans-approved consensus group.

Immediate Service Level: Full access to normal traffic available almost immediately (following the earthquake).
Repairable Damage: Damage that can be repaired with a minimum risk of losing functionality.
Limited Service Level: Limited access (reduced lanes, light emergency traffic) possible with in days. Full service restoration within months.
Significant Damage: A minimum risk of collapse, but damage that would require closure for repairs.

Note: Above performance criteria and definitions have been modified slightly in the proposed provisions for California Bridges (ACT-32, 1996) and the U.S. Bridges (ATC-18, 1997) and Caltrans (1999) MTD 20-1 (920).

37.4 Performance-Based Design Approaches

37.4.1 Caltrans Practice

Since 1989, the design criteria specified in Caltrans BDS [26] and several internal design manuals [20,25,27] have been updated continuously to reflect recent research findings and development in the field of seismic bridge design. Caltrans has been shifting toward a displacement-based design approach emphasizing capacity design. In 1994 Caltrans established the seismic performance criteria listed in Table 37.4. A bridge is categorized as an "important" or "ordinary" bridge. Project-specific two-level seismic design procedures for important bridges, such as the R-14/I-5 Interchange replacement [16], the San Francisco–Oakland Bay Bridge (SFOBB) [17], and the Benicia-Martinez Bridge [28], are required and have been developed. These performance-based seismic design criteria include site-specific ARS curves, ground motions, and specific design procedures to reflect the desired performance of these structures. For ordinary bridges, only one-level safety-evaluation design is required. The following section briefly discusses the newly developed seismic design methodology for ordinary bridges.

37.4.2 New Caltrans Seismic Design Methodology (MTD 20-1, 1999)

To improve Caltrans seismic design practice and consolidate new research findings, ATC-32 recommendations [18] and the state-of-the-art knowledge gained from the recent extensive seismic bridge design, Caltrans engineers have been developing the Seismic Design Methodology [20] and the Seismic Design Criteria (SDC) [43] for ordinary bridges.

Ordinary Bridge Category

An ordinary bridge can be classified as a "standard" or "nonstandard" bridge. An nonstandard bridge may feature irregular geometry and framing (multilevel, variable width, bifurcating, or highly horizontally curved superstructures, different structure types, outriggers, unbalanced mass and/or stiffness, high skew) and unusual geologic conditions (soft soil, moderate to high liquefaction potential, and proximity to an earthquake fault). A standard bridge does not contain nonstandard features. The performance criteria and the service and damage levels are shown in Table 37.4.

Basic Seismic Design Concept

The objective of seismic design is to ensure that all structural components have sufficient strength and/or ductility to prevent collapse— a limit state where additional deformation will potentially render a bridge incapable of resisting its self-weight during a maximum credible earthquake (MCE). Collapse is usually characterized by structural material failure and/or instability in one or more components.

Ductility is defined as the ratio of ultimate deformation to the deformation at first yield and is the predominant measure of structural ability to dissipate energy. Caltrans takes advantage of ductility and postelastic strength and does not design ordinary bridges to remain elastic during design earthquakes because of economic constraints and the uncertainties in predicting future seismic demands. Seismic deformation demands should not exceed structural deformation capacity or energy-dissipating capacity. Ductile behavior can be provided by inelastic actions either through selected structural members and/or through protective systems— seismic isolations and energy dissipation devices. Inelastic actions should be limited to the predetermined regions that can be easily inspected and repaired following an earthquake. Because the inelastic response of a concrete superstructure is difficult to inspect and repair and the superstructure damage may cause the bridge to be in an unserviceable condition, inelastic behavior on most bridges should preferably be located in columns, pier walls, backwalls, and wingwalls (see Figure 38.1).

To provide an adequate margin of strength between ductile and nonductile failure modes, capacity design is achieved by providing overstrength against seismic load in superstructure and foundations. Components not explicitly designed for ductile performance should be designed to remain essentially elastic; i.e., response in concrete components should be limited to minor cracking or limited to force demands not exceeding the strength capacity determined by current Caltrans SDC, and response in steel components should be limited to force demands not exceeding the strength capacity determined by current Caltrans SDC.

Displacement-Based Design Approach

The objective of this approach is to ensure that the structural system and its individual components have enough capacity to withstand the deformation imposed by the design earthquake. Using displacements rather than forces as a measurement of earthquake damage allows a structure to fulfill the required functions.

In a displacement-based analysis, proportioning of the structure is first made based on strength and stiffness requirements. The appropriate analysis is run and the resulting displacements are compared with the available capacity which is dependent on the structural configuration and rotational capacity of plastic hinges and can be evaluated by inelastic static push-over analysis (see Chapter 36). This procedure has been used widely in seismic bridge design in California since 1994. Alternatively, a target displacement could be specified, the analysis performed, and then design strength and stiffness determined as end products for a structure [29,30]. In displacement-based design, the designer needs to define criteria clearly for acceptable structural deformation based on postearthquake performance requirements and the available deformation capacity. Such criteria are based on many factors, including structural type and importance.

Seismic Demands on Structural Components

For ordinary bridges, safety-evaluation ground motion shall be based on deterministic assessment corresponding to the MCE, the largest earthquake which is capable of occurring based on current geologic information. The ARS curves (Figure 37.5) developed by ATC-32 are adopted as standard horizontal ARS curves in conjunction with the peak rock acceleration from the Caltrans Seismic Hazard Map 1996 to determine the horizontal earthquake forces. Vertical acceleration should be considered for bridges with nonstandard structural components, unusual site conditions, and/or close proximity to earthquake faults and can be approximated by an equivalent static vertical force applied to the superstructure.

For structures within 15 km of an active fault, the spectral ordinates of the appropriate standard ARS curve should be increased by 20%. For long-period structures ($T \geq 1.5$ s) on deep soil sites (depth of alluvium ≥ 75 m) the spectral ordinates of the appropriate standard ARS curve should be increased by 20% and the increase applies to the portion of the curves with periods greater than 1.5 s.

Displacement demands should be estimated from a linear elastic response spectra analysis of bridges with effective component stiffness. The effective stiffness of ductile components should represent the actual secant stiffness of the component near yield. The effective stiffness should include the effects of concrete cracking, reinforcement, and axial load for concrete components; residual stresses, out-of-straightness, and axial load for steel components; the restraints of the surrounding soil for pile shafts. Attempts should be made to design bridges with dynamic characteristics (mass and stiffness) so that the fundamental period falls within the region between 0.7 and 3 s where the equal displacement principle applies. It is also important that displacement demands also include the combined effects of multidirectional components of horizontal acceleration (for example, 30% rules).

For short-period bridges, linear elastic analysis underestimates displacement demands. The inability to predict displacements of a linear analysis accurately can be overcome by designing the bridge to perform elastically, multiplying the elastic displacement by an amplification factor, or using seismic isolation and energy dissipation devices to limit seismic response. For long-period ($T > 3$ s) bridges, a linear elastic analysis generally overestimates displacements and linear elastic displacement response spectra analysis should be used.

Force demands for essentially elastic components adjacent to ductile components should be determined by the joint–force equilibrium considering plastic hinging capacity of the ductile component multiplied by an overstrength factor. The overstrength factor should account for the variations in material properties between adjacent components and the possibility that the actual strength of the ductile components exceeds its estimated plastic capacity. Force demands calculated from a linear elastic analysis should not be used.

Seismic Capacity of Structural Components

Strength and deformation capacity of a ductile flexural element should be evaluated by moment–curvature analysis (see Chapters 36 and 38). Strength capacity of all components should be based on the most probable or expected material properties, and anticipated damages. The impact of the second-order P-Δ and P-δ effects on the capacity of all members subjected to combined bending and compression should be considered. Components may require re-design if the P-Δ and P-δ effects are significant.

Displacement capacity of a bridge system should be evaluated by a static push-over analysis (see Chapter 36). The rotational capacity of all plastic hinges should be limited to a safe performance level. The plastic hinge regions should be designed and detailed to perform with minimal strength degradation under cyclic loading.

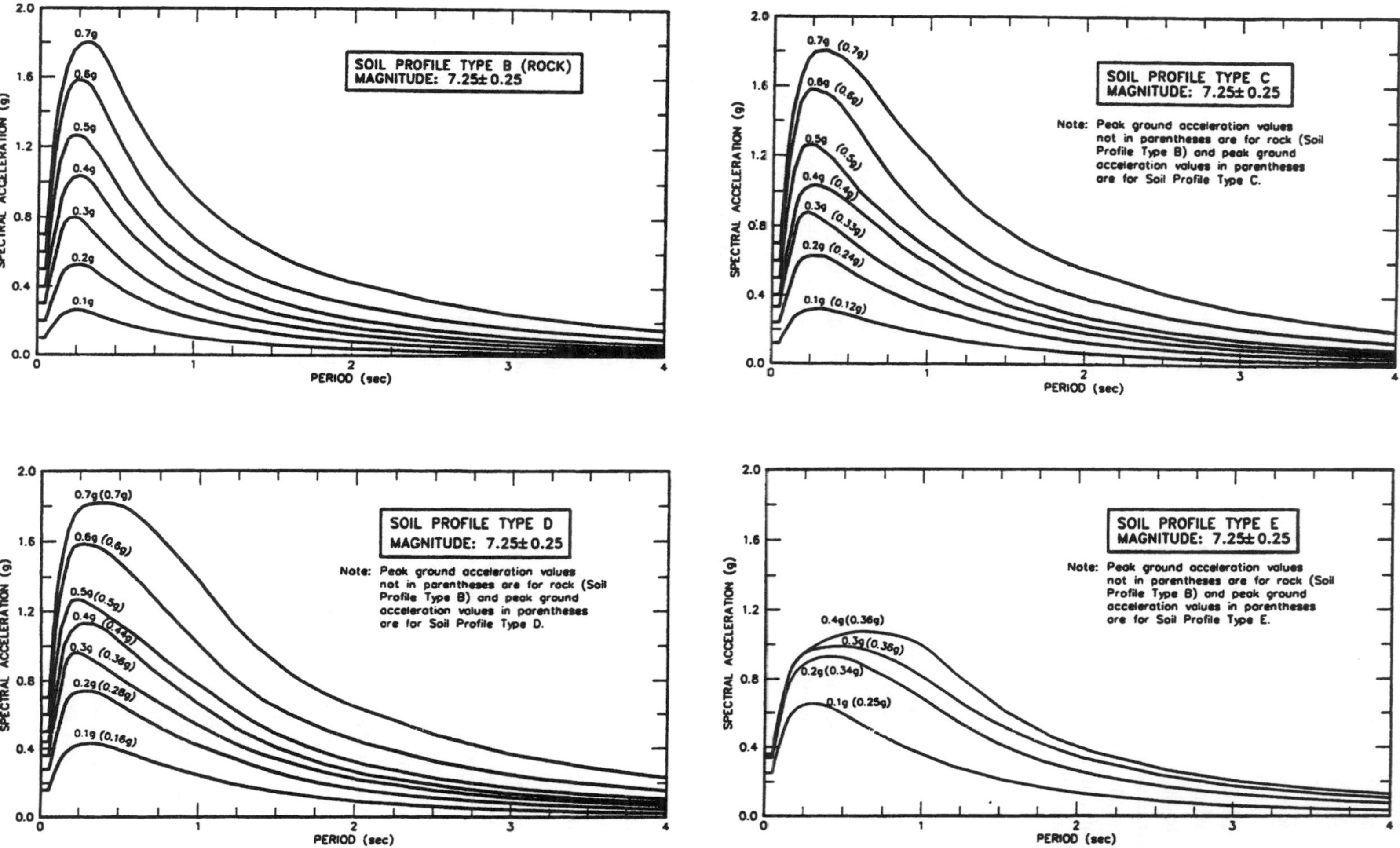

FIGURE 37.5 ATC-32 recommended ARS curves.

Seismic Design Practice

- Bridge type, component selection, member dimensions, and aesthetics should be investigated to reduce the seismic demands to the greatest extent possible. Aesthetics should not be the primary reason for producing undesirable frame and component geometry.
- Simplistic analysis models should be used for initial assessment of structural behavior. The results of more-sophisticated models should be checked for consistency with the results obtained from the simplistic models. The rotational and translational stiffness of abutments and foundations modeled in the seismic analysis must be compatible with their structural and geotechnical capacity. The energy dissipation capacity of the abutments should be considered for bridges whose response is dominated by the abutments.
- The estimated displacement demands under design earthquake should not exceed the global displacement capacity of the structure and the local displacement capacity of any of its individual components.
- Adjacent frames should be proportioned to minimize the differences in the fundamental periods and skew angles, and to avoid drastic changes in stiffness. All bridge frames must meet the strength and ductility requirements in a stand-alone condition. Each frame should provide a well-defined load path with predetermined plastic hinge locations and utilize redundancy whenever possible.
- For concrete bridges, structural components should be proportioned to direct inelastic damage into the columns, pier walls, and abutments. The superstructure should have sufficient overstrength to remain essentially elastic if the columns/piers reach their most probable plastic moment capacity. The superstructure-to-substructure connection for nonintegral caps may be designed to fuse prior to generating inelastic response in the superstructure. The girders, bent caps, and columns should be proportioned to minimize joint stresses. Moment-resisting connections should have sufficient joint shear capacity to transfer the maximum plastic moments and shears without joint distress.
- For steel bridges, structural components should be generally designed to ensure that inelastic deformation only occur in the specially detailed ductile substructure elements. Inelastic behavior in the form of controlled damage may be permitted in some of the superstructure components, such as the cross frames, end diaphragms, shear keys, and bearings. The inertial forces generated by the deck must be transferred to the substructure through girders, trusses, cross frames, lateral bracings, end diaphragms, shear keys, and bearings. As an alternative, specially designed ductile end-diaphragms may be used as structural mechanism fuses to prevent damage in other parts of structures.
- Initial sizing of columns should be based on slenderness ratios, bent cap depth, compressive stress ratio, and service loads. Columns should demonstrate dependable post-yield-displacement capacity without an appreciable loss of strength. Thrust–moment–curvature (P–M–Φ) relationships should be used to optimize the performance of a column under service and seismic loads. Concrete columns should be well proportioned, moderately reinforced, and easily constructed. Abrupt changes in the cross section and the capacity of columns should be avoided. Columns must have sufficient rotation capacity to achieve the target displacement ductility requirements.
- Steel multicolumn bents or towers should be designed as ductile moments-resisting frames (MRF) or ductile braced frames such as concentrically braced frames (CBF) and eccentrically braced frames (EBF). For components expected to behave inelastically, elastic buckling (local compression and shear, global flexural, and lateral torsion) and fracture failure modes should be avoided. All connections and joints should preferably be designed to remain essentially elastic. For MRFs, the primary inelastic deformation should preferably be columns. For CBFs, diagonal members should be designed to yield when members are in tension and to buckle inelastically when they are in compression. For EBFs, a short beam segment designated as a *link* should be well designed and detailed.

TABLE 37.5 ATC-32 Minimum Required Analysis

Bridge Type		Functional Evaluation	Safety Evaluation
Ordinary Bridge	Type I	None required	Equivalent static analysis or elastic dynamic analysis
	Type II	None required	Elastic dynamic analysis
Important Bridge	Type I	Equivalent static analysis or elastic dynamic analysis	
	Type II	Elastic dynamic analysis	Elastic dynamic analysis or inelastic static analysis or inelastic dynamic analysis

- Force demands on the foundation should be based on the most probable plastic capacity of the columns/piers with an appropriate amount of overstrength. Foundation elements should be designed to remain essentially elastic. Pile shaft foundations may experience limited inelastic deformation when they are designed and detailed in a ductile manner.
- The ability of an abutment to resist bridge seismic forces should be based on its structural capacity and the soil resistance that can be reliably mobilized. Skewed abutments are highly vulnerable to damage. Skew angles at abutments should be reduced, even at the expense of increasing the bridge length.
- Necessary restrainers and sufficient seat width should be provided between adjacent frames at all intermediate expansion joints, and at the seat-type abutments to eliminate the possibility of unseating during a seismic event.

37.4.3 ATC Recommendations

ATC-32 Recommendations to Caltrans

The Caltrans seismic performance criteria shown in Table 37.4 provide the basis for development of the ATC-32 recommendations [18]. The major changes recommended for the Caltrans BDS are as follows:

- The importance of relative (rather than absolute) displacement in the seismic performance of bridges is emphasized.
- Bridges are classified as either "important or ordinary." Structural configurations are divided into Type I, simple (similar to regular bridges), and Type II, complex (similar to irregular bridges). For important bridges, two-level design (safety evaluation and function evaluation) approaches are recommended. For ordinary bridges, a single-level design (safety evaluation) is recommended. Minimum analyses required are shown in Table 37.5.
- The proposed family of site-dependent design spectra (which vary from the current Caltrans curves) are based on four of six standard sites defined in a ground motion workshop [31].
- Vertical earthquake design loads may be taken as two thirds of the horizontal load spectra for typical sites not adjacent to active faults.
- A force-based design approach is retained, but some of the inherent shortcomings have been overcome by using new response modification factors and modeling techniques which more accurately estimate displacements. Two new sets of response modification factors Z (Figure 38.4b) are recommended to represent the response of limited and full ductile structural components. Two major factors are considered in the development of the new Z factors: the relationship between elastic and inelastic response is modeled as a function of the natural period of the structure and the predominate period of the ground motion; the distribution of elastic and inelastic deformation within a structural component is a function of its component geometry and framing configuration.

- *P*-Δ effects should be included using inelastic dynamic analysis unless the following relation is satisfied:

$$\frac{V_o}{W} \geq 4\frac{\delta_u}{H} \tag{37.2}$$

  where V_o is base shear strength of the frame obtained from plastic analysis; W is the dead load; δ_u is maximum design displacement; and H is the height of the frame. The inequality in Eq. (37.2) is recommended to keep bridge columns from being significantly affected by *P*-Δ moments.
- A adjustment factor, R_d, is recommended to adjust the displacement results from an elastic dynamic analysis to reflect the more realistic inelastic displacements that occur during an earthquake.

$$R_d = \left(1 - \frac{1}{Z}\right)\frac{T}{T^*} + \frac{1}{Z} \geq 1 \tag{37.3}$$

  where T is the natural period of the structure, T^* is the predominant period of ground motion, and Z is force-reduction coefficient defined in Figure 38.4b.
- Modification was made to the design of ductile elements, the design of nonductile elements using capacity design approach, and the detailing of reinforced concrete for seismic resistance based on recent research findings.
- Steel seismic design guidelines and detailing requirements are very similar to building code requirements.
- Foundation design guidelines include provisions for site investigation, determination of site stability, modeling and design of abutments and wing-walls, pile and spread footing foundations, drilled shafts, and Earth-retaining structures.

ATC-18 Recommendation to FHWA

The ATC recently reviewed current seismic design codes and specifications for highway structures worldwide and provided recommendations for future codes for bridge structures in the United States [19]. The recommendations have implemented significant changes to current specifications, most importantly the two-level design approach, but a single-level design approach is included. The major recommendations are summarized in Tables 37.6 and 37.7.

37.5 Sample Performance-Based Criteria

This section introduces performance-based criteria as a reference guide. A complete set of criteria will include consideration of postearthquake performance criteria, determination of seismic loads and load combinations, material properties, analysis methods, detailed qualitative acceptance criteria. The materials presented in this section are based on successful past experience, various codes and specifications, and state-of-the-art knowledge. Much of this section is based on the Seismic Retrofit Design Criteria developed for the SFOBB west span [17]. It should be emphasized that the sample criteria provided here should serve as a guide and are not meant to encompass all situations.

The postearthquake performance criteria depending on the importance of bridges specified in Table 37.4 are used. Two levels of earthquake loads, FEGM and SEGM, defined in Table 37.4 are required. The extreme event load combination specified by AASHTO-LRFD [23] should be considered (see Chapter 5).

TABLE 37.6 ATC-18 Recommendations for Future Bridge Seismic Code Development (Two-Level Design Approach)

Level		Lower Level Functional Evaluation	Upper Level Safety Evaluation
Performance Criteria	Ordinary bridges	Service level — immediate Damage level — repairable	Service level — limited Damage level — significant
	Important bridges	Service level — immediate Damage level — minimum	Service level — immediate Damage level — repairable
Design load		Functional evaluation ground motion	Safety evaluation ground motion
Design approach		• Continue current AASHTO seismic performance category • Adopt the two-level design approach at least for important bridges in higher seismic zones • Use elastic design principles for the lower-level design requirement • Use nonlinear analysis — deformation-based procedures for the upper-level design	
Analysis		Current elastic analysis procedures (equivalent static and multimodel)	Nonlinear static analysis
Design force	Ductile component	Remain undamaged	Have adequate ductility to meet the performance criteria
	Nonductile component	Remain undamaged	For sacrificial element — ultimate strength should be close to but larger than that required for the lower-level event For nonsacrificial element — based on elastic demands or capacity design procedure
	Foundation	Capacity design procedure — to ensure there is no damage	
Design displacement		Use the upper-level event Remain current seat width requirements Consider overall draft limits to avoid P-Δ effects on long-period structures	
Concrete and steel design		Use the capacity design procedure for all critical members	
Foundation design		• Complete geotechical analysis for both level events • Prevent structural capacity of the foundations at the lower level event • Allow damage in the upper-level event as long as it does not lead to catastrophic failure	

Functional Evaluation Ground Motion (*FEGM*): Probabilistic assessed ground motions that have a 72 ~ 250 year return period (i.e., 30 to 50% probability of exceedance during the useful life a bridge).

Safety Evaluation Ground Motion (*SEGM*): Probabilistic assessed ground motions that have a 950 or 2475 year return period (10% probability of exceedance for a design life of 100 ~ 250 years).

Immediate Service Level: Full access to normal traffic is available almost immediately (i.e., within hours) following the earthquake (It may be necessary to allow 24 h or so for inspection of the bridge).

Limited Service Level: Limited access (reduced lanes, light emergency traffic) is possible within 3 days of the earthquake. Full service restoration within months.

Minimum Damage: Minor inelastic deformation such as narrow flexural cracking in concrete and no apparent deformations.

Repairable Damage: Damage such as concrete cracking, minor spalling of cover concrete, and steel yield that can be repaired without requiring closure and replacing structural members. Permanent offsets are small.

Significant Damage: Damage such as concrete cracking, major spalling of concrete, steel yield that can be repaired only with closure, and partial or complete replacement. Permanent offset may occur without collapse.

37.5.1 Determination of Demands

Analysis Methods

For ordinary bridges, seismic force and deformation demands may be obtained by equivalent static analysis or elastic dynamic response spectrum analysis. For important bridges, the following guidelines may apply:

1. Static linear analysis should be used to determine member forces due to self-weight, wind, water currents, temperature, and live load.
2. Dynamic response spectrum analysis [32] should be used for local and regional stand-alone models and the simplified global model to determine mode shapes, periods, and initial estimates of seismic force and displacement demands. The analysis may be used on global models prior to a time history analysis to verify global behavior, eliminate modeling errors,

TABLE 37.7 ATC-18 Recommendations for Future Bridge Seismic Code Development (One-Level Approach)

Design philosophy		For lower-level earthquake, there should be only minimum damage For a significant earthquake, collapse should be prevented but significant damage may occur; damage should occur at visible locations The following addition to Item 2 is required if different response modification (R and Z) factors are used for important or ordinary bridges Item 2 as it stands would apply to ordinary bridges For important bridges, only repairable would be expected during a significant earthquake
Design load		Single-level — safety evaluation ground motion — 950 or 2475 year return period for the eastern and western portions of the U.S.
Design approach		• Continue current AASHTO seismic performance category • Use nonlinear analysis deformation-based procedures with strength and stiffness requirements being derived from appropriate nonlinear response spectra
Analysis		Nonlinear static analysis should be part of any analysis requirement At a minimum, nonlinear static analysis is required for important bridges Current elastic analysis and design procedure may be sufficient for small ordinary bridges Incorporate both current R-factor elastic procedure and nonlinear static analysis
Design Force	Ductile component	R-factor elastic design procedure or nonlinear static analysis
	Nonductile component	For sacrificial element, should be designed using a guideline that somewhat correspond to the design level of an unspecified lower-level event, for example, one half or one third of the force required for the upper-level event For nonsacrificial element, should be designed for elastic demands or capacity design procedure
	Foundation	Capacity design procedure — to ensure there is no damage
Design displacement		Maintain current seat width requirements Consider overall draft limits to avoid P-Δ effects on long-period structures
Concrete and steel design		Use the capacity design procedure for all critical members
Foundation design		• Complete geotechical analysis for the upper-level event • For nonessential bridges, a lower level (50% of the design acceleration) might be appropriate

and identify initial regions or members where inelastic behavior needs further refinement and inelastic nonlinear elements. In the analysis:

- Site-specific ARS curves should be used with 5% damping.
- Modal response should be combined using the complete quadratic combination (CQC) method and the resulting orthogonal responses should be combined using either the square root of the sum of the squares (SRSS) method or the "30%" rule as defined by AASHTO-LRFD [1994].

3. Dynamic Time History Analysis: Site-specific multisupport dynamic time histories should be used in a dynamic time history analysis [33].
   - Linear elastic dynamic time history analysis is defined as a dynamic time history analysis with consideration of geometric linearity (small displacement), linear boundary conditions, and elastic members. It should only be used to check regional and global models.
   - Nonlinear elastic dynamic time history analysis is defined as a dynamic time history analysis with consideration of geometric nonlinearity, linear boundary conditions, and elastic members. It should be used to determine areas of inelastic behavior prior to incorporating inelasticity into regional and global models.
   - Nonlinear inelastic dynamic time history analysis, level I, is defined as a dynamic time history analysis with consideration of geometric nonlinearity, nonlinear boundary conditions, inelastic elements (for example, seismic isolators and dampers), and elastic members. It should be used for final determination of force and displacement demands for existing structures in combination with static gravity, wind, thermal, water current, and live loads as specified in AASHTO-LRFD [23].

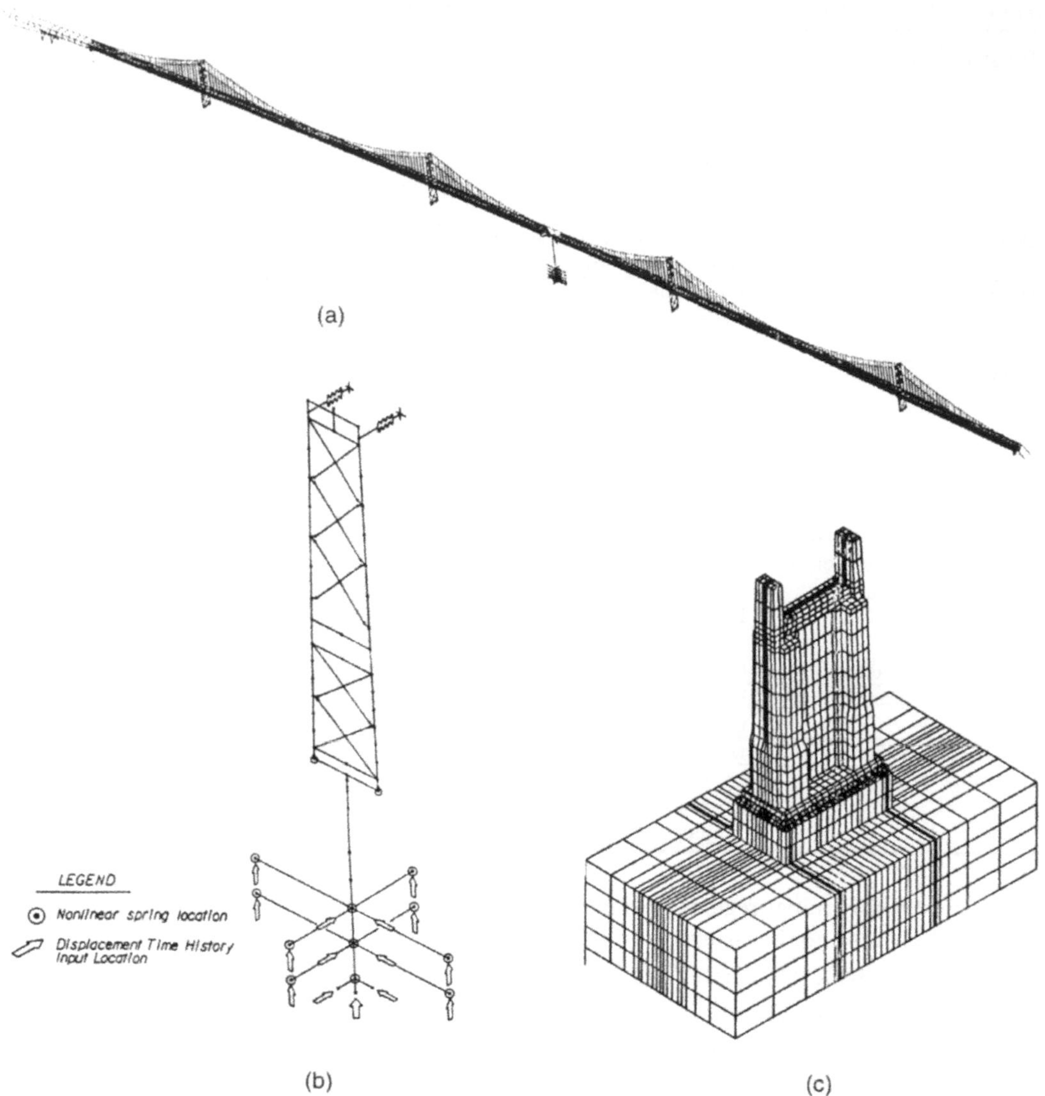

FIGURE 37.6 (a) Global, (b) Regional models for towers, and (c) local model for PW-1 for San Francisco–Oakland Bay Bridge west spans.

- Nonlinear inelastic dynamic time history analysis, level II, is defined as a dynamic time history analysis with consideration of geometric nonlinearity, nonlinear boundary conditions, inelastic elements (for example, dampers), and inelastic members. It should be used for the final evaluation of response of the structures.

Modeling Considerations

1. *Global, Regional, and Local Models*
   The global models consider overall behavior and may include simplifications of complex structural elements (Figure 37.6a). Regional models concentrate on regional behavior (Figure 37.6b). Local models (Figure 37.6c) emphasize the localized behavior, especially complex inelastic and nonlinear behavior. In regional and global models where more than one foundation location is included in the model, multisupport time history analysis should be used.

2. *Boundary Conditions*
   Appropriate boundary conditions should be included in regional models to represent the interaction between the region and the adjacent structure. The adjacent portion is not explicitly modeled but may be simplified using a combination of springs, dashpots, and lumped masses. Appropriate nonlinear elements such as gap elements, nonlinear springs, seismic response modification devices (SRMDs), or specialized nonlinear finite elements should be included where the behavior and response of the structure is sensitive to such elements.
3. Soil–Foundation–Structure Interaction
   This interaction may be considered using nonlinear or hysteretic springs in global and regional models. Foundation springs to represent the properties of the soil at the base of the structure should be included in both regional and global models (see Chapter 42).
4. Damping
   When nonlinear material properties are incorporated in the model, Rayleigh damping should be reduced (perhaps 20%) from the elastic properties.
5. Seismic Response Modification Devices
   The SRMDs should be modeled explicitly with hysteretic characteristics determined by experimental data. See Chapter 41 for a detailed discussion of this behavior.

37.5.2 Determination of Capacities

Limit States and Resistance Factors

The *limit state* is defined as that condition of a structure at which it ceases to satisfy the provisions for which it was designed. Two kinds of limit state corresponding to SEGM and FEGM specified in Table 37.4 apply for seismic design and retrofit. To account for unavoidable inaccuracies in the theory, variation in the material properties, workmanship, and dimensions, nominal strength of structural components should be modified by a resistance factor ϕ specified by AASHTO-LRFD [23] or project-specific criteria to obtain the design capacity or strength (resistance).

Nominal Strength of Structural Components

The strength capacity of structural members should be determined in accordance with specified code formula [23,26, Chapters 38 and 39], or verified with experimental and analytical computer models, or project-specific criteria [19].

Structural Deformation Capacity

Structural deformation capacity should be determined by nonlinear inelastic analysis and based on acceptable damage levels as shown in Table 37.4. The quantitative definition of the damage corresponding to different performance requirements has not been specified by the current Caltrans BDS [26], AASHTO-LRFD [23], and ATC recommendations [18,19] because of the lack of consensus. As a starting point, Table 37.8 provides a quantitative strain and ductility limit corresponding to the three damage levels.

The displacement capacity should be evaluated considering both material and geometric nonlinearities. Proper boundary conditions for various structures should be carefully considered. A static push-over analysis (see Chapter 36) may be suitable for most bridges. A nonlinear inelastic dynamic time history analysis, Level II, may be required for important bridges. The available displacement capacity is defined as the displacement corresponding to the most critical of (1) 20% load reduction from the peak load or (2) the strain limit specified in Table 37.8.

Seismic Response Modification Devices

SRMDs include energy dissipation and seismic isolation devices. Energy dissipation devices increase the effective damping of the structure, thereby reducing reaction forces and deflections. Isolation devices change the fundamental mode of vibration so that the response of the structure is lowered; however, the reduced force may be accompanied by an increased displacement.

TABLE 37.8 Damage Levels, Strain, and Ductility

	Strain		Ductility	
Damage level	Concrete	Steel	Curvature μ_ϕ	Displacement μ_Δ
Significant	ε_{cu}	ε_{sh}	8 ~ 10	4 ~ 6
Repairable	Larger $\begin{cases} 0.005 \\ \frac{2\varepsilon_{cu}}{3} \end{cases}$	Larger $\begin{cases} 0.08 \\ \frac{2\varepsilon_{y}}{3} \end{cases}$	4 ~ 6	2 ~ 4
Minimum	Larger $\begin{cases} 0.004 \\ \varepsilon_{cu} \end{cases}$	Larger $\begin{cases} 0.03 \\ 15\varepsilon_{y} \end{cases}$	2 ~ 4	1 ~ 2

ε_{cu} = ultimate concrete compression strain depending of confinement (see Chapter 36)
ε_y = yield strain of steel
ε_{sh} = hardening strain of steel
μ_ϕ = curvature ductility (ϕ_u/ϕ_y)
μ_Δ = displacement ductility (Δ_u/Δ_y) (see Chapter 36)

The properties of SRMDs should be determined by the specified testing program. References are made to AASHTO [34], Caltrans [35], and Japan Ministry of Construction (JMC) [36]. Consideration of following items should be made in the test specifications:

- Scales — at least two full-scale test specimens are required;
- Loading (including lateral and vertical) history and rate;
- Durability — design life;
- Deterioration — expected levels of strength and stiffness.

37.5.3 Performance Acceptance Criteria

To achieve the performance objectives in Table 37.4, various structural components should satisfy the acceptable demand/capacity ratios (DC_{accept},) specified in this section. The form of the equation is:

$$\frac{\text{Demand}}{\text{Capacity}} \leq DC_{\text{accept}} \tag{37.4}$$

where *demand,* in terms of factored moments, shears, and axial forces, and displacement and rotation deformations, should be determined by a nonlinear inelastic dynamic time history analysis, level I, for important bridges, and dynamic response spectrum analysis for ordinary bridges defined in Section 37.5.1, and *capacity,* in terms of factored strength and deformation capacities, should be obtained according to Section 37.5.2.

Structural Component Classifications

Structural components are classified into two categories: *critical* or *other.* It is the aim that other components may be permitted to function as *fuses* so that the critical components of the bridge system can be protected during the functionality evaluation earthquake (FEE) and the safety evaluation earthquake (SEE). As an example, Table 37.9 shows structural component classifications and their definition for a suspension bridge.

TABLE 37.9 Structural Component Classification

Component Classification	Definition	Example (SFOBB West Spans)
Critical	Components on a critical path that carry bridge gravity load directly The loss of capacity of these components would have serious consequences on the structural integrity of the bridge	Suspension cables Continuous trusses Floor beams and stringers Tower legs Central anchorage A-Frame Piers W-1 and W2 Bents A and B Caisson foundations Anchorage housings Cable bents
Other	All components other than Critical	All other components

Note: Structural components include members and connections.

Steel Structures

1. *General Design Procedure*
   Seismic design of steel members should be in accordance with the procedure shown in Figure 37.7. Seismic retrofit design of steel members should be in accordance with the procedure shown in Figure 37.8.
2. *Connections*
   Connections should be evaluated over the length of the seismic event. For connecting members with force *D*/*C* ratios larger than 1.0, 25% greater than the nominal capacity of the connecting members should be used in connection design.
3. *General Limiting Slenderness Parameters and Width–Thickness Ratios*
   For all steel members (regardless of their force *D*/*C* ratios), the slenderness parameter for axial load dominant members (λ_c) and for flexural dominant members (λ_b) should not exceed the limiting values ($0.9\lambda_{cr}$ or $0.9\lambda_{br}$ for *critical*, λ_{cr} or λ_{br} for *Others*) shown in Table 37.10.
4. *Acceptable Force* D/C *Ratios and Limiting Values*
   Acceptable force *D*/*C* ratios, DC_{accept} and associated limiting slenderness parameters and width–thickness ratios for various members are specified in Table 37.10. For all members with *D*/*C* ratios larger than 1.0, slenderness parameters and width–thickness ratios should not exceed the limiting values specified in Table 37.10. For existing steel members with *D*/*C* ratios less than 1.0, width–thickness ratios may exceed λ_r specified in Table 37.11 and AISC-LRFD [37].

The following symbols are used in Table 37.10. M_u is the factored moment demand; P_u is the vactored axial force demand; M_n is the nominal moment strength of a member; P_n is the nominal axial strength of a member; λ is the width–thickness (b/t or h/t_w) ratio of a compressive element; $\lambda_c = (KL/r\pi)\sqrt{F_y/E}$, the slenderness parameter of axial load dominant members; $\lambda_b = L/r_y$, the slenderness parameter of flexural moment dominant members; $\lambda_{cp} = 0.5$, the limiting column slenderness parameter for 90% of the axial yield load based on AISC-LRFD [37] column curve; λ_{bp} is the limiting beam slenderness parameter for plastic moment for seismic design; $\lambda_{cr} = 1.5$, the limiting column slenderness parameter for elastic buckling based on AISC-LRFD [37] column curve; λ_{br} is the limiting beam slenderness parameter for elastic lateral torsional buckling;

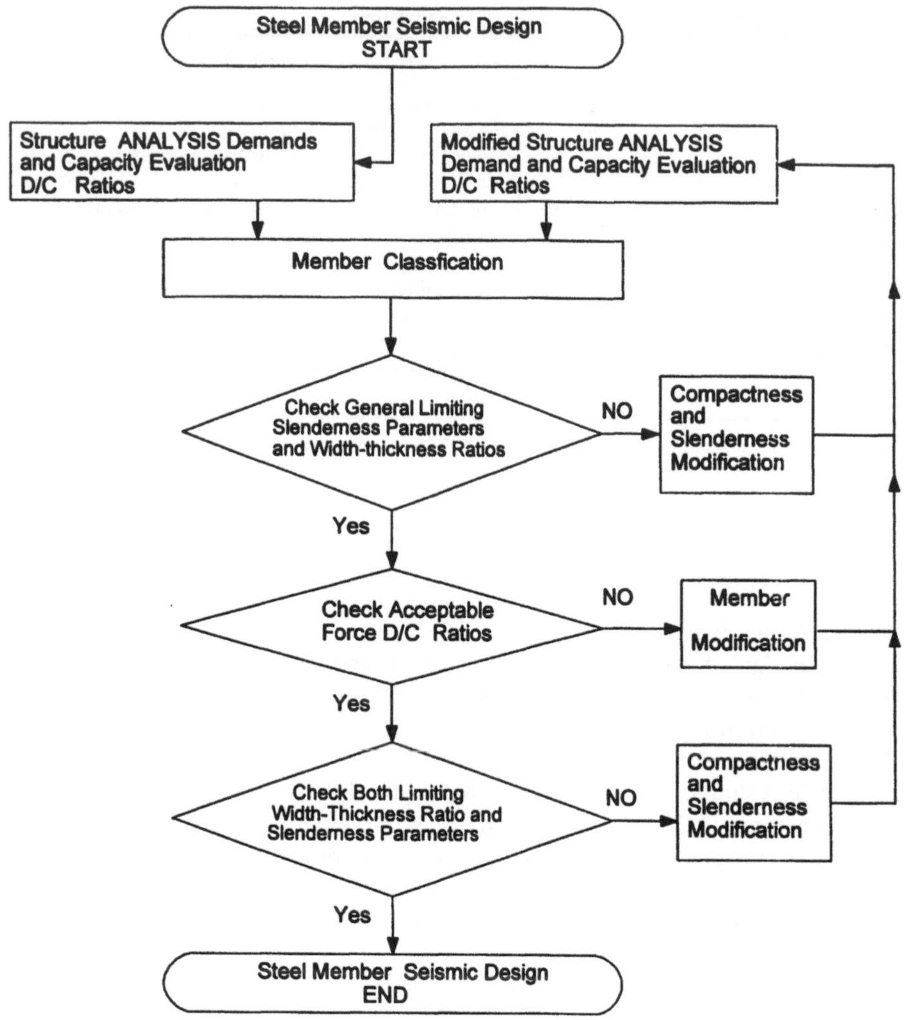

FIGURE 37.7 Steel member seismic design procedure.

$$\lambda_{br} = \begin{cases} \dfrac{57{,}000\sqrt{JA}}{M_r} & \text{for solid rectangular bars and box sections} \\[2ex] \dfrac{X_1}{F_L}\sqrt{1+\sqrt{1+X_2F_L^2}} & \text{for doubly symmetric I-shaped members and channels} \end{cases}$$

$$M_r = \begin{cases} F_L S_x & \text{for I - shaped member} \\ F_{yf} S_x & \text{for solid rectangular and box section} \end{cases}$$

$$X_1 = \frac{\pi}{S_x}\sqrt{\frac{EGJA}{2}} \qquad X_s = \frac{4C_w}{I_y}\left(\frac{S_x}{GJ}\right) \qquad F_L = \text{smaller}\begin{cases} F_{yw} \\ F_{yf} - F_r \end{cases}$$

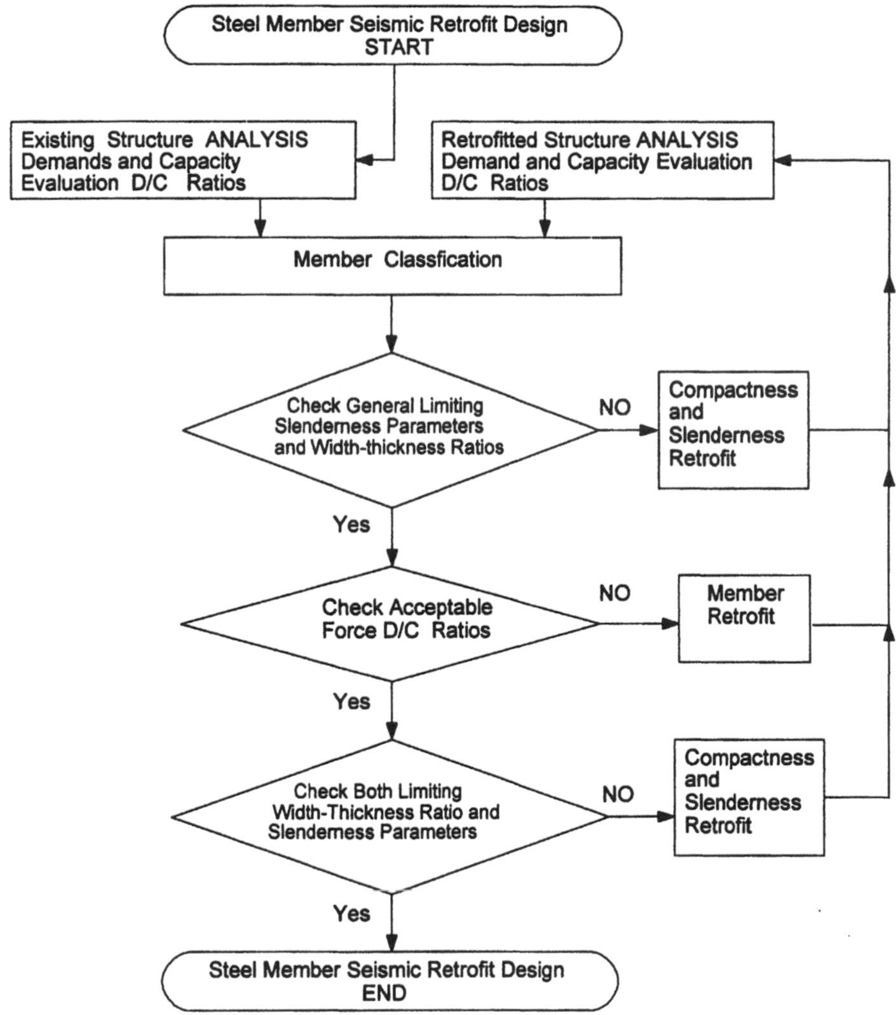

FIGURE 37.8 Steel member seismic retrofit design procedure.

where A is the cross-sectional area, in.^{2}; L is the unsupported length of a member; J is the torsional constant, in.^{4}; r is the radius of gyration, in.; r_y is the radius of gyration about minor axis, in.; F_y is the yield stress of steel; F_{yw} is the yield stress of web, ksi; F_{yf} is the yield stress of flange, ksi; E is the modulus of elasticity of steel (29,000 ksi); G is the shear modulus of elasticity of steel (11,200 ksi); S_x is the section modulus about major axis, in.^{3}; I_y is the moment of inertia about minor axis, in.^{4} and C_w is the warping constant, in.^{6} For doubly symmetric and singly symmetric I-shaped members with compression flange equal to or larger than the tension flange, including hybrid members (strong axis bending):

$$\lambda_{bp} = \begin{cases} \dfrac{\left[3600 + 2200\, M_1/M_2\right]}{F_y} & \text{for } \textit{other} \text{ members} \\ \dfrac{300}{\sqrt{F_{yf}}} & \text{for } \textit{critical} \text{ members} \end{cases} \tag{37.5}$$

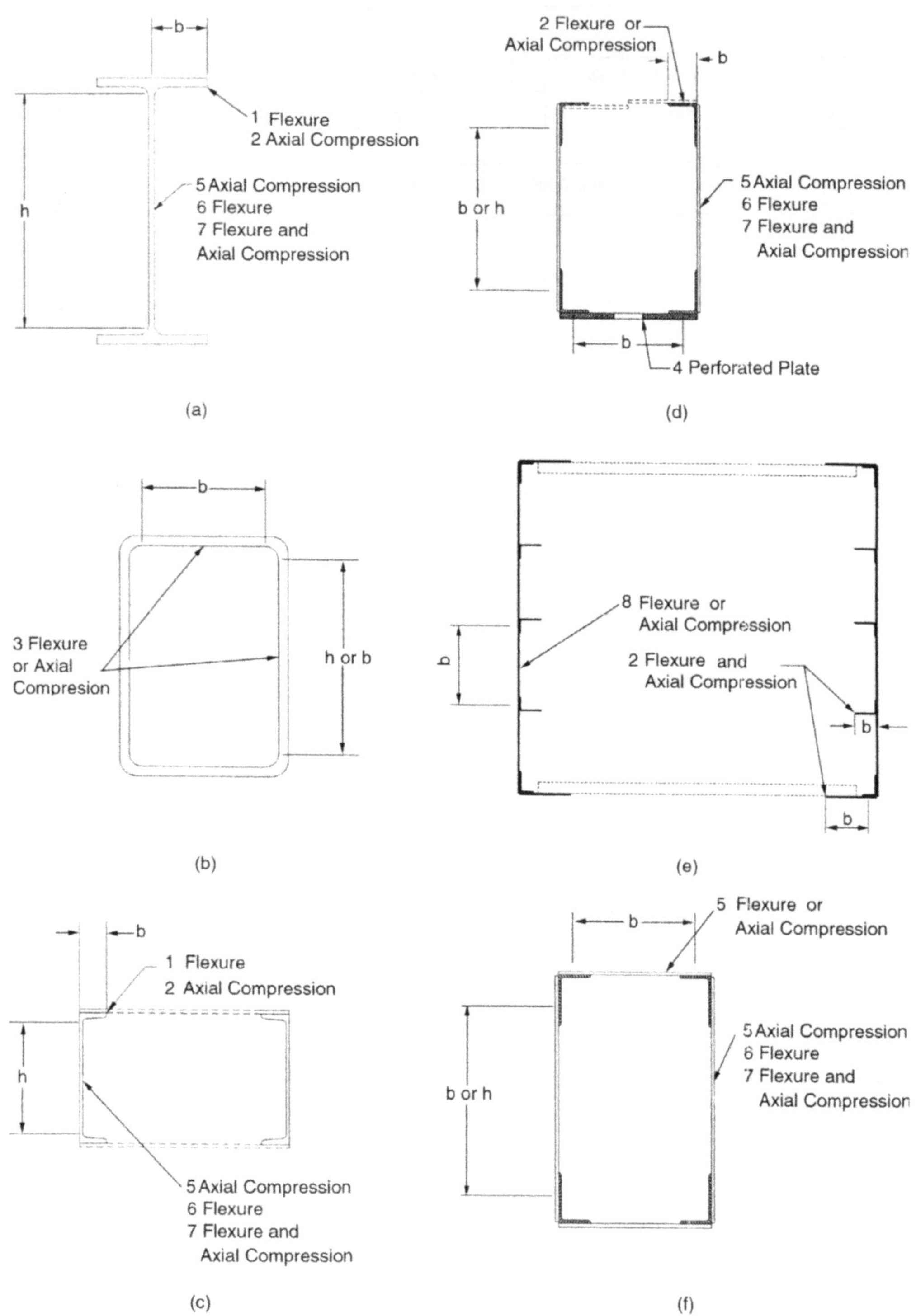

FIGURE 37.9 Typical cross sections for steel members: (a) rolled I section; (b) hollow structured tube; (c) built-up channels; (d) built-up box section; (e) longitudinally stiffened built-up box section; (f) built-up box section.

TABLE 37.10 Acceptable Force Demand/Capacity Ratios and Limiting Slenderness Parameters and Width/Thickness Ratios

Member Classification		Limiting Ratios: Slenderness Parameter (λ_c and λ_b)	Limiting Ratios: Width/Thickness λ (b/t or h/t_w)	Acceptable Force D/C Ratio DC_{accept}
Critical	Axial load dominant	$0.9\lambda_{cr}$	λ_r	$DC_r = 1.0$
	$P_u/P_n \geq M_u/M_n$	λ_{cpr}	λ_{pr}	$1.0 \sim 1.2$
		λ_{cp}	λ_p	$DC_p = 1.2$
	Flexural moment dominant	$0.9\lambda_{br}$	λ_r	$DC_r = 1.0$
	$M_u/M_n > P_u/P_n$	λ_{bpr}	λ_{pr}	$1.2 \sim 1.5$
		λ_{bp}	λ_p	$DC_p = 1.5$
Other	Axial load dominant	λ_{cr}	λ_r	$DC_r = 1.0$
	$P_u/P_n \geq M_u/M_n$	λ_{cpr}	λ_{pr}	$1.0 \sim 2.0$
		λ_{cp}	$\lambda_{p\text{-Seismic}}$	$DC_p = 2$
	Flexural moment dominant	λ_{br}	λ_r	$DC_r = 1.0$
	$M_u/M_n > P_u/P_n$	λ_{bpr}	λ_{pr}	$1.0 \sim 2.5$
		λ_{bp}	$\lambda_{p\text{-Seismic}}$	$DC_p = 2.5$

in which M_1 is larger moment at end of unbraced length of beam; M_2 is smaller moment at end of unbraced length of beam; (M_1/M_2) is positive when moments cause reverse curvature and negative for single curvature.

For solid rectangular bars and symmetric box beam (strong axis bending):

$$\lambda_{bp} = \begin{cases} \dfrac{5000 + 3000\left(M_1/M_2\right)}{F_y} \geq \dfrac{3000}{F_y} & \text{for } \textit{other} \text{ members} \\[2ex] \dfrac{3750}{M_p}\sqrt{JA} & \text{for } \textit{critical} \text{ members} \end{cases} \tag{37.6}$$

in which M_p is plastic moment (Z_xF_y); Z_x is plastic section modulus about major axis; and λ_r, λ_p, $\lambda_{p\text{-Seismic}}$ are limiting width thickness ratios specified by Table 37.11.

$$\lambda_{pr} = \begin{cases} \left[\lambda_p + \left(\lambda_r - \lambda_p\right)\left(\dfrac{DC_p - DC_{\text{accept}}}{DC_p - DC_r}\right)\right] & \text{for } \textit{critical} \text{ members} \\[2ex] \left[\lambda_{p\text{-Seismic}} + \left(\lambda_r - \lambda_{p\text{-Seismic}}\right)\left(\dfrac{DC_p - DC_{\text{accept}}}{DC_p - DC_r}\right)\right] & \text{for } \textit{other} \text{ members} \end{cases} \tag{37.7}$$

For axial load dominant members $(P_u/P_n \geq M_u/M_n)$

$$\lambda_{cpr} = \begin{cases} \lambda_{cp} + \left(0.9\lambda_{cr} - \lambda_{cp}\right)\left(\dfrac{DC_p - DC_{\text{accept}}}{DC_p - DC_r}\right) & \text{for } \textit{critical} \text{ members} \\[2ex] \lambda_{cp} + \left(\lambda_{cr} - \lambda_{cp}\right)\left(\dfrac{DC_p - DC_{\text{accept}}}{DC_p - DC_r}\right) & \text{for } \textit{other} \text{ members} \end{cases} \tag{37.8}$$

For flexural moment dominant members $(M_u/M_n > P_u/P_n)$

TABLE 37.11 Limiting Width-Thickness Ratios

No	Description of Elements	Examples	Width-Thickness Ratios	λ_r	λ_p	$\lambda_{p\text{-}Seismic}$
			Unstiffened Elements			
1	Flanges of I-shaped rolled beams and channels in flexure	Figure 37.19a Figure 37.19c	b/t	$\frac{141}{\sqrt{F_y - 10}}$	$\frac{65}{\sqrt{F_y}}$	$\frac{52}{\sqrt{F_y}}$
2	Outstanding legs of pairs of angles in continuous contact; flanges of channels in axial compression; angles and plates projecting from beams or compression members	Figure 37.19d Figure 37.19e Figure 37.19f	b/t	$\frac{95}{\sqrt{F_y}}$	$\frac{65}{\sqrt{F_y}}$	$\frac{52}{\sqrt{F_y}}$
			Stiffened Elements			
3	Flanges of square and rectangular box and hollow structural section of uniform thickness subject to bending or compression; flange cover plates and diaphragm plates between lines of fasteners or welds.	Figure 37.19b	b/t	$\frac{238}{\sqrt{F_y}}$	$\frac{190}{\sqrt{F_y}}$	$110/F_y$ (tubes) $150/F_y$ (others)
4	Unsupported width of cover plates perforated with a succession of access holes	Figure 37.19d	b/t	$\frac{317}{\sqrt{F_y}}$	$\frac{253}{\sqrt{F_y}}$	$\frac{152}{\sqrt{F_y}}$

5	All other uniformly compressed stiffened elements, i.e., supported along two edges.	Figures 37.19a, c,d,f	b/t h/t_w	$\frac{\mathbf{253}}{\sqrt{F_y}}$	$\frac{190}{\sqrt{F_y}}$	$110/\sqrt{F_y}$ (w/lacing) $150/\sqrt{F_y}$ (others)
6	Webs in flexural compression	Figures 37.19a,c,d,f	h/t_w	$\frac{\mathbf{970}}{\sqrt{F_y}}$	$\frac{\mathbf{640}}{\sqrt{F_y}}$	$\frac{\mathbf{520}}{\sqrt{F_y}}$
7	Webs in combined flexural and axial compression	Figures 37.19a,c,d,f	h/t_w	$\frac{\mathbf{970}}{\sqrt{F_y}} \times \left(1 - \frac{0.74P}{\phi_b P_y}\right)$	For $P_u \le 0.125\,\phi_b P_y$ $\frac{\mathbf{640}}{\sqrt{F_y}}\left(1 - \frac{2.75P}{\phi_b P_y}\right)$ For $P_u > 0.125\,\phi_b P_y$ $\frac{\mathbf{191}}{\sqrt{F_y}}\left(2.33 - \frac{P}{\phi_b P_y}\right) \ge \frac{\mathbf{253}}{\sqrt{F_y}}$	For $P_u \le 0.125\,\phi_b P_y$ $\frac{\mathbf{520}}{\sqrt{F_y}}\left(1 - \frac{1.54P}{\phi_b P_y}\right)$ For $P_u > 0.125\,\phi_b P_y$ $\frac{\mathbf{191}}{\sqrt{F_y}}\left(2.33 - \frac{P}{\phi_b P_y}\right) \ge \frac{\mathbf{253}}{\sqrt{F_y}}$
8	Longitudinally stiffened plates in compression	Figure 37.19e	b/t	$\frac{113\sqrt{k}}{\sqrt{F_y}}$	$\frac{95\sqrt{k}}{\sqrt{F_y}}$	$\frac{75\sqrt{k}}{\sqrt{F_y}}$

Notes:
1. Width–thickness ratios shown in **bold** are from AISC-LRFD [1993] and AISC-Seismic Provisions [1997].
2. k = buckling coefficient specified by Article 6.11.2.1.3a of AASHTO-LRFD [AASHTO, 1994]
   for $n = 1$, $k = (8I_s/bt^3)^{1/3} \le 4.0$; for $n = 2, 3, 4$, and 5, $k = (14.3I_s/bt^3n^4)^{1/3} \le 4.0$
   n = number of equally spaced longitudinal compression flange stiffeners
   I_s = moment of inertia of a longitudinal stiffener about an axis parallel to the bottom flange and taken at the base of the stiffener

$$\lambda_{bpr}\begin{cases}\lambda_{bp}+\left(0.9\lambda_{br}-\lambda_{bp}\right)\left(\dfrac{DC_p-DC_{\text{accept}}}{DC_p-DC_r}\right) & \text{for } \textit{critical} \text{ members}\\[2ex] \lambda_{bp}+\left(\lambda_{br}-\lambda_{bp}\right)\left(\dfrac{DC_p-DC_{\text{accept}}}{DC_p-DC_r}\right) & \text{for } \textit{other} \text{ members}\end{cases} \tag{37.9}$$

Concrete Structures

1. *General*
   For all concrete compression members (regardless of D/C ratios), the slenderness parameter (KL/r) should not exceed 60.

   For *critical* components, force $DC_{\text{accept}}= 1.2$ and deformation $DC_{\text{accept}} = 0.4$.
   For *other* components, force $DC_{\text{accept}}= 2.0$ and deformation $DC_{\text{accept}}= 0.67$.

2. *Beam–Column (Bent Cap) Joints*
   For concrete box-girder bridges, the beam–column (bent cap) joints should be evaluated and designed in accordance with the following guidelines [38,39]:
   a. Effective Superstructure Width: The effective width of superstructure (box girder) on either side of a column to resist longitudinal seismic moment at bent (support) should not be taken as larger than the superstructure depth.
      - The immediately adjacent girder on either side of a column within the effective superstructure width is considered effective.
      - Additional girders may be considered effective if refined bent–cap torsional analysis indicates that the additional girders can be mobilized.
   b. Minimum Bent–Cap Width: Minimum cap width outside column should not be less than $D/4$ (D is column diameter or width in that direction) or 2 ft (0.61 m).
   c. Acceptable Joint Shear Stress:
      - For existing unconfined joints, acceptable principal tensile stress should be taken as $3.5\sqrt{f_c'}$ psi $\left(0.29\sqrt{f_c'}\text{ MPa}\right)$. If the principal tensile stress demand exceeds this limiting value, the joint shear reinforcement specified in Item d should be provided.
      - For new joints, acceptable principal tensile stress should be taken as $12\sqrt{f_c'}$ psi ($1.0\sqrt{f_c'}$ MPa).
      - For existing and new joints, acceptable principal compressive stress shall be taken as f_c'.
   d. Joint Shear Reinforcement
      - Typical flexure and shear reinforcement (see Figures 37.10 and 37.11) in bent caps should be supplemented in the vicinity of columns to resist joint shear. All joint shear reinforcement should be well distributed and provided within $D/2$ from the face of column.
      - Vertical reinforcement including cap stirrups and added bars should be 20% of the column reinforcement anchored into the joint. Added bars shall be hooked around main longitudinal cap bars. Transverse reinforcement in the join region should consist of hoops with a minimum reinforcement ratio of 0.4(column steel area)/(embedment length of column bar into the bent cap)^2.
      - Horizontal reinforcement should be stitched across the cap in two or more intermediate layers. The reinforcement should be shaped as hairpins, spaced vertically at not more than 18 in. (457 mm). The hairpins should be 10% of column reinforcement. Spacing should be denser outside the column than that used within the column.
      - Horizontal side face reinforcement should be 10% of the main cap reinforcement including top and bottom steel.

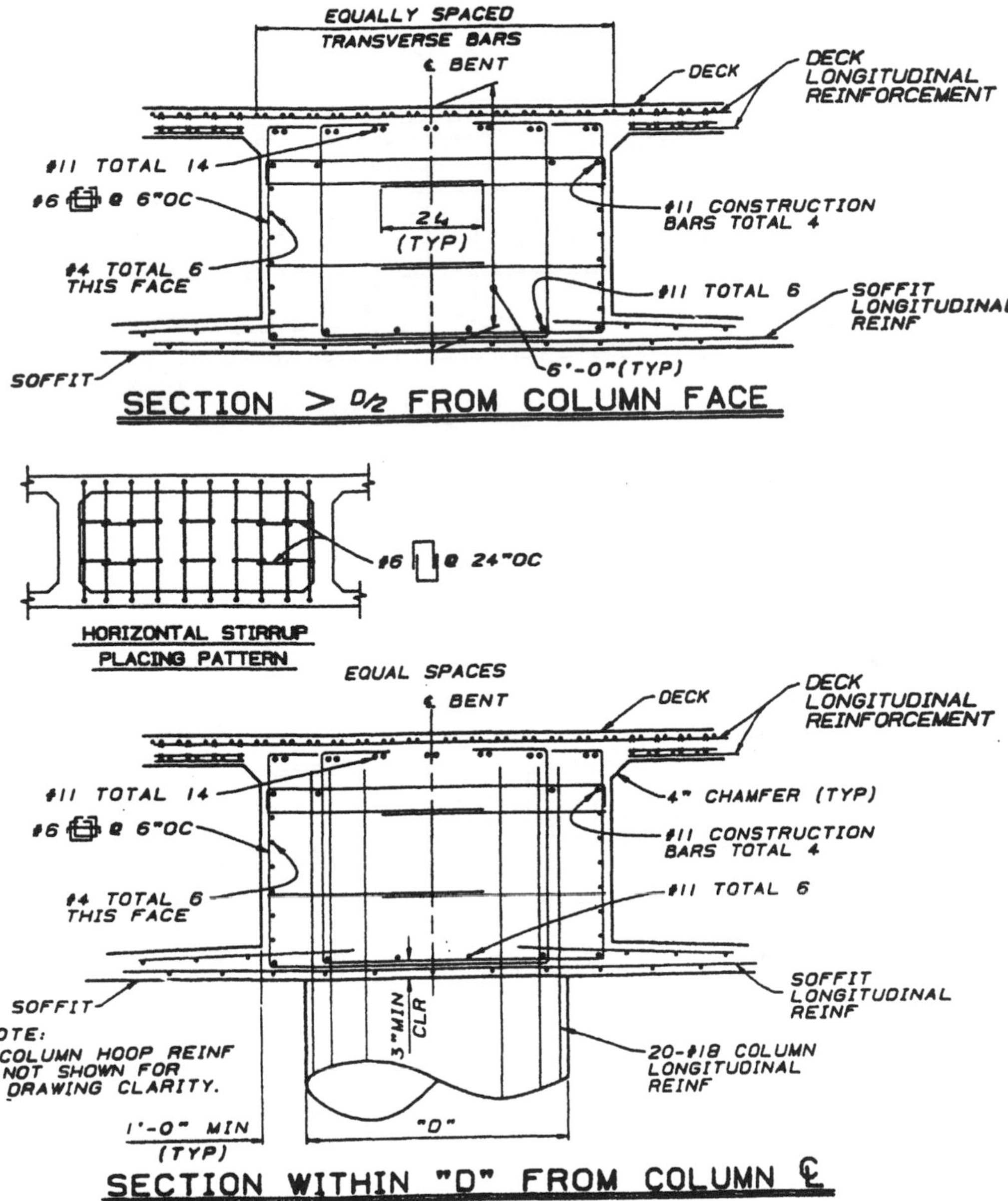

FIGURE 37.10 Example cap joint shear reinforcement — skews 0° to 20°.

- For bent caps skewed greater than 20°, the vertical J-bars hooked around longitudinal deck and bent cap steel should be 8% of column steel (see Figure 37.11). The J-bars should be alternatively 24 in. (600 mm) and 30 in. (750 mm) long and placed within a width of column dimension an either side of the column centerline.
- All vertical column bars should be extended as high as practically possible without interfering with the main cap bars.

Seismic Response Modification Devices

Analysis methods specified in Section 37.5.3 apply for determining seismic design forces and displacements on SRMDs. Properties or capacities of SRMDs should be determined by specified tests.

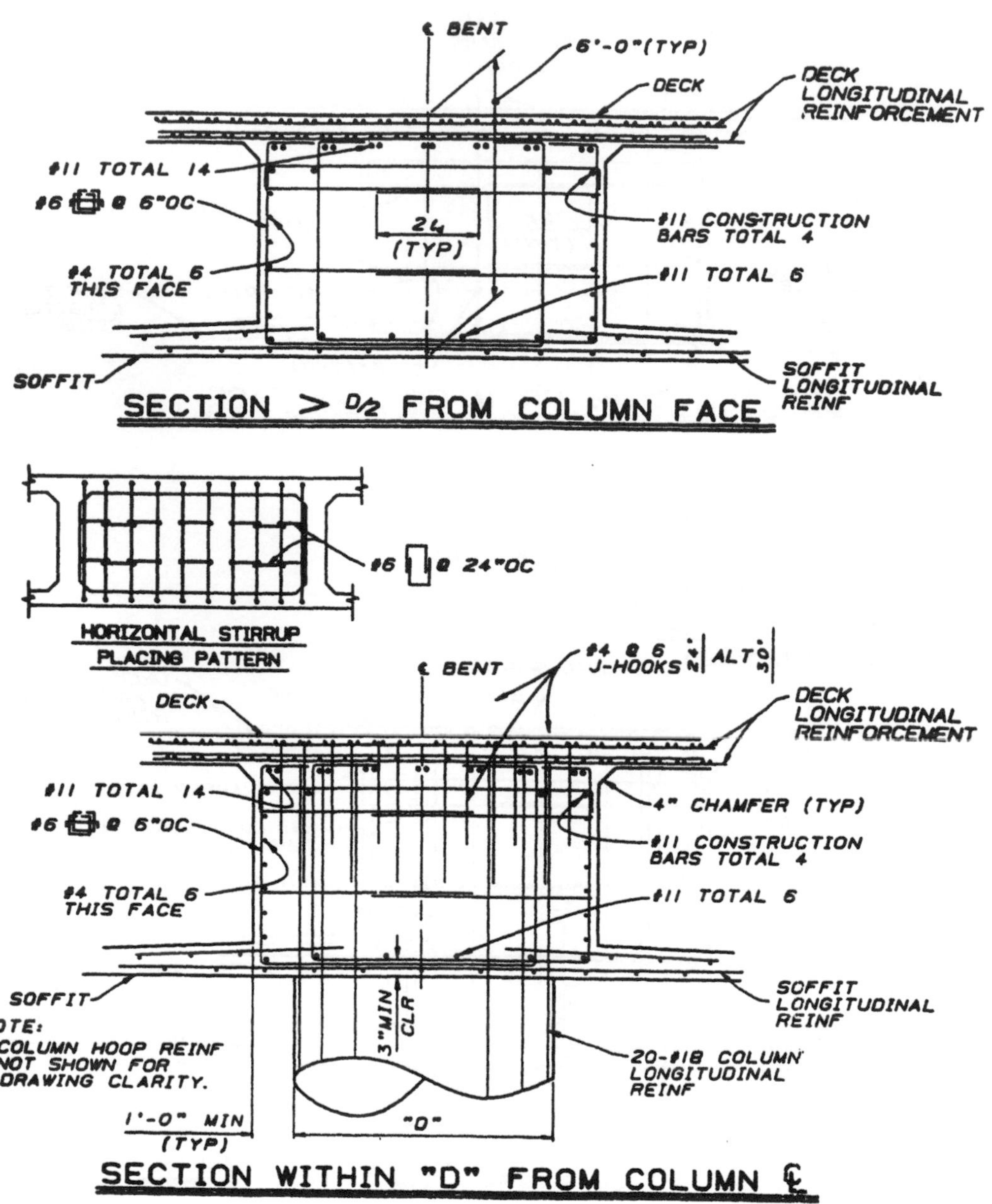

FIGURE 37.11 Example cap joint shear reinforcement — skews > 20°.

SRMDs should be able to perform their intended function and maintain their design parameters for the design life (for example, 40 years) and for an ambient temperature range (for example, from 30 to 125°F). The devices should be accessible for inspection, maintenance, and replacement. In general, SRMDs should satisfy at least the following requirements:

- Strength and stability must be maintained under increasingly large displacement. Stiffness degradation under repeated cyclic load is unacceptable.
- Energy must be dissipated within acceptable design displacement limits, for example, a limit on the maximum total displacement of the device to prevent failure, or the device can be given a displacement capacity 50% greater than the design displacement.

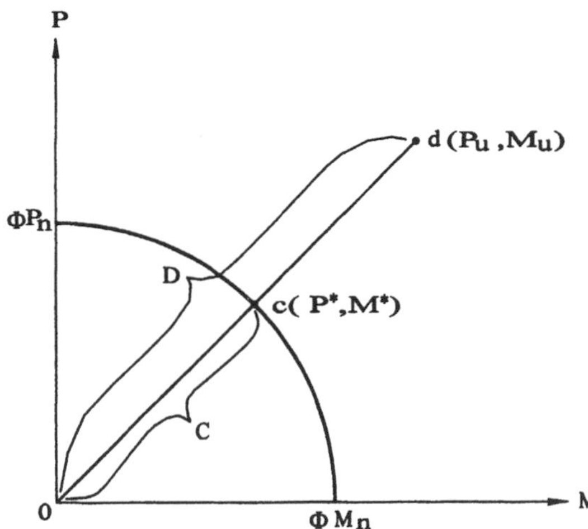

FIGURE 37.12 Definition of force D/C ratios for combined loadings.

- Heat builtup must be withstood and dissipated during "reasonable" seismic displacement time history.
- The device must survive subjected to the number of cycles of displacement expected under wind excitation during the life of the device and continue to function at maximum wind force and displacement levels for at least a given duration.

37.5.4 Acceptable Force *D/C* Ratios and Limiting Values for Structural Members

It is impossible to design bridges to withstand seismic forces elastically and the nonlinear inelastic response is expected. Performance-based criteria accept certain seismic damage in *other* components so the *critical* components will remain essentially elastic and functional after the SEE and FEE. This section presents the concept of acceptable force *D/C* ratios, limiting member slenderness parameters, and limiting width–thickness ratios, as well as expected ductility.

Definition of Force Demand/Capacity (D/C) Ratios

For members subjected to a single load, force demand is defined as a factored single force, such as factored moment, shear, or axial force. This may be obtained by a nonlinear dynamic time history analysis, level I, as specified in Section 37.5.1 and capacity is prescribed in Section 37.5.2.

For members subjected to combined loads, the force *D/C* ratio is based on the interaction. For example, for a member subjected to combined axial load and bending moment (Figure 37.12), the force demand *D* is defined as the distance from the origin point O(0, 0) to the factored force point $d(P_u, M_u)$, and capacity *C* is defined as the distance from the origin point O(0, 0) to the point $c(P^*, M^*)$ on the specified interaction surface or curve.

Ductility and Load–Deformation Curves

Ductility is usually defined as a nondimensional factor, i.e., the ratio of ultimate deformation to yield deformation [40,41]. It is normally expressed by two forms: (1) curvature ductility ($\mu_\phi = \phi_u/\phi_y$) and (2) displacement ductility ($\mu_\Delta = \Delta_u/\Delta_y$). Representing section flexural behavior, *curvature ductility* is dependent on the section shape and material properties and is based on the moment–curvature diagram. Indicating structural system or member behavior, *displacement ductility* is related to both the structural configuration and section behavior and is based on the load–displacement curve.

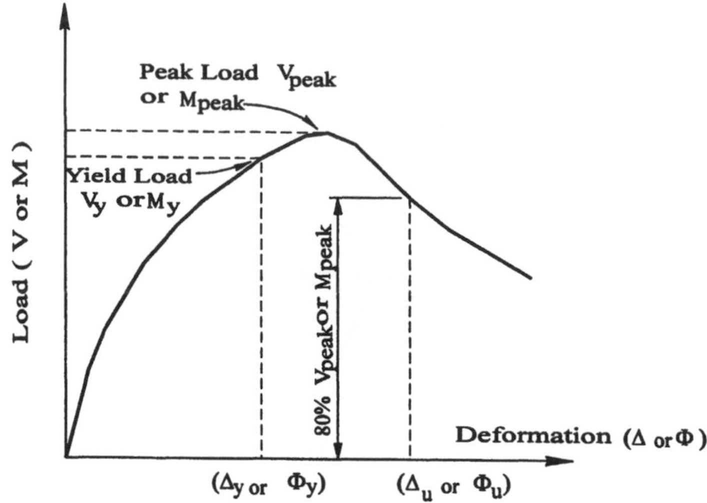

FIGURE 37.13 Load–deformation cCurves.

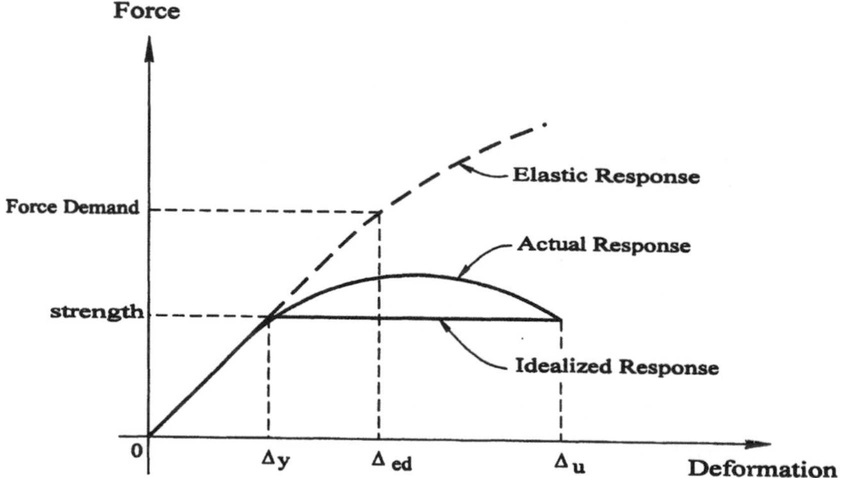

FIGURE 37.14 Response of a single-degree of freedom system.

A typical load–deformation curve, including both ascending and descending branches, is shown in Figure 37.13. The yield deformation (Δ_y or ϕ_y) corresponds to a loading state beyond which the structure responds inelastically. The ultimate deformation (Δ_u or ϕ_u) refers to the loading state at which a structural system or member can sustain without losing significant load-carrying capacity. Depending on performance requirements, it is proposed that the ultimate deformation (curvature or displacement) be defined as the most critical of (1) that deformation corresponding to a load dropping a maximum of 20% from the peak load or (2) that specified strain limit shown in Table 37.8.

Force D/C Ratios and Ductility

The following discussion will give engineers a direct measure of the seismic damage incurred by structural components during an earthquake. Figure 37.14 shows a typical load–response curve for a single-degree-of-freedom system. Displacement ductility is defined as

TABLE 37.12 Force D/C Ratio and Damage Index

Force D/C Ratio	Damage Index D_Δ	Expected System Displacement Ductility μ_Δ
1.0	No damage	No requirement
1.2	0.4	3.0
1.5	0.5	3.0
2.0	0.67	3.0
2.5	0.83	3.0

$$\mu_\Delta = \frac{\Delta_u}{\Delta_y} \tag{37.10}$$

A new term *damage index* is hereby defined as the ratio of elastic displacement demand to ultimate displacement capacity:

$$D_\Delta = \frac{\Delta_{ed}}{\Delta_u} \tag{37.11}$$

When the damage index $D_\Delta < 1/\mu_\Delta$ ($\Delta_{ed} < \Delta_y$), no damage occurs and the structure responds elastically; when $1/\mu_\Delta < D_\Delta < 1.0$, some damage occurs and the structure responds inelastically; when $D_\Delta > 1.0$, the structure collapses completely.

Based on the "equal displacement principle," the following relationship is obtained:

$$\frac{\text{Force Demand}}{\text{Force Capacity}} = \frac{\Delta_{ed}}{\Delta_y} = \mu_\Delta D_\Delta \tag{37.12}$$

It is seen from Eq. (37.12) that the force *D*/*C* ratio is related to both the structural characters in term of ductility $\mu_\Delta$and the degree of damage in terms of damage index D_Δ. Table 37.12 shows this relationship.

General Limiting Values

To ensure that important bridges have ductile load paths, general limiting slenderness parameters and width–thickness ratios are specified in Section 37.5.3.

For steel members, λ_{cr} is the limiting parameter for column elastic buckling and is taken as 1.5 from AISC-LRFD [37]; λ_{br} corresponds to beam elastic torsional buckling and is calculated by AISC-LRFD. For a *critical* member, a more strict requirement, 90% of those elastic buckling limits is proposed. Regardless of the force demand-to-capacity ratios, no members may exceed these limits. For existing steel members with *D*/*C* ratios less than 1, this limit may be relaxed. For concrete members, the general limiting parameter $KL/r = 60$ is proposed.

Acceptable Force *D*/*C* Ratios DC_{accept} for Steel Members

The acceptable force demand/capacity ratios (DC_{accept}) depends on both the structural characteristics in terms of ductility and the degree of damage acceptable to the engineer in terms of damage index D_Δ.

To ensure a member has enough inelastic deformation capacity during an earthquake and to achieve acceptable *D*/*C* ratios and energy dissipation, it is necessary, for steel member, to limit both the slenderness parameter and the width–thickness ratio within the ranges specified below.

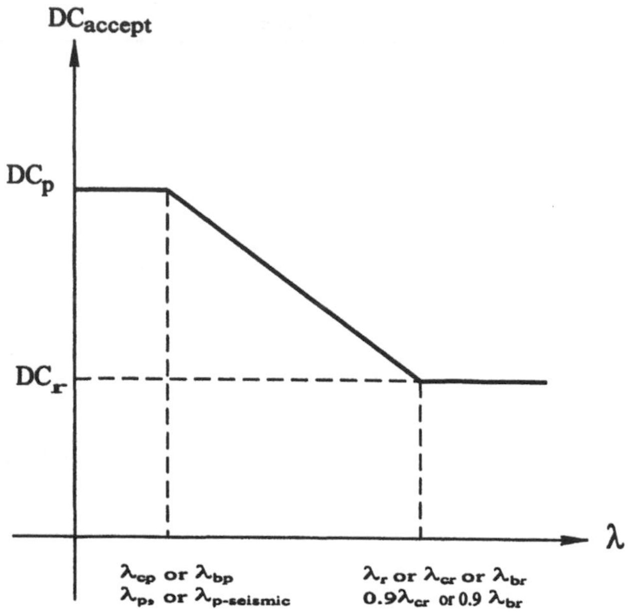

FIGURE 37.15 Acceptable D/C ratios and limiting slenderness parameters and width–thickness ratios.

Upper Bound Acceptable *D/C* Ratio DC_p

1. For *other* members, the large acceptable force *D/C* ratios ($DC_p = 2$ to 2.5) are listed in Table 37.10. The damage index is between 0.67 ~ 0.83 and more damage occurs in *other* members and great ductility is be expected. To achieve this,
   - The limiting width–thickness ratio was taken as $\lambda_{p\text{-Seismic}}$ from AISC-Seismic Provisions [42], which provides flexural ductility of 8 to 10.
   - The limiting slenderness parameters were taken as λ_{bp} for flexure-dominant members from AISC-LRFD [37], which provides flexural ductility of 8 to 10.
2. For *critical* members, small acceptable force *D/C* ratios ($DC_p = 1.2$ to 1.5) are proposed in Table 37.10, as the design purpose is to keep *critical* members essentially elastic and allow little damage (damage index values ranging from 0.4 to 0.5). Thus little member ductility is expected. To achieve this,
   - The limiting width–thickness ratio was taken as λ_p from AISC-LRFD [36], which provides a flexural ductility of at least 4.
   - The limiting slenderness parameters were taken as λ_{bp} for flexure dominant members from AISC [37], providing flexure ductility of at least 4.
3 For axial load dominant members the limiting slenderness parameter is taken as $\lambda_{cp} = 0.5$, corresponding to 90% of the axial yield load by the AISC-LRFD [37] column curve. This limit provides the potential for axial load dominant members to develop inelastic deformation.

Lower Bound Acceptable *D/C* Ratio DC_r
The lower-bound acceptable force D/C ratio $DC_{rc} = 1$ is proposed in Table 37.10. For $DC_{accept} = 1$, it is not necessary to enforce more strict limiting values for members and sections. Therefore, the limiting slenderness parameters for elastic global buckling specified in Table 27.10 and the limiting width–thickness ratios specified in Table 37.11 for elastic local buckling are proposed.

Acceptable *D/C* Ratios between Upper and Lower Bounds $DC_r < DC_{accept} < DC_p$

When acceptable force *D/C* ratios are between the upper and the lower bounds, $DC_r < DC_{accept} < DC_p$, a linear interpolation (Eqs. 37.7 to 37.9) as shown in Figure 37.15 is proposed to determine the limiting slenderness parameters and width–thickness ratios.

37.6 Summary

Seismic bridge design philosophies and current practice in the United States have been discussed. "No-collapse" based design is usually applied to ordinary bridges, and performance-based design is used for important bridges. Sample performance-based seismic design criteria are presented to bridge engineers as a reference guide. This chapter attempted to address only some of the many issues incumbent upon designers of bridges for adequate performance under seismic load. Engineers are always encouraged to incorporate to the best of their ability the most recent research findings and the most recent experimental evidence learned from past performance under real earthquakes.

References

1. ATC, Seismic Design Guidelines for Highway Bridges, Report No. ATC-6, Applied Technology Council, Redwood City, CA, 1981.
2. Housner, G. W. Competing against Time, Report to Governor George Deuknejian from the Governor's Broad of Inquiry on the 1989 Loma Prieta Earthquake, Sacramento, 1990.
3. Caltrans, The First Annual Seismic Research Workshop, Division of Structures, California Department of Transportation, Sacramento, 1991.
4. Caltrans, The Second Annual Seismic Research Workshop, Division of Structures, California Department of Transportation, Sacramento, 1993.
5. Caltrans, The Third Annual Seismic Research Workshop, Division of Structures, California Department of Transportation, Sacramento, 1994.
6. Caltrans, The Fourth Caltrans Seismic Research Workshop, Engineering Service Center, California Department of Transportation, Sacramento, 1996.
7. Caltrans, The Fifth Caltrans Seismic Research Workshop, Engineering Service Center, California Department of Transportation, Sacramento, 1998.
8. FHWA and Caltrans, *Proceedings of First National Seismic Conference on Bridges and Highways*, San Diego, 1995.
9. FHWA and Caltrans, *Proceedings of Second National Seismic Conference on Bridges and Highways*, Sacramento, 1997.
10. Kawashima, K. and Unjoh, S., The damage of highway bridges in the 1995 Hyogo-Ken Naubu earthquake and its impact on Japanese seismic design, *J. Earthquake Eng.*, 1(2), 1997, 505.
11. Park, R., Ed., Seismic design and retrofitting of reinforced concrete bridges, in *Proceedings of the Second International Workshop*, held in Queenstown, New Zealand, August, 1994.
12. Astaneh-Asl, A. and Roberts, J. Eds., *Seismic Design, Evaluation and Retrofit of Steel Bridges, Proceedings of the First U.S. Seminar,* San Francisco, 1993.
13. Astaneh-Asl, A. and Roberts, J., Ed., *Seismic Design, Evaluation and Retrofit of Steel Bridges, Proceedings of the Second U.S. Seminar,* San Francisco, 1997.
14. Housner, G.W., *The Continuing Challenge — The Northridge Earthquake of January 17, 1994,* Report to Director, California Department of Transportation, Sacramento, 1994.
15. Priestley, M. J. N., Seible, F. and Calvi, G. M., *Seismic Design and Retrofit of Bridges,* John Wiley & Sons, New York, 1996.
16. Caltrans, Design Criteria for SR-14/I-5 Replacement, California Department of Transportation, Sacramento, 1994.

17. Caltrans, San Francisco–Oakland Bay Bridge West Spans Seismic Retrofit Design Criteria, Prepared by Reno, M. and Duan, L., Edited by Duan, L., California Department of Transportation, Sacramento, 1997.
18. ATC, Improved Seismic Design Criteria for California Bridges: Provisional Recommendations, Report No. ATC-32, Applied Technology Council, Redwood City, CA, 1996.
19. Rojahn, C., et al., Seismic Design Criteria for Bridges and Other Highway Structures, Report NCEER-97-0002, National Center for Earthquake Engineering Research, State University of New York at Buffalo, Buffalo, 1997. Also refer as ATC-18, Applied Technology Council, Redwood City, CA, 1997.
20. Caltrans, Bridge Memo to Designers (20-1) — Seismic Design Methodology, California Department of Transportation, Sacramento, January 1999.
21. FHWA, Seismic Design and Retrofit Manual for Highway Bridges, Report No. FHWA-IP-87-6, Federal Highway Administration, Washington, D.C., 1987.
22. FHWA. Seismic Retrofitting Manual for Highway Bridges, Publ. No. FHWA-RD-94-052, Federal Highway Administration, Washington, D.C., 1995.
23. AASHTO, *LRFD Bridge Design Specifications*, 2nd. ed., American Association of State Highway and Transportation Officials, Washington, D.C., 1994 and 1996.
24. AASHTO, *Standard Specifications for Highway Bridges*, 16th ed., American Association of State Highway and Transportation Officials, Washington, D.C., 1996.
25. Caltrans, Bridge Memo to Designers (20-4), California Department of Transportation, Sacramento, 1995.
26. Caltrans, Bridge Design Specifications, California Department of Transportation, Sacramento, 1990.
27. Caltrans, Bridge Design Aids, California Department of Transportation, Sacramento, 1995.
28. IAI, Benicia-Martinez Bridge Seismic Retrofit — Main Truss Spans Final Retrofit Strategy Report, Imbsen and Association, Inc., Sacramento, 1995.
29. Priestley, N., Myths and fallacies in earthquake engineering — conflicts between design and reality, in *Proceedings of Tom Paulay Symposium — Recent Development in Lateral Force Transfer in Buildings*, University of California, San Diego, 1993.
30. Kowalsky, M. J., Priestley, M. J. N., and MacRae, G. A., Displacement-Based Design, Report No. SSRP-94/16, University of California, San Diego, 1994.
31. Martin, G. R. and Dobry, R., Earthquake Site Response and Seismic Code Provisions, NCEER Bulletin, Vol. 8, No. 4, National Center for Earthquake Engineering Research, Buffalo, NY, 1994.
32. Gupta, A. K. *Response Spectrum Methods in Seismic Analysis and Design of Structures*, CRC Press, Boca Raton, FL, 1992.
33. Clough, R. W. and Penzien, J., *Dynamics of Structures*, 2nd ed., McGraw-Hill, New York, 1993.
34. AASHTO, *Guide Specifications for Seismic Isolation Design*, American Association of State Highway and Transportation Officials, Washington, D.C., 1997.
35. Caltrans, Full Scale Isolation Bearing Testing Document (Draft), Prepared by Mellon, D., California Department of Transportation, Sacramento, 1997.
36. Japan Ministry of Construction (JMC), Manual of Menshin Design of Highway Bridges, (English version: EERC, Report 94/10, University of California, Berkeley), 1994.
37. AISC, *Load and Resistance Factor Design Specification for Structural Steel Buildings*, 2nd ed., American Institute of Steel Construction, Chicago, IL, 1993.
38. Zelinski, R., Seismic Design Memo, Various Topics, Preliminary Guidelines, California Department of Transportation, Sacramento, 1994.
39. Caltrans, Seismic Design Criteria for Retrofit of the West Approach to the San Francisco–Oakland Bay Bridge. Prepared by M. Keever, California Department of Transportation, Sacramento, 1996.
40. Park, R. and Paulay, T., *Reinforced Concrete Structures*, John Wiley & Sons, New York, 1975.
41. Duan, L. and Cooper, T. R., Displacement ductility capacity of reinforced concrete columns, *ACI Concrete Int.*, 17(11), 61–65, 1995.
42. AISC, *Seismic Provisions for Structural Steel Buildings*, American Institute of Steel Construction, Chicago, IL, 1991.
43. Caltrans, Seismic Design Criteria, Vers. 1.0, California Department of Transportation, Sacramento, 1999.

38

Seismic Design of Reinforced Concrete Bridges

Yan Xiao

University of Southern California

38.1 Introduction

This chapter provides an overview of the concepts and methods used in modern seismic design of reinforced concrete bridges. Most of the design concepts and equations described in this chapter are based on new research findings developed in the United States. Some background related to current design standards is also provided.

38.1.1 Two-Level Performance-Based Design

Most modern design codes for the seismic design of bridges essentially follow a two-level performance-based design philosophy, although it is not so clearly stated in many cases. The recent document ATC-32

0-8493-7434-0/00/$0.00+$.50

[2] may be the first seismic design guideline based on the two-level performance design. The two level performance criteria adopted in ATC-32 were originally developed by the California Department of Transportation [5].

The first level of design concerns control of the performance of a bridge in earthquake events that have relatively small magnitude but may occur several times during the life of the bridge. The second level of design consideration is to control the performance of a bridge under severe earthquakes that have only a small probability of occurring during the useful life of the bridge. In the recent ATC-32, the first level is defined for functional evaluation, whereas the second level is for safety evaluation of the bridges. In other words, for relatively frequent smaller earthquakes, the bridge should be ensured to maintain its function, whereas the bridge should be designed safe enough to survive the possible severe events.

Performance is defined in terms of the serviceability and the physical damage of the bridge. The following are the recommended service and damage criteria by ATC-32.

1. Service Levels:
    - *Immediate service*: Full access to normal traffic is available almost immediately following the earthquake.
    - *Limited service*: Limited access (e.g., reduced lanes, light emergency traffic) is possible within days of the earthquake. Full service is restorable within months.
2. Damage levels:
    - *Minimal damage*: Essentially elastic performance.
    - *Repairable damage*: Damage that can be repaired with a minimum risk of losing functionality.
    - *Significant damage*: A minimum risk of collapse, but damage that would require closure to repair.

The required performance levels for different levels of design considerations should be set by the owners and the designers based on the importance rank of the bridge. The fundamental task for seismic design of a bridge structure is to ensure a bridge's capability of functioning at the anticipated service levels without exceeding the allowable damage levels. Such a task is realized by providing proper strength and deformation capacities to the structure and its components.

It should also be pointed out that the recent research trend has been directed to the development of more-generalized performance-based design [3,6,8,13].

38.1.2 Elastic vs. Ductile Design

Bridges can certainly be designed to rely primarily on their strength to resist earthquakes, in other words, to perform elastically, in particular for smaller earthquake events where the main concern is to maintain function. However, elastic design for reinforced concrete bridges is uneconomical, sometimes even impossible, when considering safety during large earthquakes. Moreover, due to the uncertain nature of earthquakes, a bridge may be subject to seismic loading that well exceeds its elastic limit or strength and results in significant damage. Modern design philosophy is to allow a structure to perform inelastically to dissipate the energy and maintain appropriate strength during severe earthquake attack. Such an approach can be called ductile design, and the inelastic deformation capacity while maintaining the acceptable strength is called ductility.

The inelastic deformation of a bridge is preferably restricted to well-chosen locations (the plastic hinges) in columns, pier walls, soil behind abutment walls, and wingwalls. Inelastic action of superstructure elements is unexpected and undesirable because that damage to superstructure is difficult and costly to repair and unserviceable.

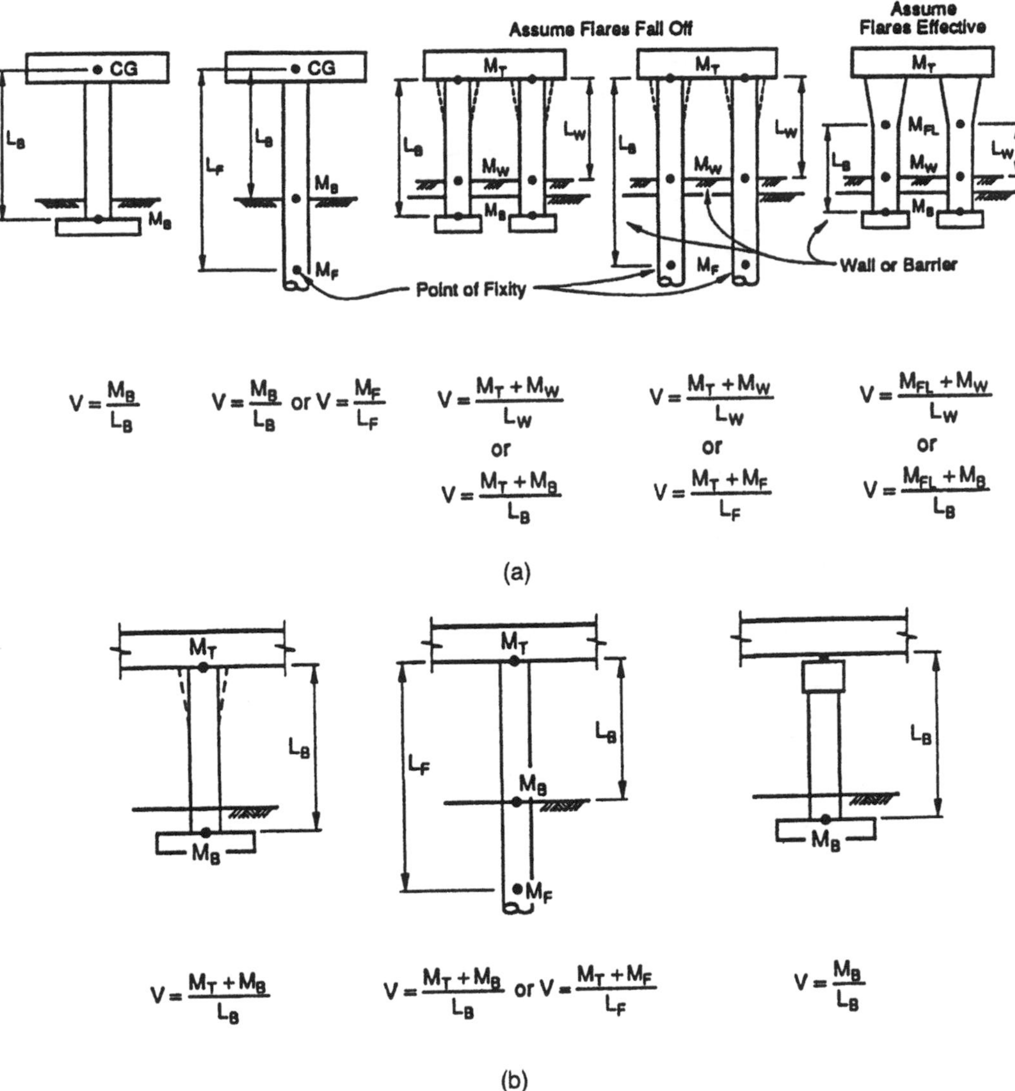

FIGURE 38.1 Potential plastic hinge locations for typical bridge bents: (a) transverse response; (b) longitudinal response. (*Source*: Caltrans, *Bridge Design Specification*, California Department of Transportation, Sacramento, June, 1990.)

38.1.3 Capacity Design Approach

The so-called capacity design has become a widely accepted approach in modern structural design. The main objective of the capacity design approach is to ensure the safety of the bridge during large earthquake attack. For ordinary bridges, it is typically assumed that the performance for lower-level earthquakes is automatically satisfied.

The procedure of capacity design involves the following steps to control the locations of inelastic action in a structure:

1. Choose the desirable mechanisms that can dissipate the most energy and identify plastic hinge locations. For bridge structures, the plastic hinges are commonly considered in columns. Figure 38.1 shows potential plastic hinge locations for typical bridge bents.

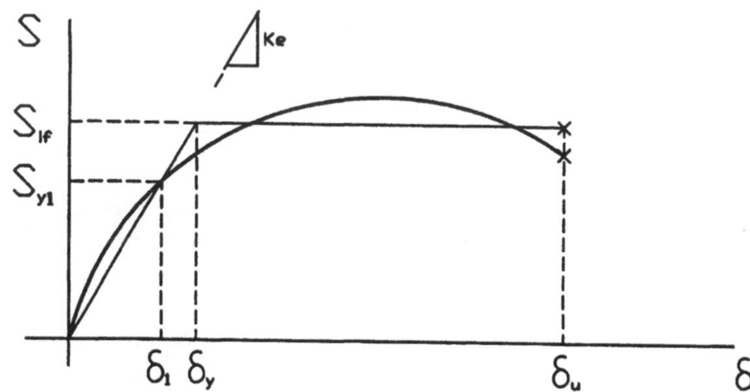

FIGURE 38.2 Idealization of column behavior.

2. Proportion structures for design loads and detail plastic hinge for ductility.
3. Design and detail to prevent undesirable failure patterns, such as shear failure or joint failure. The design demand should be based on plastic moment capacity calculated considering actual proportions and expected material overstrengths.

38.2 Typical Column Performance

38.2.1 Characteristics of Column Performance

Strictly speaking, elastic or plastic behaviors are defined for ideal elastoplastic materials. In design, the actual behavior of reinforced concrete structural components is approximated by an idealized bilinear relationship, as shown in Figure 38.2. In such bilinear characterization, the following mechanical quantities have to be defined.

Stiffness

For seismic design, the initial stiffness of concrete members calculated on the basis of full section geometry and material elasticity has little meaning, since cracking of concrete can be easily induced even under minor seismic excitation. Unless for bridges or bridge members that are expected to respond essentially elastically to design earthquakes, the effective stiffness based on cracked section is instead more useful. For example, the effective stiffness, K_e, is usually based on the cracked section corresponding to the first yield of longitudinal reinforcement,

$$K_e = S_{y1}/\delta_1 \tag{38.1}$$

where, S_{y1} and δ_1 are the force and the deformation of the member corresponding to the first yield of longitudinal reinforcement, respectively.

Strength

Ideal strength S_i represents the most feasible approximation of the "yield" strength of a member predicted using measured material properties. However, for design, such "yield" strength is conservatively assessed using *nominal strength* S_n predicted based on nominal material properties. The *ultimate* or *overstrength* represents the maximum feasible capacities of a member or a section and is predicted by taking account of all possible factors that may contribute to strength exceeding S_i or S_n. The factors include realistic values of steel yield strength, strength enhancement due to strain hardening, concrete strength increase due to confinement, strain rate, as well as actual aging, etc.

Deformation

In modern seismic design, deformation has the same importance as strength since deformation is directly related to physical damage of a structure or a structural member. Significant deformation limits are onset of cracking, onset of yielding of extreme tension reinforcement, cover concrete spalling, concrete compression crushing, or rupture of reinforcement. For structures that are expected to perform inelastically in severe earthquake, cracking is unimportant for safety design; however, it can be used as a limit for elastic performance. The first yield of tension reinforcement marks a significant change in stiffness and can be used to define the elastic stiffness for simple bilinear approximation of structural behavior, as expressed in Eq. (38.1). If the stiffness is defined by Eq. (38.1), then the yield deformation for the approximate elastoplastic or bilinear behavior can be defined as

$$\delta_y = S_{if}/S_{y1}\delta_1 \tag{38.2}$$

where, S_{yl} and δ_1 are the force and the deformation of the member corresponding to the first yield of longitudinal reinforcement, respectively; S_{if} is the idealized flexural strength for the elastoplastic behavior.

Meanwhile, the ductility factor, μ, is defined as the index of inelastic deformation beyond the yield deformation, given by

$$\mu = \delta/\delta_y \tag{38.3}$$

where δ is the deformation under consideration and δ_y is the yield deformation.

The limit of the bilinear behavior is set by an ultimate ductility factor or deformation, corresponding to certain physical events, that are typically corresponded by a significant degradation of load-carrying capacity. For unconfined member sections, the onset of cover concrete spalling is typically considered the failure. Rupture of either transverse reinforcement or longitudinal reinforcement and the crushing of confined concrete typically initiate a total failure of the member.

38.2.2 Experimentally Observed Performance

Figure 38.3a shows the lateral force–displacement hysteretic relationship obtained from cyclic testing of a well-confined column [10,11]. The envelope of the hysteresis loops can be either conservatively approximated with an elastoplastic bilinear behavior with V_{if} as the yield strength and the stiffness defined corresponding to the first yield of longitudinal steel. The envelope of the hysteresis loops can also be well simulated using a bilinear behavior with the second linear portion account for the overstrength due to strain hardening. Final failure of this column was caused by the rupture of longitudinal reinforcement at the critical sections near the column ends.

The ductile behavior shown in Figure 38.3a can be achieved by following the capacity design approach with ensuring that a flexural deformation mode to dominate the behavior and other nonductile deformation mode be prevented. As a contrary example to ductile behavior, Figure 38.3b shows a typical poor behavior that is undesirable for seismic design, where the column failed in a brittle manner due to the sudden loss of its shear strength before developing yielding, V_{if}. Bond failure of reinforcement lap splices can also result in rapid degradation of load-carrying capacity of a column.

An intermediate case between the above two extreme behaviors is shown in Figure 38.3c, where the behavior is somewhat premature for full ductility due to the fact that the tested column failed in shear upon cyclic loading after developing its yield strength but at a smaller ductility level than that shown in Figure 38.3a. Such premature behavior is also not desirable.

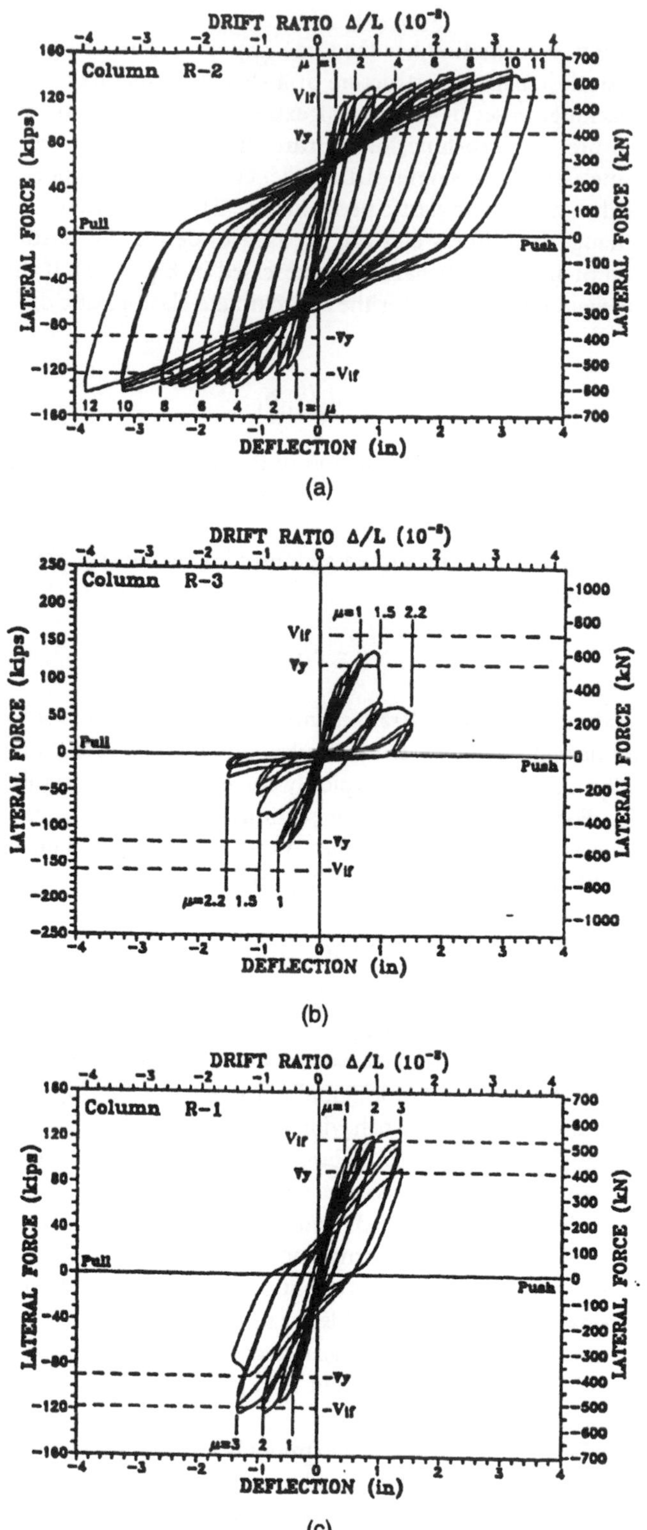

FIGURE 38.3 Typical experimental behaviors for (a) well-confined column; (b) column failed in brittle shear; (c) column with limited ductility. (*Source*: Priestley, M. J. N. et al., *ACI Struct. J.*, 91C52, 537–551, 1994. With permission.)

38.3 Flexural Design of Columns

38.3.1 Earthquake Load

For ordinary, regular bridges, the simple force design based on equivalent static analysis can be used to determine the moment demands on columns. Seismic load is assumed as an equivalent static horizontal force applied to individual bridge or frames, i.e.,

$$F_{eq} = ma_g \tag{38.4}$$

where m is the mass; a_g is the design peak acceleration depended on the period of the structure. In the Caltrans BDS [4] and the ATC-32 [2], the peak ground acceleration a_g is calculated as 5% damped elastic acceleration response spectrum at the site, expressed as ARS, which is the ratio of peak ground acceleration and the gravity acceleration g. Thus the equivalent elastic force is

$$F_{eq} = mg(\text{ARS}) = W(\text{ARS}) \tag{38.5}$$

where W is the dead load of bridge or frame.

Recognizing the reduction of earthquake force effects on inelastically responding structures, the elastic load is typically reduced by a period-dependent factor. Using the Caltrans BDS expression, the design force is found:

$$F_d = W(\text{ARS})/Z \tag{38.6}$$

This is the seismic demand for calculating the required moment capacity, whereas the capability of inelastic response (ductility) is ensured by following a capacity design approach and proper detailing of plastic hinges. Figure 38.4a and b shows the Z factor required by current Caltrans BDS and modified Z factor by ATC-32, respectively. The design seismic forces are applied to the structure with other loads to compute the member forces. A similar approach is recommended by the AASHTO-LRFD specifications.

The equivalent static analysis method is best suited for structures with well-balanced spans and supporting elements of approximately equal stiffness. For these structures, response is primarily in a single mode and the lateral force distribution is simply defined. For unbalanced systems, or systems in which vertical accelerations may be significant, more-advanced methods of analysis such as elastic or inelastic dynamic analysis should be used.

38.3.2 Fundamental Design Equation

The fundamental design equation is based on the following:

$$\phi R_n \geq R_u \tag{38.7}$$

where R_u is the strength demand; R_n is the nominal strength; and ϕ is the strength reduction factor.

38.3.3 Design Flexural Strength

Flexural strength of a member or a section depends on the section shape and dimension, amount and configuration of longitudinal reinforcement, strengths of steel and concrete, axial load magnitude, lateral confinement, etc. In most North American codes, the design flexural strength is conservatively calculated based on nominal moment capacity M_n following the ACI code recommendations [1]. The ACI approach is based on the following assumptions:

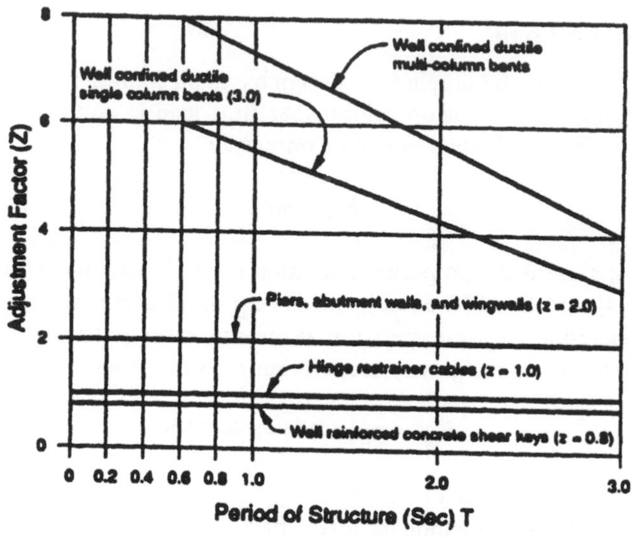

(a)

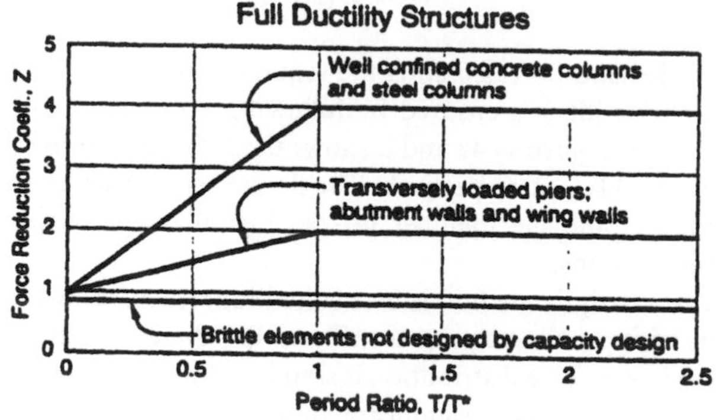

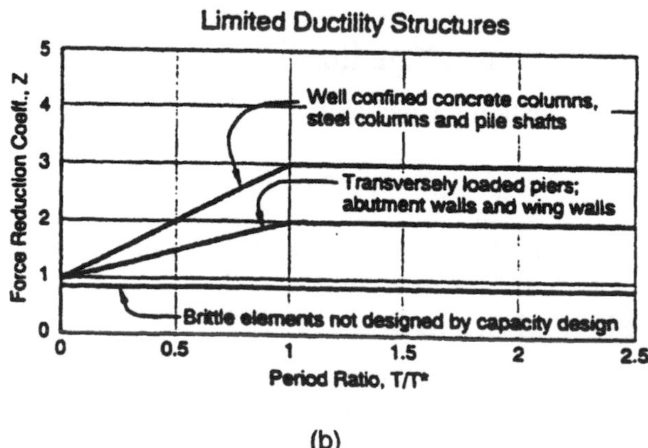

(b)

FIGURE 38.4 Force reduction coefficient Z (a) Caltrans BDS 1990; (b) ATC-32.

1. A plane section remains plane even after deformation. This implies that strains in longitudinal reinforcement and concrete are directly proportional to the distance from the neutral axis.
2. The section reaches the capacity when compression strain of the extreme concrete fiber reaches its maximum usable strain that is assumed to be 0.003.
3. The stress in reinforcement is calculated as the following function of the steel strain

$$f_s = -f_y \quad \text{for} \quad \varepsilon < -\varepsilon_y \tag{38.8a}$$

$$f_s = E_s\varepsilon \quad \text{for} \quad -\varepsilon_y \le \varepsilon \le \varepsilon_y \tag{38.8b}$$

$$f_s = f_y \quad \text{for} \quad \varepsilon > \varepsilon_y \tag{38.8c}$$

   where ε_y and f_y are the yield strain and specified strength of steel, respectively; E_s is the elastic modulus of steel.
4. Tensile stress in concrete is ignored.
5. Concrete compressive stress and strain relationship can be assumed to be rectangular, trapezoidal, parabolic, or any other shape that results in prediction of strength in substantial agreement with test results. This is satisfied by an equivalent rectangular concrete stress block with an average stress of 0.85 f_c', and a depth of $\beta_1 c$, where c is the distance from the extreme compression fiber to the neutral axis, and

$$0.65 \le \beta_1 = 0.85 - 0.05\frac{f_c' - 28}{7} \le 0.85 \quad [\, f_c' \text{ in } MPa] \tag{38.9}$$

In calculating the moment capacity, the equilibrium conditions in axial direction and bending must be used. By using the equilibrium condition that the applied axial load is balanced by the resultant axial forces of concrete and reinforcement, the depth of the concrete compression zone can be calculated. Then the moment capacity can be calculated by integrating the moment contributions of concrete and steel.

The nominal moment capacity, M_n, reduced by a strength reduction factor ϕ (typically 0.9 for flexural) is compared with the required strength, M_u, to determine the feasibility of longitudinal reinforcement, section dimension, and adequacy of material strength.

Overstrength

The calculation of the nominal strength, M_n, is based on specified minimum material strength. The actual values of steel yield strength and concrete strength may be substantially higher than the specified strengths. These and other factors such as strain hardening in longitudinal reinforcement and lateral confinement result in the actual strength of a member perhaps being considerably higher than the nominal strength. Such overstrength must be considered in calculating ultimate seismic demands for shear and joint designs.

38.3.4 Moment–Curvature Analysis

Flexural design of columns can also be carried out more realistically based on moment–curvature analysis, where the effects of lateral confinement on the concrete compression stress–strain relationship and the strain hardening of longitudinal reinforcement are considered. The typical assumptions used in the moment–curvature analysis are as follows:

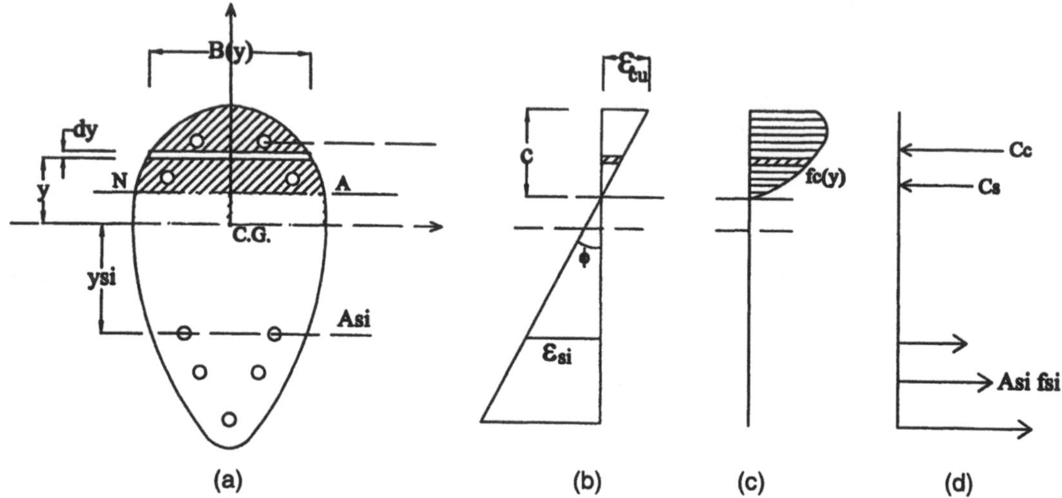

FIGURE 38.5 Moment curvature analysis: (a) generalized section; (b) strain distribution; (c) concrete stress distribution; (d) rebar forces.

1. A plane section remains plane even after deformation. This implies that strains in longitudinal reinforcement and concrete are directly proportional to the distance from the neutral axis.
2. The stress–strain relationship of reinforcement is known and can be expressed as a general function, $f_s = F_s(\varepsilon_s)$.
3. The stress–strain relationship of concrete is known and can be expressed as a general function, $f_c = F_c(\varepsilon_c)$. The tensile stress of concrete is typically ignored for seismic analysis but can be considered if the uncracked section response needs to be analyzed. The compression stress–strain relationship of concrete should be able to consider the effects of confined concrete (for example, Mander et al. [7]).
4. The resulting axial force and moment of concrete and reinforcement are in equilibrium with the applied external axial load and moment.

The procedure for moment–curvature analysis is demonstrated using a general section shown in Figure 38.5a. The distributions of strains and stresses in the cracked section corresponding to an arbitrary curvature, ϕ, are shown in Figure 38.5b, c, and d, respectively.

Corresponding to the arbitrary curvature, ϕ, the strains of concrete and steel at an arbitrary position with a distance of y to the centroid of the section can be calculated as

$$\varepsilon = \phi(y - y_c + c) \tag{38.10}$$

where y_c is the distance of the centroid to the extreme compression fiber and c is the depth of compression zone. Then the corresponding stresses can be determined using the known stress–strain relationships for concrete and steel.

Based on the equilibrium conditions, the following two equations can be established,

$$P = \sum A_{si} f_{si} + \int_A B(y) f_c(y) dy \tag{38.11}$$

$$M = \sum A_{si} f_{si} y_{si} + \int_A B(y) f_c(y) y dy \tag{38.12}$$

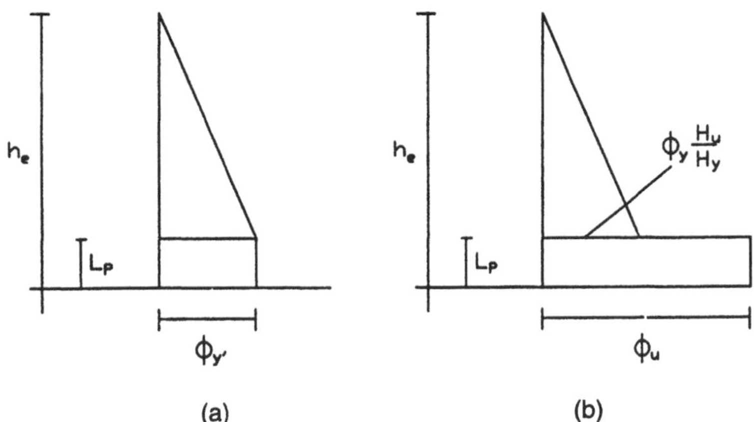

FIGURE 38.6 Idealized curvature distributions: (a) corresponding to first yield of reinforcement; (b) at ultimate flexural failure.

Using the axial equilibrium condition, the depth of the compression zone, c, corresponding to curvature, ϕ, can be determined, and then the corresponding moment, M, can be calculated. The actual computation of moment–curvature relationships is typically done by computer programs (for example, "SC-Push" [15]).

38.3.5 Transverse Reinforcement Design

In most codes, the ductility of the columns is ensured by proper detailing of transverse confinement steel. Transverse reinforcement can also be determined by a trial-and-error procedure to satisfy the required displacement member ductility levels. The lateral displacement of a member can be calculated by an integration of curvature and rotational angle along the member. This typically requires the assumption of curvature distributions along the member. Figure 38.6 shows the idealized curvature distributions at the first yield of longitudinal reinforcement and the ultimate condition for a column.

Yield Conditions: The horizontal force at the yield condition of the bilinear approximation is taken as the ideal capacity, H_{if}. Assuming linear elastic behavior up to conditions at first yield of the longitudinal reinforcement, the displacement of the column top at first yield due to flexure alone is

$$\Delta'_{yf} = \int_0^{h_e} \frac{\phi'_y}{h_e}(h_e - y)^2 dy \quad [f_y \text{ in } MPa] \tag{38.13}$$

where, ϕ'_y is the curvature at first yield and h_e is the effective height of the column. Allowing for strain penetration of the longitudinal reinforcement into the footing, the effective column height can be taken as

$$h_e = h + 0.022 f_y d_{bl} \tag{38.14}$$

where h is the column height measured from one end of a column to the point of inflexion, d_{bl} is the longitudinal reinforcement nominal diameter, and f_y is the nominal yield strength of rebar. The flexural component of yield displacement corresponding to the yield force, H_{if}, for the bilinear approximation can be found by extrapolating the first yield displacement to the ideal flexural strength, giving

$$\Delta_{yf} = \Delta'_{yf} H_{if} / H_y \tag{38.15}$$

where H_y is the horizontal force corresponding to first yield.

Ultimate Conditions: Ultimate conditions for the bilinear approximation can be taken to be the horizontal force, H_u, and displacement, Δ_{uf}, corresponding to development of an ultimate curvature, ϕ_u, based on moment-curvature analysis of the critical sections considering reinforcement strain-hardening and concrete confinement as appropriate. Ultimate curvature corresponds to the more critical ultimate concrete compression strain [7] based on an energy-balance approach, and maximum longitudinal reinforcement tensile strain, taken as $0.7\varepsilon_{su}$, where ε_{su} is the steel strain corresponding to maximum steel stress. The "ultimate" displacement is thus

$$\Delta_{uf} = \theta_p(h - 0.5L_p) + \Delta_y \frac{H_u}{H_{if}} \tag{38.16}$$

where the plastic rotation, θ_p, can be estimated as

$$\theta_p = \left(\phi_u - \phi'_y \frac{H_u}{H_y}\right) L_p \tag{38.17}$$

In Eqs. (38.16) and (38.17), L_p is the equivalent plastic hinge length appropriate to a bilinear approximation of response, given by

$$L_p = 0.08h + 0.022 f_y d_{bl} \quad [f_y \text{ in } MPa] \tag{38.18}$$

Based on the displacement check, the required transverse reinforcement within the plastic hinge region of a column to satisfy the required displacement ductility demand can be determined using a trial-and-error procedure.

Similarly, transverse reinforcement in the potential plastic hinge region of a column can be determined based on curvature ductility requirements. The ATC-32 document [2] suggests the following minimum requirement for the volumetric ratio, ρ_s, of spiral or circular hoop reinforcement within plastic hinge regions of columns based on a curvature ductility factor of 13.0 or larger.

$$\rho_s = 0.16 \frac{f'_{ce}}{f_{ye}} \left[0.5 + \frac{1.25P_e}{f'_{ce}A_g}\right] + 0.13(\rho_l - 0.01) \tag{38.19}$$

where f'_{ce} is the expected concrete compression strength taken as 1.3 f'_c ; f_{ye} is the expected steel yield strength taken as $1.1f_y$; P_e is axial load; A_g is cross-sectional area of column; and ρ_l is longitudinal reinforcement ratio. For transverse reinforcement outside the potential plastic hinge region, the volumetric ratio can be reduced to 50% of the amount given by Eq. (38.19). The length of the plastic hinge region should be the greater of (1) the section dimension in the direction considered or (2) the length of the column portion over which the moment exceeds 80% of the moment at the critical section. However, for columns with axial load ratio $P_e/ f'_{ce} A_g > 0.3$, this length should be increased by 50%. Requirements for cross ties and hoops in rectilinear sections can also be found in ATC-32 [2].

38.4 Shear Design of Columns

38.4.1 Fundamental Design Equation

As discussed previously, shear failure of columns is the most dangerous failure pattern that typically can result in the collapse of a bridge. Thus, design to prevent shear failure is of particular importance. The general design equation for shear strength can be described as

$$\phi V_n > V_u \tag{38.20}$$

where V_u is the ultimate shear demands; V_n is the nominal shear resistant; and ϕ is the strength reduction factor for shear strength.

The calculation of the ultimate shear demand, V_u, has to be based on the equilibrium of the internal forces corresponding to the maximum flexural capacity, which is calculated taking into consideration all factors for overstrength. Figure 38.1 also includes equations for determining V_u for typical bridge bents.

38.4.2 Current Code Shear Strength Equation

Following the ACI code approach [1], the shear strength of axially loaded members is empirically expressed as

$$V_n = V_c + V_s \tag{38.21}$$

In the two-term additive equations, V_c is the shear strength contribution by concrete shear resisting mechanism and V_s is the shear strength contribution by the truss mechanism provided by shear reinforcement. Concrete shear contribution, V_c, is calculated as

$$V_c = 0.166\sqrt{f_c'}\left(1 + \frac{P}{13.8A_g}\right)A_e \tag{38.22}$$

where for columns the effective shear area A_e can be taken as 80% of the cross-sectional area, A_g; P is the applied axial compression force. Note that $\sqrt{f_c'} \le 0.69$ MPa.

The contribution of truss mechanism is taken as

$$V_s = \frac{A_v f_y d}{s} \tag{38.23}$$

where A_v is the total transverse steel area within spacing s; f_y is yield strength of transverse steel; and d is the effective depth of the section.

Comparisons with existing test data indicate that actual shear strengths of columns often exceeds the design shear strength based on the ACI approach, in many cases by more than 100%.

38.4.3 Refined Shear Strength Equations

A refined shear strength equation that agrees significantly well with tests was proposed by Priestley et al. [12] in the following three-term additive expression:

$$V_n = V_c + V_s + V_a \tag{38.24}$$

where V_c and V_s are shear strength contributions by concrete shear resisting mechanism and the truss mechanism, respectively; the additional term, V_a, represents the shear resistance by the arch mechanism, provided mainly by axial compression.

$$V_c = k\sqrt{f_c'}A_e \tag{38.25}$$

and k depends on the displacement ductility factor μ_Δ, which reduces from 0.29 in MPa units (3.5 in psi) for $\mu_\Delta \le 2.0$ to 0.1 in MPa units (1.2 in psi) for $\mu_\Delta \ge 4.0$; A_e is taken as $0.8A_g$. The shear strength contribution by truss mechanism for circular columns is given by,

$$V_s = \frac{\pi}{2}\frac{A_{sp}f_y(d-c)}{s}\cot\theta \tag{38.26}$$

where A_{sp} is cross-sectional area of spiral or hoop reinforcement; d is the effective depth of the section; c is the depth of compression zone at the critical section; s is spacing of spiral or hoop; and θ is the angle of truss mechanism, taken as 30°. In general, the truss mechanism angle θ should be considered a variable for different column conditions. Note that $(d-x)\cot\theta$ in Eq. (38.26) essentially represents the length of the critical shear crack that intersects with the critical section at the position of neutral axis, as shown in Figure 38.7a [17]. Equation (38.26) is approximate for circular columns with circular hoops or spirals as shear reinforcement. As shown in Figure 38.7b, the shear contribution of circular hoops intersected by a shear crack can be calculated by integrating the components of their hoop tension forces, $A_{sp}f_s$, in the direction of the applied shear, given by [18],

$$V_s = 2\int_{c-d/2}^{d/2}\frac{2A_{sp}f_s}{sd}\cot\theta\sqrt{(d/2)^2-x^2}\,dx \tag{38.27}$$

By further assuming that all the hoops intersected by the shear crack develop yield strength, f_y, the integration results in

$$V_s = \frac{A_{sp}f_yd}{s}\cot\theta\left[\left(1-\frac{2c}{d}\right)\sqrt{\frac{c}{d}\left(1-\frac{c}{d}\right)}+\frac{1}{2}\left(1-\frac{2c}{d}\right)^2\arcsin\left(1-\frac{2c}{d}\right)+\frac{\pi}{4}\right] \tag{38.28}$$

The shear strength enhancement by axial load is considered to result from an inclined compression strut, given by

$$V_a = P\tan\alpha = \frac{D-x}{2D(M/VD)}P \tag{38.29}$$

where D is section depth or diameter; x is the compression zone depth which can be determined from flexural analysis; and (M/VD) is the shear aspect ratio.

38.5 Moment-Resisting Connection Between Column and Beam

Connections are key elements that maintain the integrity of overall structures; thus, they should be designed carefully to ensure the full transfer of seismic forces and moments. Because of their importance, complexity, and difficulty of repair if damaged, connections are typically provided with a higher degree of safety and conservativeness than column or beam members. Current Caltrans BDS and AASHTO-LRFD do not provide specific design requirements for joints, except requiring the lateral reinforcement for columns to be extended into column/footing or column/cap beam joints. A new design approach recently developed by Priestley and adopted in the ATC-32 design guidelines is summarized below.

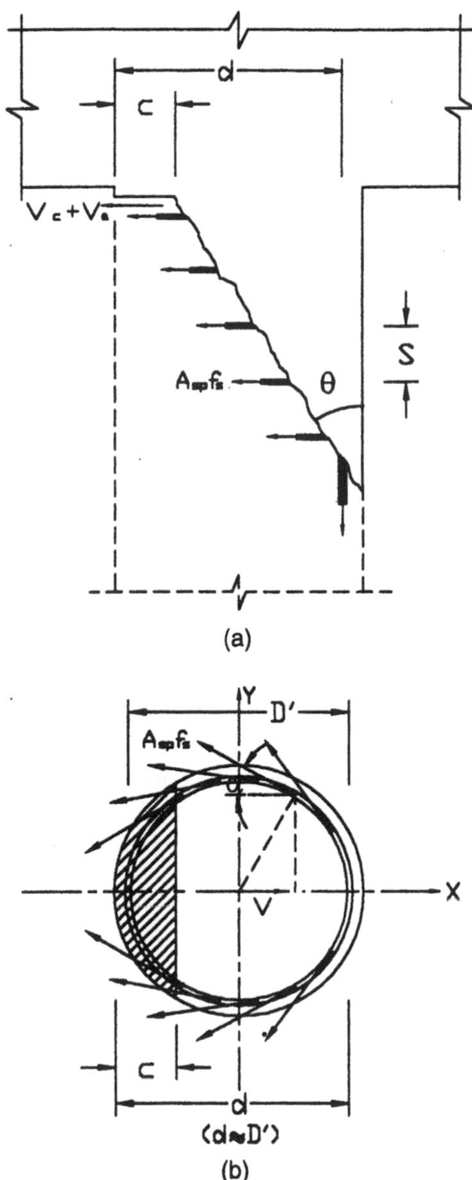

FIGURE 38.7 Shear resisting mechanism of circular hoops or spirals: (a) critical shear section; (b) stresses in hoops or spirals intersected with shear crack. (*Source*: Xiao et al. 1998. With permission.)

38.5.1 Design Forces

In moment-resisting frame structures, the force transfer typically results in sudden changes (magnitude and direction) of moments at connections. Because of the relatively small dimensions of joints, such sudden moment changes cause significant shear forces. Thus, joint shear design is the major concern of the column and beam connection, as well as that the longitudinal reinforcements of beams and columns are to be properly anchored or continued through the joint to transmit the moment. For seismic design, joint shear forces can be calculated based on the equilibrium condition using forces generated by the maximum plastic moment acting on the boundary of connections. In the following section, calculations for joint shear forces in the most common connections in bridge structures are discussed [13].

38.5.2 Design of Uncracked Joints

Joints can be conservatively designed based on elastic theory for not permitting cracks. In this approach, the principal tensile stress within a connection is calculated and compared with allowable tensile strength. The principal tensile stress, p_t, can be calculated as

$$p_t = \frac{f_h + f_v}{2} - \sqrt{\left(\frac{f_h - f_v}{2}\right)^2 + v_{hv}^2} \tag{38.30}$$

where f_h and f_v are the average axial stresses in the horizontal and vertical directions within the connection and v_{hv} is the average shear stress. In a typical joint, f_v is provided by the column axial force P_e. An average stress at the midheight of the joint should be used, assuming a 45° spread away from the boundaries of the column in all directions. The horizontal axial stress f_h is based on the mean axial force at the center of the joint, including effects of prestress, if present.

The joint shear stress, v_{hv}, can be estimated as

$$v_{hv} = \frac{M_p}{h_b h_c b_{je}} \tag{38.31}$$

where M_p is the maximum plastic moment, h_b is the beam depth, h_c is the column lateral dimension in the direction considered, and b_{je} is the effective joint width, found using a 45° spread from the column boundaries, as shown in Figure 38.8.

Based on theoretical consideration [13] and experimental observation , it is assumed that onset of diagonal cracking can be induced if the principal tensile stress exceed $0.29\sqrt{f_c'}$ MPa ($3.5\sqrt{f_c'}$ in psi units). When the principal tensile stress is less than $p_t = 0.29\sqrt{f_c'}$ MPa, the minimum amount of horizontal joint shear reinforcement capable of transferring 50% of the cracking stress resolved to the horizontal direction should be provided.

On the other hand, the principal compression stress, p_c, calculated based on the following equation should not exceed $p_c = 0.25\sqrt{f_c'}$, to prevent possible joint crushing.

$$p_c = \frac{f_h + f_v}{2} + \sqrt{\left(\frac{f_h - f_v}{2}\right)^2 + v_{hv}^2} \tag{38.32}$$

38.5.3 Reinforcement for Joint Force Transfer

Diagonal cracks are likely to be induced if the principal tensile stress exceed $0.29\sqrt{f_c'}$ MPa; thus, additional vertical and horizontal joint reinforcement is needed to ensure the force transfer within the joint. Unlike the joints in building structures where more rigorous dimension restraints exist, the joints in bridges can be reinforced with vertical and horizontal bars outside the zone where the column and beam longitudinal bars anchored to reduce the joint congestion. Figure 38.9 shows the force-resisting mechanism in a typical knee joint.

Vertical Reinforcement

On each side of the column or pier wall, the beam that is subjected to moments and shear will have vertical stirrups, with a total area $A_{jv} = 0.16A_{st}$, where A_{st} is the total area of column reinforcement anchored in the joint. The vertical stirrups shall be located within a distance $0.5D$ or $0.5h$ from the column or pier wall face, and distributed over a width not exceeding $2D$. As shown in Figure 38.9, reinforcement A_{jv} is required to provide the tie force T_s resisting the vertical component of strut $D2$. It is also clear from Figure 38.9, all the longitudinal bars contributing to the beam flexural strength should be clamped by the vertical stirrups.

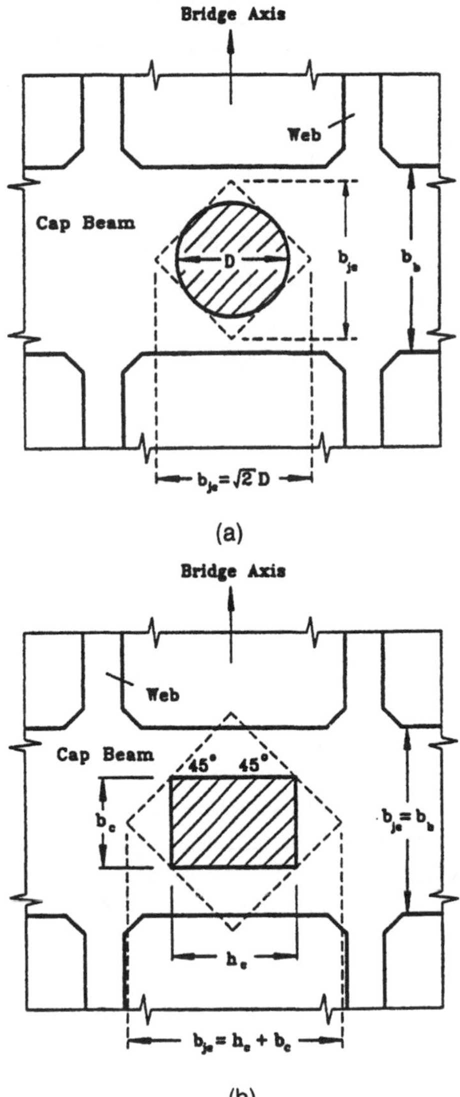

FIGURE 38.8 Effective joint width for joint shear stress calculations: (a) circular column; (b) rectangular column (Priestley et al. 1996; ATC-32, 1996).

Horizontal Reinforcement

Additional cap-beam bottom reinforcement is required to provide the horizontal resistance of the strut D2, shown in Figure 38.9. The suggested details are shown in Figure 38.9. This reinforcement may be omitted in prestressed or partially prestressed cap beams if the prestressed design force is increased by the amount needed to provide an equivalent increase in cap beam moment capacity to that provided by this reinforcement.

Hoop or Spiral Reinforcement

The hoop or spiral reinforcement is required to provide adequate confinement of the joint, and to resist the net outward thrust of struts *D1* and *D2* shown in Figure 38.9. The suggested volumetric ratio of column joint hoop or spiral reinforcement to be carried into the joint is

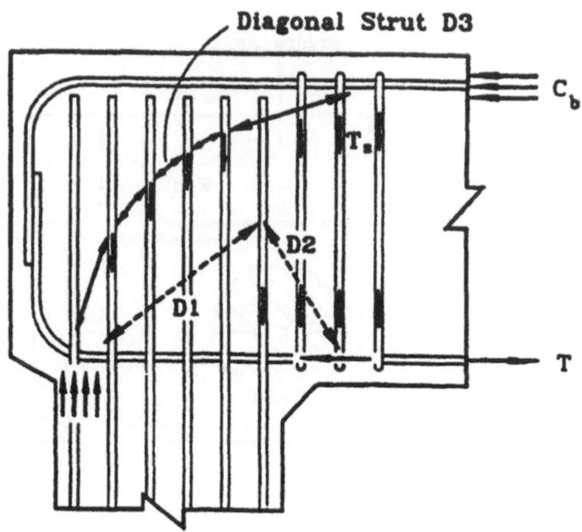

FIGURE 38.9 Joint force transfer. (*Source*: ATC-32 [2]).

$$\rho_s \geq \frac{0.4A_{st}}{l_{ac}^2} \tag{38.33}$$

where l_{ac} is the anchorage length of the column longitudinal reinforcement in the joint.

38.6 Column Footing Design

Bridge footing designs in 1950s to early 1970s were typically based on elastic analysis under relatively low lateral seismic input compared with current design provisions. As a consequence, footings in many older bridges are inadequate for resisting the actual earthquake force input corresponding to column plastic moment capacity. Seismic design for bridge structures has been significantly improved since the 1971 San Fernando earthquake. For bridge footings, a capacity design approach has been adopted by using the column ultimate flexural moment, shear, and axial force as the input to determine the required flexural and shear strength as well as the pile capacity of the footing. However, the designs for flexure and shear of footings are essentially based on a one-way beam model, which lacks experimental verification. The use of a full footing width for the design is nonconservative. In addition, despite the requirement of extending the column transverse reinforcement into the footing, there is a lack of rational consideration of column/footing joint shear in current design [16]. Based on large-scale model tests, Xiao et al. have recommended the following improved design for bridge column footings [16,18].

38.6.1 Seismic Demand

The footings are considered as under the action of column forces, due to the superimposed loads, resisted by an upward pressure exerted by the foundation materials and distributed over the area of the footing. When piles are used under footings, the upward reaction of the foundation is considered as a series of concentrated loads applied at the pile centers. For seismic design, the maximum probable moment, M_p, calculated based on actual strength with consideration of strain hardening of steel and enhancement due to confinement, with the associated axial and lateral loads are applied at the column base as the seismic inputs to the footing. Note that per current Caltrans BDS the maximum probable moment can be taken as 1.3 times the nominal moment capacity of the column, M_n, if the axial load of the column is below its balanced load, P_b. The numbers of piles and the internal forces in the footing are then determined from the seismic input. The internal moment and shear force of the footing can be determined based on the equilibrium conditions of the applied forces at the column base corresponding to the maximum moment capacity and the pile reaction forces.

38.6.2 Flexural Design

For flexural reinforcement design, the footing critical section is taken at the face of the column, pier wall, or at the edge of hinge. In case of columns that are not square or rectangular, the critical sections are taken at the side of the concentric square of equivalent area. The flexural reinforcements near top or the bottom of the footing to resist positive and negative critical moments should be calculated and placed based on the following effective footing width [16].

$$B_{feff} = B_c + 2d_f \quad \text{for rectangular columns} \tag{38.34a}$$

$$B_{feff} = D_c + 2d_f \quad \text{for circular columns} \tag{38.34b}$$

where B_{feff} is the footing effective width; B_c is the rectangular column width; d_f is the effective footing depth; and D_c is the circular column diameter. The minimum reinforcement must satisfy minimum flexural reinforcement requirements. The top reinforcement must also satisfy the requirement for shrinkage and temperature.

38.6.3 Shear Design

For shear, reinforcement of the footing should be designed against the critical shear force at the column face. As with flexure, the effective width of the footing should be used. The minimum shear reinforcement for column footings is vertical No. 5 (nominal diameter = ⅝ in. or 15.9 mm) at 12 in. (305 mm) spacing in each direction in a band between d_f, of the footing from the column surface and 6 in. (152 mm) maximum from the column reinforcement. Shear reinforcement must be hooked around the top and bottom flexure reinforcement in the footing. Inverted J stirrups with a 180° hook at the top and a 90° hook at the bottom are commonly used in California.

38.6.4 Joint Shear Cracking Check

If the footing is sufficiently large, then an uncracked joint may be designed by keeping the principal tensile stress in the joint region below the allowable cracking strength. Average principal tensile stress in the joint region can be calculated from the equivalent joint shear stress, v_{jv}, and the average vertical stress, f_a, by use of Mohr's circle for stress as

$$f_t = -f_a/2 + \sqrt{(f_a/2)^2 + v_{jv}^{\,2}} \tag{38.35}$$

It is suggested that f_a be based on the average effective axial compressive stress at middepth of the footing:

$$f_a = \frac{W_t}{A_{eff}} \tag{38.36}$$

where the effective area, A_{eff}, over which the total axial load W_t at the column base is distributed, is found from a 45-degree spread of the zone of influence as

$$A_{eff} = (B_c + d_f)(D_c + d_f) \quad \text{for rectangular column} \tag{38.37a}$$

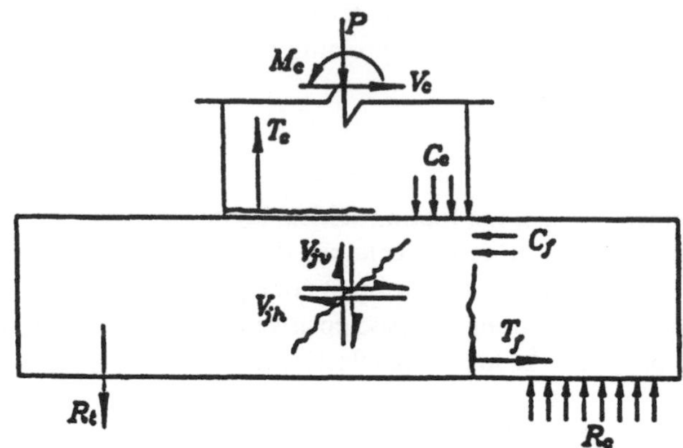

FIGURE 38.10 Shear Force in Column/Footing Joint. (*Source*: Xiao, Y. et al., *ACI Struct. J.*, 93(1), 79–94, 1996. With permission.)

$$A_{eff} = \pi(D_c + d_f)^2 / 4 \qquad \text{for circular column} \tag{38.37b}$$

where D_c is the overall section depth of rectangular column or the diameter of circular column; B_c is the section width of rectangular column; and d_f is the effective depth of the footing. As illustrated in Figure 38.10, the vertical joint shear V_{jv} can be assessed by subtracting the footing shear force due to the hold-down force, R_t, in the tensile piles and the footing self-weight, W_{ft}, outside the column tension stress resultant, from the total tensile force in the critical section of the column:

$$V_{jv} = T_c - (R_t + W_{ft}) \tag{38.38}$$

Considering an effective joint width, b_{jeff}, the average joint shear stress, v_{jv}, can be calculated as follows:

$$v_{jv} = \frac{V_{jv}}{b_{jeff} d_f} \tag{38.39}$$

where the effective joint width, b_{jeff}, can be assumed as the values given by Eq. (38.39), which is obtained based on the St. Venant 45-degree spread of influence between the tension and compression resultants in column critical section.

$$b_{jeff} = B_c + D_c \qquad \text{for rectangular column} \tag{38.40a}$$

$$b_{jeff} = \sqrt{2} D_c \qquad \text{for circular column} \tag{38.40b}$$

Joint shear distress is expected when the principal tensile stress given by Eq. (38.35) induced in the footing exceeds the direct tension strength of the concrete, which may conservatively be taken as 0.29 $\sqrt{f'_c}$ MPa (or 3.5 $\sqrt{f'_c}$ psi) [9], where f'_c is the concrete cylinder compressive strength. The minimum joint shear reinforcement should be provided even when the principal tensile stress is less than the tensile strength. This should be satisfied by simply extending the column transverse confinement into the footing.

38.6.5 Design of Joint Shear Reinforcement

When a footing cannot be prevented from joint shear cracking, additional vertical stirrups should be added around the column. For a typical column/footing designed to current standards, the assumed strut-and-tie model is shown in Figure 38.11a. The force inputs to the footing corresponding to the ultimate moment of the column critical section are the resultant tensile force, T_c, resultant compressive force, C_c, and the shear force, V_c, as shown in Figure 38.11a. The resultant tensile force, T_c, is resisted by two struts, $C1$ and $C2$, inside the column/footing joint region and a strut $C3$, outside the joint. Strut $C1$, is balanced by a horizontal tie, $T1$, provided by the transverse reinforcement of the column inside the footing and the compression zone of the critical section. Struts $C2$, and $C3$, are balanced horizontally at the intersection with the resultant tensile force, T_c. The internal strut $C2$, is supported at the compression zone of the column critical section. The external strut $C3$, transfers the forces to the ties, $T2$, $T3$, provided by the stirrups outside the joint and the top reinforcement, respectively. The forces are further transferred to the tensile piles through struts and ties, $C4$, $C5$, $C6$, and $T4$, $T5$, $T6$, $T7$. In the compression side of the footing, the resultant compressive force, C_c, and the shear force, V_c, are resisted mainly by compression struts, $C9$, $C10$, which are supported on the compression piles. It should be pointed out that the numbers and shapes of the struts and ties may vary for different column/footings.

As shown in Figure 38.11a, the column resultant tensile force, T_c, in the column/footing joint is essentially resisted by a redundant system. The joint shear design is to ensure that the tensile force, T_c, is resisted sufficiently by the internal strut and ties. The resisting system reaches its capacity, R_{ju}, when the ties, $T1$, $T2$, develop yielding. Although tie, $T3$, may also yield, it is not likely to dominate the capacity of the resisting system, since it may be assisted by a membrane mechanism near the footing face.

Assuming the inclination angles of $C1$, $C2$, and $C3$, are 45°, then the resistance can be expressed as

$$R_{ju} = (C1 + C2 + C3)\sin 45° = T1 + 2T2 \tag{38.41}$$

and at steel yielding,

$$R_{ju} = \frac{\pi}{2} n A_{sp} f_y + 2A_{jeff}\rho_{vs} f_y \tag{38.42}$$

where n is the number of layers of the column transverse reinforcement inside the footing; A_{sp} is the cross-sectional area of a hoop or spiral bar; A_{jeff} is the effective area in which the vertical stirrups are effective to resist the resultant tensile force, T_c; ρ_{vs} is the area ratio of the footing vertical stirrups.

The effective area, A_{jeff}, can be defined based on a three-dimensional crack with 45° slope around the column longitudinal bars in tension. The projection of the crack to the footing surface is shown by the shaded area in Figure 38.11b. The depth of the crack or the distance of the boundary of the shaded area in Figure 38.11b to the nearest longitudinal bar yielded in tension is assumed to be equal to the anchorage depth, d_{af}. The depth of the crack reduces from d_{af} to zero linearly if the strain of the rebar reduces from the yield strain, ε_y, to zero. Thus, A_{jeff} can be calculated as follows,

$$A_{jeff} = d_{af}(d_{af} + r_c)\arccos\left[\left(1 + \frac{\varepsilon_y}{\varepsilon_c}\right)\frac{x_n}{r_c} - 1\right] + d_{af} r_c \arccos\left(\frac{x_n}{r_c} - 1\right) \tag{38.43}$$

where d_{af} is the depth of the column longitudinal reinforcement inside footing; r_c is the radius of the centroidal circle of the longitudinal reinforcement and can be simply taken as the radius of the column section if the cover concrete is ignored; ε_y is the yield strain of the longitudinal bars; ε_c is

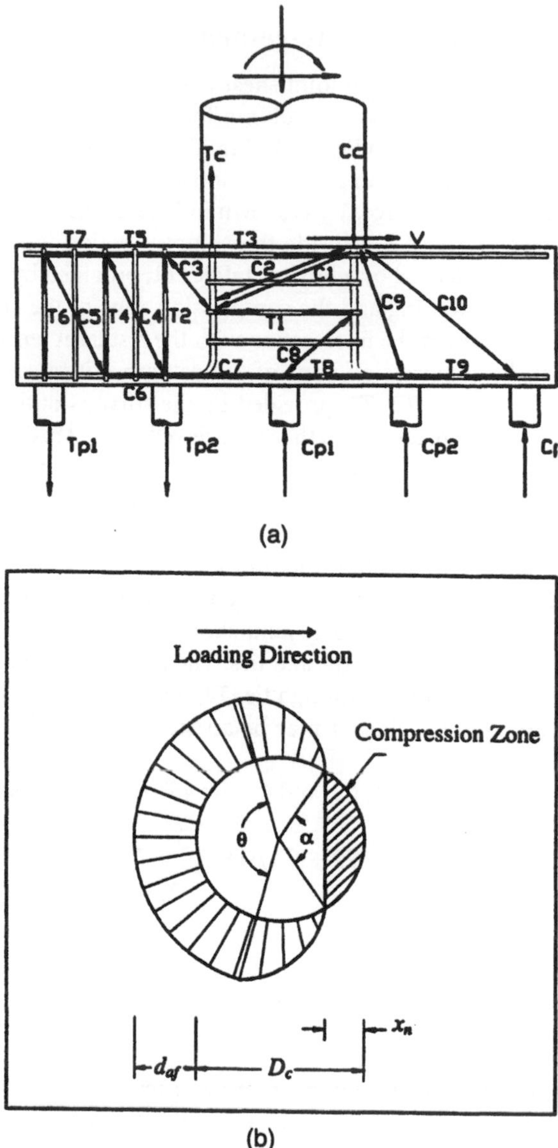

FIGURE 38.11 Column footing joint shear design: (a) force resisting mechanisms in footing; (b) effective distribution of external stirrups for joint shear resistance. (*Source*: Xiao, Y. et al. 1998. With permission.)

strain of the extreme compressive reinforcement or simply taken as the extreme concrete ultimate strain if the cover is ignored; x_n is the distance from the extreme compressive reinforcement to the neutral axis or taken as the compression zone depth, ignoring the cover.

References

1. ACI Committee 318, Building Code Requirements for Reinforced Concrete and Commentary (ACI 318-95/ACI 318R-95), American Concrete Institute, Farmington Hills, MI, 1995.
2. ATC 32, (1996), *Improved Seismic Design Criteria for California Bridges: Provisional Recommendations*, Applied Technology Council, Redwood City, CA, 1996.
3. Bertero, V. V., Overview of seismic risk reduction in urban areas: role, importance, and reliability of current U.S. seismic codes, and performance-based seismic engineering, in *Proceedings of the PRC-USA Bilateral Workshop on Seismic Codes*, Guangzhou, China, December 3–7, 1996, 10–48.

4. Caltrans, Bridge Design Specifications, California Department of Transportation, Sacramento, June, 1990.
5. Caltrans, Caltrans Response To Governor's Board of Inquiry Recommendations and Executive Order of June 2, 1990, 1994: Status Report, Roberts, J. E., California Department of Transportation, Sacramento, January, 26, 1994.
6. Jirsa, O. J. Do we have the knowledge to develop performance-based codes?" *Proceedings of the PRC-USA Bilateral Workshop on Seismic Codes,* Guangzhou, China, December 3–7, 1996, 111–118.
7. Mander, J. B., Priestley, M. J. N., and Park, R., Theoretical Stress-Strain Model for Confined Concrete, *ASCE J. Struct. Eng.,* 114(8), 1827–1849, 1988.
8. Moehle, J. P., Attempts to Introduce Modern Performance Concepts into Old Seismic Codes, in *Proceedings of the PRC-USA Bilateral Workshop on Seismic Codes,* Guangzhou, China, December 3–7, 1996, 217–230.
9. Priestley, M. J. N. and Seible, F., Seismic Assessment and Retrofit of Bridges, M. J. N. Priestley and F. Seible, Eds., University of California at San Diego, Structural Systems Research Project, Report No. SSRP-91/03, July, 1991, 418.
10. Priestley, M. J. N., Seible, F., Xiao, Y., and Verma, R., Steel jacket retrofit of short RC bridge columns for enhanced shear strength — Part 1. Theoretical considerations and test design, *ACI Struct. J.,* American Concrete Institute, 91(4), 394–405, 1994.
11. Priestley, M. J. N., Seible, F., Xiao, Y., and Verma, R., Steel jacket retrofit of short RC bridge columns for enhanced shear strength — Part 2. Experimental results, *ACI Struct. J.,* American Concrete Institute, 91(5), 537–551, 1994.
12. Priestley, M. J. N., Verma, R., and Xiao, Y., Seismic shear strength of reinforced concrete columns, *ASCE J. Struct. Eng.,* American Society of Civil Engineering, 120(8), 2310–2329, 1994.
13. Priestley, M. J. N., Seible, F., and Calvi, M., *Seismic Design and Retrofit of Bridges,* Wiley Interscience, New York, 1996, 686 pp.
14. Priestley, M. J. N., Ranzo, G., Benzoni, G., and Kowalsky, M. J., Yield Displacement of Circular Bridge Columns, in *Proceedings of the Fourth Caltrans Seismic Research Workshop,* July 9–11, 1996.
15. SC-Solution, *SC-Push 3D: Manual and Program Description,* SC-Solutions, San Jose, CA, 1995.
16. Xiao, Y., Priestley, M. J. N., and Seible, F., Seismic Assessment and Retrofit of Bridge Column Footings, *ACI Struct. J.,* 93(1), pp. 79–94, 1996.
17. Xiao, Y. and Martirossyan, A., Seismic performance of high-strength concrete columns, *ASCE J. Struct. Eng.,* 124(3), 241–251, 1998.
18. Xiao, Y., Priestley, M. J. N., and Seible, F., *Seismic Performance of Bridge Footings Designed to Current Standards,* ACI Special Publications on Earthquake Resistant Bridges, in press, 1998.
19. Xiao, Y., H. Wu, and G. R. Martin, (1998) Prefabricated composite jacketing of circular columns for enhanced shear resistance, *ASCE J. Struct. Eng.* 255–264, March, 1999.

39

Seismic Design of Steel Bridges

Chia-Ming Uang
University of California, San Diego

Keh-Chyuan Tsai
National Taiwan University

Michel Bruneau
State University of New York, Buffalo

39.1 Introduction

In the aftermath of the 1995 Hyogo-ken Nanbu earthquake and the extensive damage it imparted to steel bridges in the Kobe area, it is now generally recognized that steel bridges can be seismically vulnerable, particularly when they are supported on nonductile substructures of reinforced concrete, masonry, or even steel. In the last case, unfortunately, code requirements and guidelines on seismic design of ductile bridge steel substructures are few [12,21], and none have yet been implemented in the United States. This chapter focuses on a presentation of concepts and detailing requirements that can help ensure a desirable ductile behavior for steel substructures. Other bridge vulnerabilities common to all types of bridges, such as bearing failure, span collapses due to insufficient seat width or absence of seismic restrainers, soil liquefactions, etc., are not addressed in this chapter.

39.1.1 Seismic Performance Criteria

The American Association of State Highway and Transportation Officials (AASHTO) published both the *Standard Specifications for Highway Bridges* [2] and the *LRFD Bridge Design Specifications* [1], the latter being a load and resistance factor design version of the former, and being the preferred edition when referenced in this chapter. Although notable differences exist between the seismic

0-8493-7434-0/00/$0.00+$.50

design requirements of these documents, both state that the same fundamental principles have been used for the development of their specifications, namely:

1. Small to moderate earthquakes should be resisted within the elastic range of the structural components without significant damage.
2. Realistic seismic ground motion intensities and forces are used in the design procedures.
3. Exposure to shaking from large earthquakes should not cause collapse of all or part of the bridge. Where possible, damage that does occur should be readily detectable and accessible for inspection and repair.

Conceptually, the above performance criteria call for two levels of design earthquake ground motion to be considered. For a low-level earthquake, there should be only minimal damage. For a significant earthquake, which is defined by AASHTO as having a 10% probability of exceedance in 50 years (i.e., a 475-year return period), collapse should be prevented but significant damage may occur. Currently, the AASHTO adopts a simplified approach by specifying only the second-level design earthquake; that is, the seismic performance in the lower-level events can only be implied from the design requirements of the upper-level event. Within the content of performance-based engineering, such a one-level design procedure has been challenged [11,12].

The AASHTO also defines bridge importance categories, whereby essential bridges and critical bridges are, respectively, defined as those that must, at a minimum, remain open to emergency vehicles (and for security/defense purposes), and be open to all traffic, after the 475-year return-period earthquake. In the latter case, the AASHTO suggests that critical bridges should also remain open to emergency traffic after the 2500-year return-period event. Various clauses in the specifications contribute to ensure that these performance criteria are implicitly met, although these may require the engineer to exercise considerable judgment. The special requirements imposed on essential and critical bridges are beyond the scope of this chapter.

39.1.2 The *R* Factor Design Procedure

AASHTO seismic specification uses a response modification factor, *R*, to compute the design seismic forces in different parts of the bridge structure. The origin of the *R* factor design procedure can be traced back to the ATC 3-06 document [9] for building design. Since requirements in seismic provisions for member design are directly related to the *R* factor, it is worthwhile to examine the physical meaning of the *R* factor.

Consider a structural response envelope shown in Figure 39.1. If the structure is designed to respond elastically during a major earthquake, the required elastic force, Q_e, would be high. For economic reasons, modern seismic design codes usually take advantage of the inherent energy dissipation capacity of the structure by specifying a design seismic force level, Q_s, which can be significantly lower than Q_e:

$$Q_s = \frac{Q_e}{R} \tag{39.1}$$

The energy dissipation (or ductility) capacity is achieved by specifying stringent detailing requirements for structural components that are expected to yield during a major earthquake. The design seismic force level Q_s is the first significant yield level of the structure, which corresponds to the level beyond which the structural response starts to deviate significantly from the elastic response. Idealizing the actual response envelope by a linearly elastic–perfectly plastic response shown in Figure 39.1, it can be shown that the *R* factor is composed of two contributing factors [64]:

$$R = R_\mu \Omega? \tag{39.2}$$

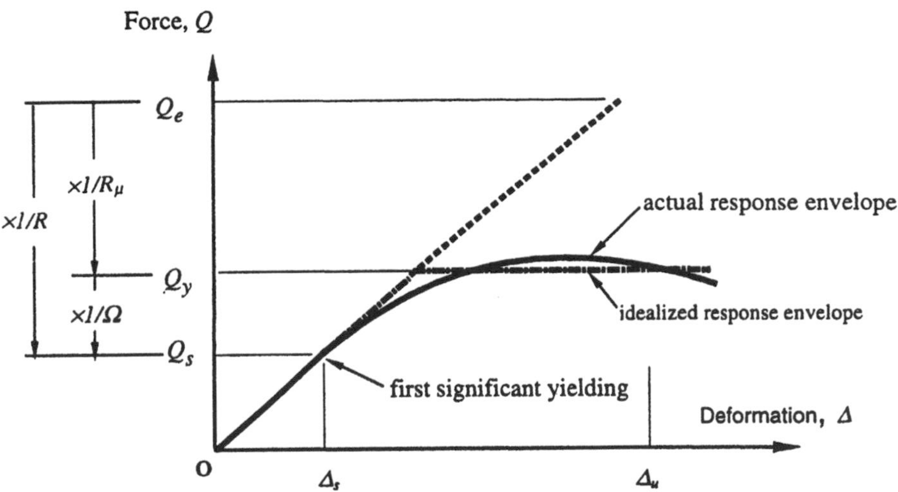

FIGURE 39.1 Concept of response modification factor, *R*.

TABLE 39.1 Response Modification Factor, *R*

Substructure	*R*	Connections	*R*
Single columns	3	Superstructure to abutment	0.8
Steel or composite steel and concrete pile bents		Columns, piers, or pile bents to cap beam or superstructure	1.0
a. Vertical piles only	5		
b. One or more batter piles	3		
Multiple column bent	5	Columns or piers to foundations	1.0

Source: AASHTO, *Standard Specifications for Seismic Design of Highway Bridges*, AASHTO, Washington, D.C., 1992.

The ductility reduction factor, R_μ, accounts for the reduction of the seismic force level from Q_e to Q_y. Such a force reduction is possible because ductility, which is measured by the ductility factor $\mu\,\Delta_u / \Delta_y$), is built into the structural system. For single-degree-of-freedom systems, relationships between μ and R_μ have been proposed (e.g., Newmark and Hall [43]).

The structural overstrength factor, Ω, in Eq. (39.2) accounts for the reserve strength between the seismic resistance levels Q_y and Q_s. This reserve strength is contributed mainly by the redundancy of the structure. That is, once the first plastic hinge is formed at the force level Q_s, the redundancy of the structure would allow more plastic hinges to form in other designated locations before the ultimate strength, Q_y, is reached. Table 39.1 shows the values of *R* assigned to different substructure and connection types. The AASHTO assumes that cyclic inelastic action would only occur in the substructure; therefore, no *R* value is assigned to the superstructure and its components. The table shows that the *R* value ranges from 3 to 5 for steel substructures. A multiple column bent with well detailed columns has the highest value (= 5) of *R* due to its ductility capacity and redundancy. The ductility capacity of single columns is similar to that of columns in multiple column bent; however, there is no redundancy and, therefore, a low *R* value of 3 is assigned to single columns.

Although modern seismic codes for building and bridge designs both use the *R* factor design procedure, there is one major difference. For building design [42], the *R* factor is applied at the system level. That is, components designated to yield during a major earthquake share the same *R* value, and other components are proportioned by the capacity design procedure to ensure that these components remain in the elastic range. For bridge design, however, the *R* factor is applied at the component level. Therefore, different *R* values are used in different parts of the same structure.

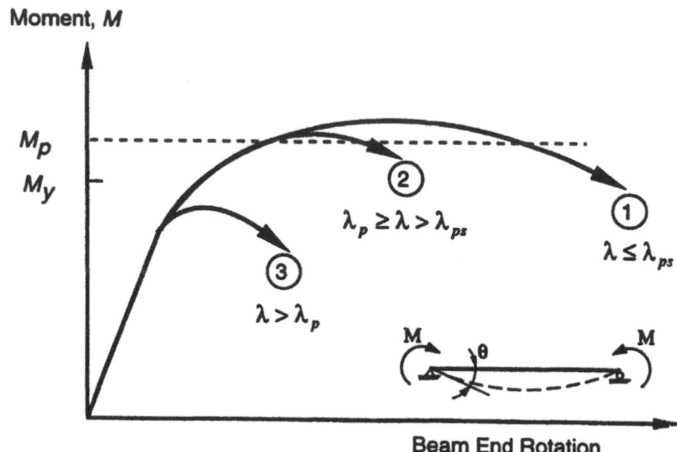

FIGURE 39.2 Effect of beam slenderness ratio on strength and deformation capacity. (Adapted from Yura et al., 1978.)

39.1.3 Need for Ductility

Using an R factor larger than 1 implies that the ductility demand must be met by designing the structural component with stringent requirements. The ductility capacity of a steel member is generally governed by instability. Considering a flexural member, for example, instability can be caused by one or more of the following three limit states: flange local buckling, web local buckling, and lateral-torsional buckling. In all cases, ductility capacity is a function of a slenderness ratio, λ. For local buckling, λ is the width–thickness ratio; for lateral-torsional buckling, λ is computed as L_b/r_y, where L_b is the unbraced length and r_y is the radius of gyration of the section about the buckling axis. Figure 39.2 shows the effect of λ on strength and deformation capacity of a wide-flanged beam. Curve 3 represents the response of a beam with a noncompact or slender section; both its strength and deformation capacity are inadequate for seismic design. Curve 2 corresponds to a beam with "compact" section; its slenderness ratio, λ, is less than the maximum ratio λp for which a section can reach its plastic moment, M_p, and sustain moderate plastic rotations. For seismic design, a response represented by Curve 1 is needed, and a "plastic" section with λ less than λps is required to deliver the needed ductility.

Table 39.2 shows the limiting width–thickness ratios λp and λps for compact and plastic sections, respectively. A flexural member with λ not exceeding λp can provide a rotational ductility factor of at least 4 [74], and a flexural member with λ less than λps is expected to deliver a rotation ductility factor of 8 to 10 under monotonic loading [5]. Limiting slenderness ratios for lateral-torsional buckling are presented in Section 39.2.

39.1.4 Structural Steel Materials

AASHTO M270 (equivalent to ASTM A709) includes grades with a minimum yield strength ranging from 36 to 100 ksi (see Table 39.3). These steels meet the AASHTO Standards for the mandatory notch toughness and weldability requirements and hence are prequalified for use in welded bridges.

For ductile substructure elements, steels must be capable of dissipating hysteretic energy during earthquakes, even at low temperatures if such service conditions are expected. Typically, steels that have $F_y < 0.8F_u$ and can develop a longitudinal elongation of 0.2 mm/mm in a 50-mm gauge length prior to failure at the expected service temperature are satisfactory.

TABLE 39.2 Limiting Width-Thickness Ratios

Description of Element	Width-Thickness Ratio	λ_p	λ_{ps}
Flanges of I-shaped rolled beams, hybrid or welded beams, and channels in flexure	b/t	$65/(\sqrt{F_y})$	$52/(\sqrt{F_y})$
Webs in combined flexural and axial compression	h/tw	for $P_u/\phi bPy \le 0.125$: $\frac{640}{\sqrt{F_y}}\left(1-\frac{2.75P_u}{\phi_b P_y}\right)$ for $P_u/\phi_b P_y > 0.125$: $\frac{191}{\sqrt{F_y}}\left(2.33-\frac{P_u}{\phi_b P_y}\right) \ge \frac{253}{\sqrt{F_y}}$	for $Pu/\phi bPy > 0.125$: $\frac{520}{\sqrt{F_y}}\left(1-\frac{1.54P_u}{\phi_b P_y}\right)$ for $P_u/\phi_b P_y > 0.125$: $\frac{191}{\sqrt{F_y}}\left(2.33-\frac{P_u}{\phi_b P_y}\right) \ge \frac{253}{\sqrt{F_y}}$
Round HSS in axial compression or flexure	D/t	$\frac{2070}{F_y}$	$\frac{1300}{F_y}$
Rectangular HSS in axial compression or flexure	b/t	$\frac{190}{\sqrt{F_y}}$	$\frac{110}{\sqrt{F_y}}$

Note: F_y in ksi, $\phi_b = 0.9$.

Source: AISC, *Seismic Provisions for Structural Steel Buildings*, AISC, Chicago, IL, 1997.

TABLE 39.3 Minimum Mechanical Properties of Structural Steel

AASHTO Designation	M270 Grade 36	M270 Grade 50	M270 Grade 50W	M270 Grade 70W	M270 Grades 100/100W	
Equivalent ASTM designation	A709 Grade 36	A709 Grade 50	A709 Grade 50W	A709 Grade 70W	A709 Grade 100/100W	
Minimum yield stress (ksi)	36	50	50	70	100	90
Minimum tensile stress (ksi)	58	65	70	90	110	100

Source: AASHTO, *Standard Specification for Highway Bridges*, AASHTO, Washington, D.C., 1996.

39.1.5 Capacity Design and Expected Yield Strength

For design purposes, the designer is usually required to use the minimum specified yield and tensile strengths to size structural components. This approach is generally conservative for gravity load design. However, this is not adequate for seismic design because the AASHTO design procedure sometimes limits the maximum force acting in a component to the value obtained from the adjacent yielding element, per a capacity design philosophy. For example, steel columns in a multiple-column bent can be designed for an R value of 5, with plastic hinges developing at the column ends. Based on the weak column–strong beam design concept (to be presented in Section 39.2), the cap beam and its connection to columns need to be designed elastically (i.e., $R = 1$, see Table 39.1). Alternatively, for bridges classified as seismic performance categories (SPC) C and D, the AASHTO recommends that, for economic reasons, the connections and cap beam be designed for the maximum forces capable of being developed by plastic hinging of the column or column bent; these forces will often be significantly less than those obtained using an R factor of 1. For that purpose, recognizing the possible overstrength from higher yield strength and strain hardening, the AASHTO [1] requires that the column plastic moment be calculated using 1.25 times the nominal yield strength.

Unfortunately, the widespread brittle fracture of welded moment connections in steel buildings observed after the 1994 Northridge earthquake revealed that the capacity design procedure mentioned

TABLE 39.4 Expected Steel Material Strengths (SSPC 1994)

Steel Grade	A36	A572 Grade 50
No. of Sample	36,570	13,536
Yield Strength (COV)	49.2 ksi (0.10)	57.6 ksi (0.09)
Tensile Strength (COV)	68.5 ksi (0.07)	75.6 ksi (0.08)

COV: coefficient of variance.

Source: SSPC, *Statistical Analysis of Tensile Data for Wide Flange Structural Shapes*, Structural Shapes Producers Council, Washington, D.C., 1994.

above is flawed. Investigations that were conducted after the 1994 Northridge earthquake indicate that, among other factors, material overstrength (i.e., the actual yield strength of steel is significantly higher than the nominal yield strength) is one of the major contributing factors for the observed fractures [52].

Statistical data on material strength of AASHTO M270 steels is not available, but since the mechanical characteristics of M270 Grades 36 and 50 steels are similar to those of ASTM A36 and A572 Grade 50 steels, respectively, it is worthwhile to examine the expected yield strength of the latter. Results from a recent survey [59] of certified mill test reports provided by six major steel mills for 12 consecutive months around 1992 are briefly summarized in Table 39.4. Average yield strengths are shown to greatly exceed the specified values. As a result, relevant seismic provisions for building design have been revised. The AISC Seismic Provisions [6] use the following formula to compute the expected yield strength, *Fye*, of a member that is expected to yield during a major earthquake:

$$F_{ye} = R_y F_y \tag{39.3}$$

where *Fy* is the specified minimum yield strength of the steel. For rolled shapes and bars, R_y should be taken as 1.5 for A36 steel and 1.1 for A572 Grade 50 steel. When capacity design is used to calculate the maximum force to be resisted by members connected to yielding members, it is suggested that the above procedure also be used for bridge design.

39.1.6 Member Cyclic Response

A typical cyclic stress–strain relationship of structural steel material is shown in Figure 39.3. When instability are excluded, the figure shows that steel is very ductile and is well suited for seismic applications. Once the steel is yielded in one loading direction, the Bauschinger effect causes the steel to yield earlier in the reverse direction, and the clearly defined yield plateau disappears in subsequent cycles. Where instability needs to be considered, the Bauschinger effect may affect the cyclic strength of a steel member.

Consider an axially loaded steel member first. Figure 39.4 shows the typical cyclic response of an axially loaded tubular brace. The initial buckling capacity can be predicted reliably using the tangent modulus concept [47]. The buckling capacity in subsequent cycles, however, is reduced due to two factors: (1) the Bauschinger effect, which reduces the tangent modulus, and (2) the increased out-of-straigthness as a result of buckling in previous cycles. Such a reduction in cyclic buckling strength needs to be considered in design (see Section 39.3).

For flexural members, repeated cyclic loading will also trigger buckling even though the width–thickness ratios are less than the λps limits specified in Table 39.2. Figure 39.5 compares the cyclic response of two flexural members with different flange *b/t* ratios [62]. The strength of the beam having a larger flange width–thickness ratio degrades faster under cyclic loading as local buckling develops. This justifies the need for more stringent slenderness requirements in seismic design than those permitted for plastic design.

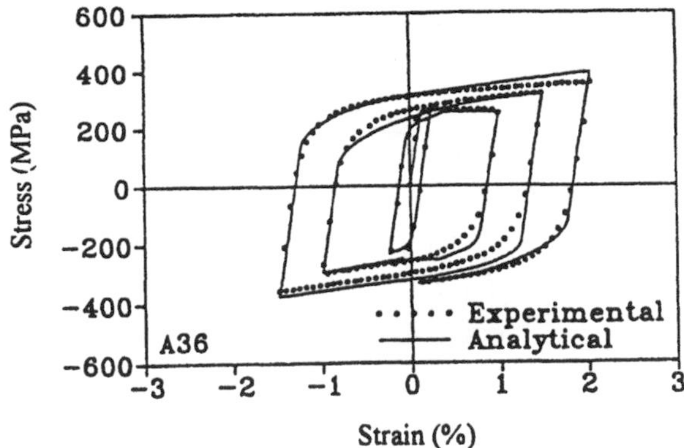

FIGURE 39.3 Typical cyclic stress–strain relationship of structural steel.

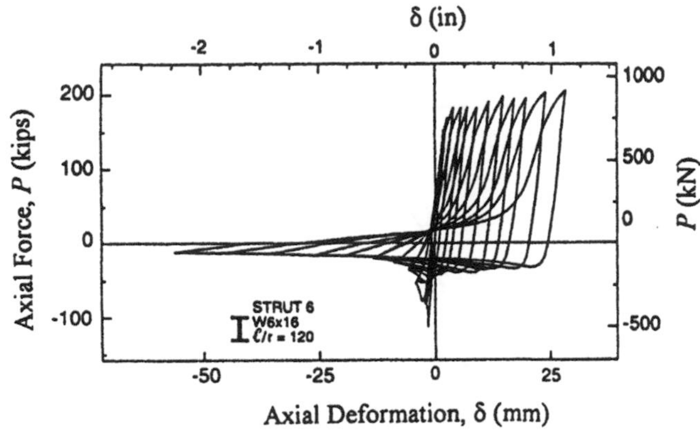

FIGURE 39.4 Cyclic response of an axially loaded member. (Source: Popov, E. P. and Black, W., *J. Struct. Div. ASCE*, 90(ST2), 223-256, 1981. With permission.)

39.2 Ductile Moment-Resisting Frame (MRF) Design

39.2.1 Introduction

The prevailing philosophy in the seismic resistant design of ductile frames in buildings is to force plastic hinging to occur in beams rather than in columns in order to better distribute hysteretic energy throughout all stories and to avoid soft-story-type failure mechanisms. However, for steel bridges such a constraint is not realistic, nor is it generally desirable. Steel bridges frequently have deep beams which are not typically compact sections, and which are much stiffer flexurally than their supporting steel columns. Moreover, bridge structures in North America are generally "single-story" (single-tier) structures, and all the hysteretic energy dissipation is concentrated in this single story. The AASHTO [3] and CHBDC [21] seismic provisions are, therefore, written assuming that columns will be the ductile substructure elements in moment frames and bents. Only the CHBDC, to date, recognizes the need for ductile detailing of steel substructures to ensure that the performance objectives are met when an R value of 5 is used in design [21]. It is understood that extra care would be needed to ensure the satisfactory ductile response of multilevel steel frame bents since these are implicitly not addressed by these specifications. Note that other recent design recommendations [12] suggest that the designer can choose to have the primary energy dissipation mechanism occur in either the beam–column panel zone or the column, but this approach has not been implemented in codes.

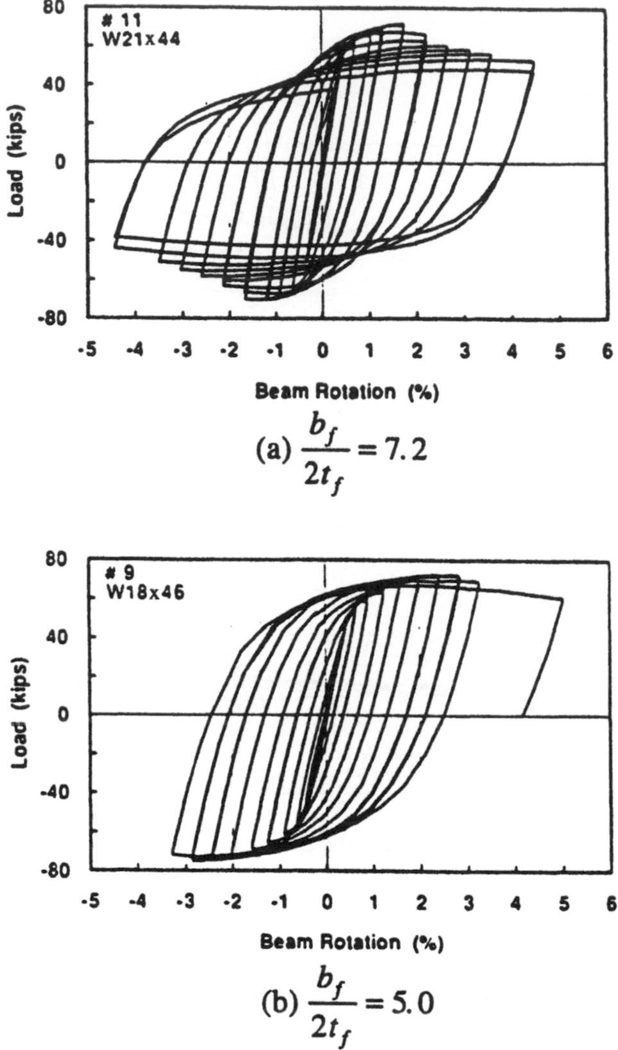

FIGURE 39.5 Effect of beam flange width–thickness ratio on strength degradation. (a) $b_f/2t_f = 7.2$; (b) $b_f/2t_f = 5.0$.

Some detailing requirements are been developed for elements where inelastic deformations are expected to occur during an earthquake. Nevertheless, lessons learned from the recent Northridge and Hyogo-ken Nanbu earthquakes have indicated that steel properties, welding electrodes, and connection details, among other factors, all have significant effects on the ductility capacity of welded steel beam–column moment connections [52]. In the case where the bridge column is continuous and the beam is welded to the column flange, the problem is believed to be less severe as the beam is stronger and the plastic hinge will form in the column [21]. However, if the bridge girder is continuously framed over the column in a single-story frame bent, special care would be needed for the welded column-to-beam connections.

Continuous research and professional developments on many aspects of the welded moment connection problems are well in progress and have already led to many conclusions that have been implemented on an interim basis for building constructions [52,54]. Many of these findings should be applicable to bridge column-to-beam connections where large inelastic demands are likely to

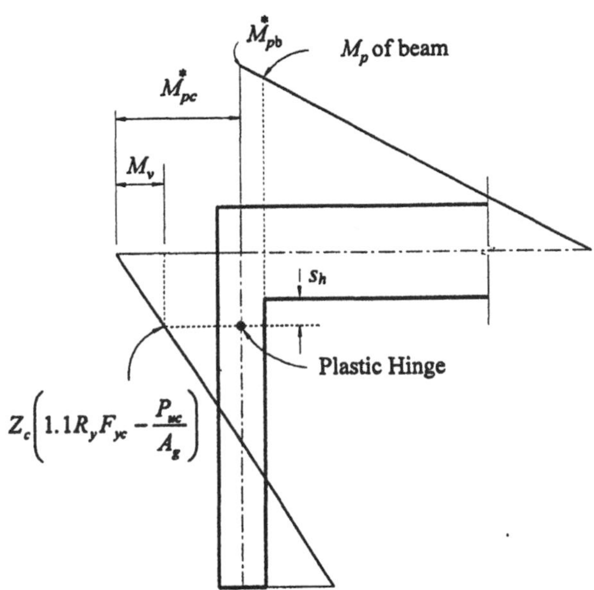

FIGURE 39.6 Location of plastic hinge.

develop in a major earthquake. The following sections provide guidelines for the seismic design of steel moment-resisting beam–column bents.

39.2.2 Design Strengths

Columns, beams, and panel zones are first designed to resist the forces resulting from the prescribed load combinations; then capacity design is exercised to ensure that inelastic deformations only occur in the specially detailed ductile substructure elements. To ensure a weak-column and strong-girder design, the beam-to-column strength ratio must satisfy the following requirement:

$$\frac{\Sigma M^*_{pb}}{\Sigma M^*_{pc}} \geq 1.0 \tag{39.4}$$

where $\Sigma\ M^*_{pb}$ is the sum of the beam moments at the intersection of the beam and column centerline. It can be determined by summing the projections of the nominal flexural strengths, M_p ($= Z_bF_y$, where Z_b is the plastic section modulus of the beam), of the beams framing into the connection to the column centerline. The term $\Sigma\ M^*_{pc}$ is the sum of the expected column flexural strengths, reduced to account for the presence of axial force, above and below the connection to the beam centerlines. The term $\Sigma\ M^*_{pc}$ can be approximated as $\Sigma\,[Z_c(1.1R_yF_{yc}-P_{uc}/A_g)+M_v]$, where A_g is the gross area of the column, P_{uc} is the required column compressive strength, Z_c is the plastic section modulus of the column, F_{yc} is the minimum specified yield strength of the column. The term M_v is to account for the additional moment due to shear amplification from the actual location of the column plastic hinge to the beam centerline (Figure 39.6). The location of the plastic hinge is at a distance s_h from the edge of the reinforced connection. The value of s_h ranges from one quarter to one third of the column depth as suggested by SAC [54].

To achieve the desired energy dissipation mechanism, it is rational to incorporate the expected yield strength into recent design recommendations [12,21]. Furthermore, it is recommended that the beam–column connection and the panel zone be designed for 125% of the expected plastic bending moment capacity, $Z_c(1.1R_yF_{yc} - P_{uc}/A_g)$, of the column. The shear strength of the panel zone, V_n, is given by

$$V_n = 0.6F_y d_c t_p \tag{39.5}$$

where d_c is the overall column depth and t_p is the total thickness of the panel zone including doubler plates. In order to prevent premature local buckling due to shear deformations, the panel zone thickness, t_p, should conform to the following:

$$t_p \geq \frac{d_z + w_z}{90} \tag{39.6}$$

where d_z and w_z are the panel zone depth and width, respectively.

Although weak panel zone is permitted by the AISC [6] for building design, the authors, however, prefer a conservative approach in which the primary energy dissipation mechanism is column hinging.

39.2.3 Member Stability Considerations

The width–thickness ratios of the stiffened and unstiffened elements of the column section must not be greater than the λ_{ps} limits given in Table 39.2 in order to ensure ductile response for the plastic hinge formation. Canadian practice [21] requires that the factored axial compression force due to the seismic load and gravity loads be less than $0.30A_gF_y$ (or twice that value in lower seismic zones). In addition, the plastic hinge locations, near the top and base of each column, also need to be laterally supported. To avoid lateral-torsional buckling, the unbraced length should not exceed $2500r_y/F_y$ [6].

39.2.4 Column-to-Beam Connections

Widespread brittle fractures of welded moment connections in building moment frames that were observed following the 1994 Northridge earthquake have raised great concerns. Many experimental and analytical studies conducted after the Northridge earthquake have revealed that the problem is not a simple one, and no single factor can be made fully responsible for the connection failures. Several design advisories and interim guidelines have already been published to assist engineers in addressing this problem [52,54]. Possible causes for the connection failures are presented below.

1. As noted in Section 39.1.5, the mean yield strength of A36 steel in the United States is substantially higher than the nominal yield value. This increase in yield strength combined with the cyclic strain hardening effect can result in a beam moment significantly higher than its nominal strength. Considering the large variations in material strength, it is questionable whether the bolted web-welded flange pre-Northridge connection details can reliably sustain the beam flexural demand imposed by a severe earthquake.
2. Recent investigations conducted on the properties of weld metal have indicated that the E70T-4 weld metal which was typically used in many of the damaged buildings possesses low notch toughness [60]. Experimental testing of welded steel moment connections that were conducted after the Northridge earthquake clearly demonstrated that notch-tough electrodes are needed for seismic applications. Note that the bridge specifications effectively prohibit the use of E70T-4 electrode.
3. In a large number of connections, steel backing below the beam bottom flange groove weld has not been removed. Many of the defects found in such connections were slag inclusions of a size that should have been rejected per AWS D1.1 if they could have been detected during

the construction. The inclusions were particularly large in the middle of the flange width where the weld had to be interrupted due to the presence of the beam web. Ultrasonic testing for welds behind the steel backing and particularly near the beam web region is also not very reliable. Slag inclusions are equivalent to initial cracks, which are prone to crack initiation at a low stress level. For this reason, the current steel building welding code [13] requires that steel backing of groove welds in cyclically loaded joints be removed. Note that the bridge welding code [14] has required the removal of steel backing on welds subjected to transverse tensile stresses.

4. Steel that is prevented from expanding or contracting under stress can fail in a brittle manner. For the most common type of groove welded flange connections used prior to the Northridge earthquake, particularly when they were executed on large structural shapes, the welds were highly restrained along the length and in the transverse directions. This precludes the welded joint from yielding, and thus promotes brittle fractures [16].
5. Rolled structural shapes or plates are not isotropic. Steel is most ductile in the direction of rolling and least ductile in the direction orthogonal to the surface of the plate elements (i.e., through-thickness direction). Thicker steel shapes and plates are also susceptible to lamellar tearing [4].

After the Northridge earthquake, many alternatives have been proposed for building construction and several have been tested and found effective to sustain cyclic plastic rotational demand in excess of 0.03 rad. The general concept of these alternatives is to move the plastic hinge region into the beam and away from the connection. This can be achieved by either strengthening the beam near the connection or reducing the strength of the yielding member near the connection. The objective of both schemes is to reduce the stresses in the flange welds in order to allow the yielding member to develop large plastic rotations. The minimum strength requirement for the connection can be computed by considering the expected maximum bending moment at the plastic hinge using statics similar to that outlined in Section 39.2.2. Capacity-enhancement schemes which have been widely advocated include cover plate connections [26] and bottom haunch connections. The demand-reduction scheme can be achieved by shaving the beam flanges [22,27,46,74]. Note that this research and development was conducted on deep beam sections without the presence of an axial load. Their application to bridge columns should proceed with caution.

39.3 Ductile Braced Frame Design

Seismic codesfor bridge design generally require that the primary energy dissipation mechanism be in the substructure. Braced frame systems, having considerable strength and stiffness, can be used for this purpose [67]. Depending on the geometry, a braced frame can be classified as either a concentrically braced frame (CBF) or an eccentrically braced frame (EBF). CBFs can be found in the cross-frames and lateral-bracing systems of many existing steel girder bridges. In a CBF system, the working lines of members essentially meet at a common point (Figure 39.7). Bracing members are prone to buckle inelastically under the cyclic compressive overloads. The consequence of cyclic buckling of brace members in the superstructure is not entirely known at this time, but some work has shown the importance of preserving the integrity of end-diaphragms [72]. Some seismic design recommendations [12] suggest that cross-frames and lateral bracing, which are part of the seismic force-resisting system in common slab-on-steel girder bridges, be designed to remain elastically under the imposed load effects. This issue is revisited in Section 39.5.

In a manner consistent with the earthquake-resistant design philosophy presented elsewhere in this chapter, modern CBFs are expected to undergo large inelastic deformation during a severe earthquake. Properly proportioned and detailed brace members can sustain these inelastic deformations and dissipate hysteretic energy in a stable manner through successive cycles of compression buckling and tension yielding. The preferred strategy is, therefore, to ensure that plastic deformation

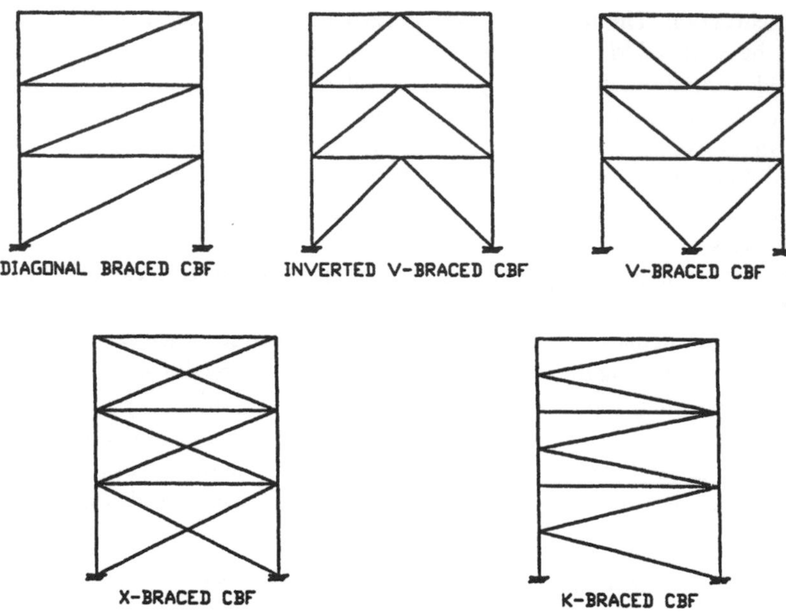

FIGURE 39.7 Typical concentric bracing configurations.

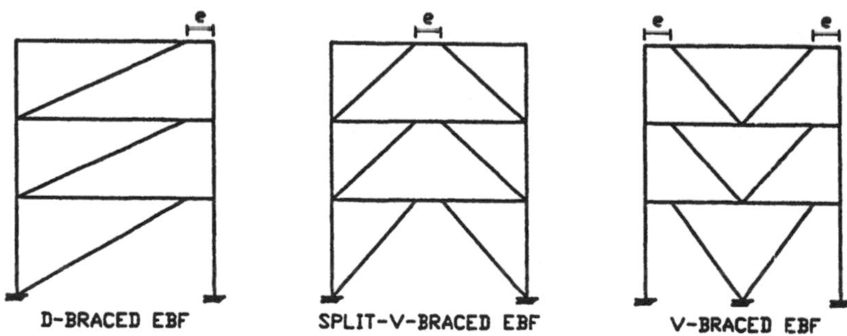

FIGURE 39.8 Typical eccentric bracing configurations.

only occur in the braces, allowing the columns and beams to remain essentially elastic, thus maintaining the gravity load-carrying capacity during a major earthquake. According to the AISC Seismic Provisions [6], a CBF can be designed as either a special CBF (SCBF) or an ordinary CBF (OCBF). A large value of R is assigned to the SCBF system, but more stringent ductility detailing requirements need to be satisfied.

An EBF is a system of columns, beams, and braces in which at least one end of each bracing member connects to a beam at a short distance from its beam-to-column connection or from its adjacent beam-to-brace connection (Figure 39.8). The short segment of the beam between the brace connection and the column or between brace connections is called the link. Links in a properly designed EBF system will yield primarily in shear in a ductile manner. With minor modifications, the design provisions prescribed in the AISC Seismic Provisions for EBF, SCBF, and OCBF can be implemented for the seismic design of bridge substructures.

Current AASHTO seismic design provisions [3] do not prescribe the design seismic forces for the braced frame systems. For OCBFs, a response modification factor, R, of 2.0 is judged appropriate. For EBFs and SCBFs, an R value of 4 appears to be conservative and justifiable by examining the ductility reduction factor values prescribed in the building seismic design recommendations [57].

For CBFs, the emphasis in this chapter is placed on SCBFs, which are designed for better inelastic performance and energy dissipation capacity.

39.3.1 Concentrically Braced Frames

Tests have shown that, after buckling, an axially loaded member rapidly loses compressive strength under repeated inelastic load reversals and does not return to its original straight position (see Figure 39.4). CBFs exhibit the best seismic performance when both yielding in tension and inelastic buckling in compression of their diagonal members contribute significantly to the total hysteretic energy dissipation. The energy absorption capability of a brace in compression depends on its slenderness ratio (KL/r) and its resistance to local buckling. Since they are subjected to more stringent detailing requirements, SCBFs are expected to withstand significant inelastic deformations during a major earthquake. OCBFs are designed to higher levels of design seismic forces to minimize the extent of inelastic deformations. However, if an earthquake greater than that considered for design occurs, structures with SCBF could be greatly advantaged over the OCBF, in spite of the higher design force level considered in the latter case.

Bracing Members

Postbuckling strength and energy dissipation capacity of bracing members with a large slenderness ratio will degrade rapidly after buckling occurs [47]. Therefore, many seismic codes require the slenderness ratio (KL/r) for the bracing member be limited to $720/\sqrt{F_y}$, where F_y is in ksi. Recently, the AISC Seismic Provisions (1997) [6] have relaxed this limit to $1000/\sqrt{F_y}$ for bracing members in SCBFs. This change is somewhat controversial. The authors prefer to follow the more stringent past practice for SCBFs. The design strength of a bracing member in axial compression should be taken as $0.8\,\phi_c P_n$, where ϕ_c is taken as 0.85 and P_n is the nominal axial strength of the brace. The reduction factor of 0.8 has been prescribed for CBF systems in the previous seismic building provisions [6] to account for the degradation of compressive strength in the postbuckling region. The 1997 AISC Seismic Provisions have removed this reduction factor for SCBFs. But the authors still prefer to apply this strength reduction factor for the design of both SCBFs and OCBFs. Whenever the application of this reduction factor will lead to a less conservative design, however, such as to determine the maximum compressive force a bracing member imposes on adjacent structural elements, this reduction factor should not be used.

The plastic hinge that forms at midspan of a buckled brace may lead to severe local buckling. Large cyclic plastic strains that develop in the plastic hinge are likely to initiate fracture due to low-cycle fatigue. Therefore, the width–thickness ratio of stiffened or unstiffened elements of the brace section for SCBFs must be limited to the values specified in Table 39.2. The brace sections for OCBFs can be either compact or noncompact, but not slender. For brace members of angle, unstiffened rectangular, or hollow sections, the width–thickness ratios cannot exceed λ_{ps}.

To provide redundancy and to balance the tensile and compressive strengths in a CBF system, it is recommended that at least 30% but not more than 70% of the total seismic force be resisted by tension braces. This requirement can be waived if the bracing members are substantially oversized to provide essentially elastic seismic response.

Bracing Connections

The required strength of brace connections (including beam-to-column connections if part of the bracing system) should be able to resist the lesser of:

1. The expected axial tension strength ($= R_y F_y A_g$) of the brace.
2. The maximum force that can be transferred to the brace by the system.

In addition, the tensile strength of bracing members and their connections, based on the limit states of tensile rupture on the effective net section and block shear rupture, should be at least equal to the required strength of the brace as determined above.

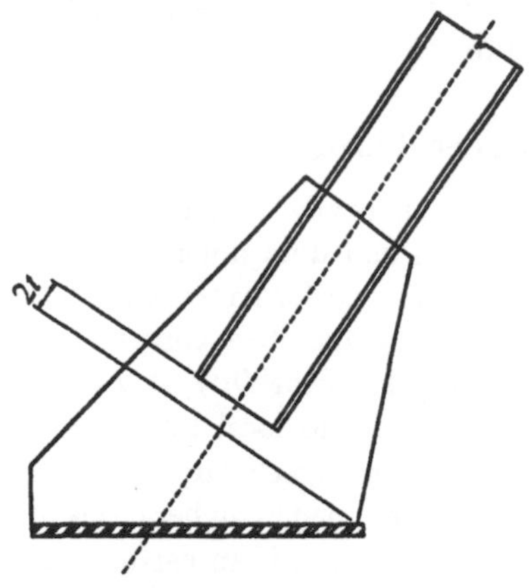

t = gusset plate thickness

FIGURE 39.9 Plastic hinge and free length of gusset plate.

End connections of the brace can be designed as either rigid or pin connection. For either of the end connection types, test results showed that the hysteresis responses are similar for a given *KL/r* [47]. When the brace is pin-connected and the brace is designed to buckle out of plane, it is suggested that the brace be terminated on the gusset a minimum of two times the gusset thickness from a line about which the gusset plate can bend unrestrained by the column or beam joints [6]. This condition is illustrated in Figure 39.9. The gusset plate should also be designed to carry the design compressive strength of the brace member without local buckling.

The effect of end fixity should be considered in determining the critical buckling axis if rigid end conditions are used for in-plane buckling and pinned connections are used for out-of-plane buckling. When analysis indicates that the brace will buckle in the plane of the braced frame, the design flexural strength of the connection should be equal to or greater than the expected flexural strength ($= 1.1R_yM_p$) of the brace. An exception to this requirement is permitted when the brace connections (1) meet the requirement of tensile rupture strength described above, (2) can accommodate the inelastic rotations associated with brace postbuckling deformations, and (3) have a design strength at least equal to the nominal compressive strength ($= A_gF_y$) of the brace.

Special Requirements for Brace Configuration

Because braces meet at the midspan of beams in V-type and inverted-V-type braced frames, the vertical force resulting from the unequal compression and tension strengths of the braces can have a considerable impact on cyclic behavior. Therefore, when this type of brace configuration is considered for SCBFs, the AISC Seismic Provisions require that:

1. A beam that is intersected by braces be continuous between columns.
2. A beam that is intersected by braces be designed to support the effects of all the prescribed tributary gravity loads assuming that the bracing is not present.
3. A beam that is intersected by braces be designed to resist the prescribed force effects incorporating an unbalanced vertical seismic force. This unbalanced seismic load must be substituted for the seismic force effect in the load combinations, and is the maximum unbalanced vertical force applied to the beam by the braces. It should be calculated using a minimum of P_y for the brace in tension and a maximum of 0.3 $\phi_c P_n$ for the brace in compression. This requirement ensures that the beam will not fail due to the large unbalanced force after brace buckling.
4. The top and bottom flanges of the beam at the point of intersection of braces must be adequately braced; the lateral bracing should be designed for 2% of the nominal beam flange strength ($= F_yb_ft_{bf}$).

For OCBFs, the AISC Seismic Provisions waive the third requirement. But the brace members need to be designed for 1.5 times the required strength computed from the prescribed load combinations.

Columns

Based on the capacity design principle, columns in a CBF must be designed to remain elastic when all braces have reached their maximum tension or compression capacity considering an overstrength factor of $1.1R_y$. The AISC Seismic Provisions also require that columns satisfy the λps requirements (see Table 39.2). The strength of column splices must be designed to resist the imposed load effects. Partial penetration groove welds in the column splice have been experimentally observed to fail in a brittle manner [17]. Therefore, the AISC Seismic Provisions require that such splices in SCBFs be designed for at least 200% of the required strength, and be constructed with a minimum strength of 50% of the expected column strength, R_yF_yA, where A is the cross-sectional area of the smaller column connected. The column splice should be designed to develop both the nominal shear strength and 50% of the nominal flexural strength of the smaller section connected. Splices should be located in the middle one-third of the clear height of the column.

39.3.2 Eccentrically Braced Frames

Research results have shown that a well-designed EBF system possesses high stiffness in the elastic range and excellent ductility capacity in the inelastic range [25]. The high elastic stiffness is provided by the braces and the high ductility capacity is achieved by transmitting one brace force to another brace or to a column through shear and bending in a short beam segment designated as a "link." Figure 39.8 shows some typical arrangements of EBFs. In the figure, the link lengths are identified by the letter *e*. When properly detailed, these links provide a reliable source of energy dissipation. Following the capacity design concept, buckling of braces and beams outside of the link can be prevented by designing these members to remain elastic while resisting forces associated with the fully yielded and strain-hardened links. The AISC Seismic Provisions (1997) [6] for the EBF design are intended to achieve this objective.

Links

Figure 39.10 shows the free-body diagram of a link. If a link is short, the entire link yields primarily in shear. For a long link, flexural (or moment) hinge would form at both ends of the link before the "shear" hinge can be developed. A short link is desired for an efficient EBF design. In order to ensure stable yielding, links should be plastic sections satisfying the width–thickness ratios $\lambda\ ps$ given in Table 39.2. Doubler plates welded to the link web should not be used as they do not perform as intended when subjected to large inelastic deformations. Openings should also be avoided as they adversely affect the yielding of the link web. The required shear strength, V_u, resulting from the prescribed load effects should not exceed the design shear strength of the link, $\phi\ V_n$, where ϕ = 0.9. The nominal shear strength of the link is

$$V_n = \min\ \{V_p,\ 2M_p/e\} \tag{39.7}$$

$$V_p = 0.60F_yA_w \tag{39.8}$$

where $A_w = (d - 2t_f)t_w$

A large axial force in the link will reduce the energy dissipation capacity. Therefore, its effect shall be considered by reducing the design shear strength and the link length. If the required link axial strength, *Pu*, resulting from the prescribed seismic effect exceeds $0.15P_y$, where $P_y = A_gF_y$, the following additional requirements should be met:

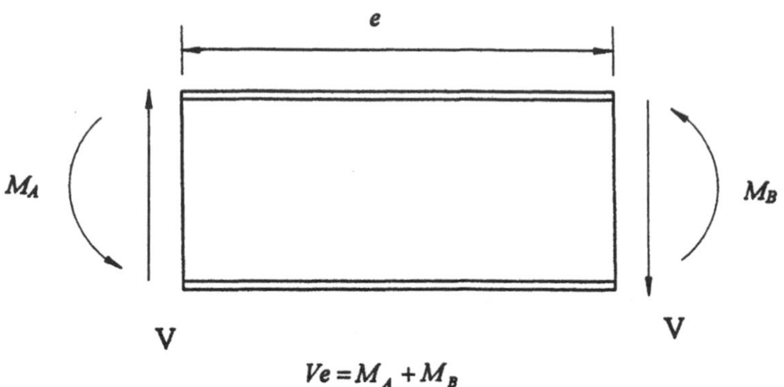

FIGURE 39.10 Static equilibrium of link.

1. The link design shear strength, ϕV_n, should be the lesser of ϕVpa or $2\phi M_{Pa}/e$, where V_{pa} and M_{Pa} are the reduced shear and flexural strengths, respectively:

$$V_{pa} = V_p \sqrt{1-(P_u / P_y)^2} \tag{39.9}$$

$$M_{Pa} = 1.18M_p[1\text{-}P_u/P_y)] \tag{39.10}$$

2. The length of the link should not exceed:

$$[1.15 - 0.5\,\rho' (A_w/A_g)]1.6M_p/V_p \quad \text{for } \rho' (A_w/A_g) \geq 0.3 \tag{39.11}$$

$$1.6M_p/V_p \quad \text{for } \rho' (A_w/A_g) < 0.3 \tag{39.12}$$

where $\rho' = P_u/V_u$.

The link rotation angle, γ, is the inelastic angle between the link and the beam outside of the link. The link rotation angle can be conservatively determined assuming that the braced bay will deform in a rigid–plastic mechanism. The plastic mechanism for one EBF configuration is illustrated in Figure 39.11. The plastic rotation is determined using a frame drift angle, θ_p, computed from the maximum frame displacement. Conservatively ignoring the elastic frame displacement, the plastic frame drift angle is $\theta_p = \delta / h$, where δ is the maximum displacement and h is the frame height.

Links yielding in shear possess a greater rotational capacity than links yielding in bending. For a link with a length of $1.6M_p/V_p$ or shorter (i.e., shear links), the link rotational demand should not exceed 0.08 rad. For a link with a length of $2.6M_p/V_p$ or longer (i.e., flexural links), the link rotational angle should not exceed 0.02 rad. A straight-line interpolation can be used to determine the link rotation capacity for the intermediate link length.

Link Stiffeners

In order to provide ductile behavior under severe cyclic loading, close attention to the detailing of link web stiffeners is required. At the brace end of the link, full-depth web stiffeners should be provided on both sides of the link web. These stiffeners should have a combined width not less than $(b_f - 2t_w)$, and a thickness not less than $0.75t_w$ nor ⅜ in. (10 mm), whichever is larger, where b_f and t_w are the link flange width and web thickness, respectively. In order to delay the link web or flange buckling, intermediate link web stiffeners should be provided as follows.

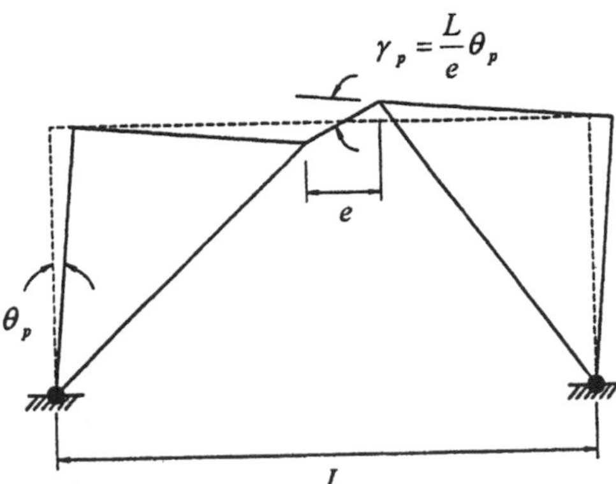

FIGURE 39.11 Energy dissipation mechanism of an eccentric braced frame.

1. In shear links, the spacing of intermediate web stiffeners depends on the magnitude of the link rotational demand. For links of lengths $1.6M_p/V_p$ or less, the intermediate web stiffener spacing should not exceed $(30t_w - d/5)$ for a link rotation angle of 0.08 rad, or $(52tw - d/5)$ for link rotation angles of 0.02 rad or less. Linear interpolation should be used for values between 0.08 and 0.02 rad.
2. Flexural links having lengths greater than $2.6M_p/V_p$ but less than $5M_p/V_p$ should have intermediate stiffeners at a distance from each link end equal to 1.5 times the beam flange width. Links between shear and flexural limits should have intermediate stiffeners meeting the requirements of both shear and flexural links. If link lengths are greater than $5M_p/V_p$, no intermediate stiffeners are required.
3. Intermediate stiffeners shall be full depth in order to react effectively against shear buckling. For links less than 25 in. deep, the stiffeners can be on one side only. The thickness of one-sided stiffeners should not be less than tw or ⅜ in., whichever is larger, and the width should not be less than $(b_f/2) - t_w$.
4. Fillet welds connecting a link stiffener to the link web should have a design strength adequate to resist a force of $A_{st}F_y$, where A_{st} is the area of the stiffener. The design strength of fillet welds connecting the stiffener to the flange should be adequate to resist a force of $A_{st}F_y/4$.

Link-to-Column Connections

Unless a very short shear link is used, large flexural demand in conjunction with high shear can develop at the link-to-column connections [25,63]. In light of the moment connection fractures observed after the Northridge earthquake, concerns have been raised on the seismic performance of link-to-column connections during a major earthquake. As a result, the AISC Seismic Provisions (1997) [6] require that the link-to-column design be based upon cyclic test results. Tests should follow specific loading procedures and results demonstrate an inelastic rotation capacity which is 20% greater than that computed in design. To avoid link-to-column connections, it is recommended that configuring the link between two braces be considered for EBF systems.

Lateral Support of Link

In order to assure stable behavior of the EBF system, it is essential to provide lateral support at both the top and bottom link flanges at the ends of the link. Each lateral support should have a design strength of 6% of the expected link flange strength ($= R_yF_yb_ft_f$).

Diagonal Brace and Beam outside of Link

Following the capacity design concept, diagonal braces and beam segments outside of the link should be designed to resist the maximum forces that can be generated by the link. Considering

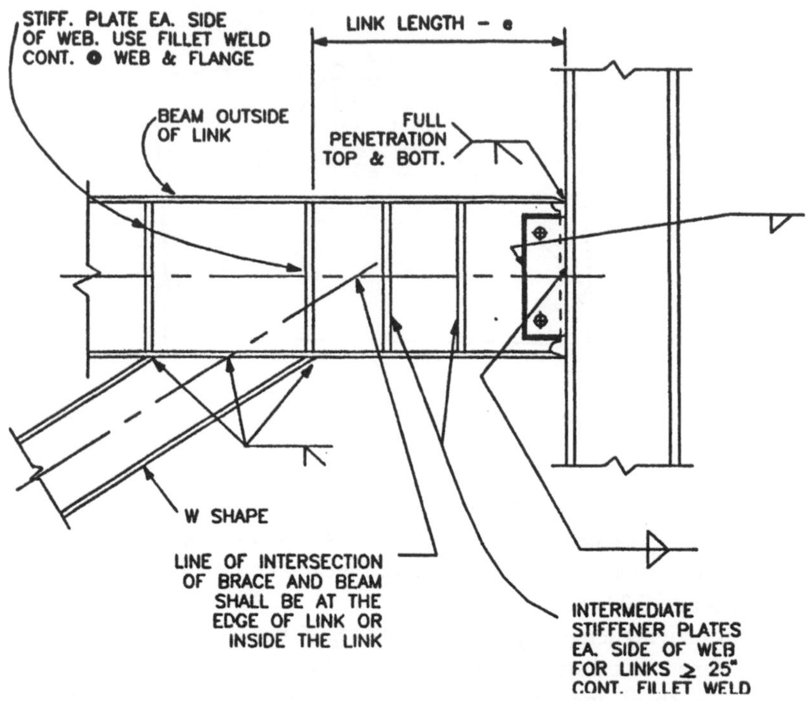

FIGURE 39.12 Diagonal brace fully connected to link. (Source: *AISC, Seismic Provisions for Structural Steel Buildings*; AISC, Chicago, IL, 1992. With permission.)

the strain-hardening effects, the required strength of the diagonal brace should be greater than the axial force and moment generated by 1.25 times the expected nominal shear strength of the link, R_yV_n.

The required strength of the beam outside of the link should be greater than the forces generated by 1.1 times the expected nominal shear strength of the link. To determine the beam design strength, it is permitted to multiply the beam design strength by the factor R_y. The link shear force will generate axial force in the diagonal brace. For most EBF configurations, the horizontal component of the brace force also generates a substantial axial force in the beam segment outside of the link. Since the brace and the beam outside of the link are designed to remain essentially elastic, the ratio of beam or brace axial force to link shear force is controlled primarily by the geometry of the EBF. This ratio is not much affected by the inelastic activity within the link; therefore, the ratio obtained from an elastic analysis can be used to scale up the beam and brace axial forces to a level corresponding to the link shear force specified above.

The link end moment is balanced by the brace and the beam outside of the link. If the brace connection at the link is designed as a pin, the beam by itself should be adequate to resist the entire link end moment. If the brace is considered to resist a portion of the link end moment, then the brace connection at the link should be designed as fully restrained. If used, lateral bracing of the beam should be provided at the beam top and bottom flanges. Each lateral bracing should have a required strength of 2% of the beam flange nominal strength, $F_yb_ft_f$. The required strength of the diagonal brace-to-beam connection at the link end of the brace should be at least the expected nominal strength of the brace. At the connection between the diagonal brace and the beam at the link end, the intersection of the brace and the beam centerlines should be at the end of the link or in the link (Figures 39.12 and 39.13). If the intersection of the brace and beam centerlines is located outside of the link, it will increase the bending moment generated in the beam and brace. The width–thickness ratio of the brace should satisfy λ_p specified in Table 39.2.

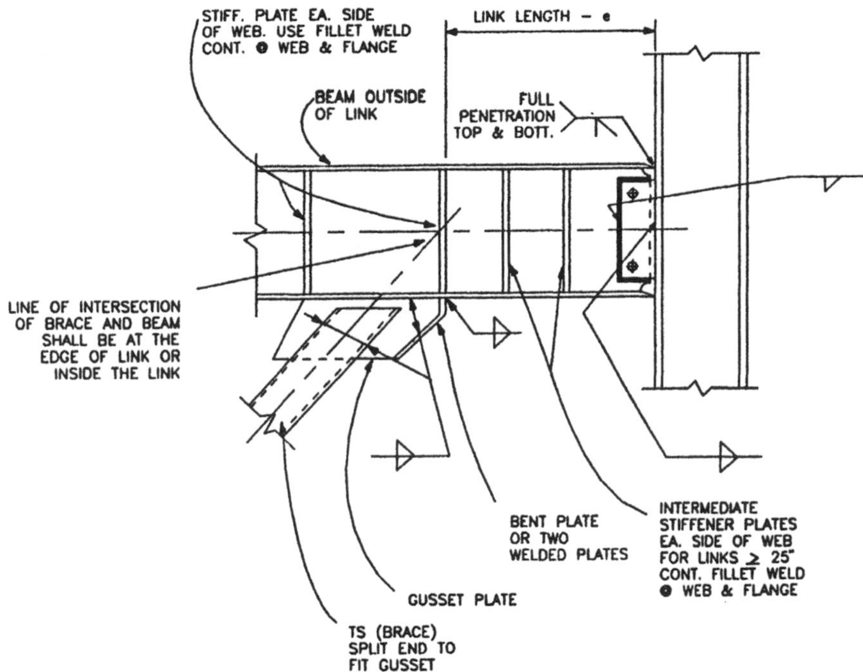

FIGURE 39.13 Diagonal brace pin-connected to link. (Source: AISC, *Seismic Provisions for Structural Steel Buildings*, AISC, Chicago, IL, 1992. With permission.)

Beam-to-Column Connections

Beam-to-column connections away from the links can be designed as simple shear connections. However, the connection must have a strength adequate to resist a rotation about the longitudinal axis of the beam resulting from two equal and opposite forces of at least 2% of the beam flange nominal strength, computed as $F_y b_f t_f$, and acting laterally on the beam flanges.

Required Column Strength

The required column strength should be determined from the prescribed load combinations, except that the moments and the axial loads introduced into the column at the connection of a link or brace should not be less than those generated by the expected nominal strength of the link, $RyVn$, multiplied by 1.1 to account for strain hardening. In addition to resisting the prescribed load effects, the design strength and the details of column splices must follow the recommendations given for the SCBFs.

39.4 Stiffened Steel Box Pier Design

39.4.1 Introduction

When space limitations dictate the use of a smaller-size bridge piers, steel box or circular sections gain an advantage over the reinforced concrete alternative. For circular or unstiffened box sections, the ductile detailing provisions of the AISC Seismic Provisions (1997) [6] or CHBDC [21] shall apply, including the diameter-to-thickness or width-to-thickness limits. For a box column of large dimensions, however, it is also possible to stiffen the wall plates by adding longitudinal and transverse stiffeners inside the section.

Design provisions for a stiffened box column are not covered in either the AASHTO or AISC design specifications. But the design and construction of this type of bridge pier has been common

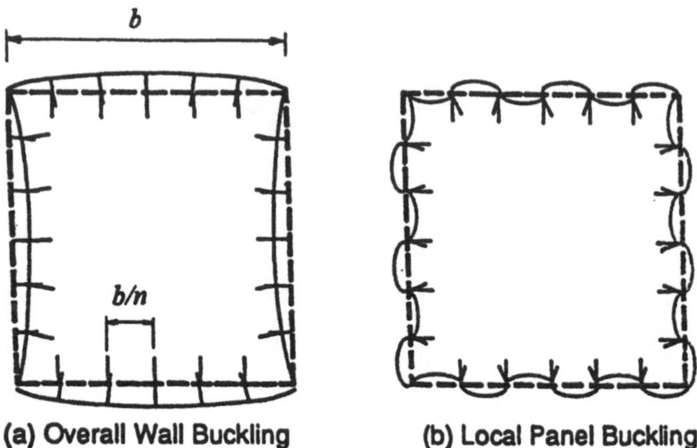

FIGURE 39.14 Buckling modes of box column with multiple stiffeners. (Source: Kawashima, K. et al., in *Stability and Ductility on Steel Structures under Cyclic Loading*; Fukumoto, Y. and G. Lee, Eds., CRC, Boca Raton, FL, 1992. With permission.)

in Japan for more than 30 years. In the sections that follow, the basic behavior of stiffened plates is briefly reviewed. Next, design provisions contained in the Japanese Specifications for Highway Bridges [31] are presented. Results from an experimental investigation, conducted prior to the 1995 Hyogo-ken Nanbu earthquake in Japan, on cyclic performance of stiffened box piers are then used to evaluate the deformation capacity. Finally, lessons learned from the observed performance of this type of piers from the Hyogo-ken Nanbu earthquake are presented.

39.4.2 Stability of Rectangular Stiffened Box Piers

Three types of buckling modes can occur in a stiffened box pier. First, the plate segments between the longitudinal stiffeners may buckle, the stiffeners acting as nodal points (Figure 39.14b). In this type of "panel buckling," buckled waves appear on the surface of the piers, but the stiffeners do not appreciably move perpendicularly to the plate. Second, the entire stiffened box wall can globally buckle (Figure 39.14a). In this type of "wall buckling," the plate and stiffeners move together perpendicularly to the original plate plane. Third, the stiffeners themselves may buckle first, triggering in turn other buckling modes.

In Japan, a design criterion was developed following an extensive program of testing of stiffened steel plates in the 1960s and 1970s [70]; the results of this testing effort are shown in Figure 39.15, along with a best-fit curve. The slenderness parameter that defines the abscissa in the figure deserves some explanation. Realizing that the critical buckling stress of plate panels between longitudinal stiffeners can be obtained by the well-known result from the theory of elastic plate buckling:

$$F_{cr} = \frac{k_o \pi^2 E}{12(1-\nu^2)(b/nt)^2} \tag{39.13}$$

a normalized panel slenderness factor can be defined as

$$R_p = \sqrt{\frac{F_y}{F_{cr}}} = \left(\frac{b}{nt}\right)\sqrt{\frac{12\ (1-\nu^2)\ F_y}{k_o\ \ \pi^2 E}} \tag{39.14}$$

where b and t are the stiffened plate width and thickness, respectively, n is the number of panel spaces in the plate (i.e., one more than the number of internal longitudinal stiffeners across the

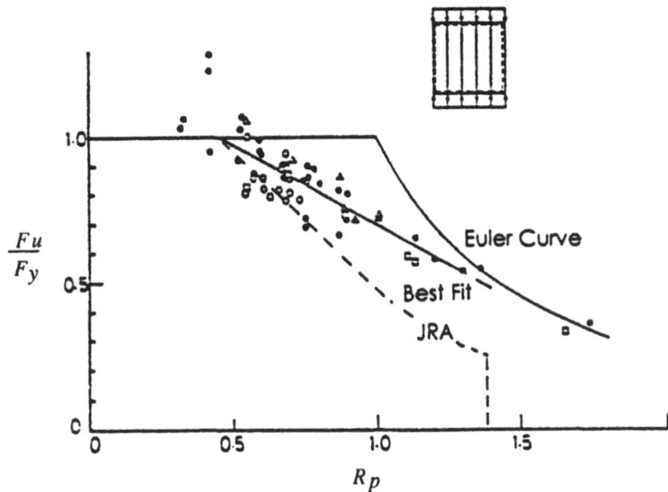

FIGURE 39.15 Relationship between buckling stress and R_p of stiffened plate.

plate), E is Young's modulus, ν is Poisson's ratio (0.3 for steel), and k_o (= 4 in. this case) is a factor taking into account the boundary conditions. The Japanese design requirement for stiffened plates in compression was based on a simplified and conservative curve obtained from the experimental data (see Figure 39.15):

$$\frac{F_u}{F_y} = 1.0 \qquad \text{for } R_p \le 0.5$$
$$\frac{F_u}{F_y} = 1.5 - R_P \qquad \text{for } 0.5 < R_P \le 1.0 \tag{39.15}$$
$$\frac{F_u}{F_y} = \frac{0.5}{R_P^2} \qquad \text{for } R_P > 1.0$$

where Fu is the buckling strength.

Note that values of F_u/F_y less than 0.25 are not permitted. This expression is then converted into the allowable stress format of the Japanese bridge code, using a safety factor of 1.7. However, as allowable stresses are magnified by a factor of 1.7 for load combinations which include earthquake effects, the above ultimate strength expressions are effectively used.

The point $R_P = 0.5$ defines the theoretical boundary between the region where the yield stress can be reached prior to local buckling ($R_P < 0.5$), and vice versa ($R_P \ge 0.5$). For a given steel grade, Eq. (39.14) for $R_P = 0.5$ corresponds to a limiting b/nt ratio:

$$\left(\frac{b}{nt}\right)_o = \frac{162}{\sqrt{F_y}} \tag{39.16}$$

and for a given plate width, b, the "critical thickness," t_o, is

$$t_o = \frac{b\sqrt{F_y}}{162n} \tag{39.17}$$

where F_y is in ksi. That is, for a stiffened box column of a given width, using a plate thicker than t_o will ensure yielding prior to panel buckling.

To be able to design the longitudinal stiffeners, it is necessary to define two additional parameters: the stiffness ratio of a longitudinal stiffener to a plate, γ_l, and the corresponding area ratio, δ_l. As the name implies:

$$\gamma_l = \frac{\text{stiffener flexural rigidity}}{\text{plate flexural rigidity}} = \frac{EI_l}{bD} = \frac{12(1-\nu^2)\,I_l}{bt^3} = \frac{10.92\,I_l}{b\,t^3} \approx \frac{11\,I_l}{b\,t^3} \tag{39.18}$$

where I_l is the moment of inertia of the T-section made up of a longitudinal stiffener and the effective width of the plate to which it connects (or, more conservatively and expediently, the moment of inertia of a longitudinal stiffener taken about the axis located at the inside face of the stiffened plate). Similarly, the area ratio is expressed as

$$\delta_l = \frac{\text{stiffener axial rigidity}}{\text{plate axial rigidity}} = \frac{A_l}{bt} \tag{39.19}$$

where A_l is the area of a longitudinal stiffener.

Since the purpose of adding stiffeners to a box section is partly to eliminate the severity of wall buckling, there exists an "optimum rigidity," γ_l^*, of the stiffeners beyond which panel buckling between the stiffeners will develop before wall buckling. In principle, according to elastic buckling theory for ideal plates (i.e., plates without geometric imperfections and residual stresses), further increases in rigidity beyond that optimum would not further enhance the buckling capacity of the box pier. Although more complex definitions of this parameter exist in the literature [35], the above description is generally sufficient for the box piers of interest here. This optimum rigidity is:

$$\gamma_l^* = 4\alpha^2 n(1 + n\delta_l) - \frac{(\alpha^2+1)^2}{n} \qquad \text{for } \alpha \le \alpha_o \tag{39.20}$$

and

$$\gamma_l^* = \frac{1}{n}\left\{\left[2\,n^2\,(1 + n\delta_l) - 1\right]^2 - 1\right\} \qquad \text{for } \alpha > \alpha_o \tag{39.21}$$

where α is the aspect ratio, a/b, a being the spacing between the transverse stiffeners (or diaphragms), and the critical aspect ratio α_o is defined as

$$\alpha_o = \sqrt[4]{1 + n\,\gamma_l} \tag{39.22}$$

These expressions can be obtained by recognizing that, for plates of thickness less than t_o, it is logical to design the longitudinal stiffeners such that wall buckling does not occur prior to panel buckling and, consequently, as a minimum, be able to reach the same ultimate stress as the latter. Defining a normalized slenderness factor, R_H, for the stiffened plate:

$$R_H = \sqrt{\frac{F_y}{F_{cr}}} = \left(\frac{b}{t}\right)\sqrt{\frac{12\,(1-\nu^2)\,F_y}{k_s\,\pi^2 E}} \tag{39.23}$$

Based on elastic plate buckling theory, k_s for a stiffened plate is equal to [15]:

$$k_s = \frac{(1+\alpha^2)^2 + n\gamma_l}{\alpha^2(1+n\delta_l)} \quad \text{for } \alpha \le \alpha_o$$

$$k_s = \frac{2\left(1+\sqrt{1+n\gamma_l}\right)}{(1+n\delta_l)} \quad \text{for } \alpha > \alpha_o \tag{39.24}$$

Letting $R_H = R_P$ (i.e, both wall buckling and panel buckling can develop the same ultimate stress), the expressions for γ_l^* in Eqs. (39.20) and (39.21) can be derived. Thus, when the stiffened plate thickness, t, is less t_o, the JRA Specifications specify that either Eq. (39.20) or (39.21) be used to determine the required stiffness of the longitudinal stiffeners.

When a plate thicker than t_o is chosen, however, larger stiffeners are unnecessary since yielding will occur prior to buckling. This means that the critical buckling stress for wall buckling does not need to exceed the yield stress, which is reached by the panel buckling when $t = t_o$. The panel slenderness ratio for $t = t_o$ is

$$R_{P(t=t_o)} = \frac{b}{nt_o}\sqrt{\frac{12(1-\nu^2)F_y}{k_o\pi^2 E}} \tag{39.25}$$

Equating R_P to R_H in Eq. (39.23), the required γ_l^* can be obtained as follows:

$$\gamma_l^* = 4\alpha^2 n\left(\frac{t_o}{t}\right)^2(1+n\delta_l) - \frac{(\alpha^2+1)^2}{n} \quad \text{for } \alpha \le \alpha_o \tag{39.26}$$

and

$$\gamma_l^* = \frac{1}{n}\left\{\left[2n^2\left(\frac{t_o}{t}\right)^2(1+n\delta_l)-1\right]^2-1\right\} \quad \text{for } \alpha > \alpha_o \tag{39.27}$$

It is noteworthy that the above requirements do not ensure ductile behavior of steel piers. To achieve higher ductility for seismic application in moderate to high seismic regions, it is prudent to limit t to t_o. In addition to the above requirements, conventional slenderness limits are imposed to prevent local buckling of the stiffeners prior to that of the main member. For example, when a flat bar is used, the limiting width–thickness ratio (λ_r) for the stiffeners is $95/\sqrt{F_y}$.

The JRA requirements for the design of stiffened box columns are summarized as follows:

1. At least two stiffeners of the steel grade no less than that of the plate are required. Stiffeners are to be equally spaced so that the stiffened plate is divided into n equal intervals. To consider the beneficial effect of the stress gradient, b/nt in Eq. (39.14) can be replaced by $b/nt\varphi$, where φ is computed as

$$\varphi = \frac{\sigma_1 - \sigma_2}{\sigma_1} \tag{39.28}$$

In the above equation, σ_1 and σ_2 are the stresses at both edges of the plate; compressive stress is defined as positive, and $\sigma_1 > \sigma_2$. The value of φ is equal to 1 for uniform compression and 2 for equal and opposite stresses at both edges of the plate. Where the plastic hinge is expected to form, it is conservative to assume a φ value of 1.

2. Each longitudinal stiffener needs to have sufficient area and stiffness to prevent wall buckling. The minimum required area, in the form of an area ratio in Eq. (39.19), is

$$(\delta_l)_{\min} = \frac{1}{10n} \tag{39.29}$$

The minimum required moment of inertia, expressed in the form of stiffness ratio in Eq. (39.18), is determined as follows. When the following two requirements are satisfied, use either Eq. (39.26) for $t \geq t_o$ or Eq. (39.20) for $t < t_o$:

$$\alpha \leq \alpha_o \tag{39.30}$$

$$I_t \geq \frac{bt^3}{11}\left(\frac{1+n\gamma_l^*}{4\alpha^3}\right) \tag{39.31}$$

where *It* is the moment of inertia of the transverse stiffener, taken at the base of the stiffener. Otherwise, use either Eq. (39.27) for $t \geq t_o$ or Eq. (39.21) for $t < t_o$.

39.4.3 Japanese Research Prior to the 1995 Hyogo-ken Nanbu Earthquake

While large steel box bridge piers have been used in the construction of Japanese expressways for at least 30 years, research on their seismic resistance only started in the early 1980s. The first inelastic cyclic tests of thin-walled box piers were conducted by Usami and Fukumoto [65] as well as Fukumoto and Kusama [29]. Other tests were conducted by the Public Works Research Institute of the Ministry of Construction (e.g., Kawashima et al. [32]; MacRae and Kawashima [36]) and research groups at various universities (e.g., Watanabe et al. [69], Usami et al. [66], Nishimura et al. [45]).

The Public Work Research Institute tests considered 22 stiffened box piers of configuration representative of those used in some major Japanese expressways. The parameters considered in the investigation included the yield strength, weld size, loading type and sequence, stiffener type (flat bar vs. structural tree), and partial-height concrete infill. An axial load ranging from 7.8 to 11% of the axial yield load was applied to the cantilever specimens for cyclic testing. Typical hysteresis responses of one steel pier and one with concrete infill at the lower one-third of the pier height are shown in Figure 39.16. For bare steel specimens, test results showed that stiffened plates were able to yield and strain-harden. The average ratio between the maximum lateral strength and the predicted yield strength was about 1.4; the corresponding ratio between the maximum strength and plastic strength was about 1.2. The displacement ductility ranged between 3 and 5. Based on Eq. (39.2), the observed levels of ductility and structural overstrength imply that the response modification factor, *R*, for this type of pier can be conservatively taken as 3.5 ($\approx 1.2 \times 3$). Specimens with a γ_l / γ_l^* ratio less than 2.0 behaved in a wall-buckling mode with severe strength degradation. Otherwise, specimens exhibited local panel buckling.

Four of the 22 specimens were filled with concrete over the bottom one-third of their height. Prior to the Hyogo-ken Nanbu earthquake, it was not uncommon in Japan to fill bridge piers with concrete to reduce the damage which may occur as a result of a vehicle collision with the pier; generally the effect of the concrete infill was neglected in design calculations. It was thought prior to the testing that concrete infill would increase the deformation capacity because inward buckling of stiffened plates was inhibited.

Figure 39.17 compares the response envelopes of two identical specimens, except that one is with and the other one without concrete infill. For the concrete-filled specimen, little plate buckling was observed, and the lateral strength was about 30% higher than the bare steel specimen. Other than

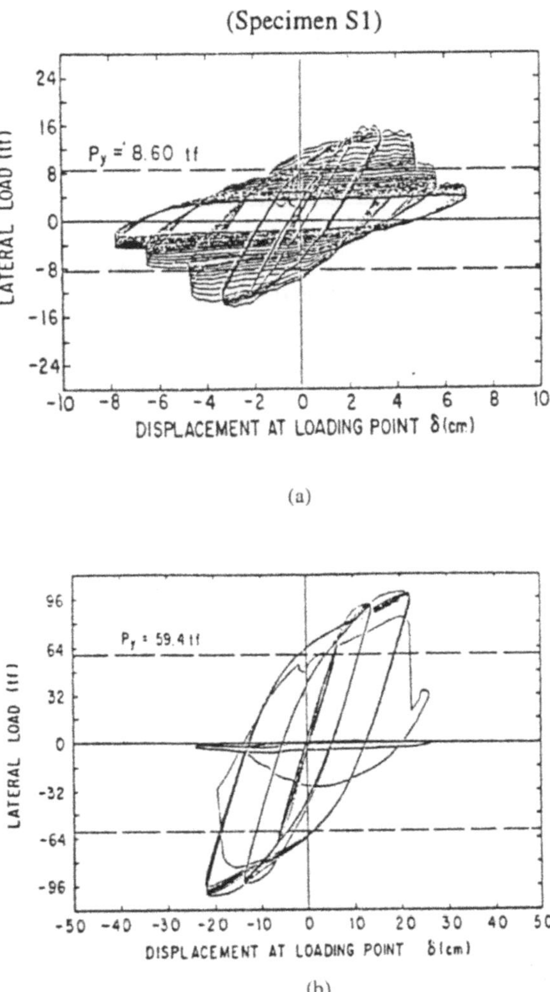

FIGURE 39.16 Hysteresis responses of two stiffened box piers (a) without concrete fill; (b) with concrete fill. (Source: Kawashima, K. et al., in *Stability and Ductility on Steel Structures under Cyclic Loading*; Fukumoto, Y. and G. Lee, Eds., CRC, Boca Raton, FL, 1992. With permission.)

exhibiting ductile buckling mode, all the concrete-filled specimens suffered brittle fracture in or around the weld at the pier base. See Figure 39.16b for a typical cyclic response. It appears that the composite effect, which not only increased the overall flexural strength but also caused a shift of the neutral axis, produced an overload to the welded joint. As a result, the ductility capacity was reduced by up to 23%. It appears from the test results that the welded connection at the pier base needs to be designed for the overstrength including the composite effect.

A large portion of the research effort in years shortly prior to the Hyogo-ken Nanbu earthquake investigated the effectiveness of many different strategies to improve the seismic performance, ductility, and energy dissipation of those steel piers [34,66,68]. Among the factors observed to have a beneficiary influence were the use of (1) a γ_l / γ_l^* ratio above 3 as a minimum, or preferably 5; (2) longitudinal stiffeners having a higher grade of steel than the box plates; (3) minimal amount of stiffeners; (4) concrete filling of steel piers; and (5) box columns having round corners, built from bend plates, and having weld seams away from the corners, thus avoiding the typically problem-prone sharp welded corners [69].

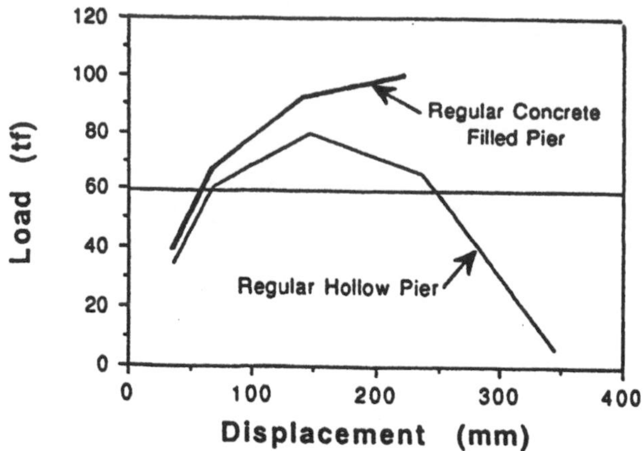

FIGURE 39.17 Effect of infill on cyclic response envelopes. (Source: Kawashima, K. et al., 1992.)

39.4.4 Japanese Research after the 1995 Hyogo-ken Nanbu Earthquake

Steel bridge piers were severely tested during the 1995 Hyogo-ken Nanbu earthquake in Japan. Recorded ground motions in the area indicated that the earthquake of a magnitude 7.2 produced a peak ground velocity of about 90 cm/s (the peak ground acceleration was about a 0.8 *g*). Of all the steel bridge piers, about 1% experienced severe damage or even collapse. But about a half of the steel bridge piers were damaged to some degree.

In addition to wall buckling and panel buckling, damage at the pier base, in the form of weld fracture or plastic elongation of anchor bolts, was also observed. Based on the observed buckling patterns, Watanabe et al. [69] suggested that the width–thickness of the wall plate needs to be further reduced, and the required stiffness of stiffeners, γ_l^*, needs to be increased by three times.

After the Hyogo-ken Nanbu earthquake, filling concrete to existing steel piers, either damaged or nondamaged ones, was suggested to be one of the most effective means to strength stiffened box piers [23,41]. Figure 39.18 shows a typical example of the retrofit. As was demonstrated in Figure 39.17, concrete infill will increase the flexural strength of the pier, imposing a higher force demand to the foundation and connections (welds and anchor bolts) at the base of the pier. Therefore, the composite effect needs to be considered in retrofit design. The capacities of the foundation and base connections also need to be checked to ensure that the weakest part of the retrofitted structure is not in these regions. Since concrete infill would force the stiffened wall plates to buckle outward, the welded joint between wall plates is likely to experience higher stresses (see Figure 39.19).

Because concrete infill for seismic retrofit may introduce several undesirable effects, alternative solutions have been sought that would enhance the deformation capacity while keeping the strength increase to a minimum. Based on the observed buckling of stiffened plates that accounted for the majority of damage to rectangular piers, Nishikawa et al. [44] postulated that local buckling is not always the ultimate limit state. They observed that bridge piers experienced limited damages when the corners of the box section remained straight, but piers were badly deformed when corner welds that fractured could not maintain the corners straight. To demonstrate their concept, three retrofit schemes shown in Figure 39.20a were verified by cyclic testing. Specimen No. 3 was retrofitted by adding stiffeners. Box corners of Specimen No. 4 were strengthened by welding angles and corner plates, while the stiffening angles of Specimen No. 5 were bolted to stiffened wall plates. Response envelopes in Figure 39.20b indicate that Specimen No. 5 had the least increase in lateral strength above the nonretrofitted Specimen No. 2, yet the deformation capacity was comparable to other retrofitted specimens.

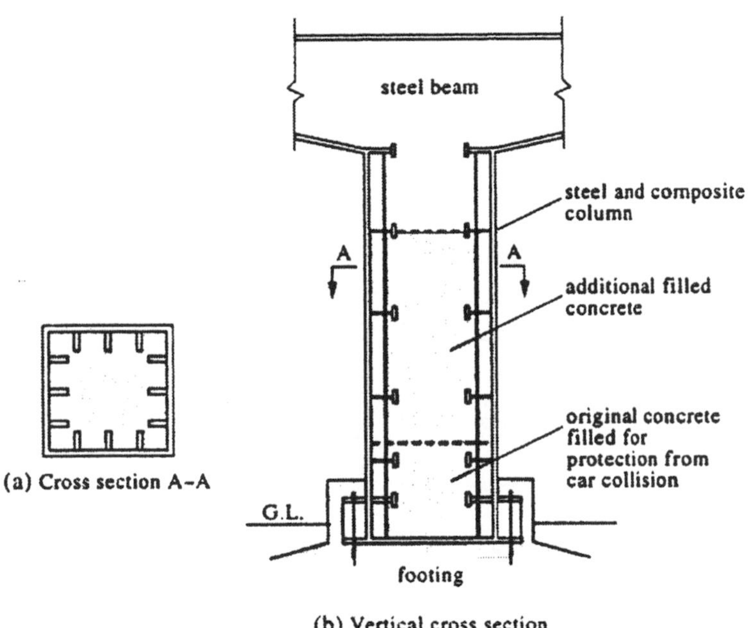

FIGURE 39.18 Retrofitted stiffened box pier with concrete infill. (Source: Fukumoto, Y. et al., in *Bridge Management*, Vol. 3, Thomas Telford, 1996. With permission.)

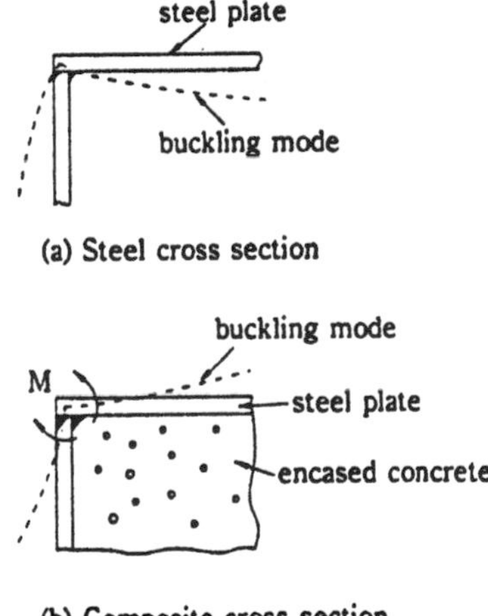

FIGURE 39.19 Effect of concrete infill on welded joint of stiffened box pier. (Source: Kitada, *Eng. Struct.*, 20(4–6), 347-354, 1998. With permission.)

39.5 Alternative Schemes

As described above, damage to substructure components such as abutments, piers, bearings, and others have proved to be of great consequence, often leading to span collapses [8,19,50]. Hence, when existing bridges are targeted for seismic rehabilitation, much attention needs to be paid to these substructure elements. Typically, the current retrofitting practice is to either strengthen or replace the existing nonductile members (e.g., ATC [10], Buckle et al. [20], Shirolé and Malik [58]),

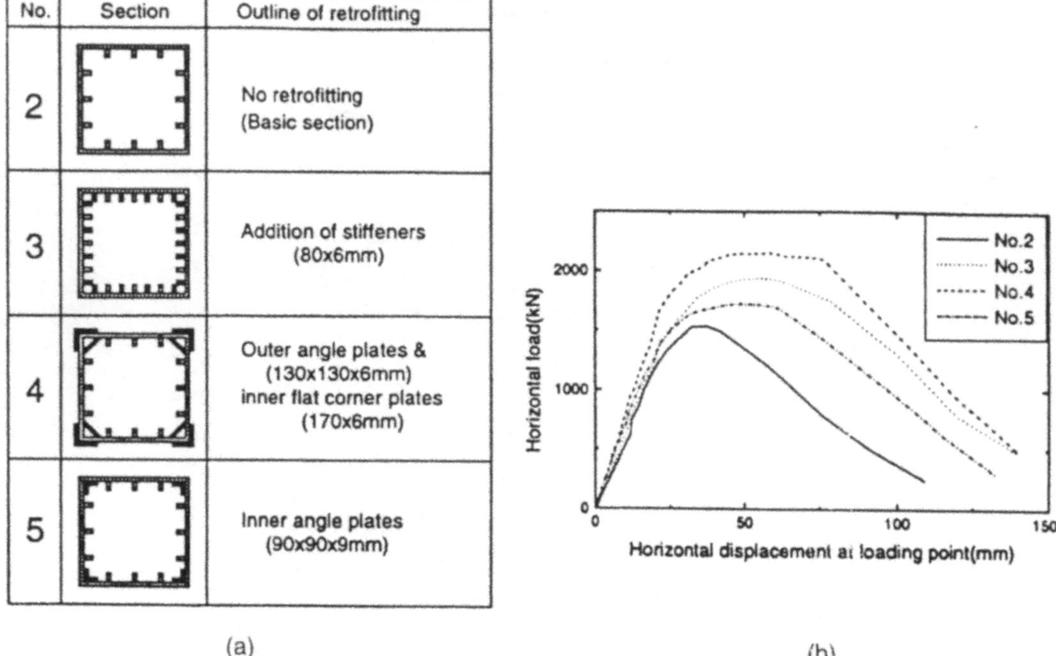

No.	Section	Outline of retrofitting
2		No retrofitting (Basic section)
3		Addition of stiffeners (80x6mm)
4		Outer angle plates & (130x130x6mm) inner flat corner plates (170x6mm)
5		Inner angle plates (90x90x9mm)

FIGURE 39.20 Seismic retrofit without concrete infill. (a) Retrofit schemes; (b) response envelopes. (Source: Nishikawa, K. et al, *Eng. Struct.*, 2062-6), 540-551, 1998. With permission.)

enhance the ductility capacity (e.g., Degenkolb [24], Priestley et al. [49]), or reduce the force demands on the vulnerable substructure elements using base isolation techniques or other structural modifications (e.g., Mayes et al. [38], Astaneh-Asl [7]). While all these approaches are proven effective, only the base isolation concept currently recognizes that seismic deficiency attributable to substructure weaknesses may be resolved by operating elsewhere than on the substructure itself. Moreover, all approaches can be costly, even base isolation in those instances when significant abutments modifications and other structural changes are needed to permit large displacements at the isolation bearings and lateral load redistribution among piers [39]. Thus, a seismic retrofit strategy that relies instead on ductile end diaphragms inserted in the steel superstructure, if effective, could provide an alternative.

Lateral load analyses have revealed the important role played by the end diaphragms in slab-on-girder steel bridges [72]. In absence of end diaphragms, girders severely distort at their supports, whether or not stiff intermediate diaphragms are present along the span. Because end diaphragms are key links along the load path for the inertia forces seismically induced at deck level, it might be possible, in some cases, to prevent damage from developing in the nonductile substructure (i.e., piers, foundation, and bearings) by replacing the steel diaphragms over abutments and piers with specially designed ductile diaphragms calibrated to yield before the strength of the substructure is reached. This objective is schematically illustrated in Figure 39.21 for slab-on-girder bridges and in Figure 39.22 for deck-truss bridges. In the latter case, however, ductile diaphragms must be inserted in the last lower lateral panels before the supports, in addition to the end diaphragms; in deck-truss bridges, seismically induced inertia forces in the transverse direction at deck level act with a sizable eccentricity with respect to the truss reaction supports, and the entire superstructure (top and lower lateral bracings, end and interior cross-frame bracings, and other lateral-load-resisting components) is mobilized to transfer these forces from deck to supports.

While conceptually simple, the implementation of ductile diaphragms in existing bridges requires consideration of many strength, stiffness, and drift constraints germane to the type of steel bridge

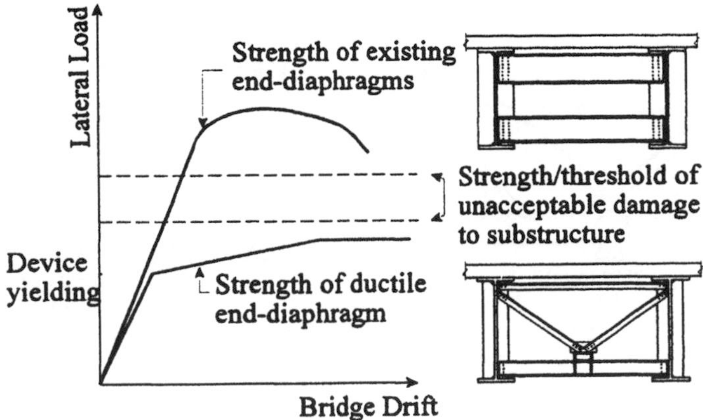

FIGURE 39.21 Schematic illustration of the ductile end-diaphragm concept.

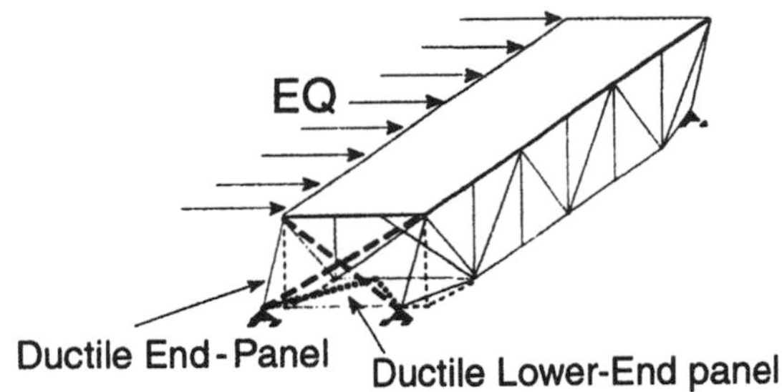

FIGURE 39.22 Ductile diaphragm retrofit concept in a deck-truss.

investigated. For example, for slab-on-girder bridges, because girders with large bearing stiffeners at the supports can contribute non-negligibly to the lateral strength of the bridges, stiff ductile diaphragms are preferred. Tests [73] confirmed that stiff welded ductile diaphragms are indeed more effective than bolted alternatives. As for deck-trusses, both upper and lower limits are imposed on the ductile diaphragm stiffnesses to satisfy maximum drifts and ductility requirements, and a systematic solution strategy is often necessary to achieve an acceptable retrofit [55,56].

Many types of systems capable of stable passive seismic energy dissipation could serve as ductile diaphragms. Among those, EBF presented in Section 39.3.2, shear panel systems (SPS) [28,40], and steel triangular-plate added damping and stiffness devices (TADAS) [61] have received particular attention in building applications. Still, to the authors' knowledge, none of these applications has been considered to date for bridge structures. This may be partly attributable to the absence of seismic design provisions in North American bridge codes. Examples of how these systems would be implemented in the end diaphragms of a typical 40-m span slab-on-girder bridge are shown in Figure 39.23. Similar implementations in deck-trusses are shown in Figure 39.24.

While the ductile diaphragm concept is promising and appears satisfactory for spans supported on stiff substructures based on the results available at the time of this writing, and could in fact be equally effective in new structures, more research is needed before common implementation is possible. In particular, large-scale experimental verification of the concept and expected behavior is desirable; parametric studies to investigate the range of substructure stiffnesses for which this retrofit strategy can be effective are also needed. It should also be noted that this concept only provides enhanced seismic resistance and substructure protection for the component of seismic excitation transverse to the bridge, and must be coupled with other devices that constrain longitudinal seismic displacements, such as simple bearings strengthening [37], rubber bumpers, and the

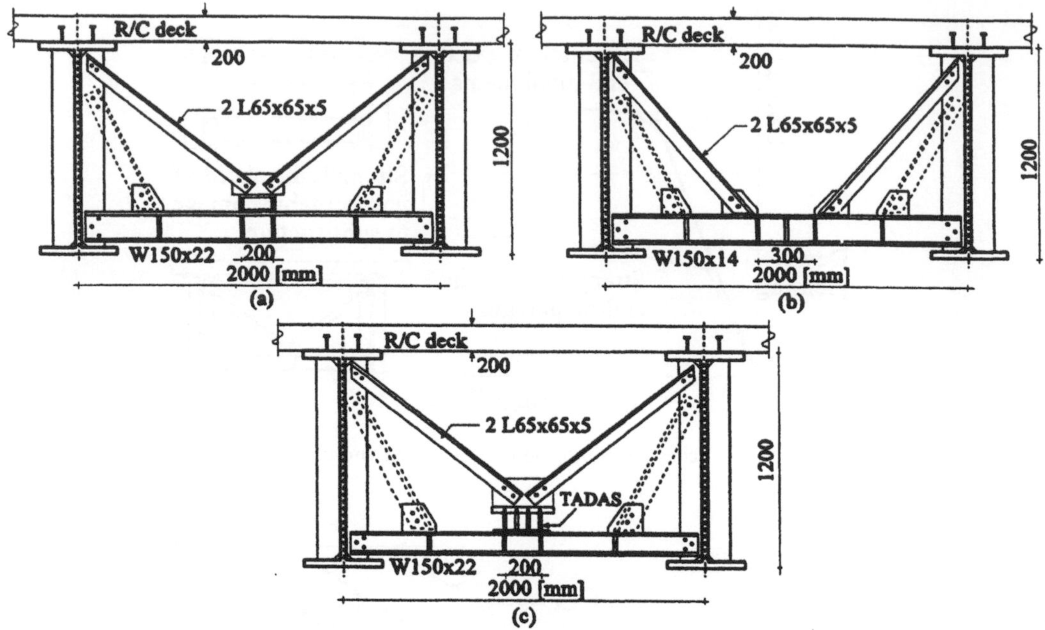

FIGURE 39.23 Ductile end diaphragm in a typical 40-m-span bridge (a) SPS; (b) EBF; (c) TADAS. (Other unbraced girders not shown; dotted members only if required for jacking purposes for nonseismic reasons).

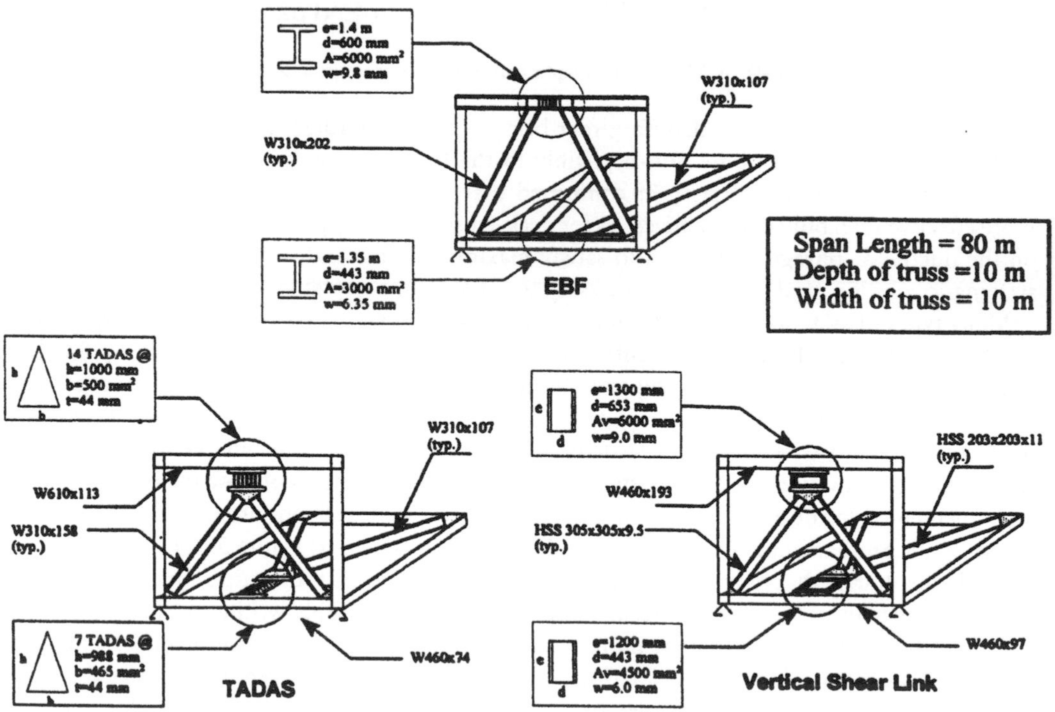

FIGURE 39.24 Examples of ductile retrofit systems at span end of deck-trusses.

like. Transportation agencies experienced in seismic bridge retrofit have indicated that deficiencies in the longitudinal direction of these bridges are typically easier to address than those in the lateral direction.

References

1. AASHTO, *LRFD Bridge Design Specifications*, American Association of State Highways and Transportaion Officials, Washington, D.C., 1994.
2. AASHTO, *Standard Specifications for Highway Bridges*, AASHTO, Washington, D.C., 1996.
3. AASHTO, *Standard Specifications for Seismic Design of Highway Bridges*, Washington, D.C., 1992.
4. AISC, Commentary on highly restrained welded connections, *Eng. J. AISC*, 10(3), 61–73, 1973.
5. AISC, *Load and Resistance Factor Design Specification for Structural Steel Buildings*, AISC, Chicago, IL, 1993.
6. AISC, *Seismic Provisions for Structural Steel Buildings*, AISC, Chicago, IL, 1992 and 1997.
7. Astaneh-Asl, A., Seismic retrofit concepts for the East Bay Crossing of the San Francisco–Oakland Bay Bridge," in *Proc. 1st U.S. Seminar on Seismic Evaluation and Retrofit of Steel Bridges*, San Francisco, CA, 1993.
8. Astaneh-Asl, A., Bolt, B., Mcmullin, K. M., Donikian, R. R., Modjtahedi, D., and Cho, S. W., Seismic Performance of Steel Bridges during the 1994 Northridge Earthquake, Report No. CE-STEEL 94/01, Berkeley, CA, 1994.
9. ATC, Tentative Provisions for the Development of Seismic Design Provisions for Buildings, Report No. ATC 3-06, Applied Technology Council, Palo Alto, CA, 1978.
10. ATC, Seismic Retrofitting Guidelines for Highway Bridges, *Report No. ATC-6-2*, Applied Technology Council, Palo Alto, CA, 1983.
11. ATC, Seismic Design Criteria for Highway Structures, *Report No. ATC-18*, Applied Technology Council, Redwood, CA,1996.
12. ATC, Improved Seismic Design Criteria for California Bridges: Provisional Recommendations, *Report No. ATC-32*, Applied Technology Council, Redwood, CA,1996.
13. AWS, *Structural Welding Code — Steel*, ANSI/AWS D1.1-98, AWS, Miami, FL, 1998.
14. AWS, *Bridge Welding Code*, ANSI/AWS D1.5-96, AWS, Miami, FL, 1996.
15. Ballio, G. and Mazzolani, F. M., *Theory and Design of Steel Structures*. Chapman and Hall, New York, 632 pp, 1983.
16. Blodgett, O. W. and Miller, D. K., Special welding issues for seismically resistant structures, in *Steel Design Handbook*, A. R. Tamboli, Ed., McGraw-Hill, New York, 1997.
17. Bruneau, M. and Mahin, S. A., Ultimate behavior of heavy steel section welded splices and design implications, *J. Struct. Eng. ASCE*, 116(8), 2214–2235, 1990.
18. Bruneau, M., Uang, C.-M., and Whittaker, A., *Ductile Design of Steel Structures*, McGraw-Hill, New York, 1997.
19. Bruneau, M., Wilson, J. W., and Tremblay, R., Performance of steel bridges during the 1995 Hyogo-ken-Nanbu (Kobe, Japan) earthquake, *Can. J. Civ. Eng.*, 23(3), 678–713, 1996.
20. Buckle, I. G., Mayes, R. L., and Button, M. R., Seismic Design and Retrofit Manual for Highway Bridges, Report No. FHWA-IP-87-6, U.S. Department of Transportation, Federal Highway Administration, 1986.
21. CHBDC, Canadian Highway Bridge Design Code, Seismic Provisions, Seismic Committee of the CHBDC, Rexdale, Ontario, Canada, 1998.
22. Chen, S. J., Yeh, C. H., and Chu, J. M., Ductile steel beam-column connections for seismic resistance, *J. Struct. Eng., ASCE*, 122(11), 1292–1299; 1996.
23. Committee on Roadway Bridges by the Hyogoken-Nanbu Earthquake, *Specifications on Retrofitting of Damaged Roadway Bridges by the Hyogoken-Nanbu Earthquake*, 1995. [in Japanese].
24. Degenkolb, O. H., Retrofitting bridges to increase seismic resistance, *J. Tech. Councils ASCE*, 104(TC1), 13–20, 1978.

25. Popov, E. P., Engelhardt, M. D., and Ricles, J. M., Eccentrically braced frames: U.S. practice, *Eng. J. AISC*, 26(2), 66–80, 1989.
26. Engelhardt, M. D. and Sabol, T., Reinforcing of steel moment connections with cover plates: benefits and limitations, *Eng. Struct.*, 20(4–6), 510–520, 1998.
27. Engelhardt, M. D., Winneburger, T., Zekany, A. J., and Potyraj, T. J., "The dogbone connection: Part II, *Modern Steel Construction, AISC*, 36(8), pp. 46–55, 1996.
28. Fehling, E., Pauli, W. and Bouwkamp, J. G., Use of vertical shear-links in eccentrically braced frames, *Proc. 10th World Conf. on Earthquake Eng.*, Madrid, Vol. 9, 1992, 4475–4479.
29. Fukumoto, Y. and Kusama, H., Cyclic bending tests of thin-walled box beams, *Proc. JSCE Struct. Eng./Earthquake Eng.*, 2(1), 117s–127s, 1985.
30. Fukumoto, Y., Watanabe, E., Kitada, T., Suzuki, I., Horie, Y., and Sakoda, H., Reconstruction and repair of steel highway bridges damaged by the Great Hanshin earthquake, in *Bridge Management*, Vol. 3, Thomas Telford, 1996, 8–16.
31. JRA, Specifications of Highway Bridges, Japan Road Association, Tokyo, Japan, 1996.
32. Kawashima, K., MacRae, G., Hasegawa, K., Ikeuchi, T., and Kazuya, O., Ductility of steel bridge piers from dynamic loading tests, in *Stability and Ductility of Steel Structures under Cyclic Loading*, Fukomoto, Y. and G. Lee, Eds., CRC Press, Boca Raton, FL, 1992.
33. Kitada, T., Ultimate strength and ductility of state-of-the-art concrete-filled steel bridge piers in Japan, *Eng. Struct.*, 20(4–6), 347–354, 1998.
34. Kitada, T., Nanjo, A., and Okashiro, S., Limit states and design methods considering ductility of steel piers for bridges under seismic load, in *Proc., 5th East Asia-Pacific Conference on Structural Engineering and Construction*, Queensland, Australia, 1995.
35. Kristek, V. and Skaloud, M., *Advanced Analysis and Design of Plated Structures, Developments in Civil Engineering*, Vol. 32, Elsevier, New York, 1991, 333 pp.
36. MacRae, G. and Kawashima, K., Estimation of the deformation capacity of steel bridge piers, in *Stability and Ductility of Steel Structures under Cyclic Loading*, Fukomoto, Y. and G. Lee, Eds., CRC Press, Boca Raton, FL, 1992.
37. Mander, J. B., Kim, D.-K., Chen, S. S., and Premus, G. J., Response of Steel Bridge Bearings to Reversed Cyclic Loading, Report No. NCEER-96-0014, State University of New York, Buffalo, 1996.
38. Mayes, R. L., Buckle, I. G., Kelly, T. E., and Jones, L. R., AASHTO seismic isolation design requirements for highway bridges, *J. Struct. Eng. ASCE*, 118(1), 284–304, 1992.
39. Mayes, R. L., Jones, D. M., Knight, R. P., Choudhury, D., and Crooks, R. S., Seismically isolated bridges come of age, *Proc., 4th Intl. Conf. on Short and Medium Span Bridges*, Halifax, Nova Scotia, 1994, 1095–1106.
40. Nakashima, M., Strain-hardening behavior of shear panels made of low-yield steel. I: Test, *J. Struct. Eng. ASCE*, 121(12), 1742–1749, 1995.
41. Nanjo, A., Horie, Y., Okashiro, S., and Imoto, I., Experimental study on the ductility of steel bridge piers, *Proc. 5th International Colloquium on Stability and Ductility of Steel Structures — SDSS '97*, Vol. 1, Nagoya, Japan, 1997, 229–236.
42. NEHRP, Recommended Provisions for the Development of Seismic Regulations for New Buildings, Federal Emergency Management Agency, Washington, D.C., 1998.
43. Newmark, N. M. and Hall, W. J., *Earthquake Spectra and Design*, EERI, 1982.
44. Nishikawa, K., Yamamoto, S., Natori, T., Terao, K., Yasunami, H., and Terada, M., Retrofitting for seismic upgrading of steel bridge columns, *Eng. Struct.*, 20(4–6), 540–551, 1998.
45. Nishimura, N., Hwang, W. S., and Fukumoto, Y., Experimental investigation on hysteretic behavior of thin-walled box beam-to-column connections, in *Stability and Ductility of Steel Structures under Cyclic Loading*, Fukomoto, Y. and G. Lee, Eds., CRC, Boca Raton, FL, 163–174, 1992 .
46. Plumier, A., The dogbone: back to the future, *Eng. J., AISC*, 34(2), 61–67, 1997.
47. Popov, E. P. and Black, W., Steel struts under severe cyclic loading, *J. Struct. Div. ASCE*, 90(ST2), 223–256, 1981.

48. Popov, E. P. and Tsai, K.-C., Performance of large seismic steel moment connections under cyclic loads, *Eng. J. AISC*, 26(2), 51–60, 1989.
49. Priestley, M. J. N., Seible, F., and Chai, Y. H., Seismic retrofit of bridge columns using steel jackets, *Proc.* 10th World Conf. on Earthquake Eng., Vol. 9, Madrid, 5285–5290, 1992.
50. Roberts, J. E., Sharing California's seismic lessons, *Modern Steel Constr., AISC*, 32(7), 32–37, 1992.
51. SAC, Interim Guidelines Advisory No. 1, Supplement to FEMA 267, Report No. FEMA 267A/SAC-96-03, SAC Joint Venture, Sacramento, CA, 1997.
52. SAC, Interim Guidelines: Evaluation, Repair, Modification, and Design of Welded Steel Moment Frame Structures, Report FEMA 267/SAC-95-02, SAC Joint Venture, Sacramento, CA, 1995.
53. SAC, Technical Report: Experimental Investigations of Beam-Column Sub-assemblages, Parts 1 and 2, Report No. SAC-96-01, SAC Joint Venture, Sacramento, CA, 1996.
54. SAC, Interim Guidelines Advisory No. 1, Supplement to FEMA 267, Report No. FEMA 267A/SAC-96-03, SAC Joint Venture, Sacramento, CA, 1997.
55. Sarraf, M. and Bruneau, M., Ductile seismic retrofit of steel deck-truss bridges. II: design applications, *J. Struct. Eng. ASCE*, 124 (11), 1263–1271,1998.
56. Sarraf, M. and Bruneau, M., Ductile seismic retrofit of steel deck-truss bridges. II: strategy and modeling, *J. Struct. Eng., ASCE*, 124 (11), 1253–1262, 1998 .
57. SEAOC, *Recommended Lateral Force Requirements and Commentary*, Seismology Committee, Structural Engineers Association of California, Sacramento, 1996.
58. Shirolé, A. M. and Malik, A. H., Seismic retrofitting of bridges in New York State," *Proc. Symposium on Practical Solutions for Bridge Strengthening and Rehabilitation*, Iowa State University, Ames, 123–131, 1993.
59. SSPC, *Statistical Analysis of Tensile Data for Wide Flange Structural Shapes*, Structural Shapes Producers Council, Washington, D.C., 1994.
60. Tide, R. H. R., Stability of weld metal subjected to cyclic static and seismic loading, *Eng. Struct.*, 20(4–6), 562–569, 1998.
61. Tsai, K. C., Chen, H. W., Hong, C. P., and Su, Y. F., Design of steel triangular plate energy absorbers for seismic-resistant construction," *Earthquake Spectra*, 9(3), 505–528, 1993.
62. Tsai, K. C. and Popov, E. P., Performance of large seismic steel moment connections under cyclic loads, *Eng. J. AISC*, 26(2), 51–60, 1989.
63. Tsai, K. C., Yang, Y. F., and Lin, J. L., Seismic eccentrically braced frames, *Int. J. Struct. Design Tall Buildings*, 2(1), 53–74, 1993.
64. Uang, C.-M., "Establishing R (or R_w) and C_d factors for building seismic provisions, *J. Struct. Eng., ASCE*, 117(1), 19–28, 1991.
65. Usami, T. and Fukumoto, Y., Local and overall buckling tests of compression members and an analysis based on the effective width concept, *Proc. JSCE*, 326, 41–50, 1982. [in Japanese].
66. Usami, T., Mizutani, S., Aoki, T., and Itoh, Y., Steel and concrete-filled steel compression members under cyclic loading, in *Stability and Ductility of Steel Structures under Cyclic Loading*, Fukomoto, Y. and G. Lee, Eds., CRC, Boca Raton, 123–138, 1992.
67. Vincent, J., Seismic retrofit of the Richmond–San Rafael Bridge, *Proc. 2nd U.S. Seminar on Seismic Design, Evaluation and Retrofit of Steel Bridges*, San Francisco, 215–232, 1996.
68. Watanabe, E., Sugiura, K., Maikawa, Y., Tomita, M., and Nishibayashi, M., Pseudo-dynamic test on steel bridge piers and seismic damage assessment, *Proc. 5th East Asia-Pacific Conference on Structural Engineering and Construction*, Queensland, Australia, 1995.
69. Watanabe, E., Sugiura, K., Mori, T., and Suzuki, I., Modeling of hysteretic behavior of thin-walled box members, in *Stability and Ductility of Steel Structures under Cyclic Loading*, Fukomoto, Y. and G. Lee, Eds., CRC Press, Boca Raton, FL, 225–236, 1992.
70. Watanabe, E., Usami, T., and Kasegawa, A., Strength and design of steel stiffened plates — a literature review of Japanese contributions, in *Inelastic Instability of Steel Structures and Structural Elements*, Y. Fujita and T. V. Galambos, Ed., U.S.–Japan Seminar, Tokyo, 1981.

71. Yura, J. A., Galambos, T. V., and Ravindra, M. K., The bending resistance of steel beams, *J. Struct. Div. ASCE*, 104(ST9), 1355–1370, 1978.
72. Zahrai, S. M. and Bruneau, M., Impact of diaphragms on seismic response of straight slab-on-girder steel bridges, *J. Struct. Eng. ASCE*, 124(8), 938–947, 1998.
73. Zahrai, S. M. and Bruneau, M, Seismic Retrofit of Steel Slab-on-Girder Bridges Using Ductile End-Diaphragms, Report No. OCEERC 98-20, Ottawa Carleton Earthquake Engineering Research Center, University of Ottawa, Ottawa, Ontario, Canada, 1998.
74. Zekioglu, A., Mozaffarian, H., Chang, K. L., Uang, C.-M., and Noel, S., Designing after Northridge, *Modern Steel Constr. AISC*, 37(3), 36–42, 1997.

40

Seismic Retrofit Practice

James Roberts
California Department of Transportation

Brian Maroney
California Department of Transportation

40.1 Introduction

Until the 1989 Loma Prieta earthquake, most of the United States had not been concerned with seismic design for bridges, although some 37 states have some level of seismic hazard and there are hundreds of bridges in these other states that have been designed to seismic criteria that are not adequate for the seismic forces and displacements that we know today. Recent earthquakes, such as the 1971 San Fernando, California; the 1976 Tangshan, China [3]; the 1989 Loma Prieta, California; the 1994 Northridge, California; and the 1995 Hyogo-ken Nanbu (Kobe), Japan, have repeatedly demonstrated the seismic vulnerability of existing bridges and the urgent need for seismic retrofit.

The California Department of Transportation (Caltrans) owns and maintains more than12,000 bridges (spans over 6 m) and some 6000 other highway structures such as culverts (spans under 6 m), pumping plants, tunnels, tubes, highway patrol inspection facilities, maintenance stations, toll plazas, and other transportation-related structures. There are about an equal number on the City and County systems. Immediately after the February 9, 1971 San Fernando earthquake, Caltrans began a comprehensive upgrading of their *Bridge Seismic Design Specifications*, construction details, and a statewide bridge seismic retrofit program to reinforce the older non-ductile bridges systematically.

The success of the bridge seismic design and retrofit program and the success of future seismic design for California bridges is based, to a large degree, on the accelerated and "problem-focused" seismic research program that has provided the bridge design community with the assurance that the new specifications and design details perform reliably and meet the performance criteria. Caltrans staff engineers, consulting firms, independent peer-review teams, and university researchers have cooperated in this program of bridge seismic design and retrofit strengthening to meet the challenge presented in the June 1990 Board of Inquiry report [4].

0-8493-7434-0/00/$0.00+$.50

This chapter discusses the bridge seismic retrofit philosophy and procedures practiced by the California bridge engineers. Issues addressed in this chapter can be of great benefit to those states and countries that are faced with seismic threats of lesser magnitude, yet have little financial support for seismic retrofitting, and much less for research and seismic detail development.

40.2 Identification and Prioritization

As part of any seismic retrofit program, the first phase should be to identify a list of specific bridges in need of retrofitting. That list of bridges also needs to be prioritized respecting which bridges pose the greatest risk to the community and therefore should be first to enter into a design phase in which a detailed analysis is completed and retrofit construction plans are completed for bidding.

In order to identify and prioritize a group of bridge projects, a type of coarse analysis must be completed. This analysis is carried out to expedite the process of achieving safety at the sites of the greatest risk. This analysis should not be confused with a detailed bridge system analysis conducted as part of the design phase. The process essentially identifies the projects that need to be addressed first. It should be recognized that it is not realistic to evaluate bridge systems to a refined degree in massive numbers simultaneously; however, it is quite possible to identify those bridges that possess the characteristics that have made bridges vulnerable, or at least more vulnerable, during past earthquakes. This coarse analysis is likely to be a collective review of databases of (1) bridge structural parameters that offer insight into the capacity of the systems to withstand earthquake loading and (2) bridge site parameters that offer insight to the potential for a site to experience threatening seismic motions. In case of many parameters to be evaluated, relative measures are possible. For example, if mass is recognized to be a characteristic that leads to poor behavior, then bridge systems can be compared quantitatively to their effective masses.

As the identification and prioritization process is well suited for high-speed computers, the process is vulnerable itself to being refined beyond its effective capacities. It is also vulnerable to errors of obvious omission because of the temptation to finalize the effort without appropriate review of the computer-generated results (i.e., never let a computer make a decision an engineer should make). The results should be reviewed carefully to check if they make engineering sense and are repeatable. In the Caltrans procedure, three separate experienced engineers reviewed each set of bridge plans and there had to be a consensus to retrofit or not. Common sense and experience are essential in this screening process.

40.2.1 Hazard

The seismic threat to a bridge structure is the potential for motions that are large enough to cause failure to occur at the bridge site. These measures of seismic threat eventually develop into the source of the demand side of the fundamental design equation. Such threats are characterized in numerous ways and presented in a variety of formats. One recognized method is to assume a deterministic approach and to recognize a single upper-bound measure of potential event magnitude for all nearby faults, assume motion characterizations for the fault sources, account for motion decay with distance from each fault, and characterize the motions at a site using a selected parameter such as spectral rock acceleration at 1 H. Alternatively, a probabilistic approach can be adopted that in a systematic manner incorporates the probabilities of numerous fault rupture scenarios and the attenuation of the motions generating the scenarios to the site. These motions then can be characterized in a variety of ways, including the additional information of a measure of the probability of occurrence. It is not economical to conduct a probabilistic ground motion study for each bridge. Size, longevity, and unusual foundations will generally determine the need.

Influences of the local geology at various sites are commonly accounted for employing various techniques. The motions can be teamed with the site response, which is often incorporated into the demand side, then called hazard. A hazard map is usually available in the bridge design specifications [1,2].

40.2.2 Structural Vulnerability

The vulnerability of a bridge system is a measure of the potential failure mechanisms of the system. To some degree, all bridge structures are ultimately vulnerable. However, judgment and reason can be applied to identify the practical vulnerabilities. Since the judgment is ideally based upon experience in observing field performances that are typically few in number, observing laboratory tests and considering/analyzing mechanisms, the judgment applied is very important and must be of high quality. Of these foundations upon which to base judgments, field observations are the most influential. The other two are more commonly used to develop or enhance understanding of the potential failure mechanisms.

Much has been learned about bridge performance in previous earthquakes. Bridge site, construction details, and structural configuration have major effects on bridge performance during an earthquake. Local site conditions amplify strong ground motions and subsequently increase the vulnerability of bridges on soft soil sites. The single-column-supported bridges were deemed more vulnerable because of lack of redundancy, based on experience in the 1971 San Fernando earthquake. Structural irregularity (such as expansion joints and C-bents) can cause stress concentration and have catastrophic consequences. Brittle elements with inadequate details always limit their ability to deform inelastically. A comprehensive discussion of earthquake damages to bridges and causes of the damage is presented in Chapter 34.

A designer's ability to recognize potential bridge system vulnerabilities is absolutely essential. A designer must have a conceptual understanding of the behavior of the system in order to identify an appropriate set of assumptions to evaluate or analyze the design elements.

40.2.3 Risk Analysis

A conventional risk analysis produces a probability of failure or survival. This probability is derived from a relationship between the load and resistance sides of a design equation. Not only is an approximate value for the absolute risk determined, but relative risks can be obtained by comparing determined risks of a number of structures. Such analyses generally require vast collections of data to define statistical distributions for all or at least the most important elements of some form of analysis, design, and/or decision equations. The acquisition of this information can be costly if obtainable at all. Basically, this procedure is to execute an analysis, evaluate both sides of the relevant design equation, and define and evaluate a failure or survival function. All of the calculations are carried out taking into account the statistical distribution of every equation component designated as a variable throughout the entire procedure.

To avoid such a large, time-consuming investment in resources and to obtain results that could be applied quickly to the retrofit program, an alternative, level-one risk analysis can be used. The difference between a conventional and level-one risk analysis is that in a level-one analysis judgments take the place of massive data supported statistical distributions.

The level-one risk analysis procedure can be summarized in the following steps:

1. Identify major faults with high event probabilities (priority-one faults)

Faults believed to be the sources of future significant seismic events should be identified by a team of seismologists and engineers. Selection criteria include location, geologic age, time of last displacement (late quarternary and younger), and length of fault (10 km min.). Each fault recognized in this step is evaluated for style, length, dip, and area of faulting in order to estimate potential earthquake magnitude. Faults are then placed in one of three categories: minor (ignored for the purposes of this project), priority two (mapped and evaluated but unused for this project), or priority one (mapped, evaluated, and recognized as immediately threatening). In California, this step was carried out by consulting the California Division of Mines and Geology and the recent U.S. Geological Survey studies.

2. **Develop average attenuation relationships at faults identified in Step 1**

3. **Define the minimum ground acceleration capable of causing severe damage to bridge structures**

The critical (i.e., damage-causing) level of ground acceleration is determined by performing nonlinear analyses on a typical highly susceptible structure (single-column connector ramp) under varying maximum ground acceleration loads. The lowest maximum ground acceleration that requires the columns providing a ductility ratio of 1.3 may be defined as the critical level of ground acceleration. The critical ground acceleration determined in the Caltrans study was 0.5 *g*.

4. **Identify all the bridges within high-risk zones defined by the attenuation model of Step 2 and the critical acceleration boundary of Step 3**

The shortest distance from every bridge to every priority-one fault is calculated. Each distance is compared to the distance from each respective level of magnitude fault to the critical ground acceleration decremented acceleration boundary. If the distance from the fault to the bridge is less than the distance from the fault to the critical acceleration boundary, the bridge shall be determined to lie in the high-risk zone and is added to the screening list for prioritization.

5. **Prioritize the threatened bridges by summing weighted bridge structural and transportation characteristic scores**

This step constitutes the process used to prioritize the bridges within the high-risk zones to establish the order of bridges to be investigated for retrofitting. It is in this step that a risk value is assigned to each bridge. A specifically selected subset of bridge structural and transportation characteristics of seismically threatened bridges should be prepared in a database. Those characteristics were ground acceleration; route type — major or minor; average daily traffic (ADT); column design single or multiple column bents; confinement details of column (relates to age); length of bridge; skew of bridge, and availability of detour.

Normalized preweight characteristic scores from 0.0 to 1.0 are assigned based on the information stored in the database for each bridge. Scores close to 1.0 represent high-risk structural characteristics or high cost of loss transportation characteristics. The preweight scores are multiplied by prioritization weights. Postweight scores are summed to produce the assigned prioritization risk value.

Determined risk values are not to be considered exact. Due to the approximations inherent in the judgments adopted, the risks are no more accurate than the judgments themselves. The exact risk is not important. Prioritization list qualification is determined by fault proximity and empirical attenuation data, not so much by judgment. Therefore, a relatively high level of confidence is associated with the completeness of the list of threatened bridges. Relative risk is important because it establishes the order of bridges to be investigated in detail for possible need of retrofit by designers.

A number of assumptions are made in the process of developing the prioritized list of seismically threatened bridges. These assumptions are based on what is believed to be the best engineering judgment available. It seems reasonable to pursue verification of these assumptions some time in the future. Two steps seem obvious: (1) monitoring the results of the design departments retrofit analyses and (2) executing a higher-level risk analysis.

Important features of this first step are the ease and cost with which it could be carried out and the database that could be developed highlighting bridge characteristics that are associated with structures in need of retrofit. This database will be utilized to confirm the assumptions made in the retrofit program. The same database will serve as part of the statistical support of a future conventional risk analysis as suggested in the second step. The additional accuracy inherent in a higher-order risk analysis will serve to verify previous assumptions, provide very good approximations of actual structural risk, and develop or evaluate postulated scenarios for emergency responses. It is reasonable to analyze only selected structures at this level. A manual screening process may be used that includes review of "as-built" plans by at least three engineers to identify bridges with common details that appeared to need upgrading.

After evaluating the results of the 1989 Loma Prieta earthquake, Caltrans modified the risk analysis algorithm by adjusting the weights of the original characteristics and adding to the list. The additional characteristics are soil type; hinges, type and number; exposure (combination of length and ADT); height; abutment type; and type of facility crossed.

Even though additional characteristics were added and weights were adjusted, the postweight scores were still summed to arrive at the prioritization risk factor. The initial vulnerability priority lists for state and locally owned bridges were produced by this technique and retrofit projects were designed and built.

In 1992, advances were made in the Caltrans procedures to prioritize bridges for seismic retrofit and a new, more accurate algorithm was developed. The most significant improvement to the prioritization procedure is the employment of the multiattribute decision theory. This prioritization scheme incorporates the information previously developed and utilizes the important extension to a multiplicative formulation.

This multiattribute decision procedure assigns a priority rating to each bridge enabling Caltrans to decide more accurately which structures are more vulnerable to seismic activity in their current state. The prioritization rating is based on a two-level approach that separates out seismic hazard from impact and structural vulnerability characteristics. Each of these three criteria (hazard, impact, and structural vulnerability) depends on a set of attributes that have direct impact on the performance and potential losses of a bridge. Each of the criteria and attributes should be assigned a weight to show their relative importance. Consistent with previous work, a global utility function is developed for each attribute.

This new procedure provides a systematic framework for treating preferences and values in the prioritization decision process. The hierarchical nature of this procedure has the distinct advantage of being able to consider seismicity prior to assessing impact and structural vulnerability. If seismic hazard is low or nonexistent, then the values of impact and structural vulnerability are not important and the overall postweight score will be low because the latter two are added but the sum of those two are multiplied by the hazard rating. This newly developed prioritization procedure is defensible and theoretically sound. It has been approved by Caltrans Seismic Advisory Board.

Other research efforts [5–7] in conjunction with the prioritization procedure involve a sensitivity study that was performed on bridge prioritization algorithms from several states. Each procedure was reviewed in order to investigate whether or not California was neglecting any important principles. In all, 100 California bridges were selected as a sample population and each bridge was independently evaluated by each of the algorithms. The 100 bridges were selected to represent California bridges with respect to the variables of the various algorithms. California, Missouri, Nevada, Washington, and Illinois have thus far participated in the sensitivity study.

The final significant improvement to the prioritization procedure is the formal introduction of varying levels of seismicity. A preliminary seismic activity map for the state of California has been developed in order to incorporate seismic activity into the new prioritization procedure. In late 1992 the remaining bridges on the first vulnerability priority list were reevaluated using the new algorithm and a significant number of bridges changed places on the priority list but there were no obvious trends. Figure 40.1 and Table 40.1 show the new algorithm and the weighting percentages for the various factors.

40.3 Performance Criteria

Performance criteria are the design goals that the designer is striving to achieve. How do you want the structure to perform in an earthquake? How much damage can you accept? What are the reasonable

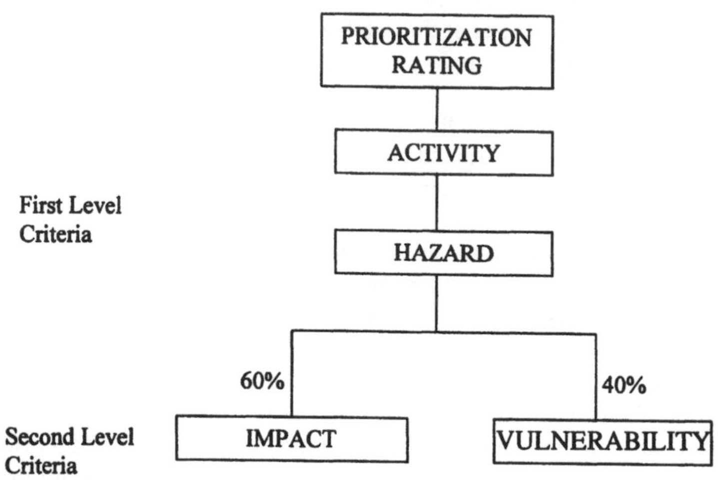

Prioritization Rating = (Activity)(Hazard)[(0.6)(Impact) + 0.4(Vulnerability)]
Activity = (Global Utility Function Value)
Hazard = Σ (Hazard Attribute Weight) (Global Utility Function Value)
Impact = Σ (Impact Attribute Weight) (Global Utility Function Value)
Vulnerability = Σ (Vulnerability Attribute Weight) (Global Utility Function Value)

FIGURE 40.1 Risk analysis — multiattribute decision procedure.

TABLE 40.1 Multi-attribute Weights

	Attributes	Weights (%)
Hazard	Soil conditions	33
	Peak rock acceleration	28
	Seismic duration	29
Impact	ADT on structure	28
	ADT under/over structure	12
	Detour length	14
	Leased air space (residential, office)	15
	Leased air space (parking, storage)	7
	RTE type on bridge	7
	Critical utility	10
	Facility crosses	7
Vulnerability	Year designed (constructed)	25
	Hinges (drop-type failure)	16.5
	Outriggers, shared column	22
	Bent redundancy	16.5
	Skew	12
	Abutment type	8

alternate routes? How do you define various levels of damage? How long do you expect for repair of various levels of damage? The form of the performance criteria can take many forms usually depending on the perspective and background of the organization presenting it. The two most common forms are functional and structural [Caltrans 1993]. The functional is the most appropriate form for the performance criteria because it refers to the justification of the existence of the structure. For example, the functional performance criteria of a bridge structure would include measures of post-earthquake capacity for traffic to flow across the bridge. Performance of the structure itself is more appropriately addressed with the structural design criteria. This would be codes, design memorandums, etc. An example of

Ground Motion at Site	Minimum Performance Level	Important Bridge Performance Level
Functional Evaluation	Immediate Service Level Repairable Damage	Immediate Service Level Minimal Damage
Safety Evaluation	Limited Service Level Significant Damage	Immediate Service Level Repairable Damage

DEFINITIONS

Immediate Service Level: Full access to normal traffic available almost immediately.
Limited Service Level: Limited access, (reduced lanes, light emergency traffic) possible within days. Full service restorable within months.

Minimal Damage: Essentially elastic performance.
Repairable Damage: Damage that can be repaired with a minimum risk of losing functionality.
Significant Damage: A minimum risk of collapse, but damage that would require closure for repair.

Important Bridge (one or more of the following items present):

- Bridge required to provide secondary life safety. (example: access to an emergency facility).
- Time for restoration of functionality after closure creates a major economic impact.
- Bridge formally designated as critical by a local emergency plan.

Safety Evaluation Ground Motion (Up to two methods of defining ground motions may be used):

- *Deterministically assessed ground motions from the maximum earthquake as defined by the Division of Mines and Geology Open-File Report 92-1 (1992).*
- *Probabilistically assessed ground motions with a long return period (approx. 1000-2000 years).*

For Important bridges both methods shall be given consideration, however the probabilistic evaluation shall be reviewed by a CALTRANS approved consensus group. For all other bridges the motions shall be based only on the deterministic evaluation. In the future, the role of the two methods for other bridges shall be reviewed by a CALTRANS approved consensus group.

Functional Evaluation Ground Motion:
Probabilistically assessed ground motions which have a 40% probability of occurring during the useful life of the bridge. The determination of this event shall be reviewed by a CALTRANS approved consensus group. **A separate Functional Evaluation is required only for Important Bridges.** All other bridges are only required to meet specified design requirements to assure Minimum Functional Performance Level compliance.

FIGURE 40.2 Seismic performance criteria for the design and evaluation of bridges.

what would be addressed in design criteria would be acceptable levels of strains in different structural elements and materials. These levels of strains would be defined to confidently avoid a defined state of failure, a deformation state associated with loss of capacity to accommodate functional performance criteria, or accommodation of relatively easy-repair.

Performance criteria must have a clear set of achievable goals, must recognize they are not independent of cost, and should be consistent with community planning. Figure 40.2 shows the seismic performance criteria for the design and evaluation of bridges in the California State Highway System [Caltrans 1993].

Once seismic performance criteria are adopted, the important issue then is to guarantee that the design criteria and construction details will provide a structure that meets that adopted performance criteria. In California a major seismic research program has been financed to physically test large-scale and full-sized models of bridge components to provide reasonable assurance to the engineering community that those details will perform as expected in a major seismic event. The current phase of that testing program involves real-time dynamic shaking on large shake tables. In addition, the Caltrans bridge seismic design specifications have been thoroughly reviewed in the ATC-32 project to ensure that they are the most up-to-date with state-of-the-art technology. On important bridges, project-based design criteria have been produced to provide guidance to the various design team members on what must be done to members to ensure the expected performance.

40.4 Retrofit Design

40.4.1 Conceptual Design

Design is the most-impacting part of the entire project. The conceptual design lays out the entire engineering challenge and sets the course for the analysis and the final detailed design. The conceptual design is sometimes referred to as type-selection or, in the case of seismic retrofit, the strategy. A seismic retrofit strategy is essentially the project engineer's plan that lays out the structural behavior to lead to the specified performance. The most important influential earthquake engineering is completed in this early phase of design. It is within this phase that "smart" engineering can be achieved (i.e., work smarter not harder). That is, type-selections or strategies can be chosen such that unreliable or unnecessary analyses or construction methods are not forced or required to be employed. When this stage of the project is completed well, a plan is implemented such that difficulties are wisely avoided when possible throughout not only the analysis, design, specification development, and construction phases, but also the remaining life of the bridge from a maintenance perspective. With such understanding, an informed decision can be made about which structural system and mechanisms should be selected and advanced in the project.

Highway multiple connector ramps on an interchange typically are supported by at least one column in the median of a busy functioning freeway. Retrofit strategies that avoid column retrofitting of the median columns have safety advantages over alternatives. Typically, columns outside the freeway traveled way can be strengthened and toughened to avoid median work and the problems of traffic handing.

On most two- and three-span shorter bridges the majority of seismic forces can be transferred into the abutments and embankments and thus reduce or entirely eliminate the amount of column retrofitting necessary. Large-diameter CIDH piles drilled adjacent to the wingwalls at abutments have been effective in resisting both longitudinal and transverse forces.

For most multiple-column bents the footing retrofits can be reduced substantially by allowing the columns to hinge at the bottom. This reduces the moments transferred into the foundations and lowers total costs. Sufficient testing on footing/pile caps and abutments has been conducted. It is found that a considerable amount of passive lateral resistance is available. Utilizing this knowledge can reduce the lateral force requirement of the structural foundations.

Continuity is extremely important and is the easiest and cheapest insurance to obtain. Well-designed monolithic structures also have the added advantage of low maintenance. Joints and bearings are some of the major maintenance problems on bridges today. If structures are not continuous and monolithic, they must be tied together at deck joints, supports, and abutments. This will prevent them from pulling apart and collapsing during an earthquake.

Ductility in the substructure elements is the second key design consideration. It is important that when you design for ductility you must be willing to accept some damage during an earthquake. The secret to good seismic design is to balance acceptable damage levels with the economics of preventing or limiting the damage. Properly designed ductile structures will perform well during an earthquake as long as the design has accounted for the displacements and controlled or provided for them at abutments and hinges. For a large majority of bridges, displacement criteria control over strength criteria in the design for seismic resistance.

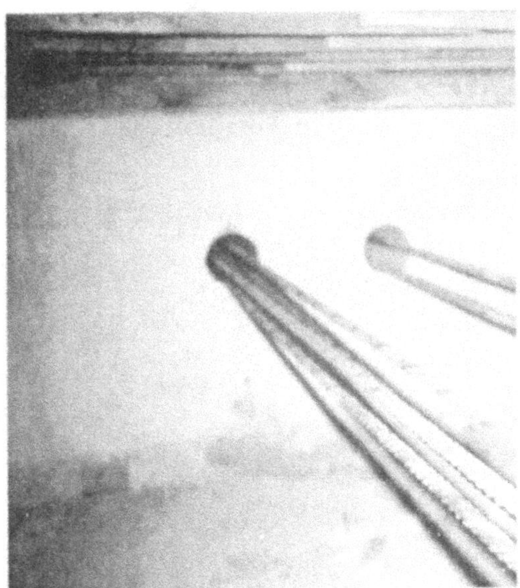

FIGURE 40.3 Hinge joint restrainer.

40.4.2 Retrofit Strategies

Designers of bridge seismic retrofit projects acquire knowledge of the bridge system, develop an understanding of the system response to potential earthquake ground motions, and identify and design modifications to the existing system that will change the expected response to one that satisfies the project performance criteria. This is accomplished by modifying any or all of the system stiffness, energy absorption, or mass characteristics. These characteristics or behavior can commonly be grouped into all structural system types, such as trusses, frames, single-column bent, shear walls, CIDH systems. This section briefly discusses various seismic retrofit strategies used in California. Chapter 43 presents more-detailed information.

Hinge Joint Restrainers

Spans dropped off from too narrow support seats and separation of expansion joints were two major causes of bridge collapse during the 1971 San Fernando earthquake. The initial phase of the Caltrans Bridge Seismic Retrofit Program involved installation of hinge and joint restrainers to prevent deck joints from separating (Figure 40.3). Included in this phase was the installation of devices to fasten the superstructure elements to the substructure in order to prevent those superstructure elements from falling off their supports (Figure 40.4). This phase was essentially completed in 1989 after approximately 1260 bridges on the California State Highway System had been retrofitted at a cost of over $55 million.

Figure 40.5 shows the installation of an external hinge extender detail that is designed to prevent the supported section of the superstructure from dropping off its support. Note the very narrow hinge details at the top of this picture, which is common on the 1960s era bridges throughout California.

The Loma Prieta earthquake of October 17, 1989 again proved the reliability of hinge and joint restrainers, but the tragic loss of life at the Cypress Street Viaduct on I-880 in Oakland emphasized the necessity to accelerate the column retrofit phase of the bridge seismic retrofit program immediately with a higher funding level for both research and implementation [8].

Confinement Jackets

The largest number of large-scale tests have been conducted to confirm the calculated ductile performance of older, nonductile bridge columns that have been strengthened by application of

FIGURE 40.4 Hold down devices for vertical acceleration.

FIGURE 40.5 External hinge extenders on Santa Monica freeway structures.

structural concrete, steel plate, prestressed strand, and fiberglass-composite jackets to provide the confinement necessary to ensure ductile performance. Since the spring of 1987 the researchers at University of California, San Diego have completed over 80 sets of tests on bridge column models [9–14]. Figure 40.6 shows reinforcement confinement for a column retrofit. Figure 40.7 shows a completed column concrete jacket retrofit. Figure 40.8 is a completed steel jacket retrofit.

Approximately 2200 of California's 12,000 bridges are located in the Los Angeles area, so it is significant to examine the damage and performance of bridges in the Northridge earthquake of January 17, 1994. About 1200 of these bridges were in an area that experienced ground accelerations greater than 0.25 *g* and several hundred were in the area that experienced ground accelerations of

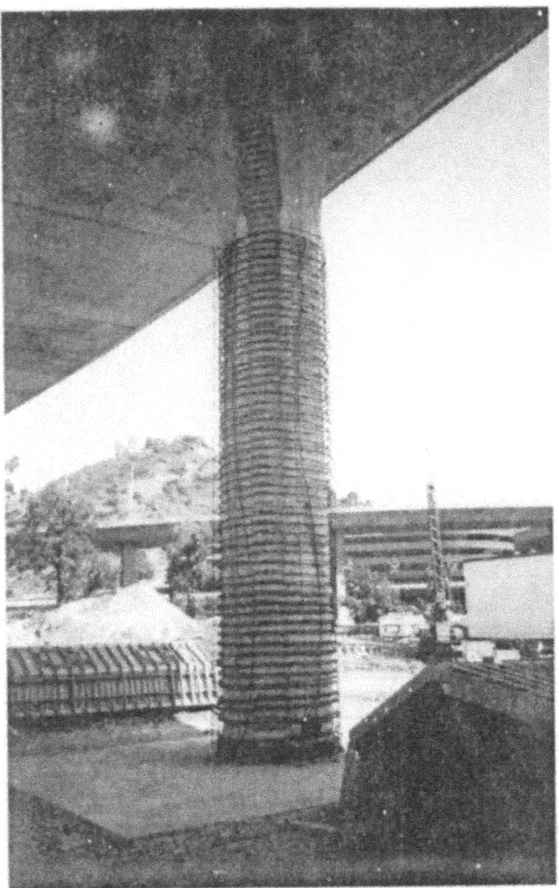

FIGURE 40.6 Reinforcement confinement retrofit.

0.50 *g*. There were 132 bridges in this area with post-San Fernando retrofit details completed and 63 with post-Loma Prieta retrofit details completed (Figure 40.9). All of these retrofitted bridges performed extremely well and most of the other bridges performed well during the earthquake; bridges constructed to the current Caltrans seismic specifications survived the earthquake with very little damage. Seven older bridges, designed for a smaller earthquake force or without the ductility of the current Caltrans design, sustained severe damage during the earthquake. Another 230 bridges suffered some damage ranging from serious problems of column and hinge damage to cracks, bearing damage, and approach settlements, but these bridges were not closed to traffic during repairs.

Link Beams

Link beams may be added to multicolumn bents to provide stiffener frame and reduce the unsupported column length. By using this development combined with other techniques, it may be possible to retrofit older, nonductile concrete columns without extensive replacement. Figure 40.10 shows link beams and installation of columns casings at the points of maximum bending and locations of anticipated plastic hinges on Santa Monica freeway structures. Half-scale models of these columns were constructed and tested under simulated seismic loading conditions to proof-test this conceptual retrofit design.

Ductile Concrete Column Details

Most concrete bridge columns designed since 1971 contain a slight increase in the main column vertical reinforcing steel and a major increase in confinement and shear reinforcing steel over the

FIGURE 40.7 Concrete jacket column retrofit.

pre-1971 designs. All new columns, regardless of geometric shape, are reinforced with one or a series of spiral-wound interlocking circular cages. The typical transverse reinforcement detail now consists of #6 (¾ in. diameter) hoops or continuous spiral at approximately 3-in. pitch over the full column height (Figure 40.11). This provides approximately eight times the confinement and shear reinforcing steel in columns than what was used in the pre-1971 nonductile designs. All main column reinforcing is continuous into the footings and superstructure. Splices are mostly welded or mechanical, both in the main and transverse reinforcing. Splices are not permitted in the plastic hinge zones. Transverse reinforcing steel is designed to produce a ductile column by confining the plastic hinge areas at the top and bottom of columns. The use of grade 60, A 706 reinforcing steel in bridges has recently been specified on all new projects.

Concrete Beam–Column–Bent Cap Details

Major advances have been made in the area of beam–column joint confinement, based on the results of research at both University of California, Berkeley, and San Diego. The performance and design criteria and structural details developed for the I-480 Terminal Separation Interchange and the I-880 replacement structures reflect the results of this research and were reported by Cooper [15]. Research is continuing at both institutions to refine the design details further to ensure ductile performance of these joints.

The concept using an integral edge beam can be used on retrofitting curved alignments, such as the Central Viaduct (U.S. 101) in downtown San Francisco and the Alemany Interchange on U.S. 101 in south San Francisco. The proofing-testing program was reported by Mahin [1991]. The concept using an independent edge beam can be used on retrofitting straight alignments. Figure 40.12 shows a graphic schematic of the proposed retrofit technique and Figure 40.13 shows the field installation of the joint reinforcement steel. Figure 40.14 shows the completed structure after retrofitting for seismic spectra that reach more than 2.0 *g* at the deck level.

FIGURE 40.8 Steel jacket column retrofit.

For outrigger bent cap under combined bending, shear, and torsion, an improved detail of column transverse reinforcement is typically continued up through the joint regions and the joints are further confined for shear and torsion resistance. The details for these joints usually require 1 to 3% confinement reinforcing steel. Thewalt and Stojadinovic [16] of University of California, Berkeley reported on this research. Figure 40.15 shows the complex joint-reinforcing steel needed to confine these joints for combined shear, bending, and torsion stresses. Design of these large joints requires use of the strut-and-tie technology to account properly for the load paths through the joint.

Steel Bridge Retrofit

Despite the fact that structural steel is ductile, members that have been designed by the pre-1972 seismic specifications must be evaluated for the seismic forces expected at the site based on earthquake magnitudes as we know them today. Typically, structural steel superstructures that had been tied to their substructures with joint and hinge restrainer systems performed well. However, we have identified many elevated viaducts and some smaller structures supported on structural steel columns that were designed prior to 1972 and that will require major retrofit strengthening for them to resist modern earthquake forces over a long period of shaking. One weak link is the older rocker bearings that will probably roll over during an earthquake. These can be replaced with modern neoprene, Teflon, pot, and base isolation bearings to ensure better performance in an earthquake. Structural steel columns can be strengthened easily to increase their toughness and ability to withstand a long period of dynamic input.

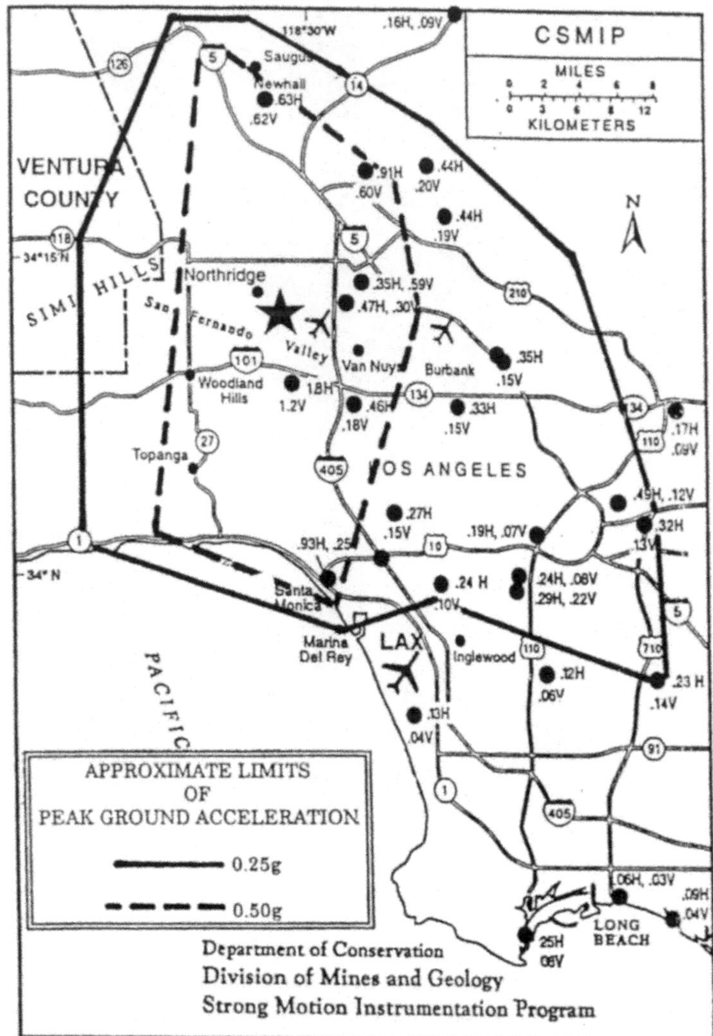

FIGURE 40.9 Peak ground acceleration zones — Northridge earthquake.

FIGURE 40.10 Steel jackets and bond beam — Santa Monica Freeway.

FIGURE 40.11 Column reinforcing steel cage.

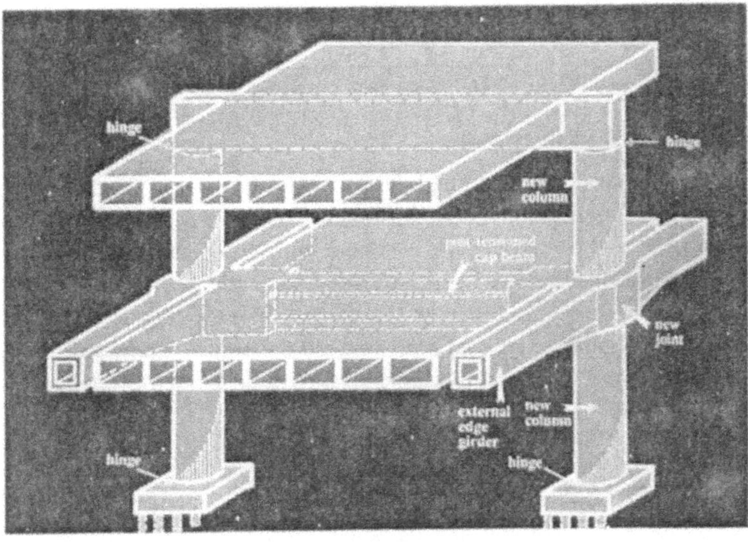

FIGURE 40.12 Graphic of edge beam retrofit scheme.

FIGURE 40.13 Field installation of joint reinforcing steel.

FIGURE 40.14 Completed retrofitted structures.

Footing and Pile Cap Modifications

Bridge column footing details established in 1980 consist of top and bottom mats of reinforcement tied together vertically by closely spaced hooked stirrups (Chapter 43). The column longitudinal rebars rest on the bottom mat, are hooked into the footing with hooks splayed outward, and are confined by spiral or hoop reinforcement between the mats. For pile foundations, the piles are reinforced and securely connected to the pile caps to resist the seismic tensile loads (Figure 40.16). The justifications for these details were widely debated, and strut-and-tie procedure seems to substantiate the need (Chapter 38). However, a proof-test of a footing with typical details performed adequately.

Seismic Isolation and Energy Dissipation Systems

Seismic isolation and supplemental energy dissipation devices have been successfully used in many bridge seismic design and retrofit projects. A detailed discussion is presented in Chapter 41. Extreme caution should be exercised when considering isolation devices. As discussed earlier, good, well-detailed, monolithic moment-resisting frames provide adequate seismic resistance without the

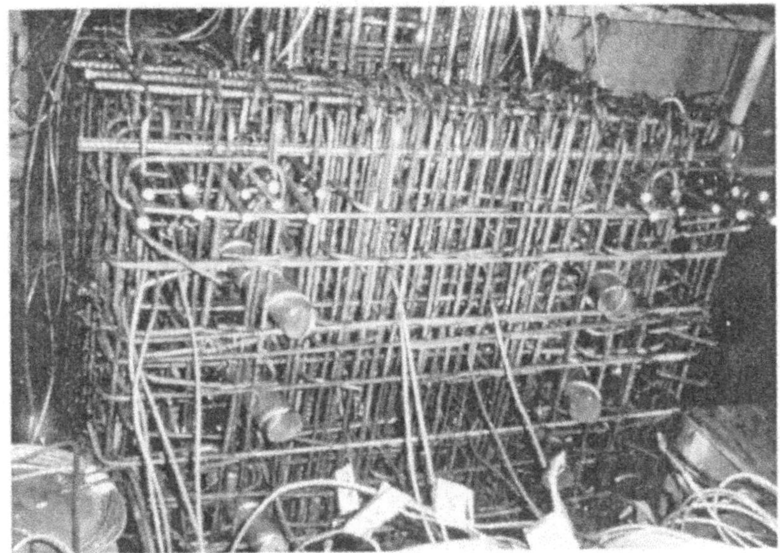

FIGURE 40.15 Reinforcing steel pattern in complex outrigger joint

FIGURE 40.16 Typical footing and pile cap modification.

inherent maintenance problems and higher initial costs. These devices, however, are excellent for replacing older, rocker bearings.

40.4.3 Analysis

Analysis is the simulation of the structure project engineer's strategy of the bridge response to the seismic motions. A good seismic design is robust and as relatively insensitive to fluctuations in ground motions as possible. Quantitative analysis is the appropriate verification of the capacity of the system and its individual subsystems being greater than the recognized demand.

The more complicated the seismic strategy, the more complicated will be the analysis. If the behavior of the system is to be nearly elastic with minor damage developed, then the analysis is

FIGURE 40.17 Completed seismic retrofit of I-5/710 interchange in Los Angeles.

likely to be simply linear-elastic analysis. However, if the behavior is likely to be complex, changing in time, with significant damage developed and loss of life or important facility loss, then the analysis is likely to be similarly complex.

As a rule of thumb, the complexity of the analysis shadows the complexity of the strategy and the importance of the bridges. However, it should be noted that a very important bridge that is being designed to behave essentially elastically will not require complex analysis. It should always be recognized that analysis serves design and is part of design. Analysis cannot be a separated form. Too many engineers confuse analysis with design. Good design combines the analysis with judgment, common sense, and use of tested details.

40.4.4 Aesthetics

The design approach to bridge architecture, whether it is a proposed new structure or seismical retrofitting of an existing structure, poses a great challenge to the design team. Successful bridge designs are created by the productive and imaginative creations of the bridge architect and bridge engineer working together. The partnership of these talents, although not recognized in many professional societies, is an essential union that has produced structures of notable fame, within immediate identity worldwide.

Why is this partnership considered so essential? There are a number of reasons. Initially, the bridge architect will research the existing structures in the geographic area with respect to the surrounding community's existing visual qualities of the structural elements and recommended materials, forms, and texture that will harmonize with rather than contrast with the built environment. Figure 40.17 shows the architecture success of seismic retrofit of I-5/710 Interchange in SLos Angeles.

The residents of most communities that possess noted historical structures are extremely proud and possessive of their inheritance. So it is incumbent upon the bridge architect to demonstrate the sensitivity that is necessary when working on modifying historically significant bridges. This process often requires presentations at community gatherings or even workshops, where the bridge architect will use a variety of presentation techniques to show how carefully the designer has seismically retrofitted that specific structure and yet preserved the original historic design. This task is by no means easy, because of the emotional attachment a community may have toward its historic fabric.

In addition to the above-noted considerations for aesthetics, the architect and engineer must also take into consideration public safety, maintenance, and constructability issues when they consider seismic retrofit ideas. In addition to aesthetics, any modifications to an existing structure must carefully take into account the other three areas that are paramount in bridge design.

Within the governmental transportation agencies and private consulting firms lies a great deal of talent in both architecture and engineering. One key to utilizing this talent is to involve the bridge architect as early as possible so the engineer can be made aware of the important community and historical issues.

40.5 Construction

Construction is a phase of any retrofit project that is often not respected to an appropriate degree by designers. This is always somewhat of a surprise as construction regularly represents 80 to 90% of the cost of a project. In the authors opinion, a good design is driven by reliable construction methods and techniques. In order to deliver a design package that will minimize construction problems, the design project engineer strives to interact with construction engineers regularly and particularly on issues involving time, limited space, heavy lifts, and unusual specifications.

As mentioned in the section covering design, legal right-of-way access and utilities are very important issues that can stop, delay, or cause tremendous problems in construction. One of the first orders of work in the construction phase is to locate and appropriately protect or relocate utilities. This usually requires a legal agreement, which requires time. A considerable cost is not uncommon. The process required varies as a function of the utility and the owner, but they always take time and money. Access right-of-way is usually available due to existing right-of-way for maintenance. If foundation extensions or additional columns are required, then additional land may need to be acquired or even greater temporary access may be necessary. This issue should be recognized in the design phase, but regularly develops into construction challenges that require significant problem solving by the construction staff. These problems can delay a project many months or even require redesign.

Safety to the traveling public and the construction personnel is always the first priority on a construction site. But most of a structure resident engineer's (SRE) time is invested in assuring the contractors' understanding and adherence to the contract documents. In order to do this well the SRE must first understand well the contract documents, including the plans, construction standard and project special specifications. Then, the SRE must understand well the plan the contractor has to construct the project in such a way as to satisfy the requirements of the contract. It is in understanding the construction plan and observing the implementation of that plan that the SRE ensures that the construction project results in a quality product that will deliver acceptable performance for the life of the structure.

As most transportation structures are in urban areas, traffic handling and safety are important elements of any retrofit project. A transportation management plan (TMP) is a necessary item to develop and maintain. Traffic safety engineers including local highway patrol or police representatives are typically involved in developing such a plan. The TMP clearly defines how and when traffic will be routed to allow the contractor working space and time to complete the required work.

Shop plans are an item that are typically addressed early in the construction phase. Shop plans are structural plans developed by the contractor for structural elements and construction procedures that are appropriately delegated to the contractor by the owner in order to allow for as competitive bids as possible. Examples of typical shop plans include prestress anchorages and steel plate strengthening details and erection procedures.

Foundation modifications have been a major component in the bridge seismic retrofit program the California Department of Transportation has undertaken since the 1989 Loma Prieta earthquake. Considerable problems have been experienced in the reconstruction of many bridge foundations. Most of the construction claim dollars leveled against the state have been associated with foundation-related issues. These problems have included as-built plans not matching actual field conditions, materials, or dimension; a lack of adequate space to complete necessary work (e.g., insufficient overhead clearance to allow for driving or placing piles); damage to existing structural components (e.g., cutting reinforcing steel while coring); splicing of reinforcing steel with couplers or welds; paint specifications and time; and unexpected changes in geologic conditions. Although these items at first appear to have little in common, each of them is founded in uncertainty. That is, the construction problem is based on a lack of information. Recognizing this, the best way to avoid such problems is as follows:

- To invest in collecting factual and specific data that can be made available to the designer and the contractor such as actual field dimensions;
- To consider as carefully as possible likely contractor space requirements given what activities the contractor will be required to conduct;
- To know and understand well the important properties of materials and structural elements that are to be placed into the structure by the contractor; and
- To conduct appropriately thorough foundation investigation which may include field testing of potential foundation systems.

The most common structural modifications to bridge structures in California have been the placement of steel shells around portions of reinforced concrete columns in order to provide or increase confinement to the concrete within the column and increase the shear strength of the column within the dimensions of the steel shell. As part of a construction project, important items to verify in a steel shell column jacket installation are the steel material properties, the placement of the steel shell, the weld material and process, the grouting of the void between the oversized steel shell and the column, and the grinding and painting of the steel shell.

Existing reinforcing steel layouts are designed for a purpose and should not be modified. In some cases they can be modified for convenience in construction. It is important that field engineers be knowledgeable in order to reject modifications to reinforcing steel layouts that could render the existing structural section inadequate.

40.6 Costs

Estimating costs for bridge seismic retrofit projects is an essential element of any retrofit program. For a program to initiate, legislation must typically be passed. As part of the legislation package, funding sources are identified, and budgets are set. The budgets are usually established from estimates. It is ironic that, typically, the word *estimate* is usually dropped in this process. Regardless of any newly assigned title of the estimate, it remains what it is — an estimate. This typical set of circumstances creates an environment in which it is essential that great care be exercised before estimates are forwarded.

The above being stated, methods have been developed to forecast retrofit costs. The most common technique is to calculate and document into a database project costs per unit deck area. When such data are nearly interpolated to similar projects with consistent parameters, this technique can realize success. This technique is better suited to program estimates rather than a specific project estimate.

TABLE 40.2 Approximate Costs of Various Pay Items of Bridge Seismic Retrofit (California, 1998)

Pay times	Approximate Cost	Notes
Access opening (deck)	$350 to $1500 per sq. ft.	
Access opening (soffit)	$400 to $750 per sq. ft.	
Restrainer cables	$3.5 to $6.6 per number	
Restrainer rods	$2.5 to $4.5 per number	
Seat extenders	$1.5 to $3.3 per number	
Steel shells for columns	$1.5 to $2.25 per lb.	
Concrete removal		
Steel removal		
Soil removal	$40 to $150 per cy	
Core concrete (6 in.)	$65 to $100 per ft.	
Concrete (bridge footing)	$175 to $420 per cy	
Concrete (bridge)	$400 to $800 per cy	
Minor concrete	$350 to $900 per cy	
Structural steel	$2.50 to $5 per lb.	
Prestressing steel	$0.80 to $1.15 per lb.	
Bar reinforcing steel	$0.50 to $1.00 per number	
Precast concrete pile (45T)	$610 to $1515 per linear ft.	
CISS piles (24 in.)	$788 to $4764 per linear ft.	
Pile shaft (48 in.)	$170 to $330 per linear ft.	
Structural backfill	$38 to $100 per cy	
Traffic lane closure (day)		
Traffic lane closure (night)		

When applied to a specific project, additional contingencies are appropriate. When an estimate for a specific project is desired, it is appropriate to evaluate the specific project parameters.

Many of the components or pay items of a seismic retrofit project when broken down to pay items are similar to new construction or widening project pay items. As a first estimate, this can be used to approximate the cost of the work crudely. Table 40.2 lists the approximate cost for various pay items in California in 1998. There certainly are exceptions to these general conditions, such as steel shells, very long coring and drilling, and pile installation in low clearance conditions.

40.7 Summary

The two most significant earthquakes in recent history that produced the best information for bridge designers were the 1989 Loma Prieta and the 1994 Northridge events. Although experts consider these to be only moderate earthquakes, it is important to note the good performance of the many bridges that had been designed for the improved seismic criteria or retrofitted with the early-era seismic retrofit details. This reasonable performance of properly designed newer and retrofitted older bridges in a moderate earthquake is significant for the rest of the United States and other countries because that knowledge can assist engineers in designing new bridges and in designing an appropriate seismic retrofit program for their older structures. Although there is a necessary concern for the "Big One" in California, especially for the performance of important structures, it must be noted that many structures that vehicle traffic can bypass need not be designed or retrofitted to the highest standards. It is also important to note that there will be many moderate earthquakes that will not produce the damage associated with a maximum event. These are the earthquake levels that should be addressed first in a multiphased retrofit strengthening program, given the limited resources that are available.

Cost–benefit analysis of retrofit details is essential to measure and ensure the effectiveness of a program. It has been the California experience that a great deal of insurance against collapse can be achieved for a reasonable cost, typically 10% of replacement cost for normal highway bridges. It is also obvious that designing for the performance criteria that provides full service immediately after a major earthquake may not be economically feasible. The expected condition of the bridge approach roadways after a major seismic event must be evaluated before large investments are made in seismic retrofitting of the bridges to the full-service criteria. There is little value to the infrastructure in investing large sums to retrofit a bridge if the approaches are not functioning after a seismic event. Roadways in the soft muds around most harbors and rivers are potentially liquefiable and will require repair before the bridges can be used.

Emerging practices on bridge seismic retrofit in the state of California was briefly presented. The excellent performance of bridges utilizing Caltrans newer design criteria and ductile details gives bridge designers an indication that these structures can withstand a larger earthquake without collapse. Damage should be expected, but it can be repaired in many cases while traffic continues to use the bridges.

References

1. AASHTO, *LRFD Bridge Design Specifications*, 2nd ed., American Association of State Highway and Transportation Officials, Washington, D.C., 1998.
2. Caltrans, Seismic Hazard Map in California, California Department of Transportation, Sacramento, CA, 1996.
3. Xie, L. L. and Housner, G. W., *The Greater Tangshan Earthquake*, Vol. I and IV, California Institute of Technology, Pasadena, CA, 1996.
4. Housner, G. W. (Chairman), Thiel, C. C. (Editor), Competing against Time, Report to Governor George Deukmejian from the Governor's Board of Inquiry on the 1989 Loma Prieta Earthquake, Publications Section, Department of General Services, State of California, Sacramento, June, 1990.
5. Maroney, B., Gates, J., and Caltrans. Seismic risk identification and prioritization in the Caltrans seismic retrofit program, in *Proceedings, 59th Annual Convention*, Structural Engineers Association of California, Sacramento, September, 1990.
6. Gilbert, A., development in seismic prioritization of California bridges, in *Proceedings: Ninth Annual US/Japan Workshop on Earthquake and Wind Design of Bridges*, Tsukuba Science City, Japan, May 1993.
7. Sheng, L. H. and Gilbert, A., California Department of Transportation seismic retrofit program-the prioritization and screening process," *Lifeline Earthquake Engineering: Proceedings of the Third U.S. Conference*, Report: Technical Council on Lifeline Earthquake Engineering, Monograph 4, American Society of Civil Engineers, New York, August, 1991.
8. Mellon, S. et al., Post earthquake investigation team report of bridge damage in the Loma Prieta earthquake, in *Proceedings: ASCE Structures Congress XI*, ASCE, Irvine, CA, April 1993.
9. Priestley, M. J. N., Seible, F., and Chai, Y. H., Flexural Retrofit of Circular Reinforced Bridge Columns by Steel Jacketing, Colret—A Computer Program for Strength and Ductility Calculation, Report No. SSRP-91/05 to the Caltrans Division of Structures, University of California at San Diego, October, 1991.
10. Priestley, M. J. N., Seible, F., and Chai, Y. H., Flexural Retrofit of Circular Reinforced Bridge Columns by Steel Jacketing, Experimental Studies, Report No. SSRP-91/05 to the Caltrans Division of Structures, University of California at San Diego, October, 1991.
11. Priestley, M. J. N. and Seible, F., Assessment and testing of column lap splices for the Santa Monica Viaduct retrofit, in *Proceedings: ASCE Structures Congress*, XI, ASCE, Irvine, California, April 1993.
12. Seible, F., Priestley, M. J. N., Latham, C. T., and Terayama, T., Full Scale Test on the Flexural Integrity of Cap/Column Connections with Number 18 Column Bars, Report No. TR-93/01 to Caltrans, University of California at San Diego, January 1993.

13. Seible, F., Priestley, M. J. N., Hamada, N., Xiao, Y., and MacRae, G. A., Rocking and Capacity Test of Model Bridge Pier, Report No. SSRP-92/06 to Caltrans, University of California at San Diego, August, 1992.
14. Seible, F., Priestley, M. J. N., Hamada, N., and Xiao, Y., Test of a Retrofitted Rectangular Column Footing Designed to Current Caltrans Retrofit Standards, Report No. SSRP-92/10 to Caltrans, University of California at San Diego, November 1992.
15. Cooper, T. R., Terminal Separation Design Criteria: A Case Study of Current Bridge Seismic Design and Application of Recent Seismic Design Research, Seismic Design and Retrofit of Bridges, *Seminar Proceedings,* Earthquake Engineering Research Center, University of California at Berkeley and California Department of Transportation, Division of Structures, Sacramento, 1992.
16. Thewalt, C. R. and Stojadinovic, B. I., Behavior and Retrofit of Outrigger Beams, Seismic Design and Retrofit of Bridges, *Seminar Proceedings,* Earthquake Engineering Research Center, University of California at Berkeley and California Department of Transportation, Division of Structures, Sacramento, 1992.

13. Seible, F., Priestley, M.J.N., Hamada, N., Xiao, Y. and MacRae, G.A. Rocking and Capacity Test of Model Bridge Pier, Report No. SSRP 92/06 to Caltrans, University of California, San Diego, August 1992.
14. Seible, F., Priestley, M.J.N., [illegible] and Xiao, Y., Test of [illegible] Footing [illegible] Retrofit Strategies [illegible] [illegible] 1994.
15. [illegible]
16. [illegible] Center, University of California, Berkeley and [illegible] Division of [illegible]
17. [illegible] Seismic Design [illegible] Center, University [illegible] of Structures.

41

Seismic Isolation and Supplemental Energy Dissipation

Rihui Zhang
California State Department of Transportation

41.1 Introduction

Strong earthquakes impart substantial amounts of energy into structures and may cause the structures to deform excessively or even collapse. In order for structures to survive, they must have the capability to dissipate this input energy through either their inherent damping mechanism or inelastic deformation. This issue of energy dissipation becomes even more acute for bridge structures because most bridges, especially long-span bridges, possess very low inherent damping, usually less than 5% of critical. When these structures are subjected to strong earthquake motions, excessive deformations can occur by relying on only inherent damping and inelastic deformation. For bridges designed mainly for gravity and service loads, excessive deformation leads to severe damage or even collapse. In the instances of major bridge crossings, as was the case of the San Francisco–Oakland

0-8493-7434-0/00/$0.00+$.50

Bay Bridge during the 1989 Loma Prieta earthquake, even noncollapsing structural damage may cause very costly disruption to traffic on major transportation arteries and is simply unacceptable.

Existing bridge seismic design standards and specifications are based on the philosophy of accepting minor or even major damage but no structural collapse. Lessons learned from recent earthquake damage to bridge structures have resulted in the revision of these design standards and a change of design philosophy. For example, the latest bridge design criteria for California [1] recommend the use of a two-level performance criterion which requires that a bridge be designed for both safety evaluation and functional evaluation design earthquakes. A safety evaluation earthquake event is defined as an event having a very low probability of occurring during the design life of the bridge. For this design earthquake, a bridge is expected to suffer limited significant damage, or immediately repairable damage. A functional evaluation earthquake event is defined as an event having a reasonable probability of occurring once or more during the design life of the bridge. Damages suffered under this event should be immediately repairable or immediate minimum for important bridges. These new criteria have been used in retrofit designs of major toll bridges in the San Francisco Bay area and in designs of some new bridges. These design criteria have placed heavier emphasis on controlling the behavior of bridge structural response to earthquake ground motions.

For many years, efforts have been made by the structural engineering community to search for innovative ways to control how earthquake input energy is absorbed by a structure and hence controlling its response to earthquake ground motions. These efforts have resulted in the development of seismic isolation techniques, various supplemental energy dissipation devices, and active structural control techniques. Some applications of these innovative structural control techniques have proved to be cost-effective. In some cases, they may be the only ways to achieve a satisfactory solution. Furthermore, with the adoption of new performance-based design criteria, there will soon come a time when these innovative structural control technologies will be the choice of more structural engineers because they offer economical alternatives to traditional earthquake protection measures.

Topics of structural response control by passive and active measures have been covered by several authors for general structural applications [2–4]. This chapter is devoted to the developments and applications of these innovative technologies to bridge structures. Following a presentation of the basic concepts, modeling, and analysis methods, brief descriptions of major types of isolation and energy dissipation devices are given. Performance and testing requirements will be discussed followed by a review of code developments and design procedures. A design example will also be given for illustrative purposes.

41.2 Basic Concepts, Modeling, and Analysis

The process of a structure responding to earthquake ground motions is actually a process involving resonance buildup to some extent. The severity of resonance is closely related to the amount of energy and its frequency content in the earthquake loading. Therefore, controlling the response of a structure can be accomplished by either finding ways to prevent resonance from building up or providing a supplemental energy dissipation mechanism, or both. Ideally, if a structure can be separated from the most-damaging energy content of the earthquake input, then the structure is safe. This is the idea behind seismic isolation. An isolator placed between the bridge superstructure and its supporting substructure, in the place of a traditional bearing device, substantially lengthens the fundamental period of the bridge structure such that the bridge does not respond to the most-damaging energy content of the earthquake input. Most of the deformation occurs across the isolator instead of in the substructure members, resulting in lower seismic demand for substructure members. If it is impossible to separate the structure from the most-damaging energy content, then the idea of using supplemental damping devices to dissipate earthquake input energy and to reduce structural damage becomes very attractive.

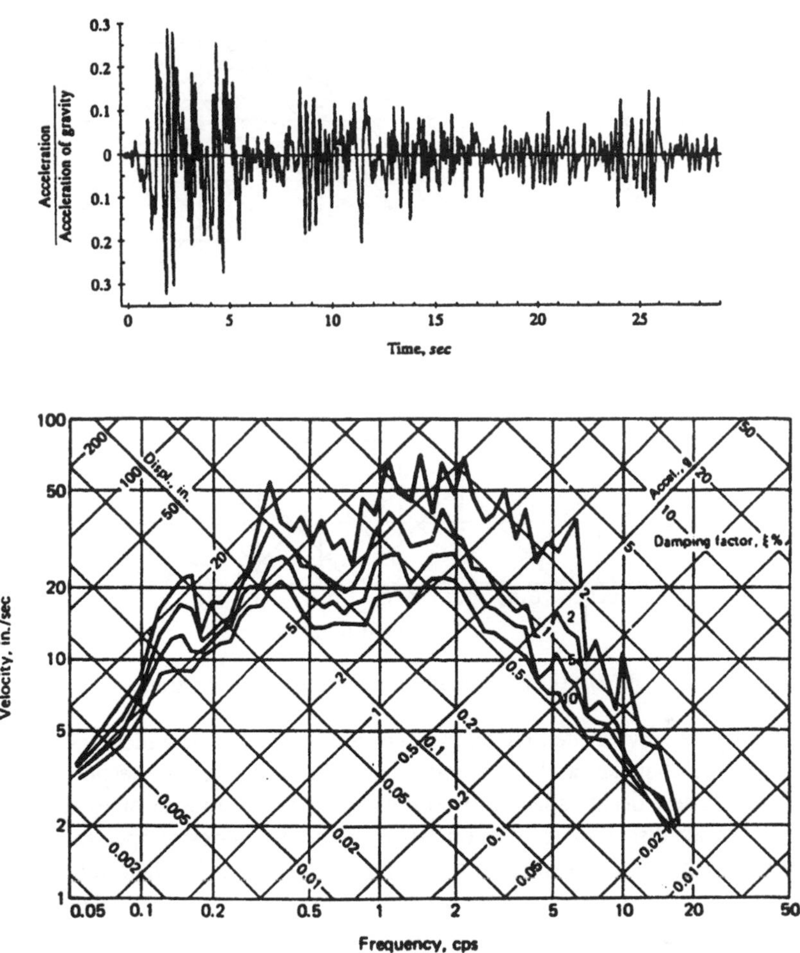

FIGURE 41.1 Acceleration time history and response spectra from El Centro earthquake, May 1940.

In what follows, theoretical basis and modeling and analysis methods will be presented mainly based on the concept of earthquake response spectrum analysis.

41.2.1 Earthquake Response Spectrum Analysis

Earthquake response spectrum analysis is perhaps the most widely used method in structural earthquake engineering design. In its original definition, an earthquake response spectrum is a plot of the maximum response (maximum displacement, velocity, acceleration) to a specific earthquake ground motion for all possible single-degree-of-freedom (SDOF) systems. One of such response spectra is shown in Figure 41.1 for the 1940 El Centro earthquake. A response spectrum not only reveals how systems with different fundamental vibration periods respond to an earthquake ground motion, when plotted for different damping values, site soil conditions and other factors, it also shows how these factors are affecting the response of a structure. From an energy point of view, response spectrum can also be interpreted as a spectrum the energy frequency contents of an earthquake.

Since earthquakes are essentially random phenomena, one response spectrum for a particular earthquake may not be enough to represent the earthquake ground motions a structure may

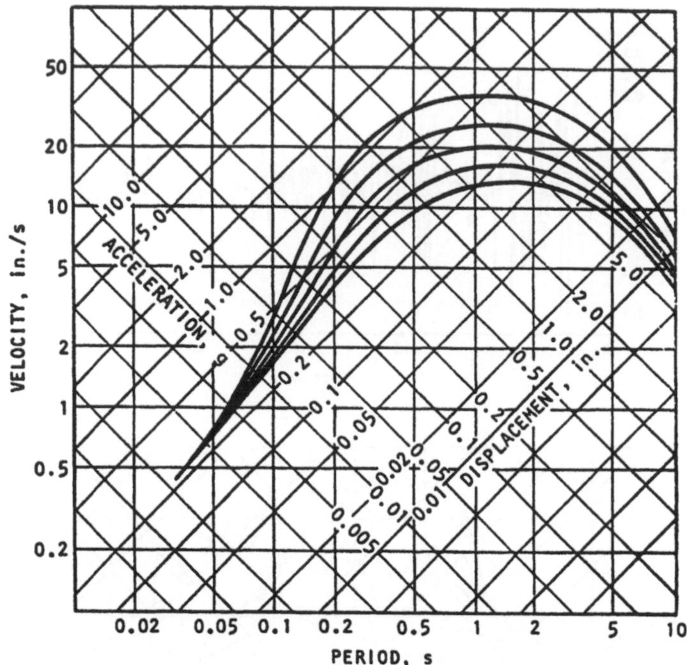

FIGURE 41.2 Example of smoothed design spectrum.

experience during its service life. Therefore, the design spectrum, which incorporates response spectra for several earthquakes and hence represents a kind of "average" response, is generally used in seismic design. These design spectra generally appear to be smooth or to consist of a series of straight lines. Detailed discussion of the construction and use of design spectra is beyond the scope of this chapter; further information can be found in References [5,6]. It suffices to note for the purpose of this chapter that design spectra may be used in seismic design to determine the response of a structure to a design earthquake with given intensity (maximum effective ground acceleration) from the natural period of the structure, its damping level, and other factors. Figure 41.2 shows a smoothed design spectrum curve based on the average shapes of response spectra of several strong earthquakes.

41.2.2 Structural Dynamic Response Modifications

By observing the response/design spectra in Figures 41.1a, it is seen that manipulating the natural period and/or the damping level of a structure can effectively modify its dynamic response. By inserting a relatively flexible isolation bearing in place of a conventional bridge bearing between a bridge superstructure and its supporting substructure, seismic isolation bearings are able to lengthen the natural period of the bridge from a typical value of less than 1 second to 3 to 5 s. This will usually result in a reduction of earthquake-induced response and force by factors of 3 to 8 from those of fixed-support bridges [7].

As for the effect of damping, most bridge structures have very little inherent material damping, usually in the range of 1 to 5% of critical. The introduction of nonstructural damping becomes necessary to reduce the response of a structure.

Some kind of a damping device or mechanism is also a necessary component of any successful seismic isolation system. As mentioned earlier, in an isolated structural system deformation mainly occurs across the isolator. Many factors limit the allowable deformation taking place across an

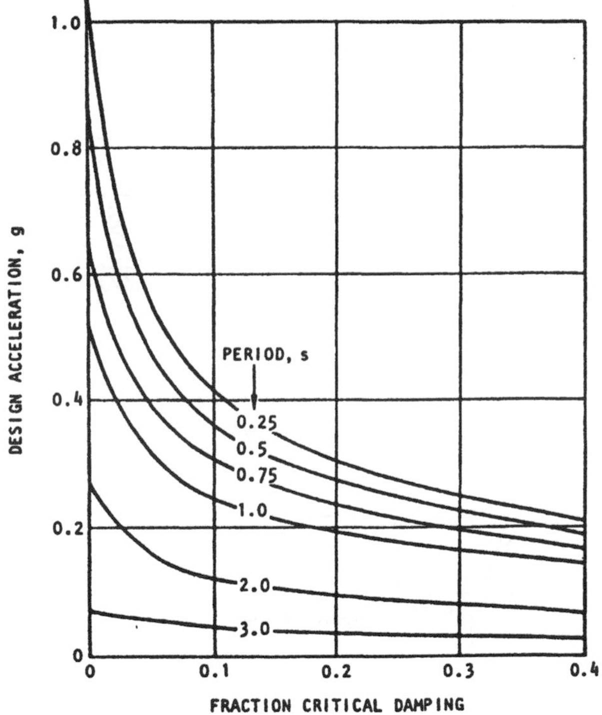

FIGURE 41.3 Effect of damping on response spectrum.

isolator, e.g., space limitation, stability requirement, etc. To control deformation of the isolators, supplemental damping is often introduced in one form or another into isolation systems.

It should be pointed out that the effectiveness of increased damping in reducing the response of a structure decreases beyond a certain damping level. Figure 41.3 illustrates this point graphically. It can be seen that, although acceleration always decreases with increased damping, its rate of reduction becomes lower as the damping ratio increases. Therefore, in designing supplemental damping for a structure, it needs to be kept in mind that there is a most-cost-effective range of added damping for a structure. Beyond this range, further response reduction will come at a higher cost.

41.2.3 Modeling of Seismically Isolated Structures

A simplified SDOF model of a bridge structure is shown in Figure 41.4. The mass of the superstructure is represented by *m*, pier stiffness by spring constant k_0, and structural damping by a viscous damping coefficient c_0. The equation of motion for this SDOF system, when subjected to an earthquake ground acceleration excitation, is expressed as:

$$m_0\ddot{x} + c_0\dot{x} + k_0 x = -m_0\ddot{x}_g \tag{41.1}$$

The natural period of motion T_0, time required to complete one cycle of vibration, is expressed as

$$T_0 = 2\pi\sqrt{\frac{m_0}{k_0}} \tag{41.2}$$

Addition of a seismic isolator to this system can be idealized as adding a spring with spring constant k_i and a viscous damper with damping coefficient c_i, as shown in Figure 41.5. The combined stiffness of the isolated system now becomes

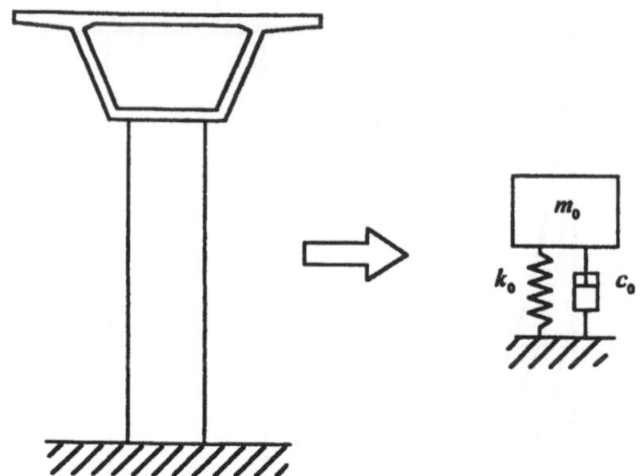

FIGURE 41.4 SDOF dynamic model.

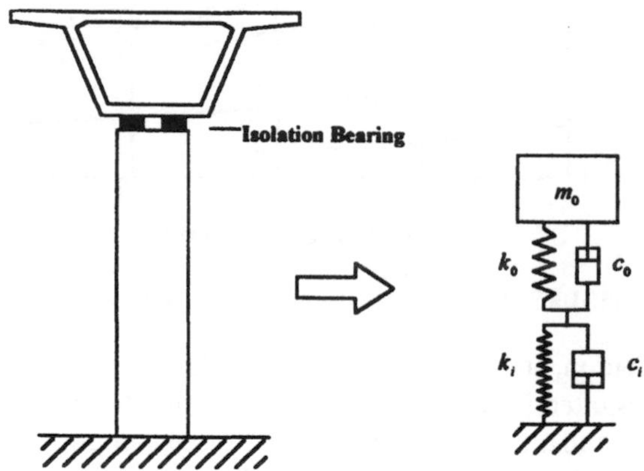

FIGURE 41.5 SDOF system with seismic isolator.

$$K = \frac{k_0 k_i}{k_0 + k_i} \tag{41.3}$$

Equation (41.1) is modified to

$$m_0 \ddot{x} + (c_0 + c_i)\dot{x} + Kx = -m_0 \ddot{x}_g \tag{41.4}$$

and the natural period of vibration of the isolated system becomes

$$T = 2\pi\sqrt{\frac{m_0}{K}} = 2\pi\sqrt{\frac{m_0(k_0 + k_i)}{k_0 k_i}} \tag{41.5}$$

When the isolator stiffness is smaller than the structural stiffness, K is smaller than k_0; therefore, the natural period of the isolated system T is longer than that of the original system. It is of interest to note that, in order for the isolator to be effective in modifying the the natural period of the structure, k_i should be smaller than k_0 to a certain degree. For example, if k_i is 50% of k_0, then T will be about 70% larger than T_0. If k_i is only 10% of k_0, then T will be more than three times of T_0.

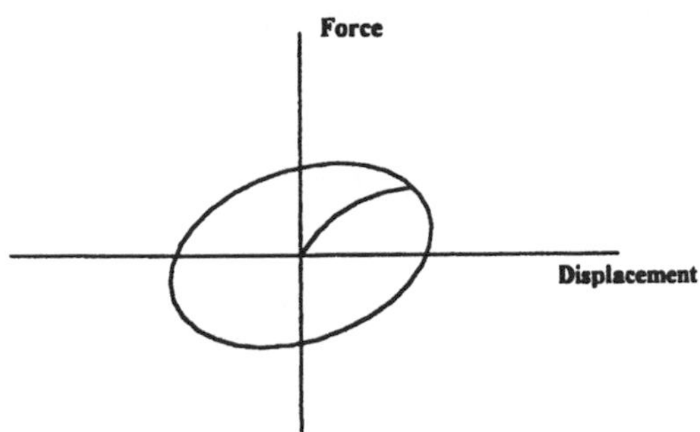

FIGURE 41.6 Generic damper hysteresis loops.

More complex structural systems will have to be treated as multiple-degree-of-freedom (MDOF) systems; however, the principle is the same. In these cases, spring elements will be added to appropriate locations to model the stiffness of the isolators.

41.2.4 Effect of Energy Dissipation on Structural Dynamic Response

In discussing energy dissipation, the terms *damping* and *energy dissipation* will be used interchangeably. Consider again the simple SDOF system used in the previous discussion. In the theory of structural dynamics [8], critical value of damping coefficient c_c is defined as the amount of damping that will prevent a dynamic system from free oscillation response. This critical damping value can be expressed in terms of the system mass and stiffness:

$$c_c = 2\sqrt{m_0 k_o} \tag{41.6}$$

With respect to this critical damping coefficient, any amount of damping can now be expressed in a relative term called damping ratio ξ, which is the ratio of actual system damping coefficient over the critical damping coefficient. Thus,

$$\xi = \frac{c_0}{c_c} = \frac{c_0}{2\sqrt{m_0 k_0}} \tag{41.7}$$

Damping ratio is usually expressed as a percentage of the critical. With the use of damping ratio, one can compare the amount of damping of different dynamic systems.

Now consider the addition of an energy dissipation device. This device generates a force $f(x, \dot{x})$ that may be a function of displacement or velocity of the system, depending on the energy dissipation mechanism. Figure 41.6 shows a hysteresis curve for a generic energy dissipation device. Equation (41.1) is rewritten as

$$x + \frac{c_0}{m_0} x + \frac{k_0}{m_0} x + \frac{f(x, x)}{m_0} = -x_g \tag{41.8}$$

There are different approaches to modeling the effects damping devices have on the dynamic response of a structure. The most accurate approach is linear or nonlinear time history analysis by modeling the true behavior of the damping device. For practical applications, however, it will often be accurate enough to represent the effectiveness of a damping mechanism by an equivalent viscous damping ratio. One way to define the equivalent damping ratio is in terms of energy E_d dissipated by the device in one cycle of cyclic motion over the maximum strain energy E_{ms} stored in the structure [8]:

$$\xi_{eq} = \frac{E_d}{4_\pi E_{ms}} \tag{41.9}$$

For a given device, E_d can be found by measuring the area of the hysteresis loop. Equation (41.9) can now be rewritten by introducing damping ratio ξ_0 and ξ_{eq}, in the form

$$\ddot{x} + 2\sqrt{\frac{k_0}{m}}\left(\xi + \xi_{eq}\right)\dot{x} + \frac{k_0}{m_0}x = -\ddot{x}_g \tag{41.10}$$

This concept of equivalent viscous damping ratio can also be generalized to use for MDOF systems by considering ξ_{eq} as modal damping ratio and E_d and E_{ms} as dissipated energy and maximum strain energy in each vibration mode [9]. Thus, for the *i*th vibration mode of a structure, we have

$$\xi^i_{eq} = \frac{E^i_d}{4\pi E^i_{ms}} \tag{41.11}$$

Now the dynamic response of a structure with supplemental damping can be solved using available linear analysis techniques, be it linear time history analysis or response spectrum analysis.

41.3 Seismic Isolation and Energy Dissipation Devices

Many different types of seismic isolation and supplemental energy dissipation devices have been developed and tested for seismic applications over the last three decades, and more are still being investigated. Their basic behaviors and applications for some of the more widely recognized and used devices will be presented in this section.

41.3.1 Elastomeric Isolators

Elastomeric isolators, in their simplest form, are elastomeric bearings made from rubber, typically in cylindrical or rectangular shapes. When installed on bridge piers or abutments, the elastomeric bearings serve both as vertical bearing devices for service loads and lateral isolation devices for seismic load. This requires that the bearings be stiff with respect to vertical loads but relatively flexible with respect to lateral seismic loads. In order to be flexible, the isolation bearings have to be made much thicker than the elastomeric bearing pads used in conventional bridge design. Insertion of horizontal steel plates, as in the case of steel reinforced elastomeric bearing pads, significantly increases vertical stiffness of the bearing and improves stability under horizontal loads. The total rubber thickness influences essentially the maximum allowable lateral displacement and the period of vibration.

For a rubber bearing with given bearing area *A*, shear modulus *G*, height *h*, allowable shear strain γ, shape factor *S*, and bulk modulus *K*, its horizontal stiffness and period of vibration can be expressed as

$$K = \frac{GA}{h} \tag{41.12}$$

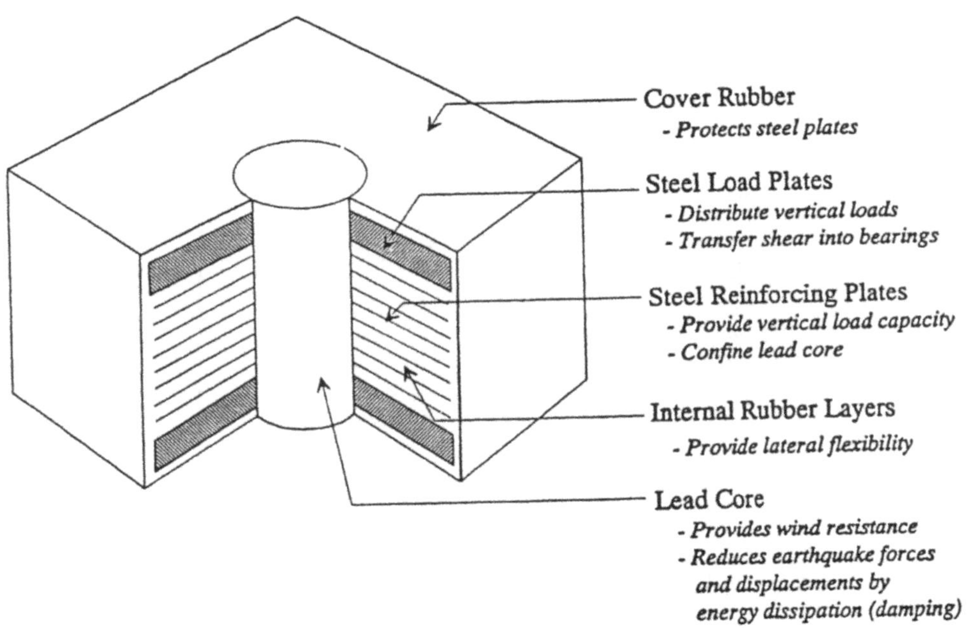

FIGURE 41.7 Typical construction of a lead core rubber bearing.

$$T_b = 2\pi\sqrt{\frac{M}{K}} = 2\pi\sqrt{\frac{Sh\gamma A'}{Ag}} \tag{41.13}$$

where A' is the overlap of top and bottom areas of a bearing at maximum displacement. Typical values for bridge elastomeric bearing properties are $G = 1$ MPa (145 psi), $K = 200$ MPa (290 psi), $\gamma = 0.9$ to 1.4, $S = 3$ to 40. The major variability lies in S, which is a function of plan dimension and rubber layer thickness.

One problem associated with using pure rubber bearings for seismic isolation is that the bearing could easily experience excessive deformation during a seismic event. This will, in many cases, jeopardize the stability of the bearing and the superstructure it supports. One solution is to add an energy dissipation device or mechanism to the isolation bearing. The most widely used energy dissipation mechanism in elastomeric isolation bearing is the insertion of a lead core at the center of the bearing. Lead has a high initial shear stiffness and relatively low shear yielding strength. It essentially has elastic–plastic behavior with good fatigue properties for plastic cycles. It provides a high horizontal stiffness for service load resistance and a high energy dissipation for strong seismic load, making it ideal for use with elastomeric bearings.

This type of lead core elastomeric isolation, also known as lead core rubber bearing (LRB), was developed and patented by the Dynamic Isolation System (DIS). The construction of a typical lead core elastomeric bearing is shown in Figure 41.7. An associated hysteresis curve is shown in Figure 41.8. Typical bearing sizes and their load bearing capacities are given in Table 41.1 [7].

Lead core elastomeric isolation bearings are the most widely used isolation devices in bridge seismic design applications. They have been used in the seismic retrofit and new design in hundreds of bridges worldwide.

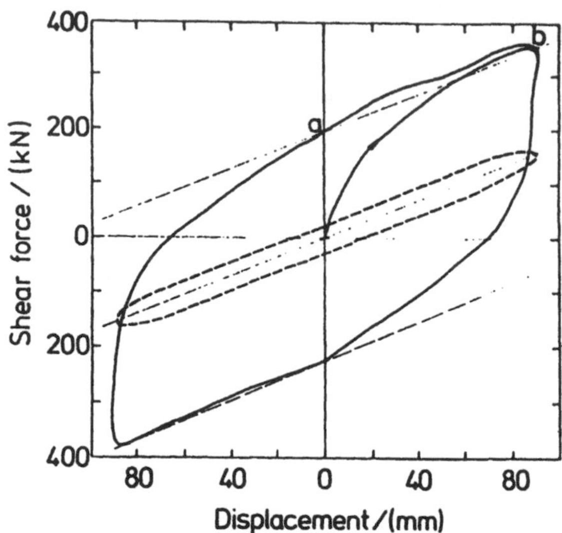

FIGURE 41.8 Hysteresis loops of lead core rubber bearing.

41.3.2 Sliding Isolators

Sliding-type isolation bearings reduce the force transferred from superstructure to the supporting substructure when subject to earthquake excitations by allowing the superstructure to slide on a low friction surface usually made from stainless steel-PTFE. The maximum friction between the sliding surfaces limits the maximum force that can be transferred by the bearing. The friction between the surfaces will also dissipate energy. A major concern with relying only on simple sliding bearings for seismic application is the lack of centering force to restore the structure to its undisplaced position together with poor predictability and reliability of the response. This can be addressed by combining the slider with spring elements or, as in the case of friction pendulum isolation (FPI) bearings, by making the sliding surface curved such that the self-weight of the structure will help recenter the superstructure. In the following, the FPI bearings by Earthquake Protection Systems (EPS) will be presented as a representative of sliding-type isolation bearings.

The FPI bearing utilizes the characteristics of a simple pendulum to lengthen the natural period of an isolated structure. Typical construction of an FPI bearing is shown in Figure 41.9. It basically consists of a slider with strength-bearing spherical surface and a treated spherical concave sliding surface housed in a cast steel bearing housing. The concave surface and the surface of the slider have the same radius to allow a good fit and a relatively uniform pressure under vertical loads. The operation of the isolator is the same regardless of the direction of the concave surface. The size of the bearing is mainly controlled by the maximum design displacement.

The concept is really a simple one, as illustrated in Figure 41.10. When the superstructure moves relative to the supporting pier, it behaves like a simple pendulum. The radius, R, of the concave surface controls the isolator period,

$$T = 2\pi\sqrt{\frac{R}{g}} \tag{41.14}$$

where g is the acceleration of gravity. The fact that the isolator period is independent of the mass of the supported structure is an advantage over the elastomeric isolators because fewer factors are involved in selecting an isolation bearing. For elastomeric bearings, in order to lengthen the period

TABLE 41.1 Total Dead Plus Live-Load Capacity of Square DIS Bearings (kN)

Plan Size		Bonded Area	Rubber Layer Thickness, mm			
W(mm)	*B*(mm)	(mm^2)	6.5	9.5	12.5	19
229	229	52,258	236	160	125	85
254	254	64,516	338	227	173	120
279	279	78,064	463	311	236	165
305	305	92,903	614	414	311	214
330	330	109,032	796	534	405	276
356	356	126,451	1,010	676	512	351
381	381	145,161	1,263	845	641	436
406	406	165,161	1,552	1,041	783	529
432	432	186,451	1,882	1,259	952	641
457	457	209,032	2,255	1,508	1,139	770
483	483	232,903	2,678	1,793	1,348	912
508	508	258,064	3,149	2,104	1,583	1,068
533	533	284,516	3,674	2,455	1,846	1,241
559	559	312,257	4,252	2,842	2,135	1,437
584	584	341,290	4,888	3,265	2,455	1,650
610	610	371,612	5,582	3,727	2,802	1,882
635	635	403,225	6,343	4,234	3,185	2,135
660	660	436,128	7,170	4,786	3,598	2,411
686	686	470,322	8,064	5,382	4,043	2,713
711	711	505,805	9,029	6,027	4,528	3,034
737	737	542,580	10,070	6,721	5,048	3,380
762	762	580,644	11,187	7,464	5,609	3,754
787	787	619,999	12,383	8,264	6,205	4,154
813	813	660,644	13,660	9,118	6,845	4,581
838	838	702,579	15,025	10,026	7,530	5,040
864	864	745,805	16,480	10,995	8,255	5,524
889	889	790,321	18,023	12,023	9,029	6,040
914	914	836,127	19,660	13,117	9,848	6,587

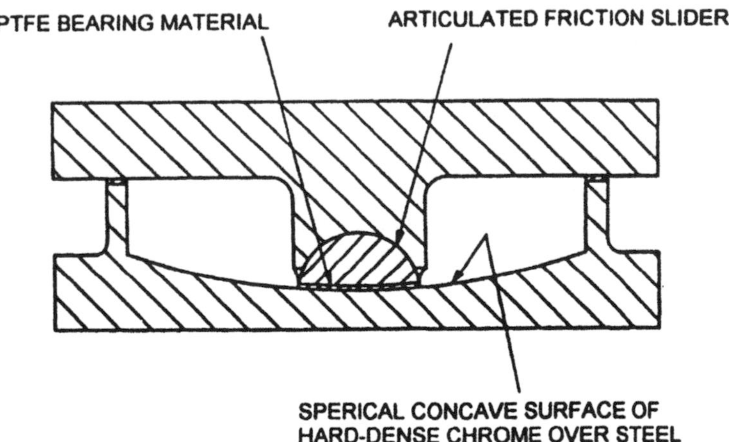

FIGURE 41.9 Typical construction of a FPI.

of an isolator without varying the plan dimensions, one has to increase the height of the bearing which is limited by stability requirement. For FPI bearings, one can vary the period simply by changing the radius of the concave surface. Another advantage the FPI bearing has is high vertical load-bearing capacity, up to 30 million lb (130,000 kN) [10].

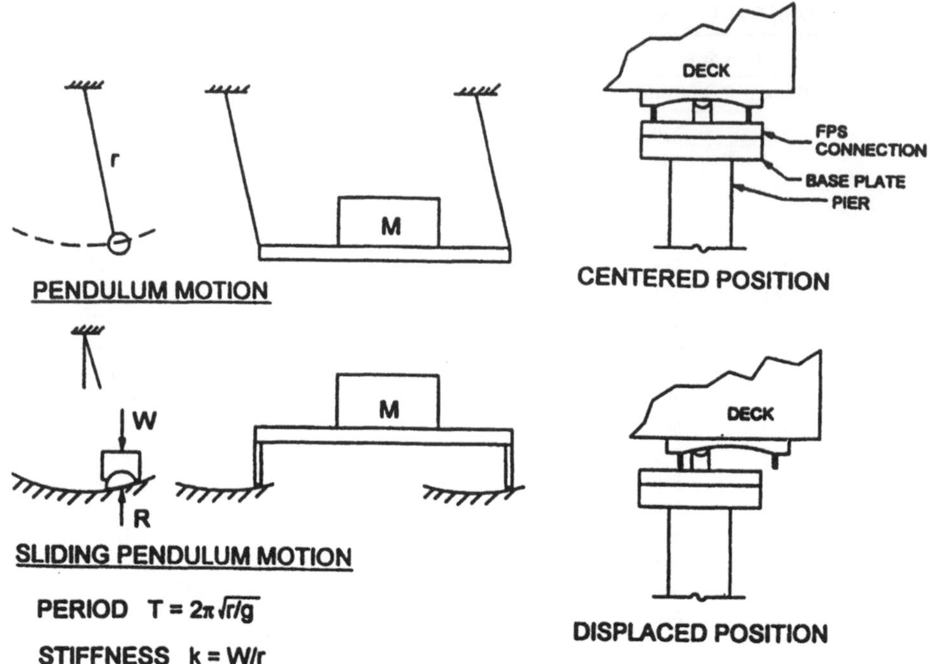

FIGURE 41.10 Basic operating principle of FPI.

The FPI system behaves rigidly when the lateral load on the structure is less than the friction force, which can be designed to be less than nonseismic lateral loads. Once the lateral force exceeds this friction force, as is the case under earthquake excitation, it will respond at its isolated period. The dynamic friction coefficient can be varied in the range of 0.04 to 0.20 to allow for different levels of lateral resistance and energy dissipation.

The FPI bearings have been used in several building seismic retrofit projects, including the U.S. Court of Appeals Building in San Francisco and the San Francisco Airport International Terminal. The first bridge structure to be isolated by FPI bearings is the American River Bridge in Folsom, California. Figure 41.11 shows one of the installed bearings on top of the bridge pier. The maximum designed bearing displacement is 250 mm, and maximum vertical load is about 16,900 kN. The largest bearings have a plan dimension of 1150 × 1150 mm. The FPI bearings will also be used in the Benicia–Martinez Bridge in California when construction starts on the retrofit of this mile-long bridge. The bearings designed for this project will have a maximum plan dimension of 4500 × 4500 mm to accommodate a maximum designed displacement of 1200 mm [11].

41.3.3 Viscous Fluid Dampers

Viscous fluid dampers, also called hydraulic dampers in some of the literature, typically consist of a piston moving inside the damper housing cylinder filled with a compound of silicone or oil. Figure 41.12 shows typical construction of a Taylor Device's viscous fluid damper and its corresponding hysteresis curve. As the piston moves inside the damper housing, it displaces the fluid which in turn generates a resisting force that is proportional to the exponent of the velocity of the moving piston, i.e.,

$$F = cV^k \tag{41.15}$$

FIGURE 41.11 A FPI bearing installed on a bridge pier.

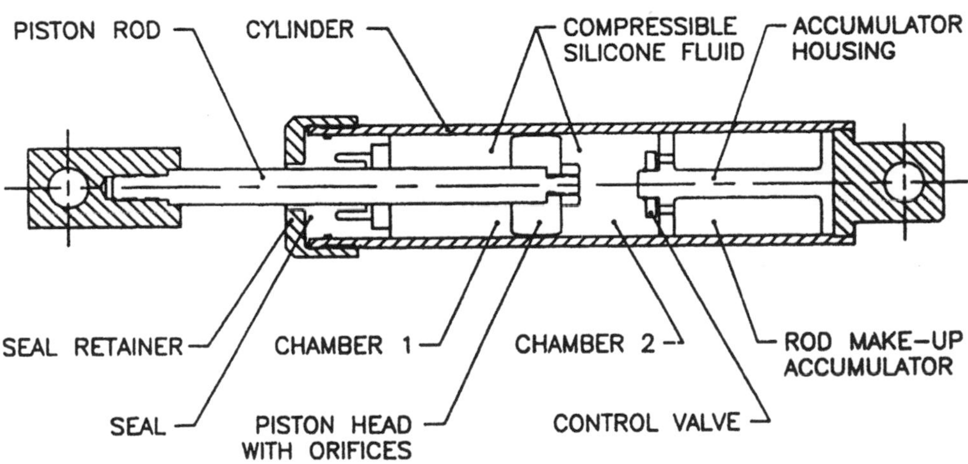

FIGURE 41.12 Typical construction of a taylor devices fluid viscous damper.

where c is the damping constant, V is the velocity of the piston, and k is a parameter that may be varied in the range of 0.1 to 1.2, as specified for a given application. If k equals 1, we have a familiar linear viscous damping force. Again, the effectiveness of the damper can be represented by the amount of energy dissipated in one complete cycle of deformation:

$$E_d = \int F dx \tag{41.16}$$

The earlier applications of viscous fluid dampers were in the vibration isolation of aerospace and defense systems. In recent years, theoretical and experimental studies have been performed in an effort to apply the viscous dampers to structure seismic resistant design [4,12]. As a result, viscous

TABLE 41.2 Fluid Viscous Damper Dimension Data (mm)

Model	A	B	C	D	E	F
100 kips (445 kN)	3327	191	64	81	121	56
200 kips (990 kN)	3353	229	70	99	127	61
300 kips (1335 kN)	3505	292	76	108	133	69
600 kips (2670 kN)	3937	406	152	191	254	122
1000 kips (4450 kN)	4216	584	152	229	362	122
2000 kips (9900 kN)	4572	660	203	279	432	152

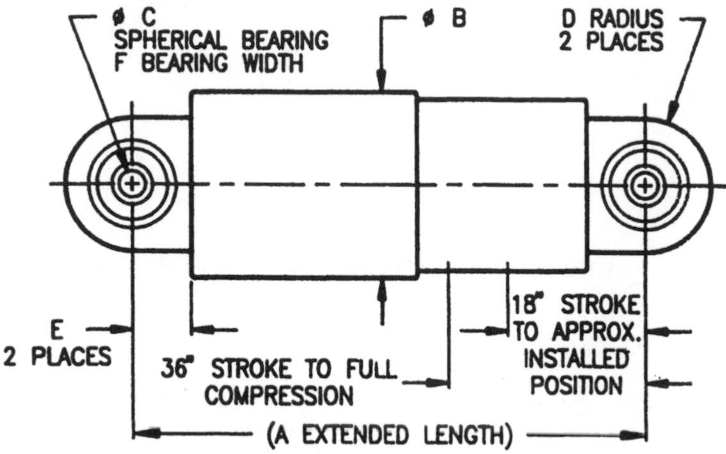

FIGURE 41.13 Viscous damper dimension.

dampers have found applications in several seismic retrofit design projects. For example, they have been considered for the seismic upgrade of the Golden Gate Bridge in San Francisco [13], where viscous fluid dampers may be installed between the stiffening truss and the tower to reduce the displacement demands on wind-locks and expansion joints. The dampers are expected to reduce the impact between the stiffening truss and the tower. These dampers will be required to have a maximum stroke of about 1250 mm, and be able to sustain a peak velocity of 1880 mm/s. This requires a maximum force output of 2890 kN.

Fluid viscous dampers are specified by the amount of maximum damping force output as shown in Table 41.2 [14]. Also shown in Table 41.2 are dimension data for various size dampers that are typical for bridge applications. The reader is referred to Figure 41.13 for dimension designations.

41.3.4 Viscoelastic Dampers

A typical viscoelastic damper, as shown in Figure 41.14, consists of viscoelastic material layers bonded with steel plates. Viscoelastic material is the general name for those rubberlike polymer materials having a combined feature of elastic solid and viscous liquid when undergoing deformation. Figure 41.14 also shows a typical hysteresis curve of viscoelastic dampers. When the center plate moves relative to the two outer plates, the viscoelastic material layers undergo shear deformation. Under a sinusoidal cyclic loading, the stress in the viscoelastic material can be expressed as

$$\sigma = \gamma_0 \left(G' \sin \omega t + G'' \cos \omega t \right) \tag{41.17}$$

where γ_0 represents the maximum strain, G' is shear storage modulus, and G'' is the shear loss modulus, which is the primary factor determining the energy dissipation capability of the viscoelastic material.

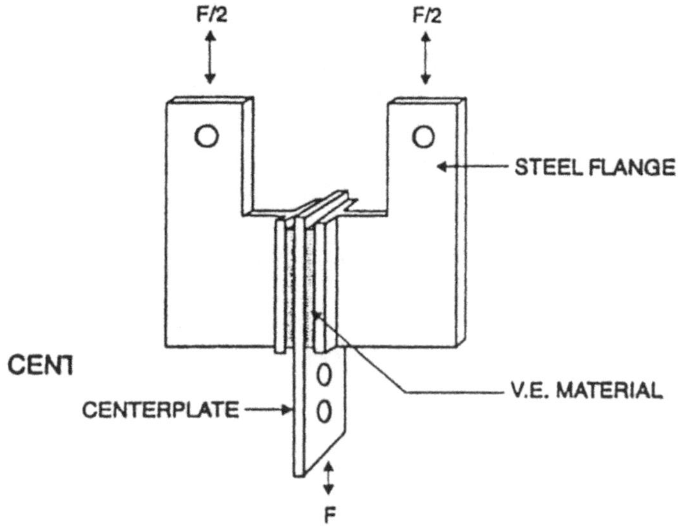

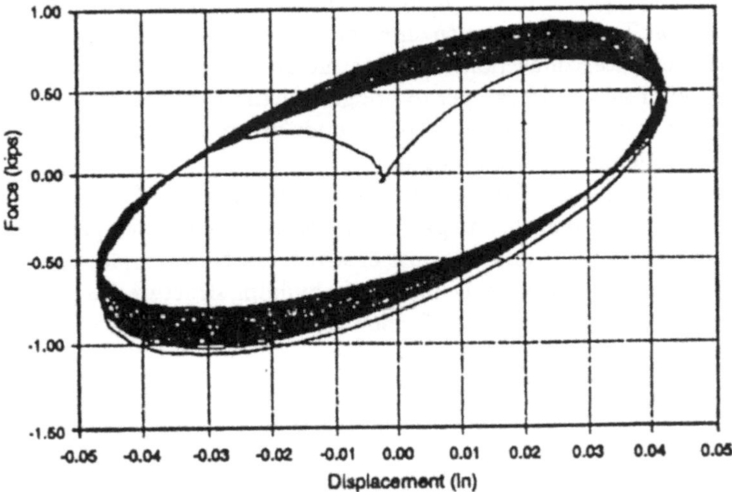

FIGURE 41.14 Typical viscoelastic damper and its hysteresis loops.

After one complete cycle of cyclic deformation, the plot of strain vs. stress will look like the hysteresis shown in Figure 41.14. The area enclosed by the hysteresis loop represents the amount of energy dissipated in one cycle per unit volume of viscoelastic material:

$$e_d = \pi\gamma_0^2 G' \tag{41.18}$$

The total energy dissipated by viscoelastic material of volume V can be expressed as

$$E_d = \pi\gamma_0^2 G'' V \tag{41.19}$$

The application of viscoelastic dampers to civil engineering structures started more than 20 years ago, in 1968, when more than 20,000 viscoelastic dampers made by the 3M Company were installed in the twin-frame structure of the World Trade Center in New York City to help resist wind load.

The application of viscoelastic dampers to civil engineering structures started more than 20 years ago, in 1968, when more than 20,000 viscoelastic dampers made by the 3M Company were installed in the twin-frame structure of the World Trade Center in New York City to help resist wind load. In the late 1980s, theoretical and experimental studies were first conducted for the possibility of applying viscoelastic dampers for seismic applications [9,15]. Viscoelastic dampers have since received increased attention from researchers and practicing engineers. Many experimental studies have been conducted on scaled and full-scale structural models. Recently, viscoelastic dampers were used in the seismic retrofit of several buildings, including the Santa Clara County Building in San Jose, California. In this case, viscoelastic dampers raised the equivalent damping ratio of the structure to 17% of critical [16].

41.3.5 Other Types of Damping Devices

There are several other types of damping devices that have been studied and applied to seismic resistant design with varying degrees of success. These include metallic yield dampers, friction dampers, and tuned mass dampers. Some of them are more suited for building applications and may be of limited effectiveness to bridge structures.

Metallic Yield Damper. Controlled use of sacrificial metallic energy dissipating devices is a relatively new concept [17]. A typical device consists of one or several metallic members, usually made of mild steel, which are subjected to axial, bending, or torsional deformation depending on the type of application. The choice between different types of metallic yield dampers usually depends on location, available space, connection with the structure, and force and displacement levels. One possible application of steel yield damper to bridge structures is to employ steel dampers in conjunction with isolation bearings. Tests have been conducted to combine a series of cantilever steel dampers with PTFE sliding isolation bearing.

Friction Damper. This type of damper utilizes the mechanism of solid friction that develops between sliding surfaces to dissipate energy. Several types of friction dampers have been developed for the purpose of improving seismic response of structures. For example, studies have shown that slip joints with friction pads placed in the braces of a building structure frame significantly reduced its seismic response. This type of braced friction dampers has been used in several buildings in Canada for improving seismic response [4,18].

Tuned Mass Damper. The basic principle behind tuned mass dampers (TMD) is the classic dynamic vibration absorber, which uses a relatively small mass attached to the main mass via a relatively small stiffness to reduce the vibration of the main mass. It can be shown that, if the period of vibration of the small mass is tuned to be the same as that of the disturbing harmonic force, the main mass can be kept stationary. In structural applications, a tuned mass damper may be installed on the top floor to reduce the response of a tall building to wind loads [4]. Seismic application of TMD is limited by the fact that it can only be effective in reducing vibration in one mode, usually the first mode.

41.4 Performance and Testing Requirements

Since seismic isolation and energy dissipation technologies are still relatively new and often the properties used in design can only be obtained from tests, the performance and test requirements are critical in effective applications of these devices. Testing and performance requirements, for the most part, are prescribed in project design criteria or construction specifications. Some nationally recognized design specifications, such as AASHO *Guide Specifications for Seismic Isolation Design* [19], also provide generic testing requirements.

Almost all of the testing specified for seismic isolators or energy dissipation devices require tests under static or simple cyclic loadings only. There are, however, concerns about how well will properties obtained from these simple loading tests correlate to behaviors under real earthquake loadings. Therefore, a major earthquake simulation testing program is under way. Sponsored by the Federal Highway Administration and the California Department of Transportation, manufacturers of isolation and energy dissipation devices were invited to provide their prototype products for testing under earthquake loadings. It is hoped that this testing program will lead to uniform guidelines for prototype and verification testing as well as design guidelines and contract specifications for each of the different systems. The following is a brief discussion of some of the important testing and performance requirements for various systems.

41.4.1 Seismic Isolation Devices

For seismic isolation bearings, performance requirements typically specify the maximum allowable lateral displacements under seismic and nonseismic loadings, such as thermal and wind loads; horizontal deflection characteristics such as effective and maximum stiffnesses; energy dissipation capacity, or equivalent damping ratio; vertical deflections; stability under vertical loads; etc. For example, the AASHTO *Guide Specifications for Seismic Isolation Design* requires that the design and analysis of isolation system prescribed be based on prototype tests and a series of verification tests as briefly described in the following:

Prototype Tests:

I. Prototype tests need to be performed on two full-size specimens. These tests are required for each type and size similar to that used in the design.
II. For each cycle of tests, the force–deflection and hysteresis behavior of the specimen need to be recorded.
III. Under a vertical load similar to the typical average design dead load, the specimen need to be tested for
   A. Twenty cycles of lateral loads corresponding to the maximum nonseismic loads;
   B. Three cycles of lateral loading at displacements equaling 25, 50, 75, 100, and 125% of the total design displacement;
   C. Not less than 10 full cycles of loading at the total design displacement and a vertical load similar to dead load.
IV. The stability of the vertical load-carrying element needs to be demonstrated by one full cycle of displacement equaling 1.5 times the total design displacement under dead load plus or minus vertical load due to seismic effect.

System Characteristics Tests:

I. The force–deflection characteristics need to be based on cyclic test results.
II. The effective stiffness of an isolator needs to be calculated for each cycle of loading as

$$k_{\text{eff}} = \frac{F_p - F_n}{\Delta_p - \Delta_n} \tag{41.20}$$

where F_p and F_n are the maximum positive and negative forces, respectively, and Δ_p and Δ_n are the maximum positive and negative displacements, respectively.

III. The equivalent viscous damping ratio ξ of the isolation system needs to be calculated as

$$\xi = \frac{\text{Total Area}}{4\pi \sum \frac{kd^2}{2}} \tag{41.21}$$

where Total Area shall be taken as the sum of areas of the hysteresis loops of all isolators; the summation in the denominator represents the total strain energy in the isolation system.

In order for a specimen to be considered acceptable, the results of the tests should show positive incremental force-carrying capability, less than a specified amount of variation in effective stiffness between specimens and between testing cycles for any given specimen. The effective damping ratio also needs to be within certain range [19].

41.4.2 Testing of Energy Dissipation Devices

As for energy dissipation devices, there have not been any codified testing requirements published. The Federal Emergency Management Agency 1994 NEHRP Recommended Provisions for Seismic Regulation for New Buildings contain an appendix that addresses the use of energy dissipation systems and testing requirements [20]. There are also project-specific testing requirements and proposed testing standards by various damper manufacturers.

Generally speaking, testing is needed to obtain appropriate device parameters for design use. These parameters include the maximum force output, stroke distance, stiffness, and energy dissipation capability. In the case of viscous dampers, these are tested in terms of damping constant C, exponential constant, maximum damping force, etc. Most of the existing testing requirements are project specific. For example, the technical requirements for viscous dampers to be used in the retrofit of the Golden Gate Bridge specify a series of tests to be carried out on model dampers [13,21]. Prototype tests were considered to be impractical because of the limitation of available testing facilities. These tests include cyclic testing of model dampers to verify their constitutive law and longevity of seals and a drop test of model and prototype dampers to help relate cyclic testing to the behavior of the actual dampers. Because the tests will be on model dampers, some calculations will be required to extrapolate the behavior of the prototype dampers.

41.5 Design Guidelines and Design Examples

In the United States, design of seismic isolation for bridges is governed by the *Guide Specifications for Seismic Isolation Design* (hereafter known as "Guide Specifications") published by AASHTO in 1992. Specifications for the design of energy dissipation devices have not been systematically developed, while recommended guidelines do exist for building-type applications.

In this section, design procedure for seismic isolation design and a design example will be presented mainly based on the AASHTO Guide Specifications. As for the design of supplemental energy dissipation, an attempt will be made to summarize some of the guidelines for building-type structures and their applicability to bridge applications.

41.5.1 Seismic Isolation Design Specifications and Examples

The AASHTO Guide Specifications were written as a supplement to the AASHTO *Standard Specifications for Highway Bridges* [22] (hereafter known as "Standard Specifications"). Therefore, the seismic performance categories and site coefficients are identical to those specified in the Standard Specifications. The response modification factors are the same as in the Standard Specifications except that a reduced R factor of 1.5 is permitted for essentially elastic design when the design intent of seismic isolation is to eliminate or significantly reduce damage to the substructure.

General Requirements

There are two interrelated parts in designing seismic isolation devices for bridge applications. First of all, isolation bearings must be designed for all nonseismic loads just like any other bearing devices. For example, for lead core rubber isolation bearings, both the minimum plan size and the thickness of individual rubber layers are determined by the vertical load requirement. The minimum isolator

TABLE 41.3 Damping Coefficient B

Damping Ratio (ξ)	$\leq$ 2%	5%	10%	20%	30%
B	0.8	1.0	1.2	1.5	1.7

Source: AASHTO, *Guide Specification for Seismic Isolation Design*, Washington, D.C., 1991. With permission.

height is controlled by twice the displacement due to combined nonseismic loads. The minimum diameter of the lead core is determined by the requirement to maintain elastic response under combined wind, brake, and centrifugal forces. Similar requirements can also be applied to other types of isolators. In addition to the above requirements, the second part of seismic isolation design is to satisfy seismic safety requirements. The bearing must be able to support safely the vertical loads at seismic displacement. This second part is accomplished through the analysis and design procedures described below.

Methods of Analysis

The Guide Specifications allow treatment of energy dissipation in isolators as equivalent viscous damping and stiffness of isolated systems as effective linear stiffness. This permits both the single and multimodal methods of analysis to be used for seismic isolation design. Exceptions to this are isolated systems with damping ratios greater than 30% and sliding type of isolators without a self-centering mechanism. Nonlinear time history analysis is required for these cases.

Single-Mode Spectral Analysis

In this procedure, equivalent static force is given by the product of the elastic seismic force coefficient C_s and dead load W of the superstructure supported by isolation bearings, i.e.,

$$F = C_s W \tag{41.22}$$

$$C_s = \frac{\sum k_{\text{eff}} \times d_i}{W} \tag{41.23}$$

$$C_s = \frac{AS_i}{T_e B} \tag{41.24}$$

where

$\sum k_{\text{eff}}$ = the sum of the effective linear stiffness of all bearings supporting the superstructure

$d_i = \dfrac{10AS_iT_e}{B}$ = displacement across the isolation bearings

A = the acceleration coefficient

B = the damping coefficient given in Table 41.3

$T_e = \sqrt{\dfrac{W}{g\sum k_{\text{eff}}}}$ = the period of vibration

The equivalent static force must be applied independently to the two orthogonal axes and combined per the procedure of the standard specifications. The effective linear stiffness should be calculated at the design displacement.

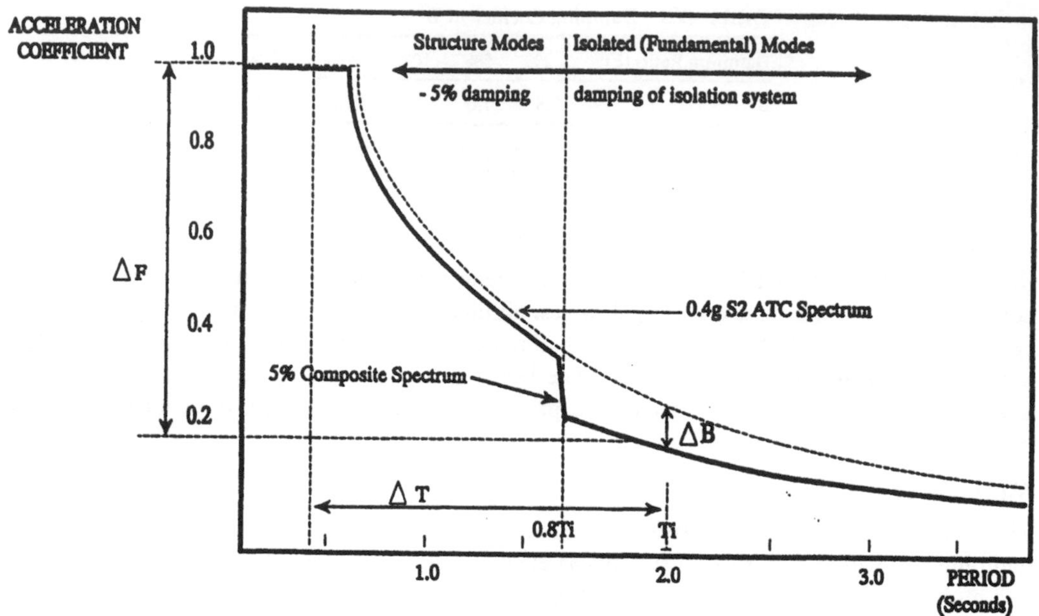

FIGURE 41.15 Modified input response spectrum.

Response Spectrum Analysis

This procedure is the same as specified in the Standard Specifications using the 5% damping ground motion response spectra with the following modifications:

1. The isolation bearings are represented by their effective stiffness values.
2. The response spectrum is modified to include the effect of higher damping of the isolated system. This results in a reduction of the response spectra values for the isolated modes. For all the other modes, the 5% damping response spectra should be used.

A typical modified response spectrum is shown in Figure 41.15.

Time History Analysis

As mentioned earlier, time history analysis is required for isolation systems with high damping ratio (>30%) or non-self-centering isolation systems. The isolation systems need to be modeled using nonlinear force–deflection characteristics of the isolator obtained from tests. Pairs of ground acceleration time history recorded from different events should be selected. These acceleration time histories should be frequency-scaled to match closely the appropriate response spectra for the site. Recommended methods for scaling are also given in the Guide Specifications. At least three pairs of time histories are required by the code. Each pair should be simultaneously applied to the model. The maximum response should be used for the design.

Design Displacement and Design Force

It is necessary to know and limit the maximum displacement of an isolation system resulting from seismic loads and nonseismic service loads for providing adequate clearance and design structural elements. The Guide Specifications require that the total design displacement be the greater of 50% of the elastomer shear strain in an elastomeric bearing system and the maximum displacement resulted from the combination of loads specified in the Standard Specifications.

Design forces for a seismically isolated bridge are obtained using the same load combinations as given for a conventionally designed bridge. Connection between superstructure and substructure shall be designed using force $F = k_{eff} d_i$. Columns and piers should be designed for the maximum force that may be developed in the isolators. The foundation design force needs not to exceed the elastic force nor the force resulted from plastic hinging of the column.

Other Requirements

It is important for an isolation system to provide adequate rigidity to resist frequently occurring wind, thermal, and braking loads. The appropriate allowed lateral displacement under nonseismic loads is left for the design engineer to decide. On the lateral restoring force, the Guide Specifications require a restoring force that is 0.25W greater than the lateral force at 50% of the design displacement. For systems not configured to provide a restoring force, more stringent vertical stability requirements have to be met.

The Guide Specifications recognize the importance of vertical stability of an isolated system by requiring a factor of safety not less than three for vertical loads in its undeformed state. A system should also be stable under the dead load plus or minus the vertical load due to seismic load at a horizontal displacement of 1.5 times the total design displacement. For systems without a lateral restoring force, this requirement is increased to three times the total design displacement.

Guidelines for Choosing Seismic Isolation

What the Guide Specifications do not cover are the conditions under which the application of seismic isolation becomes necessary or most effective. Still, some general guidelines can be drawn from various literatures and experiences as summarized below.

One factor that favors the use of seismic isolation is the level of acceptable damage to the bridge. Bridges at critical locations need to stay open to traffic following a seismic event with no damage or minor damages that can be quickly repaired. This means that the bridges are to be essentially designed elastically. The substructure pier and foundation cost could become prohibitive if using conventional design. The use of seismic isolation may be an economic solution for these bridges, if not the only solution. This may apply to both new bridge design and seismic upgrade of existing bridges.

Sometimes, it is desirable to reduce the force transferred to the superstructure, as in the case of seismic retrofit design of the Benicia–Martinez Bridge, in the San Francisco Bay, where isolation bearings were used to limit the forces in the superstructure truss members [11].

Another factor to consider is the site topography of the bridge. Irregular terrain may result in highly irregular structure configurations with significant pier height differences. This will result in uneven seismic force distributions among the piers and hence concentrated ductility demands. Use of seismic isolation bearings will make the effective stiffness and expected displacement of piers closer to each other resulting in a more even force distribution [23].

For seismic upgrading of existing bridges, isolation bearings can be an effective solution for understrength piers, insufficient girder support length, and inadequate bearings.

In some cases, there may not be an immediate saving from the use of seismic isolation over a conventional design. Considerations need to be given to a life-cycle cost comparison because the use of isolation bearings generally means much less damages, and hence lower repair costs in the long run.

Seismic Isolation Design Example

As an example, a three-span continuous concrete box-girder bridge structure, shown in Figure 41.16, will be used here to demonstrate the seismic isolation design procedure. Material and structure properties are also given in Figure 41.16. The bridge is assumed to be in a high seismic area with an acceleration coefficient A of 0.40, soil profile Type II, $S = 1.2$. For simplicity, let us use the single mode spectral analysis method for the analysis of this bridge. Assuming that the isolation bearings will be designed to provide an equivalent viscous damping of 20%, with a damping coefficient, B, of 1.5. The geometry and section properties of the bridge are taken from the worked example in the Standard Specifications with some modifications.

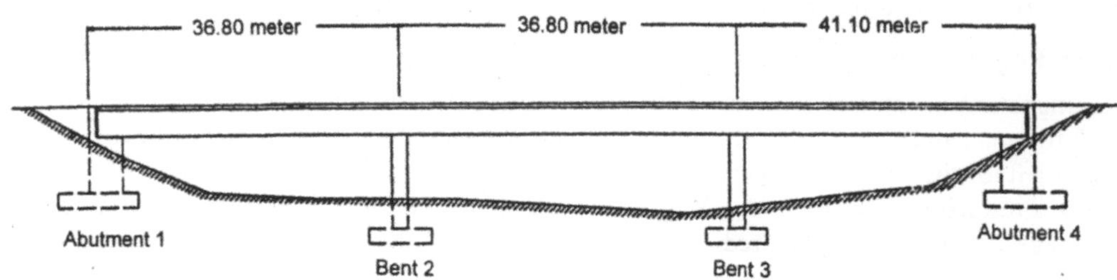

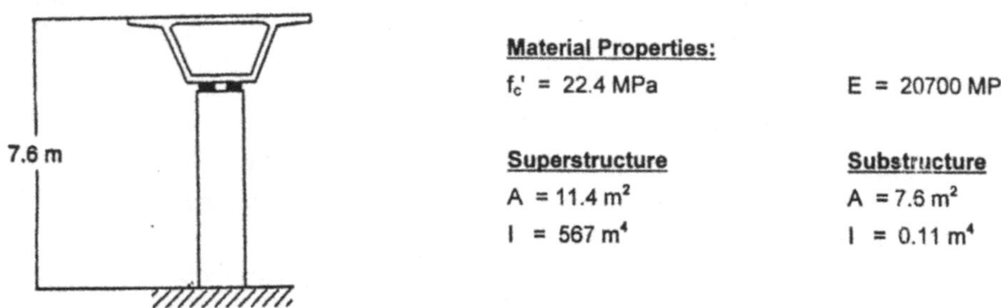

FIGURE 41.16 Example three-span bridge structure.

Force Analysis

Maximum tributary mass occurs at Bent 3, with a mass of 123 ft² × 150 lb/ft³ × 127.7 ft = 2356 kips (1,065,672 kg). Consider earthquake loading in the longitudinal direction. For fixed top of column support, the stiffness $k_0 = (12\ \mathrm{EI})/H^3 = 12 \times 432{,}000 \times 39/25^3 = 12{,}940$ kips/ft (189 kN/mm). This results in a fixed support period

$$T_0 = 2\pi\sqrt{\frac{W}{k_0 g}} = 2\pi\sqrt{\frac{2356}{12940 \times 32.3}} = 0.47\text{s}$$

The corresponding elastic seismic force

$$F_0 = C_s W = \frac{1.2AS}{T^{2/3}} = \frac{1.2 \times 0.4 \times 1.2}{0.47^{2/3}} W = 0.95W = 2238\ \text{kip}\ \ (9955\ \text{kN})$$

Now, let us assume that, with the introduction of seismic isolation bearings at the top of the columns, the natural period of the structure becomes 2.0 s, and damping $B = 1.5$. From Eqs. (41.22) and (41.24), the elastic seismic force for the isolated system,

$$F_i = C_s W = \frac{AS_i}{T_e B} W = \frac{0.4 \times 1.2}{2.0 \times 1.5} W = 0.16W = 377\ \text{kips}\ \ (1677\ \text{kN})$$

Displacement across the isolation bearing

$$d = \frac{10AS_i T_e}{B} = \frac{10 \times 0.4 \times 1.2 \times 2.0}{1.5} = 6.4\ \text{in.}\ (163\ \text{mm})$$

TABLE 41.4 Seismic Isolation Design Example Results

T_e (s)	0.5	1.0	1.5	2.0	2.5	3.0
k_{eff} (kips/in.)	306.48	76.62	34.05	19.16	12.26	8.51
(kN/mm)	(53.67)	(13.42)	(5.96)	(3.35)	(2.15)	(1.49)
d (in.)	1.60	3.20	4.80	6.40	8.00	9.60
(mm)	(40.64)	(81.28)	(121.92)	(162.56)	(203.20)	(243.84)
C_s	0.64	0.32	0.21	0.16	0.13	0.11
F_i (kip)	1507.84	753.92	502.61	376.96	301.57	251.31
(kN)	(6706.87)	(3353.44)	(2235.62)	(1676.72)	(1341.37)	(1117.81)

Table 41.4 examines the effect of isolation period on the elastic seismic force. For an isolated period of 0.5 s, which is approximately the same as the fixed support structure, the 30% reduction in elastic seismic force represents basically the effect of the added damping of the isolation system.

Isolation Bearing Design

Assume that four elastomeric (lead core rubber) bearings are used at each bent for this structure. Vertical local due to gravity load is $P = 2356/4 = 589$ kips (2620 kN). We will design the bearings such that the isolated system will have a period of 2.5 s.

$$T_e = \sqrt{\frac{W}{g\sum k_{\text{eff}}}}$$

and

$$k_{\text{eff}} = 4\left(\frac{GA}{T}\right)$$

where T is the total thickness of the elastomer. We have

$$\frac{GA}{T} = 3.06 \text{ kip/in. } (0.54 \text{ kN/mm})$$

Assuming a shear modulus $G = 145$ psi (1.0 MPa) and bearing thickness of $T = 18$ in. (457 mm) with thickness of each layer t_i equaling 0.5 in . This gives a bearing area $A = 380$ in^2 (245,070 mm^2). Hence, a plan dimension of 19.5 × 19.5 in. (495 × 495 mm).

Check shape factor:

$$S = \frac{ab}{2t_i(a+b)} = \frac{19.5 \times 19.5}{2 \times 0.5(19.5+19.5)} \qquad \text{OK}$$

Shear strain in the elastomer is the critical characteristic for the design of elastomeric bearings. Three shear strain components make up the total shear strain; these are shear strains due to vertical compression, rotation, and horizontal shear deformation. In the Guide Specifications, the shear strain due to compression by vertical load is given by

$$\gamma_c = \frac{3SW}{2A_r G(1+2kS^2)}$$

where $A_r = 19.5 \times (19.5\text{ in.}–8.0\text{ in.}) = 224.3\text{ in.}^2$ is the reduced bearing area representing the effective bearing area when undergoing horizontal displacement. In this case horizontal displacement is 8.0 in. For the purpose of presenting a simple example, an approximation of the previous expression can be used:

$$\gamma_c = \frac{\sigma}{GS} = \frac{589 \times 1000}{224.3 \times 145 \times 9.75} = 1.85$$

Shear strain due to horizontal shear deformation

$$\gamma_s = \frac{d}{T} = \frac{8\text{ in.}}{18\text{ in.}} = 0.44$$

and shear strain due to rotation

$$\gamma_r = \frac{B^2\theta}{2t_i T} = \frac{19.5^2 \times 0.01}{2 \times 0.5 \times 18} = 0.21$$

The Guide Specifications require that the sum of all three shear strain components be less than 50% of the ultimate shear strain of the elatomer, or 5.0, whichever is smaller. In this example, the sum of all three shear strain components equals 2.50 < 5.0.

In summary, we have designed four elastomeric bearings at each bent with a plan dimension of 19.5 × 19.5 in. (495 × 495 mm) and 36 layers of 0.5 in. elastomer with $G = 145$ psi (1 MPa).

41.5.2 Guidelines for Energy Dissipation Devices Design

There are no published design guidelines or specifications for application of damping devices to bridge structures. Several recommended guidelines for application of dampers to building structures have been in development over the last few years [20,24,25]. It is hoped that a brief summary of these developments will be beneficial to bridge engineers.

General Requirements

The primary function of an energy dissipation device in a structure is to dissipate earthquake-induced energy. No special protection against structural or nonstructural damage is sought or implied by the use of energy dissipation systems.

Passive energy dissipation systems are classified as displacement-dependent, velocity-dependent, or other. The fluid damper and viscoelastic damper as discussed in Section 41.3 are examples of the velocity-dependent energy dissipation system. Friction dampers are displacement-dependent. Different models need to be used for different classes of energy dissipation systems. In addition to increasing the energy dissipation capacity of a structure, energy dissipation systems may also alter the structure stiffness. Both damping and stiffness effects need to be considered in designing energy dissipation systems.

Analysis Procedures

The use of linear analysis procedures is limited to viscous and viscoelastic energy dissipation systems. If nonlinear response is likely or hysteretic or other energy dissipaters are to be analyzed and designed, nonlinear analysis procedure must be followed. We will limit our discussion to linear analysis procedure.

Similar to the analysis of seismic isolation systems, linear analysis procedures include three methods: linear static, linear response spectrum, and linear time history analysis.

When using the linear static analysis method, one needs to make sure that the structure, exclusive of the dampers, remains elastic, that the combined structure damper system is regular, and that effective damping does not exceed 30%. The earthquake-induced displacements are reduced due to equivalent viscous damping provided by energy dissipation devices. This results in reduced base shears in the building structure.

The acceptability of the damped structure system should be demonstrated by calculations such that the sum of gravity and seismic loads at each section in each member is less than the member or component capacity.

The linear dynamic response spectrum procedure is used for more complex structure systems, where structures are modeled as MDOF systems. Modal response quantities are reduced based on the amount of equivalent modal damping provided by supplemental damping devices.

Detailed System Requirements

Other factors that need to be considered in designing supplemental damping devices for seismic applications are environmental conditions, nonseismic lateral loads, maintenance and inspection, and manufacturing quality control.

Energy dissipation devices need to be designed with consideration given to environmental conditions including aging effect, creep, and ambient temperature. Structures incorporated with energy-dissipating devices that are susceptible to failure due to low-cycle fatigue should resist the prescribed design wind forces in the elastic range to avoid premature failure. Unlike conventional construction materials that are inspected on an infrequent basis, some energy dissipation hardware will require regular inspections. It is, therefore, important to make these devices easily accessible for routine inspection and testing or even replacement.

41.6 Recent Developments and Applications

The last few years have seen significantly increased interest in the application of seismic isolation and supplemental damping devices. Many design and application experiences have been published. A shift from safety-only-based seismic design philosophy to a safety-and-performance-based philosophy has put more emphasis on limiting structural damage by controlling structural seismic response. Therefore, seismic isolation and energy dissipation have become more and more attractive alternatives to traditional design methods. Design standards are getting updated with the new development both in theory and technology. While the Guide Specifications referenced in this chapter addresses mainly elastomeric isolation bearing, new design specifications under development and review will include provisions for more types of isolation devices [26].

41.6.1 Practical Applications of Seismic Isolation

Table 41.5 lists bridges in North America that have isolation bearings installed. This list, as long as it looks, is still not complete. By some estimates, there have been several hundred isolated bridges worldwide and the number is growing. The Earthquake Engineering Research Center (EERC) at the University of California, Berkeley keeps a complete listing of the bridges with isolation and energy dissipation devices. Table 41.5 is based on information available from the EERC Internet Web site.

41.6.2 Applications of Energy Dissipation Devices to Bridges

Compared with seismic isolation devices, the application of energy dissipation devices as an independent performance improvement measure is lagging behind. This is due, in part, to the lack of code development and limited applicability of the energy dissipation devices to bridge-type structures as discussed earlier. Table 41.6 gives a list of bridge structures with supplemental damping devices against seismic and wind loads. This table is, again, based on information available from the EERC Internet Web site.

TABLE 41.5 Seismically-Isolated Bridges in North America

Bridge	Location		Owner	Engineer	Bridge Description	Bearing Type	Design Criteria
Dog River Bridge, New, 1992	AL	Mobile Co.	Alabama Hwy = 2E Dept.	Alabama Hwy. Dept.	Three-span cont. steel plate girders	LRB (DIS/Furon)	AASHTO Category A
Deas Slough Bridge, Retrofit, 1990	BC	Richmond (Hwy. 99 over Deas Slough),	British Columbia Ministry of Trans. & Hwys.	PBK Eng. Ltd.	Three-span cont. riveted haunched steel plate girders	LRB (DIS/Furon)	AASHTO $A = 0.2g$, Soil profile, Type III
Burrard Bridge Main Spans, Retrofit, 1993	BC	Vancouver (Burrard St. over False Cr.),	City of Vancouver	Buckland & Taylor Ltd.	Side spans are simple span deck trusses; center span is a Pratt through truss	LRB (DIS/Furon)	AASHTO $A = 0.21g$, Soil profile, Type I
Queensborough Bridge, Retrofit, 1994	BC	New Westminster (over N. arm of Fraser River),	British Columbia Ministry of Trans. & Hwys.	Sandwell Eng.	High-level bridge, three-span cont. haunched steel plate girders; two-girder system with floor beams	LRB (DIS/Furon)	AASHTO $A = 0.2g$, Soil profile, Type I
Roberts Park Overhead, New, 1996	BC	Vancouver (Deltaport Extension over BC Rail tracks)	Vancouver Port Corp.	Buckland & Taylor Ltd.	Five-span continuous curved steel plate girders, three girder lines	LRB	AASHTO $A = 0.26g$, Soil profile, Type II
Granville Bridge, Retrofit, 1996	BC	Vancouver, Canada	—	—	—	FIP	—
White River Bridge, 1997 (est.)	YU	Yukon, Canada	Yukon Trans. Services	—	—	FPS	—
Sierra Pt. Overhead, Retrofit, 1985	CA	S. San Francisco (U.S. 101 over S.P. Railroad)	Caltrans	Caltrans	Longitudinal steel plate girders, trans. steel plate bent cap girders	LRB (DIS/Furon)	Caltrans $A = 0.6g$, 0 to 10 ft alluvium
Santa Ana River Bridge, Retrofit, 1986	CA	Riverside	MWDSC	Lindvall, Richter & Assoc.	Three 180 ft simple span through trusses, 10 steel girder approach spans	LRB (DIS/Furon)	ATC $A = 0.4g$, Soil profile, Type II
Eel River Bridge, Retrofit, 1987	CA	Rio Dell (U.S. 101 over Eel River)	Caltrans	Caltrans	Two 300 ft steel through truss simple spans	LRB (DIS/Furon)	Caltrans $A = 0.5g$, < 150 ft alluvium
Main Yard Vehicle Access Bridge, Retrofit, 1987	CA	Long Beach (former RR bridge over Long Beach Freeway)	LACMTA	W. Koo & Assoc., Inc.	Two 128 ft simple span steel through plate girders, steel floor beams, conc. deck	LRB (DIS/Furon)	Caltrans $A = 0.5g$, 10 to 80 ft alluvium
All-American Canal Bridge, Retrofit, 1988	CA	Winterhaven, Imperial Co. (I-8 over All-American Canal)	Caltrans	Caltrans	Cont. steel plate girders (replacing former steel deck trusses)	LRB (DIS/Furon)	Caltrans $A = 0.6g$, >150 ft alluvium
Carlson Boulevard Bridge, New, 1992	CA	Richmond (part of 23rd St. Grade Separation Project)	City of Richmond	A-N West, Inc.	Simple span multicell conc. box girder	LRB (DIS/Furon)	Caltrans $A = 0.7g$, 80 to 150 ft alluvium
Olympic Boulevard Separation, New, 1993	CA	Walnut Creek (part of the 24/680 Reconstruction Project)	Caltrans	Caltrans	Four-span cont. steel plate girders	LRB (DIS/Furon)	Caltrans $A = 0.6g$, 10 to 80 ft alluvium

Alemany Interchange, Retrofit, 1994	CA	I-280/U.S. 101 Interchange, San Francisco	Caltrans	PBQD	Single and double deck viaduct, R.C. box girders and cols., 7-cont. units	LRB (DIS/Furon)	Caltrans $A = 0.5g$, 10 to 80 ft alluvium
Route 242/I-680 Separation, Retrofit, 1994	CA	Concord (Rte. 242 SB over I-680)	Caltrans	HDR Eng., Inc.	8 ft-deep cont. prestressed conc. box girder	LRB (DIS/Furon)	Caltrans $A = 0.53g$, 80 to 150 ft alluvium
Bayshore Boulevard Overcrossing, Retrofit, 1994	CA	San Francisco (Bayshore Blvd. over U.S. 101)	Caltrans	Winzler and Kelly	Continuous welded steel plate girders	LRB (DIS/Furon)	Caltrans $A = 0.53g$, 0 to 10 ft alluvium
1st Street over Figuero, Retrofit, 1995	CA	Los Angeles	City of Los Angeles	Kercheval Engineers	Continuous steel plate girders with tapered end spans	LRB	Caltrans $A = 0.6g$, 0 to 10 ft alluvium
Colfax Avenue over L.A. River, Retrofit, 1995	CA	Los Angeles	City of Los Angeles	Kercheval Engineers	Deck truss center span flanked by short steel beam spans	LRB (DIS)	Caltrans $A = 0.5g$, 10 to 80 ft alluvium
Colfax Avenue over L.A. River, Retrofit, 1995	CA	Los Angeles	City of Los Angeles	—	—	Eradiquake (RJ Watson)	—
3-Mile Slough, Retrofit, 1997 (est.)	CA	—	Caltrans	—	—	LRB (Skellerup)	—
Rio Vista, Retrofit, 1997 (est.)	CA	—	Caltrans	—	—	LRB (Skellerup)	—
Rio Mondo Bridge, Retrofit, 1997 (est.)	CA	—	Caltrans	—	—	FPS (EPS)	—
American River Bridge City of Folsom, New, 1997 (est.)	CA	Folsom	City of Folsom	-HDR	Ten-span, 2-frame continuous concrete box girder bridge	FPS (EPS)	Caltrans $A = 0.5g$, 10 to 80 ft alluvium
GGB North Viaduct, Retrofit, 1998 (est.)	CA	—	GGBHTD	—	—	LRB	—
Benicia–Martinez Bridge Retrofit, 1998 (est.)	CA	—	Caltrans	—	—	FPS (EPS)	—
Coronado Bridge, Retrofit, 1998 (est.)	CA	—	Caltrans	—	—	HDR (not selected)	—
Saugatuck River Bridge, Retrofit, 1994	CT	Westport (I-95 over Saugatuck R.)	ConnDOT	H.W. Lochner, Inc.	Three cont. steel plate girder units of 3, 4, and 3 spans	LRB (DIS/Furon)	AASHTO $A = 0.16g$, Soil profile, Type II
Lake Saltonstall Bridge, New, 1995	CT	E. Haven & Branford (I-95 over Lake Saltonstall)	ConnDOT	Steinman Boynton Gronquist & Birdsall	Seven-span cont. steel plate girders	LRB (DIS/Furon)	AASHTO $A = 0.15g$, Soil profile, Type III
RT 15 Viaduct, 1996	CT	Hamden	ConnDOT	Boswell Engineers	—	EradiQuake (RJ Watson)	—
Sexton Creek Bridge, New, 1990	IL	Alexander Co. (IL Rte. 3 over Sexton Creek)	ILDOT	ILDOT	Three-span cont. steel plate girders	LRB (DIS/Furon)	AASHTO $A = 0.2g$, Soil profile, Type III

TABLE 41.5 (continued) Seismically-Isolated Bridges in North America

Bridge	Location		Owner	Engineer	Bridge Description	Bearing Type	Design Criteria
Cache River Bridge, Retrofit, 1991	IL	Alexander Co. (IL Rte. 3 over Cache R. Diversion Channel)	ILDOT	ILDOT	Three-span cont. steel plate girders	LRB (DIS/Furon)	AASHTO $A = 0.2g$, Soil profile, Type III
Route 161 Bridge, New, 1991	IL	St. Clair Co.	ILDOT	Hurst-Rosche Engrs., Inc.	Four-span cont. steel plate girders	LRB (DIS/Furon)	AASHTO $A = 0.14g$, Soil profile, Type III
Poplar Street East Approach, Bridge #082-0005, Retrofit, 1992	IL	E. St. Louis (carrying I-55/70/64 across Mississippi R.)	ILDOT	Sverdrup Corp. & Hsiong Assoc.	Two dual steel plate girder units supported on multicol. or wall piers; piled foundations	LRB (DIS/Furon)	AASHTO $A = 0.12g$, Soil profile, Type III
Chain-of-Rocks Road over FAP 310, New, 1994	IL	Madison Co.	ILDOT	Oates Assoc.	Four-span cont. curved steel plate girders	LRB (DIS/Furon)	AASHTO $A = 0.13g$, Soil profile, Type III
Poplar Street East Approach, Roadway B, New, 1994	IL	E. St. Louis	ILDOT	Sverdrup Corp.	Three-, four- and five-span cont. curved steel plate girder units	LRB (DIS/Furon)	AASHTO $A = 0.12g$, Soil profile, Type III
Poplar Street East Approach, Roadway C, New, 1995	IL	E. St. Louis	ILDOT	Sverdrup Corp.	Three-, four- and five-span cont. curved steel plate girder units	LRB (DIS/Furon)	AASHTO $A = 0.12g$, Soil profile, Type III
Poplar Street Bridge, Retrofit, 1995	IL	—	ILDOT	—	—	—	—
RT 13 Bridge, 1996	IL	Near Freeburg	ILDOT	Casler, Houser & Hutchison	—	EradiQuake (RJ Watson)	—
Wabash River Bridge, New, 1991	IN	Terra Haute, Vigo Co. (U.S.-40 over Wabash R = 2E)	INDOT	Gannett Flemming	Seven-span cont. steel girders	LRB (DIS/Furon)	AASHTO $A = 0.1g$, Soil profile, Type II
US-51 over Minor Slough, New, 1992	KY	Ballard Co.	KTC	KTC	Three 121 ft simple span prestressed conc. I girders with cont. deck	LRB (DIS/Furon)	AASHTO $A = 0.25g$, Soil profile, Type II
Clays Ferry Bridge, Retrofit, 1995	KY	I-75 over Kentucky R.	KTC	KTC	Five-span cont. deck truss, haunched at center two piers	LRB (DIS/Furon)	AASHTO $A = 0 =$ 2E1g, Soil profile, Type I
Main Street Bridge, Retrofit, 1993	MA	Saugus (Main St. over U.S. Rte 1)	MHD	Vanasse Hangen Brustlin, Inc.	Two-span cont. steel beams with conc. deck	LRB (DIS/Furon)	AASHTO $A = 0.17g$, Soil profile, Type I
Neponset River Bridge, New, 1994	MA	New Old Colony RR over Neponset R. between Boston and Quincy	MBTA	Sverdrup Corp.	Simple span steel through girders; double-track ballasted deck	LRB (DIS/Furon)	AASHTO $A = 0.15g$, Soil profile, Type III

South Boston Bypass Viaduct, New, 1994	MA	S. Boston	MHDCATP	DRC Consult., Inc.	Conc. deck supported with three trapez. steel box girders; 10-span cont. unit with two curved trapez. steel box girders	LRB (DIS/Furon)	AASHTO $A = 0.17g$, Soil profile, Type III
South Station Connector, New, 1994	MA	Boston	MBTA	HNTB	Curved, trapezoidal steel ox girders.	LRB (DIS)	AASHTO $A = 0.18g$, Soil profile, Type III
North Street Bridge No. K-26, Retrofit, 1995	MA	Grafton (North Street over Turnpike)	MTA	The Maguire Group Inc.	Steel beams, two-span continuous center unit flanked by simple spans.	LRB (DIS)	AASHTO $A = 0.17g$, Soil profile, Type II
Old Westborough Road Bridge, Retrofit, 1995	MA	Grafton	MTA	The Maguire Group Inc.	Steel beams, two-span continuous center unit flanked by simple spans.	LRB (DIS)	AASHTO $A = 0.17g$, Soil profile, Type I
Summer Street Bridge, Retrofit, 1995	MA	Boston (over Fort Point Channel)	MHD	STV Group	Six-span continuous steel beams	LRB (DIS)	AASHTO $A = 0.17g$, Soil profile, Type III
West Street over I-93, Retrofit, 1995	MA	Wilmington	MHD	Vanesse Hangen Brustlin,	Four-span continuous steel beams with concrete deck.	LRB (DIS)	AASHTO $A = 0.17g$, Soil profile, Type I
Park Hill over Mass. Pike (I-90), 1995	MA	Millbury	Mass Turnpike	Purcell Assoc./HNTB	—	EradiQuake (RJ Watson)	—
RT 6 Swing Bridge, 1995	MA	New Bedford	MHD	Lichtenstein	—	EradiQuake (RJ Watson)	—
Mass Pike (I-90) over Fuller & North Sts., 1996	MA	Ludlow	Mass Turnpike	Maguire/HNTB	—	EradiQuake (RJ Watson)	—
Endicott Street over RT 128 (I-95), 1996	MA	Danvers	MHD	Anderson Nichols	—	EradiQuake (RJ Watson)	—
I-93 Mass Ave. Interchange, 1996	MA	S. Boston (Central Artery (I-93)/Tunnel (I-90))	MHD	Ammann & Whitney	—	HDR (SEP, formerly Furon)	—
Holyoke/South Hadley Bridge, 1996	MA	South Hadley, MA (Reconstruct over Conn. River & Canal St.)	MHD	Bayside Eng. Assoc., Inc.	—	LRB, NRB (SEP, formerly Furon)	—
NB I-170 Bridge, New, 1991	MO	St. Louis (Metrolink Light Rail over NB I-170)	BSDA	Booker Assoc., Inc. and Horner & Shifrin	Two-span cont. steel box girder flanked by short span steel box girders	LRB (DIS/Furon)	AASHTO $A = 0.1g$, Soil profile, Type I
Ramp 26 Bridge, New, 1991	MO	St. Louis (Metrolink Light Rail over Ramp 26)	BSDA	Booker Assoc., Inc. and Horner & Shifrin	Four-span cont. haunched conc. box girder	LRB (DIS/Furon)	AASHTO $A = 0.1g$, Soil profile, Type I
Springdale Bridge, New, 1991	MO	St. Louis (Metrolink Light Rail over Springdale Rd.)	BSDA	Booker Assoc., Inc. and Horner & Shifrin	Three-span cont. haunched conc. box girder	LRB (DIS/Furon)	AASHTO $A = 0.1g$, Soil profile, Type I
SB I-170/EB I-70 Bridge, New, 1991	MO	St. Louis (Metrolink Light Rail over SB I-170/EB I-70)	BSDA	Booker Assoc., Inc. and Horner & Shifrin	Simple span steel box girder, cont. haunched conc. box girder; cont. curved steel box girder	LRB (DIS/Furon)	AASHTO $A = 0.1g$, Soil profile, Type I

TABLE 41.5 (continued) Seismically-Isolated Bridges in North America

Bridge	Location	Owner	Engineer	Bridge Description	Bearing Type	Design Criteria
UMSL Garage Bridge, New, 1991	MO St. Louis (Metrolink Light Rail over access to UMSL garage)	BSDA	Booker Assoc., Inc. and Horner & Shifrin	Three-span cont. haunched conc. box girder	LRB (DIS/Furon)	AASHTO $A = 0.1g$, Soil profile, Type I
East Campus Drive, Bridge New, 1991	MO St. Louis (Metrolink Light Rail over E. Campus Dr.)	BSDA	Booker Assoc., Inc. and Horner & Shifrin	Four-span cont. haunched conc. box girder	LRB (DIS/Furon)	AASHTO $A = 0.1g$, Soil profile, Type I
Geiger Road Bridge, New, 1991	MO St. Louis (Metrolink Light Rail over Geiger Rd.)	BSDA	Booker Assoc., Inc. and Horner & Shifrin	Equal cont. units: one tangent, one curved, four-span haunched conc. box girder	LRB (DIS/Furon)	AASHTO $A = 0.1g$, Soil profile, Type I
Hidalgo–San Rafael Distributor, New, 1995	MX Mexico (north of Mexico City)	MTB	Dr. Melchor Rodriguez Caballero	Multispan continuous curved steel box girder.	LRB (DIS/Furon)	AASHTO $A = 0.48g$, Soil profile, Type II
Relocated NH Route 85 over NH Route 101, New, 1992	NH Exeter-Stratham, Rockingham Co.	NHDOT	Webster-Martin, Inc.	Two-span cont. steel plate girders	LRB (DIS/Furon)	AASHTO $A = 0.15g$, Soil profile, Type I
Everett Turnpike over Nashua River & Canal, 1994	NH Nashua	NHDOT	Fay Spofford & Thorndike	—	Eradiquake (RJ Watson)	—
Squamscott River Bridge, New, 1992	NH Exeter (Relocated NH Rte. 101 over Squamscott R.)	NHDOT	Webster-Martin, Inc.	Six-span cont. steel plate girders	LRB (DIS/Furon)	AASHTO $A = 0.15g$, Soil profile, Type III
Pine Hill Road over Everett Turnpike, New, 1994	NH Nashua	NHDOT	Costello Lomasney & de Napoli, Inc.	Two-span cont. steel plate girders	LRB (DIS/Furon)	AASHTO $A = 0.15g$, Soil profile, Type I
I-93 over Fordway Ext., 1997	NH Derry	NHDOT	Clough Habour	—	Eradiquake (RJ Watson)	—
Pequannock River Bridge, New, 1991	NJ Morris & Passaic Co. (I-287 over Pequannock R.,)	NJDOT	Goodkind & O'Dea, Inc.	Three cont. steel plate girder units of 2, 3, and 3 spans	LRB (DIS/Furon)	AASHTO $A = 0.12g$, Soil profile, Type II
Foundry Street Overpass 106.68, Retrofit, 1993	NJ Newark (NJ Tpk. over Foundry St.)	NJTPA	Frederick R. Harris, Inc.	Simple span steel beams and conc. deck	LRB (DIS/Furon)	AASHTO $A = 0.18g$, Soil profile, Type II
Wilson Avenue Overpass W105.79SO, Retrofit, 1994	NJ Newark (NJ Tpk. NSO-E over Wilson Ave.)	NJTPA	Frederick R. Harris, Inc.	Steel beams, three simple spans	LRB (DIS/Furon)	AASHTO $A = 0.18g$, Soil profile, Type I

Conrail Newark Branch Overpass E106.57, Retrofit, 1994	NJ	Newark (NJ Tpk. NB over Conrail-Newark Branch)	NJTPA	Gannett-Fleming, Inc.	Steel plate girders, four simple spans	LRB (DIS/Furon)	AASHTO $A = 0.18g$, Soil profile, Type II
Wilson Avenue Overpass E105.79SO, Retrofit, 1994	NJ	Newark (NJ Tpk. Relocated E-NSO & W-NSO over Wilson Ave.)	NJTPA	Frederick R. Harris, Inc.	Steel beams, three simple spans	LRB (DIS/Furon)	AASHTO $A = 0.18g$, Soil profile, Type I
Relocated E-NSO Overpass W106.26A, New, 1994	NJ	Newark (NJ Tpk. E-NSO ramp)	NJTPA	Frederick Harris, Inc.	Steel plate girders, cont. units of five and four spans	LRB (DIS/Furon)	AASHTO $A = 0.18g$, Soil profile, Type II
Berry's Creek Bridge, Retrofit, 1995	NJ	E. Rutherford (Rte. 3 over Berry's Cr. and NJ Transit)	NJDOT	Goodkind and O'Dea, Inc.	Cont. steel plate girders; units of three, four, three, and three spans	LRB (Furon)	AASHTO $A = 0.18g$, Soil profile, Type II
Conrail Newark Branch Overpass W106.57, Retrofit, 1995	NJ	Newark (NJ Tpk. Rd. NSW over Conrail-Newark Branch & access rd.)	NJTPA	Frederick R. Harris, Inc.	Steel beams, six simple spans	LRB (DIS)	AASHTO $A = 0.18g$, Soil profile, Type I
Norton House Bridge, Retrofit, 1996	NJ	Pompton Lakes Borough and Wayne Township, Passaic County	NJDOT	A.G. Lichtenstein & Assoc.	Three-span continuous steel beams	LRB (DIS)	AASHTO $A = 0.18g$, Soil profile, Type II
Tacony-Palmyra Approaches, 1996	NJ	Palmyra, NJ	Burlington County Bridge Comm.	Steinman/Parsons Engineers	—	LRB (SEP, formerly Furon)	—
Rt. 4 over Kinderkamack Rd., 1996	NJ	Hackensack, NJ (Widening & Bridge Rehabilitation)	NJDOT	A.G. Lichtenstein & Assoc.	—	LRB, NRB (SEP)	—
Baldwin Street/Highland Avenue, 1996	NJ	Glen Ridge, NJ Bridge over Conrail	NJDOT	A.G. Lichtenstein & Asso.	—	LRB NRB (SEP, formerly Furon)	—
I-80 Bridges B764E & W, Retrofit, 1992	NV	Verdi, Washoe Co. (I-80 over Truckee R. and a local roadway)	NDOT	NDOT	Simple span composite steel plate girders or rolled beams	LRB (DIS/Furon)	AASHTO $A = 0 =$ 2E37g, Soil profile, Type I
West Street Overpass, Retrofit, 1991	NY	Harrison, Westchester Co. (West St. over I-95 New England Thwy.)	NYSTA	N.H. Bettigole, P.C.	Four simple span steel beam structures	LRB (DIS/Furon)	AASHTO $A = 0.19g$, Soil profile, Type III
Aurora Expressway Bridge, Retrofit, 1993	NY	Erie Co. (SB lanes of Rte. 400 Aurora Expy. over Cazenovia Cr.)	NYSDOT	NYSDOT	Cont. steel beams with conc. deck	LRB (DIS/Furon)	AASHTO $A = 0.19g$, Soil profile, Type III
Mohawk River Bridge, New, 1994	NY	Herkimer	NYSTA	Steinman Boynton Gronquist & Birdsall	Three-span haunched riveted steel plate girders; simple span riveted steel plate girders or rolled beams	LRB (DIS/Furon)	AASHTO $A = 0.19g$, Soil profile, Type II

TABLE 41.5 (continued) Seismically-Isolated Bridges in North America

Bridge	Location		Owner	Engineer	Bridge Description	Bearing Type	Design Criteria
Moodna Creek Bridge, Retrofit, 1994	NY	Orange County (NYST over Moodna Cr. at MP52.83)	NYSTA	Ryan Biggs Assoc., Inc.	Three simple spans; steel plate girder center span; rolled beam side spans	LRB (DIS/Furon)	AASHTO $A = 0.15g$, Soil profile, Type II
Conrail Bridge, New, 1994	NY	Herkimer (EB and WB rdwys. of NYST over Conrail, Rte. 5, etc.)	NYSTA	Steinman Boynton Gronquist & Birdsall	Four-span cont. curved haunched welded steel plate girders.	LRB (DIS/Furon)	AASHTO $A = 0.19g$, Soil profile, Type II
Maxwell Ave. over I-95, 1995	NY	Rye	NYS Thruway Authority	Casler Houser & Hutchison	—	EradiQuake (RJ Watson)	—
JFK Terminal One Elevated Roadway, New, 1996	NY	JFK International Airport, New York City	Port Authority of New York & New Jersey	STV Group	Continuous and simple span steel plate girders	LRB	AASHTO $A = 0.19g$, Soil profile, Type III
Buffalo Airport Viaduct, 1996	NY	Buffalo	NFTA	Lu Engineers	—	EradiQuake (RJ Watson)	—
Yonkers Avenue Bridge, 1997	NY	Yonkers	NY DOT	Voilmer & Assoc.	—	EradiQuake (RJ Watson)	—
Clackamas Connector, New, 1992	OR	Milwaukie (part of Tacoma St. Interchange)	ODOT	ODOT	Eight-span cont=2E post-tensioned conc. trapez. box girder	LRB (DIS/Furon)	AASHTO $A = 0.29g$, Soil profile, Type III
Hood River Bridges, 1995	OR	Hood River, OR	ODOT	ODOT	—	NRB (Furon)	—
Marquam Bridge, Retrofit, 1995	OR	—	ODOT	—	—	FIP	—
Hood River Bridge, Retrofit, 1996	OR	Hood River, OR	ODOT	ODOT	—	FIP	—
Toll Plaza Road Bridge, New, 1990	PA	Montgomery Co. (Approach to toll plaza over Hwy. LR145)	PTC	CECO Assoc., Inc.	176 ft simple span composite steel plate girder	LRB (DIS/Furon)	AASHTO $A = 0.1g$, Soil profile, Type II
Montebella Bridge Relocation, 1996	PR	Puerto Rico	P.R. Highway Authority	Walter Ruiz & Assoc.	—	LRB, NRB (SEP, formerly Furon)	—
Blackstone River Bridge, New, 1992	RI	Woonsocket	RIDOT	R.A. Cataldo & Assoc.	Four-span cont. composite steel plate girders	LRB (DIS/Furon)	AASHTO $A = 0.1g$, Soil profile, Type II
Providence Viaduct, Retrofit, 1992	RI	Rte. I-95, Providence	RIDOT	Maguire Group	Five-span steel plate girders/hunched steel plate girder units	LRB (DIS/Furon)	AASHTO $A = 0.32g$, Soil profile, Type III
Seekonk River Bridge, Retrofit, 1995	RI	Pawtuckett (I-95 over Seekonk River)	RIDOT	A.G Lichenstein & Assoc.	Haunched steel, two-girder floor beam construction.	LRB (DIS)	AASHTO $A = 0.32g$, Soil profile, Type I

I-295 to Rt. 10, 1996	RI	Warwick/Cranston (Bridges 662 & 663)	RIDOT	Commowealth Engineers & Consultants	—	LRB (SEP, formerly Furon)	—
Chickahominy River Bridge, New, 1996	VA	Hanover-Hennico County Line (US1 over Chickahominy River)	VDOT	Alpha Corp.	Simple span prestress concrete I-girders with continuous deck.	LRB (DIS)	AASHTO $A = 0.13g$, Soil profile, Type I
Ompompanoosuc River Bridge, Retrofit, 1992	VT	Rte. 5, Norwich	VAT	VAT	Three-span cont. steel plate girders	LRB (DIS/Furon)	AASHTO $A = 0.25g$, Soil profile, Type III
Cedar River Bridge New, 1992	WA	Renton (I-405 over Cedar R. and BN RR)	WSDOT	WSDOT	Four-span cont. steel plate girders	LRB (DIS/Furon)	AASHTO $A = 0.25g$, Soil profile, Type II
Lacey V. Murrow Bridge, West Approach, Retrofit, 1992	WA	Seattle (Approach to orig. Lake Washington Floating Br.)	WSDOT	Arvid Grant & Assoc., Inc.	Cont. conc. box girders; cont. deck trusses; simple span tied arch	LRB (DIS/Furon)	AASHTO $A = 0.25g$, Soil profile, Type II
Coldwater Creek Bridge No. 11, New, 1994	WA	SR504 (Mt. St. Helens Hwy.) over Coldwater Lake Outlet	WSDOT	WSDOT	Three-span cont. steel plate girders	LRB (DIS/Furon)	AASHTO $A = 0.55g$, Soil profile, Type I
East Creek Bridge No. 14, New, 1994	WA	SR504 (Mt. St. Helens Hwy.) over East Cr.	WSDOT	WSDOT	Three-span cont. steel plate girders	LRB (DIS)	AASHTO $A = 0.55g$, Soil profile, Type I
Home Bridge, New, 1994	WA	Home (Key Penninsula Highway over Von Geldem Cove)	Pierce Co. Public Works/Road Dept.	Pierce Co. Public WorksDept.	Prestressed concrete girders; simple spans; continuous for live load.	LRB (DIS)	AASHTO $A = 0.25g$, Soil profile, Type II
Duwamish River Bridge, Retrofit, 1995	WA	Seattle (I-5 over Duwamish River)	WSDOT	Exceltech	Cont. curved steel plate girder unit flanked by curved concrete box girder end spans	LRB (DIS)	AASHTO $A = 0.27g$, Soil profile, Type II

TABLE 41.6 Bridges in North America with Supplemental Damping Devices

Bridge	Location	Type and Number of Dampers	Year	Notes
San Francisco–Oakland Bay Bridge	San Francisco, CA	Viscous dampers Total: 96	1998 (design)	Retrofit of West Suspension spans. 450~650 kips force output, 6~22 in. strokes
Gerald Desmond Bridge	Long Beach, CA	Viscous dampers (Enidine) Total: 258	1996	Retrofit, 258 × 50 kip shock absorbers, 6 in. stroke
Cape Girardeau Bridge	Cape Girardeau, MO	Viscous dampers (Taylor)	1997	New construction of a cable-stayed bridge; Dampers used to control longitudinal earthquake movement while allowing free thermal movement.
The Golden Gate Bridge	San Francisco, CA	Viscous dampers (to be det.) Total: 40	1999 (est.)	Retrofit, 40 × 650 kip nonlinear dampers, ± 24 in.
Santiago Creek Bridge	California	Viscous dampers (Enidine)	1997 (est.)	New construction; dampers at abutments for energy dissipation in longitudinal direction
Sacramento River Bridge at Rio Vista	Rio Vista, CA	Viscous dampers (Taylor)	1997 (est.)	Retrofit; eight dampers used to control uplift of lift-span towers
Vincent Thomas Bridge	Long Beach, CA	Viscous dampers (to be det.) Total: 16	—	Retrofit, 8 × 200 kip and 8 × 100 kip linear dampers, ± 12 in.
Montlake Bridge	Seattle, WA	Viscous dampers (Taylor)	1996	Protection of new bascule leafs from runaway
West Seattle Bridge	Seattle, WA	Viscous dampers (Taylor)	1990	Deck isolation for swing bridge.

41.7 Summary

An attempt has been made to introduce the basic concepts of seismic isolation and supplemental energy dissipation, their history, current developments, applications, and design-related issues. Although significant strides have been made in terms of implementing these concepts to structural design and performance upgrade, it should be mentioned that these are emerging technologies and advances are being made constantly. With more realistic prototype testing results being made available to the design community of seismic isolation and supplemental energy dissipation devices from the FHWA/Caltrans testing program, significant improvement in code development will continuously make design easier and more standardized.

Acknowledgments

The author would like to express his deepest gratitude to Professor T. T. Soong, State University of New York at Buffalo, for his careful, thorough review, and many valuable suggestions. The author is also indebted to Dr. Lian Duan, California State Department of Transportation, for his encouragement, patience, and valuable input.

References

1. Applied Technology Council, *Improved Seismic Design Criteria for California Bridges: Provisional Recommendations*, ATC-32, Applied Technology Council, Redwood City, 1996.
2. Housner, G. W., Bergman, L. A., Caughey, T. K., Chassiakos, A. G., Claus, R. O., Masri, S. F., Seklton, R. E., Soong, T. T., Spencer, B. F., and Yao, J. T. P., Structural control: past, present, and future, *J. Eng. Mech. ASCE*, 123(9), 897–971.
3. Soong, T. T., and Dargush, G. F., Passive energy dissipation and active control, in *Structure Engineering Handbook*, Chen, W. F., Ed., CRC Press, Boca Raton, FL, 1997.
4. Soong, T. T. and Dargush, G. F., *Passive Energy Dissipation Systems in Structural Engineering*, John Wiley & Sons, New York, 1997.
5. Housner, G. W. and Jennings, P. C., *Earthquake Design Criteria*, Earthquake Engineering Research Institute, 1982.
6. Newmark, N. M. and Hall, W. J., Procedures and Criteria for Earthquake Resistant Design, *Building Practice for Disaster Mitigation*, Department of Commerce, Feb. 1973.
7. Dynamic Isolation Systems, *Force Control Bearings for Bridges — Seismic Isolation Design*, Rev. 4, Lafayette, CA, Oct. 1994.
8. Clough, R. W. and Penzien, J., *Dynamics of Structures*, 2nd ed., McGraw-Hill, New York, 1993.
9. Zhang, R., Soong, T. T., and Mahmoodi, P., Seismic response of steel frame structures with added viscoelastic dampers, *Earthquake Eng. Struct. Dyn.*, 18, 389–396, 1989.
10. Earthquake Protection Systems, *Friction Pendulum Seismic Isolation Bearings*, Product Technical Information, Earthquake Protection Systems, Emeriville, CA, 1997.
11. Liu, D. W, Nobari, F. S., Schamber, R. A., and Imbsen, R. A., Performance based seismic retrofit design of Benicia–Martinez bridge, in *Proceedings, National Seismic Conference on Bridges and Highways*, Sacramento, CA, July 1997.
12. Constantinou, M. C. and Symans, M. D., Seismic response of structures with supplemental damping, *Struct. Design Tall Buildings*, 2, 77–92, 1993.
13. Ingham, T. J., Rodriguez, S., Nader, M. N., Taucer, F., and Seim, C., Seismic retrofit of the Golden Gate Bridge, in *Proceedings, National Seismic Conference on Bridges and Highways*, Sacramento, CA, July 1997.
14. Taylor Devices, Sample Technical Specifications for Viscous Damping Devices, Taylor Devices, Inc., North Tonawanda, NY, 1996.
15. Lin, R. C., Liang, Z., Soong, T. T., and Zhang, R., An Experimental Study of Seismic Structural Response with Added Viscoelastic Dampers, Report No. NCEER-88-0018, National Center For Earthquake Engineering Research, State University of New York at Buffalo, February, 1988.
16. Crosby, P., Kelly, J. M., and Singh, J., Utilizing viscoelastic dampers in the seismic retrofit of a thirteen story steel frame building, *Structure Congress*, XII, Atlanta, GA, 1994.
17. Skinner, R. I., Kelly, J. M., and Heine, A. J., Hysteretic dampers for earthquake resistant structures, *Earthquake Eng. Struct. Dyn.*, 3, 297–309, 1975.
18. Pall, A. S. and Pall, R., Friction-dampers used for seismic control of new and existing buildings in Canada, in *Proceedings ATC 17-1 Seminar on Isolation, Energy Dissipation and Active Control*, San Francisco, 1992.
19. AASHTO, *Guide Specifications for Seismic Isolation Design*, American Association of State Highway and Transportation Officials, Washington, D.C., June 1991.
20. FEMA, NEHRP Recommended Provisions for Seismic Regulations for New Buildings, 1994 ed., Federal Emergency Management Agency, Washington, D.C., May 1995.
21. EERC/Berkeley, Pre-qualification Testing of Viscous Dampers for the Golden Gate Bridge Seismic Rehabilitation Project, A Report to T. Y. Lin International, Inc., Report No. EERCL/95-03, Earthquake Engineering Research Center, University of California at Berkeley, December 1995.

22. AASHTO, *Standard Specifications for Highway Bridges*, 16th ed., American Association of State Highway and Transportaion Officials, Washington, D.C., 1996.
23. Priestley, M. J. N, Seible, F., and Calvi, G. M., *Seismic Design and Retrofit of Bridges*, John Wiley an&& Sons, New York, 1996.
24. Whittaker, A., Tentative General Requirements for the Design and Construction of Structures Incorporating Discrete Passive Energy Dissipation Devices, ATC-15-4, Redwood City, CA, 1994.
25. Applied Technology Council, BSSC Seismic Rehabilitation Projects, ATC-33.03 (Draft), Redwood City, CA, 1997. .
26. AASHTO-T3 Committee Task Group, *Guide Specifications for Seismic Isolation Design*, T3 Committee Task Group Draft Rewrite, America Association of State Highway and Transportation Officials, Washington, D.C., May, 1997.

42

Soil–Foundation–Structure Interaction

Wen-Shou Tseng
International Civil Engineering Consultants, Inc.

Joseph Penzien
International Civil Engineering Consultants, Inc.

42.1 Introduction

Prior to the 1971 San Fernando, California earthquake, nearly all damages to bridges during earthquakes were caused by ground failures, such as liquefaction, differential settlement, slides, and/or spreading; little damage was caused by seismically induced vibrations. Vibratory response considerations had been limited primarily to wind excitations of large bridges, the great importance of which was made apparent by failure of the Tacoma Narrows suspension bridge in the early 1940s, and to moving loads and impact excitations of smaller bridges.

The importance of designing bridges to withstand the vibratory response produced during earthquakes was revealed by the 1971 San Fernando earthquake during which many bridge structures collapsed. Similar bridge failures occurred during the 1989 Loma Prieta and 1994 Northridge, California earthquakes, and the 1995 Kobe, Japan earthquake. As a result of these experiences, much has been done recently to improve provisions in seismic design codes, advance modeling and analysis

0-8493-7434-0/00/$0.00+$.50

procedures, and develop more effective detail designs, all aimed at ensuring that newly designed and retrofitted bridges will perform satisfactorily during future earthquakes.

Unfortunately, many of the existing older bridges in the United States and other countries, which are located in regions of moderate to high seismic intensity, have serious deficiencies which threaten life safety during future earthquakes. Because of this threat, aggressive actions have been taken in California, and elsewhere, to retrofit such unsafe bridges bringing their expected performances during future earthquakes to an acceptable level. To meet this goal, retrofit measures have been applied to the superstructures, piers, abutments, and foundations.

It is because of this most recent experience that the importance of coupled soil–foundation–structure interaction (SFSI) on the dynamic response of bridge structures during earthquakes has been fully realized. In treating this problem, two different methods have been used (1) the "elastodynamic" method developed and practiced in the nuclear power industry for large foundations and (2) the so-called empirical *p–y* method developed and practiced in the offshore oil industry for pile foundations. Each method has its own strong and weak characteristics, which generally are opposite to those of the other, thus restricting their proper use to different types of bridge foundation. By combining the models of these two methods in series form, a hybrid method is reported herein which makes use of the strong features of both methods, while minimizing their weak features. While this hybrid method may need some further development and validation at this time, it is fundamentally sound; thus, it is expected to become a standard procedure in treating seismic SFSI of large bridges supported on different types of foundation.

The subsequent sections of this chapter discuss all aspects of treating seismic SFSI by the elastodynamic, empirical *p–y*, and hybrid methods, including generating seismic inputs, characterizing soil–foundation systems, conducting force–deformation demand analyses using the substructuring approach, performing force–deformation capacity evaluations, and judging overall bridge performance.

42.2 Description of SFSI Problems

The broad problem of assessing the response of an engineered structure interacting with its supporting soil or rock medium (hereafter called soil medium for simplicity) under static and/or dynamic loadings will be referred here as the soil–structure interaction (SSI) problem. For a building that generally has its superstructure above ground fully integrated with its substructure below, reference to the SSI problem is appropriate when describing the problem of interaction between the complete system and its supporting soil medium. However, for a long bridge structure, consisting of a superstructure supported on multiple piers and abutments having independent and often distinct foundation systems which in turn are supported on the soil medium, the broader problem of assessing interaction in this case is more appropriately and descriptively referred to as the soil–foundation–structure interaction (SFSI) problem. For convenience, the SFSI problem can be separated into two subproblems, namely, a soil–foundation interaction (SFI) problem and a foundation–structure interaction (FSI) problem. Within the context of SFSI, the SFI part of the total problem is the one to be emphasized, since, once it is solved, the FSI part of the total problem can be solved following conventional structural response analysis procedures. Because the interaction between soil and the foundations of a bridge makes up the core of an SFSI problem, it is useful to review the different types of bridge foundations that may be encountered in dealing with this problem.

42.2.1 Bridge Foundation Types

From the perspective of SFSI, the foundation types commonly used for supporting bridge piers can be classified in accordance with their soil-support configurations into four general types: (1) spread footings, (2) caissons, (3) large-diameter shafts, and (4) slender-pile groups. These types as described separately below are shown in Figure 42.1.

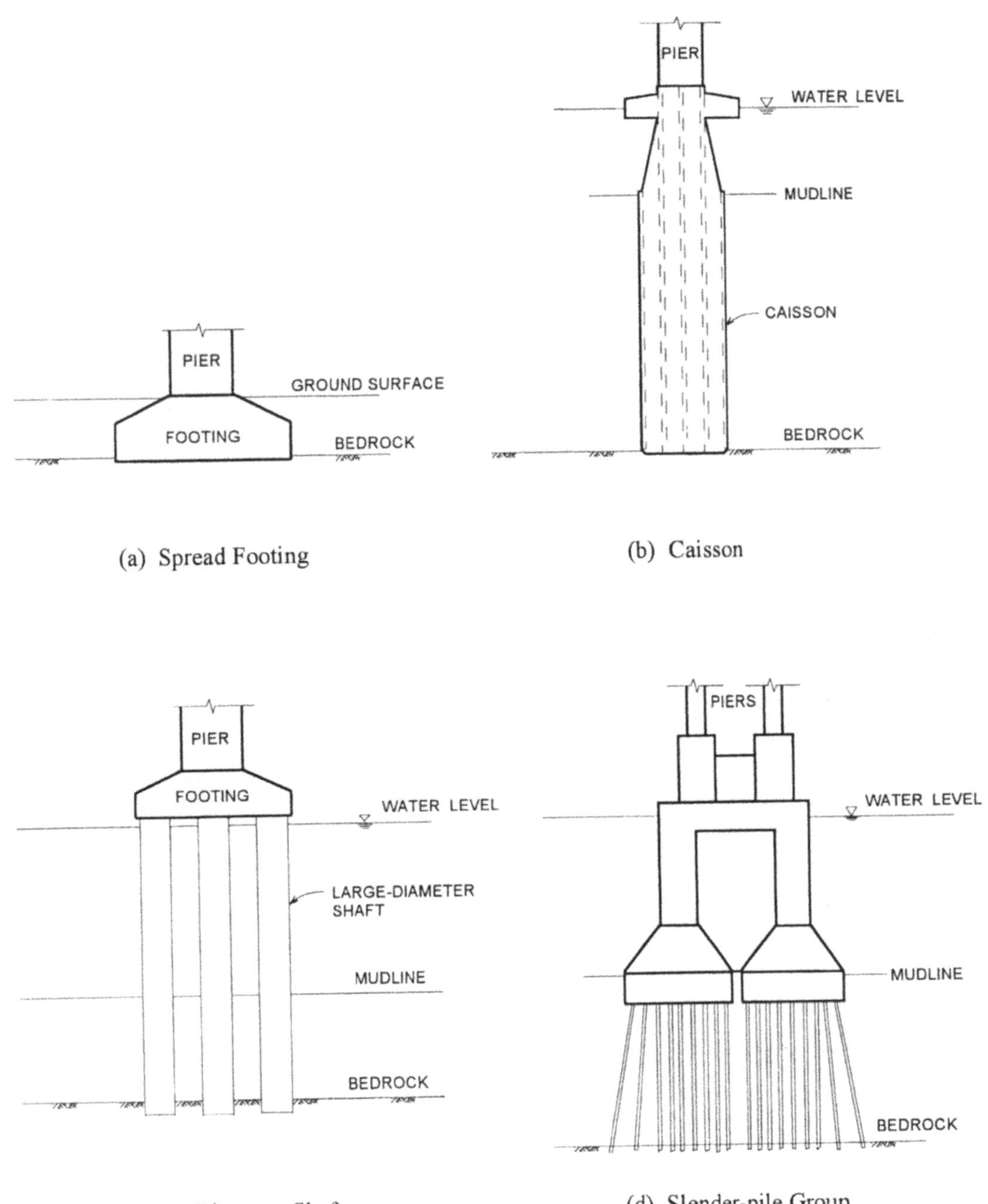

FIGURE 42.1 Bridge foundation types: (a) spread footing; (b) caisson; (c) large-diameter shafts; and (d) slender-pile group.

Spread Footings

Spread footings bearing directly on soil or rock are used to distribute the concentrated forces and moments in bridge piers and/or abutments over sufficient areas to allow the underlying soil strata to support such loads within allowable soil-bearing pressure limits. Of these loads, lateral forces are resisted by a combination of friction on the foundation bottom surface and passive soil pressure on its embedded vertical face. Spread footings are usually used on competent soils or rock which have high allowable bearing pressures. These foundations may be of several forms, such as (1) isolated footings, each supporting a single column or wall pier; (2) combined footings, each supporting two or more closely spaced bridge columns; and (3) pedestals which are commonly used for supporting steel bridge columns where it is desirable to terminate the structural steel above grade for corrosion protection. Spread footings are generally designed to support the superimposed forces and moments without uplifting or sliding. As such, inelastic action of the soils supporting the footings is usually not significant.

Caissons

Caissons are large structural foundations, usually in water, that will permit dewatering to provide a dry condition for excavation and construction of the bridge foundations. They can take many forms to suit specific site conditions and can be constructed of reinforced concrete, steel, or composite steel and concrete. Most caissons are in the form of a large cellular rectangular box or cylindrical shell structure with a sealed base. They extend up from deep firm soil or rock-bearing strata to above mudline where they support the bridge piers. The cellular spaces within the caissons are usually flooded and filled with sand to some depth for greater stability. Caisson foundations are commonly used at deep-water sites having deep soft soils. Transfer of the imposed forces and moments from a single pier takes place by direct bearing of the caisson base on its supporting soil or rock stratum and by passive resistance of the side soils over the embedded vertical face of the caisson. Since the soil-bearing area and the structural rigidity of a caisson is very large, the transfer of forces from the caisson to the surrounding soil usually involves negligible inelastic action at the soil–caisson interface.

Large-Diameter Shafts

These foundations consist of one or more large-diameter, usually in the range of 4 to 12 ft (1.2 to 3.6 m), reinforced concrete cast-in-drilled-hole (CIDH) or concrete cast-in-steel-shell (CISS) piles. Such shafts are embedded in the soils to sufficient depths to reach firm soil strata or rock where a high degree of fixity can be achieved, thus allowing the forces and moments imposed on the shafts to be safely transferred to the embedment soils within allowable soil-bearing pressure limits and/or allowable foundation displacement limits. The development of large-diameter drilling equipment has made this type of foundation economically feasible; thus, its use has become increasingly popular. In actual applications, the shafts often extend above ground surface or mudline to form a single pier or a multiple-shaft pier foundation. Because of their larger expected lateral displacements as compared with those of a large caisson, a moderate level of local soil nonlinearities is expected to occur at the soil–shaft interfaces, especially near the ground surface or mudline. Such nonlinearities may have to be considered in design.

Slender-Pile Groups

Slender piles refer to those piles having a diameter or cross-sectional dimensions less than 2 ft (0.6 m). These piles are usually installed in a group and provided with a rigid cap to form the foundation of a bridge pier. Piles are used to extend the supporting foundations (pile caps) of a bridge down through poor soils to more competent soil or rock. The resistance of a pile to a vertical load may be essentially by point bearing when it is placed through very poor soils to a firm soil stratum or rock, or by friction in case of piles that do not achieve point bearing. In real situations, the vertical resistance is usually achieved by a combination of point bearing and side friction. Resistance to lateral loads is achieved by a combination of soil passive pressure on the pile cap, soil resistance around the piles, and flexural resistance of the piles. The uplift capacity of a pile is generally governed by the soil friction or cohesion acting on the perimeter of the pile. Piles may be installed by driving or by casting in drilled holes. Driven piles may be timber piles, concrete piles with or without prestress, steel piles in the form of pipe sections, or steel piles in the form of structural shapes (e.g., H shape). Cast-in-drilled-hole piles are reinforced concrete piles installed with or without steel casings. Because of their relatively small cross-sectional dimensions, soil resistance to large pile loads usually develops large local soil nonlinearities that must be considered in design. Furthermore, since slender piles are normally installed in a group, mutual interactions among piles will reduce overall group stiffness and capacity. The amounts of these reductions depend on the pile-to-pile spacing and the degree ofsoil nonlinearity developed in resisting the loads.

42.2.2 Definition of SFSI Problem

For a bridge subjected to externally applied static and/or dynamic loadings on the aboveground portion of the structure, the SFSI problem involves evaluation of the structural performance (demand/capacity ratio) of the bridge under the applied loadings taking into account the effect of SFI. Since in this case the ground has no initial motion prior to loading, the effect of SFI is to provide the foundation–structure system with a flexible boundary condition at the soil–foundation interface location when static loading is applied and a compliant boundary condition when dynamic loading is applied. The SFI problem in this case therefore involves (1) evaluation of the soil–foundation interface boundary flexibility or compliance conditions for each bridge foundation, (2) determination of the effects of these boundary conditions on the overall structural response of the bridge (e.g., force, moment, or deformation) demands, and (3) evaluation of the resistance capacity of each soil–foundation system that can be compared with the corresponding response demand in assessing performance. That part of determining the soil–foundation interface boundary flexibilities or compliances will be referred to subsequently in a gross term as the "foundation stiffness or impedance problem"; that part of determining the structural response of the bridge as affected by the soil–foundation boundary flexibilities or compliances will be referred to as the "foundation–structure interaction problem"; and that part of determining the resistance capacity of the soil–foundation system will be referred to as the "foundation capacity problem."

For a bridge structure subjected to seismic conditions, dynamic loadings are imposed on the structure. These loadings, which originate with motions of the soil medium, are transmitted to the structure through its foundations; therefore, the overall SFSI problem in this case involves, in addition to the foundation impedance, FSI, and foundation capacity problems described above, the evaluation of (1) the soil forces acting on the foundations as induced by the seismic ground motions, referred to subsequently as the "seismic driving forces," and (2) the effects of the free-field ground-motion-induced soil deformations on the soil–foundation boundary compliances and on the capacity of the soil–foundation systems. In order to evaluate the seismic driving forces on the foundations and the effects of the free-field ground deformations on compliances and capacities of the soil–foundation systems, it is necessary to determine the variations of free-field motion within the ground regions which interact with the foundations. This problem of determining the free-field ground motion variations will be referred to herein as the "free-field site response problem." As will be shown later, the problem of evaluating the seismic driving forces on the foundations is equivalent to determining the "effective or scattered foundation input motions" induced by the free-field soil motions. This problem will be referred to here as the "foundation scattering problem."

Thus, the overall SFSI problem for a bridge subjected to externally applied static and/or dynamic loadings can be separated into the evaluation of (1) foundation stiffnesses or impedances, (2) foundation–structure interactions, and (3) foundation capacities. For a bridge subjected to seismic ground motion excitations, the SFSI problem involves two additional steps, namely, the evaluation of free-field site response and foundation scattering. When solving the total SFSI problem, the effects of the nonzero soil deformation state induced by the free-field seismic ground motions should be evaluated in all five steps mentioned above.

42.2.3 Demand vs. Capacity Evaluations

As described previously, assessing the seismic performance of a bridge system requires evaluation of SFSI involving two parts. One part is the evaluation of the effects of SFSI on the seismic-response demands within the system; the other part is the evaluation of the seismic force and/or deformation capacities within the system. Ideally, a well-developed methodology should be one that is capable of solving these two parts of the problem concurrently in one step using a unified suitable model for the system. Unfortunately, to date, such a unified method has not yet been developed. Because of the complexities of a real problem and the different emphases usually demanded of the solutions for the two parts, different solution strategies and methods of analysis are warranted for solving these two parts of the overall SFSI problem. To be more specific, evaluation on the demand side of the problem is concerned with the overall SFSI system behavior which is controlled by the mass, damping (energy dissipation), and stiffness properties, or, collectively, the impedance properties, of the entire system; and, the solution must satisfy the dynamic equilibrium and compatibility conditions of the global system. This system behavior is not sensitive, however, to approximations made on local element behavior; thus, its evaluation does not require sophisticated characterizations of the detailed constitutive relations of its local elements. For this reason, evaluation of demand has often been carried out using a linear or equivalent linear analysis procedure. On the contrary, evaluation of capacity must be concerned with the extreme behavior of local elements or subsystems; therefore, it must place emphasis on the detailed constitutive behaviors of the local elements or subsystems when deformed up to near-failure levels. Since only local behaviors are of concern, the evaluation does not have to satisfy the global equilibrium and compatibility conditions of the system fully. For this reason, evaluation of capacity is often obtained by conducting nonlinear analyses of detailed local models of elements or subsystems or by testing of local members, connections, or sub-assemblages, subjected to simple pseudo-static loading conditions.

Because of the distinct differences between effective demand and capacity analyses as described above, the analysis procedures presented subsequently differentiate between these two parts of the overall SFSI problem.

42.3 Current State-of-the-Practice

The evaluation of SFSI effects on bridges located in regions of high seismicity has not received as much attention as for other critical engineered structures, such as dams, nuclear facilities, and offshore structures. In the past, the evaluation of SFSI effects for bridges has, in most cases, been regarded as a part of the bridge foundation design problem. As such, emphasis has been placed on the evaluation of load-resisting capacities of various foundation systems with relatively little attention having been given to the evaluation of SFSI effects on seismic-response demands within the complete bridge system. Only recently has formal SSI analysis methodologies and procedures, developed and applied in other industries, been adopted and applied to seismic performance evaluations of bridges [1], especially large important bridges [2,3].

Even though the SFSI problems for bridges pose their own distinct features (e.g., multiple independent foundations of different types supported in highly variable soil conditions ranging from hard to very soft), the current practice is to adopt, with minor modifications, the same methodologies and procedures developed and practiced in other industries, most notably, the nuclear power and offshore oil industries. Depending upon the foundation type and its soil-support condition, the procedures currently being used in evaluating SFSI effects on bridges can broadly be classified into two main methods, namely, the so-called elastodynamic method that has been developed and practiced in the nuclear power industry for large foundations, and the so-called empirical *p–y* method that has been developed and practiced in the offshore oil industry for pile foundations. The bases and applicabilities of these two methods are described separately below.

42.3.1 Elastodynamic Method

This method is based on the well-established elastodynamic theory of wave propagation in a linear elastic, viscoelastic, or constant-hysteresis-damped elastic half-space soil medium. The fundamental element of this method is the constitutive relation between an applied harmonic point load and the corresponding dynamic response displacements within the medium called the dynamic Green's functions. Since these functions apply only to a linear elastic, visoelastic, or constant-hysteresis-damped elastic medium, they are valid only for linear SFSI problems. Since application of the elastodynamic method of analysis uses only mass, stiffness, and damping properties of an SFSI system, this method is suitable only for global system response analysis applications. However, by adopting the same equivalent linearization procedure as that used in the seismic analysis of free-field soil response, e.g., that used in the computer program SHAKE [4], the method has been extended to one that can accommodate global soil nonlinearities, i.e., those nonlinearities induced in the free-field soil medium by the free-field seismic waves [5].

Application of the elastodynamic theory to dynamic SFSI started with the need for solving machine–foundation vibration problems [6]. Along with other rapid advances in earthquake engineering in the 1970s, application of this theory was extended to solving seismic SSI problems for building structures, especially those of nuclear power plants [7–9]. Such applications were enhanced by concurrent advances in analysis techniques for treating soil dynamics, including development of the complex modulus representation of dynamic soil properties and use of the equivalent linearization technique for treating ground-motion-induced soil nonlinearities [10–12]. These developments were further enhanced by the extensive model calibration and methodology validation and refinement efforts carried out in a comprehensive large-scale SSI field experimental program undertaken by the Electric Power Research Institute (EPRI) in the 1980s [13]. All of these efforts contributed to advancing the elastodynamic method of SSI analysis currently being practiced in the nuclear power industry [5].

Because the elastodynamic method of analysis is capable of incorporating mass, stiffness, and damping characteristics of each soil, foundation, and structure subsystem of the overall SFSI system, it is capable of capturing the dynamic interactions between the soil and foundation subsystems and between the foundations and structure subsystem; thus, it is suitable for seismic demand analyses. However, since the method does not explicitly incorporate strength characteristics of the SFSI system, it is not suitable for capacity evaluations.

As previously mentioned in Section 42.2.1, there are four types of foundation commonly used for bridges: (1) spread footings, (2) caissons, (3) large-diameter shafts, and (4) slender-pile groups. Since only small local soil nonlinearities are induced at the soil–foundation interfaces of spread footings and caissons, application of the elastodynamic method of seismic demand analysis of the complete SFSI system is valid. However, the validity of applying this method to large-diameter shaft foundations depends on the diameter of the shafts and on the amplitude of the imposed loadings. When the shaft diameter is large so that the load amplitudes produce only small local soil nonlinearities, the method is reasonably valid. However, when the shaft diameter is relatively small, the larger-amplitude loadings will produce local soil nonlinearities sufficiently large to require that the method be modified as discussed subsequently. Application of the elastodynamic method to slender-pile groups is usually invalid because of the large local soil nonlinearities which develop near the pile boundaries. Only for very low amplitude loadings can the method be used for such foundations.

42.3.2 Empirical "*p-y*" Method

This method was originally developed for the evaluation of pile–foundation response due to lateral loads [14–16] applied externally to offshore structures. As used, it characterizes the lateral soil resistance per unit length of pile, p, as a function of the lateral displacement, y. The p–y relation is generally developed on the basis of an empirical curve which reflects the nonlinear resistance of the local soil surrounding the pile at a specified depth (Figure 42.2). Construction of the curve depends mainly on soil material strength parameters, e.g., the friction angle, ϕ, for sands and cohesion, c, for clays at the specified depth. For shallow soil depths where soil surface effects become important, construction of these curves also depends on the local soil failure mechanisms, such as failure by a passive soil resistance wedge. Typical p–y curves developed for a pile at different soil depths are shown in Figure 42.3. Once the set of p–y curves representing the soil resistances at discrete values

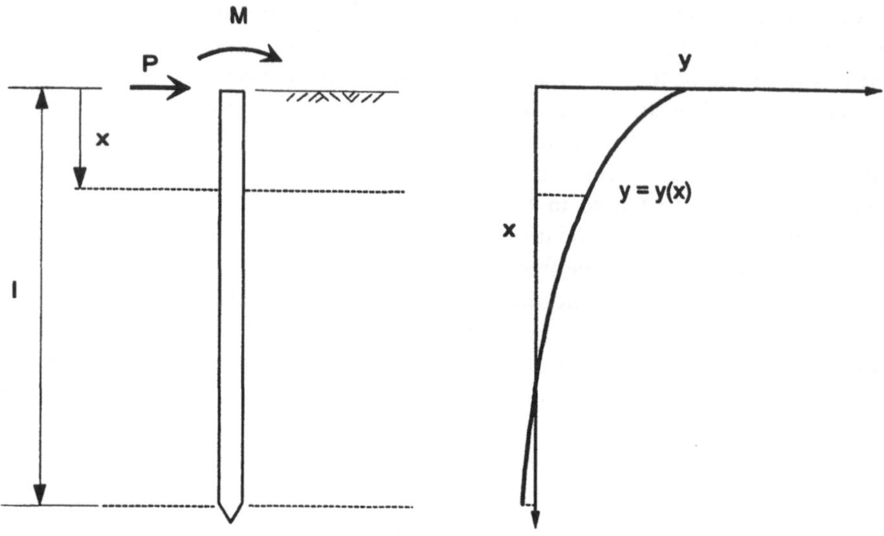

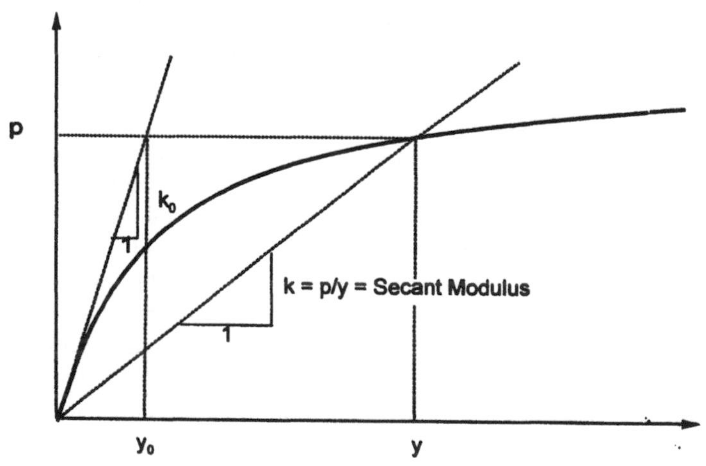

FIGURE 42.2 Empirical *p–y* curves and secant modulus.

of depth along the length of the pile has been constructed, evaluation of pile response under a specified set of lateral loads is accomplished by solving the problem of a beam supported laterally on discrete nonlinear springs. The validity and applicability of this method are based on model calibrations and correlations with field experimental results [15,16].

Based on the same model considerations used in developing the *p–y* curves for lateral response analysis of piles, the method has been extended to treating the axial resistance of soils to piles per unit length of pile, *t*, as a nonlinear function of the corresponding axial displacement, z, resulting in the so-called axial *t–z* curve, and treating the axial resistance of the soils at the pile tip, *Q*, as a

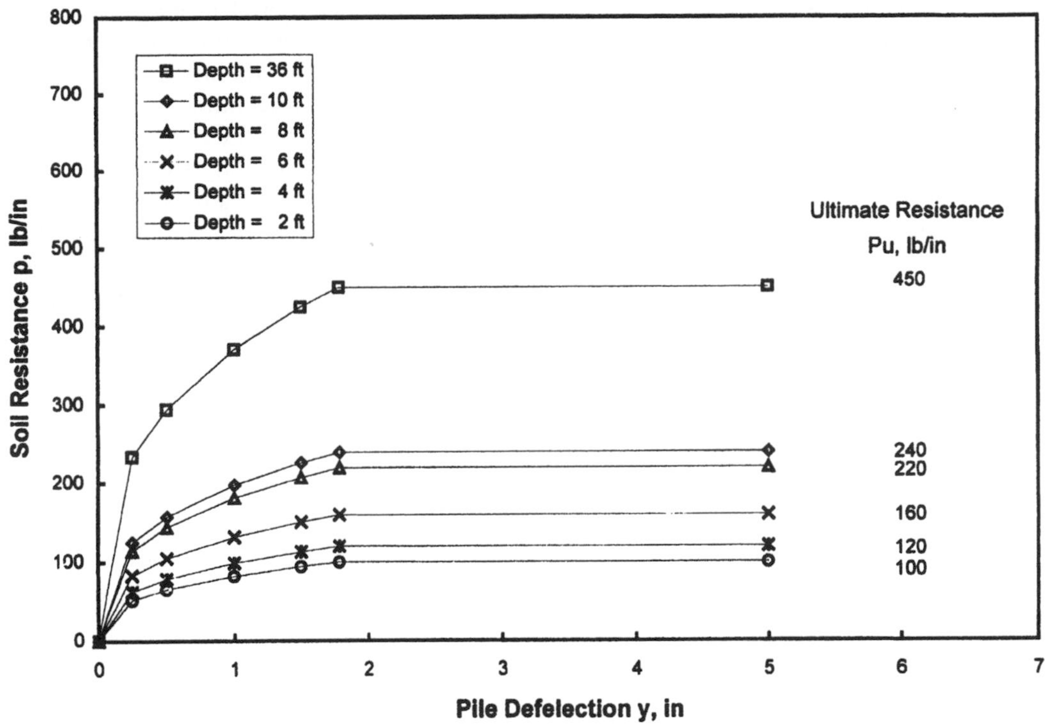

FIGURE 42.3 Typical *p–y* curves for a pile at different depths.

nonlinear function of the pile tip axial displacement, *d*, resulting in the so-called *Q–d* curve. Again, the construction of the *t–z* and *Q–d* curves for a soil-supported pile is based on empirical curvilinear forms and the soil strength parameters as functions of depth. By utilizing the set of *p–y*, *t–z*, and *Q–d* curves developed for a pile foundation, the response of the pile subjected to general three-dimensional (3-D) loadings applied at the pile head can be solved using the model of a 3-D beam supported on discrete sets of nonlinear lateral *p–y*, axial *t–z*, and axial *Q–d* springs. The method as described above for solving a soil-supported pile foundation subjected to applied loadings at the pile head is referred to here as the empirical *p–y* method, even though it involves not just the lateral *p–y* curves but also the axial *t–z* and *Q–d* curves for characterizing the soil resistances.

Since this method depends primarily on soil-resistance strength parameters and does not incorporate soil mass, stiffness, and damping characteristics, it is, strictly speaking, only applicable for capacity evaluations of slender-pile foundations and is not suitable for seismic demand evaluations because, as mentioned previously, a demand evaluation for an SFSI system requires the incorporation of the mass, stiffness, and damping properties of each of the constituent parts, namely, the soil, foundation, and structure subsystems.

Even though the *p–y* method is not strictly suited to demand analyses, it is current practice in performing seismic-demand evaluations for bridges supported on slender-pile group foundations to make use of the empirical nonlinear *p–y*, *t–z*, and *Q–d* curves in developing a set of equivalent linear lateral and axial soil springs attached to each pile at discrete elevations in the foundation. The soil–pile systems developed in this manner are then coupled with the remaining bridge structure to form the complete SFSI system for use in a seismic demand analysis. The initial stiffnesses of the equivalent linear *p–y*, *t–z*, and *Q–d* soil springs are based on secant moduli of the nonlinear *p–y*, *t–y*, and *Q–d* curves, respectively, at preselected levels of lateral and axial pile displacements, as shown schematically in Figure 42.2. After completing the initial demand analysis, the amplitudes of pile displacement are compared with the corresponding preselected amplitudes to check on their mutual compatibilities. If incompatibilities exist, the initial set of equivalent linear stiffnesses is adjusted and a second demand analysis is performed. Such iterations continue until reasonable compatibility is achieved. Since soil inertia and damping properties are not included in the above-described demand analysis procedure, it must be considered approximate; however, it is reasonably valid when the nonlinearities in the soil resistances become so large that the inelastic components of soil deformations adjacent to piles are much larger than the corresponding elastic components. This condition is true for a slender-pile group foundation subjected to relatively large amplitude pile-head displacements. However, for a large-diameter shaft foundation, having larger soil-bearing areas and higher shaft stiffnesses, the inelastic components of soil deformations may be of the same order or even smaller than the elastic components, in which case, application of the empirical *p–y* method for a demand analysis as described previously can result in substantial errors.

42.4 Seismic Inputs to SFSI System

The first step in conducting a seismic performance evaluation of a bridge structure is to define the seismic input to the coupled soil–foundation–structure system. In a design situation, this input is defined in terms of the expected free-field motions in the soil region surrounding each bridge foundation. It is evident that to characterize such motions precisely is practically unachievable within the present state of knowledge of seismic ground motions. Therefore, it is necessary to use a rather simplistic approach in generating such motions for design purposes. The procedure most commonly used for designing a large bridge is to (1) generate a three-component (two horizontal and vertical) set of accelerograms representing the free-field ground motion at a "control point" selected for the bridge site and (2) characterize the spatial variations of the free-field motions within each soil region of interest relative to the control motions.

The control point is usually selected at the surface of bedrock (or surface of a firm soil stratum in case of a deep soil site), referred to here as "rock outcrop," at the location of a selected reference pier; and the free-field seismic wave environment within the local soil region of each foundation is assumed to be composed of vertically propagating plane shear (*S*) waves for the horizontal motions and vertically propagating plane compression (*P*) waves for the vertical motions. For a bridge site consisting of relatively soft topsoil deposits overlying competent soil strata or rock, the assumption of vertically propagating plane waves over the depth of the foundations is reasonably valid as confirmed by actual field downhole array recordings [17].

The design ground motion for a bridge is normally specified in terms of a set of parameter values developed for the selected control point which include a set of target acceleration response spectra (ARS) and a set of associated ground motion parameters for the design earthquake, namely (1) magnitude, (2) source-to-site distance, (3) peak ground (rock-outcrop) acceleration (PGA), velocity (PGV), and displacement (PGD), and (4) duration of strong shaking. For large important bridges, these parameter values are usually established through regional seismic investigations coupled with site-specific seismic hazard and ground motion studies, whereas, for small bridges, it is customary to establish these values based on generic seismic study results such as contours of regional PGA values and standard ARS curves for different general classes of site soil conditions.

For a long bridge supported on multiple piers which are in turn supported on multiple foundations spaced relatively far apart, the spatial variations of ground motions among the local soil regions of the foundations need also be defined in the seismic input. Based on the results of analyses using actual earthquake ground motion recordings obtained from strong motion instrument arrays, such as the El Centro differential array in California and the SMART-1 array in Taiwan, the spatial variations of free-field seismic motions have been characterized using two parameters: (1) apparent horizontal wave propagation velocity (speed and direction) which controls the first-order spatial variations of ground motion due to the seismic wave passage effect and (2) a set of horizontal and vertical ground motion "coherency functions" which quantifies the second-order ground motion variations due to scattering and complex 3-D wave propagation [18]. Thus, in addition to the design ground motion parameter values specified for the control motion, characterizing the design seismic inputs to long bridges needs to include the two additional parameters mentioned above, namely, (1) apparent horizontal wave velocity and (2) ground motion coherency functions; therefore, the seismic input motions developed for the various pier foundation locations need to be compatible with the values specified for these two additional parameters.

Having specified the design seismic ground motion parameters, the steps required in establishing the pier foundation location-specific seismic input motions for a particular bridge are

1. Develop a three-component (two horizontal and vertical) set of free-field rock-outcrop motion time histories which are compatible with the design target ARS and associated design ground motion parameters applicable at a selected single control point location at the bridge site (these motions are referred to here simply as the "response spectrum compatible time histories" of control motion).
2. Generate response-spectrum-compatible time histories of free-field rock-outcrop motions at each bridge pier support location such that their coherencies relative to the corresponding components of the response spectrum compatible motions at the control point and at other pier support locations are compatible with the wave passage parameters and the coherency functions specified for the site (these motions are referred to here as "response spectrum and coherency compatible motions).
3. Carry out free-field site response analyses for each pier support location to obtain the time-histories of free-field soil motions at specified discrete elevations over the full depth of each foundation using the corresponding response spectrum and coherency compatible free-field rock-outcrop motions as inputs.

In the following sections, procedures will be presented for generating the set of response spectrum compatible rock-outcrop time histories of motion at the control point location and for generating the sets of response spectrum and coherency compatible rock-outcrop time histories of motion at all pier support locations, and guidelines will be given for performing free-field site response analyses.

42.4.1 Free-Field Rock-Outcrop Motions at Control-Point Location

Given a prescribed set of target ARS and a set of associated design ground motion parameters for a bridge site as described previously, the objective here is to develop a three-component set of time histories of control motion that (1) provides a reasonable match to the corresponding target ARS and (2) has time history characteristics reasonably compatible with the other specified associated ground motion parameter values. In the past, several different procedures have been used for developing rock-outcrop time histories of motion compatible with a prescribed set of target ARS. These procedures are summarized as follows:

1. *Response Spectrum Compatibility Time History Adjustment Method* [19–22] — This method as generally practiced starts by selecting a suitable three-component set of initial or "starting" accelerograms and proceeds to adjust each of them iteratively, using either a time-domain [21,23] or a frequency-domain [19,20,22] procedure, to achieve compatibility with the specified target ARS and other associated parameter values. The time-domain adjustment procedure usually produces only small local adjustments to the selected starting time histories, thereby producing response spectrum compatible time histories closely resembling the initial motions. The general "phasing" of the seismic waves in the starting time history is largely maintained while achieving close compatibility with the target ARS: minor changes do occur, however, in the phase relationships. The frequency-domain procedure as commonly used retains the phase relationships of an initial motion, but does not always provide as close a fit to the target spectrum as does the time-domain procedure. Also, the motion produced by the frequency-domain procedure shows greater visual differences from the initial motion.

2. *Source-to-Site Numerical Model Time History Simulation Method* [24–27] — This method generally starts by constructing a numerical model to represent the controlling earthquake source and source-to-site transmission and scattering functions, and then accelerograms are synthesized for the site using numerical simulations based on various plausible fault-rupture scenarios. Because of the large number of time history simulations required in order to achieve a "stable" average ARS for the ensemble, this method is generally not practical for developing a complete set of time histories to be used directly; rather it is generally used to supplement a set of actual recorded accelerograms, in developing site-specific target response spectra and associated ground motion parameter values.
3. *Multiple Actual Recorded Time History Scaling Method* [28,29] — This method starts by selecting multiple 3-component sets (generally ≥7) of actual recorded accelerograms which are subsequently scaled in such a way that the average of their response spectral ordinates over the specified frequency (or period) range of interest matches the target ARS. Experience in applying this method shows that its success depends very much on the selection of time histories. Because of the lack of suitable recorded time histories, individual accelerograms often have to be scaled up or down by large multiplication factors, thus raising questions about the appropriateness of such scaling. Experience also indicates that unless a large ensemble of time histories (typically >20) are selected, it is generally difficult to achieve matching of the target ARS over the entire spectral frequency (or period) range of interest.
4. *Connecting Accelerogram Segments Method* [55] — This method produces a synthetic time history by connecting together segments of a number of actual recorded accelerograms in such a way that the ARS of the resulting time history fits the target ARS reasonably well. It generally requires producing a number of synthetic time histories to achieve acceptable matching of the target spectrum over the entire frequency (or period) range of interest.

At the present time, Method 1 is considered most suitable and practical for bridge engineering applications. In particular, the time-domain time history adjustment procedure which produces only local time history disturbances has been applied widely in recent applications. This method as developed by Lilhanand and Tseng [21] in 1988, which is based on earlier work by Kaul [30] in 1978, is described below.

The time-domain procedure for time history adjustment is based on the inherent definition of a response spectrum and the recognition that the times of occurrence of the response spectral values for the specified discrete frequencies and damping values are not significantly altered by adjustments of the time history in the neighborhoods of these times. Thus, each adjustment, which is made by adding a small perturbation, $\delta a(t)$, to the selected initial or starting acceleration time history, $a(t)$, is carried out in an iterative manner such that, for each iteration, i, an adjusted acceleration time history, $a_i(t)$, is obtained from the previous acceleration time history, $a_{(i-1)}(t)$, using the relation

$$a_i(t) = a_{(i-1)}(t) + \delta a_i(t) \tag{42.1}$$

The small local adjustment, $\delta a_i(t)$, is determined by solving the integral equation

$$\delta Ri(\omega_j, \beta_k) = \int_0^{t_{jk}} \delta a_i(\tau) h_{jk}(t_{jk} - \tau) d\tau \tag{42.2}$$

which expresses the small change in the acceleration response value $\delta R_i(\omega_j, \beta_k)$ for frequency ω_j and damping β_k resulting from the local time history adjustment $\delta a_i(t)$. This equation makes use of the acceleration unit–impulse response function $h_{jk}(t)$ for a single-degree-of-freedom oscillator having a natural frequency ω_j and a damping ratio β_k. Quantity t_{jk} in the integral represents the time at which its corresponding spectral value occurs, and τ is a time lag.

By expressing $\delta a_i(t)$ as a linear combination of impulse response functions with unknown coefficients, the above integral equation can be transformed into a system of linear algebraic equations that can easily be solved for the unknown coefficients. Since the unit–impulse response functions decay rapidly due to damping, they produce only localized perturbations on the acceleration time history. By repeatedly applying the above adjustment, the desired degree of matching between the response spectra of the modified motions and the corresponding target spectra is achieved, while, in doing so, the general characteristics of the starting time history selected for adjustment are preserved.

Since this method of time history modification produces only local disturbances to the starting time history, the time history phasing characteristics (wave sequence or pattern) in the starting time history are largely maintained. It is therefore important that the starting time history be selected carefully. Each three-component set of starting accelerograms for a given bridge site should preferably be a set recorded during a past seismic event that has (1) a source mechanism similar to that of the controlling design earthquake, (2) a magnitude within about ±0.5 of the target controlling earthquake magnitude, and (3) a closest source-to-site distance within 10 km of the target source-to-site distance. The selected recorded accelerograms should have their PGA, PGV, and PGD values and their strong shaking durations within a range of ±25% of the target values specified for the bridge site and they should represent free-field surface recordings on rock, rocklike, or a stiff soil site; no recordings on a soft site should be used. For a close-in controlling seismic event, e.g., within about 10 km of the site, the selected accelerograms should contain a definite velocity pulse or the so-called fling. When such recordings are not available, Method 2 described previously can be used to generate a starting set of time histories having an appropriate fling or to modify the starting set of recorded motions to include the desired directional velocity pulse.

Having selected a three-component set of starting time histories, the horizontal components should be transformed into their principal components and the corresponding principal directions should be evaluated [31]. These principal components should then be made response spectrum compatible using the time-domain adjustment procedure described above or the standard frequency-domain adjustment procedure[20,22,32]. Using the latter procedure, only the Fourier amplitude spectrum, not the phase spectrum, is adjusted iteratively.

The target acceleration response spectra are in general identical for the two horizontal principal components of motion; however, a distinct target spectrum is specified for the vertical component. In such cases, the adjusted response spectrum compatible horizontal components can be oriented horizontally along any two orthogonal coordinate axes in the horizontal plane considered suitable for structural analysis applications. However, for bridge projects that have controlling seismic events with close-in seismic sources, the two horizontal target response spectra representing motions along a specified set of orthogonal axes are somewhat different, especially in the low-frequency (long-period) range; thus, the response spectrum compatible time histories must have the same definitive orientation. In this case, the generated three-component set of response spectrum compatible time histories should be used in conjunction with their orientation. The application of this three-component set of motions in a different coordinate orientation requires transforming the motions to the new coordinate system. It should be noted that such a transformation of the components will generally result in time histories that are not fully compatible with the original target response spectra. Thus, if response spectrum compatibility is desired in a specific coordinate orientation (such as in the longitudinal and transverse directions of the bridge), target response spectra in the specific orientation should be generated first and then a three-component set of fully response spectrum compatible time histories should be generated for this specific coordinate system.

As an example, a three-component set of response spectrum compatible time histories of control motion, generated using the time-domain time history adjustment procedure, is shown in Figure 42.4.

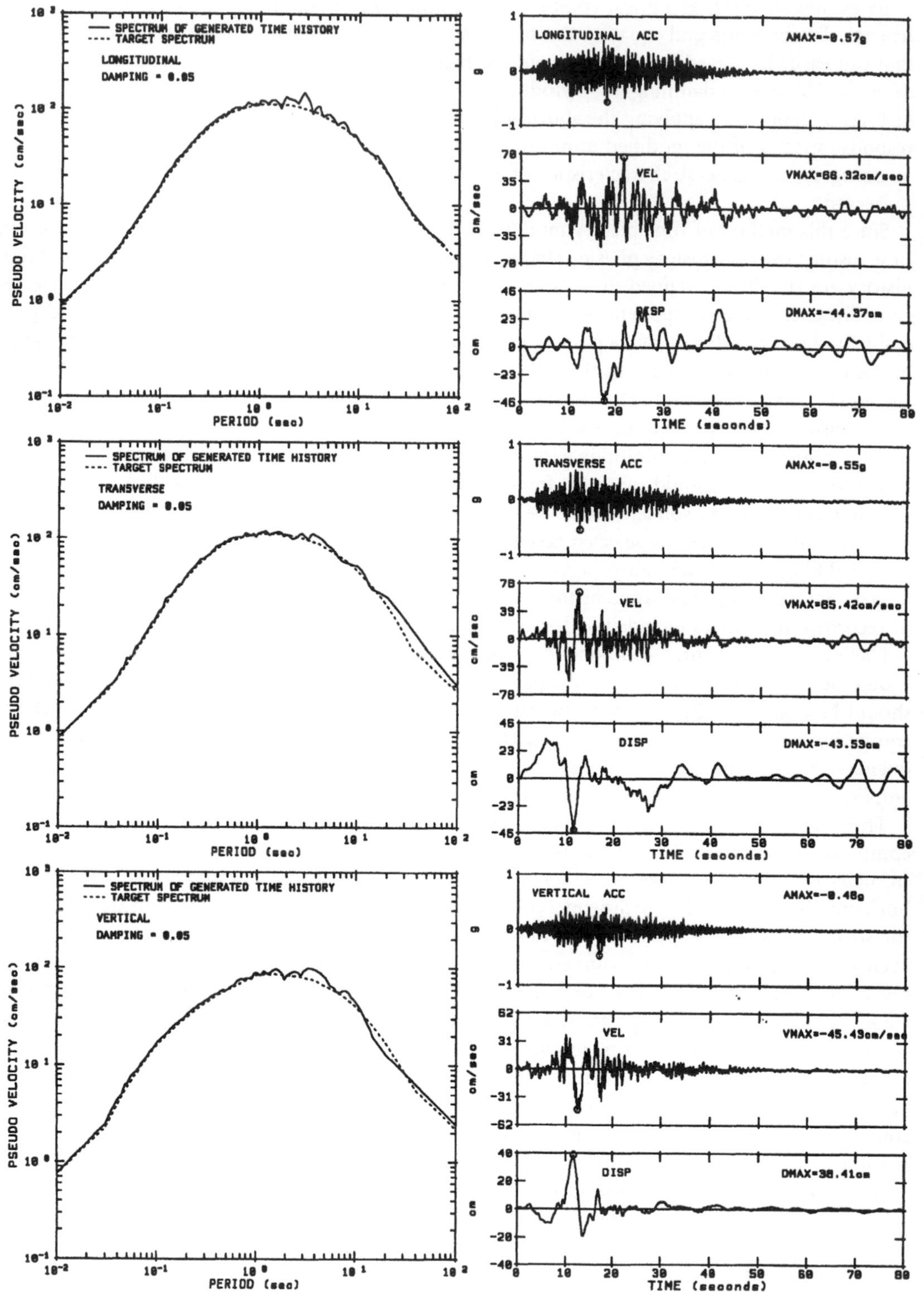

FIGURE 42.4 Examples of a three-component set of response spectrum compatible time histories of control motion.

42.4.2 Free-Field Rock-Outcrop Motions at Bridge Pier Support Locations

As mentioned previously, characterization of the spatial variations of ground motions for engineering purposes is based on a set of wave passage parameters and ground motion coherency functions. The wave passage parameters currently used are the apparent horizontal seismic wave speed, V, and its direction angle θ relative to an axis normal to the longitudinal axis of the bridge. Studies of strong- and weak-motion array data including those in California, Taiwan, and Japan show that the apparent horizontal speed of S-waves in the direction of propagation is typically in the 2 to 3 km/s range [18,33]. In applications, the apparent wave-velocity vector showing speed and direction must be projected along the bridge axis giving the apparent wave speed in that direction as expressed by

$$V_{\text{bridge}} = \frac{V}{\sin\theta} \tag{42.3}$$

To be realistic, when θ becomes small, a minimum angle for θ, say, 30°, should be used in order to account for waves arriving in directions different from the specified direction.

The spatial coherency of the free-field components of motion in a single direction at various locations on the ground surface has been parameterized by a complex coherency function defined by the relation

$$\Gamma_{ij}(i\omega) = \frac{S_{ij}(i\omega)}{\sqrt{S_{ii}(\omega)}\sqrt{S_{jj}(\omega)}} \qquad i, j = 1, 2, \ldots, n \text{ locations} \tag{42.4}$$

in which $S_{ij}(i\omega)$ is the smoothed complex cross-power spectral density function and $S_{ii}(\omega)$ and $S_{jj}(\omega)$ are the smoothed real power spectral density (PSD) functions of the components of motion at locations i and j. The notation $i\omega$ in the above equation is used to indicate that the coefficients $S_{ij}(i\omega)$ are complex valued (contain both real and imaginary parts) and are dependent upon excitation frequency ω. Based on analyses of strong-motion array data, a set of generic coherency functions for the horizontal and vertical ground motions has been developed [34]. These functions for discrete separation distances between locations i and j are plotted against frequency in Figure 42.5.

Given a three-component set of response spectrum compatible time histories of rock-outcrop motions developed for the selected control point location and a specified set of wave passage parameters and "target" coherency functions as described above, response spectrum compatible and coherency compatible multiple-support rock-outcrop motions applicable to each pier support location of the bridge can be generated using the procedure presented below. This procedure is based on the "marching method" developed by Hao et al. [32] in 1989 and extended by Tseng et al. [35] in 1993.

Neglecting, for the time being, ground motion attenuation along the bridge axis, the components of rock-outcrop motions at all pier support locations in a specific direction have PSD functions which are common with the PSD function $S_o(\omega)$ specified for the control motion, i.e.,

$$S_{ii}(\omega) = S_{jj}(\omega) = S_o(\omega) = |u_o(i\omega)|^2 \tag{42.5}$$

where $u_o(i\omega)$ is the Fourier transform of the corresponding component of control motion, $u_o(t)$. By substituting Eq. (42.5) into Eq. (42.4), one obtains

$$S_{ij}(i\omega) = \Gamma_{ij}(i\omega)\, S_o(\omega) \tag{42.6}$$

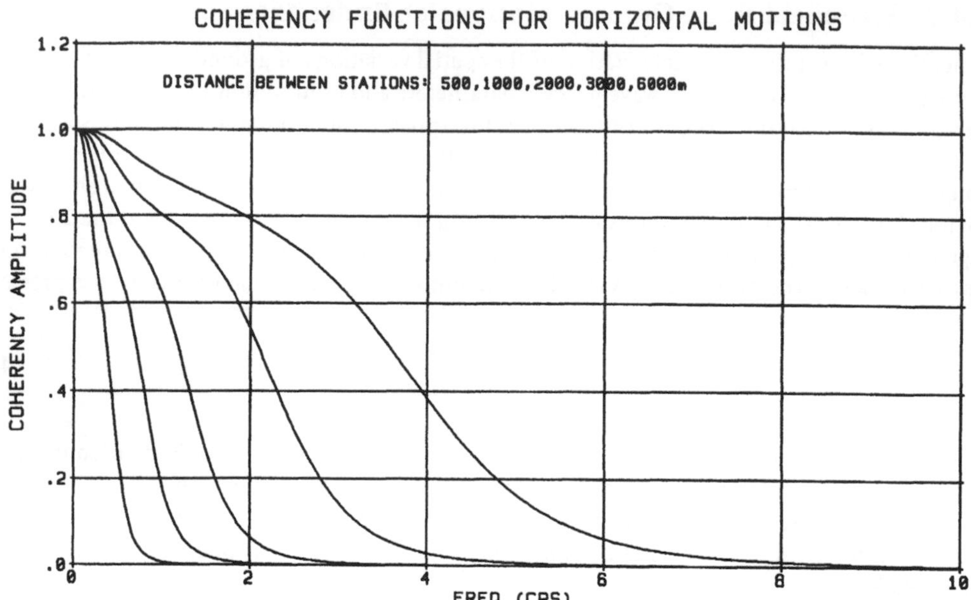

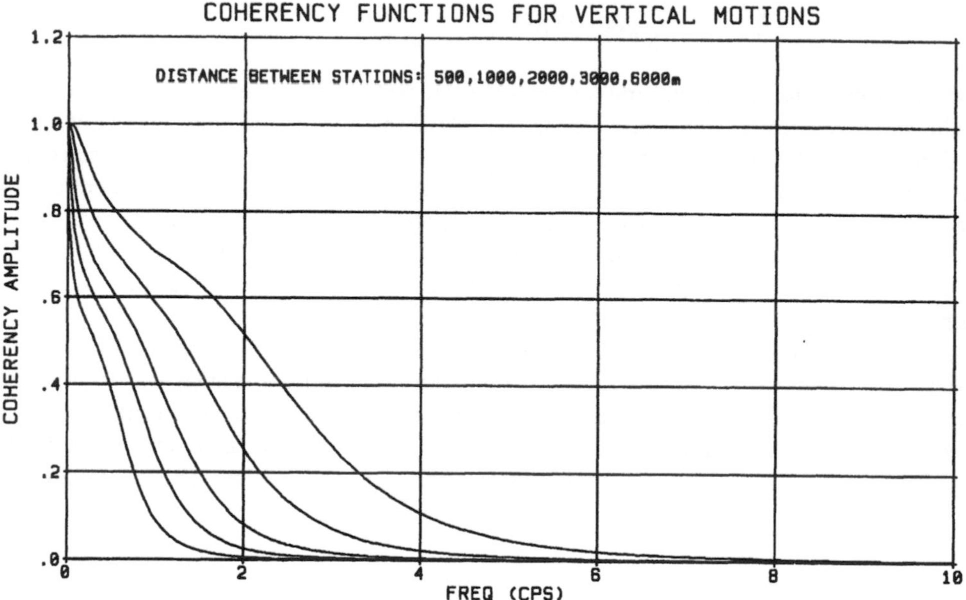

FIGURE 42.5 Example of coherency functions of frequency at discrete separation distances.

which can be rewritten in a matrix form for all pier support locations as follows:

$$S(i\omega) = \boldsymbol{\Gamma}(i\omega)\ S_o(\omega) \tag{42.7}$$

Since, by definition, the coherency matrix $\boldsymbol{\Gamma}(i\omega)$ is an Hermitian matrix, it can be decomposed into a complex conjugate pair of lower and upper triangular matrices $\boldsymbol{L}(i\omega)$ and $\boldsymbol{L}^*(i\omega)^T$ as expressed by

$$\boldsymbol{\Gamma}(i\omega) = \boldsymbol{L}(i\omega)\ \boldsymbol{L}^*(i\omega)^T \tag{42.8}$$

in which the symbol * denotes complex conjugate. In proceeding, let

$$u(i\omega) = L(i\omega)\eta_{\phi_i}(i\omega)u_o(i\omega) \tag{42.9}$$

in which $u(\omega)$ is a vector containing components of motion $u_i(\omega)$ for locations, $i = 1, 2, \ldots, n$; and, $\eta_{\phi_i}(i\omega) = \{e^{i\phi_i(\omega)}\}$ is a vector containing unit amplitude components having random-phase angles $\phi_i(\omega)$. If $\phi_i(\omega)$ and $\phi_j(\omega)$ are uniformly distributed random-phase angles, the relations

$$E[\eta_{\phi_i}(i\omega)\eta^*_{\phi_j}(i\omega)] = 0 \quad \text{if } i \neq j$$

$$E[\eta_{\phi_i}(i\omega)\eta^*_{\phi_j}(i\omega)] = 1 \quad \text{if } i = j \tag{42.10}$$

will be satisfied, where the symbol $E[\ ]$ represents ensemble average. It can easily be shown that the ensemble of motions generated using Eq. (42.9) will satisfy Eq. (42.7). Thus, if the rock-outcrop motions at all pier support locations are generated from the corresponding motions at the control point location using Eq. (42.9), the resulting motions at all locations will satisfy, on an ensemble basis, the coherency functions specified for the site. Since the matrix $L(i\omega)$ in Eq. (42.9) is a lower triangular matrix having its diagonal elements equal to unity, the generation of coherency compatible motions at all pier locations can be achieved by marching from one pier location to the next in a sequential manner starting with the control pier location.

In generating the coherency compatible motions using Eq. (42.9), the phase angle shifts at various pier locations due to the single plane-wave passage at the constant speed V_{bridge} defined by Eq. (42.3) can be incorporated into the term $\eta_{\phi_i}(i\omega)$. Since the motions at the control point location are response spectrum compatible, the coherency compatible motions generated at all other pier locations using the above-described procedure will be approximately response spectrum compatible. However, an improvement on their response spectrum compatibility is generally required, which can be done by adjusting their Fourier amplitudes but keeping their Fourier phase angles unchanged. By keeping these angles unchanged, the coherencies among the adjusted motions are not affected. Consequently, the adjusted motions will not only be response spectrum compatible, but will also be coherency compatible.

In generating the response spectrum- and coherency-compatible motions at all pier locations by the procedure described above, the ground motion attenuation effect has been ignored. For a long bridge located close to the controlling seismic source, attenuation of motion with distance away from the control pier location should be considered. This can be achieved by scaling the generated motions at various pier locations by appropriate scaling factors determined from an appropriate ground motion attenuation relation. The acceleration time histories generated for all pier locations should be integrated to obtain their corresponding velocity and displacement time histories, which should be checked to ensure against having numerically generated baseline drifts. Relative displacement time histories between the control pier location and successive pier locations should also be checked to ensure that they are reasonable. The rock-outcrop motions finally obtained should then be used in appropriate site-response analyses to develop the corresponding free-field soil motions required in conducting the SFSI analyses for each pier location.

42.4.3 Free-Field Soil Motions

As previously mentioned, the seismic inputs to large bridges are defined in terms of the expected free-field soil motions at discrete elevations over the entire depth of each foundation. Such motions must be evaluated through location-specific site-response analyses using the corresponding previously described rock-outcrop free-field motions as inputs to appropriately defined soil–bedrock models. Usually, as mentioned previously, these models are based on the assumption that the horizontal and vertical free-field soil motions are produced by upward/downward propagation of one-dimensional shear and compression waves, respectively, as caused by the upward propagation of incident waves in the underlying rock or firm soil formation. Consistent with these types of motion, it is assumed that the local soil medium surrounding each foundation consists of uniform horizontal layers of infinite lateral extent. Wave reflections and refractions will occur at all interfaces of adjacent layers, including the soil–bedrock interface, and reflections of the waves will occur at the soil surface. Computer program SHAKE [4,44] is most commonly used to carry out the above-described one-dimensional type of site-response analysis. For a long bridge having a widely varying soil profile from end to end, such site-response analyses must be repeated for different soil columns representative of the changing profile.

The cyclic free-field soil deformations produced at a particular bridge site by a maximum expected earthquake are usually of the nonlinear hysteretic form. Since the SHAKE computer program treats a linear system, the soil column being analyzed must be modeled in an equivalent linearized manner. To obtain the equivalent linearized form, the soil parameters in the model are modified after each consecutive linear time history response analysis is complete, which continues until convergence to strain-compatible parameters are reached.

For generating horizontal free-field motions produced by vertically propagating shear waves, the needed equivalent linear soil parameters are the shear modulus G and the hysteretic damping ratio β. These parameters, as prepared by Vucetic and Dobry [36] in 1991 for clay and by Sun et al. [37] in 1988 and by the Electric Power Research Institute (EPRI) for sand, are plotted in Figures 42.6 and 42.7, respectively, as functions of shear strain γ. The shear modulus is plotted in its nondimensional form G/G_{max} where G_{max} is the *in situ* shear modulus at very low strains ($\gamma \leq 10^{-4}\%$). The shear modulus G must be obtained from cyclic shear tests, while G_{max} can be obtained using $G_{max} = \rho V_s^2$ in which ρ is mass density of the soil and V_s is the *in situ* shear wave velocity obtained by field measurement. If shear wave velocities are not available, G_{max} can be estimated using published empirical formulas which correlate shear wave velocity or shear modulus with blow counts and/or other soil parameters [38–43]. To obtain the equivalent linearized values of G/G_{max} and β following each consecutive time history response analysis, values are taken from the G/G_{max} vs. γ and β vs. γ relations at the effective shear strain level defined as $\gamma_{eff} = \alpha\gamma_{max}$ in which γ_{max} is the maximum shear strain reached in the last analysis and α is the effective strain factor. In the past, α has usually been assigned the value 0.65; however, other values have been proposed (e.g., Idriss and Sun [44]). The equivalent linear time history response analyses are performed in an iterative manner, with soil parameter adjustments being made after each analysis, until the effective shear strain converges to essentially the same value used in the previous iteration [45]. This normally takes four to eight iterations to reach 90 to 95% of full convergence when the effective shear strains do not exceed 1 to 2%. When the maximum strain exceeds 2%, a nonlinear site-response analysis is more appropriate. Computer programs available for this purpose are DESRA [46], DYNAFLOW [47], DYNAID [48], and SUMDES [49].

For generating vertical free-field motions produced by vertically propagating compression waves, the needed soil parameters are the low-strain constrained elastic modulus $E_p = \rho V_p^2$, where V_p is the compression wave velocity, and the corresponding damping ratio. The variations of these soil parameters with compressive strain have not as yet been well established. At the present time, vertical site-response analyses have generally been carried out using the low-strain constrained elastic moduli, E_p, directly and the strain-compatible damping ratios obtained from the horizontal response analyses, but limited to a maximum value of 10%, without any further strain-compatibility iterations. For soils submerged in water, the value of E_p should not be less than the compression wave velocity of water.

Having generated acceleration free-field time histories of motion using the SHAKE computer program, the corresponding velocity and displacement time histories should be obtained through

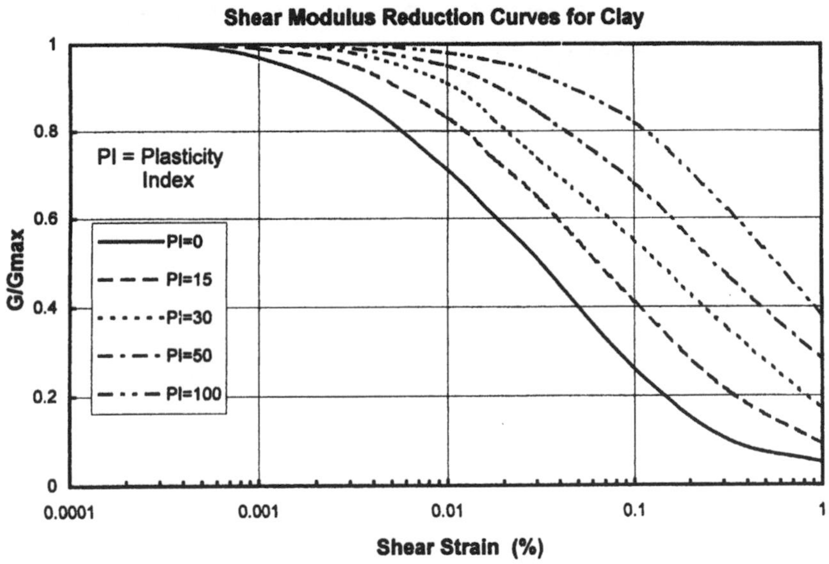

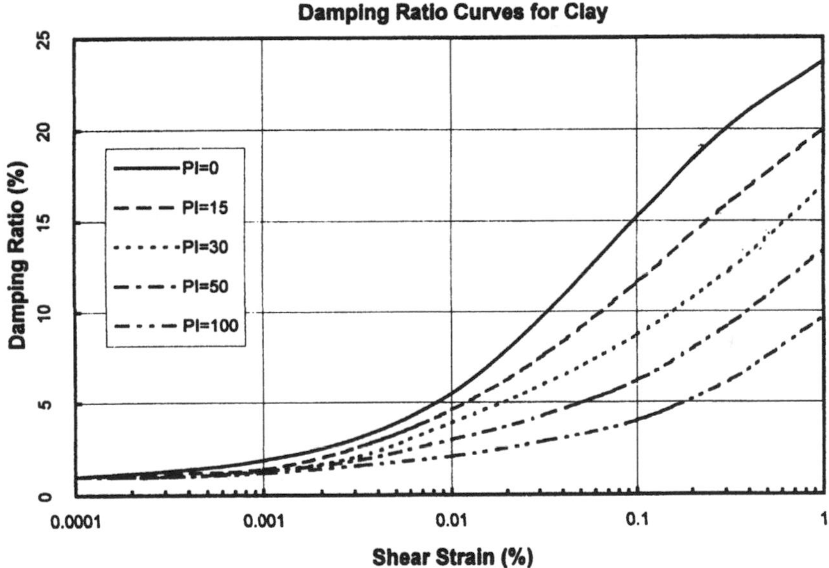

FIGURE 42.6 Equivalent linear shear modulus and hysteretic damping ratio as functions of shear strain for clay. (Source: Vucetic, M. and Dobry, R., *J. Geotech. Eng. ASCE,* 117(1), 89-107, 1991. With permission.)

single and double integrations of the acceleration time histories. Should unrealistic drifts appear in the displacement time histories, appropriate corrections should be applied. Should such drifts appear in a straight-line fashion, it usually indicates that the durations specified for Fourier transforming the recorded accelerograms are too short; thus, increasing these durations will usually correct the problem. If the baseline drifts depart significantly from a simple straight line, this tends to indicate that the analysis results may be unreliable; in which case, they should be carefully checked before being used. Time histories of free-field relative displacement between pairs of pier locations should also be generated and then be checked to judge the reasonableness of the results obtained.

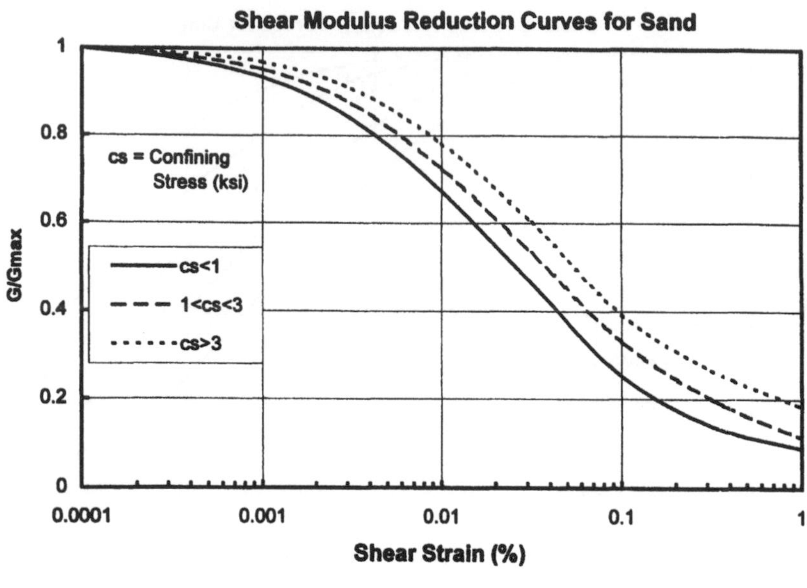

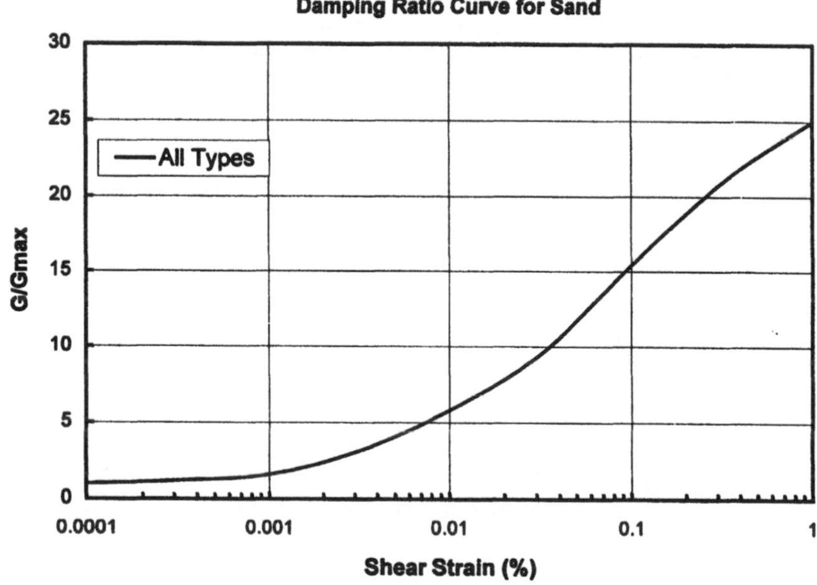

FIGURE 42.7 Equivalent linear shear modulus and hysteretic damping ratio as functions of shear strain for sand. (Source: Sun, J. I. et al., Reort No. UBC/EERC-88/15, Earthquake Engineer Research Center, University of California, Berkeley, 1988.)

42.5 Characterization of Soil–Foundation System

The core of the dynamic SFSI problem for a bridge is the interaction between its structure–foundation system and the supporting soil medium, which, for analysis purposes, can be considered to be a full half-space. The fundamental step in solving this problem is to characterize the constitutive relations between the dynamic forces acting on each foundation of the bridge at its interface boundary with the soil and the corresponding foundation motions, expressed in terms of the displacements, velocities, and accelerations. Such forces are here called the soil–foundation interaction forces. For a bridge subjected to externally applied loadings, such as dead, live, wind, and wave loadings, these SFI forces are functions of the foundation motions only; however, for a bridge subjected to seismic loadings, they are functions of the free-field soil motions as well.

Let h be the total number of degrees of freedom (DOF) of the bridge foundations as defined at their soil–foundation interface boundaries; $\boldsymbol{u}_h(t)$, $\dot{\boldsymbol{u}}_h(t)$, and $\ddot{\boldsymbol{u}}_h(t)$ be the corresponding foundation displacement, velocity, and acceleration vectors, respectively; and $\bar{\boldsymbol{u}}_h(t)$, $\dot{\bar{\boldsymbol{u}}}_h(t)$, and $\ddot{\bar{\boldsymbol{u}}}_h(t)$ be the free-field soil displacement, velocity, and acceleration vectors in the h DOF, respectively; and let $\boldsymbol{f}_h(t)$ be the corresponding SFI force vector. By using these notations, characterization of the SFI forces under seismic conditions can be expressed in the general vectorial functional form:

$$\boldsymbol{f}_h(t) = \Im_h\,(\boldsymbol{u}_h(t),\ \dot{\boldsymbol{u}}_h(t),\ \ddot{\boldsymbol{u}}_h(t),\ \bar{\boldsymbol{u}}_h(t)\ \ \dot{\bar{\boldsymbol{u}}}_h(t),\ \ddot{\bar{\boldsymbol{u}}}_h(t)) \tag{42.11}$$

Since the soils in the local region immediately surrounding each foundation may behave nonlinearly under imposed foundation loadings, the form of $\Im_h$ is, in general, a nonlinear function of displacements $\boldsymbol{u}_h(t)$ and $\bar{\boldsymbol{u}}_h(t)$ and their corresponding velocities and accelerations.

For a capacity evaluation, the nonlinear form of $\Im_h$ should be retained and used directly for determining the SFI forces as functions of the foundation and soil displacements. Evaluation of this form should be based on a suitable nonlinear model for the soil medium coupled with appropriate boundary conditions, subjected to imposed loadings which are usually much simplified compared with the actual induced loadings. This part of the evaluation will be discussed further in Section 42.8.

For a demand evaluation, the nonlinear form of $\Im_h$ is often linearized and then transformed to the frequency domain. Letting $\boldsymbol{u}_h(i\omega)$, $\dot{\boldsymbol{u}}_h(i\omega)$, $\ddot{\boldsymbol{u}}_h(i\omega)$, $\bar{\boldsymbol{u}}_h(i\omega)$, $\dot{\bar{\boldsymbol{u}}}_h(i\omega)$, $\ddot{\bar{\boldsymbol{u}}}_h(i\omega)$, and $\boldsymbol{f}_h(i\omega)$ be the Fourier transforms of $\boldsymbol{u}_h(t)$, $\dot{\boldsymbol{u}}_h(t)$, $\ddot{\boldsymbol{u}}_h(t)$, $\bar{\boldsymbol{u}}_h(t)$, $\dot{\bar{\boldsymbol{u}}}_h(t)$, $\ddot{\bar{\boldsymbol{u}}}_h(t)$, and $\boldsymbol{f}_h(t)$, respectively, and making use of the relations

$$\dot{\boldsymbol{u}}_h(i\omega) = i\omega\,\boldsymbol{u}_h(i\omega)\,; \qquad \ddot{\boldsymbol{u}}_h(i\omega) = -\omega^2\,\boldsymbol{u}_h(i\omega)$$

and

$$\dot{\bar{\boldsymbol{u}}}_h(i\omega) = i\omega\,\bar{\boldsymbol{u}}_h(i\omega)\,; \qquad \ddot{\bar{\boldsymbol{u}}}_h(i\omega) = -\omega^2\,\bar{\boldsymbol{u}}_h(i\omega)\,, \tag{42.12}$$

Equation (42.11) can be cast into the more convenient form:

$$\boldsymbol{f}_h(i\omega) = \Im_h\,(\,\boldsymbol{u}_h(i\omega), \bar{\boldsymbol{u}}_h(i\omega)\,) \tag{42.13}$$

To characterize the linear functional form of $\Im_h$, it is necessary to solve the dynamic boundary-value problem for a half-space soil medium subjected to force boundary conditions prescribed at the soil–foundation interfaces. This problem is referred to here as the "soil impedance" problem, which is a part of the foundation impedance problem referred to earlier in Section 42.2.2.

In linearized form, Eq. (42.13) can be expressed as

$$\boldsymbol{f}_h(i\omega) = \boldsymbol{G}_{hh}(i\omega)\ \ \{\boldsymbol{u}_h(i\omega) - \bar{\boldsymbol{u}}_h(i\omega)\} \tag{42.14}$$

in which $\boldsymbol{f}_h(i\omega)$ represents the force vector acting on the soil medium by the foundation and the matrix $\boldsymbol{G}_{hh}(i\omega)$ is a complex, frequency-dependent coefficient matrix called here the "soil impedance matrix."

Define a force vector $\bar{\boldsymbol{f}}_h(i\omega)$ by the relation

$$\bar{\boldsymbol{f}}_h(i\omega) = \boldsymbol{G}_{hh}(i\omega)\ \ \bar{\boldsymbol{u}}_h(i\omega) \tag{42.15}$$

This force vector represents the internal dynamic forces acting on the bridge foundations at their soil–foundation interface boundaries resulting from the free-field soil motions when the foundations are held fixed, i.e., $\boldsymbol{u}_h(i\omega) = \mathbf{0}$. The force vector $\bar{\boldsymbol{f}}_h(i\omega)$ as defined in Eq. (42.15) is the "seismic driving force" vector mentioned previously in Section 42.2.2. Depending upon the type of bridge foundation, the characterization of the soil impedance matrix $\boldsymbol{G}_{hh}(i\omega)$ and associated free-field soil input motion vector $\bar{\boldsymbol{u}}_h(i\omega)$ for demand analysis purposes may be established utilizing different soil models as described below.

42.5.1 Elastodynamic Model

As mentioned in Section 42.3.1, for a large bridge foundation such as a large spread footing, caisson, or single or multiple shafts having very large diameters, for which the nonlinearities occurring in the local soil region immediately adjacent to the foundation are small, the soil impedance matrix $\boldsymbol{G}_{hh}(i\omega)$ can be evaluated utilizing the dynamic Green's functions (dynamic displacements of the soil medium due to harmonic point-load excitations) obtained from the solution of a dynamic boundary-value problem of a linear damped-elastic half-space soil medium subjected to harmonic point loads applied at each of the h DOF on the soil–foundation interface boundaries. Such solutions have been obtained in analytical form for a linear damped-elastic continuum half-space soil medium by Apsel [50] in 1979. Because of complexities in the analytical solution, dynamic Green's functions have only been obtained for foundations having relatively simple soil–foundation interface geometries, e.g., rectangular, cylindrical, or spherical soil–foundation interface geometries, supported in simple soil media. In practical applications, the dynamic Green's functions are often obtained in numerical forms based on a finite-element discretization of the half-space soil medium and a corresponding discretization of the soil–foundation interface boundaries using a computer program such as SASSI [51], which has the capability of properly simulating the wave radiation boundary conditions at the far field of the half-space soil medium. The use of finite-element soil models to evaluate the dynamic Green's functions in numerical form has the advantage that foundations having arbitrary soil–foundation interface geometries can be easily handled; it, however, suffers from the disadvantage that the highest frequency, i.e., cutoff frequency, of motion for which a reliable solution can be obtained is limited by size of the finite element used for modeling the soil medium.

Having evaluated the dynamic Green's functions using the procedure described above, the desired soil impedance matrix can then be obtained by inverting, frequency-by-frequency, the "soil compliance matrix," which is the matrix of Green's function values evaluated for each specified frequency ω. Because the dynamic Green's functions are complex valued and frequency dependent, the coefficients of the resulting soil impedance matrix are also complex-valued and frequency dependent. The real parts of the soil impedance coefficients represent the dynamic stiffnesses of the soil medium which also incorporate the soil inertia effects; the imaginary parts of the coefficients represent the energy losses resulting from both soil material damping and radiation of stress waves into the far-field soil medium. Thus, the soil impedance matrix as developed reflects the overall dynamic characteristics of the soil medium as related to the motion of the foundation at the soil–foundation interfaces.

Because of the presence of the foundation excavation cavities in the soil medium, the vector of free-field soil motions $\bar{\boldsymbol{u}}_h(i\omega)$ prescribed at the soil–foundation interface boundaries has to be derived from the seismic input motions of the free-field soil medium without the foundation excavation cavities as described in Section 42.4. The derivation of the motion vector $\bar{\boldsymbol{u}}_h(i\omega)$ requires the solution of a dynamic boundary-value problem for the free-field half-space soil medium having foundation excavation cavities subjected to a specified seismic wave input such that the resulting solution satisfies the traction-free conditions at the surfaces of the foundation excavation cavities. Thus, the resulting seismic response motions, $\bar{\boldsymbol{u}}_h(i\omega)$, reflect the effects of seismic wave scattering due to the presence of the cavities. These motions are, therefore, referred to here as the "scattered free-field soil input motions."

The effects of seismic wave scattering depend on the relative relation between the characteristic dimension, ℓ_f, of the foundation and the specific seismic input wave length, λ, of interest, where

$\lambda = 2\pi V_s/\omega$ or $2\pi V_p/\omega$ for vertically propagating plane shear or compression waves, respectively; V_s and V_p are, as defined previously, the shear and compression wave velocities of the soil medium, respectively. If the input seismic wave length λ is much longer than the characteristic length ℓ_f, the effect of wave scattering will be negligible; on the other hand, when $\lambda \leq \ell_f$, the effect of wave scattering will be significant. Since the wave length λ is a function of the frequency of input motion, the effect of wave scattering is also frequency dependent. Thus, it is evident that the effect of wave scattering is much more important for a large bridge foundation, such as a large caisson or a group of very large diameter shafts, than for a small foundation having a small characteristic dimension, such as a slender-pile group; it can also be readily deduced that the scattering effect is more significant for foundations supported in soft soil sites than for those in stiff soil sites.

The characterization of the soil impedance matrix utilizing an elastodynamic model of the soil medium as described above requires soil material characterization constants which include (1) mass density, ρ; (2) shear and constrained elastic moduli, G and E_p (or shear and compression wave velocities, V_s and V_p); and (3) constant-hysteresis damping ratio, β. As discussed previously in Section 42.4.3, the soil shear modulus decreases while the soil hysteresis damping ratio increases as functions of soil shear strains induced in the free-field soil medium due to the seismic input motions. The effects of these so-called global soil nonlinearities can be easily incorporated into the soil impedance matrix based on an elastodynamic model by using the free-field-motion-induced strain-compatible soil shear moduli and damping ratios as the soil material constants in the evaluation of the dynamic Green's functions. For convenience of later discussions, the soil impedance matrix, $\boldsymbol{G}_{hh}(i\omega)$, characterized using an elastodynamic model will be denoted by the symbol $\boldsymbol{G}^e_{hh}(i\omega)$.

42.5.2 Empirical *p–y* Model

As discussed in Section 42.3.2, for a slender-pile group foundation for which soil nonlinearities occurring in the local soil regions immediately adjacent to the piles dominate the behavior of the foundation under loadings, the characterization of the soil resistances to pile deflections has often relied on empirically derived *p–y* curves for lateral resistance and *t–z* and *Q–d* curves for axial resistance. For such a foundation, the characterization of the soil impedance matrix needed for demand analysis purposes can be made by using the secant moduli derived from the nonlinear *p–y*, *t–z*, and *Q–d* curves, as indicated schematically in Figure 42.2. Since the development of these empirical curves has been based upon static or pseudo-static test results, it does not incorporate the soil inertia and material damping effects. Thus, the resulting soil impedance matrix developed from the secant moduli of the *p–y*, *t–z*, and *Q–d* curves reflects only the static soil stiffnesses but not the soil inertia and soil material damping characteristics. Hence, the soil impedance matrix so obtained is a real-valued constant coefficient matrix applicable at the zero frequency ($\omega = 0$); it, however, is a function of the foundation displacement amplitude. This matrix is designated here as $\boldsymbol{G}^s_{hh}(0)$ to differentiate it from the soil impedance matrix $\boldsymbol{G}^e_{hh}(i\omega)$ defined previously. Thus, Eq. (42.14) in this case is given by

$$\boldsymbol{f}_h(i\omega) = \boldsymbol{G}^s_{hh}(0)\{\boldsymbol{u}_h(i\omega) - \bar{\boldsymbol{u}}_h(i\omega)\} \tag{42.16}$$

where $\boldsymbol{G}^s_{hh}(0)$ depends on the amplitudes of the relative displacement vector $\Delta\boldsymbol{u}_h(i\omega)$ defined by

$$\Delta\boldsymbol{u}_h(i\omega) = \boldsymbol{u}_h(i\omega) - \bar{\boldsymbol{u}}_h(i\omega) \tag{42.17}$$

As mentioned previously in Section 42.3.2, the construction of the *p–y*, *t–z*, and *Q–d* curves depends only on the strength parameters but not on the stiffness parameters of the soil medium; thus, the effects of global soil nonlinearities on the dynamic stiffnesses of the soil medium, as caused by soil shear modulus decrease and soil-damping increase as functions of free-field-motion-induced soil shear strains, cannot be incorporated into the soil impedance matrix developed from these curves. Furthermore, since these curves are developed on the basis of results from field tests in which there are no free-field ground-motion-induced soil deformations, the effects of such global soil nonlinearities on the soil strength characterization parameters and hence the *p–y*, *t–z*, and *Q–d* curves cannot be incorporated.

Because of the small cross-sectional dimensions of slender piles, the seismic wave-scattering effect due to the presence of pile cavities is usually negligible; thus, the scattered free-field soil input motions $\bar{u}_h(i\omega)$ in this case are often taken to be the same as the free-field soil motions when the cavities are not present.

42.5.3 Hybrid Model

From the discussions in the above two sections, it is clear that characterization of the SFI forces for demand analysis purposes can be achieved using either an elastodynamic model or an empirical *p–y* model for the soil medium, each of which has its own merits and deficiencies. The elastodynamic model is capable of incorporating soil inertia, damping (material and radiation), and stiffness characteristics, and it can incorporate the effects of global soil nonlinearities induced by the free-field soil motions in an equivalent linearized manner. However, it suffers from the deficiency that it does not allow for easy incorporation of the effects of local soil nonlinearities. On the contrary, the empirical *p–y* model can properly capture the effects of local soil nonlinearities in an equivalent linearized form; however, it suffers from the deficiencies of not being able to simulate soil inertia and damping effects properly, and it cannot treat the effects of global soil nonlinearities. Since the capabilities of the two models are mutually complementary, it is logical to combine the elastodynamic model with the empirical *p–y* model in a series form such that the combined model has the desired capabilities of both models. This combined model is referred to here as the "hybrid model."

To develop the hybrid model, let the relative displacement vector, $\Delta \boldsymbol{u}_h(i\omega)$, between the foundation displacement vector $\boldsymbol{u}_h(i\omega)$ and the scattered free-field soil input displacement vector $\bar{\boldsymbol{u}}_h(i\omega)$, as defined by Eq. (42.17), be decomposed into a component representing the relative displacements at the soil–foundation interface boundary resulting from the elastic deformation of the global soil medium outside of the soil–foundation interface, designated as $\Delta \boldsymbol{u}_h^e(i\omega)$, and a component representing the relative displacements at the same boundary resulting from the inelastic deformations of the local soil regions adjacent the foundation, designated as $\Delta \boldsymbol{u}_h^i(i\omega)$; thus,

$$\Delta \boldsymbol{u}_h(i\omega) = \Delta \boldsymbol{u}_h^i(i\omega) + \Delta \boldsymbol{u}_h^e(i\omega) \tag{42.18}$$

Let $\boldsymbol{f}_h^e(i\omega)$ represent the elastic force vector which can be characterized in terms of the elastic relative displacement vector $\boldsymbol{u}_h^e(i\omega)$ using the elastodynamic model, in which case

$$\boldsymbol{f}_h^e(i\omega) = \boldsymbol{G}_{hh}^e(i\omega)\Delta \boldsymbol{u}_h^e(i\omega) \tag{42.19}$$

where $\boldsymbol{G}_{hh}^e(i\omega)$ is the soil impedance matrix as defined previously in Section 42.5.1, which can be evaluated using an elastodynamic model. Let $\boldsymbol{f}_h^i(i\omega)$ represent the inelastic force vector which is assumed to be related to $\Delta \boldsymbol{u}_h^i(i\omega)$ by the relation

$$\boldsymbol{f}_h^i(i\omega) = \boldsymbol{G}_{hh}^i(i\omega)\Delta \boldsymbol{u}_h^i(i\omega) \tag{42.20}$$

The characterization of the matrix $\boldsymbol{G}_{hh}^i(i\omega)$ can be accomplished by utilizing the soil secant stiffness matrix $\boldsymbol{G}_{hh}^s(0)$ developed from the empirical *p–y* model by the procedure discussed below.

Solving Eqs. (42.19) and (42.20) for $\Delta \boldsymbol{u}_h^e(i\omega)$ and $\Delta \boldsymbol{u}_h^i(i\omega)$, respectively, substituting these relative displacement vectors into Eq. (42.18), and making use of the force continuity condition that $f_h^e(i\omega) = f_h^i(i\omega)$, since the elastodynamic model and the inelastic local model are in series, one obtains

$$f_h(i\omega) = \left\{[G_{hh}^i(i\omega)]^{-1} + [G_{hh}^e(i\omega)]^{-1}\right\}^{-1} \ \Delta u_h(i\omega) \tag{42.21}$$

Comparing Eq. (42.14) with Eq. (42.21), one finds that by using the hybrid model, the soil impedance matrix is given by

$$G_{hh}(i\omega) = \left\{[G_{hh}^i(i\omega)]^{-1} + [G_{hh}^e(i\omega)]^{-1}\right\}^{-1} \tag{42.22}$$

Since the soil impedance matrix $G_{hh}^s(i\omega)$ is formed by the static secant moduli of the nonlinear *p–y*, *t–z*, and *Q–d* curves when $\omega = 0$, Eq. (42.22) becomes

$$G_{hh}^s(0) = \left\{[G_{hh}^i(0)]^{-1} + [G_{hh}^e(0)]^{-1}\right\}^{-1} \tag{42.23}$$

where $G_{hh}^s(0)$ is the soil stiffness matrix derived from the secant moduli of the nonlinear *p–y*, *t–z*, and *Q–d* curves. Solving Eq. (42.23), for $G_{hh}^i(0)$ gives

$$G_{hh}{}^i(0) = \left\{[G_{hh}^s(0)]^{-1} - [G_{hh}^e(0)]^{-1}\right\}^{-1} \tag{42.24}$$

Thus, Eq. (42.22) can be expressed in the form

$$G_{hh}(i\omega) = \left\{[G_{hh}^i(0)]^{-1} + [G_{hh}^e(i\omega)]^{-1}\right\}^{-1} \tag{42.25}$$

From Eq. (42.25), it is evident that when $\Delta u_h{}^i(i\omega) << \Delta u_h{}^e(i\omega)$, $G_{hh}(i\omega) \rightarrow G_{hh}^e(i\omega)$; however, when $\Delta u_h{}^i(i\omega) >> \Delta u_h{}^e(i\omega)$, $G_{hh}(i\omega) \rightarrow G_{hh}^i(0) \rightarrow G_{hh}^s(0)$. Thus, the hybrid model represented by this equation converges to the elastodynamic model when the local inelastic soil deformations are relatively small, as for the case of a large footing, caisson, or very large diameter shaft foundation, whereas it converges to the empirical *p–y* model when the local inelastic soil deformations are relatively much larger, as for the case of a slender-pile group foundation. For a moderately large diameter shaft foundation, the local inelastic and global elastic soil deformations may approach a comparable magnitude; in which case, the use of a hybrid model to develop the soil impedance matrix as described above can properly represent both the global elastodynamic and local inelastic soil behaviors.

As local soil nonlinearities are induced by the relative displacements between the foundation and the scattered free-field soil input motions, they do not affect the scattering of free-field soil motions due to the traction-free conditions present at the surface of the foundation cavities. Therefore, in applying the hybrid model described above, the scattered free-field soil input motion vector $\bar{u}_h(i\omega)$ should still be derived using the elastodynamic model described in Section 42.5.1.

42.6 Demand Analysis Procedures

42.6.1 Equations of Motion

The seismic response of a complete bridge system involves interactions between the structure and its supporting foundations and between the foundations and their surrounding soil media. To develop the equations of motion governing the response of this system in discrete (finite-element) form, let s denote the number of DOF in the structure, excluding its f DOF at the structure/foundation interface locations, and let g denote the number of DOF in the foundations, also excluding the f DOF but including the h DOF at all soil–foundation interfaces as defined in Section 42.5. Corresponding with those DOF, let vectors $\boldsymbol{u}_s(t)$, $\boldsymbol{u}_f(t)$, and $\boldsymbol{u}_g(t)$ contain the total displacement time histories of motion at the DOF s, f, and g, respectively.

Linear Modeling

Since the soil medium surrounding all foundations is continuous and of infinite extent, a rigorous model of a complete bridge system must contain stiffness and damping coefficients which are dependent upon the excitation (or response) frequencies. Such being the case, the corresponding equations of motion of the complete system having n DOF ($n = s + f + g$) must rigorously be represented in the frequency domain.

Considering the coupled structure–foundation system as a free-free (no boundary constraints) system having externally applied forces $-\boldsymbol{f}_h(t)$ acting in the h DOF, its equations of motion can be expressed in the frequency-domain form:

$$\begin{bmatrix} \boldsymbol{D}_{ss}(i\omega) & \boldsymbol{D}_{sf}(i\omega) & \boldsymbol{0} \\ \boldsymbol{D}_{sf}^T(i\omega) & \boldsymbol{D}_{ff}(i\omega) & \boldsymbol{D}_{fg}(i\omega) \\ \boldsymbol{0} & \boldsymbol{D}_{sf}^T(i\omega) & \boldsymbol{D}_{gg}(i\omega) \end{bmatrix} \begin{Bmatrix} \boldsymbol{u}_s(i\omega) \\ \boldsymbol{u}_f(i\omega) \\ \boldsymbol{u}_g(i\omega) \end{Bmatrix} = \begin{Bmatrix} \boldsymbol{0} \\ \boldsymbol{0} \\ \boldsymbol{f}_g(i\omega) \end{Bmatrix} \tag{42.26}$$

in which $\boldsymbol{u}_s(i\omega)$, $\boldsymbol{u}_f(i\omega)$, $\boldsymbol{u}_g(i\omega)$, and $\boldsymbol{f}_g(i\omega)$ are the Fourier transforms of vectors $\boldsymbol{u}_s(t)$, $\boldsymbol{u}_f(t)$, $\boldsymbol{u}_g(t)$, and $\boldsymbol{f}_g(t)$, respectively; and matrices $\boldsymbol{D}_{ij}(i\omega)$, $i, j = s, f, g$, are the corresponding impedance (dynamic stiffness) matrices. The g components in vectors $\boldsymbol{u}_g(i\omega)$ and $\boldsymbol{f}_g(i\omega)$ are ordered such that their last h components make up vectors $\boldsymbol{u}_h(i\omega)$ and $-\boldsymbol{f}_h(i\omega)$, respectively, with all other components being equal to zero.

For a viscously damped linear structure–foundation system, the impedance matrices $\boldsymbol{D}_{ij}(i\omega)$ are of the form:

$$\boldsymbol{D}_{ij}(i\omega) = \boldsymbol{K}_{ij} + i\omega\boldsymbol{C}_{ij} - \omega^2\boldsymbol{M}_{ij} \quad i, j = s, f, g \tag{42.27}$$

in which $\boldsymbol{K}_{ij}$, $\boldsymbol{C}_{ij}$, and $\boldsymbol{M}_{ij}$ are the standard stiffness, damping, and mass matrices, respectively, which would appear in the equations of motion of the system if expressed in the time domain. For a constant-hysteresis-damped linear system, the impedance matrices are given by

$$\boldsymbol{D}_{ij}(i\omega) = \boldsymbol{K}_{ij}^* - \omega^2\boldsymbol{M}_{ij} \quad i, j = s, f, g \tag{42.28}$$

in which $\boldsymbol{K}_{ij}^*$ is a complex stiffness matrix obtained by assembling individual finite-element matrices $\boldsymbol{K}^{*(m)}$ of the form

$$\boldsymbol{K}^{*(m)} \equiv \left\{1 - 2(\beta^{(m)})^2 + 2i\beta^{(m)}\sqrt{1-(\beta^{(m)})^2}\right\}\boldsymbol{K}^{(m)} \doteq (1 + 2i\beta^{(m)})\boldsymbol{K}^{(m)} \tag{42.29}$$

where $\boldsymbol{K}^{(m)}$ denotes the standard elastic stiffness matrix for finite element m as used in the assembly process to obtain matrix $\boldsymbol{K}_{ij}$ and $\beta^{(m)}$ is a damping ratio specified appropriately for the material used in finite-element m [56].

The hysteretic form of damping represented in Eq. (42.28) is the more appropriate form to use for two reasons: (1) it is easy to accommodate different damping ratios for the different materials used in the system and (2) the resulting modal damping is independent of excitation (or response) frequency ω, consistent with test evidence showing that reai damping is indeed essentially independent of this frequency. As noted by the form of Eq. (42.27), viscous damping is dependent upon frequency ω, contrary to test results; thus, preference should definitely be given to the use of hysteretic damping for linear systems which can be solved in the frequency domain. Hysteretic damping is unfortunately incompatible with solutions in the time domain.

Vector $-\boldsymbol{f}_h(i\omega)$, which makes up the last h components in force vector $\boldsymbol{f}_g(i\omega)$ appearing in Eq. (42.26), represents, as defined in Section 42.5, the internal SFI forces at the soil–foundation interfaces when the entire coupled soil–foundation–structure system is responding to the free-field soil input motions. Therefore, to solve the SFSI problem, this vector must be characterized in terms of the foundation displacement vector $\boldsymbol{u}_h(i\omega)$ and the scattered free-field soil displacement vector $\bar{\boldsymbol{u}}_h(i\omega)$. As discussed previously in Section 42.5, for demand analysis purposes, this vector can be linearized to the form

$$-\boldsymbol{f}_h(i\omega) = \boldsymbol{G}_{hh}(i\omega)\{\bar{\boldsymbol{u}}_h(i\omega) - \boldsymbol{u}_h(i\omega)\} \tag{42.30}$$

in which $-\boldsymbol{f}_h(i\omega)$ represents the force vector acting on the foundations from the soil medium and $\boldsymbol{G}_{hh}(i\omega)$ is the soil impedance matrix which is complex valued and frequency dependent.

Substituting Eq. (42.30) into Eq. (42.26), the equations of motion of the complete bridge system become

$$\begin{bmatrix} \boldsymbol{D}_{ss}(i\omega) & \boldsymbol{D}_{sf}(i\omega) & \boldsymbol{0} \\ \boldsymbol{D}_{sf}^T(i\omega) & \boldsymbol{D}_{ff}(i\omega) & \boldsymbol{D}_{fg}(i\omega) \\ \boldsymbol{0} & \boldsymbol{D}_{fg}^T(i\omega) & [\boldsymbol{D}_{gg}(i\omega)+\boldsymbol{G}_{gg}(i\omega)] \end{bmatrix} \begin{Bmatrix} \boldsymbol{u}_s(i\omega) \\ \boldsymbol{u}_f(i\omega) \\ \boldsymbol{u}_g(i\omega) \end{Bmatrix} = \begin{Bmatrix} \boldsymbol{0} \\ \boldsymbol{0} \\ \bar{\boldsymbol{f}}_g(i\omega) \end{Bmatrix} \tag{42.31}$$

in which

$$\boldsymbol{G}_{gg}(i\omega) = \begin{bmatrix} \boldsymbol{0} & \boldsymbol{0} \\ \boldsymbol{0} & \boldsymbol{G}_{hh}(i\omega) \end{bmatrix}; \quad \bar{\boldsymbol{f}}_g(i\omega) = \begin{Bmatrix} \boldsymbol{0} \\ \bar{\boldsymbol{f}}_h(i\omega) \end{Bmatrix} \tag{42.32}$$

Vector $\bar{\boldsymbol{f}}_h(i\omega)$ is the free-field soil "seismic driving force" vector defined by Eq. (42.15), in which the free-field soil displacements in vector $\bar{\boldsymbol{u}}_h(i\omega)$ result from scattering of incident seismic waves propagating to the bridge site as explained previously in Section 42.5.

Nonlinear Modeling

When large nonlinearities develop in the structure–foundation subsystem during a seismic event, evaluation of its performance requires nonlinear modeling and analysis in the time domain. In this case, the standard linear equations of motion of the complete system as expressed by

$$\begin{bmatrix} \boldsymbol{M}_{ss} & \boldsymbol{M}_{sf} & \boldsymbol{0} \\ \boldsymbol{M}_{sf}^T & \boldsymbol{M}_{ff} & \boldsymbol{M}_{fg} \\ \boldsymbol{0} & \boldsymbol{M}_{fg}^T & \boldsymbol{M}_{gg} \end{bmatrix} \begin{Bmatrix} \ddot{\boldsymbol{u}}_s(t) \\ \ddot{\boldsymbol{u}}_f(t) \\ \ddot{\boldsymbol{u}}_g(t) \end{Bmatrix} + \begin{bmatrix} \boldsymbol{C}_{ss} & \boldsymbol{C}_{sf} & \boldsymbol{0} \\ \boldsymbol{C}_{sf}^T & \boldsymbol{C}_{ff} & \boldsymbol{C}_{fg} \\ \boldsymbol{0} & \boldsymbol{C}_{fg}^T & \boldsymbol{C}_{gg} \end{bmatrix} \begin{Bmatrix} \dot{\boldsymbol{u}}_s(t) \\ \dot{\boldsymbol{u}}_f(t) \\ \dot{\boldsymbol{u}}_g(t) \end{Bmatrix} + \begin{bmatrix} \boldsymbol{K}_{ss} & \boldsymbol{K}_{sf} & \boldsymbol{0} \\ \boldsymbol{K}_{sf}^T & \boldsymbol{K}_{ff} & \boldsymbol{K}_{fg} \\ \boldsymbol{0} & \boldsymbol{K}_{fg}^T & \boldsymbol{K}_{gg} \end{bmatrix} \begin{Bmatrix} \boldsymbol{u}_s(t) \\ \boldsymbol{u}_f(t) \\ \boldsymbol{u}_g(t) \end{Bmatrix} = \begin{Bmatrix} \boldsymbol{0} \\ \boldsymbol{0} \\ \boldsymbol{f}_g(t) \end{Bmatrix} \tag{42.33}$$

must be modified appropriately to characterize the nonlinearities for use in a step-by-step numerical solution. Usually, it is the third term on the left-hand side of this equation that must be modified to represent the nonlinear hysteretic force–deformation behavior taking place in the individual finite elements of the system. The second term in this equation, representing viscous damping forces, is usually retained in its linear form with the full viscous damping matrix C being expressed in the Rayleigh form

$$C = \alpha_R M + \beta_R K \tag{42.34}$$

in which M and K are the full mass and elastic-stiffness matrices shown in Eq. (42.33) and α_R and β_R are constants assigned numerical values which will limit the modal damping ratios to levels within acceptable bounds over a range of modal frequencies dominating the seismic response.

For a time-domain solution of Eq. (42.33) in its modified nonlinear form, all parameters in the equation must be real (no imaginary parts) and frequency independent. It remains therefore to modify the soil impedance matrix $G_{hh}(i\omega)$ so that when introduced into Eq. (42.30), the inverse Fourier transform of $-f_h(i\omega)$ to the time domain will yield a vector $-f_h(t)$ having no frequency-dependent parameters. To accomplish this objective, separate $G_{hh}(i\omega)$ into its real and imaginary parts in accordance with

$$G_{hh}(i\omega) = G_{hh}^R(\omega) + iG_{hh}^I(\omega) \tag{42.35}$$

in which $G_{hh}^R(\omega)$ and $G_{hh}^I(\omega)$ are real functions of ω. Then approximate these functions using the relations

$$G_{hh}^R(\omega) \doteq \overline{K}_{hh} - \omega^2 \overline{M}_{hh}\,; \quad G_{hh}^I(\omega) \doteq \omega \overline{C}_{hh} \tag{42.36}$$

where the real constants in matrices $\overline{K}_{hh}$, $\overline{M}_{hh}$, and $\overline{C}_{hh}$ are assigned numerical values to provide best fits to the individual frequency-dependent functions in matrices $G_{hh}^R(\omega)$ and $G_{hh}^I(\omega)$ over the frequency range of major influence on seismic response. Typically, applying these best fits to the range $0 < \omega < 4\pi$ radians/second, corresponding to the range $0 < f < 2$ Hz, where $f = \omega/2\pi$, is adequate for most large bridges. In this fitting process, it is sufficient to treat $\overline{M}_{hh}$ as a diagonal matrix, thus affecting only the diagonal functions in matrix $G_{hh}^R(\omega)$. The reason for selecting the particular frequency-dependent forms of Eqs. (42.36) is that when they are substituted into Eq. (42.35), which in turn is substituted into Eq. (42.30), the resulting expression for $f_h(i\omega)$ can be Fourier transformed to the time domain yielding

$$-f_h(t) = \overline{K}_{hh}\{\overline{u}_h(t) - u_h(t)\} + \overline{C}_{hh}\{\dot{\overline{u}}_h(t) - \dot{u}_h(t)\} + \overline{M}_{hh}\{\ddot{\overline{u}}_h(t) - \ddot{u}_h(t)\} \tag{42.37}$$

Substituting $-f_h(t)$ given by this equation for the last h components in vector $f_g(t)$, with all other components in $f_g(t)$ being equal to zero, and then substituting the resulting vector $f_g(t)$ into Eq. (42.33) gives

$$\begin{bmatrix} M_{ss} & M_{sf} & 0 \\ M_{sf}^T & M_{ff} & M_{fg} \\ 0 & M_{fg}^T & [M_{gg} + \overline{M}_{gg}] \end{bmatrix} \begin{Bmatrix} \ddot{u}_s(t) \\ \ddot{u}_f(t) \\ \ddot{u}_g(t) \end{Bmatrix} + \begin{bmatrix} C_{ss} & C_{sf} & 0 \\ C_{sf}^T & C_{ff} & C_{fg} \\ 0 & C_{fg}^T & [C_{gg} + \overline{C}_{gg}] \end{bmatrix} \begin{Bmatrix} \dot{u}_s(t) \\ \dot{u}_f(t) \\ \dot{u}_g(t) \end{Bmatrix} + \tag{42.38}$$

$$\begin{bmatrix} K_{ss} & K_{sf} & 0 \\ K_{sf}^T & K_{ff} & K_{fg} \\ 0 & K_{fg}^T & [K_{gg} + \overline{K}_{gg}] \end{bmatrix} \begin{Bmatrix} u_s(t) \\ u_f(t) \\ u_g(t) \end{Bmatrix} = \begin{Bmatrix} 0 \\ 0 \\ \overline{K}_{gg}\overline{u}_g(t) + \overline{C}_{gg}\dot{\overline{u}}_g(t) + \overline{M}_{gg}\ddot{\overline{u}}_g(t) \end{Bmatrix}$$

in which

$$\overline{M}_{gg} = \begin{bmatrix} 0 & 0 \\ 0 & \overline{M}_{hh} \end{bmatrix}; \quad \overline{K}_{gg} = \begin{bmatrix} 0 & 0 \\ 0 & \overline{K}_{hh} \end{bmatrix}; \quad \overline{C}_{gg} = \begin{bmatrix} 0 & 0 \\ 0 & \overline{C}_{hh} \end{bmatrix} \tag{42.39}$$

showing that no frequency-dependent parameters remain in the equations of motion, thus allowing the standard time-domain solution procedure to be used for solving them. Usually, the terms $\overline{C}_{gg}\dot{\overline{u}}_g(t)$ and $\overline{M}_{gg}\ddot{\overline{u}}_g(t)$ on the right-hand side of Eq. (42.38) have small effects on the solution of this equation; however, the importance of their contributions should be checked. Having modified the third term on the left-hand side of Eq. (42.38) to its nonlinear hysteretic form, the complete set of coupled equations can be solved for displacements $u_s(t)$, $u_f(t)$, $u_g(t)$ using standard step-by-step numerical integration procedures.

42.6.2 Solution Procedures

One-Step Direct Approach

In this approach, the equations of motion are solved directly in their coupled form. If the system is treated as being fully linear (or equivalent linear), the solution can be carried out in the frequency domain using Eq. (42.31). In doing so, the complete set of complex algebraic equations is solved separately for discrete values of ω over the frequency range of interest yielding the corresponding sets of displacement vectors $u_s(i\omega)$, $u_f(i\omega)$, and $u_g(i\omega)$. Having obtained these vectors for the discrete values of ω, they are inverse Fourier transformed to the time domain giving vectors $u_s(t)$, $u_f(t)$, $u_g(t)$. The corresponding time histories of internal forces and/or deformations in the system can then be obtained directly using standard finite-element procedures.

If the structure–foundation subsystem is modeled as a nonlinear system, the solution can be carried out in the time domain using Eq. (42.38). In this case, the coupled nonlinear equations of motion are solved using standard step-by-step numerical integration procedures.

This one-step direct approach is simple and straightforward to implement for a structural system supported on a single foundation, such as a building. However, for a long, multiple-span bridge supported on many independent foundations, a very large system of equations and an associated very large number of seismic free-field inputs in vector $\overline{u}_g\ (i\omega)$ result, making the solution computationally impractical, especially when large nonlinearities are present in the equations of motion. In this case, it is desirable to simplify the problem by finding separate solutions to a set of smaller problems and then combine the solutions in steps so as to achieve the desired end result. The multiple-step substructuring approach described subsequently is ideally suited for this purpose.

Multiple-Step Substructuring Approach

For long bridges supported on multiple foundations, the support separation distances are sufficiently large so that each foundation subsystem can be treated as being independent of the others; therefore, the soil impedance matrix for each foundation will be uncoupled from those of the other foundations. In this case, to simplify the overall problem, each foundation subsystem can be analyzed separately to obtain a boundary impedance matrix called the foundation impedance matrix and a consistent boundary force vector called the foundation driving-force vector, both of which are associated with the DOF at its structure–foundation interface. Having obtained the foundation impedance matrix and associated driving force vector for each foundation subsystem, all such matrices and vectors can be combined into the equations of motion for the total structure as a free-free system, resulting in $(s + f)$ DOF present in the structure–foundation subsystem rather than the $(s + f + g)$ DOF present in the complete soil–structure–foundation system. This reduced set of equations having $(s + f)$ DOF can be solved much more efficiently than solving the equations for the complete system having $(s + f + g)$ DOF as required by the one-step direct approach.

Referring to Eq. (42.31), it is seen that the linear equations of motion for each independent foundation system j can be expressed in the frequency-domain form:

$$\begin{bmatrix} \boldsymbol{D}^j_{ff}(i\omega) & \boldsymbol{D}^j_{fg}(i\omega) \\ \boldsymbol{D}^j_{fg}(i\omega)^T & [\boldsymbol{D}^j_{gg}(i\omega)+\boldsymbol{G}^j_{gg}(i\omega)] \end{bmatrix} \begin{Bmatrix} \boldsymbol{u}^j_f(i\omega) \\ \boldsymbol{u}^j_g(i\omega) \end{Bmatrix} = \begin{Bmatrix} \boldsymbol{0} \\ \overline{\boldsymbol{f}}^j_g(i\omega) \end{Bmatrix} \tag{42.40}$$

in which

$$\overline{\boldsymbol{f}}^j_g(i\omega) = \boldsymbol{G}^j_{gg}(i\omega)\overline{\boldsymbol{u}}^j_g(i\omega) \tag{42.41}$$

Solving the second of Eqs. (42.40) for $\boldsymbol{u}^j_g(i\omega)$ gives

$$\boldsymbol{u}^j_g(i\omega) = \left[\boldsymbol{D}^j_{gg}(i\omega)+\boldsymbol{G}^j_{gg}(i\omega)\right]^{-1}\left[-\boldsymbol{D}^j_{fg}(i\omega)^T\boldsymbol{u}^j_f(i\omega)+\overline{\boldsymbol{f}}^j_g(i\omega)\right] \tag{42.42}$$

Substituting this equation into the first of Eqs. (42.40) yields

$$\left[\boldsymbol{D}^j_{ff}(i\omega)+\boldsymbol{F}^j_{ff}(i\omega)\right]\boldsymbol{u}^j_f(i\omega) = \overline{\boldsymbol{f}}^j_f(i\omega) \tag{42.43}$$

where

$$\boldsymbol{F}^j_{ff}(i\omega) \equiv -\boldsymbol{D}^j_{fg}(i\omega)[\boldsymbol{D}^j_{gg}(i\omega)+\boldsymbol{G}^j_{gg}(i\omega)]^{-1}\boldsymbol{D}^j_{fg}(i\omega)^T \tag{42.44}$$

$$\overline{\boldsymbol{f}}^j_f(i\omega) \equiv -\boldsymbol{D}^j_{fg}(i\omega)[\boldsymbol{D}^j_{gg}(i\omega)+\boldsymbol{G}^j_{gg}(i\omega)]^{-1}\overline{\boldsymbol{f}}^j_g(i\omega) \tag{42.45}$$

Matrix $\boldsymbol{F}^j_{ff}(i\omega)$ and vector $\overline{\boldsymbol{f}}^j_f(i\omega)$ will be referred to here as the foundation impedance matrix and its associated foundation driving-force vector, respectively, for the *j*th foundation. For convenience, a foundation motion vector $\overline{\boldsymbol{u}}^j_f(i\omega)$ will now be defined as given by

$$\overline{\boldsymbol{u}}^j_f(i\omega) \equiv \boldsymbol{F}^j_{ff}(i\omega)^{-1}\overline{\boldsymbol{f}}^j_f(i\omega) \tag{42.46}$$

so that the driving-force vector $\overline{\boldsymbol{f}}^j_f(i\omega)$ can be expressed in the form

$$\overline{\boldsymbol{f}}^j_f(i\omega) = \boldsymbol{F}^j_{ff}(i\omega)\overline{\boldsymbol{u}}^j_f(i\omega) \tag{42.47}$$

The motion vector $\overline{\boldsymbol{u}}^j_f(i\omega)$ given by Eq. (42.46) will be referred to subsequently as the "effective (scattered) foundation input motion" vector. Conceptually, this is the vector of foundation motions which, when multiplied by the foundation impedance matrix $\boldsymbol{F}^j_{ff}(i\omega)$, yields the foundation driving-force vector $\overline{\boldsymbol{f}}^j_f(i\omega)$ resulting from the prescribed scattered free-field soil input motions contained in vector $\overline{\boldsymbol{u}}^j_g(i\omega)$.

Combining Eqs. (42.43) for all foundation subsystems with the equations of motion for the complete free-free structure subsystem yields the desired reduced matrix equation of motion for the entire structure–foundation system in the linear form

$$\begin{bmatrix} \boldsymbol{D}_{ss}(i\omega) & \boldsymbol{D}_{sf}(i\omega) \\ \boldsymbol{D}_{sf}(i\omega)^T & [\boldsymbol{D}^s_{ff}(i\omega)+\boldsymbol{F}_{ff}(i\omega)] \end{bmatrix} \begin{Bmatrix} \boldsymbol{u}_s(i\omega) \\ \boldsymbol{u}_f(i\omega) \end{Bmatrix} = \begin{Bmatrix} \boldsymbol{0} \\ \overline{\boldsymbol{f}}_f(i\omega) \end{Bmatrix} \tag{42.48}$$

in which $D_{ss}(i\omega)$ and $D_{sf}(i\omega)$ are given by Eqs. (42.27) and (42.28) directly, $D^s_{ff}(i\omega)$ is that part of $D_{ff}(i\omega)$ given by these same equations as contributed by the structure only, and

$$\bar{f}_f(i\omega) = F_{ff}(i\omega)\bar{u}_f(i\omega) \tag{42.49}$$

The solution of Eq. (42.48) for discrete values of ω over the frequency range of interest gives the desired solutions for $u_s(i\omega)$ and $u_f(i\omega)$. To obtain the corresponding solution $u^j_g(i\omega)$ for each foundation subsystem *j*, a backsubstitution is required. This is done by substituting the solution $u^j_f(i\omega)$ for each foundation subsystem *j* into Eq. (42.42) and computing the corresponding response motions in vector $u^j_g(i\omega)$. This step will be called the "foundation feedback" analysis.

When large nonlinearities develop in the structure during a seismic event, the reduced equations of motion representing the coupled structure–foundation system must be expressed in the time domain. To do so, consider the structure alone as a free-free linear system having externally applied forces $f_f(t)$ acting in the *f* DOF. The equations of motion for this system can be expressed in the frequency domain form:

$$\begin{bmatrix} D_{ss}(i\omega) & D_{sf}(i\omega) \\ D_{sf}(i\omega)^T & D^s_{ff}(i\omega) \end{bmatrix} \begin{Bmatrix} u_s(i\omega) \\ u_f(i\omega) \end{Bmatrix} = \begin{Bmatrix} 0 \\ f_f(i\omega) \end{Bmatrix} \tag{42.50}$$

in which $f_f(i\omega)$ is the Fourier transform of vector f_f(t). If Eq. (42.50) is to represent the coupled structure–foundation system, then $f_f(i\omega)$ must satisfy the relation

$$f_f(i\omega) = F_{ff}(i\omega)\left\{\bar{u}_f(i\omega) - u_f(i\omega)\right\} \tag{42.51}$$

in which matrix $F_{ff}(i\omega)$ is an assembly of the individual foundation impedance matrices $F^j_{ff}(i\omega)$ given by Eq. (42.44) for all values of *j* and vector $\bar{u}_f(i\omega)$ is the corresponding complete foundation-motion vector containing all individual vectors $\bar{u}^j_f(i\omega)$ given by Eq. (42.46).

Equation (42.50) can be converted to the time-domain form:

$$\begin{bmatrix} M_{ss} & M_{sf} \\ M^T_{sf} & M^s_{ff} \end{bmatrix} \begin{Bmatrix} \ddot{u}_s(t) \\ \ddot{u}_f(t) \end{Bmatrix} + \begin{bmatrix} C_{ss} & C_{sf} \\ C^T_{sf} & C^s_{ff} \end{bmatrix} \begin{Bmatrix} \dot{u}_s(t) \\ \dot{u}_f(t) \end{Bmatrix} + \begin{bmatrix} K_{ss} & K_{sf} \\ K^T_{sf} & K^s_{ff} \end{bmatrix} \begin{Bmatrix} u_s(t) \\ u_f(t) \end{Bmatrix} = \begin{Bmatrix} 0 \\ f_f(t) \end{Bmatrix} \tag{42.52}$$

in which K^s_{ff}, C^s_{ff}, and M^s_{ff} are the standard stiffness, damping, and mass matrices contributed by the structure only (no contributions from the foundation) and $f_f(t)$ is the inverse Fourier transform of $f_f(i\omega)$ given by Eq. (42.51). In order for $f_f(t)$ to have no frequency-dependent parameters, as required by a time-domain solution, matrix $\mathrm{F}_{ff}(i\omega)$ should be separated into its real and imaginary parts in accordance with

$$F_{ff}(i\omega) = F^R_{ff}(\omega) + iF^I_{ff}(\omega) \tag{42.53}$$

in which $F^R_{ff}(\omega)$ and $F^I_{ff}(\omega)$ can be approximated using the relations

$$F^R_{ff}(\omega) \doteq (\bar{K}_{ff} - \omega^2\bar{M}_{ff}); \quad F^I_{ff}(\omega) \doteq \omega\bar{C}_{ff} \tag{42.54}$$

where the real constants in matrices $\bar{K}_{ff}$, $\bar{M}_{ff}$, and $\bar{C}_{ff}$ are assigned numerical values to provide best fits to the individual frequency-dependent functions in matrices $F^R_{ff}(\omega)$ and $F^I_{ff}(\omega)$ over the frequency range of major influence on seismic response; usually the range $0 < \omega < 4\pi$ rad/s. is adequate for large bridges. In this fitting process, it is sufficient to treat $\bar{M}_{ff}$ as a diagonal matrix, thus affecting only the diagonal functions in matrix $F^R_{ff}(\omega)$.

Substituting Eqs. (42.54) into Eq. (42.53) and the resulting Eq. (42.53) into Eq. (42.51), this latter equation can be inverse Fourier transformed, giving

$$f_t(t) = \bar{K}_{ff}\{\bar{u}_f(t) - u_f(t)\} + \bar{C}_{ff}\{\dot{\bar{u}}_f(t) - \dot{u}_f(t)\} + \bar{M}_{ff}\{\ddot{\bar{u}}_f(t) - \ddot{u}_f(t)\} \tag{42.55}$$

which when introduced into Eq. (42.50) yields the desired reduced linear equations of motion in the time-domain form

$$\begin{bmatrix} M_{ss} & M_{sf} \\ M_{sf}^T & [M_{ff}^s + \bar{M}_{ff}] \end{bmatrix} \begin{Bmatrix} \ddot{u}_s(t) \\ \ddot{u}_f(t) \end{Bmatrix} + \begin{bmatrix} C_{ss} & C_{sf} \\ C_{sf}^T & [C_{ff}^s + \bar{C}_{ff}] \end{bmatrix} \begin{Bmatrix} \dot{u}_s(t) \\ \dot{u}_f(t) \end{Bmatrix} +$$

$$\begin{bmatrix} K_{ss} & K_{sf} \\ K_{sf}^T & [K_{ff}^s + \bar{K}_{ff}] \end{bmatrix} \begin{Bmatrix} u_s(t) \\ u_f(t) \end{Bmatrix} = \begin{bmatrix} 0 \\ \bar{K}_{ff}\bar{u}_f(t) + \bar{C}_{ff}\dot{\bar{u}}_f(t) + \bar{M}_{ff}\ddot{\bar{u}}_f(t) \end{bmatrix} \tag{42.56}$$

showing that no frequency-dependent parameters remain in the equations of motion, thus satisfying the time-domain solution requirement. Again, as explained previously, the full viscous damping matrix in this equation is usually expressed in the Rayleigh form given by Eq. (42.34) in which constants α_R and β_R are assigned numerical values to limit the modal damping ratios to levels within acceptable bounds over the range of frequencies dominating seismic response. As explained previously for Eq. (42.38), the damping and mass terms on the right-hand side of Eq. (42.56) usually have small effects on the solution; however, their importance should be checked.

Having modified the third term on the left-hand side of Eq. (42.56) to its nonlinear hysteretic form, the complete set of coupled equations can be solved for displacements $u_s(t)$ and $u_f(t)$ using standard step-by-step numerical integration procedures.

To obtain the corresponding $u_g^j(i\omega)$ for each foundation subsystem j, the previously defined foundation feedback analyses must be performed. To do so, each subvector $u_f^j(t)$ contained in vector $u_f(t)$, must be Fourier transformed to obtain $u_f^j(i\omega)$. Having these subvectors for all values of j, each one can be substituted separately into Eq. (42.42) giving the corresponding subvector $u_g^j(i\omega)$. Inverse Fourier transforming each of these subvectors yields the corresponding vectors $u_f^j(t)$ for all values of j.

42.7 Demand Analysis Examples

This section presents the results of three example solutions to illustrate applications of the demand analysis procedures described in the previous section, in particular, the multiple-step substructuring approach. These examples have been chosen from actual situations to illustrate application of the three methods of soil–foundation modeling: (1) the elastodynamic method, (2) the empirical *p–y* method, and (3) the hybrid method.

42.7.1 Caisson Foundation

The first example is chosen to illustrate application of the elastodynamic method of modeling and analysis to a deeply embedded caisson foundation of a large San Francisco Bay crossing bridge. The foundation considered is a large reinforced concrete cellular caisson, 80 ft (24.4 m) long, 176 ft (53.6 m) wide, and 282 ft (86.0 m) tall, located at a deep soil site and filled with water. The configuration of the caisson and its supporting soil profile and properties are shown in Figure 42.8. The soil properties are the shear-strain-compatible equivalent linear properties obtained from free-field site-response analyses using SHAKE with the seismic input motions prescribed at the bedrock surface

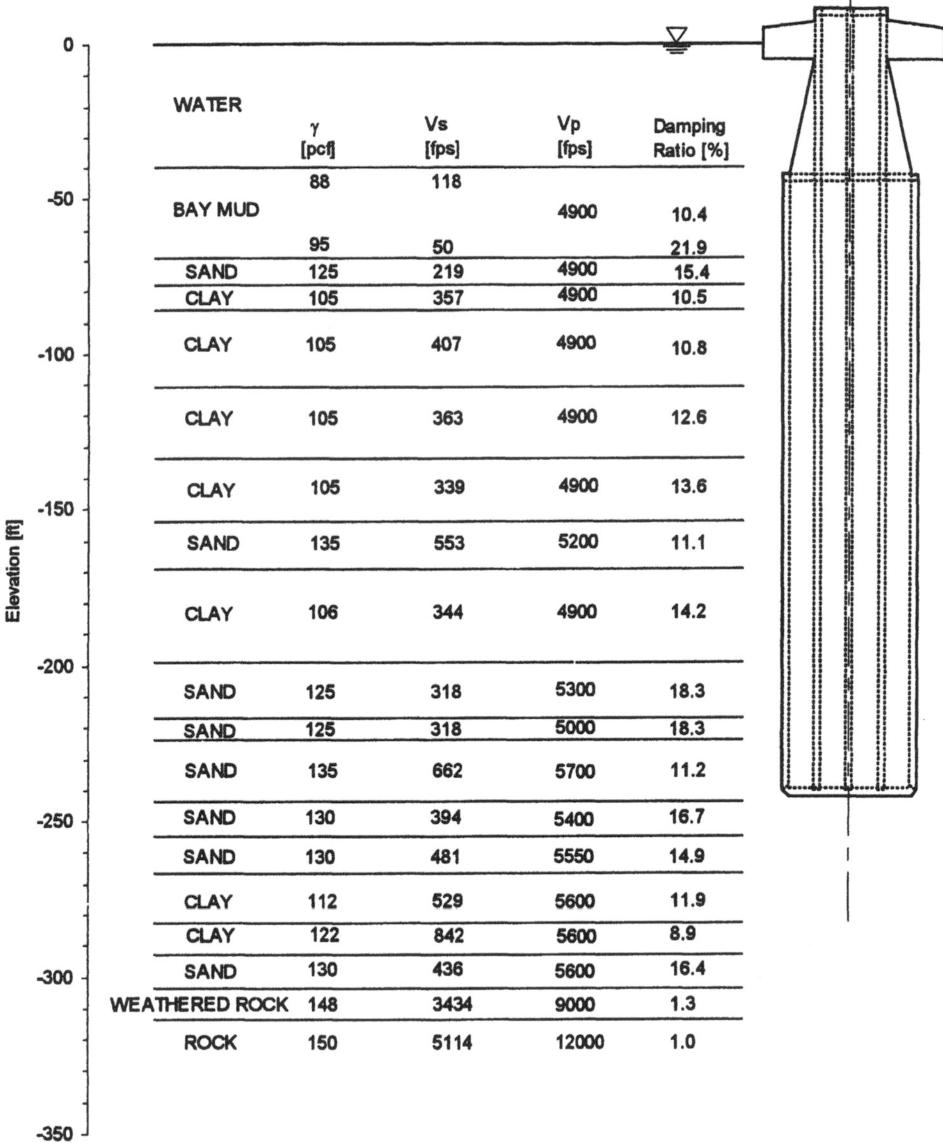

FIGURE 42.8 Configuration and soil profile and properties of the cassion foundation at its SASSI half-model.

in the form of rock-outcrop motion. Thus, these properties have incorporated stiffness degradation effects due to global soil nonlinearities induced in the free-field by the selected seismic input.

Since the caisson is deeply embedded and has large horizontal dimensions, the local soil nonlinearities that develop near the soil–caisson interface are relatively small; therefore, they were neglected in the demand analysis. The soil–caisson system was modeled using the elastodynamic method; i.e., the system was modeled by an elastic foundation structure embedded in a damped-elastic soil medium having the properties shown in Figure 42.8. This model, developed using the finite-element SASSI computer program for one quarter of the soil–caisson system, is shown in Figure 42.8. Using this model, the foundation impedance matrix, i.e., $\boldsymbol{F}_{ff}^{j}(i\omega)$ defined by Eq. (42.44), and its associated effective (scattered) foundation input motion vector, i.e., $\bar{\boldsymbol{u}}_{f}^{j}(i\omega)$ defined by Eq. (42.46), were evaluated consistent with the free-field seismic input using SASSI. The foundation impedance matrix associated with the six DOF of the node located at the top of the caisson El. 40 ft. (12.2 m) was

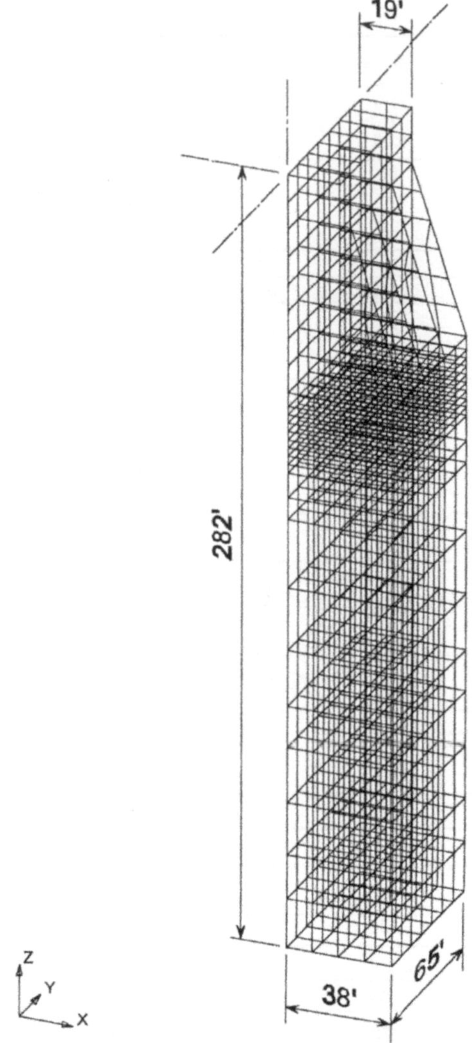

FIGURE 42.8 (continued)

computed following the procedure described in Section 42.6. The individual impedance functions in this matrix are shown in Figure 42.9. The amplitudes of the transfer functions for longitudinal response motions of the caisson relative to the corresponding seismic input motion, as computed for several elevations, are shown in Figure 42.10. The 5% damped acceleration response spectra computed for these motions are also shown in Figure 42.10 where they can be compared with the 5% damped response spectra for the corresponding seismic input motion prescribed at the bedrock level and the corresponding free-field soil motion at the mudline elevation.

As indicated in Figure 42.10, the soil–caisson interaction system alone, without pier tower and superstructure of the bridge being present, has characteristic translational and rocking mode frequencies of 0.7 and 1.4 Hz (periods 1.4 and 0.7 s), respectively. The longitudinal scattered foundation motion associated with the foundation impedance matrix mentioned above is the motion represented by the response spectrum for El. 40 ft (12.2 m) as shown in Figure 42.10.

The response spectra shown in Figure 42.10 indicate that, because of the 0.7-s translational period of the soil–caisson system, the scattered foundation motion at the top of the caisson where the bridge pier tower would be supported, exhibits substantial amplifications in the neighborhood of

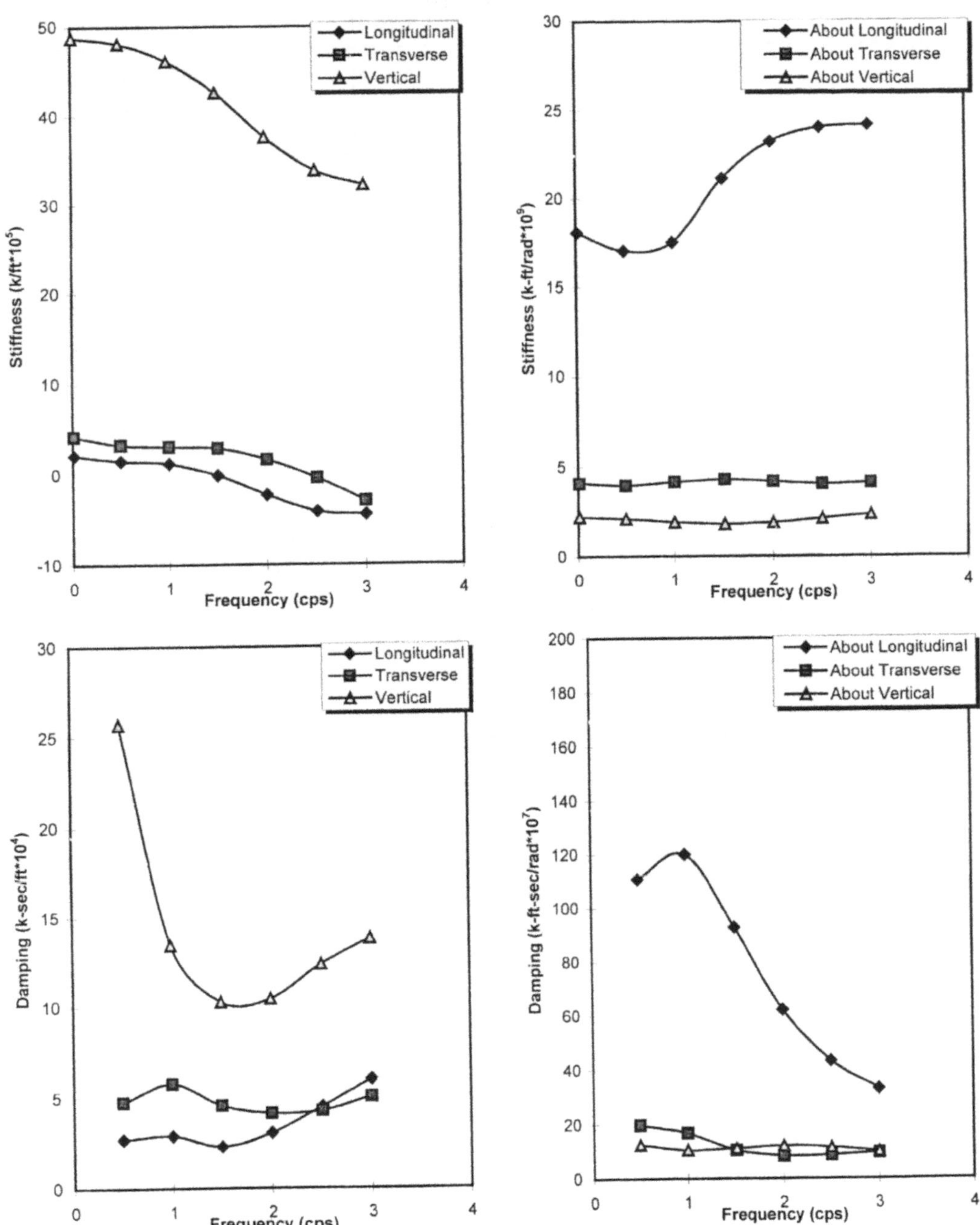

FIGURE 42.9 Foundation impedance functions at the top of the caisson considered.

this period. In the period range longer than 2.0 s, in which the major natural vibration frequencies of the bridge system are located, the spectral values for the scattered foundation input motion are seen to be smaller than the corresponding values for the free-field mudline motion. The above results point out the importance of properly modeling both the stiffness and the inertial properties of the soil–caisson system so that the resulting scattered foundation motions to be used as input to the foundation–structure system will appropriately represent the actual dynamic characteristics of the soil–caisson interaction system.

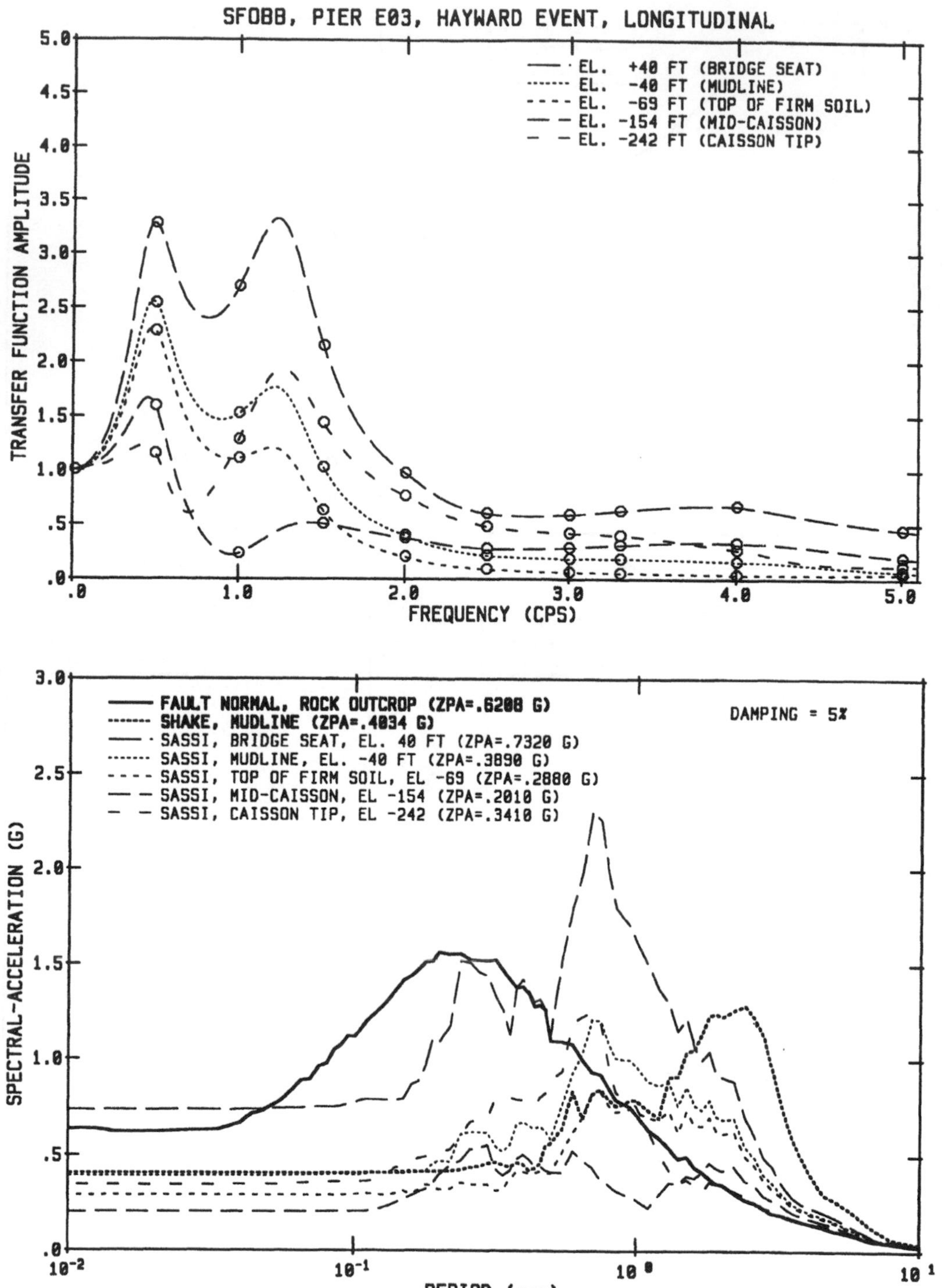

FIGURE 42.10 Transfer function amplitudes and 5% damped response spectra for scattered foundation motions of the caisson at several elevations.

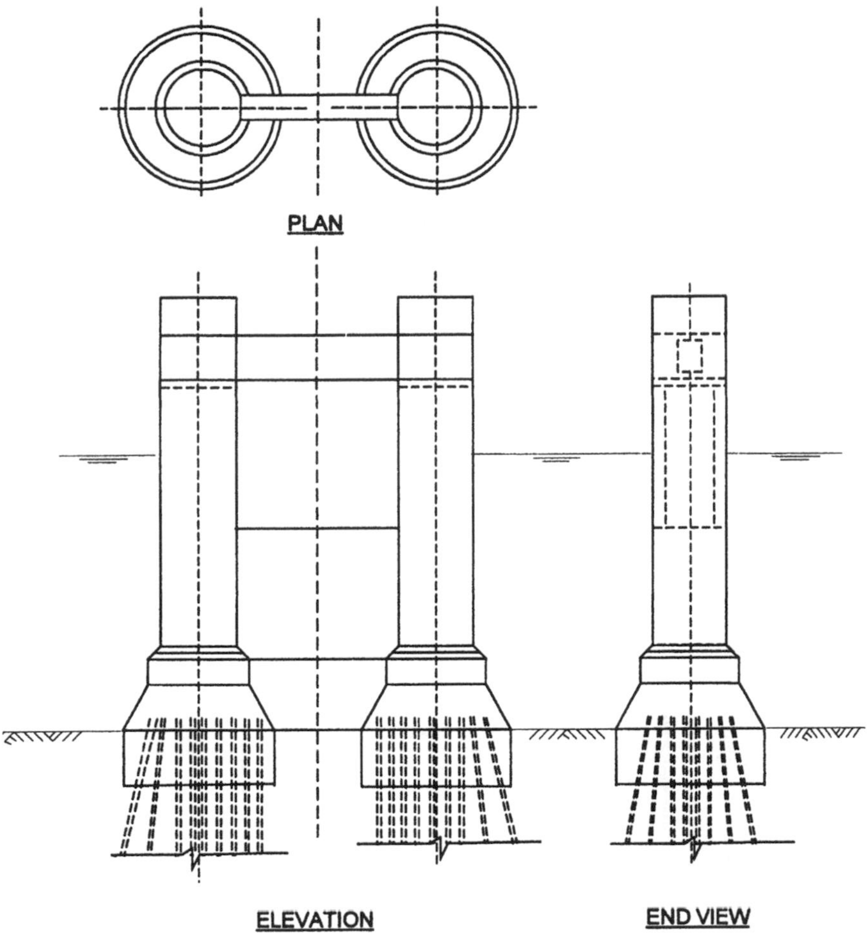

FIGURE 42.11 Configuration of the slender-pile group foundation considered.

42.7.2 Slender-Pile Group Foundation

The second example is to illustrate application of the empirical *p–y* method in a demand analysis of a slender-pile group foundation constructed at a deep soil site. The pile group foundation selected is one of 78 pier foundations of a long, water-crossing steel truss bridge. The foundation is constructed of two 24-ft (7.32-m)-diameter bell-shaped precast reinforced concrete pile caps, which are linked together by a deep cross-beam, as shown in Figure 42.11. Each bell-shaped pile cap is supported on a group of 28 steel 14BP89 H-pipes, giving a total of 56 piles supporting the combined two-bell pile cap. The piles in the outer ring and in the adjacent inner ring are battered at an angle of 4 to 1 and 6 to 1, respectively, leaving the remaining piles as vertical piles. The top ends of all piles are embedded with sufficient lengths into the concrete that fills the interior space of the bell-shaped pile caps such that these piles can be considered as fixed-head piles. The piles penetrate deep into the supporting soil medium to an average depth of 147 ft (44.8 m) below the mudline, where they encounter a thick dense sand layer. The soil profile and properties at this foundation location are shown in Figure 42.12. As indicated in this figure, the top 55 ft (16.8 m) of the site soil is composed of a 35-ft (10.7-m) layer of soft bay mud overlying a 20-ft (6.1-m) layer of loose silty sand.

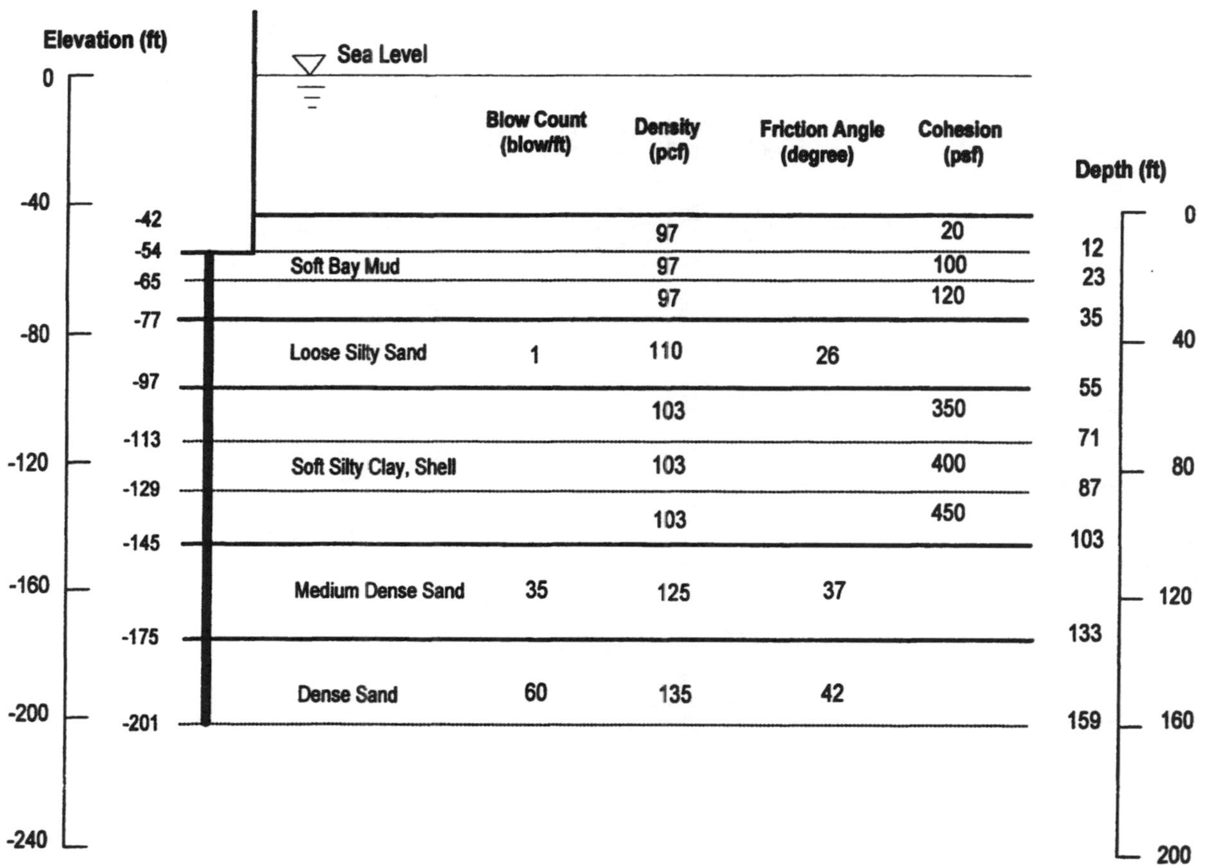

FIGURE 42.12 Soil profile and properties at the slender-pile group foundation considered.

Because of the soft topsoil layers and the slender piles used, the foundation under seismic excitations is expected to undergo relatively large foundation lateral displacements relative to the free-field soil. Thus, large local soil nonlinearities are expected to occur at the soil–pile interfaces. To model the nonlinear soil resistances to the lateral and axial deflections of the piles, the empirically derived lateral *p–y* and axial *t–z*, and the pile-tip *Q–d* curves for each pile were used. Typical *p–y* and *t–z* curves developed for the piles are shown in Figure 42.13. Using the nonlinear *p–y, t–z*, and *Q–d* curves developed, evaluation of the foundation impedance matrix, $\boldsymbol{F}_{ff}^{j}(i\omega)$, and the associated scattered foundation input motion vector, $\bar{\boldsymbol{u}}_f(i\omega)$, were obtained following the procedures described below:

1. Determine the pile group deflected shape using a nonlinear analysis program such as GROUP [52], LPIPE [53], and APILE2 [54], as appropriate, under an applied set of monotonically increasing axial and lateral forces and an overturning moment.
2. Select target levels of axial and lateral deflections at each selected soil depth corresponding to a selected target level of pile cap displacement and determine the corresponding secant moduli from the applicable nonlinear *p–y, t–z*, and *Q–d* curves.
3. Develop a model of a group of elastic beams supported on elastic axial and lateral soil springs for the pile group using the elastic properties of the piles and the secant moduli of the soil resistances obtained in Step 2 above.
4. Compute the foundation impedance matrix and associated scattered foundation input motion vector for the model developed in Step 3 using Eqs. (42.44) and (42.46).

Since the *p–y, t–z*, and *Q–d* curves represent pseudo-static force–deflection relations, the resulting foundation impedance matrix computed by the above procedure is a real (not complex) frequency-independent pseudo-static stiffness matrix, i.e., $\boldsymbol{F}_{ff}^{j}(i\omega)=\boldsymbol{F}_{ff}^{j}(0)$. For the pile group foundation considered in this example, the beam-on-elastic-spring model shown schematically in Figure 42.14 was used. The foundation stiffness matrix is associated with the six DOF of the nodal point located at the bottom center of the pile cap is shown in Figure 42.14. The scattered foundation motions in the longitudinal, transverse, and vertical directions associated with this foundation stiffness matrix are represented by their 5% damped acceleration response spectra shown in Figure 42.15. These spectra can be compared with the corresponding spectra for the seismic input motion prescribed at the pile tip elevation and the free-field mudline motions computed from free-field site-response analyses using SHAKE. As shown in Figure 42.15, the spectral values for the scattered pile cap motions, which would be used as input to the foundation–structure system, are lower than the spectral values for the free-field mudline motions. This result is to be expected for two reasons: (1) the soft topsoil layers present at the site are not capable of driving the pile group foundation and (2) the battered piles, acting with the vertical piles, resist lateral loads primarily through stiff axial truss action, in which case, the effective input motions at the pile cap are controlled more by the free-field soil motions at depth, where more competent soil resistances are present, than by the soil motions near the surface.

42.7.3 Large-Diameter Shaft Foundation

The third example illustrates the application of the demand analysis procedure using the hybrid method of modeling. This method is preferred for a foundation constructed of a group of large-diameter CISS or CIDH shafts. Because of the large horizontal dimensions and substantial masses associated with the shafts in this type of foundation, the dynamic interaction of the shafts with the surrounding soil medium is more appropriately modeled and analyzed using the elastodynamic method; however, because the shafts resist loadings in a manner like piles, the local soil nonlinearities present in the soil–shaft interface regions near the ground surface where soft soils are usually present may be sufficiently large that they should be explicitly considered using a method such as the empirical *p–y* method.

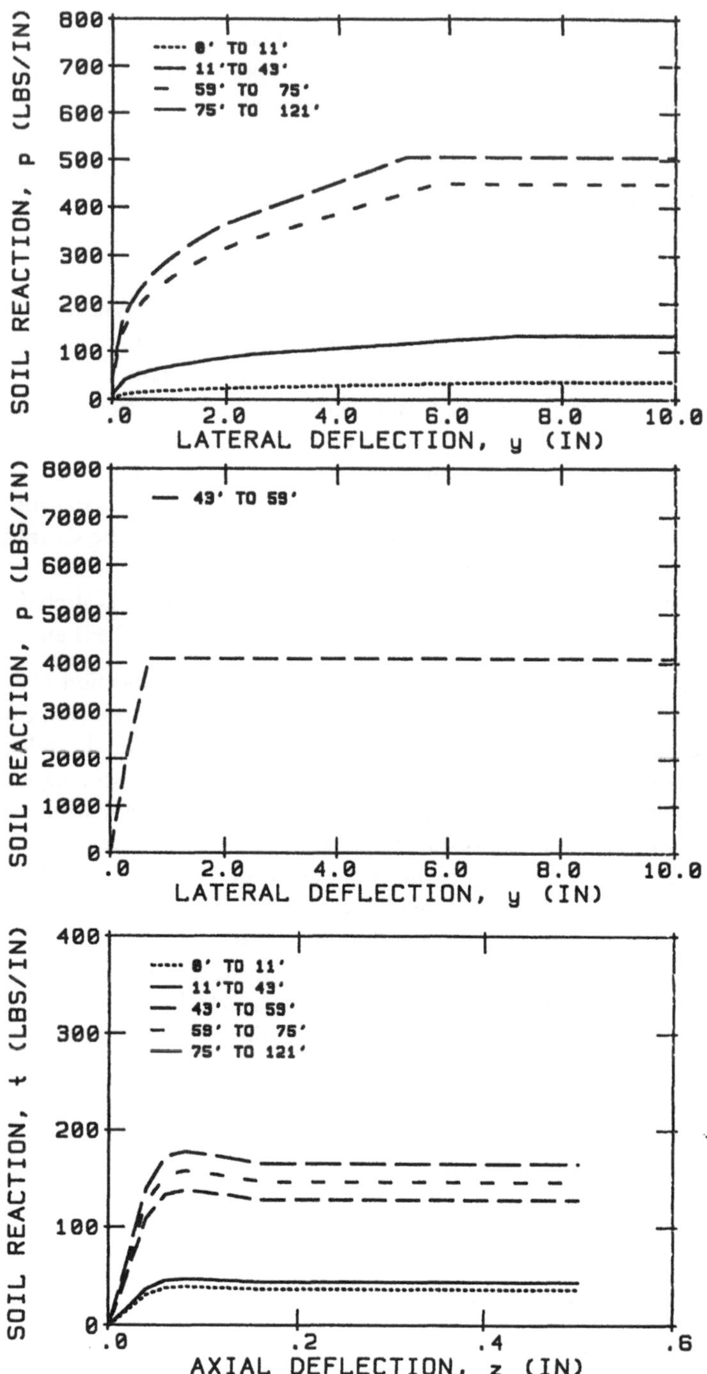

FIGURE 42.13 Typical p–y and t–z curves for the piles of the slender-pile group foundation considered.

The foundation selected for this example is composed of two 10.5-ft (3.2 m)-diameter shafts 150 ft (45.7 m) long, each consisting of a steel shell of wall thickness 1.375 in. (34.9 mm) filled with concrete. These two shafts are designed to be used as seismic retrofit shear piles for adding lateral stiffnesses and lateral load-resistance capacities to the H-pile group foundation considered in the second example discussed previously. The two shafts are to be linked to the existing pile group at

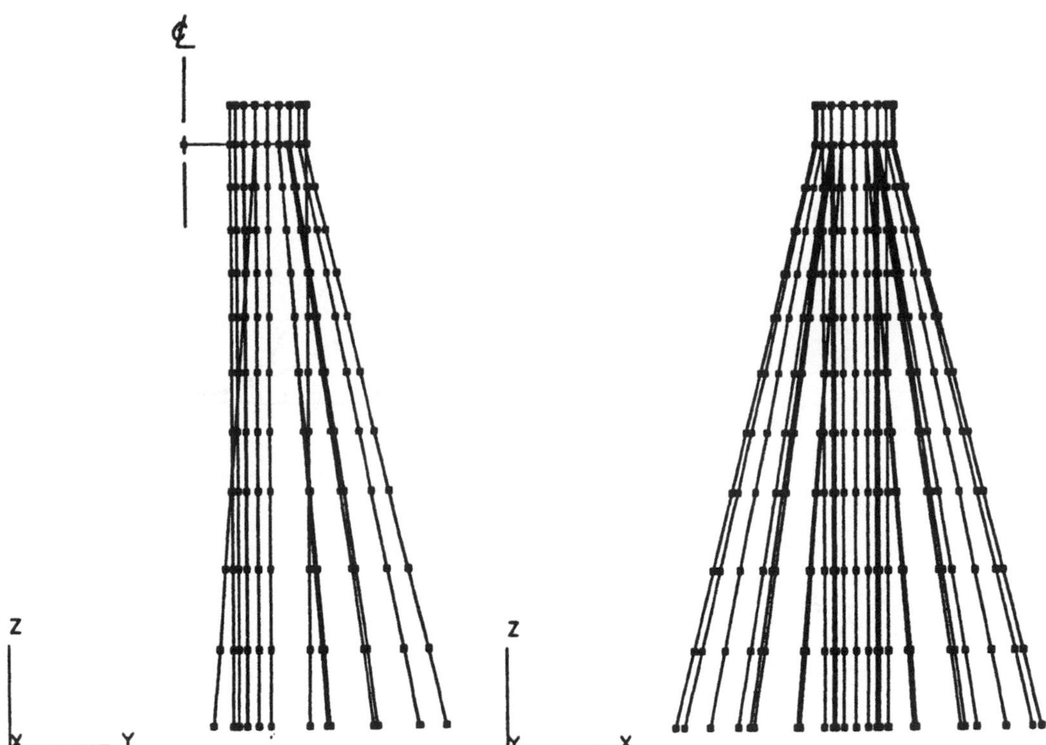

FIGURE 42.14 Beam-on-elastic-foundation half-model for the slender-pile group foundation considered.

the pile cap through a pile cap extension which permits the shafts to resist only horizontal shear loads acting on the pile cap, not axial loads and overturning moments. These shear piles have been designed to resist seismic horizontal shear loads acting on the pile head up to 3000 kips (13,344 kN) each.

To determine the foundation impedance matrix and the scattered pile cap motion vector associated with the horizontal displacements of the shafts at the pile cap, an SASSI model of one half of the soil–shaft system is developed, as shown in Figure 42.16. The soil properties used in this model are the strain-compatible properties shown in Table 42.1, which were obtained from the free-field site-response analyses using SHAKE; thus, the effects of global soil nonlinearities induced in the free-field soil by the design seismic input have been incorporated. To model the local soil nonlinearities occurring near the soil–shaft interface, three-directional (two lateral and one axial) soil springs are used to connect the beam elements representing the shafts to the soil nodes located at the boundary of the soil–shaft interfaces. The stiffnesses of these springs are derived in such a manner that they match the secant moduli of the empirical *p–y*, *t–z*, and *Q–d* curves developed for the shafts, as described previously in Section 42.5.3. Using the complete hybrid model shown in Figure 42.16, foundation compliances as functions of frequency were developed for harmonic pile-head shear loads varying from 500 (2224) to 3000 kips (13,334 kN). The results obtained are shown in Figure 42.17. It is seen that by incorporating local soil nonlinearities using the hybrid method, the resulting foundation compliance coefficients are not only frequency dependent due to the soil and shaft inertias and soil-layering effects as captured by the elastodynamic method, but they are also load–deflection amplitude dependent due to the local soil nonlinearities, as captured by the empirical *p–y* method. The shear load–deflection curves obtained at the pile head in the low-frequency range (≤1.0 Hz) are shown in Figure 42.18. The deflection curve for zero frequency, i.e., the static loading case, compares well with that obtained from a nonlinear analysis using LPILE [53], as indicated in Figure 42.18.

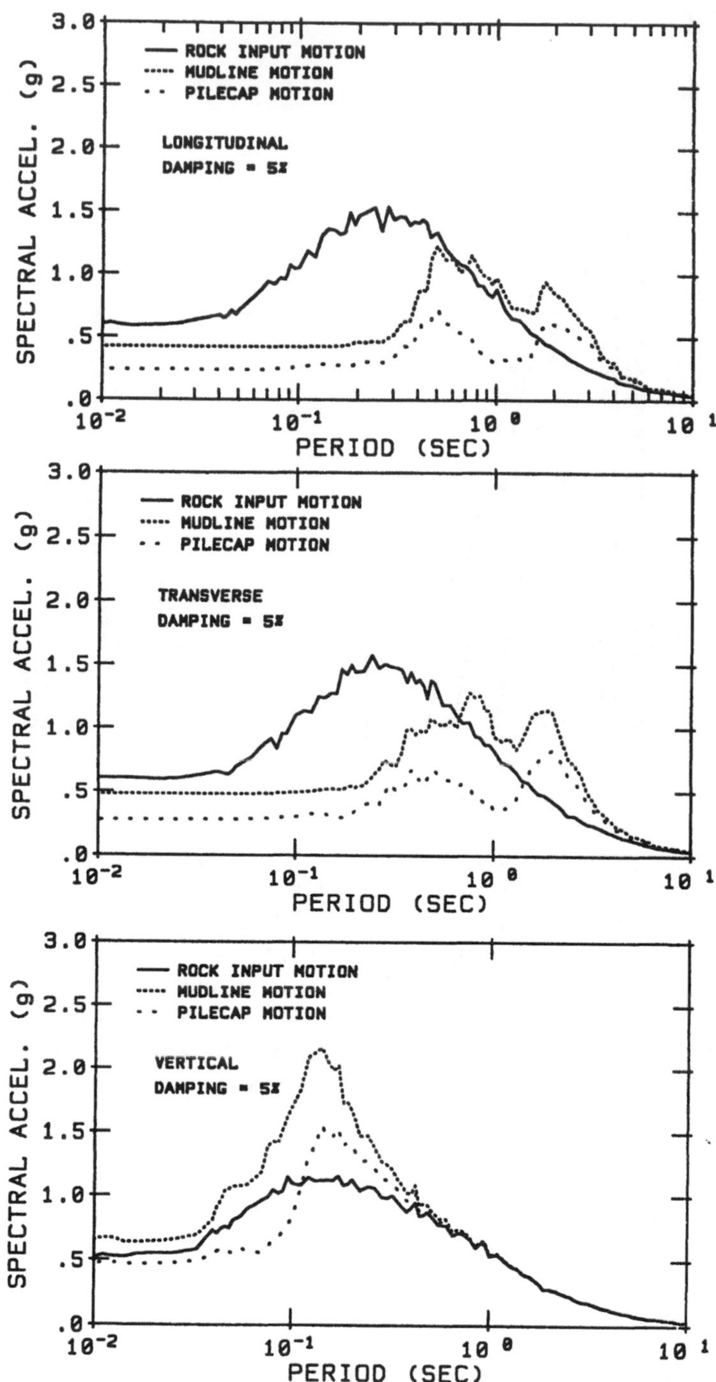

FIGURE 42.15 Comparisons of 5% damped response spectra for the rock input, mudline, and scattered pile cap motions in longitudinal, transverse, and vertical directions.

Subjecting the foundation to the design seismic input motions prescribed at the pile tip elevation and the corresponding free-field soil motions over its full depth, scattered foundation motions in the longitudinal and transverse directions of the bridge at the bottom center of the pile cap were

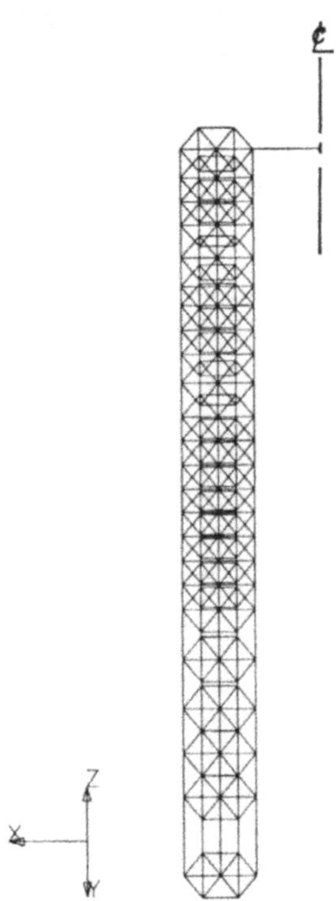

FIGURE 42.16 SASSI half-model of the large-diameter shaft foundation considered.

TABLE 42.1 Strain-Compatible Soil Properties for the Large-Diameter Shaft Foundation

El. ft (m)	Depth; ft (m)	Thickness; ft (m)	Unit Wt., $\frac{k}{ft^3}\left(\frac{kN}{m^3}\right)$	Shear Wave Velocity $\frac{ft}{s}\left(\frac{m}{s}\right)$	Shear Wave Damping Ratio	Compression Wave Velocity, $\frac{ft}{s}\left(\frac{m}{s}\right)$	Compression Wave Damping Ratio
–50 (–15.2)	0 (0.0)	10 (3.05)	0.096 (15.1)	202.1 (61.6)	0.10	4,800 (1,463)	0.09
–60 (–18.3)	10 (3.05)	10 (3.05)	0.096 (15.1)	207.5 (63.3)	0.15	5,000 (1,524)	0.10
–70 (–21.3)	20 (6.10)	10 (3.05)	0.096 (15.1)	217.7 (66.4)	0.17	5,000 (1,524)	0.10
–80 (–24.4)	30 (9.15)	20 (6.10)	0.110 (17.3)	137.5 (41.9)	0.25	4,300 (1,311)	0.10
–100 (–30.5)	50 (15.2)	10 (3.05)	0.096 (15.1)	215.7 (65.8)	0.20	4,800 (1,463)	0.10
–110 (–33.5)	60 (18.3)	20 (6.10)	0.096 (15.1)	218.4 (66.6)	0.20	4,300 (1,311)	0.10
–130 (–39.6)	80 (24.4)	20 (6.10)	0.096 (15.1)	233.0 (71.0)	0.20	4,900 (1,494)	0.10
–150 (–45.7)	100 (30.5)	20 (6.10)	0.120 (18.8)	420.4 (128.2)	0.20	5,500 (1,677)	0.10
–170 (–51.8)	120 (36.6)	10 (3.05)	0.120 (18.8)	501.0 (152.7)	0.19	6,000 (1,829)	0.10
–180 (–54.9)	130 (39.6)	10 (3.05)	0.120 (18.8)	532.7 (162.4)	0.19	5,800 (1,768)	0.10
–190 (–57.9)	140 (42.7)	20 (6.10)	0.125 (19.6)	607.2 (185.1)	0.18	5,800 (1,768)	0.10
–210 (–64.0)	160 (48.8)	20 (6.10)	0.128 (20.1)	806.9 (246.0)	0.16	5,800 (1,768)	0.10
–230 (–70.1)	180 (54.9)	10 (3.05)	0.133 (20.9)	1,374.4 (419.0)	0.11	6,400 (1,951)	0.10
–240 (–78.2)	190 (57.9)	5 (1.52)	0.140 (21.9)	2,844.9 (867.3)	0.02	12,000 (3,658)	0.02
–245 (–74.7)	195 (59.5)	halfspace	0.145 (22.8)	6,387.2 (1,947.3)	0.01	12,000 (3,658)	0.01

obtained as shown in terms of their 5% damped acceleration response spectra in Figure 42.19, where they can be compared with the corresponding response spectra for the seismic input motions and the free-field mudline motions. It is seen that, because of the substantial masses of the shafts, the spectral amplitudes of the scattered motions are higher than those of the free-field mudline motions

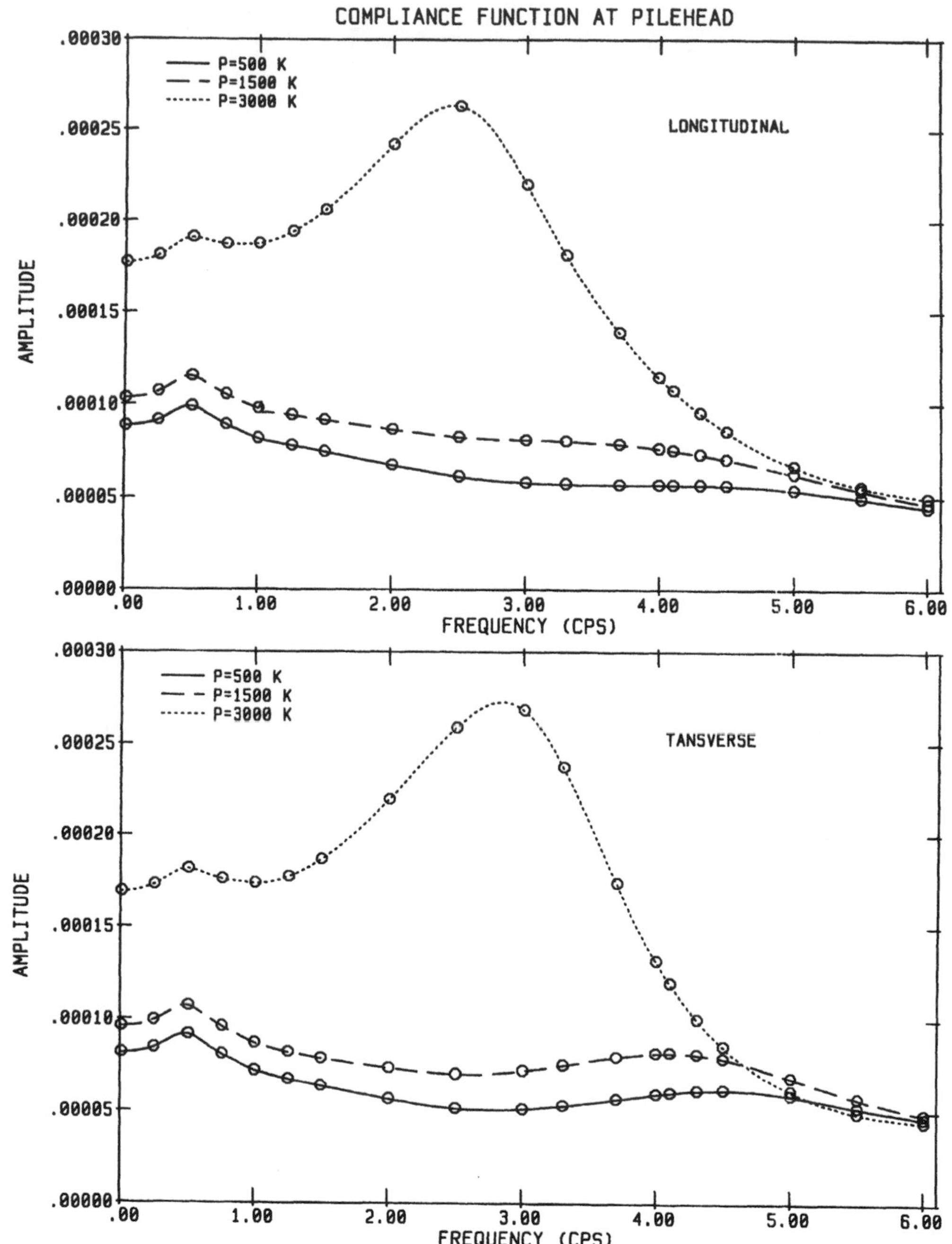

FIGURE 42.17 Foundation compliance functions at discrete values of shear load applied at the top of the shaft foundation.

for frequencies in the neighborhood of the soil–shaft system frequencies. Thus, for large-diameter shaft foundations constructed in deep, soft soil sites, it is important that the soil and shaft inertias be properly included in the SFI. Neglecting the shaft masses will result in underestimating the scattered pile cap motions in the longitudinal and transverse directions of the bridge, as represented in Figure 42.20.

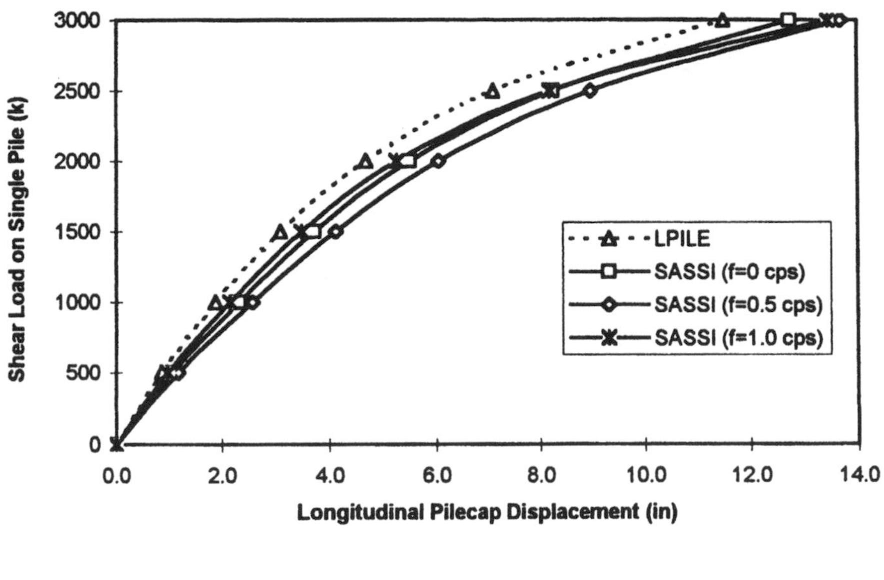

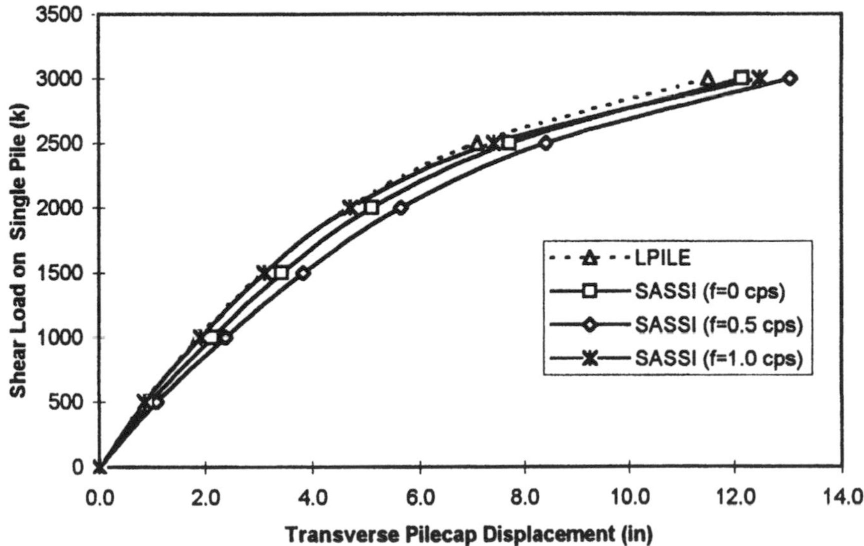

FIGURE 42.18 Typical sher load–deflection curves at several forcing frequencies.

42.8 Capacity Evaluations

The objective of the capacity evaluation is to determine the most probable levels of seismic resistance of the various elements, components, and subsystems of the bridge. The resistance capacities provided by this evaluation, along with the corresponding demands, provide the basis for judging seismic performance of the complete bridge system during future earthquakes. In the domain of SFSI as discussed here, the capacity evaluation focuses on soil–foundation systems.

For a bridge subjected to static loadings, the soil–foundation capacities of interest are the load resistances and the associated foundation deflections and settlements. Their evaluation constitutes the bulk of the traditional foundation design problem. When the bridge is subjected to oscillatory dynamic loadings, including seismic, the static capacities mentioned above are, alone, insufficient in the process of judging soil–foundation performance. In this case, it is necessary to assess the

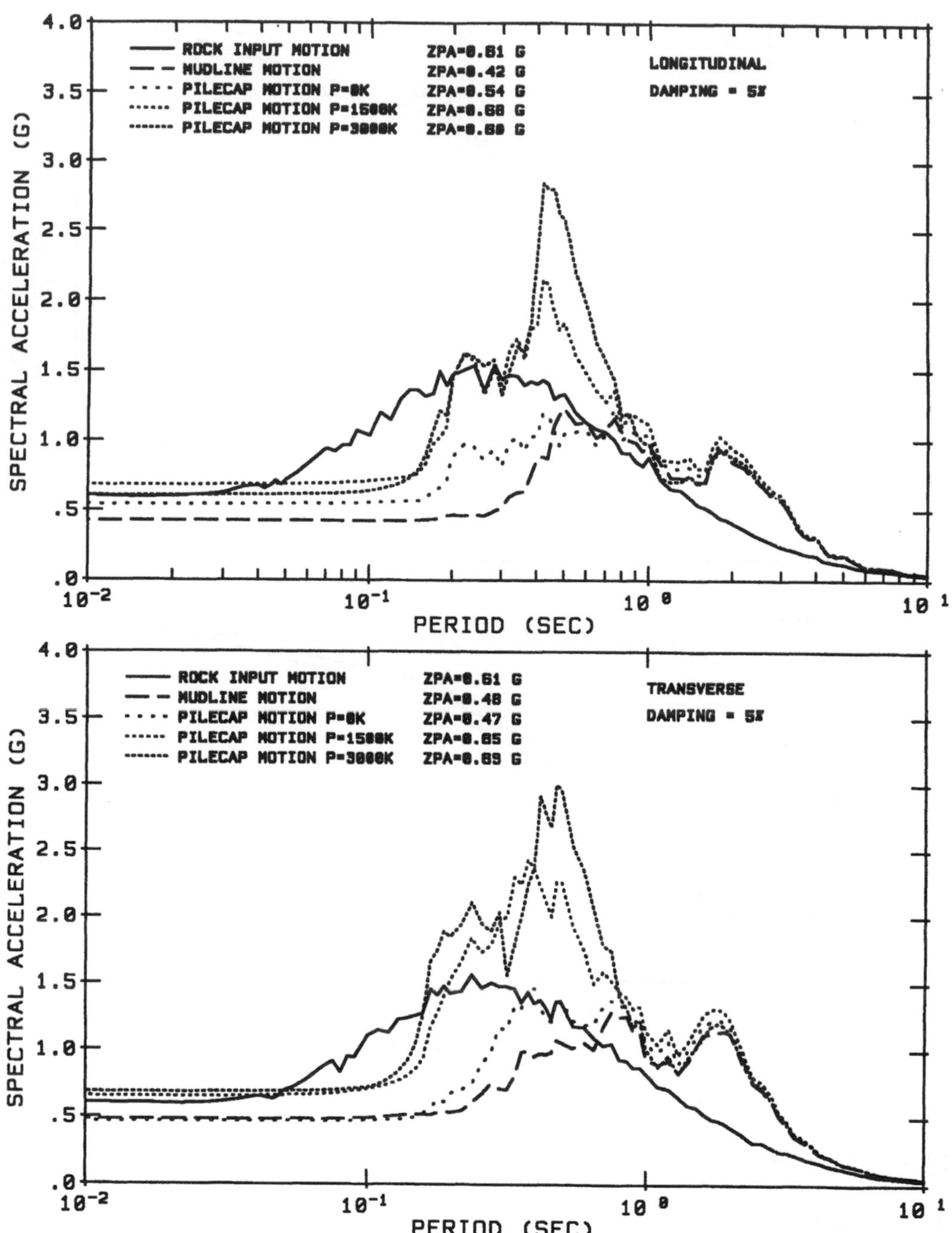

FIGURE 42.19 Comparisons of 5% damped response spectra for the longitudinal and transverse rock input, mudline, and scattered pile cap motions for the shaft foundations at several shear load levels.

entire load–deflection relationships, including their cyclic energy dissipation characteristics, up to load and/or deformation limits approaching failure conditions in the soil–foundation system. Because of the complexity of this assessment, the capacity evaluation must be simplified in order to make it practical. This is usually done by treating each soil–foundation system independently and by subjecting it to simplified pseudo-static monotonic and/or cyclic deformation-controlled step-by-step patterns of loading, referred to here as "push-over" analysis.

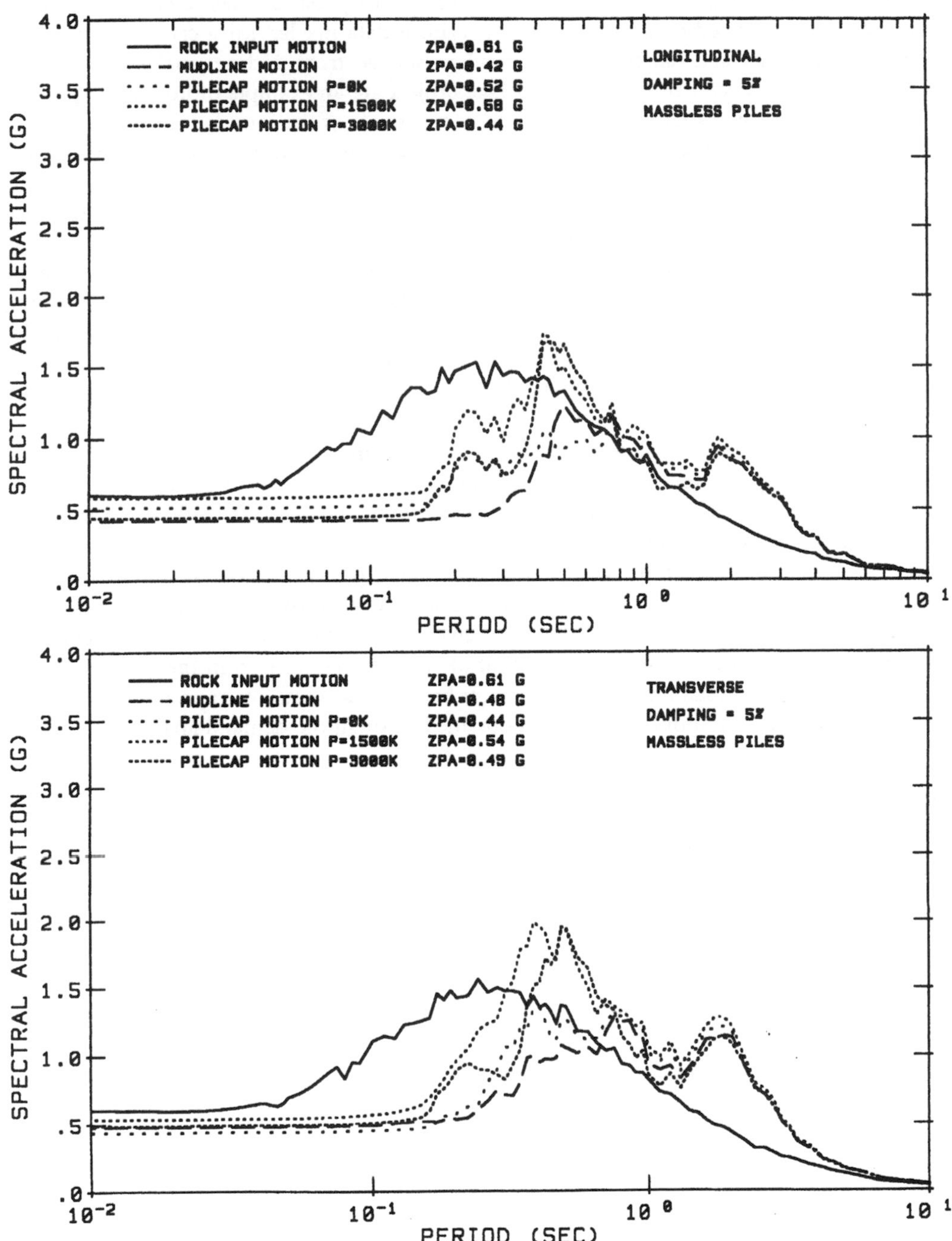

FIGURE 42.20 Comparisons of 5% damped response spectra for the longitudinal and transverse rock input, mudline, and scattered pile cap motions for the shaft foundation without masses in the shafts.

Because near-failure behavior of a soil–foundation system is involved in the capacity evaluation, it necessarily involves postelastic nonlinear behavior in the constituent components of the system, including the structural elements and connections of the foundation and its surrounding soil medium. Thus, ideally, a realistic evaluation of the capacities should be based on *in situ* tests conducted on prototypical foundation systems. Practical limitations, however, generally do not allow the conduct of such comprehensive tests. It is usually necessary, therefore, to rely solely on a combination of analysis and limited-scope *in situ* or laboratory tests of selected critical components. These tests are performed either to provide the critical data needed for a capacity analysis or to confirm the adequacy and reliability of the results obtained from such an analysis. Indicator-pile tests that have often been performed for a bridge project are an example of limited-scope testing.

In a typical push-over analysis, the structural components of the foundation are represented by appropriate nonlinear finite elements capable of representing the near-failure nonlinear features, such as plastic hinging, ductile or brittle shearing, tensile or compressive yielding and fracturing, local and global buckling, and stiffness and capacity degradations due to P-Δ effects; further, the surrounding soil medium is usually represented either by nonlinear finite elements capable of modeling the postelastic constitutive behavior of the material or by empirically derived generalized nonlinear soil springs such as those developed from the p–y, t–z, and Q–d curves used for pile foundations. Ideally, the soil–foundation model used should also be able to represent properly the important nonlinear behaviors that could develop at the soil–foundation interfaces, such as slippage, debonding, and gapping. After the model has been developed, it is then subjected to a set of suitable push-over loading programs that simulate the loading conditions imposed on the soil–foundation system by the bridge pier at its interface with the foundation.

Conducting a step-by-step push-over analysis of the model described above, one can identify load and deformation levels associated with the various failure modes in the soil–foundation system. Then, load and deformation limits can be set beyond which the performance goals set for the bridge will no longer be met. These limits can be considered the capacity limits of the foundation system.

Because large uncertainties usually exist in a capacity evaluation, the capacity limits obtained therefrom should be reduced using appropriate capacity reduction ϕ factors. Each reduction factor adopted should adequately cover the lower limit of capacity resulting from the uncertainties. The reduced capacity limits established in this manner become the allowable capacity limits for use in comparing with the corresponding demands obtained through the demand analysis.

42.9 Concluding Statements

The previous sections of this chapter discuss the various elements of a modern state-of-the-art SFSI seismic analysis for large important bridges. These elements include (1) generating the site-specific rock-outcrop motions and corresponding free-field soil motions, (2) modeling and analysis of individual soil–foundation systems to establish foundation impedances and scattered motions, (3) determining SFSI using the substructuring method of analyses, and (4) assessing overall bridge performance by comparing force–deformation demands with corresponding capacities. Without retracing the details of these elements, certain points are worthy of special emphasis, as follows:

- Best-estimate rock and soil properties should be used in the generation of free-field seismic motions, with full recognition of the variations (randomness) and uncertainties (lack of knowledge) involved.
- Likewise, best-estimate material properties should be used in modeling the foundations, piers, abutments, and superstructure, also recognizing the variations and uncertainties involved.
- In view of the above-mentioned variations and uncertainties, sensitivity analyses should be conducted to provide a sound basis for judging overall seismic performance.
- Considering the current state of development, one should clearly differentiate between the requirements of a seismic force–deformation demand analysis and the corresponding capacity evaluation. The former is concerned with global system behavior; thus, it must satisfy only global dynamic equilibrium and compatibility. The latter, however, places emphasis on the behavior of local elements, components, and subsystems, requiring that equilibrium and compatibility be satisfied only at the local level within both the elastic and postelastic ranges of deformation.

- In conducting a demand analysis, equivalent linear modeling, coupled with the substructuring method of analysis, has the advantages that (1) the results are more controllable and predictable, (2) the uncertainties in system parameters can easily be evaluated separately, and (3) the SFSI responses can be assessed at stages. These advantages lead to a high level of confidence in the results when the nonlinearities are relatively weak. However, when strong nonlinearities are present, nonlinear time history analyses should be carried out in an iterative manner so that system response is consistent with the nonlinearities.
- When strong nonlinearities are present in the overall structural system, usually in the piers and superstructure, multiple sets of seismic inputs should be used separately in conducting the demand analyses; since, such nonlinearities cause relatively large dispersions of the maximum values of critical response.
- The elastodynamic method of treating SFSI is valid for foundations having large horizontal dimensions, such as large spread footings and caissons; while the empirical *p–y* method is valid only for slender-pile foundations subjected to large-amplitude deflections. For foundations intermediate between these two classes, e.g., those using large-diameter shafts, both of these methods are deficient in predicting SFSI behavior. In this case, the hybrid method of modeling has definitive advantages, including its ability to treat all classes of foundations with reasonable validity.
- The *p–y* method of treating SFSI in both demand analyses and capacity evaluations needs further development, refinement, and validation through test results, particularly with regard to establishing realistic *p–y*, t–z, and *Q–d* curves. For seismic applications, changes in the characteristics of these curves, due to global soil nonlinearities induced by the free-field ground motions, should be assessed.
- The hybrid method of treating SFSI, while being fundamentally sound, also needs further development, refinement, and test validation to make it fully acceptable for bridge applications.
- Systematic research and development efforts, involving laboratory and field tests and analytical correlation studies, are required to advance the SFSI analysis methodologies for treating bridge foundations.

The state of the art of SFSI analysis of large bridge structures has been rapidly changing in recent years, a trend that undoubtedly will continue on into the future. The reader is therefore encouraged to take note of new developments as they appear in future publications.

Acknowledgment

The authors wish to express their sincere thanks and appreciation to Joseph P. Nicoletti and Abbas Abghari for their contributions to Sections 42.2 and 42.4, respectively.

References

1. Mylonakis, G., Nikolaou, A., and Gazetas, G., Soil-pile-bridge seismic interaction: kinematic and inertia effects. Part I: Soft soil, *Earthquake Eng. Struct. Dyn.*, 26, 337–359, 1997.
2. Tseng, W. S., Soil-foundation-structure interaction analysis by the elasto-dynamic method, in *Proc. 4th Caltrans Seismic Research Workshop*, Sacramento, July 9–11, 1996.
3. Lam, I. P. and Law, H., Soil-foundation-structure interaction — analytical considerations by empirical *p–y* method, in *Proc. 4th Caltrans Seismic Research Workshop*, Sacramento, July 9–11, 1996.
4. Schnabel, P. B., Lysmer, J., and Seed, H. B., SHAKE — A Computer Program for Earthquake Response Analysis of Horizontally Layered Sites, Report No. EERC 72–12, Earthquake Engineering Research Center, University of California, Berkeley, 1972.

5. Tseng, W. S. and Hadjian, A. H., Guidelines for Soil-Structure Interaction Analysis, EPRI NP-7395, Electric Power Research Institute, Palo Alto, CA, October 1991.
6. Richart, F. E., Hall, J. R., Jr., and Woods, R. D., *Vibrations of Soils and Foundations*, Prentice-Hall, Englewood Cliffs, NJ, 1970.
7. Veletsos, A. S. and Wei, Y. T., Lateral and rocking vibration of footings, *J. Soil Mech. Found. Div. ASCE*, 97(SM9), 1227–1249, 1971.
8. Kausel, E. and Roësset, J. M., Soil-structure interaction problems for nuclear containment structures, in *Proc. ASCE Power Division Specialty Conference*, Denver, CO, August 1974.
9. Wong, H. L. and Luco, J. E., Dynamic response of rigid foundations of arbitrary shape, *Earthquake Eng. Struct. Dyn.*, 4, 587–597, 1976.
10. Seed, H. B. and Idriss, I. M., Soil Moduli and Damping Factors for Dynamic Response Analysis, Report No. EERC 70-10, Earthquake Engineering Research Center, University of California, Berkeley, 1970.
11. Waas, G., Analysis Method for Footing Vibrations through Layered Media, Ph.D. dissertation, University of California, Berkeley, 1972.
12. Lysmer, J., Udaka, T., Tsai, C. F., and Seed, H. B., FLUSH — A Computer Program for Approximate 3-D Analysis of Soil-Structure Interaction, Report No. EERC75-30, Earthquake Engineering Research Center, University of California, Berkeley, 1975.
13. Tang, Y. K., Proceedings: EPRI/NRC/TPC Workshop on Seismic Soil-Structure Interaction Analysis Techniques Using Data from Lotung Taiwan, EPRI NP-6154, Electric Power Research Institute, Palo Alto, CA, March 1989.
14. Matlock, H. and Reese, L. C., Foundation analysis of offshore pile supported structures, in *Proc. 5th Int. Conf. on Soil Mech. and Found. Eng.*, Paris, France, July 17–22, 1961.
15. Matlock, H., Correlations for design of laterally loaded piles in soft clay, in *Proc. Offshore Technology Conference*, Paper No. OTC1204, Dallas, TX, April 22–24, 1970.
16. Reese, L. C., Cox, W. R., and Koop, F. D., Analysis of laterally loaded piles in sand, in *Proc. Offshore Technology Conference*, Paper No. OTC2080, Dallas, TX, May 6–8, 1974.
17. Chang, C. Y., Tseng, W. S., Tang, Y. K., and Power, M. S., Variations of earthquake ground motions with depth and its effects on soil–structure interaction, *Proc. Second Department of Energy Natural Phenomena Hazards Mitigation Conference*, Knoxville, Tennessee, October 3–5, 1989.
18. Abrahamson, N. A., Spatial Variation of Earthquake Ground Motion for Application to Soil-Structure Interaction, Report No. TR-100463, Electric Power Research Institute, Palo Alto, CA, March 1992.
19. Gasparini, D. and Vanmarcke, E. H., SIMQKE: A program for Artificial Motion Generation, Department of Civil Engineering, Massachusetts Institute of Technology, Cambridge, 1976.
20. Silva, W. J. and Lee, K., WES RASCAL Code for Synthesizing Earthquake Ground Motions, State-of-the-Art for Assessing Earthquake Hazards in United States, Report 24, Army Engineers Waterway Experimental Station, Miscellaneous Paper 5-73-1, 1987.
21. Lilhanand, K. and Tseng, W. S., Development and application of realistic earthquake time histories compatible with multiple-damping design response spectra, in *Proc. 9th World Conference of Earthquake Engineers*, Tokyo-Kyoto, Japan, August 2–9, 1988.
22. Bolt, B. A. and Gregor, N. J., Synthesized Strong Ground Motions for the Seismic Condition Assessment of Eastern Portion of the San Francisco Bay Bridge, Report No. EERC 93-12, Earthquake Engineering Research Center, University of California, Berkeley, 1993.
23. Abrahamson, N. A., Nonstationary spectral matching, *Seismol. Res. Lett.*, 63(1), 1992.
24. Boore D. and Akinson, G., Stochastic prediction of ground motion and spectral response parameters at hard-rock sites in eastern North America, *Bull. Seismol. Soc. Am.*, 77, 440–467, 1987.
25. Sommerville, P. G. and Helmberger, D. V., Modeling earthquake ground motion at close distance, in *Proc. EPRI/Stanford/USGS Workshop on Modeling Earthquake Ground Motion at Close Distances*, August 23, 1990.

26. Papageorgiou, A. S. and Aki, K., A specific barrier model to the quantitative description of inhomogeneous faulting and the prediction of strong motion, *Bull. Seismol. Soc. Am.*, Vol. 73, 693–722, 953–978, 1983.
27. Bolt, B. A., Ed., *Seismic Strong Motion Synthetics*, Academic Press, New York, 1987.
28. U.S. Nuclear Regulatory Commission, *Standard Review Plan*, Section 3.71., Revision 2, Washington, D.C., August, 1989.
29. International Conference of Building Officials, *Uniform Building Code*, Whittier, CA, 1994.
30. Kaul, M. K., Spectrum-consistent time-history generation, *J. Eng. Mech. Div. ASCE*, 104(EM4), 781, 1978.
31. Penzien, J. and Watabe, M., Simulation of 3-dimensional earthquake ground motions, *J. Earthquake Eng. Struct. Dyn.*, 3(4), 1975.
32. Hao, H., Oliviera, C. S., and Penzien, J., Multiple-station ground motion processing and simulation based on SMART-1 array data, *Nuc. Eng. Des.*, III(6), 2229–2244, 1989.
33. Chang, C. Y., Power, M. S., Idriss, I. M., Sommerville, P. G., Silva, W., and Chen, P. C., Engineering Characterization of Ground Motion, Task II: Observation Data on Spatial Variations of Earthquake Ground Motion, NUREG/CR-3805, Vol. 3, U.S. Nuclear Regulatory Commission, Washington, D.C., 1986.
34. Abrahamson, N. A., Schneider, J. F., and Stepp, J. C., Empirical spatial coherency functions for application to soil-structure interaction analyses, *Earthquake Spectra*, 7, 1, 1991.
35. Tseng, W. S., Lilhanand, K., and Yang, M. S., Generation of multiple-station response-spectra-and-coherency-compatible earthquake ground motions for engineering applications, in *Proc. 12th Int. Conference on Struct. Mech. in Reactor Technology*, Paper No. K01/3, Stuttgart, Germany, August 15–20, 1993.
36. Vucetic, M. and Dobry, R., Effects of soil plasticity on cyclic response, *J. Geotech. Eng. ASCE*, 117(1), 89–107, 1991.
37. Sun, J. I., Golesorkhi, R., and Seed, H. B., Dynamic Moduli and Damping Ratios for Cohesive Soils, Report No. UBC/EERC-88/15, Earthquake Engineer Research Center, University of California, Berkeley, 1988.
38. Hardin, B. O. and Drnevich, V., Shear modulus and damping in soils: design equations and curves, *J. Soil Mech. Found. Div. ASCE*, 98(7), 667–691, 1972.
39. Seed, H. B., and Idriss, I. M., Soil Moduli and Damping Factors for Dynamic Response Analyses, Report No. EERC 70-10, Earthquake Engineering Research Center, University of California, Berkeley, 1970.
40. Seed, H. B., Wong, R. T., Idriss, I. M., and Tokimatsu, K., Moduli and Damping Factors for Dynamic Analyses of Cohesionless Soils, Report No. UCB/EERC-84/14, Earthquake Engineering Research Center, University of California, Berkeley, 1984.
41. Hardin, B. O. and Black, W. L., Vibration modulus of normally consolidated clay, *J. Soil Mech. Found. Div. ASCE*, 94(2), 353–369, 1968.
42. Hardin, B. O., The nature of stress-strain behavior for soils, *Proc. ASEC Geotech. Eng. Div. Specialty Conference on Earthquake Eng. and Soil Dyn.*, Vol. 1, 3-90, 1978.
43. Dickenson, S. E., Dynamic Response of Soft and Deep Cohesive Soils During the Loma Prieta earthquake of October 17, 1989, Ph.D. dissertation, University of California, Berkeley, 1994.
44. Idriss, I. M. and Sun, J. I., User's Manual for SHAKE 91, Center for Geotechnical Modeling, University of California, Davis, 1992.
45. Seed, H. B. and Idriss, I. M., Influence of soil conditions on ground motions during earthquakes, *J. Soil Mech. Found. Div.*, 95(SM1), 99–138, 1969.
46. Lee, M. K. W. and Finn, W. D. L., DESRA-2: Dynamic Effective Stress Response Analysis of Soil Deposits with Energy Transmitting Boundary Including Assessment of Liquification Potential, Report No. 38, Soil Mechanics Series, Department of Civil Engineering, University of British Columbia, Vancouver, 1978.

47. Prevost, J. H. DYNAFLOW: A Nonlinear Transient Finite Element Analysis Program, Department of Civil Engineering and Operations Research, Princeton University, Princeton, NJ, 1981; last update, 1993.
48. Prevost, J. H., DYNAID: A Computer Program for Nonlinear Seismic Site Response Analysis, Report No. NCEER-89-0025, National Center for Earthquake Engineering Research, Buffalo, New York, 1989.
49. Li, X. S., Wang, Z. L., and Shen, C. K., SUMDES, A Nonlinear Procedure for Response Analysis of Horizontally-Layered Sites, Subjected to Multi-directional Earthquake Loading, Department of Civil Engineering, University of California, Davis, 1992.
50. Apsel, R. J., Dynamic Green's Function for Layered Media and Applications to Boundary Value Problems, Ph.D. thesis, University of California, San Diego, 1979.
51. Lysmer, J., Tabatabaie-Raissai, M., Tajirian, F., Vahdani, S., and Ostadan, F., SASSI — A System for Analysis of Soil-Structure Interaction, Report No. UCB/GT/81-02, Department of Civil Engineering, University of California, Berkeley, 1981.
52. Reese, L. C., Awoshirka, K., Lam, P. H. F., and Wang, S. T., Documentation of Computer Program GROUP — Analysis of a Group of Piles Subjected to Axial and Lateral Loading, Ensoft, Inc., Austin, TX, 1990.
53. Reese, L. C. and Wang, S. T., Documentation of Computer Program LPILE — A Program for Analysis of Piles and Drilled Shafts under Lateral Loads," Ensoft, Inc., Austin, TX, 1989; latest update, Version 3.0, May 1997.
54. Reese, L. C. and Wang, S. T., Documentation of Computer Program APILE2 — Analysis of Load Vs. Settlement for an Axially Loaded Deep Foundation, Ensoft, Inc., Austin, TX, 1990.
55. Seed, H. B. and Idriss, I. M., Rock Motion Accelerograms for High-Magnitude Earthquakes, Report No. EERC 69-7, Earthquake Engineering Research Center, University of California, Berkeley, 1969.
56. Clough, R. W. and Penzien, J., *Dynamics of Structures,* 2nd ed., McGraw-Hill, New York, 1993.

43

Seismic Retrofit Technology

Kevin I. Keady
California Department of Transportation

Fadel Alameddine
California Department of Transportation

Thomas E. Sardo
California Department of Transportation

43.1 Introduction

Upon completion of the seismic analysis and the vulnerability study for existing bridges, the engineer must develop a retrofit strategy to achieve the required design criteria. Depending on the importance of structures, there are two levels of retrofit. For ordinary structures, a lower level of retrofit may be implemented. The purpose of this level of retrofit is to prevent collapse. With this level of retrofit, repairable damage is generally expected after a moderate earthquake. Following a major earthquake, extensive damage is expected and replacement of structures may be necessary. For important structures, a higher level of retrofit could be required at a considerably higher cost. The purpose of this level of retrofit is not only to prevent collapse, but also to provide serviceability after a major earthquake.

There are two basic retrofit philosophies for a concrete bridge. The first is to force plastic hinging into the columns and keep the superstructure elastic. This is desirable because columns can be more easily inspected, retrofitted, and repaired than superstructures. The second is to allow plastic hinging in the superstructure provided that ductility levels are relatively low and the vertical shear load-carrying capacity is maintained across the hinge. This is desirable when preventing hinging in the superstructure is either prohibitively expensive or not possible. In other words, this strategy is permissible provided that the hinge in the superstructure does not lead to collapse. To be conservative, the contribution of concrete should be ignored and the steel stirrups need to be sufficient to carry 1.5 times the dead-load shear reaction if hinging is allowed in the superstructure.

There are two basic retrofit philosophies for steel girder bridges. The first is to let the bearings fail and take retrofit measures to ensure that the spans do not drop off their seats and collapse. In this scenario, the bearings act as a "fuse" by failing at a relatively small seismic force and thus protecting the substructure from being subjected to any potential larger seismic force. This may be

0-8493-7434-0/00/$0.00+$.50

TABLE 43.1 Seismic Performance Criteria

Ground Motion at Site	Minimum performance level	Important bridge performance level
Functional Evaluation	Immediate service level repairable damage	Immediate service level minimal damage
Safety Evaluation	Limited service level significant damage	Limited service level repairable damage

the preferred strategy if the fusing force is low enough such that the substructure can survive with little or no retrofit. The second philosophy is to make sure that the bearings do not fail. It implies that the bearings transfer the full seismic force to the substructure and retrofitting the substructure may be required. The substructure retrofit includes the bent caps, columns or pier walls, and foundations. In both philosophies, a superstructure retrofit is generally required, although the extent is typically greater with the fixed bearing scheme.

The purpose of this chapter is to identify potential vulnerabilities to bridge components and suggest practical retrofit solutions. For each bridge component, the potential vulnerabilities will be introduced and retrofit concepts will be presented along with specific design considerations.

43.2 Analysis Techniques for Bridge Retrofit

For ordinary bridges, a dynamic modal response spectrum analysis is usually performed under the input earthquake loading. The modal responses are combined using the complete quadratic combination (CQC) method. The resulting orthogonal responses are then combined using the 30% rule. Two cases are considered when combining orthogonal seismic forces. Case 1 is the sum of forces due to transverse loading plus 30% of forces due to longitudinal loading. Case 2 is the sum of forces due to longitudinal loading plus 30% of forces due to transverse loading.

A proper analysis should consider abutment springs and trusslike restrainer elements. The soil foundation structure interaction should be considered when deemed important. Effective properties of all members should be used. Typically, two dynamic models are utilized to bound the assumed nonlinear response of the bridge: a "tension model" and a "compression model." As the bridge opens up at its joints, it pulls on the restrainers. In contrast, as the bridge closes up at its joints, its superstructure elements go into compression.

For more important bridges, a nonlinear time history analysis is often required. This analysis can be of uniform support excitation or of multiple supports excitation depending on the length of the bridge and the variability of the subsurface condition.

The input earthquake loading depends on the type of evaluation that is considered for the subject bridge. Table 43.1 shows the seismic performance criteria for the design and evaluation of bridges developed by the California Department of Transportation [1]. The safety evaluation response spectrum is obtained using:

1. Deterministic ground motion assessment using maximum credible earthquake, or
2. Probabilistically assessed ground motion with a long return period.

The functional evaluation response spectrum is derived using probabilistically assessed ground motions which have a 60% probability of not being exceeded during the useful life of the bridge. A separate functional evaluation is usually required only for important bridges.

With the above-prescribed input earthquake loading and using an elastic dynamic multimodal response spectrum analysis, the displacement demand can be computed. The displacement capacity of various bents may be then calculated using two-dimensional or three-dimensional nonlinear static push-over analysis with strain limits associated with expected damage at plastic hinge locations. When performing a push-over analysis, a concrete stress–strain model that considers effects of transverse confinement, such as Mander's model, and a steel stress–strain curve are used for considering material nonlinearity [2]. Limiting the concrete compressive strain to a magnitude smaller than the confined concrete ultimate compressive strain and the steel strain to a magnitude

TABLE 43.2 Strain Limits

	Significant Damage	Repairable Damage	Minimal Damage
Concrete strain limit ε_c	ε_{cu}	2/3 * ε_{cu}	The greater of 1/3 * ε_{cu} or 0.004
Grade 430 bar #29 to #57 steel strain limit ε_s	0.09	0.06	0.03
Grade 280 bar #29 to #57 steel strain limit ε_s	0.12	0.08	0.03
Grade 430 bar #10 to #25 steel strain limit ε_s	0.12	0.08	0.03
Grade 280 bar #10 to #25 steel strain limit ε_s	0.16	0.10	0.03

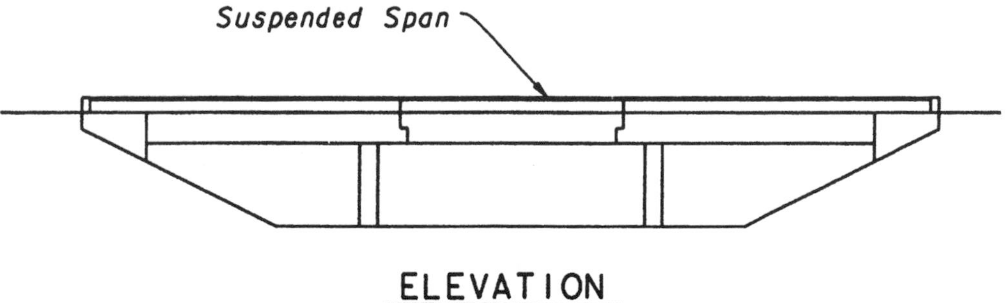

FIGURE 43.1 Suspended span.

smaller than steel rupture strain results in lesser curvature of the cross section under consideration. Smaller curvatures are usually associated with smaller cracks in the plastic hinge region.

Table 43.2 shows the general guidelines on strain limits that can be considered for a target level of damage in a plastic hinge zone. These limits are applied for the ultimate concrete strain ε_{cu} and the ultimate strain in the reinforcing steel ε_{su}. The ultimate concrete strain can be computed using a concrete confinement model such as Mander's model. It can be seen that for a poorly confined section, the difference between minimal damage and significant damage becomes insignificant. With displacement demands and displacement capacities established, the demand-to-capacity ratios can be computed showing adequacy or inadequacy of the subject bridge.

43.3 Superstructure Retrofits

Superstructures can be categorized into two different categories: concrete and steel. After the 1971 San Fernando, California earthquake, the primary failure leading to collapse was identified as unseating of superstructures at the expansion joints and abutments, a problem shared by both types of superstructures. Other potential problems that may exist with steel superstructures are weak cross-bracing and/or diaphragms. Concrete bridge superstructures have the potential to form plastic hinges during a longitudinal seismic response which is largely dependent upon the amount of reinforcement used and the way it is detailed.

43.3.1 Expansion Joints and Hinges

During an earthquake, adjacent bridge frames will often vibrate out of phase, causing two types of displacement problems. The first type is a localized damage caused by the frames pounding together at the hinges. Generally, this localized damage will not cause bridge collapse and is therefore not a major concern. The second type occurs when the hinge joint separates, possibly allowing the adjacent spans to become unseated if the movement is too large. Suspended spans (i.e., two hinges within one span) are especially vulnerable to becoming unseated (Figure 43.1).

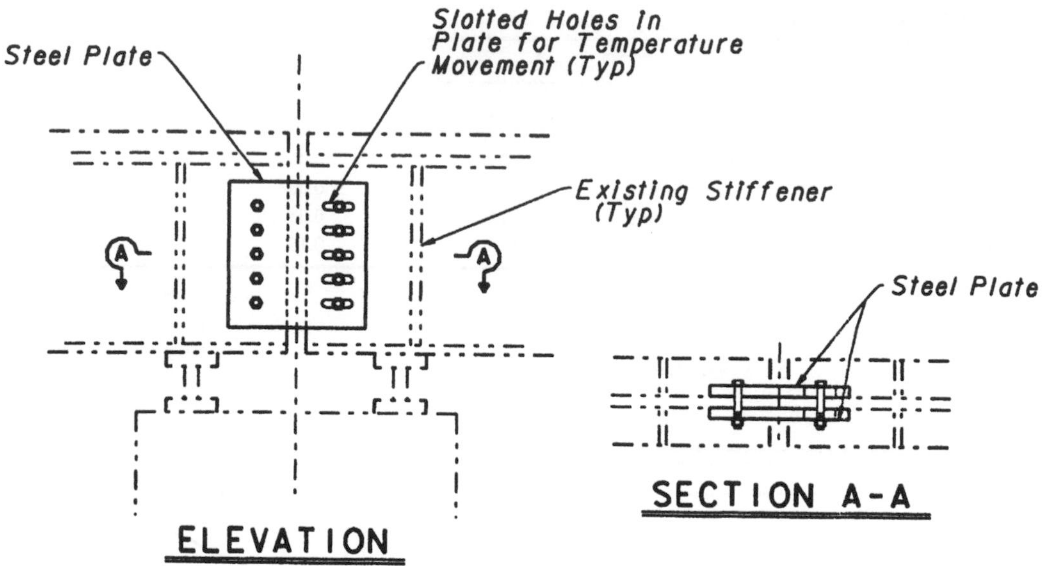

FIGURE 43.2 Steel girder hinge plate retrofit.

43.3.1.1 Simply Supported Girders

The most common problem for simply supported structures is girders falling off their seats due to a longitudinal response. If the seismic force on the structure is large enough to fail the bearings, then the superstructure becomes vulnerable to unseating at the supports.

There are several ways of retrofitting simply supported steel girders and/or precast concrete girders. The most common and traditional way is to use cable restrainers, since the theory is fundamentally the same for both types of girders. For more about cable restrainers, refer to Section 43.3.1.3. Care should be taken when designing the cables to intrude as little as possible on the vertical clearance between the girders and the roadway. Note that the cable retrofit solution for simply supported girders can be combined with a cap seat extension if expected longitudinal displacements are larger than the available seat width.

Another possible solution for steel girders is to make the girders continuous over the bents by tying the webs together with splice plates (Figure 43.2). The splice plate should be designed to support factored dead-load shears assuming the girder becomes unseated. The splice plate is bolted to the girder webs and has slotted or oversized holes to allow for temperature movement. This retrofit solution will usually work for most regular and straight structures, but not for most irregular structures. Any situation where the opposing girders do not line up will not work. For example, bridges that vary in width or are bifurcated may have different numbers of girders on opposite sides of the hinge. Bridges that are curved may have the girders at the same location but are kinked with respect to each other. In addition, many structures may have physical restrictions such as utilities, bracing, diaphragms, stiffeners, etc. which need to be relocated in order for this strategy to work.

43.3.1.2 Continuous Girders with In-Span Hinges

For continuous steel girders, the hinges are typically placed near the point of zero moment which is roughly at 20% of the span length. These hinges can be either seat type as shown in Figure 43.3 or hanger type as shown in Figure 43.4. The hanger-type hinges are designed for vertical dead and live loads. These loads are typically larger than forces that can be imparted onto the hanger bar from a longitudinal earthquake event, and thus retrofitting the hanger bar is generally unnecessary. Hanger-type hinges typically have more seismic resistance than seat-type hinges but may still be subjected to seismic damage. Hanger bars are tension members that are vulnerable to differential

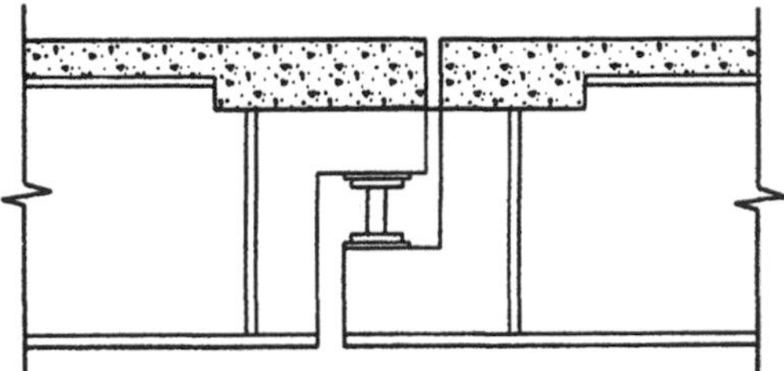

FIGURE 43.3 Seat-type hinge.

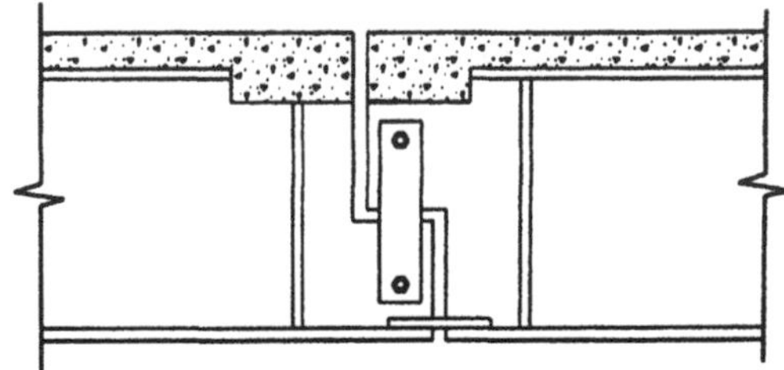

FIGURE 43.4 Hanger-type hinge.

transverse displacement on either side of the hinge. The differential displacement between the girders causes the hanger bars to go into bending plus tension. These hinges often have steel bars or angles that bear against the opposite web, or lugs attached to the flanges, which were designed to keep the girders aligned transversely for wind forces. These devices are usually structurally inadequate and are too short to be effective with even moderate seismic shaking. Consideration should be given to replacing them or adding supplemental transverse restrainers [3]. Cross-bracing or diaphragms on both sides of the hinge may have to be improved in conjunction with the transverse restrainer.

It can generally be assumed that any seat-type hinge used with steel girders will need additional transverse, longitudinal, and vertical restraint in even moderately severe seismic areas [3].

Continuous concrete box girders typically have in-span-type hinges. These hinge seats were typically 150 to 200 mm, while some were even less on many of the older bridges. Because of the localized damage that occurs at hinges (i.e., spalling of the concrete, etc.), the actual length of hinge seat available is much less than the original design. Therefore, a means of providing a larger hinge seat and/or tying the frames together is necessary.

43.3.1.3 Restrainers

Restrainers are used to tie the frames together, limiting the relative displacements from frame to frame and providing a load path across the joint. The main purpose is to prevent the frames from falling off their supports. There are two basic types of restrainers, cables and rods. The choice between cables and rods is rather arbitrary, but some factors to consider may be structure period, flexibility, strength of hinge/bent diaphragm, tensile capacity of the superstructure, and, to some degree, the geometry of the superstructure.

There are various types of longitudinal cable restraining devices as shown in Figures 43.5 to 43.7. Cable restraining units, such as the ones shown in Figures 43.5 and 43.6 generally have an advantage over high-strength rods because of the flexibility with its usage for varying types of superstructures. For simply supported girders, cables may be anchored to a bracket mounted to the underside of the girder flange and wrapped around the bent cap and again anchored as shown in Figure 43.8. This is the preferred method. Another possibility is to have the cables anchored to a bracket mounted to the bottom flange and simply attached to an opposing bracket on the other side of the hinge, as

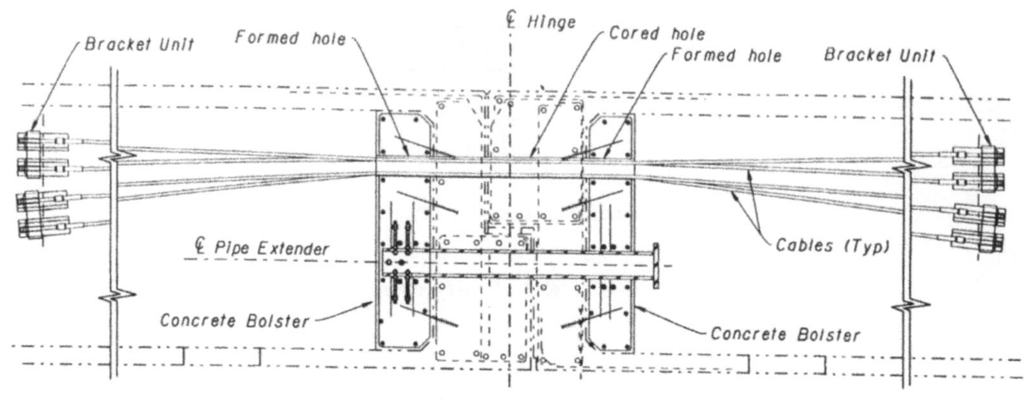

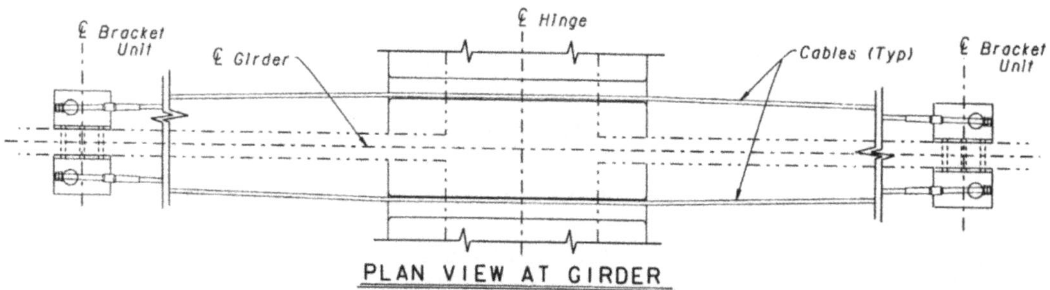

FIGURE 43.5 Type 1 hinge restrainer.

shown in Figure 43.9. The latter example is generally used for shorter bridges with a larger seat area at the bent cap or in situations where vertical clearances may be limited. Moreover, cables have the advantage of using a variety of lengths, since the anchorage devices can be mounted anywhere along the girders, whether it be steel or concrete, in addition to being anchored to the nearest bent cap or opposite side of the hinge diaphragm. For example, if the restrainer is relatively short, this may shorten the period, possibly increasing the demand to the adjacent bridge frames. Therefore, for this example, it may be desirable to lengthen the restrainer keeping the force levels to within the capacities of the adjacent segments [4]. On the other hand, if the restrainer is too long, unseating can thus occur, and an additional means of extending the seat length becomes necessary.

High-strength rods are another option for restricting the longitudinal displacements and can be used with short seats without the need for seat extenders. Unlike cables, when high-strength rods are used, shear keys or pipes are generally used in conjunction since rods can be sheared with transverse movements at hinge joints. Geometry may be a limiting factor when using high-strength rods. For example, if a box-girder bridge is shallow, it may not be possible to install a long rod through a narrow access opening. For both cables and rods, the designer needs to consider symmetry when locating restraining devices.

43.3.1.4 Pipe Seat Extenders

When a longer restraining device is preferred, increased longitudinal displacements will result and may cause unseating. It is therefore necessary to incorporate pipe seat extenders to be used in conjunction with longer restrainers when unseating will result. A 200 mm (8 in.), XX strong pipe is used for the pipe seat extender which is placed in 250 mm (10 in.) cored (and formed) hole (Figure 43.10). A 250-mm cored hole allows vertical jacking if elastomeric pads are present and replacement is required after the pad fails. Longitudinal restraining devices (namely, cables and rods) must be strain compatible with the seismic deflections imposed upon the hinge joint. In other

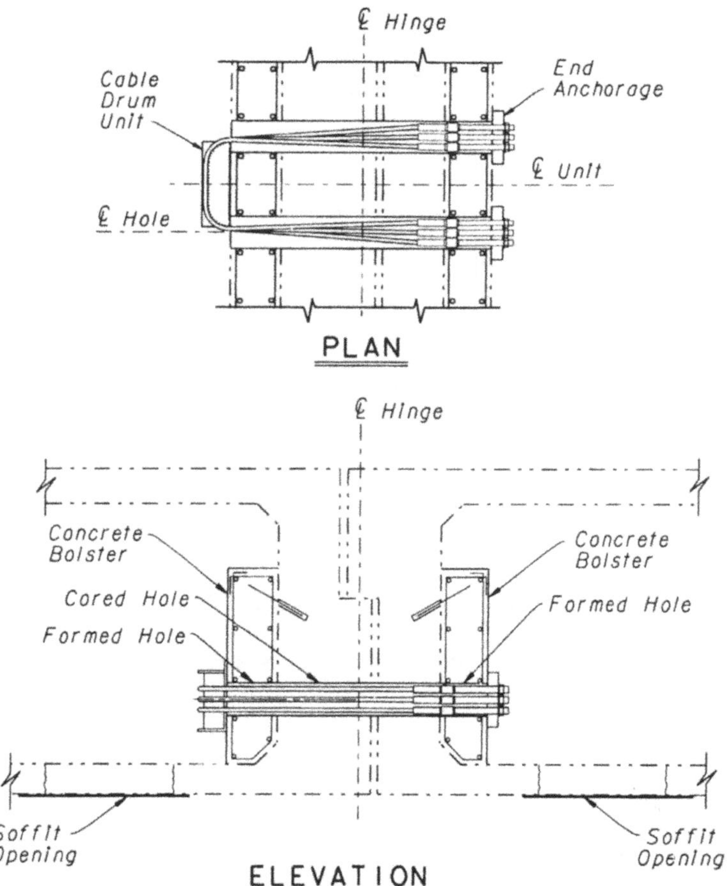

FIGURE 43.6 C-1 hinge restrainer.

words, if the longitudinal restrainers were too short, the device would have yielded long before the pipe seat extender was mobilized, deeming the longitudinal restrainers useless. To limit the number of cored holes in the diaphragm, a detail has been developed to place the restrainer cables through the pipes as shown in Figure 43.11. The pipes are not only used for vertical load-carrying capacity, but can also be used successfully as transverse shear keys.

43.3.2 Steel Bracing and Diaphragms

Lateral stiffening between steel girders typically consists of some type of cross-bracing system or channel diaphragm. These lateral bracing systems are usually designed to resist wind loads, construction loads, centrifugal force from live loads and seismic loads. The seismic loads prescribed by older codes were a fraction of current code seismic loads and, in some cases, may not have controlled bracing design. In fact, in many cases, the lateral bracing system is not able to withstand the "fusing" forces of the bearing capacity and/or shear key capacity. As a result, bracing systems may tend to buckle and, if channel diaphragms are not full depth of the girder, the webs could cripple. In general, the ideal solution is to add additional sets of bracing, stiffeners, and/or full-depth channel diaphragms as close to the bearings as physically possible.

Retrofit solutions chosen will depend on space restrictions. New bracing or diaphragms must be placed to not interfere with existing bracing, utilities, stiffeners, cable restrainers and to leave enough access for maintenance engineers to inspect the bearings. Skewed bents further complicate space

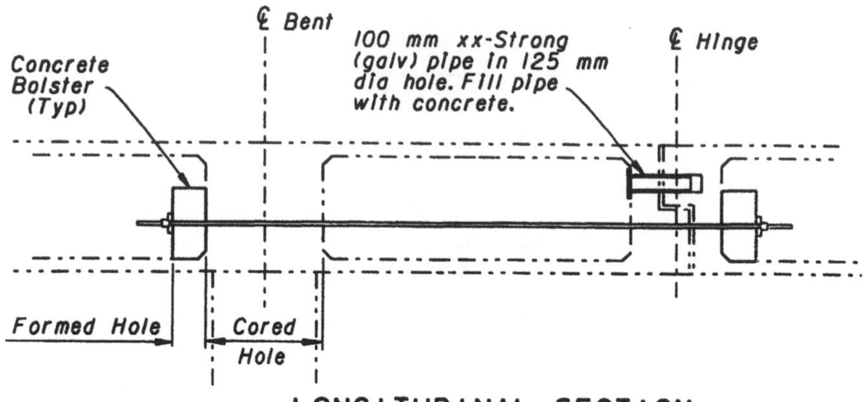

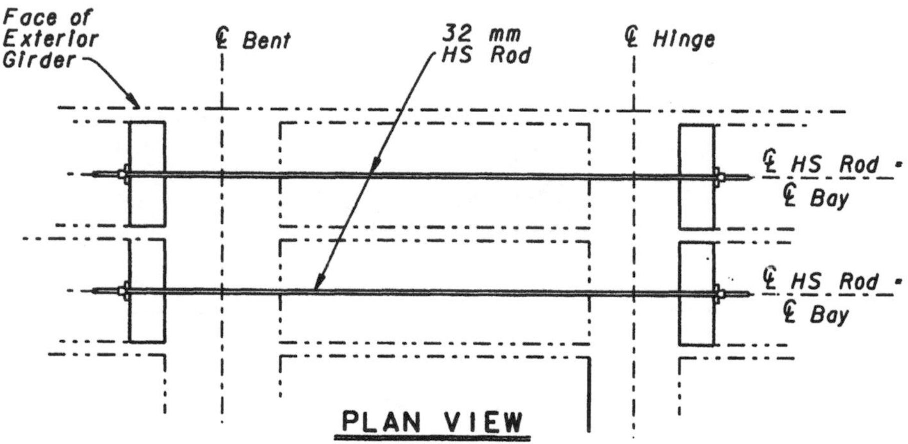

FIGURE 43.7 HS rod restrainer.

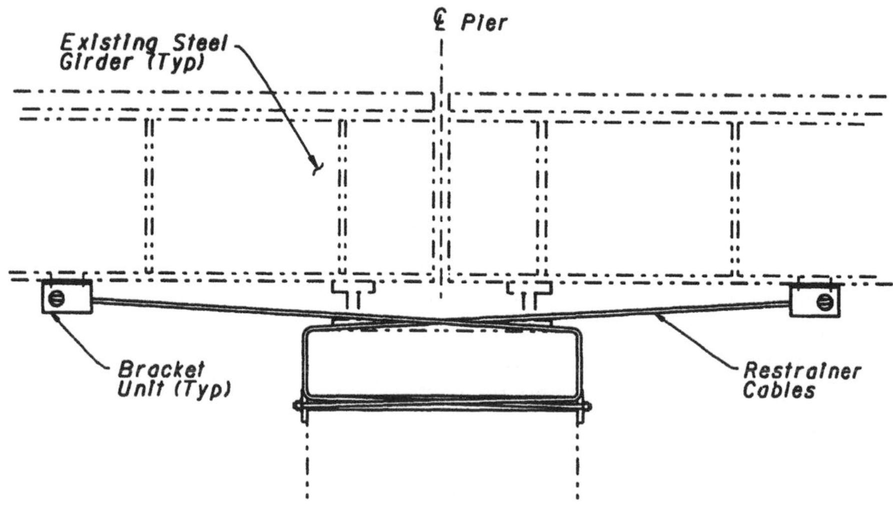

FIGURE 43.8 Cable restrainer through bent cap.

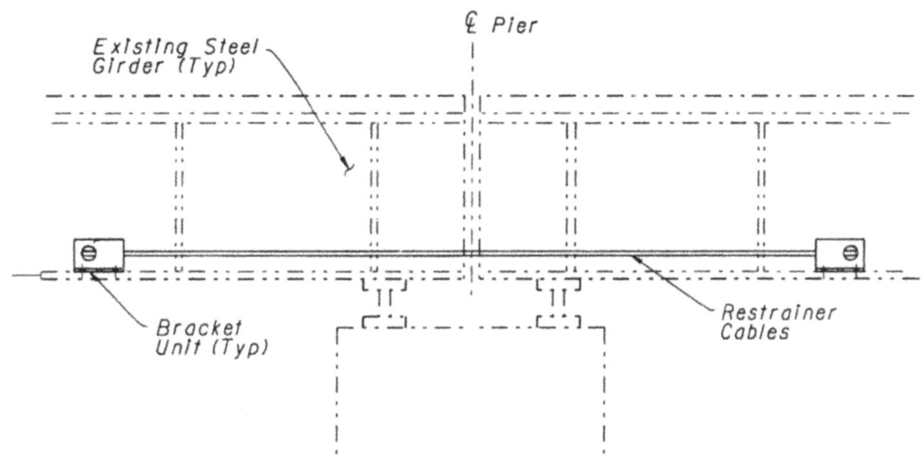

FIGURE 43.9 Girder-to-girder cable restrainer.

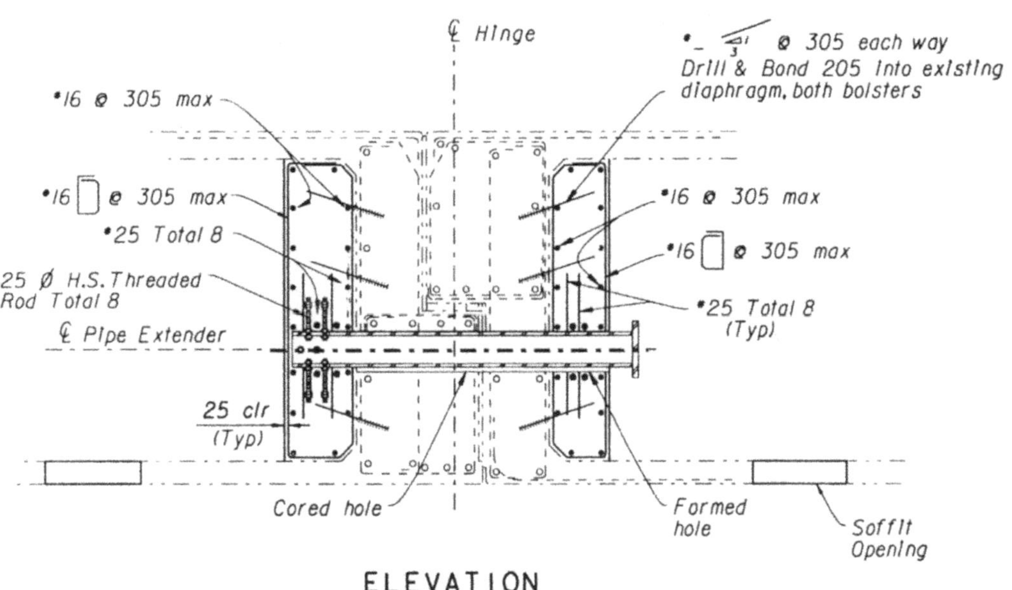

FIGURE 43.10 Pipe seat extender.

restrictions. When choosing a retrofit solution, the engineer must keep in mind that the retrofit will be constructed while the structure is carrying traffic. Stresses in a bracing member tend to cycle under live load which makes it difficult for the engineer to assess actual member stresses. As a result, any retrofit solution that requires removing and replacing existing members is not recommended. In addition, careful consideration should be given to bolted vs. welded connections. As previously mentioned, structures are under live loads during the retrofit operation and thus members are subjected to cyclic stresses. It may be difficult to achieve a good-quality weld when connecting to a constantly moving member. If bolted connections are used, preference should be given to end-bearing connections over friction connections. Friction connections have more stringent surface preparation requirements which are difficult and expensive to achieve in the field, as is the case with lead-based paint.

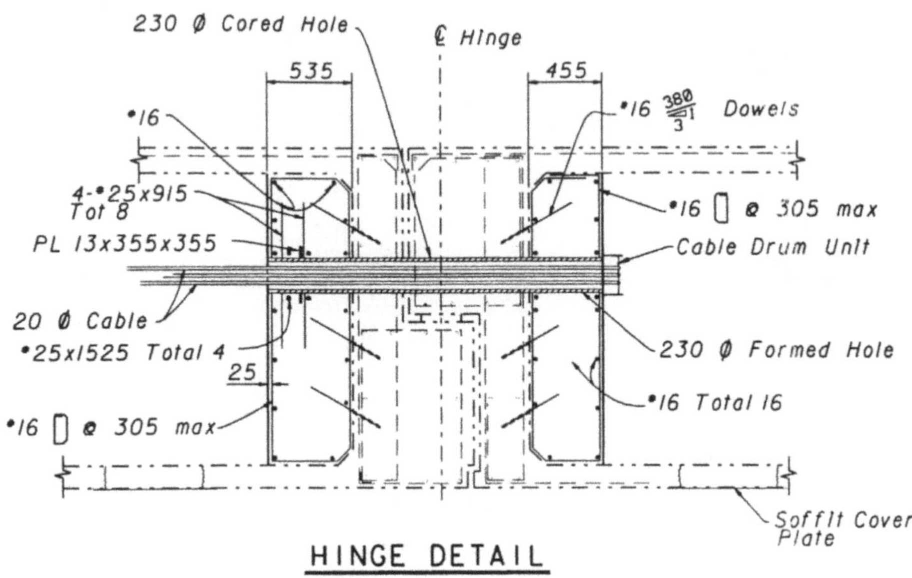

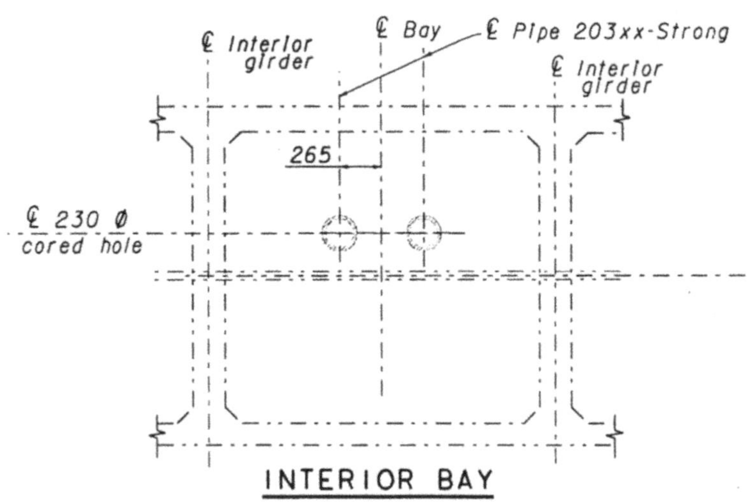

FIGURE 43.11 Restrainers/pipe seat extender.

43.3.3 Concrete Edge Beams

Edge beams are used to enhance the longitudinal capacity of a concrete bridge. These beams link consecutive bents together outside the existing box structure. In the United States, edge beams have been used to retrofit double-deck structures with long outriggers.

In a single-level bridge structure, outriggers are vulnerable in torsion under longitudinal excitation. Two retrofit alternatives are possible. The first alternative is to strengthen the outrigger cap while maintaining torsional and flexural fixity to the top of the column. The second alternative is to pin the top of the column; thus reducing the torsional demand on the vulnerable outrigger cap. Using the second alternative, the column bottom fixity needs to be ensured by means of a full footing retrofit.

In a double-deck structure, pinning the connection between the lower level outrigger cap and the column is not possible since fixity at that location is needed to provide lateral support to the upper deck. In situations where the lower deck is supported on a long outrigger cap, the torsional softening of that outrigger may lead to loss of lateral restraint for the upper-deck column. This weakness can be remedied by using edge beams to provide longitudinal lateral restraint. The edge beams need to be stiff and strong enough to ensure plastic hinging in the column and reduce torsional demand on the lower-deck outrigger cap.

43.4 Substructure Retrofits

Most earthquake damage to bridge structures occurs at the substructure. There are many types of retrofit schemes to increase the seismic capacity of existing bridges and no one scheme is necessarily more correct than another. One type of retrofit scheme may be to encase the columns and add an overlay to footings. Another might be to attract forces into the abutments and out of the columns and footings. Listed below are some different concepts to increase the capacities of individual members of the substructure.

43.4.1 Concrete Columns

Bridge columns constructed in the United States prior to 1971 are generally deficient in shear, flexure, and/or lateral confinement. Stirrups used were typically #13 bars spaced at 300 mm on center (#4 bars at 12 in.) for the entire column length including the regions of potential plastic hinging. Typically, the footings were constructed with footing dowels, or starter bars, with the longitudinal column reinforcement lapped onto the dowels. As the force levels in the column approach yield, the lap splice begins to slip. At the onset of yielding, the lap splice degrades into a pin-type condition and within the first few cycles of inelastic bending, the load-carrying capacity degrades significantly. This condition can be used to allow a "pin" to form and avoid costly footing repairs. Various methods have been successfully used to both enhance the shear capacity and ductility by increasing the lateral confinement of the plastic hinge zone for bridge columns with poor transverse reinforcement details. Following is a list of these different types and their advantages and/or disadvantages.

43.4.1.1 Column Casings

The theory behind any of the column casing types listed below is to enhance the ductility, shear, and/or flexural capacity of an existing reinforced concrete column and, in some cases, to limit the radial dilating strain within the plastic hinge zone. Because of the lap splice detail employed in older columns, one of the issues facing column retrofitting is to maintain fixity at the column base. The lateral confining pressure developed by the casing is capable of limiting the radial dilating strain of the column, enough to "clamp" the splice together, preventing any slippage from occurring. Tests have shown that limiting the radial dilating strains to less than 0.001, the lap splice will remain fixed and is capable of developing the full plastic moment capacity of the section [5]. Contrary to limiting the radial strains is to permit these strains to take place (i.e., radial dilating strains greater than 0.001) allowing a "pin" to form while providing adequate confinement throughout the plastic hinge region.

Steel Casings

There are three types of retrofit schemes that are currently employed to correct the problems of existing columns through the use of steel jacketing. The first type is typically known as a class F type of column casing retrofit, as shown in Figure 43.12. This type of casing is fully grouted and is placed the full height of the column. It is primarily used for a column that is deficient in shear and flexure. It will limit the radial dilating strain to less than 0.001, effectively fixing the lap splice from

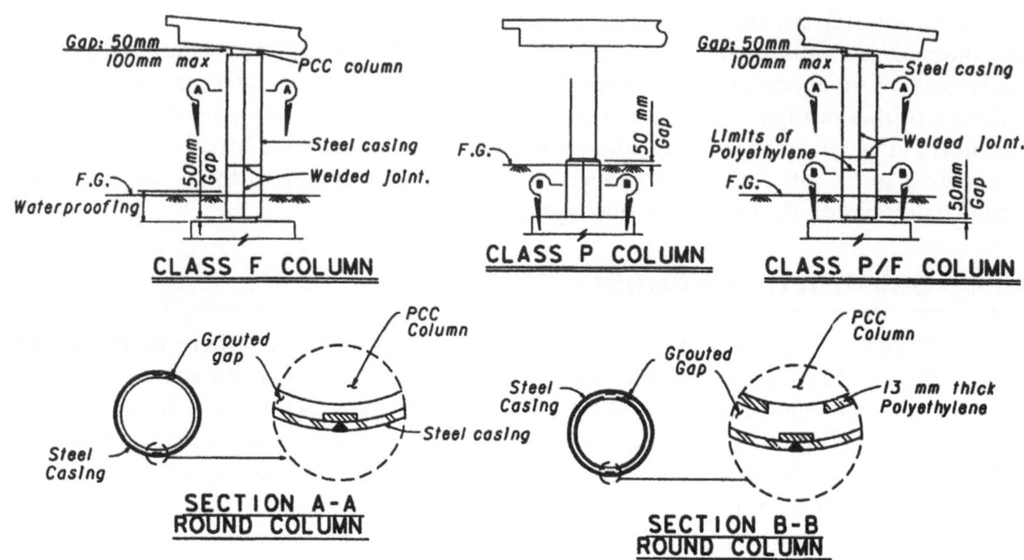

FIGURE 43.12 Steel column casings.

slipping. The lateral confining pressure for design is taken as 2.068 MPa (300 psi) and, when calculated for a 13-mm-thick steel casing with A36 steel, is equivalent to a #25 bar (#8 bar) at a spacing of about 38 mm. It can therefore be seen that the confinement, as well as the shear, is greatly enhanced. The allowable displacement ductility ratio for a class F column casing is typically 6, allowing up to 8 in isolated locations. It has been tested well beyond this range; however, the allowable ductility ratio is reduced to prohibit fracturing of the longitudinal bars and also to limit the level of load to the footing.

The second type of casing is a class P type and is only a partial height casing; therefore, it does not help a column that is deficient in shear. As can be seen from Figure 43.12, the main difference between a class F and a class P type column casing is the layer of polyethylene between the grout and the column. This will permit the column to dilate outward, allowing the strain to exceed 0.001, forming a pin at the base of the column. It should be noted that the casing is still required in this condition to aid in confining the column. The limit of this type of retrofit is typically taken as 1.5 times the columns diameter or to where the maximum moment has decreased to 75%.

The third type of steel column casing is a combination of the first two and, hence, is known as a class P/F, also shown in Figure 43.12. It is used like a class P casing, but for a column with a shear deficiency. All of these casings can be circular (for circular or square columns) or oblong (for rectangular columns). If a rectangular column is deficient in shear only, it is sometimes permitted to use flat steel plates if the horizontal clearances are limited and it is not possible to fit an oblong column in place. For aesthetic purposes, the class P casing may be extended to full height in highly visible areas. It can be unsightly if an oblong casing is only partial height. It is important to mention that the purpose of the 50-mm gap at the ends of the column is to prevent the casing from bearing against the supporting member acting as compression reinforcement, increasing the flexural capacity of the column. This would potentially increase the moments and shears into the footing and/or bent cap under large seismic loads.

Concrete Casings

When retrofitting an unusually shaped column without changing the aesthetic features of the geometry of the column, a concrete casing may be considered as an alternative. Existing columns are retrofitted by placing hoops around the outer portion of the column and then drilling and bonding bars into the column to enclose the hoops. The reinforcement is then encased with concrete,

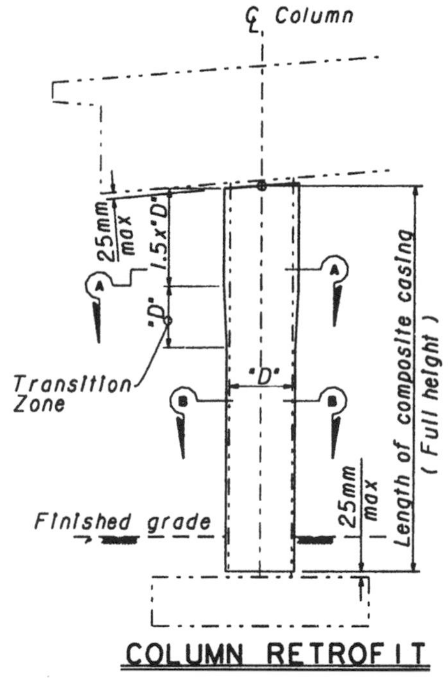

FIBER VOLUME = 35% min

E-GLASS		
ROUND COLUMN		
COLUMN DIA	t1(min)	t2(min)
305mm	5mm	5mm
610mm	9mm	5mm
915mm	13mm	7mm
1220mm	17mm	9mm
1525mm	22mm	11mm
1830mm	25mm	13mm

Note A:
Epoxy Resin-Glass Composite

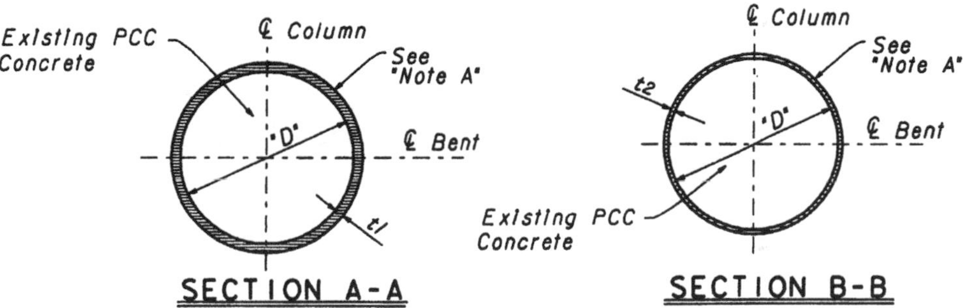

FIGURE 43.13 Advanced composite column casing.

thus maintaining the original shape of the column. The design of a concrete jacket follows the requirements of a new column. Although this method increases the shear and flexural capacities of the column and provides additional confinement without sacrificing aesthetics, it is labor intensive and therefore can be costly.

Advanced Composite Casings

Recently, there has been significant research and development using advanced composites in bridge design and retrofit. Similar to steel casings, advanced composite casings increase the confinement and shear capacity of existing concrete bridge columns. This type of column retrofit has proved to be competitive with steel casings when enhancing column shear capacity and may also provide an economic means of strengthening bridge columns (Figure 43.13). However, currently, composites are not economic when limiting lap splice slippage inside expected plastic hinge zones. The advantage of using some types of composite casings is that the material can be wrapped to the column without changing its geometric shape. This is important when aesthetics are important or lateral clearances at roadways are limited.

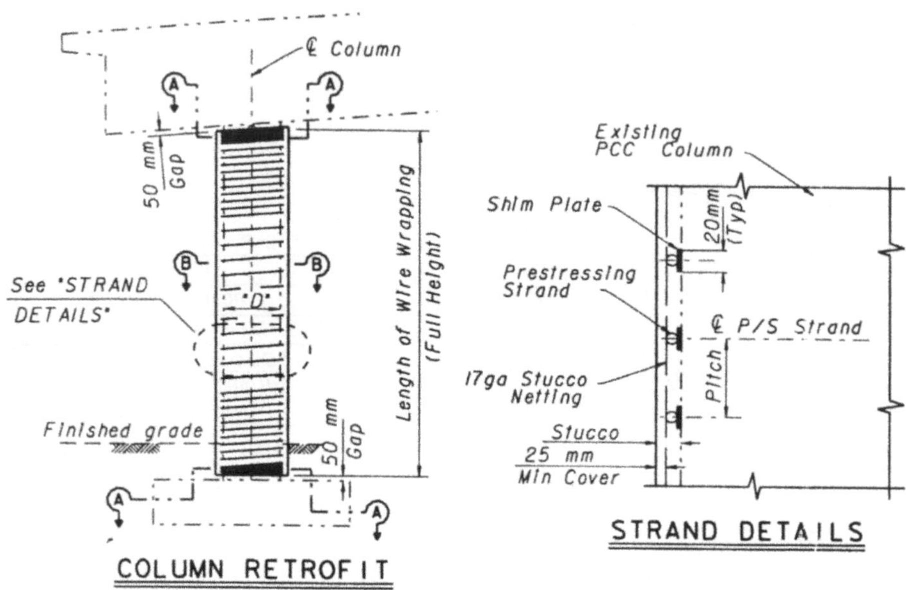

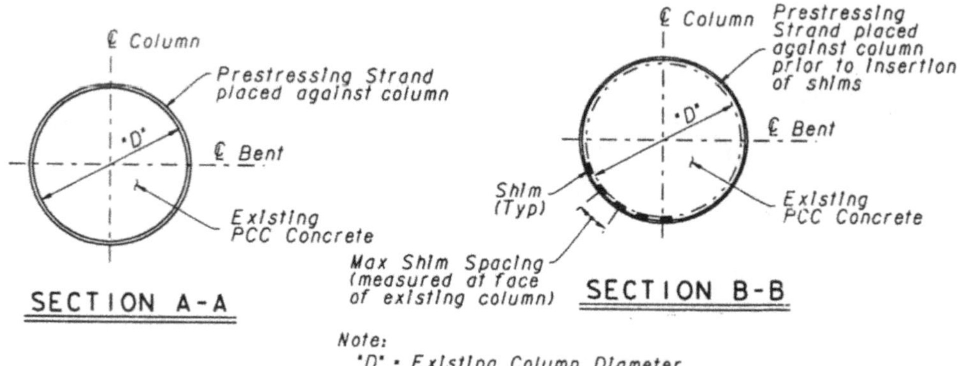

FIGURE 43.14 Column wire wrap.

Wire Wrap Casings

Another type of system that was recently approved by the California Department of Transportation is a "wire wrap" system. It consists of a prestressing strand hand wrapped onto a column; then wedges are placed between the strand and the column, effectively prestressing the strand and actively confining the column, as shown in Figure 43.14. The advantage of this type of system is that, like the advanced composites, it can be wrapped to any column without changing the geometric shape. Its basic disadvantage is that it is labor intensive and currently can only be applied to circular columns.

43.4.1.2 In-Fill Walls

Reinforced concrete in-fill walls may also be used as an alternative for multicolumn bridge bents, as shown in Figure 43.15. This has two distinct advantages: it increases the capacity of the columns in the transverse direction and limits the transverse displacements. By limiting the displacements

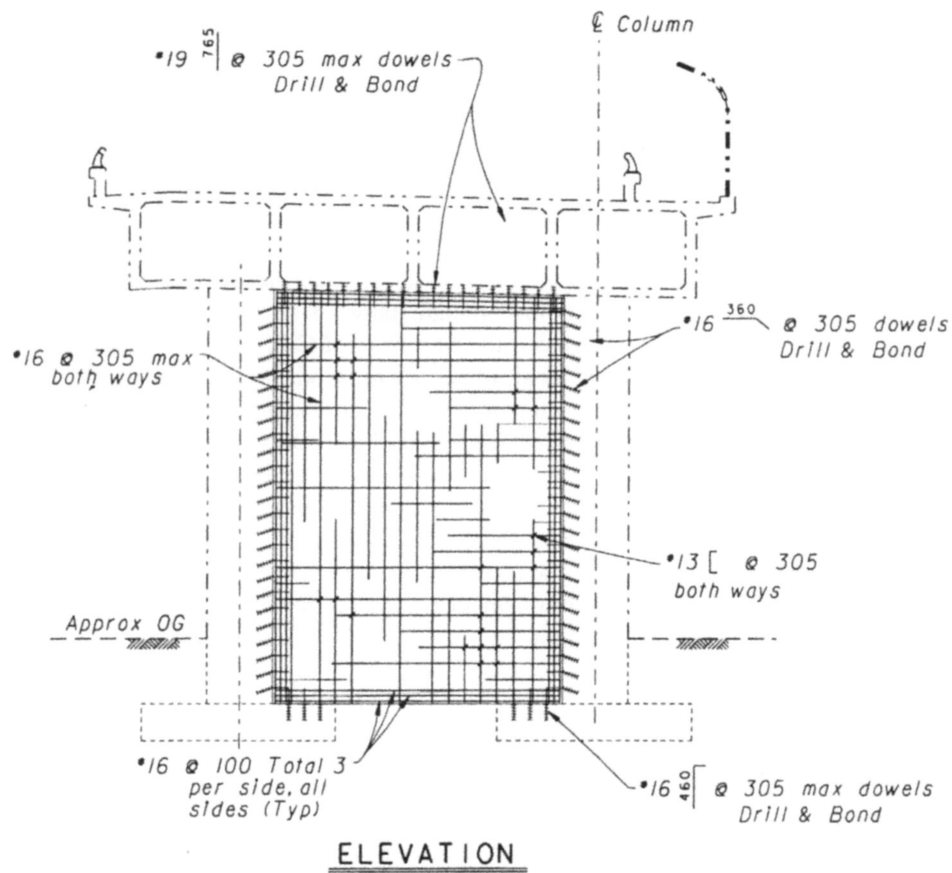

FIGURE 43.15 In-fill wall.

transversely, the potential for plastic hinge formation in the bent cap is eliminated. Therefore, cost may prove to be less than some other retrofit alternatives mentioned earlier. It is important to note that the in-fill wall is not effective in the longitudinal capacity of bridge bents with little or no skew.

43.4.1.3 Link Beams

Link beams are used to enhance the transverse capacity of a concrete bent. The placement of a link beam over a certain height above ground level determines its function.

A link beam can be placed just below the soffit level and acts as a substitute to a deficient existing bent cap. The main function of this kind of a link beam is to protect the existing superstructure and force hinging in the column.

In other cases, a link beam is placed somewhere between the ground level and the soffit level in order to tune the transverse stiffness of a particular bent. This type of retrofit can be encountered in situations where a box superstructure is supported on bents with drastically unequal stiffnesses. In this case, the center of mass and the center of rigidity of the structure are farther apart. This eccentricity causes additional displacement on the outer bents which can lead to severe concentrated ductility demands on just a few bents of the subject bridge. This behavior is not commonly preferred in seismically resistant structures and the use of link beams in this case can reduce the eccentricity between the center of mass and the center of rigidity. This structural tuning is important in equally distributing and reducing the net ductility demands on all the columns of the retrofitted bridge.

43.4.2 Pier Walls

Until recently, pier walls were thought to be more vulnerable to seismic attack than columns. However, extensive research performed at the University of California, Irvine has proved otherwise [6]. Pier walls, nevertheless, are not without their problems. The details encompass poor confinement and lap splices, similar to that of pre-1971 bridge columns. Pier walls are typically designed and analyzed as a shear wall about the strong axis and as a column about the weak axis. The shear strength of pier walls in the strong direction is usually not a concern and one can expect a shear stress of about $0.25\sqrt{f_c'}$ (MPa). For the weak direction, the allowable demand displacement ductility ratio in existing pier walls is 4.0. Similar to columns, many older pier walls were also built with a lap splice at the bottom. If the lap splice is long enough, fixity will be maintained and the full plastic moment can be developed. Tests conducted at the University of California, Irvine, have shown that lap splices 28 times the diameter of the longitudinal bar to be adequate [6]. However, a lap splice detail with as little as 16 bar diameters will behave in the same manner as that of a column with an inadequate lap splice and may slip, forming a pin condition. Because of the inherent flexibility of a pier wall about its weak axis, the method of retrofit for this type of lap splice is a plate with a height two times the length of the splice, placed at the bottom of the wall. The plate thickness is not as critical as the bolt spacing. It is generally recommended that the plate be 25 mm thick with a bolt spacing equal to that of the spacing of the main reinforcement, only staggered (not to exceed approximately 355 mm). If additional confinement is required for the longer lap splice, the plate height may be equal to the lap splice length.

43.4.3 Steel Bents

Most steel bents encountered in older typical bridges can be divided in two groups. One group contains trestle bents typically found in bridges spanning canyons, and the second group contains open-section built-up steel columns. The built-up columns are typically I-shaped sections which consist of angles and plates bolted or riveted together. The second group is often found on small bridges or elevated viaducts.

Trestle steel bents are commonly supported on pedestals resting on rock or relatively dense foundation. In general, the truss members in these bents have very large slenderness ratios which lead to very early elastic buckling under low-magnitude earthquake loading. Retrofitting of this type of bent consists basically of balancing between member strengthening and enhancing the tensile capacity of the foundation and keeping connection capacities larger than member capacities. In many situations, foundation retrofit is not needed where bent height is not large and a stable rocking behavior of the bent can be achieved. Strengthening of the members can be obtained by increasing the cross-sectional area of the truss members or reducing the unsupported length of the members.

Figure 43.16 shows the retrofit of Castro Canyon bridge in Monterey County, California. The bent retrofit consists of member strengthening and the addition of a reinforced concrete block around the bent-to-pedestal connection. In this bridge, the pedestals were deeply embedded in the soil which added to the uplift capacity of the foundation.

For very tall trestle bents (i.e., 30 m high), foundation tie-downs, in addition to member strengthening, might be needed in order to sustain large overturning moments. Anchor bolts for base plates supported on top of pedestals are usually deficient. Replacement of these older bolts with high-strength bolts or the addition of new bolts can be done to ensure an adequate connection capable of developing tension and shear strength. The addition of new bolts can be achieved by coring through the existing base plate and pedestal or by enlarging the pedestal with a concrete jacket surrounding the perimeter bolts. The use of sleeved anchor bolts is desirable to induce some flexibility into the base connection.

The second group of steel bents contains open section built-up columns. These members may fail because of yielding, local buckling, or lateral torsional buckling. For members containing a single I-shaped section, lateral torsional buckling typically governs. Retrofit of this type of column

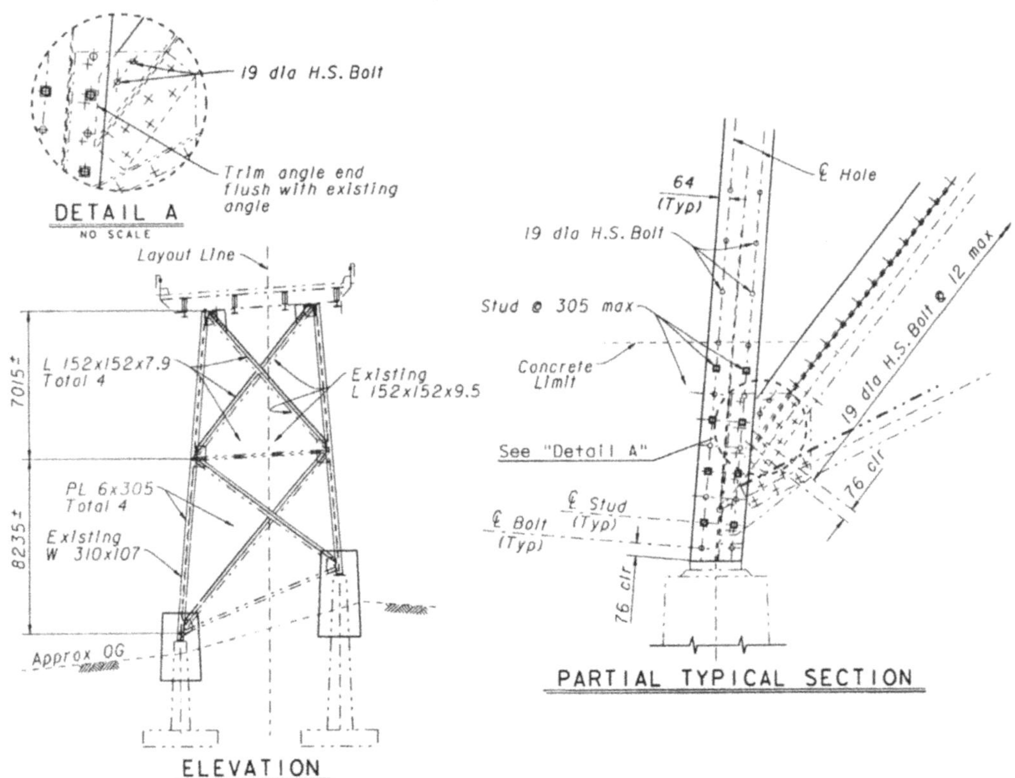

FIGURE 43.16 Trestle bent retrofit.

consists of enclosing the section by bolting channel sections to the flanges. Figure 43.17 shows this type of retrofit. Installation of these channels is made possible by providing access through slotted holes. These holes are later covered by tack welding plates or left open. For larger members with an open section as seen in Figure 43.18, retrofit consists of altering the existing cross section to a multicell box section.

The seismic behavior of the multicell box is quite superior to an open section of a single box. These advantages include better torsional resistance and a more ductile postelastic behavior. In a multicell box, the outside plates sustain the largest deformation. This permits the inside plates to remain elastic in order to carry the gravity load and prevent collapse during an earthquake. To maintain an adequate load path, the column base connection and the supporting foundation should be retrofitted to ensure the development of the plastic hinge just above the base connection. This requires providing a grillage to the column base as seen in Figure 43.18 and a footing retrofit to ensure complete load path.

43.4.4 Bearing Retrofit

Bridge bearings have historically been one of the most vulnerable components in resisting earthquakes. Steel rocker bearings in particular have performed poorly and have been damaged by relatively minor seismic shaking. Replacement of any type of bearing should be considered if failure will result in collapse of the superstructure. Bearing retrofits generally consist of replacing steel rocker-type bearings with elastomeric bearings. In some cases, where a higher level of serviceability is required, base isolation bearings may be used as a replacement for steel bearings. For more information on base isolation, see the detailed discussion in Chapter 41. Elastomeric bearings are preferred over steel rockers because the bridge deck will only settle a small amount when the bearings

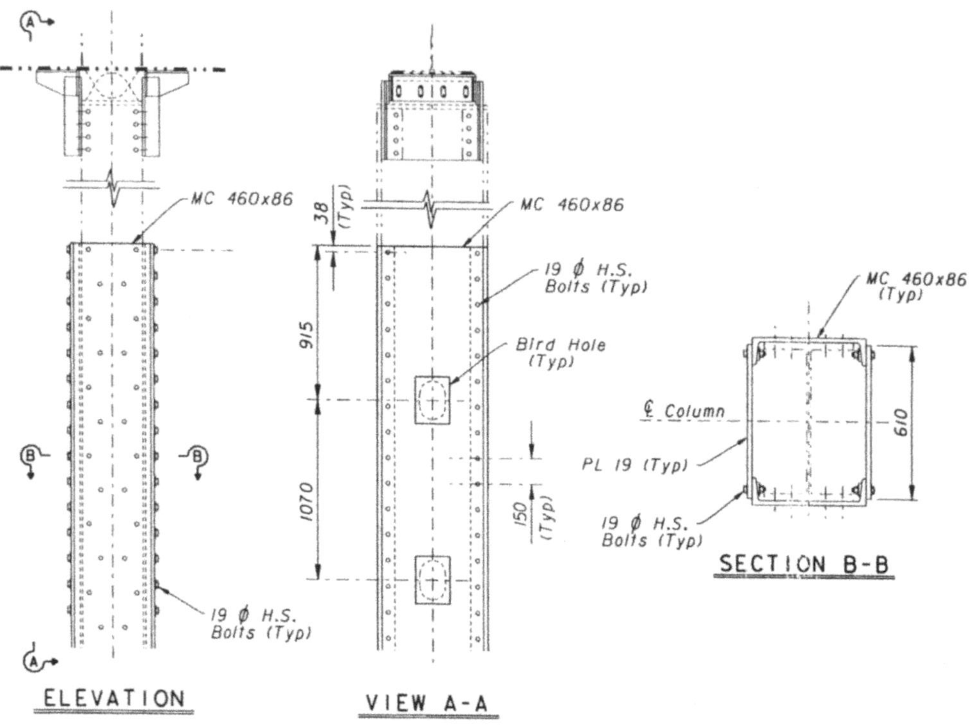

FIGURE 43.17 I-shaped steel column retrofit.

fail, whereas the deck will settle several inches when rockers fail. Elastomeric bearings also have more of a base isolation effect than steel rockers. Both types of bearings may need catchers, seat extenders, or some other means of providing additional support to prevent the loss of a span. Although elastomeric bearings perform better than steel rockers, it is usually acceptable to leave the existing rockers in place since bearings replacement is more expensive than installing catchers to prevent collapse during an earthquake.

43.4.5 Shear Key Retrofit

The engineer needs to consider the ramifications of a shear key retrofit. The as-built shear keys may have been designed to "fuse" at a certain force level. This fusing will limit the amount of force transmitted to the substructure. Thus, if the shear keys are retrofitted and designed to be strong enough to develop the plastic capacity of the substructure, this may require a more expensive substructure retrofit. In many cases, it is rational to let the keys fail to limit forces to the substructure and effectively isolate the superstructure. Also note that superstructure lateral bracing system retrofits will also have to be increased to handle increased forces from a shear key retrofit. In many cases, the fusing force of the existing shear keys is large enough to require a substructure retrofit. In these situations, new or modified shear keys need to be constructed to be compatible with the plastic capacity of the retrofitted substructure. There are other situations that may require a shear key retrofit. Transverse movements may be large enough so that the external girder displaces beyond the edge of the bent cap and loses vertical support. For a multiple-girder bridge, it is likely that the side of the bridge may be severely damaged and the use of a shoulder or lane will be lost, but traffic can be routed over a portion of the bridge with few or no emergency repairs. This is considered an acceptable risk. On the other hand, if the superstructure of a two- or three-girder bridge is displaced transversely so that one line of girders loses its support, the entire bridge may collapse. Adequate transverse restraint, commonly in the form of shear keys, should be provided.

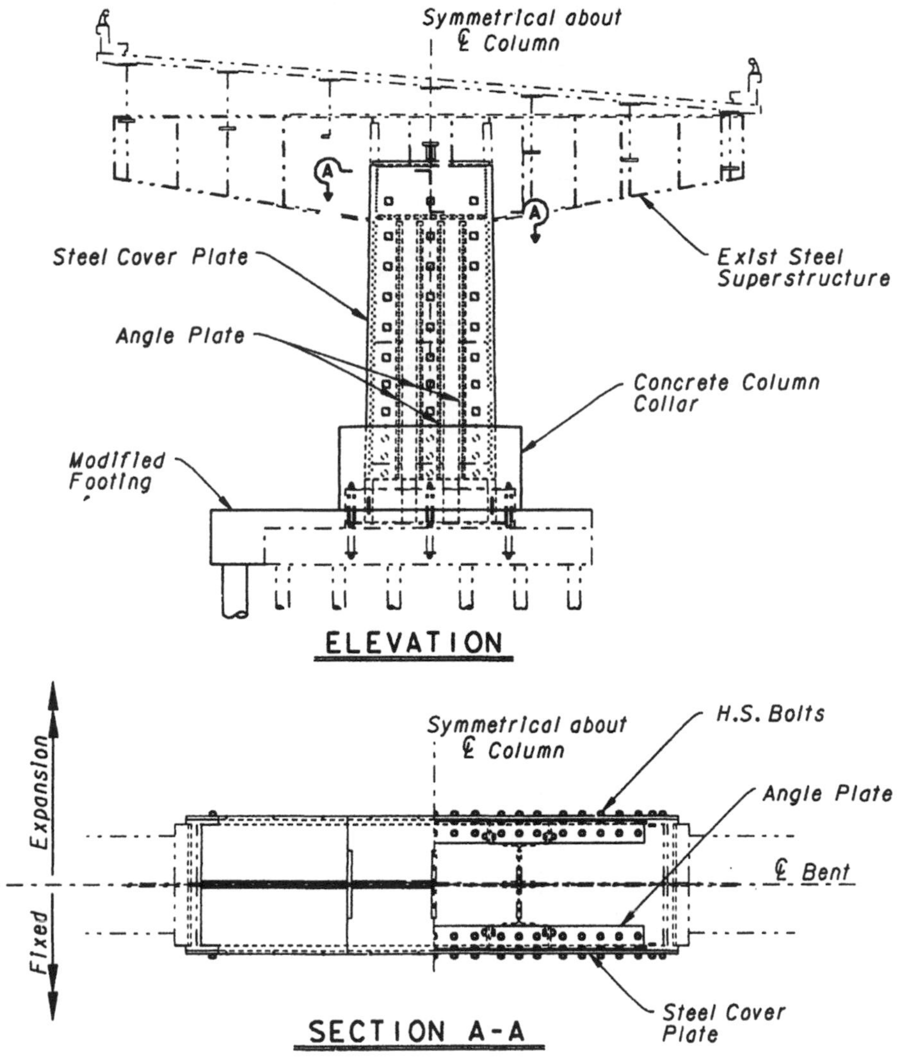

FIGURE 43.18 Open-section steel column retrofit.

43.4.6 Cap Beam Retrofit

There are several potential modes of failure associated with bent caps. Depending on the type of bent cap, these vulnerabilities could include bearing failures, shear key failures, inadequate seat widths, and cap beam failures. Table 43.3 lists several types of bent caps and their associated potential vulnerabilities.

Cap beam modes of failure may include flexure, shear, torsion, and joint shear. Prior to the 1989 Loma Prieta earthquake, California, there was very little emphasis placed on reinforcement detailing of bent cap beams for lateral seismic loads in the vicinity of columns. As a result, cap beams supported on multiple columns were not designed and detailed to handle the increased moment and shear demands that result from lateral transverse framing action. In addition, the beam–column joint is typically not capable of developing the plastic capacity of the column and thus fails in joint shear. For cap beams supported by single columns, although they do not have framing action in the transverse direction and are not subjected to moment and shear demands that are in addition

TABLE 43.3 Potential Bent Cap Vulnerabilities

Cap Type	Bearings	Shear Keys	Seat Width	Cap Beam				
				Moment	Shear	Torsion	Joint Shear	Bolted Cap/Col Connection
Concrete drop cap — single column bent	x	x	x				x	
Concrete drop cap — multicolumn bents	x	x	x	x	x		x	
Integral concrete cap — single column bent							x	
Integral concrete cap — multicolumn bent				x	x	x	x	
Inverted T — simple support for dead load, continuous for live load — single col. bent		x					x	
Inverted T — simple support for dead load, continuous for live load — multi col. bent		x					x	
Inverted T — simple support for both dead load, live load — single column bent	x	x	x				x	
Inverted T — simple support for both dead load and live load — multiple column bent	x	x	x	x	x	x	x	
Steel bent cap — single column bent	x	x	x					x
Steel bent cap — ulticolumn bent	x	x	x					x
Integral outrigger bent						x	x	
Integral C bent						x	x	

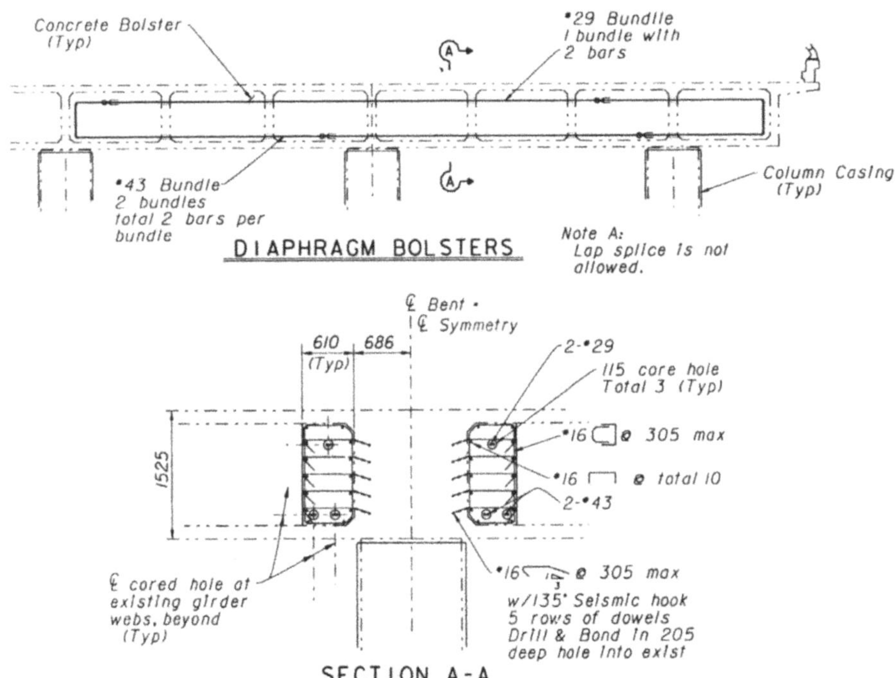

FIGURE 43.19 Bent cap retrofit.

to factored vertical loads, joint shear must still be considered as a result of longitudinal seismic response. In these situations, retrofit of the superstructure is not common since single-column bents are typically fixed to the footing and fixity at top of the column is not necessary.

Retrofit solutions that address moment and shear deficiencies typically include adding a bolster to the existing cap (Figure 43.19). Additional negative and positive moment steel can be placed on the top and bottom faces of the bolsters as required to force plastic hinging into the columns. These bolsters will also contain additional shear stirrups and steel dowels to ensure a good bond with the existing cap for composite action. Prestressing can also be included in the bolsters. In fact, prestressing has proved to be an effective method to enhance an existing cap moment, shear, and joint shear capacity. This is particularly true for bent caps that are integral with the superstructure.

Special consideration should be given to the detailing of bolsters. The engineer needs to consider bar hook lengths and bending radii to make sure that the stirrups will fit into the bolsters. Although, the philosophy is to keep the superstructure elastic and take all inelastic action in the columns, realistically, the cap beams will have some yield penetration. Thus, the new bolster should be detailed to provide adequate confinement for the cap to guarantee ductile behavior. This suggests that bolsters should not be just doweled onto the existing bent cap. There should be a continuous or positive connection between the bolsters through the existing cap. This can be achieved by coring through the existing cap. The hole pattern should be laid out to miss the existing top and bottom steel of the cap. It is generally difficult to miss existing shear stirrups so the engineer should be conservative when designing the new stirrups by not depending on the existing shear steel. The steel running through the existing cap that connects the new bolsters can be continuous stirrups or high-strength rods which may or may not be prestressed.

The cap retrofit is much easier with an exposed cap but can be done with integral caps. In order to add prestressing or new positive and negative steel and to add dowels to make sure that the bolsters in adjacent bays are continuous or monolithic, the existing girders have to be cored. Care must be taken to avoid the main girder steel and/or prestressing steel.

Torsion shall be investigated in situations where the superstructure, cap beam, and column are monolithic. In these situations, longitudinal loads are transferred from the superstructure into the columns through torsion of the cap beam. Superstructures supported on cap beams with bearings are unlikely to cause torsional problems. Torsion is mainly a problem in outriggers connected to columns with top fixed ends. However, torsion can also exist in bent cap beams susceptible to softening due to longitudinal displacements. This softening is initiated when top or especially bottom longitudinal reinforcement in the superstructure is not sufficient to sustain flexural demands due to the applied plastic moment of the column. Retrofit solutions should ensure adequate member strength along the load path from superstructure to column foundation.

In general, the philosophy of seismic design is to force column yielding under earthquake loads. In the case of an outrigger, the cap beam torsional nominal yield capacity should be greater than the column flexural plastic moment capacity. Torsion reinforcement should be provided in addition to reinforcement required to resist shear, flexure, and axial forces. Torsion reinforcement consists of closed stirrups, closed ties, or spirals combined with transverse reinforcement for shear, and longitudinal bars combined with flexural reinforcement. Lap-spliced stirrups are considered ineffective in outriggers, leading to a premature torsional failure. In such cases, closed stirrups should not be made up of pairs of U-stirrups lapping one another. Where necessary, mechanical couplers or welding should be used to develop the full capacity of torsion bars. When plastic hinging cannot be avoided in the superstructure, the concrete should be considered ineffective in carrying any shear or torsion. Regardless where plastic hinging occurs, the outrigger should be proportioned such that the ultimate torsional moment does not exceed four times the cracking torque. Prestressing should not be considered effective in torsion unless bonded in the member. Unbonded reinforcement, however, can be used to supply axial load to satisfy shear friction demands to connect outrigger caps to columns and superstructure. Bonded tendons should not be specified in caps where torsional yielding will occur. Designers must consider effects of the axial load in caps due to transverse column plastic hinging when satisfying shear and torsion demands.

43.4.7 Abutments

Abutments are generally classified into two types: seat type and monolithic. The monolithic type of abutment is commonly used for shorter bridges, whereas longer bridges typically use a seat type. Contrary to the seat-type abutment, the monolithic abutment has the potential for heavy damage. This is largely due to the fact that the designer has more control through the backwall design. The backwall behaves as a fuse to limit any damage to the piles. However, since this damage is not a collapse mechanism, it is therefore considered to be acceptable damage. Additionally, the monolithic abutment has proven itself to perform very well in moderate earthquakes, sustaining little or no damage. Some typical problems encountered in older bridges are

- Insufficient seat length for seat-type abutments;
- Large gallery, or gap, between the backwall and superstructure end diaphragm;
- Insufficient longitudinal and/or transverse shear capacity;
- Weak end diaphragms at monolithic abutments.

Following are some of the more common types of retrofits used to remedy the abutment problems mentioned above.

43.4.7.1 Seat Extenders

Seat extenders at abutments and drop caps generally consist of additional concrete scabbed onto the existing face (Figures 43.20 and 43.21). The design of seat extenders that are attached to existing abutment or bent cap faces should be designed like a corbel. When designing the connecting steel between the new seat extender and the existing concrete, shear friction for vertical loads should be considered. Tensile forces caused by friction should also be considered when the girder moves in

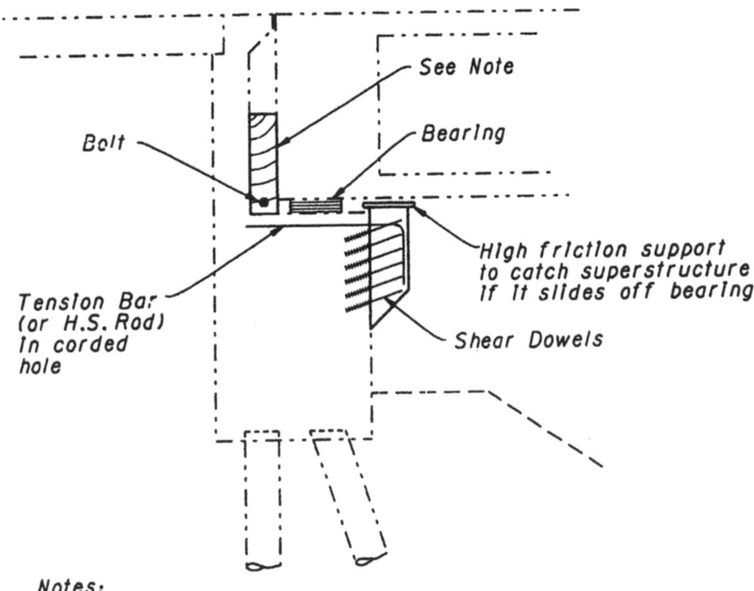

Notes:

Gap filler to mobilize backwall and embankment soil. Filler can be:

A) Steel or hardwood strips inserted by slipping them horizontally in space above seat and rotating to vertical. Bolt together.

B) Fill space with concrete through top use polystyrene between new concrete and bridge superstructure requires traffic control.

Hardwood filler (Option A) shown in sketch above.

FIGURE 43.20 Seat extender at abutment.

the longitudinal direction and pulls the new concrete away from the existing bent cap or abutment. Compression strut and bearing loads under the girder also need to be considered. Note that the face of the existing concrete should be intentionally roughened before the new concrete is placed to ensure a good bond.

If bearing failure results in the superstructure dropping 150 mm or more, catchers could be added to minimize the drop. Catchers generally are designed to limit the superstructure drop to 50 mm and can provide additional seat width. In other words, catchers are basically seat extenders that are detailed to reduce the amount the superstructure is allowed to drop. The design procedure is similar for both seat extenders and catchers. In some cases, an elastomeric bearing pad is placed on top of the catcher to provide a landing spot for the girder after bearing failure. The friction factor for concrete on an elastomeric pad is less than for concrete on concrete so the tension force in the corbel could possibly be reduced. One special consideration for catchers is to make sure to leave enough room to access the bearing for inspection and replacement.

43.4.7.2 Fill galleries with timber, concrete or steel

Some seat-type abutments typically have a gallery, or a large gap, between the superstructures end diaphragm and the backwall. It is important to realize that the columns must undergo large deformations before the soil can be mobilized behind the abutment if this gap is not filled. Therefore, as a means of retrofit, the gallery is filled with concrete, steel, or timber to engage the backwall and, hence, the soil (Figure 43.22). However, timber is a potential fire hazard and in some parts of the

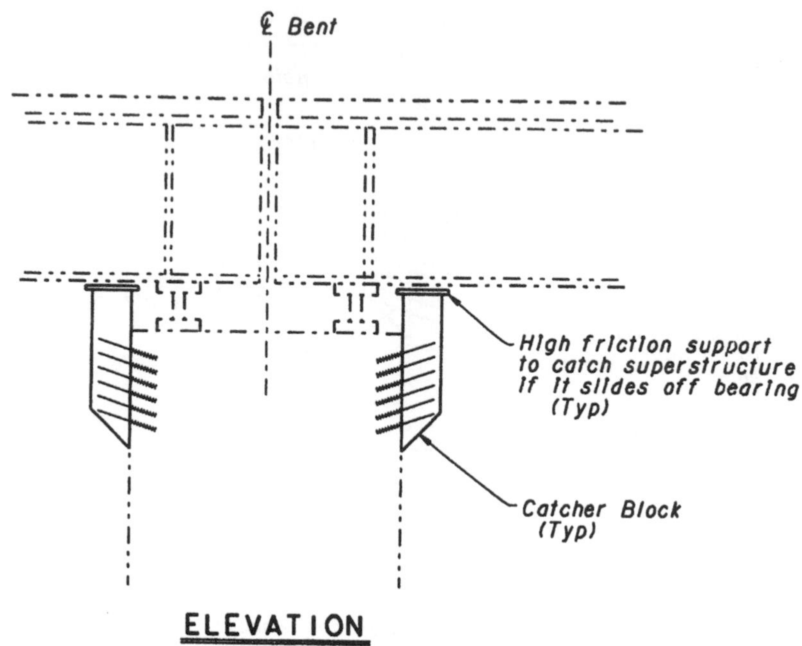

FIGURE 43.21 Seat extender at bent cap.

United States may be susceptible to termite attack. When filling this gap, the designer should specify the expected thermal movements, rather than the required thickness. This prevents any problems that may surface if the backwall is not poured straight.

43.4.7.3 L Brackets on Superstructure Soffit

Similar in theory to filling the gallery behind the backwall is adding steel angles (or brackets) to the flanges of steel I-girders, as seen in Figure 43.23. These brackets act as "bumpers" that transfer the longitudinal reaction from the superstructure into the abutment, and then into the soil.

43.4.7.4 Shear Keys, Large CIDH Piles, Anchor Slabs, and Vertical Pipes

For shorter bridges, an effective retrofit scheme is to attract the forces away from the columns and footings and into the abutments. This usually means modifying and/or strengthening the abutment, thereby "locking up" the abutment, limiting the displacements and, hence, attracting most of the loads. Although this type of retrofit is more effective in taking the load out of columns for shorter structures, the abutments still may require strengthening, in addition to retrofitting the columns, for longer bridges.

Methods that are intended to mobilize the abutment and the soil behind the abutment may consist of vertical pipes, anchor piles, seismic anchor slabs, or shear keys as shown in Figures 43.24 and 43.25. For heavily skewed or curved bridges, anchor piles or vertical pipes are generally the preferred method due to the added complication from geometry. For instance, as the bridge rotates away from the abutment, there is nothing to resist this movement. By adding an anchor pile at the acute corners of the abutment, the rotation is prohibited and the anchor pile then picks up the load.

43.4.7.5 Catchers

When the superstructure is founded on tall bearings at the abutments, the bearings, as mentioned earlier, are susceptible to damage. If tall enough, the amount of drop the superstructure would undergo can significantly increase the demands at the bents. To remedy this, catcher blocks are constructed next to the bearings to "catch" the superstructure and limit the amount of vertical displacement.

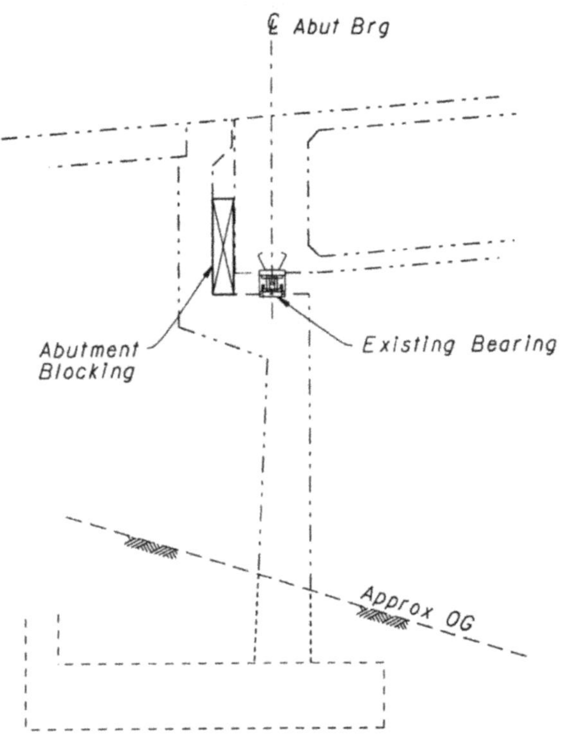

FIGURE 43.22 Abutment blocking.

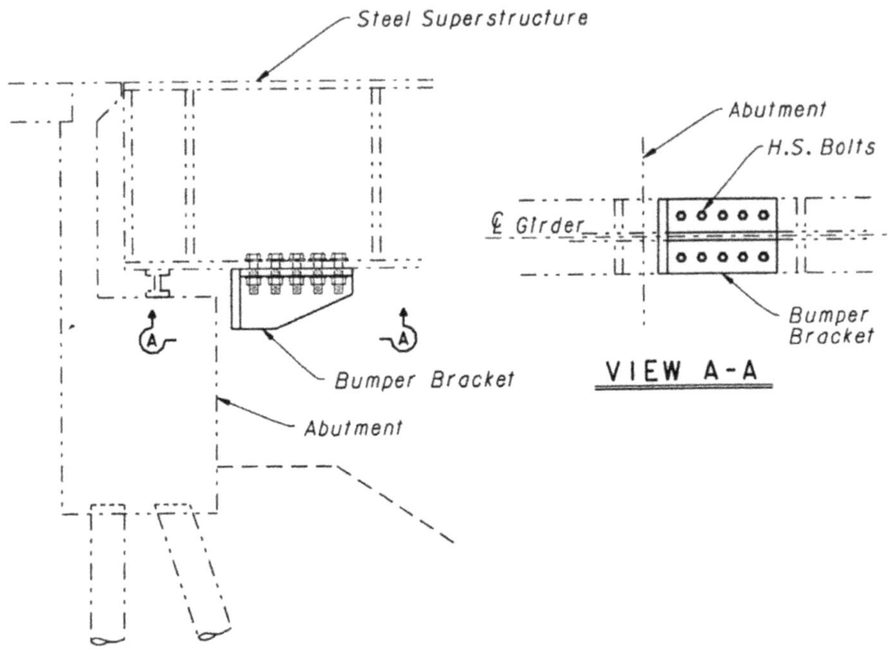

FIGURE 43.23 Bumper bracket at abutment.

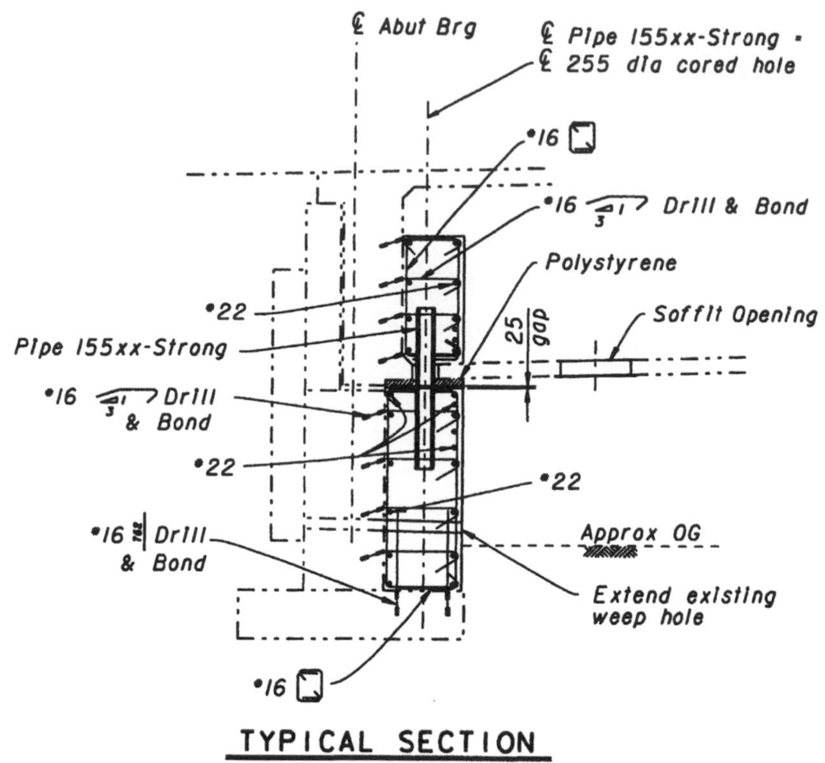

FIGURE 43.24 Vertical pipe at abutment.

43.4.8 Foundations

Older footings have many vulnerabilities that can lead to failure. The following is a list of major weaknesses encountered in older footings:

- Lack of top mat reinforcement and shear reinforcement;
- Inadequate development of tension pile capacity;
- Inadequate size for development of column plastic moment.

Footings can be lumped into two categories:

I. Spread footings resting on relatively dense material or footings resting on piles with weak tension connection to the footing cap. This latter group is treated similarly to spread footings since a strategy can be considered to ignore the supporting piles in tension.
II. Footings with piles that act in tension and compression.

In general, retrofit of footings supporting columns with a class P type casing is not needed; retrofit of footings supporting columns with class F casing is needed to develop the ultimate demand forces from the column. Typically, complete retrofit of one bent per frame including the column and the footing is recommended. However, retrofit in multicolumn bents can often be limited to columns because of common pin connections to footings. Footing retrofit is usually avoided on multicolumn bridges by allowing pins at column bases as often as possible. Pins can be induced by allowing lap splices in main column bars to slip, or by allowing continuous main column bars to cause shear cracking in the footing.

For category I footings supporting low- to medium-height single columns, rocking behavior of the bent should be investigated for stability and the footing capacity can be compared against the

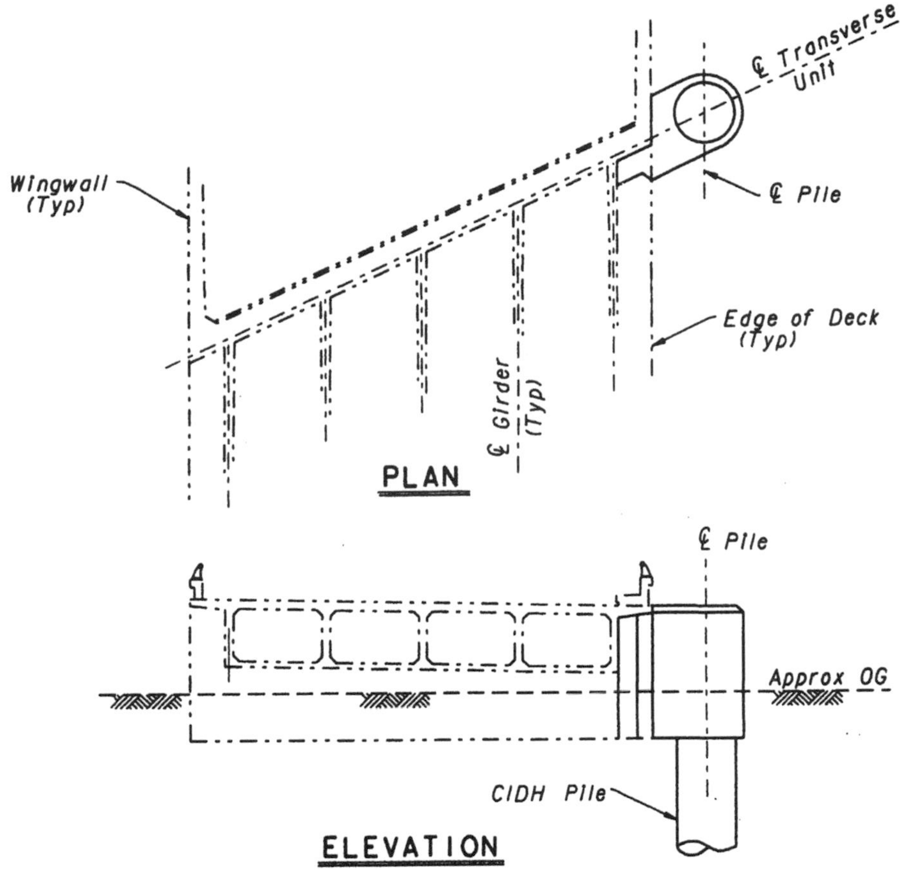

FIGURE 43.25 Anchor pile at abutment.

resulting forces from the rocking analysis. These forces can be of lesser magnitude than forces induced by column plastic hinging. Typically, retrofits for this case consist of adding an overlay to enhance the footing shear capacity or even widening of the footing to gain a larger footprint for stability and increase of the flexural moment capacity. The new concrete is securely attached to the old footing. This is done by chipping away at the concrete around the existing reinforcement and welding or mechanically coupling the new reinforcement to the old one. Holes are then drilled and the dowels are bonded between the faces of the old and new concrete as shown in Figure 43.26.

For category I footings supporting tall, single columns, rocking behavior of the bent can lead to instability, and some additional piles might be needed to provide stability to the tall bent. This type of modification leads to increased shear demand and increased tension demand on the top fiber of the existing footing that requires addition of a top mat reinforcement (Figure 43.26). The top mat is tied to the existing footing with dowels, and concrete is placed over the new piles and reinforcement. Where high compressive capacity piles are added, reinforcement with an extension hook is welded or mechanically coupled to existing bottom reinforcement. The hook acts to confine the concrete in the compression block where the perimeter piles are under compression demand.

When tension capacity is needed, the use of standard tension/compression piles is preferred to the use of tie-downs. In strong seismic events, large movements in footings are associated with tie-downs. Generally, tie-downs cannot be prestressed to reduce movements without overloading existing piles in compression. The tie-down movements are probably not a serious problem with short columns where P–Δ effects are minimal. Also, tie-downs should be avoided where groundwater could affect the quality of installation.

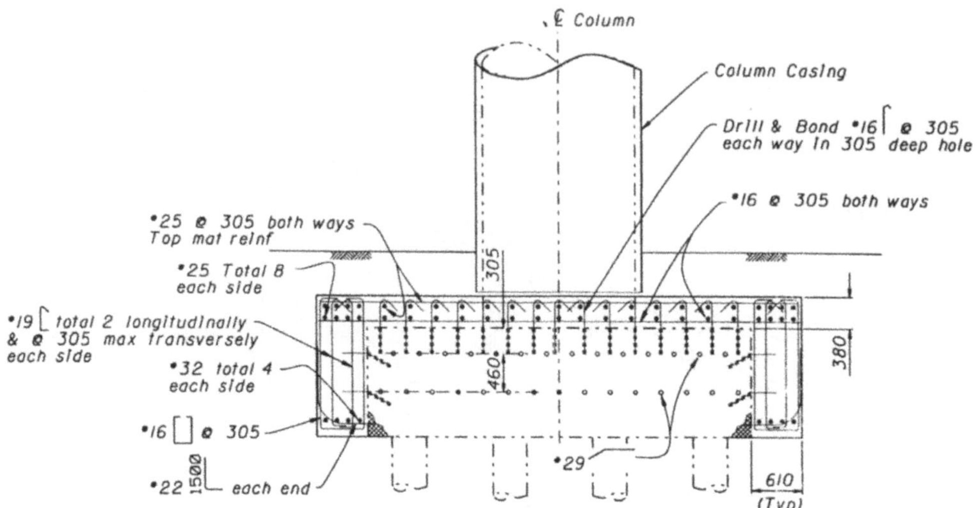

FIGURE 43.26 Widening footing retrofit.

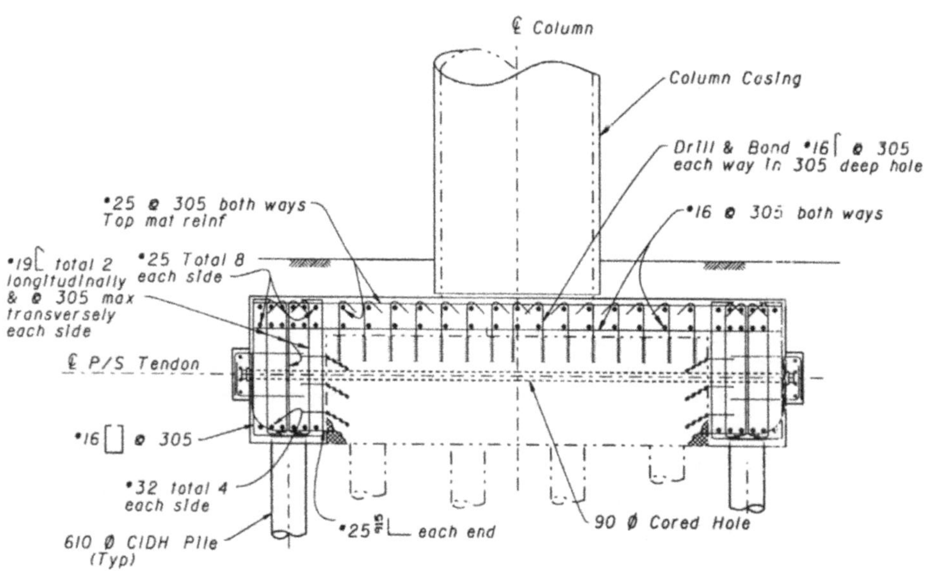

FIGURE 43.27 Footing retrofit using prestressing.

For category II footings, the ability of existing piles to cause tension on the top fiber of the footing where no reinforcement is present can lead to footing failure. Therefore, adding a top overlay in conjunction with footing widening might be necessary.

In sites where soft soil exists, the use of larger piles (600 mm and above) may be deemed necessary. These larger piles may induce high flexural demands requiring additional capacity from the bottom reinforcement. In this situation, prestressing of the footing becomes an alternative solution since it enhances the footing flexural capacity in addition to confining the concrete where perimeter piles act in compression (Figure 43.27). This retrofit is seldom used and is considered a last recourse.

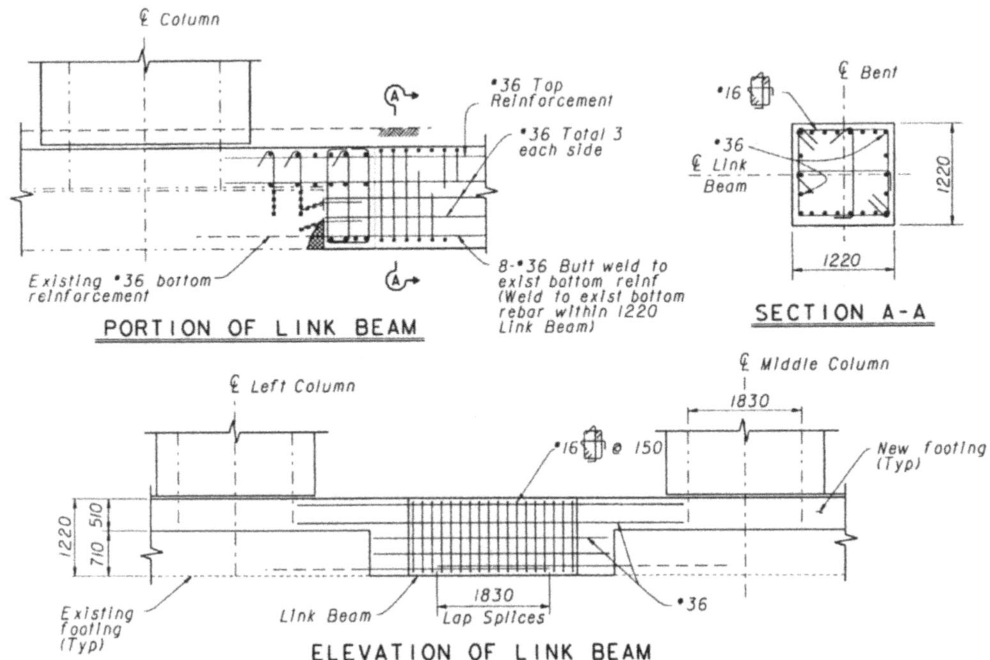

FIGURE 43.28 Link beam footing retrofit.

A rare but interesting situation occurs when a tall multicolumn bent has pinned connections to the superstructure instead of the usual monolithic connections and is resting on relatively small spread footings. This type of bent is quite vulnerable under large overturning moments and the use of link beams, as shown in Figure 43.28, is considered economical and sufficient to provide adequate stability and load transfer mechanism in a seismic event.

43.5 Summary

The seismic-resistant retrofit design of bridges has been evolving dramatically in the last decade. Many of the retrofit concepts and details discussed in this chapter have emerged as a result of research efforts and evaluation of bridge behavior in past earthquakes. This practice has been successfully tested in relatively moderate earthquakes but has not yet seen the severe test of a large-magnitude earthquake. The basic philosophy of current seismic retrofit technology in the U.S. is to prevent collapse by providing sufficient seat for displacement to take place or by allowing ductility in the supporting members. The greatest challenge to this basic philosophy will be the next big earthquake. This will serve as the utmost test to current predictions of earthquake demands on bridge structures.

References

1. Department of Transportation, State of California, Bridge Design Specifications, unpublished revision, Sacramento, 1997.
2. Mander, J. B., Priestley, M. J. N., and Park, R., *Theoretical stress-strain model for confined concrete, J. Struct. Eng. ASCE,* 114(8), 1804–1826,
3. Department of Transportation, State of California, Caltrans Memo to Designers, Sacramento, May 1994.

4. Department of Transportation, State of California, Caltrans Bridge Design Aids, Sacramento, October 1989.
5. Priestley, M. J. N. and Seible, F., *Seismic Assessment and Retrofit of Bridges*, Department of Applied Mechanics and Engineering Sciences, University of California, San Diego, La Jolla, July 1991, Structural Systems Research Project SSRP — 91/03.
6. Haroun, M. A., Pardoen, G. C., Shepherd, R., Haggag, H. A., and Kazanjy, R. P., *Cyclic Behavior of Bridge Pier Walls for Retrofit*, Department of Civil and Environmental Engineering, University of California, Irvine, Irvine, December 1993, Research Technical Agreement RTA No. 59N974.

44

Seismic Design Practice in Japan

Shigeki Unjoh
Public Works Research Institute

Nomenclature

The following symbols are used in this chapter. The section number in parentheses after definition of a symbol refers to the section where the symbol first appears or is defined.

a space of tie reinforcement (Section 44.4.4)
A_{CF} sectional area of carbon fiber (Figure 44.19)
A_h area of tie reinforcements (Section 44.4.4)
A_w sectional area of tie reinforcement (Section 44.4.4)
b width of section (Section 44.4.4)
c_B coefficient to evaluate effective displacement (Section 44.4.7)
c_B modification coefficient for clearance (Section 44.4.11)
c_{df} modification coefficient (Section 44.4.2)
c_c modification factor for cyclic loading (Section 44.4.4)
c_D modification coefficient for damping ratio (Section 44.4.6)
c_e modification factor for scale effect of effective width (Section 44.4.4)
c_E modification coefficient for energy-dissipating capability (Section 44.4.7)
c_P coefficient depending on the type of failure mode (Section 44.4.2)
c_{pt} modification factor for longitudinal reinforcement ratio (Section 44.4.4)

0-8493-7434-0/00/$0.00+$.50

c_R	factor depending on the bilinear factor r (Section 44.4.2)
c_W	corrective coefficient for ground motion characteristics (Section 44.4.9)
c_Z	modification coefficient for zone (Section 44.4.3)
d	effective width of tie reinforcements (Section 44.4.4)
d	height of section (Section 44.4.4)
D	a width or a diameter of a pier (Section 44.4.4)
D_E	coefficient to reduce soil constants according to F_L value (Section 44.4.11)
E_c	elastic modules of concrete (Section 44.4.4)
E_{CF}	elastic modulus of carbon fiber (Figure 44.19)
E_{des}	gradient at descending branch (Section 44.4.4)
F_L	liquefaction resistant ratio (Section 44.4.9)
$F(u)$	restoring force of a device at a displacement u (Section 44.4.7)
h	height of a pier (Section 44.4.4)
h_B	height of the center of gravity of girder from the top of bearing (Figure 44.13)
h_B	equivalent damping of a Menshin device (Section 44.4.7)
h_i	damping ratio of ith mode (Section 44.4.6)
h_{ij}	damping ratio of jth substructure in ith mode (Section 44.4.6)
h_{Bi}	damping ratio of ith damper (Section 44.4.7)
h_{Pi}	damping ratio of ith pier or abutment (Section 44.4.7)
h_{Fui}	damping ratio of ith foundation associated with translational displacement (Section 44.4.7)
$h_{F\theta i}$	damping ratio of ith foundation associated with rotational displacement(Section 44.4.7)
H	distance from a bottom of pier to a gravity center of a deck (Section 44.4.7)
H_0	shear force at the bottom of footing (Figure 44.12)
I	importance factor (Section 44.5.2)
k_{hc}	lateral force coefficient (Section 44.4.2)
k_{hc}	design seismic coefficient for the evaluation of liquefaction potential (Section 44.4.9)
k_{hc0}	standard modification coefficient (Section 44.4.3)
k_{hcm}	lateral force coefficient in Menshin design (Section 44.4.7)
k_{he}	equivalent lateral force coefficient (Section 44.4.2)
k_{hem}	equivalent lateral force coefficient in Menshin design (Section 44.4.7)
k_{hp}	lateral force coefficient for a foundation (Section 44.4.2)
k_j	stiffness matrix of jth substructure (Section 44.4.6)
K	stiffness matrix of a bridge (Section 44.4.6)
K_B	equivalent stiffness of a Menshin device (Section 44.4.7)
K_{Pi}	equivalent stiffness of ith pier or abutment (Section 44.4.7)
K_{Fui}	translational stiffness of ith foundation (Section 44.4.7)
$K_{F\theta i}$	rotational stiffness of ith foundation (Section 44.4.7)
L	shear stress ratio during an earthquake (Section 44.4.9)
L_A	redundancy of a clearance (Section 44.4.11)
L_E	clearance at an expansion joint (Section 44.4.11)
L_P	plastic hinge length of a pier (Section 44.4.4)
M_0	moment at the bottom of footing (Figure 44.12)
P_a	lateral capacity of a pier (Section 44.4.2)
P_s	shear capacity in consideration of the effect of cyclic loading (Section 44.4.4)
P_{s0}	shear capacity without consideration of the effect of cyclic loading (Section 44.4.4)
P_u	bending capacity (Section 44.4.2)
r	bilinear factor defined as a ratio between the first stiffness (yield stiffness) and the second stiffness (postyield stiffness) of a pier (Section 44.4.2)
r_d	modification factor of shear stress ratio with depth (Section 44.4.9)
R	dynamic shear strength ratio (Section 44.4.9)
R	priority (Section 44.5.2)
R_D	dead load of superstructure (Section 44.4.11)
R_{heq} and R_{veq}	vertical reactions caused by the horizontal seismic force and vertical force (Section 44.4.11)
R_L	cyclic triaxial strength ratio (Section 44.4.9)
R_U	design uplift force applied to the bearing support (Section 44.4.11)
s	space of tie reinforcements (Section 44.4.4)
S	earthquake force (Section 44.5.2)
S_c	shear capacity shared by concrete (Section 44.4.4)
S_I and S_{II}	acceleration response spectrum for Type-I and Type-II ground motions (Section 44.4.6)

S_{I0} and S_{II0} standard acceleration response spectrum for Type-I and Type-II ground motions (Section 44.4.6)
S_E seat length (Section 44.4.11)
S_{EM} minimum seat length (cm) (Section 44.4.11)
S_s shear capacity shared by tie reinforcements (Section 44.4.4)
T natural period of fundamental mode (Table 44.3)
ΔT difference of natural periods (Section 44.4.11)
T_1 and T_2 natural periods of the two adjacent bridge systems (Section 44.4.11)
u_B design displacement of isolators (Section 44.4.7)
u_{Be} effective design displacement (Section 44.4.7)
u_{Bi} design displacement of *i*th Menshin device (Section 44.4.7)
u_G relative displacement of ground along the bridge axis (Section 44.4.11)
u_R relative displacement (cm) developed between a superstructure and a substructure (Section 44.4.11)
V_0 vertical force at the bottom of footing (Figure 44.12)
V_T structural factor (Section 44.5.2)
V_{RP1} design specification (Section 44.5.2)
V_{RP2} pier structural factor (Section 44.5.2)
V_{RP3} aspect ratio (Section 44.5.2)
V_{MP} steel pier factor (Section 44.5.2)
V_{FS} unseating device factor (Section 44.5.2)
V_F foundation factor (Section 44.5.2)
w_v weighting factor on structural members (Section 44.5.2)
W equivalent weight (Section 44.4.2)
W elastic strain energy (Section 44.4.7)
W_P weight of a pier (Section 44.4.2)
W_U weight of a part of superstructure supported by the pier (Section 44.4.2)
ΔW energy dissipated per cycle (Section 44.4.7)
α safety factor (Section 44.4.4)
α, β coefficients depending on shape of pier (Section 44.4.4)
α_m safety factor used in Menshin design (Section 44.4.7)
δ_y yield displacement of a pier (Section 44.4.2)
δ_R residual displacement of a pier after an earthquake (Section 44.4.2)
δ_{Ra} allowable residual displacement of a pier (Section 44.4.2)
δ_u ultimate displacement of a pier (Section 44.4.4)
ε_c strain of concrete (Section 44.4.4)
ε_{cc} strain at maximum strength (Section 44.4.4)
ε_G ground strain induced during an earthquake along the bridge axis (Section 44.4.11)
ε_s strain of reinforcements (Section 44.4.4)
ε_{sy} yield strain of reinforcements (Section 44.4.4)
θ angle between vertical axis and tie reinforcement (Section 44.4.4)
θ_{pu} ultimate plastic angle (Section 44.4.4)
μ_a allowable displacement ductility factor of a pier (Section 44.4.2)
μ_m allowable ductility factor of a pier in Menshin design (Section 44.4.7)
μ_R response ductility factor of a pier (Section 44.4.2)
ρ_s tie reinforcement ratio (Section 44.4.4)
σ_c stress of concrete (Section 44.4.4)
σ_{cc} strength of confined concrete (Section 44.4.4)
σ_{CF} stress of carbon fiber (Figure 44.19)
σ_{ck} design strength of concrete (Section 44.4.4)
σ_s stress of reinforcements (Section 44.4.4)
σ_{sy} yield strength of reinforcements (Section 44.4.4)
σ_v total loading pressure (Section 44.4.9)
σ_v' effective loading pressure (Section 44.4.9)
τ_c shear stress capacity shared by concrete (Section 44.4.4)
ϕ_{ij} mode vector of *j*th substructure in *i*th mode (Section 44.4.6)
ϕ_i mode vector of a bridge in *i*th mode (Section 44.4.6)
ϕ_y yield curvature of a pier at bottom (Section 44.4.4)
ϕ_u ultimate curvature of a pier at bottom (Section 44.4.4)

44.1 Introduction

Japan is one of the most seismically disastrous countries in the world and has often suffered significant damage from large earthquakes. More than 3000 highway bridges have suffered damage since the 1923 Kanto earthquake. The earthquake disaster prevention technology for highway bridges has been developed based on such bitter damage experiences. Various provisions for designing bridges have been developed to prevent damage due to the instability of soils such as soil liquefaction. Furthermore, design detailings including unseating prevention devices are implemented. With progress in improving seismic design provisions, damage to highway bridges caused by the earthquakes has been decreasing in recent years.

However, the Hyogo-ken Nanbu earthquake of January 17, 1995 caused destructive damage to highway bridges. Collapse and near collapse of superstructures occurred at nine sites, and other destructive damage occurred at 16 sites [1]. The earthquake revealed that there are a number of critical issues to be revised in the seismic design and seismic retrofit of bridges [2,3].

This chapter presents technical developments for seismic design and seismic retrofit of highway bridges in Japan. The history of the earthquake damage and development of the seismic design methods is first described. The damage caused by the 1995 Hyogo-ken Nanbu earthquake, the lessons learned from the earthquake, and the seismic design methods introduced in the 1996 *Seismic Design Specifications for Highway Bridges* are then described. Seismic performance levels and design methods as well as ductility design methods for reinforced concrete piers, steel piers, foundations, and bearings are described. Then the history of the past seismic retrofit practices is described. The seismic retrofit program after the Hyogo-ken-Nanbu earthquake is described with emphasis on the seismic retrofit of reinforced concrete piers as well as research and development on the seismic retrofit of existing highway bridges.

44.2 History of Earthquake Damage and Development of Seismic Design Methods

A year after the 1923 Great Kanto earthquake, consideration of the seismic effect in the design of highway bridges was initiated. The Civil Engineering Bureau of the Ministry of Interior promulgated "The Method of Seismic Design of Abutments and Piers" in 1924. The seismic design method has been developed and improved through bitter experience in a number of past earthquakes and with progress of technical developments in earthquake engineering. Table 44.1 summarizes the history of provisions in seismic design for highway bridges.

In particular, the seismic design method was integrated and upgraded by compiling the "Specifications for Seismic Design of Highway Bridges" in 1971. The design method for soil liquefaction and unseating prevention devices was introduced in the Specifications. It was revised in 1980 and integrated as "Part V: Seismic Design" in Design Specifications of Highway Bridges. The primitive check method for ductility of reinforced concrete piers was included in the reference of the Specifications. It was further revised in 1990 and ductility check of reinforced concrete piers, soil liquefaction, dynamic response analysis, and design detailings were prescribed. It should be noted here that the detailed ductility check method for reinforced concrete piers was first introduced in the 1990 Specifications.

However, the Hyogo-ken Nanbu earthquake of January 17, 1995, exactly 1 year after the Northridge earthquake, California, caused destructive damage to highway bridges as described earlier. After the earthquake the Committee for Investigation on the Damage of Highway Bridges Caused by the Hyogo-ken Nanbu Earthquake (chairman, Toshio Iwasaki, Executive Director, Civil Engineering Research Laboratory) was established in the Ministry of Construction to investigate the damage and to identify the factors that caused the damage.

TABLE 44.1 History of Seismic Design Methods

		1926 Details of Road Structure (draft) Road Law, MIA	1939 Design Specifications of Steel Highway Bridges (draft) MIA	1956 Design Specifications of Steel Highway Bridges, MOC	1964 Design Specifications of Substructures (Pile Foundations), MOC	1964 Design Specifications of Steel Highway Bridges. MOC	1966 Design Specifications of Substructures (Survey and Design), MOC	1968 Design Specifications of Substructures (Piers and Direct Foundations), MOC	1970 Design Specifications of Substructures (Caisson Foundations), MOC	1971 Specifications for Seismic Design of Highway Bridges, MOC	1972 Design Specifications of Substructures (Cast-in-Piles), MOC	1975 Design Specifications of Substructures (Pile Foundations), MOC	1980 Design Specifications of Highway Bridges, MOC	1990 Design Specifications of Highway Bridges, MOC
Seismic loads	Seismic coefficient	Largest seismic loads Varied dependent on the site	$k_h = 0.2$	$k_h = 0.1$–0.35 Varied dependent on the site and ground condition					Standardization of seismic coefficient provision of modified seismic coefficient method	$k_h = 0.1$–0.3			Revision of application range of modified seismic coefficient method	$k_h = 0.1$–0.3 Integration of seismic coefficient method and modified one.
	Dynamic earth pressure	Equations proposed by Mononobe and Okabe were supposed to be used					Provision of dynamic earth pressure							
	Dynamic hydraulic pressure	Less effect on piers except high piers in deep water					Provision of hydraulic pressure			Provision of dynamic hydraulic pressure				
Reinforced concrete column	Bending at bottom	Supposed to be designed in a similar way provided in current design Specifications						Provisions of Definite Design Method						
	Shear	Less effect on RC piers except those with smaller section area such as RC frame and hollow section						Check of shear strength						Provision of definite design method, decreasing of allowable shear stress
	Termination of Main Reinforcement at Midheight												Elongation of anchorage length of terminated reinforcement at midheight	
	Bearing capacity for lateral force							Less effect on RC piers with larger section area					Ductility check	Check for bearing capacity for lateral force
Footing								Provisions of definite design method (designed as a cantilever plate)					Provisions of effective width and check of shear strength	
Pile foundation		Bearing capacity in vertical direction was supposed to be checked			Provisions of Definite Design Method (bearing capacity in vertical and horizontal directions)							Special Condition (Foundation on Slope, Consolidation Settlement, Lateral Movement)		Provisions of Design Details for Pile Head
Direct foundation		Stability (overturning and slip) was supposed to be checked						Provisions of Definite Design Method (bearing capacity, stability analysis)						
Caisson foundation					Supposed to be designed in a similar way provided in Design Specification of Caisson Foundation of 1969				Provisions of Definite Design Method					
Soil Liquefaction										Provisions of soil layers of which bearing capacity shall be ignored in seismic design			Provisions of evaluation method of soil liquefaction and the treatment in seismic design	Consideration of effect of fine sand content
Bearing support	Bearing support			Provisions of Design Methods for steel bearing supports (bearing, roller, anchor bolt)						Provision of transmitting method of seismic load at bearing				
	Devices preventing falling-off of superstructure							Provision of bearing seat length S		Provisions of stopper at movable bearings, devices for preventing superstructure from falling (seat length S, connection of adjacent decks)			Provisions of stopper at movable bearings, devices for preventing superstructure from falling (seat length S_E devices)	

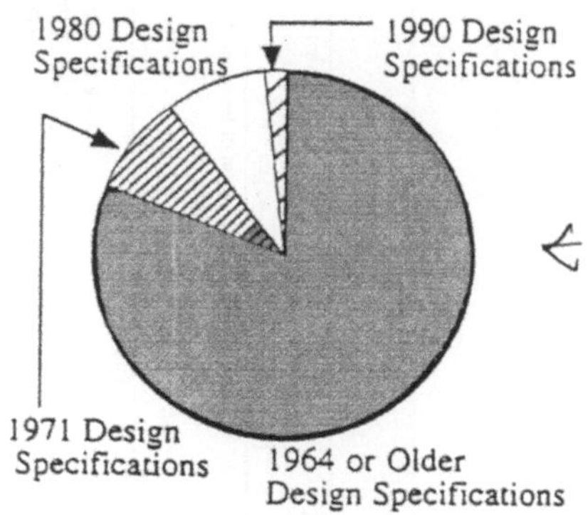

FIGURE 44.1 Design specifications referred to in design of Hanshin Expressway [2].

On February 27, 1995, the Committee approved the "Guide Specifications for Reconstruction and Repair of Highway Bridges Which Suffered Damage Due to the Hyogo-ken Nanbe Earthquake," [4], and the Ministry of Construction announced on the same day that the reconstruction and repair of the highway bridges which suffered damage in the Hyogo-ken Nanbu earthquake should be made by the Guide Specifications. It was decided by the Ministry of Construction on May 25, 1995 that the Guide Specifications should be tentatively used in all sections of Japan as emergency measures for seismic design of new highway bridges and seismic strengthening of existing highway bridges until the Design Specifications of Highway Bridges is revised.

In May, 1995, the Special Sub-Committee for Seismic Countermeasures for Highway Bridges (chairman, Kazuhiko Kawashima, Professor of the Tokyo Institute of Technology) was established in the Bridge Committee (chairman, Nobuyuki Narita, Professor of the Tokyo Metropolitan University), Japan Road Association, to draft the revision of the Design Specifications of Highway Bridges. The new Design Specifications of Highway Bridges [5,6] was approved by the Bridge Committee, and issued by the Ministry of Construction on November 1, 1996.

44.3 Damage of Highway Bridges Caused by the Hyogo-ken Nanbu Earthquake

The Hyogo-ken Nanbu earthquake was the first earthquake to hit an urban area in Japan since the 1948 Fukui earthquake. Although the magnitude of the earthquake was moderate (M7.2), the ground motion was much larger than anticipated in the codes. It occurred very close to Kobe City with shallow focal depth.

Damage was developed at highway bridges on Routes 2, 43, 171, and 176 of the National Highway, Route 3 (Kobe Line) and Route 5 (Bay Shore Line) of the Hanshin Expressway, and the Meishin and Chugoku Expressways. Damage was investigated for all bridges on national highways, the Hanshin Expressway, and expressways in the area where destructive damage occurred. The total number of piers surveyed reached 3396 [1]. Figure 44.1 shows Design Specifications referred to in the design of the 3396 highway bridges. Most of the bridges that suffered damage were designed according to the 1964 Design Specifications or the older Design Specifications. Although the seismic design methods have been improved and amended several times since 1926, only a requirement for lateral force coefficient was provided in the 1964 Design Specifications or the older Specifications.

Figure 44.2 compares damage of piers (bridges) on the Route 3 (Kobe Line) and Route 5 (Bay Shore Line) of the Hanshin Expressway. Damage degree was classified as A_s (collapse), A (nearly collapse), B (moderate damage), C (damage of secondary members), and D (minor or no damage). Substructures on Route 3 and Route 5 were designed with the 1964 Design Specifications and the 1980 Design Specifications, respectively. It should be noted in this comparison that the intensity of

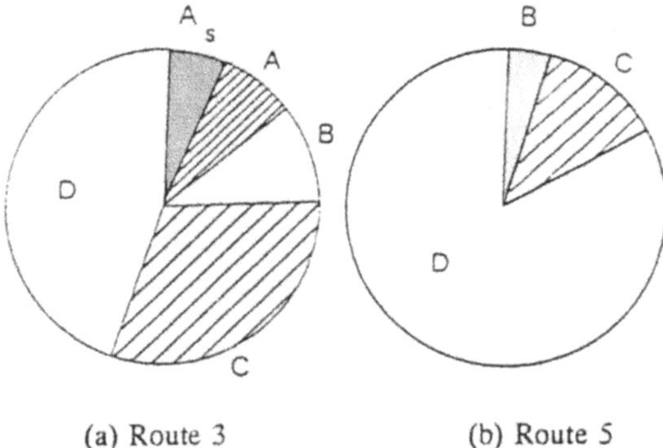

(a) Route 3 (b) Route 5

FIGURE 44.2 Comparison of damage degree between Route 3 (a) and Route 5 (b) (As: collapse, A: near collapse, B: moderate damage, C: damage of secondary members, D: minor or no damage) [2].

ground shaking in terms of response spectra was smaller at the Bay Area than the narrow rectangular area where JMA seismic intensity was VII (equivalent to modified Mercalli intensity of X-XI). Route 3 was located in the narrow rectangular area, while Route 5 was located in the Bay Area. Keeping in mind such differences in ground motion, it is apparent in Figure 44.2 that about 14% of the piers on Route 3 suffered As or A damage while no such damage was developed in the piers on Route 5.

Although damage concentrated on the bridges designed with the older Design Specifications, it was thought that essential revision was required even in the recent Design Specifications to prevent damage against destructive earthquakes such as the Hyogo-ken Nanbu earthquake. The main modifications were as follows:

1. To increase lateral capacity and ductility of all structural components in which seismic force is predominant so that ductility of a total bridge system is enhanced. For such purpose, it was required to upgrade the "Check of Ductility of Reinforced Concrete Piers," which has been used since 1990, to a "ductility design method" and to apply the ductility design method to all structural components. It should be noted here that "check" and "design" are different; the check is only to verify the safety of a structural member designed by another design method, and is effective only to increase the size or reinforcements if required, while the design is an essential procedure to determine the size and reinforcements.
2. To include the ground motion developed at Kobe in the earthquake as a design force in the ductility design method.
3. To specify input ground motions in terms of acceleration response spectra for dynamic response analysis more actively.
4. To increase tie reinforcements and to introduce intermediate ties for increasing ductility of piers. It was decided not to terminate longitudinal reinforcements at midheight to prevent premature shear failure, in principle.
5. To adopt multispan continuous bridges for increasing number of indeterminate of a total bridge system.
6. To adopt rubber bearings for absorbing lateral displacement between a superstructure and substructures and to consider correct mechanism of force transfer from a superstructure to substructures.
7. To include the Menshin design (seismic isolation).
8. To increase strength, ductility, and energy dissipation capacity of unseating prevention devices.
9. To consider the effect of lateral spreading associated with soil liquefaction in design of foundations at sites vulnerable to lateral spreading.

TABLE 44.2 Seismic Performance Levels

<table>
<tr><th colspan="2" rowspan="2">Type of Design Ground Motions</th><th colspan="2">Importance of Bridges</th><th colspan="2">Design Methods</th></tr>
<tr><th>Type-A (Standard Bridges)</th><th>Type-B (Important Bridges)</th><th>Equivalent Static Lateral Force Methods</th><th>Dynamic Analysis</th></tr>
<tr><td colspan="2">Ground motions with high probability to occur</td><td colspan="2">Prevent Damage</td><td>Seismic coefficient method</td><td rowspan="3">Step by Step analysis
or
Response spectrum analysis</td></tr>
<tr><td rowspan="2">Ground motions with low probability to occur</td><td>Type I (plate boundary earthquakes)</td><td rowspan="2">Prevent critical damage</td><td rowspan="2">Limited damage</td><td rowspan="2">Ductility design method</td></tr>
<tr><td>Type II (Inland earthquakes)</td></tr>
</table>

44.4 1996 Seismic Design Specifications of Highway Bridges

44.4.1 Basic Principles of Seismic Design

The 1995 Hyogo-ken Nanbu earthquake, the first earthquake to be considered that such destructive damage could be prevented due to the progress of construction technology in recent years, provided a large impact on the earthquake disaster prevention measures in various fields. Part V: Seismic Design of the Design Specifications of Highway Bridges (Japan Road Association) was totally revised in 1996, and the design procedure moved from the traditional seismic coefficient method to the ductility design method. The revision was so comprehensive that the past revisions of the last 30 years look minor.

A major revision of the 1996 Specifications is the introduction of explicit two-level seismic design consisting of the seismic coefficient method and the ductility design method. Because Type I and Type II ground motions are considered in the ductility design method, three design seismic forces are used in design. Seismic performance for each design force is clearly defined in the Specifications.

Table 44.2 shows the seismic performance level provided in the 1996 Design Specifications. The bridges are categorized into two groups depending on their importance: standard bridges (Type A bridges) and important bridges (Type B bridges). The seismic performance level depends on the importance of the bridge. For moderate ground motions induced in earthquakes with a high probability of occurrence, both A and B bridges should behave in an elastic manner without essential structural damage. For extreme ground motions induced in earthquakes with a low probability of occurrence, Type A bridges should prevent critical failure, whereas Type B bridges should perform with limited damage.

In the ductility design method, two types of ground motions must be considered. The first is the ground motions that could be induced in plate boundary-type earthquakes with a magnitude of about 8. The ground motion at Tokyo in the 1923 Kanto earthquake is a typical target of this type of ground motion. The second is the ground motion developed in earthquakes with magnitude of about 7 to 7.2 at very short distance. Obviously, the ground motions at Kobe in the Hyogo-ken Nanbu earthquake is a typical target of this type of ground motion. The first and the second ground motions are called Type I and Type II ground motions, respectively. The recurrence time of Type II ground motion may be longer than that of Type I ground motion, although the estimation is very difficult.

The fact that lack of near-field strong motion records prevented serious evaluation of the validity of recent seismic design codes is important. The Hyogo-ken Nanbu earthquake revealed that the history of strong motion recording is very short, and that no near-field records have yet been measured by an earthquake with a magnitude on the order of 8. It is therefore essential to have sufficient redundancy and ductility in a total bridge system.

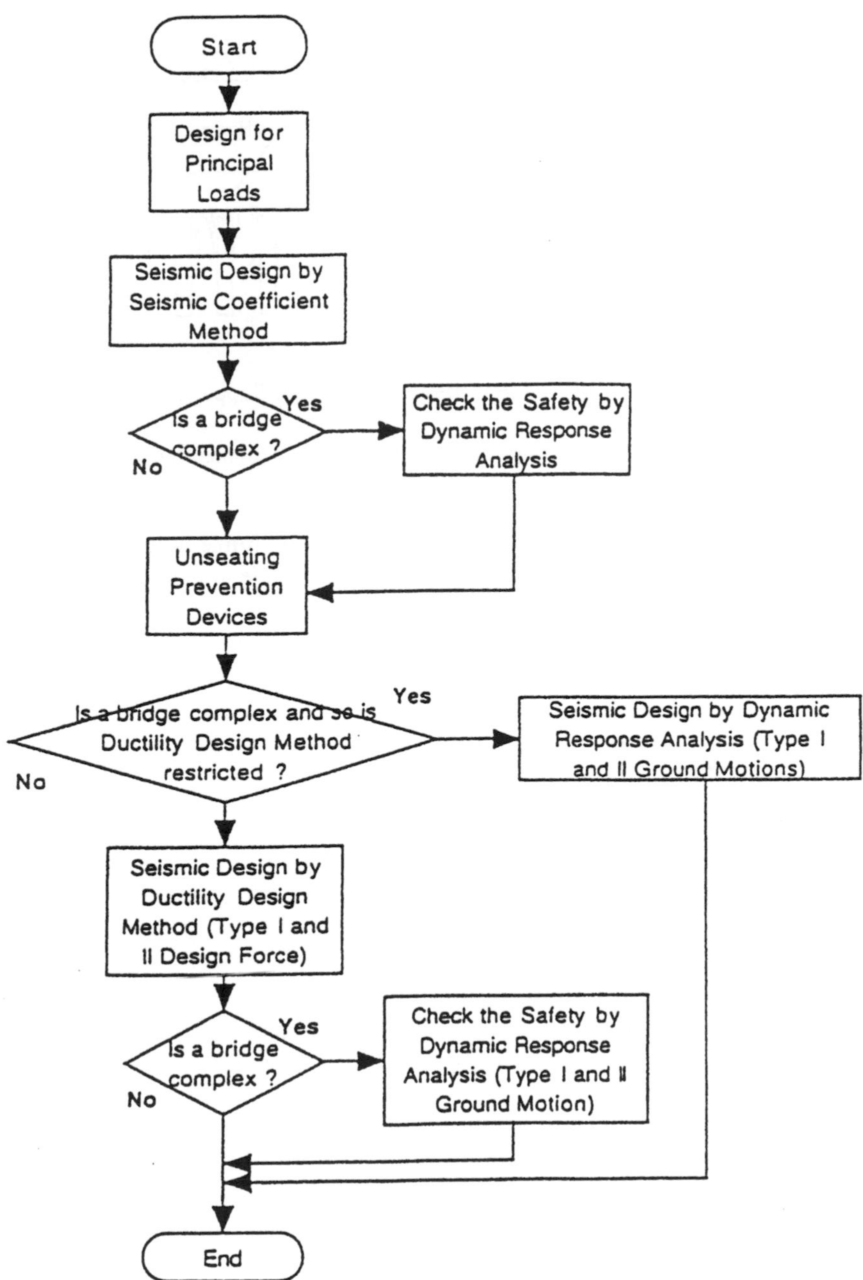

FIGURE 44.3 Flowchart of seismic design.

44.4.2 Design Methods

Bridges are designed by both the seismic coefficient method and the ductility design method as shown in Figure 44.3. In the seismic coefficient method, a lateral force coefficient ranging from 0.2 to 0.3 has been used based on the allowable stress design approach. No change has been made since the 1990 Specifications in the seismic coefficient method.

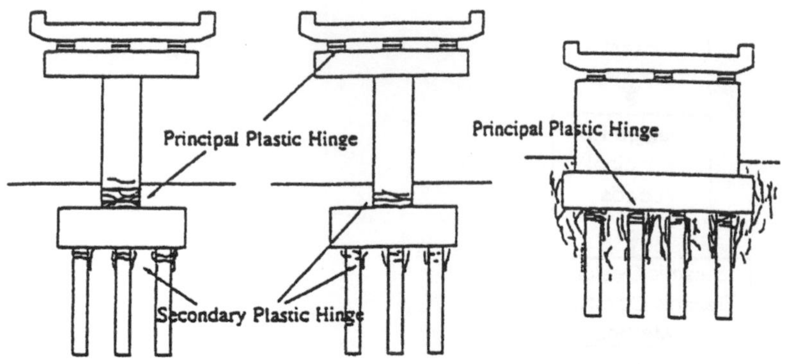

FIGURE 44.4 Location of primary plastic hinge. (a) Conventional design; (b) Menshin design; (c) bridge supported by a wall-type pier.

In the ductility design method, assuming a principal plastic hinge is formed at the bottom of pier as shown in Figure 44.4a and that the equal energy principle applies, a bridge is designed so that the following requirement is satisfied:

$$P_a > k_{he} W \tag{44.1}$$

where

$$k_{he} = \frac{k_{hc}}{\sqrt{2\mu_a - 1}} \tag{44.2}$$

$$W = W_U + c_P W_P \tag{44.3}$$

in which P_a = lateral capacity of a pier, k_{he} = equivalent lateral force coefficient, W = equivalent weight, k_{hc} = lateral force coefficient, μ_a = allowable displacement ductility factor of a pier, W_U = weight of a part of superstructure supported by the pier, W_P = weight of a pier, and c_P = coefficient depending on the type of failure mode. The c_P is 0.5 for a pier in which either flexural failure or shear failure after flexural cracks are developed, and 1.0 is for a pier in which shear failure is developed. The lateral capacity of a pier P_a is defined as a lateral force at the gravity center of a superstructure.

In Type B bridges, residual displacement developed at a pier after an earthquake must be checked as

$$\delta_R < \delta_{Ra} \tag{44.4}$$

where

$$\delta_R = c_R \left(\mu_R - 1\right)\left(1 - r\right)\delta_y \tag{44.5}$$

$$\mu_R = 1/2 \left\{ \left(k_{hc} \cdot W / P_a\right)^2 + 1 \right\} \tag{44.6}$$

in which δ_R = residual displacement of a pier after an earthquake, δ_{Ra} = allowable residual displacement of a pier, r = bilinear factor defined as a ratio between the first stiffness (yield stiffness) and the second stiffness (postyield stiffness) of a pier, c_R = factor depending on the bilinear factor r, μ_R = response ductility factor of a pier, and δ_y = yield displacement of a pier. The δ_{Ra} should be 1/100 of the distance between the bottom of a pier and the gravity center of a superstructure.

In a bridge with complex dynamic response, the dynamic response analysis is required to check the safety of a bridge after it is designed by the seismic coefficient method and the ductility design method. Because this is only for a check of the design, the size and reinforcements of structural members once determined by the seismic coefficient method and the ductility design methods may be increased if necessary. It should be noted, however, that under the following conditions in which the ductility design method is not directly applied, the size and reinforcements can be determined based on the results of a dynamic response analysis as shown in Figure 44.3. Situations when the ductility design method should not be directly used include:

1. When principal mode shapes that contribute to bridge response are different from the ones assumed in the ductility design methods
2. When more than two modes significantly contribute to bridge response
3. When principal plastic hinges form at more than two locations, or principal plastic hinges are not known where to be formed
4. When there are response modes for which the equal energy principle is not applied

In the seismic design of a foundation, a lateral force equivalent to the ultimate lateral capacity of a pier P_u is assumed to be a design force as

$$k_{hp} = c_{df}\, P_u / W \tag{44.7}$$

in which k_{hp} = lateral force coefficient for a foundation, c_{df} = modification coefficient (= 1.1), and W = equivalent weight by Eq. (44.3). Because the lateral capacity of a wall-type pier is very large in the transverse direction, the lateral seismic force evaluated by Eq. (44.7) in most cases becomes excessive. Therefore, if a foundation has sufficiently large lateral capacity compared with the lateral seismic force, the foundation is designed assuming a plastic hinge at the foundation and surrounding soils as shown in Figure 44.4c.

44.4.3 Design Seismic Force

Lateral force coefficient k_{hc} in Eq. (44.2) is given as

$$k_{hc} = c_z \cdot k_{hc0} \tag{44.8}$$

in which c_Z = modification coefficient for zone, and is 0.7, 0.85, and 1.0 depending on the zone, and k_{hc0} = standard modification coefficient. Table 44.3 and Figure 44.5 show the standard lateral force coefficients k_{hc0} for Type I and Type II ground motions. Type I ground motions have been used since 1990 (1990 Specifications), while Type II ground motions were newly introduced in the 1996 Specifications. It should be noted here that the k_{hc0} at stiff site (Group I) has been assumed smaller than the k_{hc0} at moderate (Group II) and soft soil (Group III) sites in Type I ground motions as well as the seismic coefficients used for the seismic coefficient method. Type I ground motions were essentially estimated from an attenuation equation for response spectra that is derived from a statistical analysis of 394 components of strong motion records. Although the response spectral accelerations at short natural period are larger at stiff sites than at soft soil sites, the tendency has not been explicitly included in the past. This was because damage has been more developed at soft sites than at stiff sites. To consider such a fact, the design force at stiff sites is assumed smaller than

TABLE 44.3 Lateral Force Coefficient k_{hc0} in the Ductility Design Method

Soil Condition	Lateral Force Coefficient k_{hc0}		
	Type I Ground Motion		
Group I (stiff)		$k_{hc0} = 0.7$ for $T \leqq 1.4$	$k_{hc0} = 0.876T^{2/3}$ for $T > 1.4$
Group II (moderate)	$k_{hc0} = 1.51T^{1/3}$ ($k_{hc0} \geqq 0.7$) for $T < 0.18$	$k_{hc0} = 0.85$ for $0.18 \leqq T \leqq 1.6$	$k_{hc0} = 1.16T^{2/3}$ for $T > 1.6$
Group III (soft)	$k_{hc0} = 1.51T^{1/3}$ ($k_{hc0} \geqq 0.7$) for $T < 0.29$	$k_{hc0} = 1.0$ for $0.29 \leqq T \leqq 2.0$	$k_{hc0} = 1.59T^{2/3}$ for $T > 2.0$
	Type II Ground Motion		
Group I (stiff)	$k_{hc0} = 4.46T^{2/3}$ for $T \leqq 0.3$	$k_{hc0} = 2.00$ for $0.3 \leqq T \leqq 0.7$	$k_{hc0} = 1.24T^{4/3}$ for $T > 0.7$
Group II (moderate)	$k_{hc0} = 3.22T^{2/3}$ for $T < 0.4$	$k_{hc0} = 1.75$ for $0.4 \leqq T \leqq 1.2$	$k_{hc0} = 2.23T^{4/3}$ for $T > 1.2$
Group III (soft)	$k_{hc0} = 2.38T^{2/3}$ for $T < 0.5$	$k_{hc0} = 1.50$ for $0.5 \leqq T \leqq 1.5$	$k_{hc0} = 2.57T^{4/3}$ for $T > 1.5$

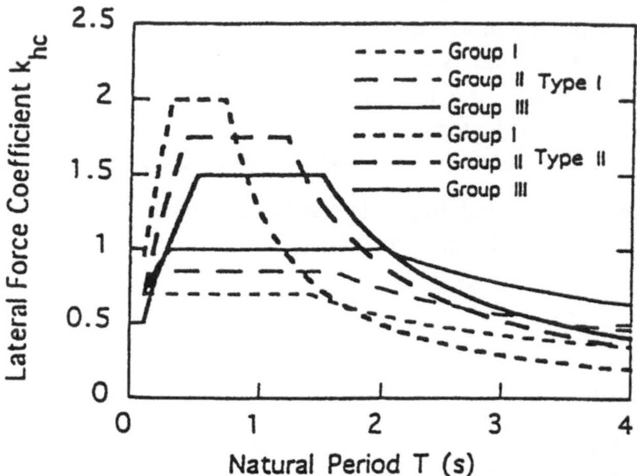

FIGURE 44.5 Type I and Type II ground motions in the ductility design method.

that at soft sites even at short natural period. However, being different from such a traditional consideration, Type II ground motions were determined by simply taking envelopes of response accelerations of major strong motions recorded at Kobe in the Hyogo-ken Nanbu earthquake.

Although the acceleration response spectral intensity at short natural period is higher in Type II ground motions than in Type I ground motions, the duration of extreme accelerations excursion is longer in Type I ground motions than Type II ground motions. As will be described later, such a difference of the duration has been taken into account to evaluate the allowable displacement ductility factor of a pier.

44.4.4 Ductility Design of Reinforced Concrete Piers

44.4.4.1 Evaluation of Failure Mode

In the ductility design of reinforced concrete piers, the failure mode of the pier is evaluated as the first step. Failure modes are categorized into three types based on the flexural and shear capacities of the pier as

1. $P_u \leqq P_s$ bending failure
2. $P_s \leq P_u \leqq P_{s0}$ bending to shear failure
3. $P_{s0} \leq P_u$ shear failure

in which P_u = bending capacity, P_s = shear capacity in consideration of the effect of cyclic loading, and P_{s0} = shear capacity without consideration of the effect of cyclic loading.

The ductility factor and capacity of the reinforced concrete piers are determined according to the failure mode as described later.

44.4.4.2 Displacement Ductility Factor

Th allowable displacement ductility factor of a pier μ_a in Eq. (44.2) is evaluated as

$$\mu_a = 1 + \frac{\delta_u - \delta_y}{\alpha \delta_y} \tag{44.9}$$

in which α = safety factor, δ_y = yield displacement of a pier, and δ_u = ultimate displacement of a pier. As well as the lateral capacity of a pier P_a in Eq. (44.1), the δ_y and δ_u are defined at the gravity center of a superstructure. In a reinforced concrete single pier as shown in Figure 44.4a, the ultimate displacement δ_u is evaluated as

$$\delta_u = \delta_y + \left(\phi_u - \phi_y\right) L_P \left(h - L_P/2\right) \tag{44.10}$$

in which ϕ_y = yield curvature of a pier at bottom, ϕ_u = ultimate curvature of a pier at bottom, h = height of a pier, and L_P = plastic hinge length of a pier. The plastic hinge length is given as

$$L_p = 0.2\,h - 0.1\,D\left(0.1\,D \leqq L_p \leqq 0.5D\right) \tag{44.11}$$

in which D is a width or a diameter of a pier.

The yield curvature ϕ_y and ultimate curvature ϕ_u in Eq. (44.10) are evaluated assuming a stress–strain relation of reinforcements and concrete as shown in Figure 44.6. The stress σ_c – strain ε_c relation of concrete with lateral confinement is assumed as

$$\sigma_c = \begin{cases} E_c \varepsilon_c \left(1 - \dfrac{1}{n}\left(\dfrac{\varepsilon_c}{\varepsilon_{cc}}\right)^{n-1}\right) & \left(0 \leqq \varepsilon_c \leqq \varepsilon_{cc}\right) \\ \sigma_{cc} - E_{des}\left(\varepsilon_c - \varepsilon_{cc}\right) & \left(\varepsilon_{cc} < \varepsilon_c \leqq \varepsilon_{cu}\right) \end{cases} \tag{44.12}$$

$$n = \frac{E_c \varepsilon_{cc}}{E_c \varepsilon_{cc} - \sigma_{cc}} \tag{44.13}$$

in which σ_{cc} = strength of confined concrete, E_c = elastic modules of concrete, ε_{cc} = strain at maximum strength, and E_{des} = gradient at descending branch. In Eq. (44.12), σ_{cc}, ε_{cc}, and E_{des} are determined as

$$\sigma_{cc} = \sigma_{ck} + 3.8\alpha\rho_s\sigma_{sy} \tag{44.14}$$

$$\varepsilon_{cc} = 0.002 + 0.033\,\beta\frac{\rho_s\sigma_{sy}}{\sigma_{ck}} \tag{44.15}$$

$$E_{des} = 11.2\frac{\sigma_{ck}^2}{\rho_s\sigma_{sy}} \tag{44.16}$$

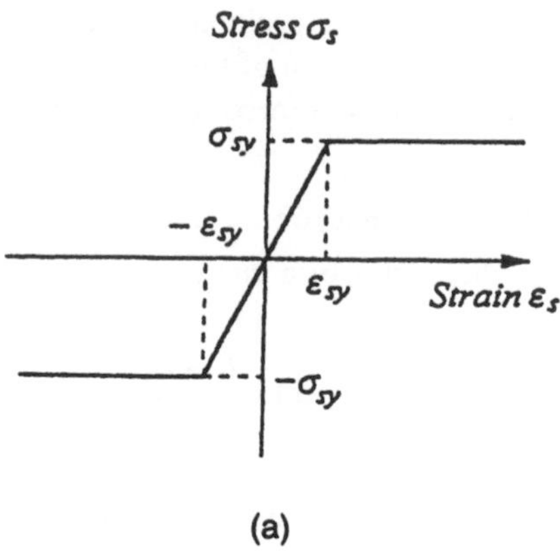

(a)

(b) Concrete

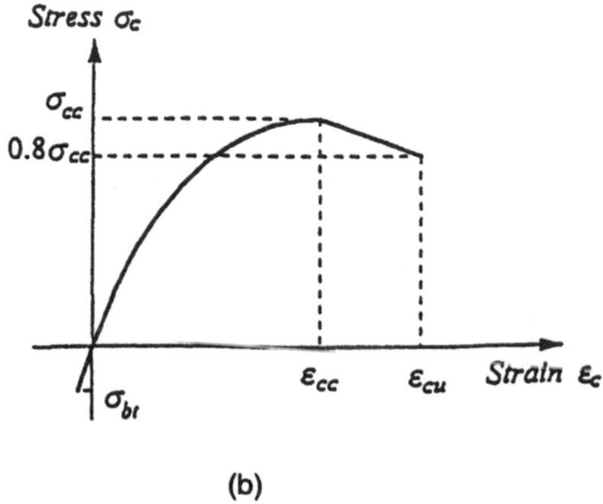

(b)

FIGURE 44.6 Stress and strain relation of confined concrete and reinforcing bars. (a) Steel (b) concrete.

in which σ_{ck} = design strength of concrete, σ_{sy} = yield strength of reinforcements, α and β = coefficients depending on shape of pier (α = 1.0 and β = 1.0 for a circular pier, and α = 0.2 and β = 0.4 for a rectangular pier), and ρ_s = tie reinforcement ratio defined as

$$\rho_s = \frac{4A_h}{sd} \leqq 0.018 \tag{44.17}$$

in which A_h = area of tie reinforcements, s = space of tie reinforcements, and d = effective width of tie reinforcements.

The ultimate curvature ϕ_u is defined as a curvature when concrete strain at longitudinal reinforcing bars in compression reaches an ultimate strain ε_{cu} defined as

$$\varepsilon_{cu} = \begin{cases} \varepsilon_{cc} & \text{for Type I ground motions} \\ \varepsilon_{cc} + \dfrac{0.2\sigma_{cc}}{E_{des}} & \text{for Type II ground motions} \end{cases} \tag{44.18}$$

It is important to note that the ultimate strain ε_{cu} depends on the types of ground motions; the ε_{cu} for Type II ground motions is larger than that for Type I ground motions. Based on a loading test,

TABLE 44.4 Safety Factor α in Eq. 44.9

Type of Bridges	Type I Ground Motion	Type II Ground Motion
Type B	3.0	1.5
Type A	2.4	1.2

TABLE 44.5 Modification Factor on Scale Effect for Shear Capacity Shared by Concrete

Effective Width of Section d (m)	Coefficient c_c
$d \leqq 1$	1.0
$d = 3$	0.7
$d = 5$	0.6
$d \geqq 10$	0.5

it is known that a certain level of failure in a pier such as a sudden decrease of lateral capacity occurs at smaller lateral displacement in a pier subjected to a loading hysteresis with a greater number of load reversals. To reflect such a fact, it was decided that the ultimate strain ε_{cu} should be evaluated by Eq. (44.18), depending on the type of ground motions. Therefore, the allowable ductility factor μ_a depends on the type of ground motions; the μ_a is larger in a pier subjected to Type II ground motions than a pier subjected to Type I ground motions.

It should be noted that the safety factor α in Eq. (44.9) depends on the type of bridges as well as the type of ground motions as shown in Table 44.4. This is to preserve higher seismic safety in the important bridges, and to take account of the difference of recurrent time between Type I and Type II ground motions.

44.4.4.3 Shear Capacity

Shear capacity of reinforced concrete piers is evaluated by a conventional method as

$$P_s = S_c + S_s \tag{44.19}$$

$$S_c = c_c c_e c_{pt} \tau_c bd \tag{44.20}$$

$$S_s = \frac{A_w \sigma_{sy} d(\sin\theta + \cos\theta)}{1.15a} \tag{44.21}$$

in which P_s = shear capacity; S_c = shear capacity shared by concrete; S_s = shear capacity shared by tie reinforcements, τ_c = shear stress capacity shared by concrete; c_c = modification factor for cyclic loading (0.6 for Type I ground motions; 0.8 for Type II ground motions); c_e = modification factor for scale effect of effective width; c_{pt} = modification factor for longitudinal reinforcement ratio; b and d = width and height of section, A_w = sectional area of tie reinforcement; σ_{sy} = yield strength of tie reinforcement, θ = angle between vertical axis and tie reinforcement, and a = space of tie reinforcement.

The modification factor on the scale effect of effective width, c_e, was based on experimental study of loading tests of beams with various effective heights and was newly introduced in the 1996 Specifications. Table 44.5 shows the modification factor on scale effect.

44.4.4.4 Arrangement of Reinforcement

Figure 44.7 shows a suggested arrangement of tie reinforcement. Tie reinforcement should be deformed bars with a diameter equal or larger than 13 mm, and it should be placed in most bridges

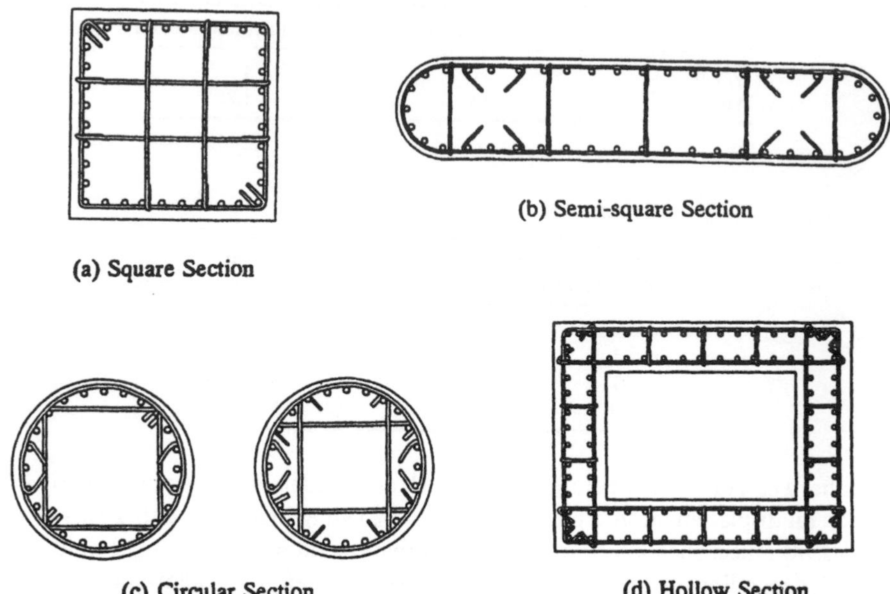

FIGURE 44.7 Confinement of core concrete by tie reinforcement. (a) Square section; (b) semisquare section; (c) circular section; (d) hollow section.

at a distance of no longer than 150 mm. In special cases, such as bridges with pier height taller than 30 m, the distance of tie reinforcement may be increased at height so that pier strength should not be sharply decreased at the section. Intermediate ties should be also provided with the same distance with the ties to confine the concrete. Space of the intermediate ties should be less than 1 m.

44.4.4.5 Two-Column Bent

To determine the ultimate strength and ductility factor for two-column bents, it is modeled as a frame model with plastic hinges at both ends of a lateral cap beam and columns as shown in Figure 44.8. Each elastic frame member has the yield stiffness which is obtained based on the axial load by the dead load of the superstructure and the column. The plastic hinge is assumed to be placed at the end part of a cap beam and the top and bottom part of each column. The plastic hinges are modeled as spring elements with a bilinear moment–curvature relation. The location of plastic hinges is half the distance of the plastic hinge length off from the end edge of each member, where the plastic hinge length L_P is assumed to be Eq. (44.11).

When the two-column bent is subjected to lateral force in the transverse direction, axial force developed in the beam and columns is affected by the applied lateral force. Therefore, the horizontal force–displacement relation is obtained through the static push-over analysis considering axial force *N*/moment *M* interaction relation. The ultimate state of each plastic hinge is obtained by the ultimate plastic angle θ_{pu} as

$$\theta_{pu} = \left(\phi_u/\phi_y - 1\right)L_P\phi_y \tag{44.22}$$

in which ϕ_u = ultimate curvature and ϕ_y = yield curvature.

The ultimate state of the whole two-bent column is determined so that all four plastic hinges developed reach the ultimate plastic angle.

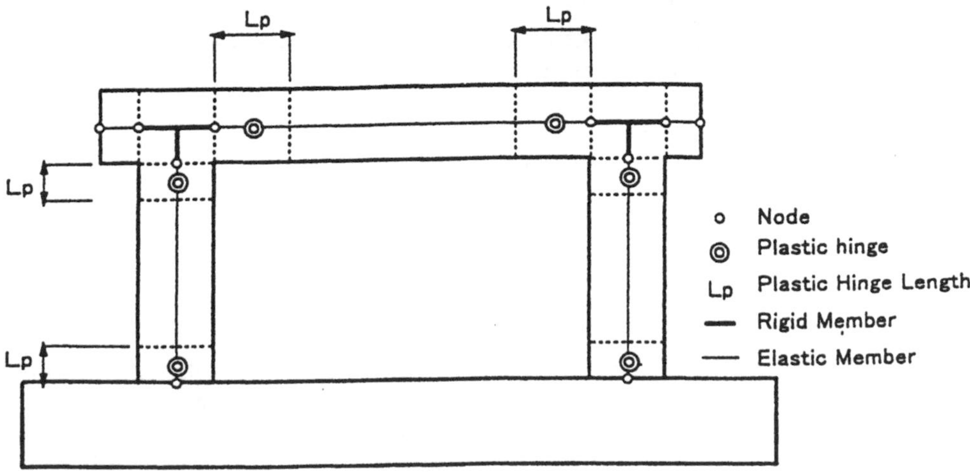

FIGURE 44.8 Analytical idealization of a two-column bent.

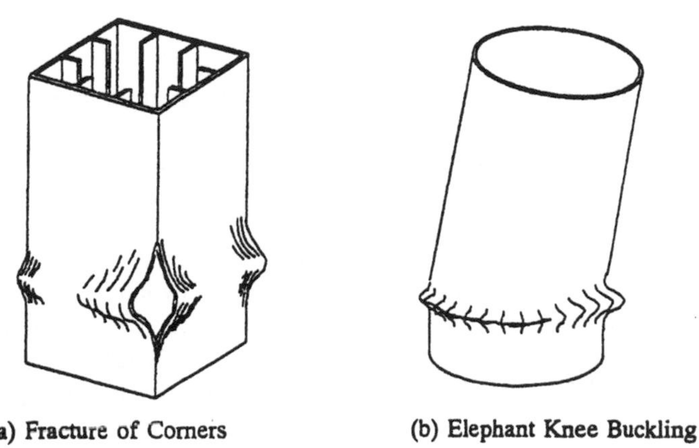

(a) Fracture of Corners **(b) Elephant Knee Buckling**

FIGURE 44.9 Typical brittle failure modes of steel piers. (a) Fracture of corners; (b) elephant knee buckling.

44.4.5 Ductility Design of Steel Piers

44.4.5.1 Basic Concept

To improve seismic performance of a steel pier, it is important to avoid specific brittle failure modes. Figure 44.9 shows the typical brittle failure mode for rectangular and circular steel piers. The following are the countermeasures to avoid such brittle failure modes and to improve seismic performance of steel piers:

1. Fill the steel column with concrete.
2. Improve structural parameters related to buckling strength.
   - Decrease the width–thickness ratio of stiffened plates of rectangular piers or the diameter–thickness ratio of steel pipes;
   - Increase the stiffness of stiffeners;
   - Reduce the diaphragm spacing;
   - Strengthen corners using the corner plates;
3. Improve welding section at the corners of rectangular section
4. Eliminate welding section at the corners by using round corners.

44.4.5.2 Concrete-Infilled Steel Pier

In a concrete-infilled steel pier, the lateral capacity P_a and the allowable displacement ductility factor μ_a in Eqs. (44.1) and (44.2) are evaluated as

$$P_a = P_y + \frac{P_u - P_y}{\alpha} \tag{44.23}$$

$$\mu_a = \left(1 + \frac{\delta_u - \delta_y}{\alpha\delta_y}\right)\frac{P_u}{P_a} \tag{44.24}$$

in which P_y and P_u = yield and ultimate lateral capacity of a pier; δ_y and δ_u = yield and ultimate displacement of a pier; and α = safety factor (refer to Table 44.4). The P_a and the μ_a are evaluated idealizing that a concrete-infilled steel pier resists flexural moment and shear force as a reinforced concrete pier. It is assumed in this evaluation that the steel section is idealized as reinforcing bars and that only the steel section resists axial force. A stress vs. strain relation of steel and concrete as shown in Figure 44.10 is assumed. The height of infilled concrete has to be decided so that bucking is not developed above the infilled concrete.

44.4.5.3 Steel Pier without Infilled Concrete

A steel pier without infilled concrete must be designed with dynamic response analysis. Properties of the pier need to be decided based on a cyclic loading test. Arrangement of stiffness and welding at corners must be precisely evaluated so that brittle failure is avoided.

44.4.6 Dynamic Response Analysis

Dynamic response analysis is required in bridges with complex dynamic response to check the safety factor of the static design. Dynamic response analysis is also required as a "design" tool in the bridges for which the ductility design method is not directly applied. In dynamic response analysis, ground motions which are spectral-fitted to the following response spectra are used;

$$S_I = c_Z \cdot c_D \cdot S_{I0} \tag{44.25}$$

$$S_{II} = c_Z \cdot c_D \cdot S_{II0} \tag{44.26}$$

in which S_I and S_{II} = acceleration response spectrum for Type I and Type II ground motions; S_{I0} and S_{II0} = standard acceleration response spectrum for Type I and Type II ground motions, respectively; c_Z = modification coefficient for zone, refer to Eq. (44.8); and c_D = modification coefficient for damping ratio given as

$$c_D = \frac{1.5}{40h_i + 1} + 0.5 \tag{44.27}$$

Table 44.6 and Figure 44.11 show the standard acceleration response spectra (damping ratio h = 0.05) for Type I and Type II ground motions.

It is recommended that at least three ground motions be used per analysis and that an average be taken to evaluate the response.

In dynamic analysis, modal damping ratios should be carefully evaluated. To determine the modal damping ratios, a bridge may be divided into several substructures in which the energy-dissipating mechanism is essentially the same. If one can specify a damping ratio of each substructure for a given mode shape, the modal damping ratio for the ith mode, h_i, may be evaluated as

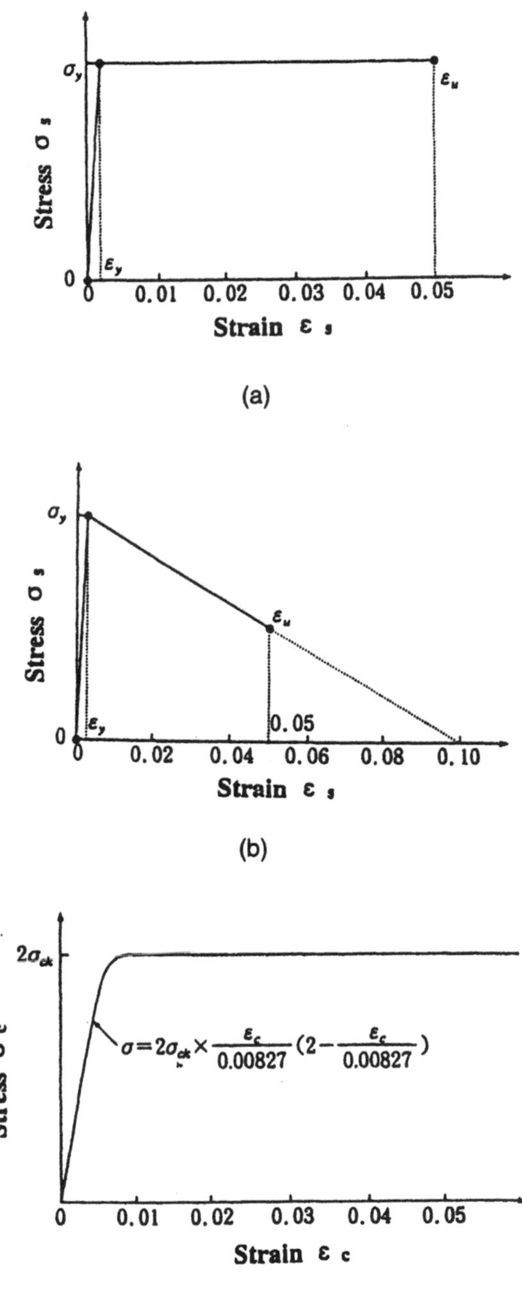

FIGURE 44.10 Stress–strain relation of steel and concrete. (a) Steel (tension); (b) steel (compression); (c) concrete.

$$h_i = \frac{\sum_{j=1}^{n} \phi_{ij}^T \cdot h_{ij} \cdot K_j \cdot \phi_{ij}}{\Phi_i^T \cdot K \cdot \Phi_i} \tag{44.28}$$

TABLE 44.6 Standard Acceleration Response Spectra

Soil Condition	Response Acceleration S_{10} (gal = cm/s²)		
	Type I Response Spectra S_{10}		
Group I	$S_{10} = 700$ for $T_i = \leqq 1.4$	$S_{10} = 980/T_i$ for $T_i > 1.4$	
Group II	$S_{10} = 1505T_i^{1/3}$ ($S_{10} \geqq 700$) for $T_i < 0.18$	$S_{10} = 850$ for $0.18 \leqq T_i \leqq 1.6$	$S_{10} = 1360/T_i$ for $T_i > 1.6$
Group III	$S_{10} = 1511T_i^{1/3}$ ($S_{10} \geqq 700$) for $T_i < 0.29$	$S_{10} = 1000$ for $0.29 \leqq T_i \leqq 2.0$	$S_{10} = 2000/T_i$ for $T_i > 2.0$
	Type II Response Spectra S_{110}		
Group I	$S_{110} = 4463T_i^{2/3}$ for $T_i \leqq 0.3$	$S_{110} = 2000$ for $0.3 \leqq T_i \leqq 0.7$	$S_{110} = 1104/T_i^{5/3}$ for $T_i > 0.7$
Group II	$S_{110} = 3224T_i^{2/3}$ for $T_i < 0.4$	$S_{110} = 1750$ for $0.4 \leqq T_i \leqq 1.2$	$S_{110} = 2371/T_i^{5/3}$ for $T_i > 1.2$
Group III	$S_{110} = 2381T_i^{2/3}$ for $T_i < 0.5$	$S_{110} = 1500$ for $0.5 \leqq T_i \leqq 1.5$	$S_{110} = 2948T_i^{5/3}$ for $T_i > 1.5$

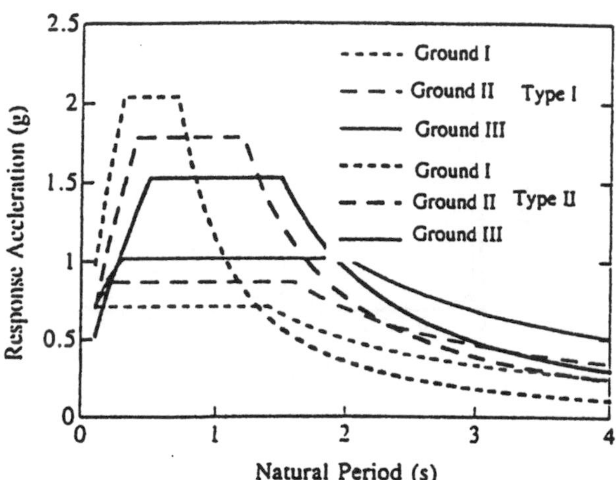

FIGURE 44.11 Type I and Type II standard acceleration response spectra.

TABLE 44.7 Recommended Damping Ratios for Major Structural Components

Structural Components	Elastic Response		Nonlinear Response	
	Steel	Concrete	Steel	Concrete
Superstructure	0.02 ~ 0.03	0.03 ~ 0.05	0.02 ~ 0.03	0.03 ~ 0.05
Rubber bearings	0.02		0.02	
Menshin bearings	Equivalent damping ratio by Eq. 44.26		Equivalent damping ratio by Eq. 44.46	
Substructures	0.03 ~ 0.05	0.05 ~ 0.1	0.1 ~ 0.2	0.12 ~ 0.2
Foundations	0.1 ~ 0.3		0.2 ~ 0.4	

in which h_{ij} = damping ratio of the jth substructure in the ith mode, ϕ_{ij} = mode vector of the jth substructure in the ith mode, k_j = stiffness matrix of the jth substructure, K = stiffness matrix of a bridge, and Φ_i = mode vector of a bridge in the ith mode, which is given as

$$\Phi_i^T = \left\{\phi_{i1}^T, \phi_{i2}^T, \ldots, \phi_{in}^T\right\} \tag{44.29}$$

Table 44.7 shows recommended damping ratios for major structural components.

TABLE 44.8 Modification Coefficient for Energy Dissipation Capability

Damping Ratio for First Mode h	Coefficient c_E
$h < 0.1$	1.0
$0.1 \leqq h < 0.12$	0.9
$0.12 \leqq h < 0.15$	0.8
$h \geqq 0.15$	0.7

44.4.7 Menshin (Seismic Isolation) Design

44.4.7.1 Basic Principle

Implementation of Menshin bridges should be carefully chosen from the point of view not only of seismic performance but also of function for traffic and maintenance, based on the advantage and disadvantage of increasing natural period. The Menshin design should not be adopted in the following situations:

1. Sites vulnerable to loss of bearing capacity due to soil liquefaction and lateral spreading;
2. Bridges supported by flexible columns;
3. Soft soil sites where potential resonance with surrounding soils could be developed by increasing the fundamental natural period; and
4. Bridges with uplift force at bearings.

It is suggested that the design be made with an emphasis on an increase of energy-dissipating capability and a distribution of lateral force to as many substructures as possible. To concentrate the hysteretic deformation not at piers, but at bearings, the fundamental natural period of a Menshin bridge should be about two times or more longer than the fundamental natural period of the same bridge supported by conventional bearings. It should be noted that an elongation of natural period aiming to decrease the lateral force should not be attempted.

44.4.7.2 Design Procedure

Menshin bridges are designed by both the seismic coefficient method and the ductility design method. In the seismic coefficient method, no reduction of lateral force from the conventional design is made.

In the ductility design method, the equivalent lateral force coefficient k_{hem} in the Menshin design is evaluated as

$$k_{hem} = \frac{k_{hcm}}{\sqrt{2\mu_m - 1}} \tag{44.30}$$

$$k_{hcm} = c_E \cdot k_{hc} \tag{44.31}$$

in which k_{hcm} = lateral force coefficient in Menshin design, μ_m = allowable ductility factor of a pier, c_E = modification coefficient for energy-dissipating capability (refer to Table 44.8), and k_{hc} = lateral force coefficient by Eq. (44.8). Because the k_{hc} is the lateral force coefficient for a bridge supported by conventional bearings, Eq. (44.31) means that the lateral force in the Menshin design can be reduced, as much as 30%, by the modification coefficient c_E depending on the modal damping ratio of a bridge.

Modal damping ratio of a menshin bridge h for the fundamental mode is computed as Eq. (44.32). In Eq. (44.32), h_{Bi} = damping ratio of the ith damper, h_{Pi} = damping ratio of the ith pier or abutment, h_{Fui} = damping ratio of the ith foundation associated with translational displacement, $h_{F\phi i}$ = damping ratio of the ith foundation associated with rotational displacement, K_{Pi} = equivalent stiffness of the ith pier or abutment, K_{Fui} = translational stiffness of the ith foundation, $K_{F\phi i}$ = rotational stiffness of the ith foundation, u_{Bi} = design displacement of the ith Menshin device, and H = distance from the bottom of a pier to a gravity center of a deck.

In the Menshin design, the allowable displacement ductility factor of a pier μ_m in Eq. (44.30) is evaluated by

$$h = \frac{\sum K_{Bi} \cdot u_{Bi}^2 \left(h_{Bi} + \frac{h_{Pi} \cdot K_{Bi}}{K_{Pi}} + \frac{h_{Fui} \cdot K_{Bi}}{K_{Fui}} + \frac{h_{F\theta i} \cdot K_{Bi} \cdot H^2}{K_{Fei}} \right)}{\sum K_{Bi} \cdot u_{Bi}^2 \left(1 + \frac{K_{Bi}}{K_{Pi}} + \frac{K_{Bi}}{K_{Fui}} + \frac{K_{Bi} \cdot H^2}{K_{F\theta i}} \right)} \tag{44.32}$$

$$\mu_m = 1 + \frac{\delta_u - \delta_y}{\alpha_m \delta_y} \tag{44.33}$$

in which α_m is a safety factor used in Menshin design and is given as

$$\alpha_m = 2\alpha \tag{44.34}$$

where α is the safety factor in the conventional design (refer to Table 44.4). Equation (44.34) means that the allowable displacement ductility factor in the Menshin design μ_m should be smaller than the allowable displacement ductility factor μ_a by Eq. (44.2) in the conventional design. The reason for the smaller allowable ductility factor in the Menshin design is to limit the hysteretic displacement of a pier at the plastic hinge zone so that the principal hysteretic behavior occurs at the Menshin devices, as shown in Figure 44.4b.

44.4.7.2 Design of Menshin Devices

Simple devices that can resist extreme earthquakes must be used. The bearings have to be anchored to a deck and substructures with bolts, and should be replaceable. Clearance has to be provided between a deck and an abutment or between adjacent decks.

Isolators and dampers must be designed for a desired design displacement uB. The design displacement u_B is evaluated as

$$u_B = \frac{k_{hem} W_U}{K_B} \tag{44.35}$$

in which k_{hem} = equivalent lateral force coefficient by Eq. (44.31), K_B = equivalent stiffness, and W_U = dead weight of a superstructure. It should be noted that, because the equivalent lateral force coefficient k_{hem} depends on the type of ground motions, the design displacement u_B also depends on the same.

The equivalent stiffness K_B and the equivalent damping ratio h_B of a Menshin device are evaluated as

$$K_B = \frac{F(u_{Be}) - F(-u_{Be})}{2u_{Be}} \tag{44.36}$$

$$h_B = \frac{\Delta W}{2\pi W} \tag{44.37}$$

$$u_{Be} = c_B \cdot u_B \tag{44.38}$$

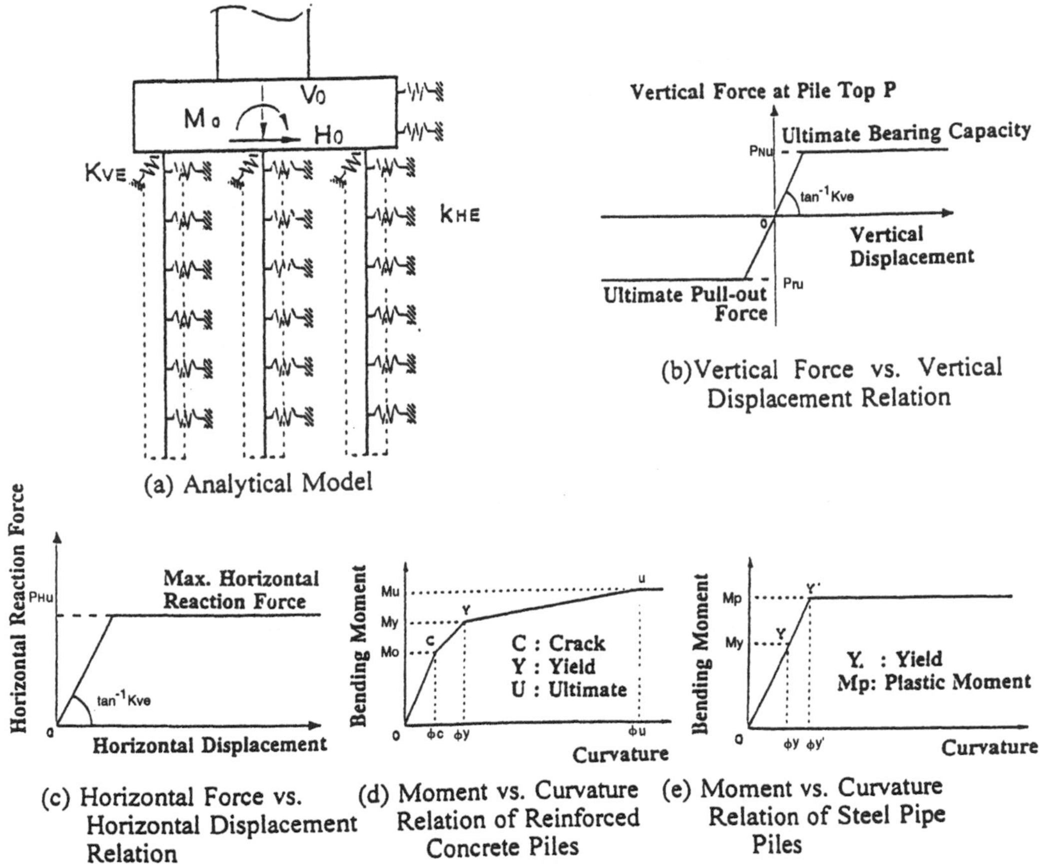

FIGURE 44.12 Idealized nonlinear model of a pile foundation. (a) Analybical model; (b) vertical force vs. vertical displacement relation; (c) horizontal force vs. horizontal displacement relation; (d) moment vs. curvature relation of reinforced concrete piles; (e) moment vs. curvature relation of steel pipe piles.

in which $F(u)$ = restoring force of a device at a displacement u, u_{Be} = effective design displacement, ΔW = energy dissipated per cycle, W = elastic strain energy, and c_B = coefficient to evaluate effective displacement (= 0.7).

44.4.8 Design of Foundations

The evaluation methods of ductility and strength of foundations such as pile foundations and caisson foundations were newly introduced in the 1996 Specifications.

For a pile foundation, a foundation should be so idealized that a rigid footing is supported by piles which are supported by soils. The flexural strength of a pier defined by Eq. (44.7) is to be applied as a seismic force to foundations at the bottom of the footing together with the dead-weight superstructure, pier, and soils on the footing. Figure 44.12 shows the idealized nonlinear model of a pile foundation. The nonlinearity of soils and piles is considered in the analysis.

The safety of the foundation is to be checked so that (1) the foundation does not reach its yield point; (2) if the primary nonlinearity is developed in the foundations, the response displacement is less than the displacement ductility limit; and (3) the displacement developed in the foundation is less than the allowable limit. The allowable ductility and the allowable limit of displacement were noted as 4 in displacement ductility, 40 cm in horizontal displacement, and 0.025 rad in rotation angle.

For a caisson-type foundation, the foundation should be modeled as a reinforced concrete column that is supported by soil spring model; the safety is checked in the same way as the pile foundations.

44.4.9 Design against Soil Liquefaction and Liquefaction~Induced Lateral Spreading

44.4.9.1 Estimation of Liquefaction Potential

Since the Hyogo-ken Nanbu earthquake of 1995 caused liquefaction even at coarse sand or gravel layers which had been regarded as invulnerable to liquefication, a gravel layer was included in the soil layers that require liquefaction potential estimation. Soil layers that satisfy the following conditions are estimated to be potential liquefaction layers:

1. Saturated soil layer which is located within 20 m under the ground surface and in which the groundwater level is less than 10 m deep;
2. Soil layer in which fine particle content ratio FC is equal or less than 35% or the plasticity index I_P is equal to or less than 15;
3. Soil layer in which mean grain size D_{50} is equal or less than 10 mm and 10% grain size D_{10} is equal or less than 1 mm.

Liquefaction potential is estimated by the safety factor against liquefaction F_L as

$$F_L = R/L \tag{44.39}$$

where, F_L = liquefaction resistant ratio, R = dynamic shear strength ratio, and L = shear–stress ratio during an earthquake. The dynamic shear strength ratio R may be expressed as

$$R = c_W R_L \tag{44.40}$$

where c_W = corrective coefficient for ground motion characteristics (1.0 for Type I ground motions, 1.0 to 2.0 for Type II ground motions), and R_L = cyclic triaxial strength ratio. The cyclic triaxial strength ratio was estimated by laboratory tests with undisturbed samples by the *in situ* freezing method.

The shear–stress ratio during an earthquake may be expressed as

$$L = r_d k_{hc}\, \sigma_v / \sigma'_v \tag{44.41}$$

where r_d = modification factor shear–stress ratio with depth, k_{hc} = design seismic coefficient for the evaluation of liquefaction potential, σ_v = total loading pressure, σ'_v = effective loading pressure.

It should be noted here that the design seismic coefficient for the evaluation of liquefaction potential k_{hc} ranges from 0.3 to 0.4 for Type I ground motions, and from 0.6 to 0.8 for Type II ground motions.

44.4.9.2 Design Treatment of Liquefaction for Bridge Foundations

When liquefaction occurs, the strength and the bearing capacity of a soil decreases. In the seismic design of highway bridges, soil constants of a sandy soil layer which is judged liable to liquefy are reduced according to the F_L value. The reduced soil constants are calculated by multiplying the coefficient D_E in Table 44.9 to the soil constants estimated on an assumption that the soil layer does not liquefy.

44.4.9.3 Design Treatment of Liquefaction-Induced Ground Flow for Bridge Foundations

The influence of liquefaction-induced ground flow was included in the revised Design Specifications in 1996. The case in which ground flow that may affect bridge seismicity is likely to occur is generally that the ground is judged to be liquefiable and is exposed to biased Earth pressure, for example, the ground behind a seaside protection wall. The effect of liquefaction-induced ground flow is

TABLE 44.9 Reduction Coefficient for Soil Constants Due to Soil Liquefaction

Range of F_L	Depth from the Present Ground Surface x (m)	Dynamic Shear Strength Ratio R	
		$R \leqq 0.3$	$0.3 < R$
$F_L \leqq 1/3$	$0 \leqq x \leqq 10$	0	1/6
	$10 < x \leqq 20$	1/3	1/3
$1/3 < F_L \leqq 2/3$	$0 \leqq x \leqq 10$	1/3	2/3
	$10 < x \leqq 20$	2/3	2/3
$2/3 < F_L \leqq 1$	$0 \leqq x \leqq 10$	2/3	1
	$10 < x \leqq 20$	1	1

considered as the static force acting on a structure. This method premises that the surface soil is of the nonliquefiable and liquefiable layers, and the forces equivalent to the passive Earth pressure and 30% of the overburden pressure are applied to the structure in the nonliquefiable layer and liquefiable layer, respectively.

The seismic safety of a foundation is checked by confirming that the displacement at the top of foundation caused by ground flow does not exceed an allowable value, in which a foundation and the ground are idealized as shown in Figure 44.12. The allowable displacement of a foundation may be taken as two times the yield displacement of a foundation. In this process, the inertia force of structure is not necessary to be considered simultaneously, because the liquefaction-induced ground flow may take place after the principal ground motion.

44.4.10 Bearing Supports

The bearings are classified into two groups: Type A bearings resisting the seismic force considered in the seismic coefficient method, and Type B bearings resisting the seismic force of Eq. (44.2). Seismic performance of Type B bearings is, of course, much higher than that of Type A bearings. In Type A bearings, a displacement-limiting device, which will be described later, has to be coinstalled in both longitudinal and transverse directions, while it is not required in Type B bearings. Because of the importance of bearings as one of the main structural components, Type B bearings should be used in Menshin bridges.

The uplift force applied to the bearing supports is specified as

$$R_U = R_D - \sqrt{R_{heq}^2 + R_{veq}^2} \tag{44.42}$$

in which R_U = design uplift force applied to the bearing support, R_D = dead load of superstructure, R_{heq} and R_{veq} are vertical reactions caused by the horizontal seismic force and vertical force, respectively. Figure 44.13 shows the design forces for the bearing supports.

44.4.11 Unseating Prevention Systems

Unseating prevention measures are required for highway bridges. Unseating prevention systems consist of enough seat length, a falling-down prevention device, a displacement-limiting device, and a settlement prevention device. The basic requirements are as follows:

1. The unseating prevention systems have to be so designed that unseating of a superstructure from its supports can be prevented even if unpredictable failures of structural members occur;
2. Enough seat length must be provided and a falling-down prevention device must be installed at the ends of a superstructure against longitudinal response. If Type A bearings are used, a displacement-limiting device has to be further installed at not only the ends of a superstructure but at each intermediate support in a continuous bridge; and

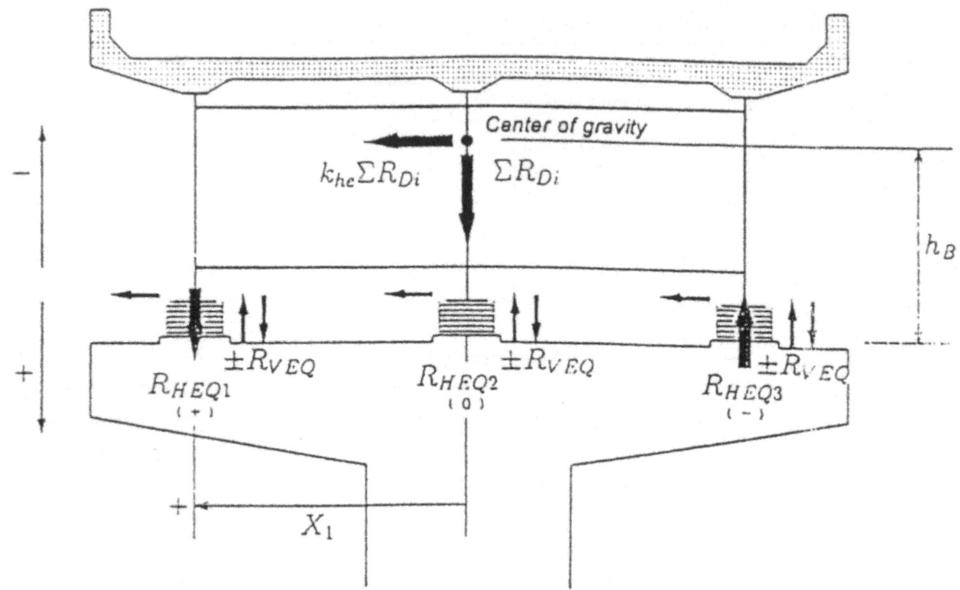

FIGURE 44.13 Design forces for bearing supports.

3. If Type A bearings are used, a displacement-limiting device is required at each support against transverse response. The displacement-limiting device is not generally required if Type B bearings are used. But, even if Type B bearings are adopted, it is required in skewed bridges, curved bridges, bridges supported by columns with narrow crests, bridges supported by few bearings per pier, and bridges constructed at sites vulnerable to lateral spreading associated with soil liquefaction.

The seat length S_E is evaluated as

$$S_E = u_R + u_G \geqq S_{EM} \tag{44.43}$$

$$S_{EM} = 70 + 0.5l \tag{44.44}$$

$$u_G = 100 \cdot \varepsilon_G \cdot L \tag{44.45}$$

in which u_R = relative displacement (cm) developed between a superstructure and a substructure subjected to a seismic force equivalent to the equivalent lateral force coefficient k_{hc} by Eq. (44.2); u_G = relative displacement of ground along the bridge axis; S_{EM} = minimum seat length (cm); ε_G = ground strain induced during an earthquake along the bridge axis, which is 0.0025, 0.00375, and 0.005 for Group I, II, and III sites, respectively; L = distance that contributes to the relative displacement of ground (m); and 1 = span length (m). If two adjacent decks are supported by a pier, the larger span length should be l in evaluating the seat length.

In the Menshin design, in addition to the above requirements, the following considerations have to be made.

1. To prevent collisions between a deck and an abutment or between two adjacent decks, enough clearance must be provided. The clearance between those structural components S_B should be evaluated as

$$S_B = \begin{cases} u_B + L_A & \text{between a deck and an abutment} \\ c_B \cdot u_B + L_A & \text{between two adjacent decks} \end{cases} \tag{44.46}$$

TABLE 44.10 Modification Coefficient for Clearance c_B

$\Delta T/T_1$	c_B
$0 \leqq \Delta T/T_1 < 0.1$	1
$0.1 \leqq \Delta T/T_1 < 0.8$	$\sqrt{2}$
$0.8 \leqq \Delta T/T_1 \leqq 1.0$	1

in which u_B = design displacement of Menshin devices (cm) by Eq. (44.39), L_A = redundancy of a clearance (generally ±1.5 cm), and c_B = modification coefficient for clearance (refer to Table 44.10). The modification coefficient c_B was determined based on an analysis of the relative displacement response spectra. It depends on a difference of natural periods $\Delta T = T_1 - T_2$ ($T_1 > T_2$), in which T_1 and T_2 represent the natural period of the two adjacent bridge systems.

2. The clearance at an expansion joint L_E is evaluated as

$$L_E = u_B + L_A \tag{44.47}$$

in which u_B = design displacement of Menshin devices (cm) by Eq. (44.39), and L_A = redundancy of a clearance (generally ±1.5 cm).

44.5 Seismic Retrofit Practices for Highway Bridges

44.5.1 Past Seismic Retrofit Practices

The Ministry of Construction has conducted seismic evaluations of highway bridges throughout the country five times since 1971 as a part of the comprehensive earthquake disaster prevention measures for highway facilities. Seismic retrofit for vulnerable highway bridges had been successively made based on the seismic evaluations. Table 44.11 shows the history of past seismic evaluations [7,8].

The first seismic evaluation was made in 1971 to promote earthquake disaster prevention measures for highway facilities. The significant damage of highway bridges caused by the 1971 San Fernando earthquake in the United States triggered the seismic evaluation. Highway bridges with span lengths longer than or equal to 5 m on all systems of national expressways and highways were evaluated. Attention was paid to detect deterioration such as cracks of reinforced concrete structures, tilting, sliding, settlement, and scouring of foundations. Approximately 18,000 highway bridges in total were evaluated and approximately 3200 bridges were found to require retrofit.

Following the first, seismic evaluations had been subsequently made in 1976, 1979, 1986, and 1991 with gradually expanding highways and evaluation items. The seismic evaluation in 1986 was made with the increase of social needs to ensure seismic safety of highway traffic after the damage caused by the Urakawa-oki earthquake in 1982 and the Nihon-kai-chubu earthquake in 1983. The highway bridges with span lengths longer than or equal to 15 m on all systems of national expressways, national highways and principal local highways, and overpasses were evaluated. The evaluation items included deterioration, unseating prevention devices, strength of substructures, and stability of foundations. Approximately 40,000 bridges in total were evaluated and approximately 11,800 bridges were found to require retrofit. The latest seismic evaluation was made in 1991. The number of highways to be evaluated has increased from the number evaluated in 1986. Approximately 60,000 bridges in total were evaluated and approximately 18,000 bridges were found to require retrofit. Through a series of seismic retrofit works, approximately 32,000 bridges were retrofitted by the end of 1994.

TABLE 44.11 Past Seismic Evaluations of Highway Bridges

			Number of Bridges		
Year	Highways Inspected	Inspection Items	Inspected	Require Strengthening	Strenghtened
1971	All sections of national expressways and national highways, and sections of others (bridge length ≧ 5m)	1. Deterioration 2. Bearing seat length *S* for bridges supported by bent piles	18,000	3,200	1,500
1976	All sections of national expressways and national highways, and sections of others (Bridge Length ≧ 15m or Overpass Bridges)	1. Deterioration of substructures, bearing supports, and girders/slabs 2. Bearing seat length *S* and devices for preventing falling-off of superstructure	25,000	7,000	2,500
1979	All sections of national expressways, national highways, and principallocal highways, and sections of others (bridge length ≧ 15 m or overpass bridges)	1. Deterioration of substructures and bearing supports 2. Devices for preventing falling-off of superstructure 3. Effect of soil liquefaction 4. Bearing capacity of soils and piles 5. Strength of RC piers 6. Vulnerable foundations (bent pile and RC frame on two independent caisson founcations)	35,000	16,000	13,000
1986	All sections of national expressways, national highways and principal local highways, and sections of others (bridge length ≧ 15 m or overpass bridges)	1. Deterioration of substructures, bearing supports, and concrete girders 2. Devices for preventing falling-off of superstructure 3. Effect of soil liquefaction 4. Strength of RC piers (bottom of piers and termination zone of main reinforcement) 5. Bearing capacity of piles 6. Vulnerable foundations (bent piles and RC frame on two independent caisson foundations)	40,000	11,800	8,000
1991	All sections of national expressways, national highways and principal local highways, and sections of others (bridge length ≧ 15 m or overpass bridges)	1. Deterioration of substructures, bearing supports, and concrete girders 2. Devices for preventing falling-off of superstructure 3. Effect of soil liquifaction 4. Strength of RC piers (piers and termination zone of main reinforcement) 5. Vulnerable foundations (bent piles and RC frame on two independent caisson foundations)	60,000	18,000	7,000 (as of the end of 1994)

Note: Number of bridges inspected, number of bridges that required strengthening, and number of bridges strengthened are approximate numbers.

TABLE 44.12 Application of the Guide Specifications

Types of Roads and Bridges	Double Deckers, Overcrossings on Roads and Railways, Extremely Important Bridges from Disaster Prevention and Road Network	Others
Expressways, urban expressways, designated urban expressway, Honshu–Shikoku Bridges, designated national highways	Apply all items, in principle	Apply all items, in principle
Nondesignated national highways, prefectural roads, city, town, and village roads	Apply all items, in principle	Apply partially, in principle

The seismic evaluations in 1986 and 1991 were made based on a statistical analysis of bridges damaged and undamaged in the past earthquakes [9]. Because the collapse of bridges tends to develop because of excessive relative movement between the superstructure and the substructures and the failure of substructures associated with inadequate strength, the evaluation was made based on both the relative movement and the strength of the substructure.

Emphasis had been placed on installing unseating prevention devices in the past seismic retrofit. Because the installation of the unseating prevention devices was being completed, it had become important to promote strengthening of those substructures with inadequate strength and lateral stiffness.

44.5.2 Seismic Retrofit after the Hyogo-ken Nanbu Earthquake

44.5.2.1 Reference for Applying Guide Specifications to New Highway Bridges and Seismic Retrofit of Existing Highway Bridges

After the 1995 Hyogo-ken Nanbu earthquake, the "Part V: Seismic Design" of the "Design Specifications of Highway Bridges" (Japan Road Association) was completely revised in 1996 as discussed in the previous sections.

Because most of the substructures designed and constructed before 1971 do not meet the current seismic requirements, it is urgently needed to study the level of seismic vulnerability requiring retrofit. Upgrading the reliability of predictions of possible failure modes in future earthquakes is also very important. Since the seismic retrofit of substructures requires more cost, it is necessary to develop and implement effective and inexpensive retrofit measures and to design methods to provide for the next event.

For increasing seismic safety of the highway bridges that suffered damage by the Hyogo-ken Nanbu earthquake, various new drastic changes were tentatively introduced in the "Guide Specifications for Reconstruction and Repair of Highway Bridges Which Suffered Damage Due to the Hyogo-ken Nanbu Earthquake." Although intensified review of design could be made when it was applied to the bridges only in the Hanshin area, it may not be so easy for field design engineers to follow up the new Guide Specifications when the Guide Specifications is used for seismic design of all new highway bridges and seismic strengthening of existing highway bridges. Based on such demand, the "Reference for Applying the Guide Specifications to New Bridges and Seismic Strengthening of Existing Bridges" [10] was issued on June 30, 1995 by the Sub-Committee for Seismic Countermeasures for Highway Bridges, Japan Road Association.

The Reference classified the application of the Guide Specifications as shown in Table 44.12 based on the importance of the roads. All items of the Guide Specifications are applied for bridges on extremely important roads, while some items which prevent brittle failure of structural components are applied for bridges on important roads. For example, for bridges on important roads, the items for Menshin design, tie reinforcements, termination of longitudinal reinforcements, type of bearings, unseating prevention devices and countermeasures for soil liquefaction are applied, while the remaining items such as the design force, concrete-infilled steel bridges, and ductility check for foundations, are not applied.

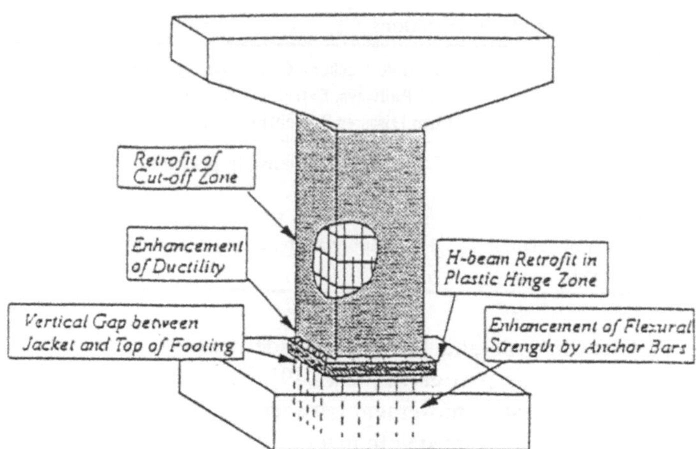

FIGURE 44.14 Seismic retrofit of reinforced concrete piers by steel jacket with controlled increase of flexural strength.

Because damage was concentrated in single reinforced concrete piers/columns with small concrete sections, a seismic retrofit program has been initiated for those columns that were designed according to the pre-1980 Design Specifications, at extremely important bridges such as bridges on expressways, urban expressways, and designated highway bridges, and also double-deckers and overcrossings, etc. which significantly affect highway functions once damaged. In the 3-year program, approximately 30,000 piers will be evaluated and retrofitted. Unseating devices also should be installed for these extremely important bridges.

The main purpose of the seismic retrofit of reinforced concrete columns is to increase their shear strength, in particular in piers with termination of longitudinal reinforcements without enough anchoring length. This increases the ductility of columns, because premature shear failure can be avoided.

However, if only ductility of piers is increased, residual displacement developed at piers after an earthquake may increase. Therefore, the flexural strength should also be increased. However, the increase of flexural strength of piers tends to increase the seismic force transferred from the piers to the foundations. It was found from an analysis of various types of foundations that failure of the foundations by increasing the seismic force may not be significant if the increasing rate of the flexural strength of piers is less than two. It is therefore suggested to increase the flexural strength of piers within this limit so that it does not cause serious damage to foundations.

For such requirements, seismic strengthening by steel jackets with controlled increase of flexural strength was suggested [10, 11]. This uses a steel jacket surrounding the existing columns as shown in Figure 44.14. Epoxy resin or nonshrinkage concrete mortar are injected between the concrete surface and the steel jacket. A small gap is provided at the bottom of piers between the steel jacket and the top of the footing. This prevents excessive increase in the flexural strength.

To increase the flexural strength of columns in a controlled manner, anchor bolts are provided at the bottom of the steel jacket. They are drilled into the footing. By selecting an appropriate number and size of the anchor bolts, the degree of increase of the flexural strength of piers may be controlled. The gap is required to trigger the flexural failure at the bottom of columns. A series of loading tests are being conducted at the Public Works Research Institute to check the appropriate gap and number of anchor bolts. Table 44.13 shows a tentatively suggested thickness of steel jackets and size and number of anchor bolts. They are for reinforced concrete columns with a/b less than 3, in which a and b represent the width of a column in transverse and longitudinal direction, respectively. The size and number of anchor bolts were evaluated so that the increasing rate of flexural strength of columns is less than about 2.

TABLE 44.13 Tentative Retrofit Method by Steel Jacketing

Column/Piers	Steel Jackets	Anchor Bolts
$a/b \leqq 2$	SM400, $t = 9$ mm	SD295, D35 ctc 250 mm
$2 < a/b \leqq 3$ Column supporting lateral force of a continuous girder through fixed bearing and with $a/b \leqq 3$	SM400, $t = 12$ mm	

Conventional reinforced concrete jacketing methods are also applied for the retrofit of reinforced concrete piers, especially for piers that require an increase of strength. It should be noted here that the increase of the strength of the pier should be carefully designed in consideration with the strength of foundations and footings.

44.5.2.2 Research and Development on Seismic Evaluation and Retrofit of Highway Bridges

Prioritization Concept for Seismic Evaluation

The 3-year retrofit program was completed in the 1997 fiscal year. In the program, the single reinforced concrete piers/columns with small concrete section which were designed by the pre-1980 Design Specifications on important highways have been evaluated and retrofitted and other bridges with wall-type piers, steel piers, and frame piers, and so on, as well as the bridges on the other highways, should be evaluated and retrofitted if required in the next retrofit program. Since there are approximately 200,000 piers, it is required to develop prioritization methods and methods to evaluate vulnerability for the intentional retrofit program.

Figure 44.15 shows the simple flowchart to prioritize the retrofit work to bridges. The importance of the highway, structural factors, member vulnerability (reinforced concrete piers, steel piers, unseating prevention devices, foundations) are the factors to be considered for prioritization.

Priority R of each bridge may be evaluated by Eq. (44.48).

$$R = I \cdot S \cdot V_T \cdot w_v \cdot \left[f\left(V_{RP1}, V_{RP2}, V_{RP3}\right), V_{MP}, V_{FS}, V_F \right] \times 100 \quad (44.48)$$

$$f\left(V_{RP1}, V_{RP2}, V_{RP3}\right) = V_{RP1} \cdot V_{RP2} \cdot V_{RP3} \quad (44.49)$$

in which R = priority, I = importance factor, S = earthquake force, V_T = structural factor, w_v = weighting factor on structural members, V_{RP1} = design specification, V_{RP2} pier structural factor, V_{RP3} = aspect ratio, V_{MP} = steel pier factor, V_{FS} = unseating device factor, and V_F = foundation factor. Each item and category with a weighting number is tentatively shown in Table 44.14. If this prioritization method is to applied the bridges damaged during the Hyogo-ken Nanbu earthquake, the categorization number is given as shown in Table 44.14.

Seismic Retrofit of Wall-Type Piers

The steel-jacketing method as described in the above was applied for reinforced concrete with circular section or rectangular section of $a/b < 3$. It is required to develop the seismic retrofit method for a wall-type pier. The confinement of concrete was provided by a confinement beam such as the H-shaped steel beam for rectangular piers. However, since the size of the confinement beam becomes very large, the confinement may be provided by other measures, such as intermediate anchors for a wall-type pier.

The seismic retrofit concept for a wall-type pier is the same as that for rectangular piers. It is important to increase the flexural strength and ductility capacity with the appropriate balance. Generally, the longitudinal reinforcement ratio is smaller than that for rectangular piers; therefore, the flexural strength is smaller. Thus, it is essential to increase the flexural strength appropriately. Since the longitudinal

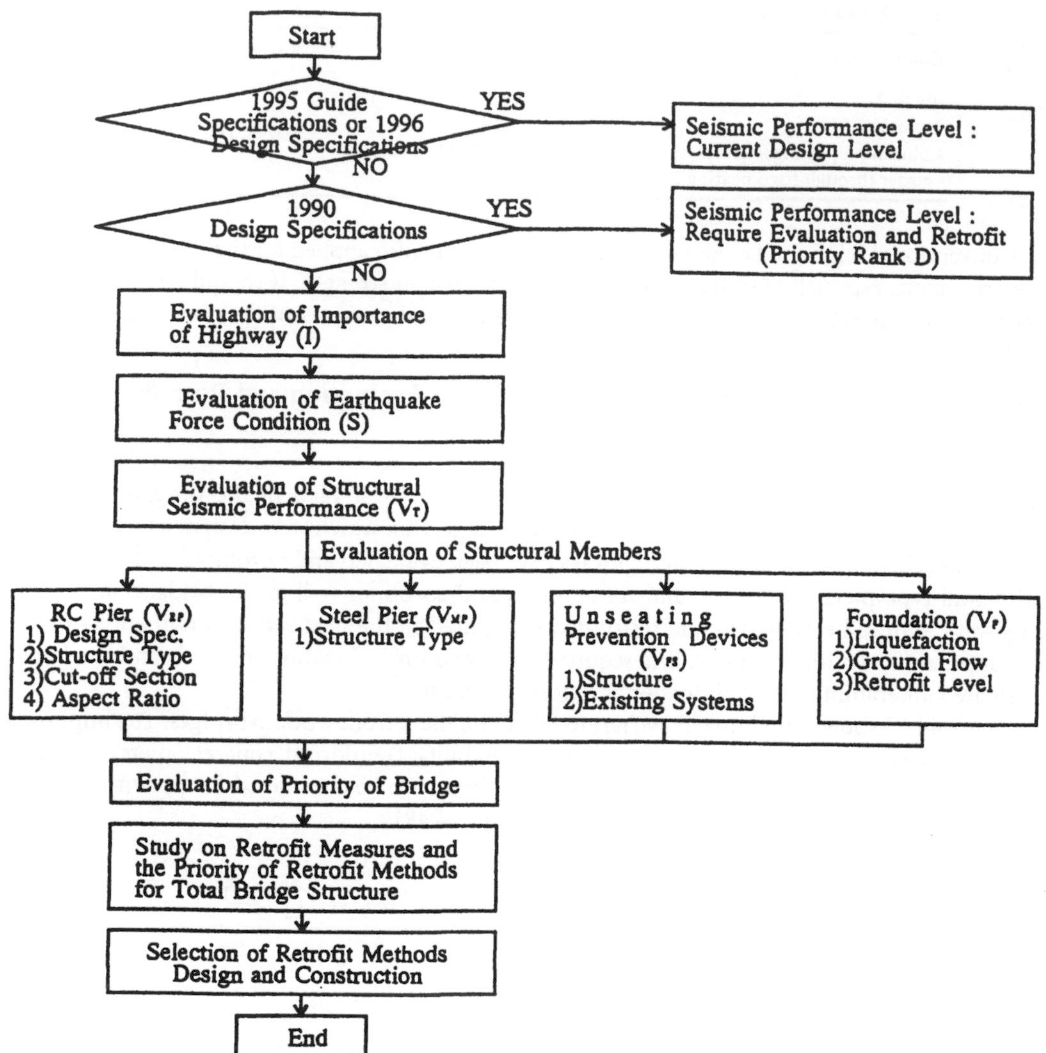

FIGURE 44.15 Prioritization concept of seismic retrofit of highway bridges.

reinforcement was generally terminated at midheight without appropriate anchorage length, it is also important to strengthen both the flexural and shear strength midheight section.

Figure 44.16 shows the possible seismic retrofit method for wall-type piers. To increase the flexural strength, the additional reinforcement by rebars or anchor bars are fixed to the footing. The number of reinforcements is designed to give the necessary flexural strength. It should be noted here that anchoring of additional longitudinal reinforcement is controlled to develop plastic hinge to the bottom of pier rather than the midheight section with termination of longitudinal reinforcement. And the increase of strength should be carefully designed considering the effect on the foundations and footings. The confinement in the plastic hinge zone is provided by steel bars for prestressed concrete or rebars which were installed inside of the column section.

Seismic Retrofit of Two-Column Bents

During the Hyogo-ken Nanbu earthquake, some two-column bents were damaged in the longitudinal and transverse directions. The strength and ductility characteristics of the two-column bents have been studied and the analysis and design method was introduced in the 1996 Design Specifications [12].

TABLE 44.14 Example of Prioritization Factors for Seismic Retrofit of Highway Bridges

Item	Category	Evaluation Point
Importance of highway (I)	1. Emergency routes	1.0
	2. Overcrossing with emergency routes	0.9
	3. Others	0.6
Earthquake force (S)	1. Ground condition Type I	1.0
	2. Ground condition Type II	0.9
	3. Ground condition Type III	0.8
Structural factor (V_T)	1. Viaducts	1.0
	2. Supported by abutments at both ends	0.5
Weighting factor on structural members (V_i)	1. Reinforced concrete pier	1.0
	2. Steel pier	0.95
	3. Unseating prevention devices	0.9
	4. Foundation	0.8
Reinforced concrete pier 1. Design specification (V_{RP1})	1. Pre-1980 Design Specifications	1.0
	2. Post-1980 Design Specifications	0.7
2. Pier structure (V_{RP2})	1. Single column	1.0
	2. Wall-type column	0.8
	3. Two-column bent	0.7
3. Aspect ratio (V_{RP3})	1. $h/D \leqq 3$	1.0
	2. $3 < h/D < 4$ with cutoff section	0.9
	3. $H/D \geqq 4$ with cutoff section	0.9
	4. $3 < h/D < 4$ without cutoff section	0.7
	5. $H/D \geqq 4$ without cutoff section	0.7
Steel pier (V_{MP})	1. Single column	1.0
	2. Frame structure	0.8
Unseating prevention devices (V_{FS})	1. Without unseating devices	1.0
	2. With one device	0.9
	3. With two devices	0.8
Foundations (V_F)	1. Vulnerable to Ground Flow (without unseating devices)	1.0
	2. Vulnerable to Ground Flow	0.9
	3. Vulnerable to Liquefaction (without unseating devices)	0.7
	4. Vulnerable to Liquefaction	0.6
Evaluation of the priority R	1. $R \geqq 0.8$	Priority Rank A
	2. $0.7 \leqq R < 0.8$	Priority Rank B
	3. $R < 0.7$	Priority Rank C

The strength and ductility of existing two-column bents were studied both in the longitudinal and transverse directions. In the longitudinal direction, the same as a single column, it is required to increase the flexural strength and ductility with appropriate balance. In the transverse direction, the shear strength of the columns or the cap beam is generally not enough in comparison with the flexural strength.

Figure 44.17 shows the possible seismic retrofit methods for two-column bents. The concept of the retrofit is to increase flexural strength and ductility as well as shear capacity for columns and cap beams. Since axial force in the cap beam is much smaller than that in the columns, increasing the shear capacity is essential for the retrofit of the cap beam. It should be noted that since the jacketing of cap beam is difficult because of the existing bearing supports and construction space, it is required to develop more effective retrofit measures for cap beams such as application of jacketing by new materials with high modulus of elasticity and high strength and out-cable pre-stressing, etc.

Seismic Retrofit Using New Materials

Retrofit work is often restricted because construction space is limited to open the structure for public traffic, particularly for the seismic retrofit of highway bridges in urban areas [13]. Therefore, there are sites where conventional steel jacketing and reinforced concrete jacketing methods are

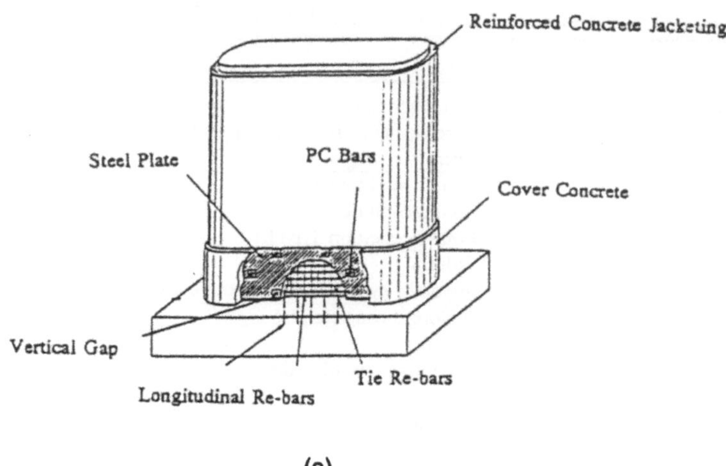

(a)

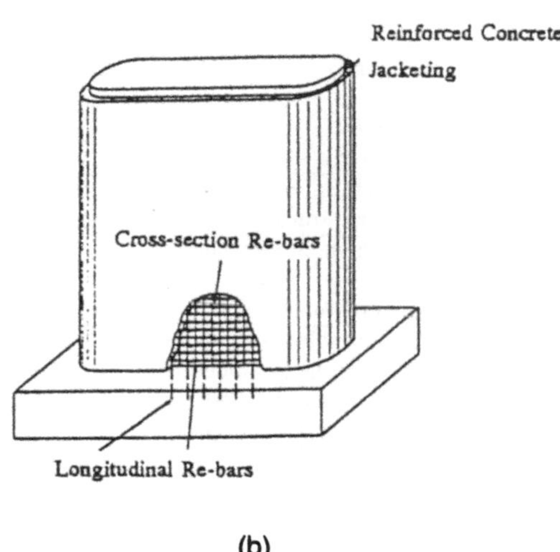

(b)

FIGURE 44.16 Seismic retrofit of wall-type piers. (a) Integrated seismic retrofit method with reinforced concrete and steel jacketing; (b) reinforced concrete jacketing.

difficult to apply. New materials such as carbon fiber sheets and aramid fiber sheets are attractive for application in the seismic retrofit of such bridges with construction restrictions as shown in Figure 44.18. The new materials such as fiber sheets are very light, do not need machines for use, and are easy to construct using glue bond as epoxy resin.

There are various studies on seismic retrofit methods using fiber sheets. Figure 44.19 shows the cooperative effect between fiber sheets and reinforcement for shear strengthening of a single reinforced concrete column. When carbon fiber sheets, which have almost the same elasticity and 10 times the failure strength as those of a reinforcing bar, are assumed to be applied, it is important to design the effects of carbon fiber sheets to achieve the required performance of seismic retrofit. In particular, strengthening of flexural, shear capacities, and ductility for reinforced concrete columns should be carefully evaluated. Based on experimental studies, it is essential to evaluate appropriately the effect of materials on the strengthening, carefully considering the material properties such as the modulus of elasticity and strength.

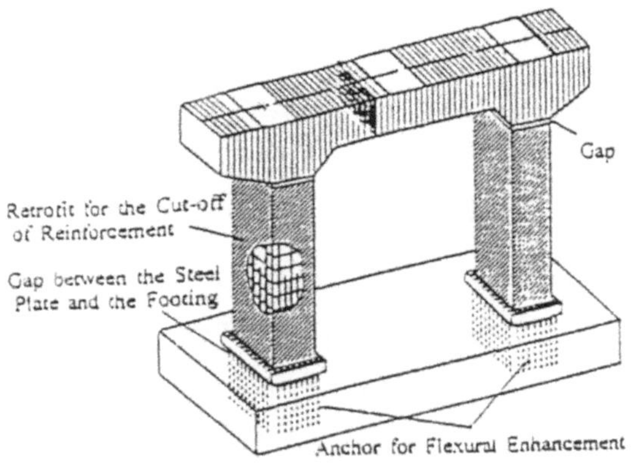

(a)

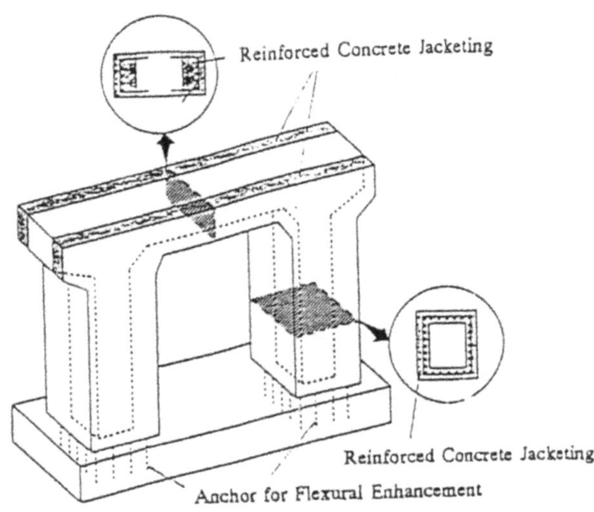

(b)

FIGURE 44.17 Seismic retrofit of two-column bents. (a) Steel jacketing; (b) reinforced concrete jacketing.

Acknowledgments

Drafting of the revised version of the "Part V: Seismic Design" of the "Design Specifications of Highway Bridges" was conducted by the Special Sub-Committee for Seismic Countermeasures for Highway Bridges" and was approved by the Bridge Committee, Japan Road Association. Dr. Kazuhiko Kawashima, Professor of the Tokyo Institute of Technology, chairman of the Special Sub-Committee and all other members of the Special Sub-Committee and the Bridge Committee are gratefully acknowledged.

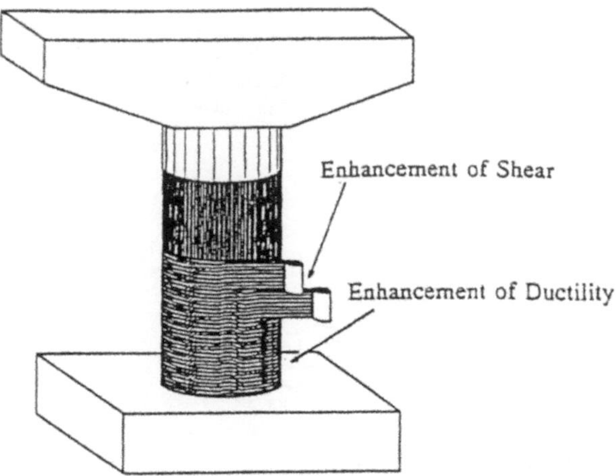

FIGURE 44.18 Application to new materials for seismic retrofit of reinforced column.

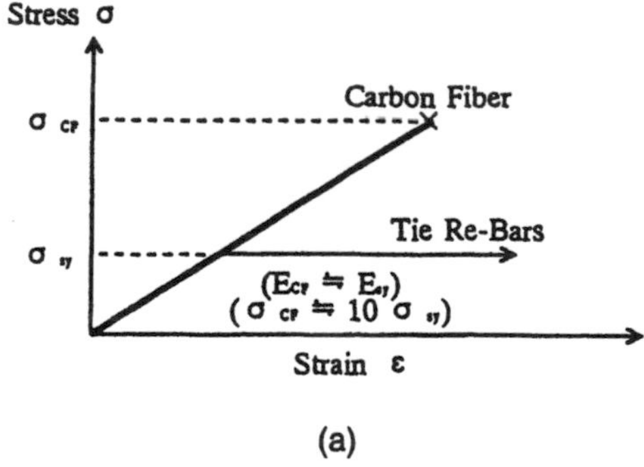

(a)

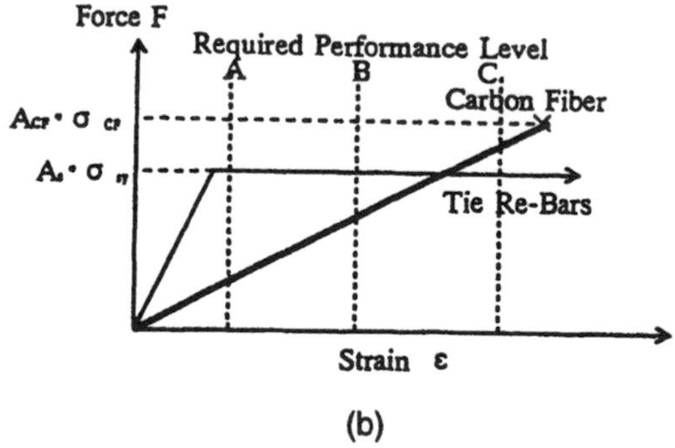

(b)

FIGURE 44.19 Cooperative effect between tie reinforcement and carbon fiber sheets. (a) Stress–strain relation; (b) force–strain relation.

References

1. Ministry of Construction, Report on the Damage of Highway Bridges by the Hyogo-ken Nanbu Earthquake, Committee for Investigation on the Damage of Highway Bridges Caused by the Hyogo-ken Nanbu Earthquake, 1995 [in Japanese].
2. Kawashima, K., Impact of Hanshin/Awaji Earthquake on Seismic Design and Seismic Strengthening of Highway Bridges, Report No. TIT/EERG 95-2, Tokyo Institute of Technology, 1995.
3. Kawashima, K. and Unjoh, S., The damage of highway bridges in the 1995 Hyogo-ken Nanbu earthquake and its impact on Japanese seismic design, *J. Earthquake Eng.*, 1(3), 1997.
4. Ministry of Construction, Guide Specifications for Reconstruction and Repair of Highway Bridges Which Suffered Damage Due to the Hyogo-ken Nanbu Earthquake, 1995 [in Japanese].
5. Japan Road Association, Design Specifications of Highway Bridges, Part I: Common Part, Part II: Steel Bridges, Part III: Concrete Bridges. Part IV: Foundations. Part V: Seismic Design, 1996 [in Japanese].
6. Kawashima, K., Nakano, M., Nishikawa, K., Fukui, J., Tamura, K., and Unjoh, S., The 1996 seismic design specifications of highway bridges, *Proceedings of the 29th Joint Meeting of U.S.–Japan Panel on Wind and Seismic Effects,* UJNR, Technical Memorandum of PWRI, No. 3524, 1997.
7. Kawashima, K., Unjoh, S., and Mukai, H., Seismic strengthening of highway bridges, in *Proceedings of the 2nd U.S.–Japan Workshop on Seismic Retrofit of Bridges,* Berkeley, January 1994, Technical Memorandum of PWRI, No. 3276.
8. Unjoh, S., Terayama, T., Adachi, Y., and Hoshikuma, J., Seismic retrofit of existing highway bridges in Japan, in *Proceedings of the 29th Joint Meeting of U.S.–Japan Panel on Wind and Seismic Effects,* UJNR, Technical Memorandum of PWRI, No. 3524, 1997.
9. Kawashima, K. and Unjoh, S., An inspection method of seismically vulnerable existing highway bridges, *Struct. Eng. Earthquake Eng.*, 7(7), *Proc. JSCE,* April 1990.
10. Japan Road Association, Reference for Applying Guided Specifications to New Highway Bridge and Seismic Strengthening of Existing Highway Bridges, June, 1995.
11. Hoshikuma. J., Otsuka, H., and Nagaya, K., Seismic retrofit of square RC Column by steel jacketing, *Proceedings of the 3rd U.S. Japan Workshop on Seismic Retrofit of Bridges,* Osaka, Japan, December 1996, Technical Memorandum of PWRI, No. 3481.
12. Terayama, T. and Otsuka, H., Seismic evaluation and retrofit of existing multi-column bents, in *Proceedings of the 3rd U.S.–Japan Workshop on Seismic Retrofit of Bridges,* Osaka, Japan, December 1996, Technical Memorandum of PWRI, No. 3481.
13. Public Works Research Center, Research Report by the Committee on Seismic Retrofit Methods Using Carbon Fiber Sheet, September 1996 [in Japanese].

References

1. Ministry of Construction, Report on the Damage of Highway Bridges [illegible] Earthquake, Committee for Investigation on the Damage of [illegible] Earthquake, 1995 (in Japanese).
2. [illegible]
3. [illegible]
4. [illegible]
5. [illegible]

Index

A

C

D

E

F

G

H

I

J

K

L

M

N

O

P

Q

R

S

T

U

V

W

X

Y

Z

Milton Keynes UK
Ingram Content Group UK Ltd.
UKHW051902071024
449327UK00025B/2053